PL.I.

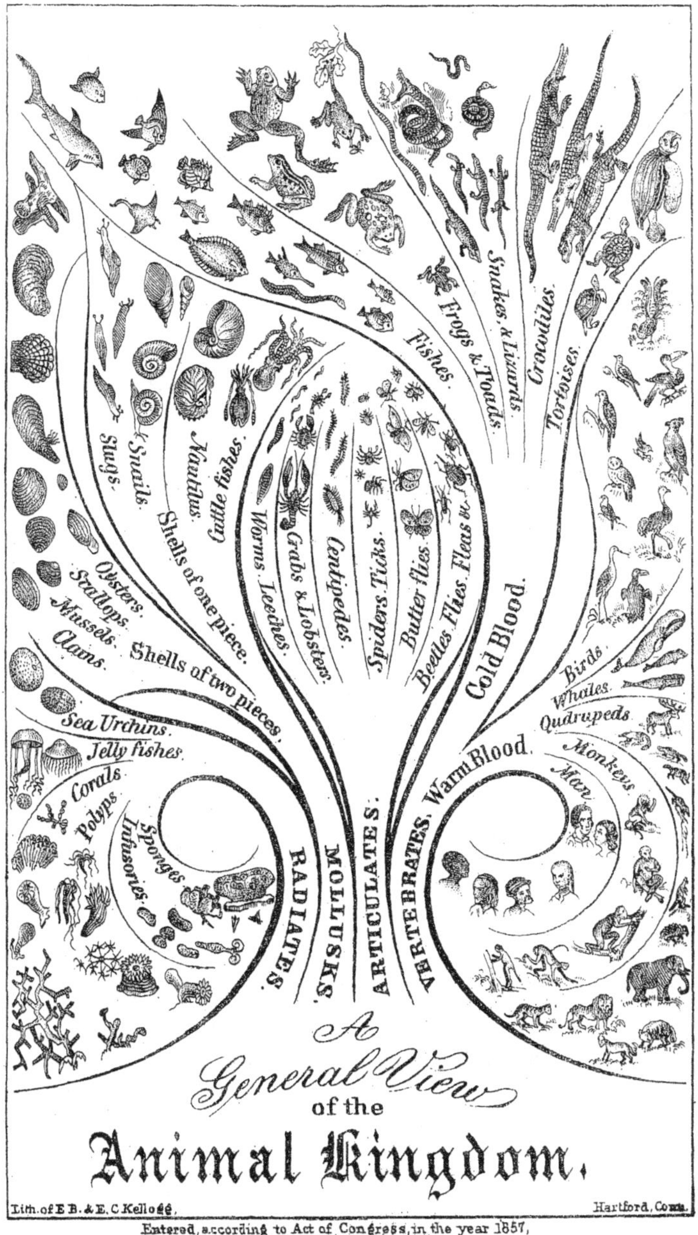

Fishes. Frogs & Toads. Snakes, & Lizards. Crocodiles. Tortoises.

Slugs. Snails. Nautilus. Cuttle fishes. Worms. Leeches. Crabs & Lobsters. Centipedes. Spiders. Ticks. Butter flies. Beetles Flies. Fleas &c.

Oysters. Scallops. Mussels. Clams. Shells of two pieces. Shells of one piece.

Cold Blood.

Birds. Whales. Qudrupeds. Monkeys. Man. Warm Blood.

Sea Urchins. Jelly fishes. Corals. Polyps. Sponges. Infusories.

RADIATES. MOLLUSKS. ARTICULATES. VERTEBRATES.

A General View of the

Animal Kingdom.

Lith. of E.B. & E.C. Kellogg, Hartford, Conn.

ZOÖLOGICAL SCIENCE

OR

NATURE IN LIVING FORMS,

ILLUSTRATED BY NUMEROUS PLATES.

ADAPTED TO ELUCIDATE THE

CHART OF THE ANIMAL KINGDOM,

BY A. M. REDFIELD,

AND

"Ask now the beasts, and they shall teach thee; and the fowls of the air, and they shall tell thee; or speak to the earth, and it shall teach thee; and the fishes of the sea shall declare unto thee. Who knoweth not in all these that the hand of the Lord hath wrought this?" (Job xii. 7.)

E. B. & E. C. KELLOGG, PUBLISHERS.
87 FULTON ST., NEW YORK.
FREDERICK COHOON, COLUMBUS, OHIO.
1858.

DEDICATION.

The following work has with my aid been prepared by an esteemed and highly competent friend, to whom I am also much indebted for valuable assistance rendered in connection with the publication of the Chart which it is adapted to elucidate.

Prepared, as it has been, with the utmost care and exactness; with un usual regard to order, and fullness of explanation as to the terms employed, I am sanguine in the belief it will every where meet with a cordial welcome as a suitable accompaniment of the Chart. Though both are capable of being used separately, each will be found to shed light upon the other.

To Teachers, to Parents and Heads of Families, to all who are lovers of Natural History and desire its advancement, I humbly but respectfully dedicate this volume and the Chart it is intended to explain and illustrate.

ANN M. REDFIELD.

PREFACE.

THE following work has been prepared as an accompaniment to the "GENERAL VIEW OF THE ANIMAL KINGDOM"—a CHART which, in the beautiful and harmonious arrangement of its several parts; its lucid and orderly classification; its brief but comprehensive statements and explanations,—presents the subject in an outline so full and consistent as to make it valuable even to the most scientific naturalist, both for convenient private reference, and as a help or guide in public lectures; while the more uninitiated, and such as are just setting out in the study of Natural History, becoming familiar with the details of the Chart, will, it is believed, desire and be prepared the better to appreciate *additional* information in relation to the subject; such information it is the aim of this work to impart.

The possessor of the Chart might have recourse to works of two kinds—one purely scientific, like those of Cuvier and others, or the works on Natural History published by State authority; the other, of a strictly popular character, in which not a single scientific or technical term is employed. The array of unexplained technical language in the former class of works, he would, perhaps, deem repulsive and discouraging; the descriptions of the latter class, he might, as related to the Chart, be often at a loss to apply correctly, though presenting to him the appearance of more interesting details than those which are found in works strictly scientific. The present volume, being a sort of meditım between these two kinds of works, is adapted to meet the exigencies of such a case. It does not give the "characters" and "descriptions" with the technicality and minuteness of the purely scientific treatise; to do this was found to be incompatible with the desired limits, as well as the general design of this publication; at the same time, it is far from ignoring these things, after the manner of some popular treatises. The "characters" of the Classes, Orders, and Families will here be found given with considerable fullness; the main or prominent ones of the genera and species are also usually given: not in all cases in a separate and formal manner, but occasionally are blended with other particulars relating to the general hâbits of animals, or interspersed with illustrative anecdote. In most, if not all cases, the reader will, from the statements made, be able to form some correct and consistent ideas as to the genera and species noticed. When more full discriminations are desired, reference can be had to other and larger works.

The medium character of this volume, and its relation to the extremely wide range of topics presented on the Chart, have increased the difficulty of preparing it within limits so restricted. To have furnished an amusing work composed chiefly or entirely of anecdotes or kindred material, would have been, comparatively, an easy task. In its present form, this work will perhaps not be unacceptable to such as are already somewhat acquainted with Natural History in its scientific aspects and relations; while others, the young especially, may, from the use of this volume, pass, by an easy transition, to the study of larger works and those more purely scientific. To TEACHERS in particular, is this volume respectfully commended. Questions are added to each section with special reference to its use in Academies and Common Schools.

It is proper to remark that this work is not published as containing the results of *original observation*, excepting to a limited extent: mainly it embodies materials newly moulded and arranged, but derived from approved standards, and some of the latest issues relating to the subjects of which it treats. The range of reference and comparison has been extensive; the results of protracted investigation are sometimes condensed into a single brief paragraph or sentence. This work will be found orderly and harmonious in several respects in which some other publications betray confusion and inconsistency; in the explanation of scientific terms, also, it is unusually full. Neither on the Chart, nor in this volume has the aim been to give *all* the different names which may have been applied by naturalists to a particular object; for this there was not room; and besides, in the case of some, such a course might have tended to confuse rather than really enlighten. Many of the pictorial illustrations are original, and with the accompanying explanations, will be found to add much to the interest and intrinsic value of the work.

It is confidently trusted that the CHART, with this explanatory volume, will be welcomed in Seminaries generally; and be accepted as valuable auxiliaries by all lovers of physical science. May they tend to create and foster widely a taste for the study of nature; and by the developments which they make, and the researches and meditations to which they lead, awaken loftier and more worthy thoughts of the Infinite Creator. M.

SYRACUSE, March 1, 1858.

CLASSIFICATION OF THE ANIMAL KINGDOM.

SUB-KINGDOMS, four: VERTEBRATES, ARTICULATES, MOLLUSKS, RADIATES.

VERTEBRATES: Grand Divisions: WARM and COLD BLOODED:

The WARM BLOODED Division includes MAMMALS and BIRDS.

1. MAMMALS, three sub-classes, nine orders.
   FIRST SUB-CLASS, UNGUICULATA, (with nails or claws.)

(1.) BIMANA, (Two-handed) Man.

(2.) QUADRUMANA (Four-handed). Three families.
Simiadæ, Apes, Baboons, Monkeys of the Old World.
Cebidæ, Monkeys (American).
Lemuridæ, Lemurs.

(3.) CARNIVORA (Flesh-eating Quadrupeds).
Sub-order, CHEIROPTERA (Hand-winged) Bats.
DIGITIGRADA (walking on the toes). Three families.
Felidæ, Cats, Lions, Tigers, &c.
Canidæ, Dogs, Wolves, &c.
Mustelidæ, Weasels, &c.
PLANTIGRADA (walking on the soles of the feet). Two families.
Ursidæ, Bears, Racoons.
Phocidæ, Seals, Walruses, &c.
TRUE INSECTIVORA (Insect-eaters). Four families.
Talpidæ, Moles.
Sorecidæ, Shrews.
Erinaceadæ, Hedge-hogs.
Tupaiadæ, Banxrings (of the Indian Archipelago).

(4.) MARSUPIALIA (Pouched Quadrupeds). Four sections.
Ovo-viviparous.
Sarcophaga (Flesh-eaters). Dasyuri.
Entomophaga (Insect-eaters). Opossums.
Carpophaga (Fruit-eaters). Phalangers.
Pöephaga (Grass-eaters). Kangaroos.
Sub-order *Rhizophaga* (Root-eaters). Wombats.
MONOTREMATA (Monotremes). Echidna and Ornithorhyncus or Water-Mole.

(5.) **Edentata** (Toothless or without front teeth). Four families.
Bradypodidæ (Slow-footed) or *Tardigrada*, Sloths.
Megatheriadæ (Great-beasts). Fossil Sloths.
Myrmecophagadæ (Ant-eaters).
Armadillidæ (Armadillos).

(6.) Rodentia (Gnawing Quadrupeds). Seven families.
Sciuridæ (Squirrels).
Muridæ (Mice & Rats).
Castoridæ (Beavers).
Hystricidæ (Porcupines).
Cavidæ(Cavies or Guinea Pigs).
Chinchillidæ (Chinchillas).
Leporidæ (Hares).

Second Sub-class, Ungulata (with hoofs).

(7.) Pachydermata (Thick-skinned Quadrupeds). Three families.
Elephantidæ (or Proboscideans,) Elephants, &c.
Suidæ, Swine, Rhinoceros, &c.
Equidæ, Horses, Zebras, &c. } Solipedes or *Solidungula* Solid-hoofed.

(8.) Ruminantia (Cud-chewing Quadrupeds). Eight Families.
Camelidæ, Camels, Llamas.
Camelopardæ, Camelopards or Giraffes.
Moschidæ, Musk-Deer.
Cervidæ, Deer or Stags.
Bovidæ, Oxen, Bison, (Buffalo,) &c.
Ovidæ, Sheep.
Capridæ, Goats.
Antilopidæ, Antelopes.

Third Sub-class, Marine Mammals.

(9.) Cetacea (Whale-tribe). Four families.
Balænidæ, Baleen or Whale-bone Whales.
Catadontidæ or *Physeteridæ*, } Blowers or Spermaceti Whales.
Delphinidæ, Dolphins, Porpoises, &c.
Manatidæ or Herbivorous Cetacea, } Manatees or Sea-Cows, &c.

II. Division of Warm Blooded Vertebrates.

Birds: *Land Birds*, five orders; *Water Birds*, two orders.

(1.) Raptores (Raveners or Birds of Prey). Three families.
Falconidæ, Falcon tribe. Three families.
Sub-families, *Aquilinæ*, Eagles.
Milvinæ, Kites.
Buteoninæ, Buzzards.
Falconinæ, Falcons.
Accipitrinæ, Hawks.
Vulturidæ, Vultures.
Strigidæ, Owls.

(2.) Insessores (Perchers). Four sub-orders. Fissirostres, Dentirostres, Conirostres, Tenuirostres.

1. Fissirostres (Cleft-bills). **Seven families.**
Caprimulgidæ, Night-jars.
Hirundinidæ, Swallows.
Meropidæ, Bee-eaters.
Todidæ, Todies.
Trogonidæ, Trogons.
Halcyonidæ or } King-fishers.
Alcedinidæ, }
Trogonidæ, Trogons.

2. Dentirostres (Toothed-bills).
Silviadæ, Warblers.
Merulidæ or } Thrushes.
Turdinidæ, }
Muscicapidæ, Fly-catchers.
Ampelidæ, Chatterers.
Laniadæ, Shrikes or Butcher Birds.

3. Conirostres (Cone-billed). Seven families.
Corvidæ, Crows, sub-fam.: ***Paradiseadæ*, Birds of Paradise.**
Sturnidæ, Starlings.
Fringillidæ, Finches.
Loxiadæ, Cross-bills.
Bucerotidæ, Hornbills.
Musophagidæ, Plaintain-eaters.

4. Tenuirostres (Thin-billed). **Five families.**
Promeropidæ or *Upupadæ*, Hoopœs.
Cinnyridæ or } Sun-birds, Honey-suckers **or Nectar Birds.**
Nectarinidæ, }
Trochilidæ, Humming-birds.
Meliphagidæ, Honey-eaters.
Certhiadæ, Creepers.

(3.) Scansores, Climbers. **Four families.**
Rhamphastidæ, Toucans.
Psittacidæ, Parrots.
Picidæ, Wood-peckers.
Cuculidæ, Cuckoos.

(4.) Rasores or Gallinæ, (Scratchers, Poultry Birds.) **Seven families.**
Columbidæ, Pigeons.
Cracidæ, Curassows.
Megapodidæ, Megapodes or Great-foots.
Phasianidæ, Pheasants, &c.
Tetraonidæ, Grouse.
Chionidæ, Sheath-bills.
Tinamidæ, Tinamous.

(5.) Cursores (Runners). One family. ***Struthionidæ*, Ostriches, &c.**

(6.) Grallatores, (Waders.) **Six families.** Aquatic Birds.
Charadriadæ, Plovers.
Ardeidæ, Herons.
Rostridæ, Spoon-bills.
Tantalidæ, Ibises.
Scolopacidæ, Snipes.
Rallidæ, Rails.

(7.) NATATORES, (Swimmers). Six families.
- *Anatidæ*, Ducks.
- *Colymbidæ*, Divers.
- *Alcidæ*, Auks.
- *Procellaridæ*, Petrels.
- *Laridæ*, Gulls.
- *Pelecanidæ*, Pelicans.

The COLD BLOODED Division includes REPTILES and FISHES.

I. REPTILES, four orders, viz. CHELONIANS (Turtles).
SAURIANS (Lizards, Crocodiles).
OPHIDIANS (Snakes).
AMPHIBIANS (Frogs, Toads, &c.)

1st. CHELONIANS, (CHELONIA,) arranged by Agassiz in his late work, as follows:

	Sub-orders.	Families.	
ORDER, TESTUDINATA.	1. *Amydæ*,	*Testudinina*,	Land Tortoises.
		Emydoidæ, *Cinosternoidæ*, *Chelydroidæ*, *Hydraspidæ*, *Chelyoidæ*, *Trionychidæ*,	Marsh and River Tortoises.
	2. *Chelonii*,	*Chelonidæ*, *Sphargidæ*,	Sea Turtles.

2d. SAURIANS.

ORDER, SAURIA.
- Fam. *Crocodilidæ*.
  - Alligators or Cäimans of America.
  - Crocodiles of the Nile.
  - Gavials of the Ganges.
  - Enaliosauria, (Fossil Fish-Lizards).
- " *Chamæleonidæ*, Chamæleons.
- " *Geckotidæ*, Geckos.
- " *Iguanidæ*, Iguanas.
- " *Varanidæ*, Varans.
- " *Teidæ*, Teguixans.
- " *Lacertidæ*, True Lizards.
- " *Chalcidæ*, Snakelike do.
- " *Scincidæ*, Scinks.

3d. OPHIDIANS. ORDER OPHIDIA.

Family	*Colubridæ*, (mostly) harmless Snakes.	Sub-order COLUBRINA.
"	*Boidæ*, Boas and Pythons.	
"	*Hydridæ*, Water (Venomous) Snakes.	
"	*Viperidæ*, Vipers, do.	Sub-order VIPERINA
"	*Crotalidæ*, Rattle Snakes (all kinds.)	

4th. AMPHIBIANS, sub-order. *Caducibranchiata*, (Gills perishable in the tadpole state.)

ORDER AMPHIBIA.
- Family *Cæciliidæ*, (Cæcilia.) Apodous or without feet.
- " *Ranidæ*, (Frogs.) } Anoura or tailless.
- " *Bufoidæ*, (Toads.) }
- " *Salamandridæ*, Salamanders, Land-Newts, Tritons, Water-Newts. } with tails
- " *Amphiumidæ*, Amphiuma, Menopoma or Mud-devil.
- Sub-order PERENNIBRANCHIATA, (with enduring gills.)
- " *Proteidæ*, (Proteus, Axolotl, Siren.)

*II. **Fishes.** Three Groups or Divisions based upon the distinctive character of the fins, viz.

- Acanthopterygii, (Spine-rayed fins.)
- Malacopterygii, (Soft-rayed fins.)
- Chondropterygii, (Cartilage-fins.)

Agassiz bases the orders upon the scales and makes them four.

- Ctenoids, (Comb-like.)
- Cycloids, (Circle-like.)
- Ganoids, (Splendor-like.)
- Placoids, (Plate-like.)

1st Order. Acanthopterygii, (Spine-rayed,) or Ctenoids.

Family
- ***Percidæ***, (Perch.)
- ***Triglidæ***, (Gurnards.)
- ***Scienidæ***, (Maigres, Sheep's-heads, Drum-fish, &c.)
- ***Sparidæ***, (Sea-Breams.)
- ***Chætontidæ***, (Chætodons, Moon-fish, Razor-fish.)
- ***Scombridæ***, (Mackerel.)
- ***Anabassidæ***. (Climbing-Perches.)
- ***Cepolidæ*** or ***Tæniadæ***, (Ribbon-fish.)
- ***Teuthidæ***, (Surgeon-fish.)
- ***Atherinidæ***, (Silver-sides.)
- ***Mugilidæ***, (Mullets.)
- ***Gobidæ***, (Gobias.)
- ***Lophidæ***, (Crested or Toad-fish.)
- ***Labridæ***, (Wrasses or Rock-fish.)

2d Order. Malacopterygii, (Soft-rayed) or Cycloids.

- Abdominales, (Ventral fins behind the pectoral.)
  - ***Siluridæ***, (Cat-fish.)
  - ***Cyprinidæ***, (Carps.)
  - ***Esocidæ***, (Pikes.)
  - ***Fistularidæ***, (Pipe-fish.)
  - ***Salmonidæ***, (Salmon.)
  - ***Clupeidæ***, (Herring.)
- Sub-brachials. (Ventral fins under the pectoral.)
  - ***Gadidæ***, (Codfish.)
  - ***Pleuronectidæ***, or ***Planidæ***, (Flat-fish.)
  - ***Echeneidæ***, (Sucking-fish.)
  - ***Cyclopteridæ***, (Lump-fish.)
- Apodes, without ventral fins.
  - ***Murænidæ*** or ***Anguillidæ***, (Eels.)
- Lophobranchia or Lophobranchii, (Tufted-gills.)
  - ***Syngnathidæ***, (Sea-horse, &c.)
- Plectognathi, (Plaited jaws)
  - ***Gymnodontidæ***, Balloon-fish.)
  - ***Balistidæ***, (File-fish.)
  - ***Ostracionidæ***, (Trunk-fish.)

---

* See Page 560.

Ganoids. *Sauridæ*, (Gar-fish, &c.)

CHONDROPTERIGII.
- PLACOIDS
  - Eleutheropomi, (gills free.)
    - *Chimæridæ*, (Sea Monsters.)
    - *Sturionidæ*, (Sturgeons.)
  - Plagiostomi, (transverse mouths.)
    - *Squalidæ*, (Sharks.)
    - *Raiidæ*, (Rays.)
  - Cyclostomi, (*Round* fleshy mouth or lip.)
    - [*Petromyzonidæ*, (Lampreys.)
  - Branchiostoma, (Gill-mouth, i. e., having *cirri*, or curled filaments in the mouth.)
    - [*Amphioxidæ*, (Lancelets.)
    - Very anomalous, and sometimes included with the Cyclostomi.

ARTICULATES. Three classes. INSECTS.
CRUSTACEANS.
WORMS OR ANNELIDANS.

I. INSECTS.

Twelve ORDERS Of TRUE INSECTS.

With biting mouths.
1. COLEOPTERA (Sheath-wings), Beetles, Hornbugs.
2. STREPSIPTERA (Twisted-wings), Wasp-flies.
3. DERMAPTERA (Skin-wings), Ear-wigs.
4. ORTHOPTERA (Strait-wings).
   Sub-orders, CURSORIA (Runners), Cockroaches.
   RAPTORIA (Graspers), Mantises.
   AMBULATORIA (Walkers), Walking Sticks.
   SALTATORIA (Leapers), Grasshoppers, Crickets, &c.
5. TRICHOPTERA (Hair-wings), Caddice-flies, &c.
6. NEUROPTERA (Nerve-wings), White Ants, Dragon-flies, &c.
7. HYMENOPTERA (Membranous-wings), Bees, Wasps, [&c.

With sucking mouths.
8. LEPIDOPTERA (Scale-wings), Moths, Butterflies, &c.
9. HEMIPTERA (Half-wings), Fruit-bugs, Bed-bugs, &c.
10. DIPTERA (Two-wings), Flies, Musquitoes, &c.
11. APHANIPTERA (Invisible or rudimental wings), Fleas, Jiggers.
12. APTERA (No wings), Lice, Lepismas.

Sometimes called Sub-classes.
13. MYRIAPODA (with innumerable feet), Thousand-legged Worms, Centipedes.
14. ARACHNIDA, Spiders, Scorpions, Ticks, Mites.

2d. CRUSTACEANS. Five orders (or sub-classes,) (Dana).
1. Decapoda (Ten-footed), Crabs, Lobsters, Shrimps.
2. Tetradecapoda (Fourteen-footed), Sow-bugs, Sand-fleas, &c.
3. Entomostraca (Shell insects), Cyclops, Daphnia, Cypris, Limulus, (Sea-Spiders), and possibly also the TRILOBITES.
4. Cirripedes (curled jointed-feet), Barnacles.
5. Rotatoria or Rotifera—Wheel Animalcules.

3d. WORMS or ANNELIDANS. Four orders.
1. Tubulibranchiata (Gills in tubes), Serpula, Vermilia, &c.
2. Dorsibranchiata (Gills on the back), Sea-Centipedes.
3. Abranchiata (without gills), Leeches & Earth-worms.
4. Entozoa, (Internal worms,) or White blooded Worms.

These sometimes resemble worms found in the other classes or orders, while differing from them as to their *locality*. They have been arranged into the following sub-orders:

(1). Nematoidea or Nematoids, Round Worms, Thread Worms, Pin Worms, Guinea Worms.
(2). Acanthocephala, (Spine-headed,) Hooked Worms.
(3). Trematoda, (from Gr. *trema*, hole, having Sucker-like openings,) Fluke Worms, &c.
(4). Cestoidea, (Gr. *Kestos*, girdle,) Tape Worms.
(5). Cystica, (Gr. *Kustis*, a bladder,) Hytatids or Bladder-like Worms.

MOLLUSKS. Two Grand Divisions:
Céphalata (with heads) or Univalves.
Acephala (without heads) or Bivalves.

I. Cephalata: Three classes or Sub-divisions, viz.

(1). Cephalopods, { Head-footed, i. e. Arms about the head. } Two sections.

Dibranchiata, (Two branchiæ,) Cuttle-fish.
Tetrabranchiata, (Four branchiæ,) Nautilus, Ammonites, &c.

(2). Pteropods, { Wing-footed, i. e., wing-like arms for swimming. } Three families.

Hyalæidæ, Hyalæa, Cleodora.
Limacinidæ, Limacina, Spiralis.
Clionidæ, Clio.

(3). Gasteropods, { Stomach-footed, i. e., Feet on the Stomach. } Divided into

(1). Pulmobranchia (Lung-like Gills). (Four families.)
Limacidæ, (from *Limax*,) Slugs.
Helicidæ, (from *Helix*,) Snails.
Auriculidæ, (from *Auricula*,) Ear-shaped Shells.
Limnæidæ, (" *Limnæa*, Aquatic-Snails.

(2). Pectinibranchia, (Comb-like gills.) (Nine families.)
Trochidæ, (from *Trochus*,) Trochi.
Turbinidæ, (from *Turbo*,) Turbines, Periwinkles.
Muricidæ, (" *Murex*,) Murices.
Strombidæ, (" *Strombus*,) Conch-Shells.
Buccinidæ, (" *Buccinum*,) Harp-Shells, Whelks.
Cypræidæ, (" *Cypræa*,) Cowries.
Conidæ, (from *Conus*,) Cones.
Volutidæ, (from *Voluta*,) Volutes, Olives, Mitres.
Capuloideæ, (from *Capula*,) Cup-shaped Shells.

(3). Branchifers, Sub-class. { Gill-bearing, i. e., breathing by gills. } Seven orders.

Tubulibranchia, (Tubular-gills,) Vermetus, Siliquaria.
Scutibranchia or *Aspidobranchia*, { Gills shielded, by the Shell. } Haliotis.
Cyclobranchia, { Gills circular, i. e., *around* the body of the animal. } Chiton, Limpet.

Tectibranchia,	Covered-gills, i. e., by the mantle.	Bulla or Bubble.
Inferobranchia,	Under-gills, i. e., under the edge of the mantle.	Phyllidia, Diphyllidia.
Nudibranchia,	Naked-gills, i. e., without Shells.	Glaucus, Doris.
Heteropoda,	Other-footed, i. e., feet different from the others.	Carinaria, Firola.

II. ACEPHALA. HEADLESS MOLLUSKS. Four orders.

(1). CONCHIFERA (Shell bearing) or LAMELLIBRANCHIA, (Plate or leaf-like gills.)

(Oyster Family) *Ostraceæ*– sub-families–*Anomiidæ*, Anomia. *Placunidæ*, Placuna. *Ostreidæ*, Ostrea. *Pectinidæ*, Pecten. *Aviculidæ*, Avicula. — Monomyaria or having one muscle.

(Fresh Water Mussels) 4 *Naiades* or *Unionidæ*, Unio, Anodon, Alasmodon.
(Salt-water do.) 4 *Mytilaceæ*, Mytilus, Modiola, Pinna, Crenella.
4 *Chamaceæ*–sub-families, *Tridacnidæ*, Tridacna. *Chamidæ*, Chama.
4 *Cardiaceæ*, Mantle closed behind. Siphons united or distinct. do. *Carditidæ*, Cardita. *Cycladidæ*, Cyclas. *Tellinidæ*, Tellina. *Lucinidæ*, Lucina. *Veneridæ*, Venus. *Crassitellidæ*, Crassitella,
Sub-order INCLUSA (inclosed, i. e., within the mantle, which has but one opening for the passage of the foot.)
Families *Mactridæ*, Mactra.
Myidæ, Mya.
Solemyidæ, Solemya.
Saxicavidæ, Saxicava.
Pandoridæ, Pandora.
Solenidæ, Solen (Razor Shell).
Pholadidæ, Pholas.
Teredinidæ, Teredo (Wood or Ship Worm.
Enclosed in a tube but not attached. *Tubicolidæ*, Aspergillum Gastrochæna. or Watering-pot.
— Dimyaria (or having two muscles.)

(2). BRACHIOPODA, Arm-footed, i. e., having two long spiral arms each side of the mouth capable of protrusion. — Terebratula, Lingula, Orbicula.

(3). TUNICATA, Coated, i. e., body enveloped in an elastic tunic or coat, —including the *Ascidians* (or Mollusks of a Leathern bottle-shape).

(4). Bryozoa, Gr. Moss-animals, i. e., largely aggregated like corallagineous Zoophytes.

(Agassiz proposes the following classification, Contributions to Nat. Hist., Vol. 1, page 185.)

1st Class. Acephala, (orders as already given.)
2d do. Gasteropoda, with three orders, Pteropoda, Heteropoda, and Gasteropoda proper.
3d do. Cephalopoda, with two orders, Tetrabranchiata and Dibranchiata.

RADIATES. Four Classes.

I. ECHINODERMS, (Gr. *Echinos*, the Sea-urchin; *derma* skin.) Four orders.
(1). HOLOTHURIDEA, (Gr. *Holothourion*,) Sea-slugs or Sea-cucumbers.
(2). ECHINIDEA, (Gr. *echinos*). Sea-urchins.
(3). ASTERIDEA (Gr. *astēr*, a Star.) Star-fish.
(4). CRINOIDEA. (Gr. *krinon*, a lily, lily-like). Encrinites.

II. ACALEPHS., (Gr. *akalēphe*, a nettle). Three orders.
(1). PULMONIGRADES, (*pulmo*, lungs; *gradior*, to advance, *i. e.*, contracting or expanding their umbrella-shaped disk, thus showing a resemblance to the motion of the lungs in breathing.
(2.) PHYSOGRADA, (Gr. *phusao*, to inflate; *gradior*, i. e., supported and moving in the water by means of one or more bladders, capable of being filled with air at the will of the animal). Hydrostatic Acalephs of Cuvier.
(3). CILIOGRADA. (*cilia*, vibratile hairs; *gradior*, i. e., moving by means of vibratile cilia disposed on the surface of the body.)

The orders are otherwise named thus:
DISCOPHORA, (Disk-bearing) Medusæ or Jelly-fish.
SIPHONOPHORA, (Siphon or Sucker-bearing, i. e., having aerial vesicles.)
CTENOPHORA, (Comb-bearing, i. e., moving by vibrating hairs resembling the teeth of a comb.)

III. PHYTOZOA or ZOOPHYTA, { (*phuton*, a plant; zōon, animal,) Plant-like animals. Two orders.
Polyps, { ACTINOIDS, (*aktin*, a ray,) Ray-like animals.
Hydroids, (*hudra*, a hydra or water-snake,) Hydra-like [animals.

IV. PROTOZOA, (*proton*, first; zōon, animal: i. e., the lowest form of organized bodies.)

[The last is a very numerous, but a very *uncertain* class. Linnæus placed them all at the end of *Worms*, and called them *Chaos*. So great is the number of the INFUSORIES that they have sometimes been arranged into Legions. Some have been transferred to the Articulates; others have been removed to the Vegetable Kingdom. Prof. Agassiz is of the opinion that the entire class will soon be dispensed with.

NOTE.

An interesting and instructive use of the "Chart of the Animal Kingdom" will be to employ the method of CLASSIFICATION, which it embodies, in tracing an individual of any species, through the successive gradations, to the Sub-Kingdom to which it belongs.

1. In the VERTEBRATES, take, for example, the Common Dog, *Canis familiaris;* and it may be traced as follows: The generic term (which is always placed before the specific, or stands alone when the specific term is omitted) is *Canis; familiaris* is the specific term. Genera are formed into families; the family name is *Canidæ;* families are formed into sub-orders or orders (the orders are in larger or capital letters); *Canidæ* belongs to the sub-order DIGITIGRADA; to the order Carnivora. Orders are formed into classes. CARNIVORA belongs to the sub-class UNGUICULATA; to the class MAMMALS. Classes (denoted by larger letters) are formed into SUB-KINGDOMS. The MAMMALS belong to the Sub-Kingdom VERTEBRATES, denoted by letters next in size to those of the "ANIMAL KINGDOM."

2. In the ARTICULATES, take the Lobster, *Astacus marinus. Marinus* denotes the species; *Astacus*, the genus—of the order (or sub-class) MALACOSTRACA, of the class CRUSTACEA, of the Sub-Kingdom ARTICULATES.

3. In the MOLLUSKS, take the Shell, *Mitra episcopalis. Episcopalis* is the name of the species: *Mitra*, of the genus. This genus belongs to the family *Volutidæ.* The family *Volutidæ* belongs to the order PECTINIBRANCHIA; this order to the class GASTEROPODS; this class to the UNIVALVES, the first grand division of the Sub-Kingdom MOLLUSKS.

4. In the RADIATES, take the Portugese Man of War, *Physalis pelagica.* The generic term is *Physalis;* the specific term, *pelagica; Physalis* belongs to the order SIPHONOPHORI, to the class ACALEPHS, to the Sub-Kingdom RADIATES.

The above are given as specimens in the several sub-kingdoms, showing the manner in which the species named in the Chart, may in conformity with the system of Classification, be followed up to their respective places.

NATURAL HISTORY.

SECTION I.

The science of Natural History is truly vast in its extent, including all bodies found on the earth, or of which its mass is composed. Its most general divisions are Mineralogy, Botany and Zoology. These divisions are founded upon the different and distinguishing characters and states of the various objects which they respectively include. Minerals are *inorganic bodies* ; they are without life, and incapable of increase or diminution except by means of some force outwardly applied. These are earth, rock, metals, &c. Organic bodies are divided into animate and inanimate. The former comprehend substances endowed with sense and motion and belong to the department of Zoology ; the latter are without the faculties of sense and motion, and included in Botany. Organized beings, whether animate or inanimate, differ from inorganic ones in having the power of reproduction, or continuing the existence of beings like themselves. Animals derive their nourishment either directly or indirectly from vegetables, of which hydrogen and carbon are the principal ingredients. The latter derive their nourishment from the soils of the earth and from the atmosphere.

In the survey of objects so numerous and possessing such varied characteristics as those of Natural History, *classification* is obviously of high importance. A union of several traits is almost always required to distinguish a single being from others around it which have some, but not all of the same traits, or have them in combination with others of which that single being is destitute. In the work of classification a number of neighboring beings are compared with each other; and their differences, which are supposed to be the least part of their formation, are made indexes of their character. The union formed by the comparison of objects which agree, but with certain differences, is called a *genus*; a union with fewer differences is called a *species*. Genera are formed into *orders*, and orders into *classes*.

The CHART of which this volume is explanatory, exhibits the "Animal Kingdom" by means of a Tree having four branches, each representing one of the four sub-kingdoms into which it is divided, viz., VERTEBRATES, ARTICULATES, MOLLUSKS and RADIATES. Each branch puts forth other branches bearing subdivisions—classes, orders, families, genera, &c., illustrated by numerous and appropriate figures, and so variously lettered and marked as to be easily distinguished. It was prepared with great labor, and in the use of much research, in order to facilitate acquisitions in the department of physical science which it delineates and with the hope of thus encouraging a more general introduction of the Study of Natural History into our Seminaries of learning, from the Common School to the College and University. "Man," said Lord Bacon, "is the minister and interpreter of NATURE."

More attention should be given in the domestic circle, and in the various schools of instruction to the business of training the young to be *observers* of nature. A fondness for the lessons and researches of natural history, implanted in the mind during the period of youth, will, in all probability, last through life, affecting favorably the entire mental development.

None should neglect the investigations to which by the "View of the Animal Kingdom," they are invited. Such investigations, it should be remembered, pertain neither to fiction nor hypothesis—but to *realities*. They seem specially adapted to man's endowments in his present state of existence; but the facts and impressions which he derived from an earnest contemplation of the works of God, memory will embalm and render immortal. "And as now the memory of home is pleasurable in proportion to the vividness and distinctness of its image; as we now attach importance to the most insignificant object around the place of our birth; as we regard with intense interest the old elm, the green lawn, the hawthorn bush, the rivulet because they are inseparably connected with our developments of *mind*, even so perhaps may we then, after millions of ages shall have elapsed, recall with increasing pleasure the physical scenery of this birth-place of our existence."

QUESTIONS ON SECTION I.

What does the science of Natural History include? What are its general divisions? What are minerals? How are organic bodies divided? Which belong to Zoology? Which to Botany? How do organic bodies

differ from inorganic? From what do animals derive their nourishment? Of what do vegetables principally consist? From what do they derive their nourishment? What is necessary to distinguish one being from another? How do you proceed in classifying objects? What is a genus? What is a species? Of what are orders and classes formed? What is the definition of genus and species at the bottom of the chart on the left hand? What are minuter differences called? Answer. Varieties. What does a generic name signify or comprehend? Ans. It comprehends all the species; *Canis*, for example, is the generic name of animals of the Dog kind, including the Fox (*Canis Vulpes*,) the Wolf (*C. Lupus*,) the Jackal (*C. aureus*,) and the domestic Dog (*C. familiaris*.) How are generic terms printed on the chart? Ans. Always larger than the common name by which the animal is known, and commencing with a capital letter. How do you distinguish the specific from the generic name? Ans. It follows the generic term in letters of the same size, and should not commence with a capital, unless it is derived from some person or place, or is sometimes used in a generic sense. Why is the name of the species often omitted on the chart? Ans. For want of room, and fear of confusing the student by crowding too much in a small space. How are the families distinguished on the chart? Ans. By their terminating in *idae*, as *mustelidae* for the Weasel Tribe, or Family. How can you distinguish the orders? Ans. They are printed in CAPITALS, and the number of orders is mentioned on the branch, as in the *Ungulata*, or hoofed *Mammals*. Are there any other divisions or distinctions on the chart? Ans. Several, as among the cud chewing some have solid horns, some are hollow, and some are entirely without horns; some shed them annually as in the deer, in others they are permanent, as in the ox or sheep. Some birds are terrestrial, others aquatic; some insects and reptiles are venomous (poisonous;) others are non-venomous, or harmless. Wherever there is room, you will find these things noticed on the branches, or as near the classes, orders or figures as practicable. Dots are often added to make the connection or relation still plainer; and where there is but small space allotted to explanation or figures, the deficiency will be remedied as we proceed. How many ranks, or grades of groups does Swainson enumerate? Ans. Nine, commencing with the highest, and terminating with the lowest assemblages. 1. Kingdom; 2. Subkingdom; 3. Class; 4. Order; 5. Tribe; 6. Family; 7. Sub-family; 8. Genus; 9. Sub-genus. Name the four great Classes, or Sub-Kingdoms from the chart.

SECTION II.

THE ANIMAL KINGDOM.

THE system of Zoology places MAN at the head of this Kingdom. As he is endowed with intellectual and moral faculties, and fitted for responsible action, there is room for doubt whether, in his pre-eminence, he should have a place among the tribes of animals. But as his being is compound, he becomes the connecting link between them and beings purely spiritual. To the former he is allied by his bodily frame with its appetites and passions; to the latter by his reason and mental susceptibilities. INSTINCT distinguishes the lower animals—truly wonderful in some of its actings as will be shown hereafter; but yet only a mere internal impulse, and incapable of improvement. The bird shows it in building its nest; the bee in constructing its cells; but both the nest and comb are made as skillfully at the first as in any subsequent trial.

There seems no occasion to mistake by referring to mineralogy or botany what properly belongs to the Animal Kingdom; and yet in such animals as the *oyster* we discern but little of the sensibility and capacity for voluntary motion which are usually adduced as characteristics of the animal tribes.

Chemistry has ascertained that the substances found both in animals and vegetables are chiefly formed of four elements, viz., carbon, hydrogen, oxygen and nitrogen. These have, therefore, been called *organic elements*. The opposite and distinctive natures of plants and animals may be seen in the functions which they perform dependently one on another. In animal respiration, the oxygen of the atmosphere is combined with the blood, forming carbonic acid gas, which is thrown off from the entire surface of the body in some animals; from the gills of those that live in water, and the lungs of those that live in air. Animals thus consume oxygen—to them it is *pabulum vitae*—the food of life. Plants, on the contrary, consume carbonic acid and give off oxygen. They thus become able to furnish animals with carbon. Animals, in their turn furnish food to plants. The excretions which they throw off, yield ammonia (consisting of hydrogen and nitrogen,) from which substance vegetables principally derive their nitrogen. The animal derives the constituents of its body from the vegetable kingdom; the plant obtains its elements from the mineral kingdom. The tissues of the plant change mineral into organic substances; those of the animal change organic substances into mineral.

A further contrast between plants and animals is presented in the effects produced upon them, respectively, by light and heat. Both of these are indispensable to the proper growth of plants. The productions found in their tissues are but the expression of the light and heat they have, as it were, appropriated. Many of the substances in this way formed, are taken as food into the systems of animals; but in them are again set free in the form of "vital animal forces."

Differences of structure also constitute an important ground of distinction between the animal and vegetable kingdoms; yet, sometimes, as in the sponge, it is only by considering to which there is the greatest *general* resemblance, it can be decided whether a particular being should be classed as an animal or vegetable.

The different methods by which they receive food, and assimilate it or convert it into their own substance, form another distinction between animals and plants. Vegetables imbibe their nourishment through their outward surface, or through their roots and leaves; but animals, for the most part, have a *stomach*, or internal cavity, into which the food is received, where it is digested, and by appropriate vessels, absorbed into the body.

The food of animals is generally in a solid state, and must be rendered fluid before it can be formed into the tissues. Taken at intervals, and stored in the stomach, it does not hinder their movements from place to place. During the intervals of its reception, it is kept in contact with the absorbent vessels. Hence, animals are said to "bear their soil about with them." The earth is called "the stomach of plants."

The *habits* and *instincts* of animals must also be considered by the zoologist in making up the account of the differences between them and plants. This is a field which affords a wide scope for comparison and research in tracing analogies between objects in many respects diverse, and one which teaches many lessons concerning the Divine wisdom and benevolence.

The chart of "the Animal Kingdom" presents a view of that branch of Natural History which is called Zoology, a term derived from the Greek *Zôon*, an animal, and *logos*, a discourse. This includes nine divisions, viz.; I. Mammalogy, which treats of the *Mammalia*, or animals that nurse their young; II. Ornithology, which relates to Birds; III. Erpetology, which includes the Natural History of Reptiles; IV. Ichthyology, which gives the Natural History of Fishes; V. Entomology, which gives the Natural History of Insects; VI. Crustaceology, which treats of Crabs, Lobsters, &c.; VII. Helminthology, which treats of Worms; VIII. Malacology, which includes Conchology, and describes soft-bodied animals, with and without shells; IX.* Actinology, which treats of radiate animals, as the Star-fish, Sea-Anemone, &c. The Animal Kingdom is divided, as on the chart, into four sub-kingdoms, viz.: Vertebrates, Articulates, Mollusks, and Radiates.

* We have ventured to introduce this new term, formed from the Greek word *aktin*, a ray, (corresponding with the Latin radius,) and *logos*, a discourse, in order to have the names of the several branches alike as to their termination and Greek derivation, though the terms actinia and actiniadæ, (generic and family,) refer distinctively to the Sea-Anemones.

The VERTEBRATES, (from the Latin *vertebra*, a joint, which comes from *vertĕre*, to turn,) have a jointed backbone, or internal bony skeleton. They are divided into WARM and COLD BLOODED; the former, including Mammals, (*Mammalia*,) and Birds, (*Aves*;) the latter, Reptiles, (*Reptilia*,) and Fishes, *Pisces*.) The Whale tribe. (*Cetacea*,) inhabiting the sea, form one order of the Mammalia.

ARTICULATES, (from the Latin *articulus*, a ring or joint,) are animals in which the body and legs are jointed, and the hardest parts are outside. These are arranged into three classes, viz.: Insects, Crustaceans, and Worms.

MOLLUSKS, (from the Latin *mollis*, soft,) are shell-fish whose nervous system is composed of several scattered masses, or ganglions, united by means of nervous threads, and whose soft bodies are generally protected by a shell.

RADIATES, (from the Latin *radius*, a ray,) are animals whose parts are disposed in the form of *rays*, tending to a common center, where the mouth is placed, as in the Star-fish.

QUESTIONS ON SECTION 2.

Who is placed at the head of the Animal Kingdom? With what is he endowed? For what is he fitted? What does his compound being constitute him? How is he allied to animals? How to spiritual beings? What guides the lower animals instead of reason? Does the bird or bee construct its last nest or comb with more skill than the first? Is there any need of mistake in referring to Mineralogy or Botany, what properly belongs to the Animal Kingdom? How is it with the Oyster? What are the four elements both in vegetables and animals? What name is given to these elements? What shows the opposite natures of plants and animals? When animals breathe, what is combined with the blood? What gas is thus formed? How is this thrown off in some animals? How in others? What is oxygen called? On what do plants live? What do they give off? What do they furnish to animals? What do animals furnish plants? What is obtained from animal excretions? What do vegetables derive from it? Whence does an animal derive the constituents of its body, and whence the plant its elementary ingredients? What is a further source of contrast between plants and animals? What additional ground of distinction is there between the animal and vegetable kingdoms? In some cases, how is it determined to which of the two a particular being belongs? What further distinction between plants and animals is referred to? How do vegetables take in their nourishment? How animals? What is said about the food of animals? What are animals said to do? What has the earth been called? What is said of the habits and instincts of animals as relates to the differences between them and plants? What benefits flow from tracing the analogies between animals and plants? Is this a wide field and what does it teach?

What does the Chart present? From what is the term ZOOLOGY derived? Of which of the three kingdoms of nature is this Chart a general view?

PL. II.

VERTEBRATES.

ARTICULATES.

MOLLUSKS.

RADIATES.

EXPLANATION OF PLATE II.

VERTEBRATES, ARTICULATES, MOLLUSKS AND RADIATES.

VERTEBRATES.

1. *Homo sapiens*, Man.
2. *Cebus*, Monkey.
3. *Camelus Dromedarius*, Dromedary.
4. *Avis*, Bird.
5. *Ciconia Alba*, White Stork.
6. *Pisces*, Fishes.
7. *Ophis*, Snake.
8. *Rana pipiens*, Bull-frog.
9. *Alligator Lucius*, Alligator.

ARTICULATES.

1. *Astacus marinus*, Lobster.
2. *Papilio*, Butterfly.
3. *Culex pipiens*, Mosquitoe.
4. *Musca domestica*, Common House Fly.
5. Larva, or Caterpillar of a Moth or Butterfly.
6. *Tettigonia verrucivora*, Spotted Grasshopper of Europe.
7. *Clerus apiarius*, Hive Beetle.
8. *Lucanus cervus*, Stag Beetle.

MOLLUSKS.

1. *Buccinum*, Whelk.
2. *Mitra Episcopalis*, Bishop's Mitre.
3. *Tridacna gigas*, Giant Tridacna.
4. *Planorbis*, Coil-shell.
5. *Siliquaria.*
6. *Nautilus umbiculatus*, Umbilicated Nautilus.
7. *Loligo vulgaris*, Common Calamary.
8. *Triton variegatus*, Variegated Triton.
9. *Physa fontinalis*, Bubble-shell.

RADIATES.

Fig. 1. *Corallum rubrum*, Red Coral.
2. *Apiocrinites rotundus.*
3. *Edwardsia vestita.*
4. *Dianæa*, a Jelly-fish, or Medusa.
5. *Tima flavilabris*, Jelly-fish.
6. *Asterias*, Star-fish.
7. *Zoanthus Solanderi*, Animal Flower, or Zoophyte.
8. *Astræa ananas*, Pine-apple Coral.

How many divisions does it include? Of what does Mammalogy treat? To what does Ornithology relate? What does Erpetology include? What does Ichthyology give? What science treats of Insects? What of Crabs, Lobsters, and Barnacles? Of what does Helminthology treat? What does Malacology include and describe? Of what does Actinology treat?

QUESTIONS ON THE CHART.

How is the Animal Kingdom divided on the Chart? To which of these four great Classes, or Sub-kingdoms, do the first four of the above nine divisions belong? Point out each division of this right hand branch. Give the name of the science pertaining to or describing each. In what particular do they all agree? Ans. In having a backbone, or spinal column. Define vertebra and give its derivation. Which are warm blooded? Which are cold blooded? How cold or warm are they? How many orders of Reptiles? How many of Fishes? How many of Mammals? Which order ranks first, and is far above all others? What is said of man, near the bottom on the right hand of the chart? What is said of his brain? What of his birth? What of his wants? How does he compare with others in regard to strength, speed, &c.? Is his reason an improvable gift? Does it supply the place of strength? What order comes next to man? How do the Quadrupeds differ from QUADRUMANA? What marine animals belong to the class MAMMALIA? In what element do they live? With what organs do they move? Is the largest living animal found in this class? What is its name and what are its uses? Which of the VERTEBRATES live in the water? Which on land? Which in the trees? Which fly? Which swim? Which crawl? Which are covered with feathers? Which with hair? Which with scales? Which are born alive, (viviparous?) Which hatched from eggs, (oviparous?) Which are entirely without limbs? Which have but two?

In which Sub-kingdom, or on what branch do you find Insects, Crustaceans, and Worms or Annelidans? From what is the name Crustaceans derived? In what do they resemble one another? Have they any internal skeleton? Where are the hardest parts? Which is the largest of all articulated animals? Ans. Lobsters. Name some of the worms on the chart. Of what use is the leech? Of what use is the earth or angle worm, (Lumbricus terrestris?) Ans. This despised creature is of great use in loosening the earth, so that air and water can pass through it freely, and in covering barren tracts of land with their worm casts, thus rendering them productive. Mention some of the Insects and Crustaceans. Name the sciences describing them. Are Insects a numerous class? Ans. They outnumber all other classes together. There are 80,000 species of the beetles alone, (order *Coleoptera.*) Here you find the *Curculio*, or weevil, deathwatch, lightning-bug, horn-bugs, &c., &c.

From what is the name of the third branch, (Mollusks,) derived? How are these soft bodies protected? How are Mollusks divided? Which have heads? Which none? To which division do snails and slugs belong? On which branch do you find Oysters and Clams? Which move about, (are free?) Which are fixed, (stationary?) Is the Oyster always attached to other substances? Ans. No. Which branch of the Mollusks are entirely aquatic, or never leave the water? Are the TUNICATA, or ASCIDIANS protected by shells? Name from the chart the largest genus of known shells. Is it a bivalve, (of two pieces,) or a univalve, (of one piece?)

Which is the fourth, last and lowest branch of the Animal Kingdom? Define Radiate. How are the parts disposed? Where is the mouth? From what is the term derived? Are they aquatic? What is said of these animals near the bottom of the chart, on the left hand? Are they less perfect of their kind than those on the right branch? Why, then, are they said to be the lowest in the scale of animal life? Which animals are always lowest in organization in the class, division, or order to which they belong? Which rank next in the ascending scale? Which rank highest of all? Which is the lowest order of land animals which nurse their young? Ans. The MONOTREMATA and MARSUPIALS. Why are the branches of the orders Marsupialia and Rodentia, (gnawers,) bent and carried around next the marine mammals? Ans. To show that though having nails, they come next the order *Cetacea*, (Whales, Dolphins, &c.,) in organization. Which is the lowest or most simply organized class of animals? How many orders does it contain? What does proto signify? Is it a well established class? What is said of it? What animals are found on the chart among the RADIATES? Which are microscopic? Which fossil? Which used as food? What is said of Sponge? Where does Agassiz class it? Are Animalcules, Infusories, and Microscopic or very minute animals common? Ans. They are dispersed like seed through all nature. Are Animalcules tenacious of life? Ans. It is so difficult to kill them that they can be repeatedly dried and kept for a long time, and will revive or come again to life, as soon as put into water.

FIRST BRANCH OF ZOOLOGY.

MAMMALOGY, (Gr. *μαμμα*, *mamma*, a breast; *λόγος*, *logos*, a discourse.)

I. GRAND DIVISION OF VERTEBRATES, (Warm-Blooded Animals.)

SECTION III.

VERTEBRATES.

(Lat. Vertebrata, possessing Vertebræ, or joints in the backbone.)

The first class of the Vertebrates consists of the Mammals, or *Mammalia*, (Gr. *Mamma*, a breast,) a term first used by Linnæus and designating all animals which nurse their young. The highest position in the Animal Kingdom is given to this class, composed as it is of beings whose faculties are the most numerous, which are most perfect in their structure and capable of the most varied movements, and whose intelligence is most largely developed. A large part of the Mammals are formed for walking: some can fly in the air, and water is the element in which others live and move. Their skeletons are all constructed after the same general plan, changed, however, and modified in certain parts or organs, to fit them for the stations which they are designed to occupy. (See Plates III. and XII.) All of them are viviparous, (born alive.) The young, as the name of the class denotes, are, for a longer or shorter time, nourished by the milk of the mother. Sometimes they are born with their eyes open, and able immediately to move about and seek their own food; but not a few of them are born with their eyes closed, and in a state of extreme helplessness.

The leading characters of the Mammalia are founded on the number and kind of their teeth, (see Plate IV.) and the construction of their hands and feet. (See Plates III. and VI.) The expertness of these animals is closely connected with the perfection of the organs of touch. The nature of their food and their digestive functions may, in great part, be inferred from the number and structure of their teeth. (See Plates III. and VI.)

They are divided into three sub-classes, viz.: UNGULATA, (lat. *ungula*, a hoof,) animals with hoofs; and UNGUICULATA, (lat. *Unguiculus*, a soft, small nail,) animals with nails or claws; and CETACEA, with fins, (Gr. *Ketos*,) a whale, or sea monster.

The Mammals are, (on the Chart,) arranged into nine orders, after the plan of Cuvier, that arrangement being deemed, on the whole, the most satisfactory The number of well established species, according to Dr. Hitchcock, is somewhat more than 2000.

The names of the nine orders are, I. BIMANA; II. QUADRUMANA; III. CARNIVORA; IV. MARSUPIALIA; V. EDENTATA; VI. RODENTIA; VII. PACHYDERMATA; VIII. RUMINANTIA; IX. CETACEA.

Some naturalists have elevated the CHEIROPTERA, the INSECTIVORA, and the MONOTREMATA to the rank of orders, making the number XII; but the first two of these are flesh-eaters, and therefore properly included among the *Carnivora*, (or the Carnassiers of Cuvier;) and the MONOTREMES, including but two genera, have such points of resemblance to the MARSUPIALIA, as justify referring them to that order.

QUESTIONS ON THE VERTEBRATES.

What is the first class of Vertebrates? Who first used the term? Who was Linnæus? Ans. An eminent Swedish naturalist. He was the author of the Linnæan, or artificial system of Botany. What does the term Mammals, or Mammalia designate? What position in the ANIMAL KINGDOM does this class occupy? Of what beings is it composed? For what are a large part of the Mammals formed? How do others of this class live and move? What is said of their skeletons? Are all able at first to move about, use their eyes, and seek their own food? Upon what are the prominent characters of the MAMMALIA founded? What distinguishes the three sub-classes into which all Mammals are divided? Spell, define and give the derivation of these words. Which have nails? Which hoofs? Which fins? Which have hair? Which live on land? Which in the water? Under how many orders are the Mammals on the chart arranged? Whose arrangement is this, and why adopted? Who was Cuvier? Ans. An eminent French naturalist who could, like Prof. Owen, of England, describe an animal by seeing a single bone, and the nature of its food, by looking at its teeth, or examining its intestines. Name the nine orders from the chart, giving examples of each. Read the explanations along the sides of the branches and limbs, as you trace them up from the root or foundation of the tree. To what rank have some naturalists elevated the CHEIROPTERA, INSECTIVORA, and MONOTREMATA? What animals on the chart belong to these sub-orders? What reason is assigned for giving them this rank?

SECTION IV.

FIRST SUB-CLASS. UNGUICULATA.

FIRST ORDER. BIMANA, (Lat. *bis*, twice; *manus*, hand; two-handed.)

MAN fills the first place in the animal series. In reality, he stands alone, sole order, genus and species. His full zoological relations are: class, VERTEBRATA; order, MAMMALIA; genus HOMO; species, SAPIENS. The position at the head of the Animal Kingdom, given to man by the great body of zoologists, is, however, objected to by some eminent naturalists, "who are not disposed to admit that because he possesses certain zoological characters which are entirely secondary and subordinate, he should be classed with brutes, when his noblest attribute, reason, destroys every vestige of affinity, and places him immeasurably above them all."*

The most prominent of the *characters* by which man is distinguished from the lower animals, are as follows:

Rational; endowed with speech; able to walk erect, two handed; having a prominent chin; four incisor (cutting) teeth above and below; and all the teeth side by side; the canine (eye) teeth of the same length as the others; the lower cutting teeth erect; a peculiar relative proportion of the thighs and arms, and wide soles to the feet.

Considering him in his higher or spiritual nature, we may name his sentiments, feelings, sympathies, internal consciousness and purposes; and the courses of action thence resulting as among his proper and essential characteristics. Even physically, he is *first* of all the living creatures on earth; not, however, in size, or in animal strength, in which respects many of the Ver-

* Zoology of New York, by Dr. De Kay.

tebrates excel him,—but in the *plan* or *model* after which he is constructed

The eagle, for example, has a more powerful vision; the hare is more keenly sensible to sound; the dog and vulture are more ready to catch the scent which is borne upon the breeze; but in man is found a nice adjustment, a "peculiar and felicitous accuracy" of the senses, which, while ministering to his enjoyment, enables him to cultivate a more thorough and pleasing acquaintance with the objects by which he is surrounded. In the power of speech, and the various exercises of this power by which he makes known his wants, his desires, and his most abstract mental conceptions; in his processes of reasoning and in his susceptibility of endlessly progressive improvement, he rises high above every other animal existence.

The several parts of the living human frame are suited to the *erect attitude* for which it is distinguished. (See Plate III.) Man's structure fits him for moving in an erect posture, and unfits him for moving with ease *in any other*. He has, however, the ability to *imitate* almost every motion but that of flight. As aids to such imitation, he possesses, when in maturity and health, sixty bones in his head, sixty in his thighs and legs, sixty-two in his arms and hands, and sixty-seven in his trunk, and he has also four hundred and thirty-four muscles. His foot is, in proportion to his whole body, larger, broader, and stronger than that of any other animal. The muscle called "*flexor longus policis pedis*," (the muscle of the great toe,) terminates in a *single* tendon, and its force is centered in the great toe, the chief point of resistance in raising the body upon the heel. In the Orang-outang, the corresponding muscle terminates in three tendons, separately and exclusively inserted in the three middle toes, to enable him to grasp an object more forcibly in climbing, and thus more fully meeting the wants of an animal that makes its home in the trees. "Surely," says Professor Owen, "it is asking too much to believe that in the course of time, these *three* muscles should, under any circumstances, become consolidated into *one*, and that one implanted in a toe to which *none* of the three tendons were before attached." The teeth, bones and muscles of the monkey decisively forbid the conclusion that he could by any ordinary natural process, ever be expanded into a MAN. Man alone is two handed; in him the faculty of opposing the thumb to the other fingers is carried to the highest perfection. In his "Bridgewater Treatise," Sir Charles Bell says: "The structure of the human hand is so much more complicated, and suited to so many different offices, we ought to define the hand as belonging

Pl. III.

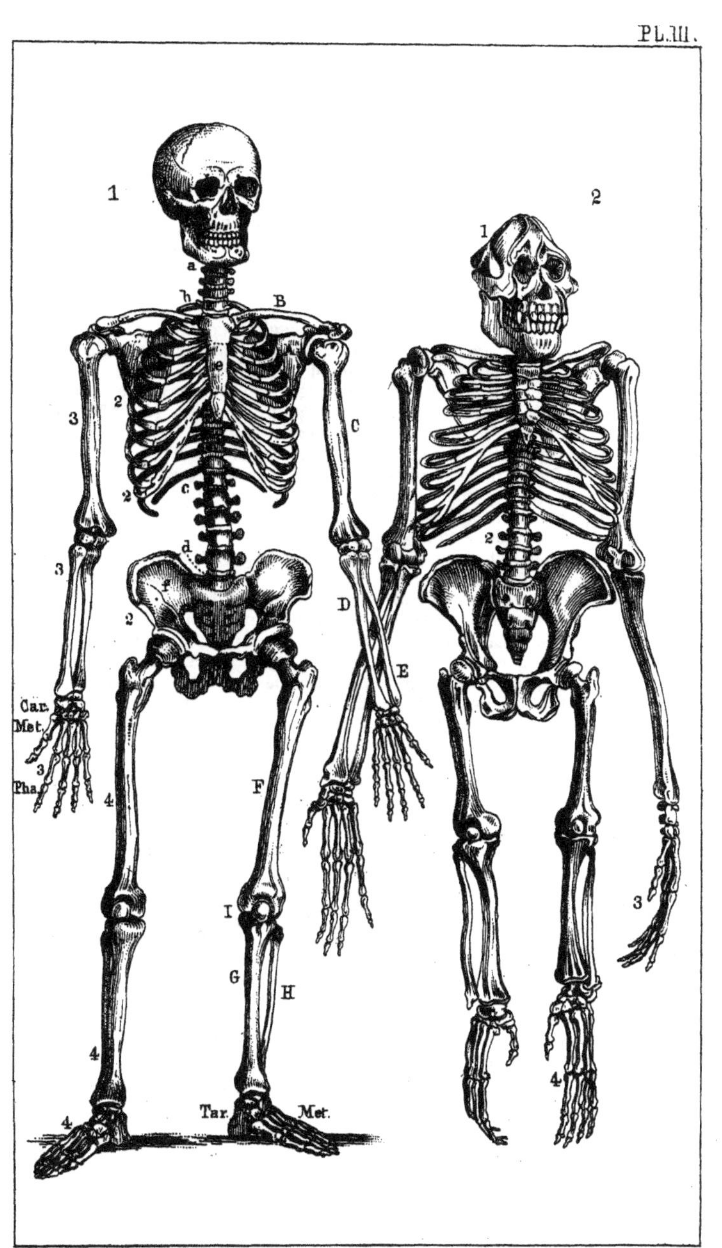

EXPLANATION OF PLATE III.

Fig. 1. The HUMAN SKELETON divided into three principal parts; the Head (1,) the Trunk (2,) and the extremities (3 and 4.) Physiologists enumerate as many as 260 bones; but some of these bones, which are separated in early life, are afterwards united, so as to admit of the following enumeration: Cranium, 8; Face, 14; Internal ears, 8; Vertebral column, 24; Chest, 26; Pelvis, 11; Upper extremities, 68; Lower extremities, 64; in the whole, 223.

1. The bones of the Skull, divided into two sets, viz., those of the Cranium, or case for the brain, and those of the Face.

2. The Trunk, composed of the Spine, or Vertebral column, extending from *a* to *d*, the Chest, including the Ribs, and Sternum or Breast-bone, (*e;*) the Pelvis; the circle of bones on which the Spine rests.

The Spine, extending from *a* to *d*, in the erect man, supports the head upon its summit, (a,) while its base rests upon the sacrum (d.) It consists of 24 bones, called Vertebræ, (Lat. *verto*, to turn,) because the trunk is turned by their motion upon each other. It is the center about which the limbs move, and the chief support of the skeleton.

The Cervical vertebræ, (the 7 bones of the neck,) extend from *a* to *b;* the middle, dorsal or back vertebræ, from *b* to *c*, and the 5 lowest or lumbar vertebræ, from *c* to *d*.

3. and 4. Are the last main divisions, consisting of the upper and lower extremities.

3. The upper extremities (the arms) consist of the scapula, A, or shoulder-blade, the Clavicle or collar-bone, B, the Humerus, or bone of the upper arm, (C,) the Ulna, (D,) situated on the inner side, and the Radius, (E,) on the outer side of the fore-arm, the Carpus, (Car;) the 8 small bones of the wrist, the 5 bones of the metacarpus between the wrist, and the bones of the fingers, (Met.,) and the bones of the fingers, called Phalanges, (Pha.,) of which the thumb has two, and the fingers three each.

4. The lower extremities, or legs, consist of the Femur or thigh bone, (F,) which is the largest bone of the body, the Tibia or shin-bone, (G,) on the front and inner part, and the Fibula, (H,) at the outer part of the leg, the Patella or knee-pan, (I,) the Tarsus, the 7 bones forming the heel and instep, (Tar.,) the metatarsus (Met.,) between the instep and the toes, and Phalanges of the toes similar in number and arrangement to those of the fingers.

Fig. 2. SKELETON OF A CHIMPANZEE. The ape that comes nearest to man. 3 and 4 show how the extremities terminating with long fingers, and a small feeble thumb set far back, adapt it for climbing rather than walking, thus differing from those organs in man.

1. The Cranium,—showing none of the fine sweep of the forehead seen in man, and indicating a small cerebral development as compared with him.

2. The Vertebral column, without the pyramidal form seen in man, and not adapted to an erect posture.

Pelvis, narrow as compared with that of man. (See description in the text.)

exclusively to man. The whole frame conforms to the hand, and acts with reference to it." The human hand is not only powerful, but exquisitely susceptible of impressions, and possesses the most delicate touch. Every finger, except the one called the ring finger, is capable of independent movements,—a power possessed by no other mammal. The thumb is lengthened so as to meet readily the tips of any of the fingers; the fingers themselves, and especially the pulpy tip at their ends, are supplied with a nervous tissue endowed with a discriminating sensibility that is peculiar to man.

"The difference in the *length* of the fingers serves a thousand purposes, adapting the hand and fingers, as in holding a rod, a switch, a sword, a hammer, a pen or pencil, engraving tool, etc., in all which a secure hold and freedom of motion are admirably combined. Nothing is more remarkable, as forming a part of the prospective design to prepare an instrument fitted for the various uses of the human hand, than the manner in which the delicate and moving apparatus of the palm and fingers is guarded. The power with which the hand grasps, as when a sailor lays hold to raise his body to the rigging, would be too great for the texture of mere tendons, nerves and vessels; they would be crushed were not every part that bears the pressure defended with a cushion of fat as elastic as that we have described in the foot of the horse and camel. To add to this purely passive defence, there is a muscle which runs across the palm, and more especially supports the cushion on its inner edge. It is this muscle which, raising the edge of the palm, adapts it to lave water, forming the cup of Diogenes."*

The brain of man, in proportion to the residue of the human system, surpasses in volume or extent that of every other mammal, as is shown by the proportion which the cavities containing the brain and face bear to each other. The size of the brain is sometimes estimated by the facial angle,† which, in the average of Europeans and their descendants on this continent, is 80°; but in the adult Chimpanzee is only 35°, and in the Orang or Satyr is, according to Professor Owen, 30°.

The blood necessary for an organ so developed as the human brain, is carried to it by arteries which do not subdivide as in

---

* Sir C. Bell's Bridgewater Treatise on the Hand.

"† The facial angle is found by drawing a line from the most prominent part of the forehead to that of the upper jaw bone, and observing the angle which it forms with another line through the external auditory canal to the base of the nose, or, (the head being in a vertical position,) with a horizontal line."

most quadrupeds, but allow of the full and free circulation which its energies require.

The fine sweep of cranium and the smooth spherical surface of the human skull, showing the volume of the interior brain, are also noticeable, as contrasting strikingly with the heavy ridges, the irregular prominences and the small capacity of the Monkey's skull. The face of the Monkey is an aid to him in procuring food, and a weapon for attack and defence; Man's face bespeaks the workings of the inner MIND. He uses his *hands* to procure his food, and naturally unarmed, protects himself with weapons which he has *manufactured*. His jaws and teeth are both as small as could consist with the preservation of life. Though at first weak and defenceless, he becomes able not only to assert his dominion over animated nature, but to make the very elements subserve his designs. No monkey or ape has ever been able to make weapons of either attack or defence; nor can he procure fire or renew it, which the lowest of the human species readily does. The most benighted Hottentot can form weapons with which he is able to destroy the ferocious lion, the swift antelope, and the wary ostrich; "he constructs for himself a hut by the side of his prey, strikes fire, fetches fuel, and dresses his meat." There seems, as Buffon has intimated, no anatomical reason why an ape should not speak; but it has no language, and cannot by the most patient labor, be taught to speak. *Articulate language,* of itself, makes a difference, vast in extent, between man and every other tribe of the *Mammalia.*

His physical system is peculiar in the readiness with which it accommodates itself to the variations of climate, and in modes of living. The Arctic explorations of Captains Ross and Parry, of Sir John Franklin, and of our own lamented Dr. Kane, have signally evinced the capacity of the human constitution for enduring with safety, the intensest cold. On the other hand, men long accustomed to the air of the temperate zones, have penetrated far into the interior of Africa, and traversed other equatorial regions, without experiencing any serious evils from the heat.

QUESTIONS ON THE ORDER BIMANA.

What is the first order? How is it spelled, defined, and from what derived? Who is at the head, or fills the first place in the animal series? What is said of him, and to what class, order, genus, and species does he belong? Are all Zoologists agreed as to the propriety of placing man with animals? What places him immeasurably above them all? What are his most prominent distinctions, or what is said of his speech, walk, chin, teeth, &c.? Contrast these with those of the inferior animals. What is said of man, physically? In what respect does he surpass all other created beings?

In what senses is he inferior, or in what way does the eagle, hare, dog, or vulture surpass him? What is found in man? What does this enable him to cultivate? What elevates him so highly above other animal existences? To what are the several parts of the human frame suited? For what does a man's structure fit him? For what does it unfit him? Has he the power of imitation? What aids this power or faculty? How many bones and muscles has he? What is said of his foot? What is said of the muscle of the great toe in man? Give its technical name. What of the corresponding muscle in the Orang Outang? What does Prof. Owen say in relation to this, and how does this bear upon the *development theory?* What do the teeth, bones, and muscles of the monkey forbid? What is said of the hands, thumbs, and fingers of man? What does Sir Charles Bell say in his "Bridgewater Treatise?" What is further said of the human hand? What of the ring finger? What of the thumb and other fingers? Of what use is the different length of the fingers? Does it evince design, or did it occur by chance? What is chance? What is the cup of Diogenes; and how is it formed? Who was Diogenes? Ans. A celebrated Cynic philosopher, of Greece, who died in great misery and indigence, B. C. 324, at the age of 96. What is said of the human brain? How is this shown? How is the size of the brain sometimes estimated? How is this angle found? What is said of the arteries supplying blood to the human brain? How does the cranium, or human skull, contrast with that of the monkey? What is said of the monkey's face? What of man's? Which bespeaks the most intelligence? For what does he use his hands? How does he protect himself? What is said of his jaws and teeth? What are monkeys unable to do? By whom are they surpassed? Is there any anatomical reason why an ape should not speak? Have they ever been taught to speak? What makes a vast difference between man and all other mammals? In what is man's physical system peculiar? What have Arctic and African explorations shown?

VARIETIES OF THE HUMAN RACE.

The variations of mankind, in respect to climate and modes of life, are connected with changes in complexion and feature, with differences in the skull, in the color and nature of the hair, etc. The divisions of the race to which these differences have given rise, are stated diversely by naturalists, some numbering more, and others fewer varieties. The Caucasian, Mongolian, and Nigritian tribes, are by some regarded as the three distinctly marked *types;* and the other varieties as but a blending of these and their peculiarities, and hence merely sub-typical.

The "Chart of the Animal Kingdom" exhibits the division of Blumenbach, the one which has commonly been made, which, separating the Malay and American varieties from the Mongolian, one of the distinctly marked types, makes the number FIVE, viz.: 1. THE EUROPEAN OR CAUCASIAN; 2. THE ASIATIC, MONGOLIAN, OR TURANIAN, of Dr. Pritchard; 3. THE MALAY OR AUSTRALIAN; 4. THE AMERICAN; 5. THE ETHIOPIAN OR AFRICAN.

1. THE CAUCASIAN VARIETY was so called because it origi-

nated among the tribes of men found in the region of the Caucasus. It is distinguished for general symmetry and regularity of outline. The head is, in the Caucasian, almost round, the face oval, the forehead much expanded, the features not very prominent. The skin is white, the hair soft, long and brown, more or less dark, and curled. The facial angle is from 80° to 90°. The entire conformation of the head shows a superior intellectual organization. In respect both to mental power, and attainments in art and science, the Caucasians have ever stood in the foremost rank.

2. THE ASIATIC OR MONGOLIAN VARIETY.—This variety is remarkable for a feminine aspect in both sexes; the color is, for the most part, pale yellow or olive; the head almost square; the facial angle 80°; the cheek bones are prominent; the face broad and flattened, and without a beard; and the hair straight and black.

3. In the MALAY or AUSTRALIAN, the color varies from a clear mahogany to dark chestnut brown; the hair is black and bushy; the beard thin; the nose broad, and the mouth wide; the forehead slightly arched; the upper jaw projecting; the eye is more sunken and piercing, and the lips less uniformly thick than in the negro.

4. THE AMERICAN VARIETY is allied to the Malay and Mongolian varieties. It includes Indians, or native Americans, Toltecans, &c. In these, the cheek bones are prominent; the face broad; the forehead low; the eyes deeply seated; the hair black and straight.

5. THE ETHIOPIAN OR BLACK VARIETY includes Negroes, Africans, Hottentots, Bushmen, (Bosjesmans,) Bochmen, (Bechuanas.) The color is black, with greater or less intensity; the lips extremely thick; the nose flat and thick; the nostrils wide; the hair black and frizzly like wool; the head narrow; the forehead convex; the face projecting; the facial angle 70°. Between this and the European or Caucasian variety, the differences are marked; but there is no character in which the contrast between the lowest negro and highest ape is not many times greater than between the same negro and the highest European. The differences in respect to *structure* between the Ethiopian and the other varieties, would not be deemed sufficient to constitute a specific character among the lowest animals.

In regard to the varieties above described, it will be seen that one of the enumerated distinctions relates to the color and nature of the hair. At a trial held in South Carolina, in which the point in dispute, property in a mulatto girl, rested on a question of race, Dr. Gibbs stated, as a curious fact resulting from microscopic observation, that in the mulatto cross the hair of one or the other parent was present, and sometimes hairs of both, but never a *mongrel* hair; that no amalgamated hair existed; that the mulatto as often had straight hair as kinky. He stated that the microscope revealed that the hair of the white race is, when transversely divided, oval; that of the Indian, circular; and that of the Negro, eccentrically elliptical with flattened edges; that of the Negro is not hair, but wool, and capable of being felted; that the coloring matter of true hair is in an internal tube, while in the negro it is in the epidermis, or scales covering the shaft of hair. In corroboration of the statement that both white and negro hair were sometimes found in the same head, a singular case was mentioned by Dr. Gibbs. He remarked that he once attended a half-breed Indian and Negro, who had straight Indian hair. He was ill and had his head shaved and blistered. On his recovery, when his hair grew out, it was negro hair, crisped and wiry.

The late Dr. Morton, of our own country, in a disquisition relative to the "Size of the Brain" in the different varieties, presents the following results:

"The ancient Egyptians, whose civilization antedates that of all other people, and whose country has been justly called 'the cradle of the arts and sciences,' have the least sized brain of any Caucasian nation, excepting the Hindoos.

The Negro brain is nine cubic inches less than the Teutonic, and three cubic inches larger than that of the ancient Egyptians.

The brain of the Australian and Hottentot falls far below that of the Negro, and measures precisely the same as the ancient Peruvian." (See Silliman's Journal.)

QUESTIONS ON THE VARIETIES OF THE HUMAN RACE.

With what are the variations of the HUMAN RACE connected? Are naturalists agreed as to the number of these varieties? What three are by some regarded as distinctly marked types? What do they consider the other varieties? How many distinct types or races are named on the chart? Whose arrangement has been followed? From what did the Caucasians derive their name? What nations belong to this variety? [See the chart.] For what are they distinguished? What are their characteristics? What does the entire conformation of the head show? What is said of their mental attainments? For what is the MONGOLIAN variety remarkable? What nations does it include? How do you describe the MALAY or AUSTRALIAN variety? Name the people or nations belonging to this variety. To which variety is the AMERICAN allied? Name the tribes or people which it includes. [See on the chart.] What are their distinguishing peculiarities?

What does the Ethiopian or black variety include? Describe their features, color, hair, &c. Is there a greater contrast between the highest European and the negro, than between the same negro and the ape? What is said as to the difference in respect to structure between the Ethiopian and the other varieties? In what respect does the hair of the Caucasian, Indian, and Negro varieties differ? What cases corroborate this curious fact? What were the results arrived at by Dr. Morton, of Philadelphia?

Obs. Here is a good opportunity for a general exercise about the people of the different varieties, the countries they inhabit, their customs, religion, degrees of civilization, &c., showing the pupil how to apply his geographical or historical knowledge.

SECTION V.

Second Order. QUADRUMANA.—FOUR HANDED.

(Lat. *quatuor*, four, and *manus*, hand.)

This order includes the *Simiadae*, (Lat. *Simia*, an ape,—ape-kind;) *Cebidae*, (Gr. κῆβος, *kebos*, a monkey,—monkey tribe;) pronounced *kebidae; Lemuridae,* (Lat. *Lemures*, ghosts,—ghost-like.)

The Simiadae are spread over the tropical regions of Asia and Africa, including the larger islands of the Indian Ocean; the Cebidae are found in South America; the Lemuridae, in Madagascar and the smaller adjacent islands.

The name "Quadrumana" is given to these animals because, while having two hands, resembling those of man, they have feet which are also formed like hands, and can grasp branches of trees. Like man, they have no natural means of defence; but they are endowed with a cunning, a quickness and agility not often equaled and never surpassed by any other quadrupeds. The peculiarities of their structure do not adapt them either to an erect or a horizontal position, but to one that is diagonal or sloping. Their great muscular strength, combined with the faculty of climbing, enables them to escape from the carnivorous quadrupeds which are found in the same forests with themselves. "Leaping from bough to bough, they pass through the most entangled forests with greater swiftness than an ordinary horse would travel on a turnpike road. The apes upon the rocks of Gibraltar, (Barbary apes, which are the only ones found in Europe,) can never be approached by the most cautious sportsmen. They climb, with the greatest facility, among frightful precipices, where neither dogs nor men can follow."*

The hand of the highest Quadrumana is greatly inferior to that of man, both in respect to its structure, and the uses for

* "Swainson's Habits and Instincts of Animals."

which it is fitted. The thumb is a mere rudiment, and in some species, entirely wanting. The fingers are very long, and fitted for hooking an object, but have but little power of separate motion among themselves; the palm, instead of being hollow, is narrow and flat, and tapers from the wrist. All of them have three sorts of teeth, like man, but the canine, (eye) teeth, are more developed in the Quadrumana than in him, and there are spaces between them and the other teeth.

The principal food of these animals is fruit, which Providence furnishes them most plentifully in tropical countries, though occasionally they prey upon the young and eggs of birds, also upon lizards and insects. When captured and domesticated, they become almost *omnivorous*, (Lat. *omnis*, all, and *voro*, to devour.) They are peculiar to tropical regions, and are useful there as tending to diminish the annoyances which might otherwise arise from the insects which they consume for food. In some countries these animals are themselves used for food, and their skins converted into leather.

The SIMIADAE include three divisions: I. The APES, *without tails;* II. the BABOONS, *with short tails* and sometimes none; III. the MONKEYS, *with tails*, which as connected with this family are adroit, agile, and restless, but usually live only two or three years. In this family, the tail has no prehensile, or grasping power. Their teeth, of which there are ten molar in each jaw, are thirty-two in number; their nostrils separated by a very narrow division. The larger portion have cheek pouches and callosities, (hard parts,) on the hind parts of the body. Of the Apes we name first the *Troglodytes*, (Gr. τρώγλη, *trogle*, a hole; δύνω, *duno*, to creep, a creeper into holes.)

This is the CHIMPANZEE, (not to be confounded with the Orang-Outang,) found rather commonly on the banks of the Gambia and Congo. It is more man-like than any other animal, especially when young. When full grown, its height is at least five feet, and according to some naturalists, six or seven. The hair is black, long and coarse, falling down on each side of the head, forming large whiskers on the cheeks; the eyes are hazel, deep set and lively; the ears large and spreading; the lips covered with a thin white beard, and large and wrinkled; the face and hands, of a dark brown color. An officer in the English navy, who saw the animal in 1838, says that in its natural state, "it mounts trees only for food or observation, has enormous strength, easily snapping boughs from trees which the united strength of two men could scarcely bend." These animals reach their full growth when between eight and nine years old. They travel in large bands, armed with sticks, which they handle with great dexterity; and sometimes are so full of courage and fury that they drive the elephant and lion from their haunts. As their name imports, they spend much of their time in holes, or rocky caves. They are very watchful, even when united in a herd; and the first one who notices the approach of a stranger, utters a long drawn cry, which resembles that of a human being in distress. This is done to notify the herd of the stranger's coming. They then immediately leave any place which would expose them to danger, and betake themselves to the bushes. It is said to be very difficult to obtain them alive, owing to a superstitious notion of the natives that they have the "power of witching."

Several young Chimpanzees have, at different times, been imported into England and the United States. These appeared to be mild and docile, but were short lived, being unable to endure the changes to which they were subjected in respect to climate and mode of living. Had they lived to full age, they would probably have manifested the ape's naturally fierce and obstinate disposition. One of them, which lived about a year in the menagerie of the British Zoological Society, is described as appearing like "an old, bent, and diminutive negro." The appearance of age was increased by its short white beard and wrinkled face, though at the time not more than two and a half years old. All its actions seemed child-like. It would "examine every object within its reach with an air so considerate and thoughtful as to create a smile on the face of the gravest spectator. When perfectly free and unconstrained, Tommy's usual mode of progression was on all fours. His feet, and particularly his heels, were broader and better adapted for the biped race than those of the Orang-Outang, and this he adopted when occasion required. He frequently indulged in a kind of rude, stamping dance; would seat himself in his swing with great good humor, when ordered to do so, stretching out his foot to some of the company to set him in motion; and interpreting your wishes and intentions from your looks, tones, and gestures, exhibited the most wonderful quickness of apprehension."

Pithecus Satyrus.

(Gr. πίθηκος, *pithecos*, ape; σατυρὸς, *saturos*, satyr.)

The Orang-Outang, or wild man, (from *Orang*, the Malay term for man, and *Outang*, wild.)

The Orang-Outang is found in the islands of Borneo and Sumatra. Though called by this name, it is less man-like than the

Chimpanzee. In the young animal, the forehead and skull appear well developed and somewhat human; in the adult, the bones of the face are so increased in size that they throw the skull backwards, which, combined in its effect with other differences, takes away the resemblance, which is seen in the young, to the human face. The arms are so long that they reach the ground, or nearly so, when the animal stands erect; and the palms of the hands show lines and *papillae*, like those of man. The ears are small; the eyes dark and round; the throat is swollen, the skin about it being loose and folded, and enveloping a double membranous sac, which connects with the larynx or wind-pipe, and becomes inflated when the animal expresses pleasure or anger. The body is stoutly built and very muscular; the belly round and protuberant; the hair is of a reddish brown hue, long and coarse. The Orang has no tail or cheek pouches. A very marked characteristic is the disproportion between the size and length of the arms, as compared with the legs, which, viewed in connexion with the long and hooked hands, indicates that the animal is, more than the Chimpanzee, formed to live on trees. Among the branches, he moves with surprising facility. By weaving these together, he constructs a sort of rude hut, which he seldom leaves, except when forced by the calls of appetite. In Borneo, the natives call the two species found there, *mias-kassar* and *mias-pappan.* Of these the latter is much the larger and more powerful, and justly named *Satyrus*, from his ugly face and disgusting callosities. Some naturalists consider the Orang of Sumatra to be a distinct species.

The Orang may be ranked as the largest of the apes. A specimen from Borneo was in height five feet ten inches, and one from Sumatra reached the enormous stature of seven feet six inches. Those animals are described by persons who have seen them in their native climes, as "leading a solitary life, more than two or three never being found together;" and as "roused from their habitual dullness by nothing but hunger or the approach of danger." Their strength is so great they can not be safely encountered except with fire-arms. A female Orang snapped a strong spear asunder, after receiving many wounds. Hence, the natives of Borneo hold these animals in especial dread, and carefully avoid them.

Hylobates, (Gr. ῞υλη, *hule*, a wood; βαίνω, *baino*, to traverse,) Long Armed Ape, or Gibbon. *H. Syndactulus*, (Gr. Συν, *Sun*, connected, together; δακτυλος, *daktulos*, a finger.)

This species of Gibbons receives the name *Syndactylus*, from having the second and third toes of the hind foot *united* by a narrow membrane the whole length of the first joint. As the generic name, ***Hylobates***, imports, this animal lives in the recesses of dense woods, (in the East Indian islands and the Malay peninsula.) The hands are extremely powerful, and so long that they reach to the heel, and their span extends from four to six feet. These greatly assist him in making his rapid movements among the trees. The fur is longer and more abundant than that of the Orangs. The animal is like the Orang in temper and manners, but much smaller, when standing upright, being but two feet four inches. It is a better walker than the Orang, but its gait is unsteady, and it frequently places its hands on the ground to assist its position. An adult male of this species was taken in 1830, but died while on its way to England. It fed on vegetables, yet eagerly accepted animal food; fowls it especially preferred. It appeared to be good tempered and affectionate; "when pleased, uttering a chirping note; when frightened or angry, uttering the loud guttural sounds of *ra, ra, ra.*" It was fond of play and became quite attached to a Papuan girl who was on board the vessel—"would sit on the capstan with its long paw around her neck, and lovingly eat biscuit with her." This Gibbon is sometimes called the SIAMANG, and is said to be celebrated for the pains which it takes to wash the face of its young, which it does with maternal faithfulness, in spite of its screams and struggles.

H. agilis. The AGILE, or SILVERY GIBBON, also called the *Ungka*, or *Oungka.*

This species is a native of Sumatra, deriving its name, *agilis*, (active,) from its remarkable activity in leaping among the branches. One of these animals, which was exhibited in London some years since, "sprang with the greatest ease through distances of twelve and eighteen feet; and when apples or nuts were thrown to her while in the air, she would catch them without discontinuing her course. She kept up a succession of springs, hardly touching the branches in her progress, continually uttering a musical but almost deafening cry. She was very tame and gentle, and would permit herself to be touched or caressed." This Gibbon is distinguished by its low forehead, as well as its activity. The color varies a good deal, according to the sex or age, but is usually brown. In the male, a white band over the eyes unites with the whitish whiskers. The hair is fine except about the neck, where it is rather woolly and curled.

BABOONS.

The most striking peculiarity of these animals is the resemblance of their head and face to those of a large dog. Their muzzles are long and truncated. They have cheek pouches, short tails and sharp claws. The malignant expression of their countenances, their gigantic strength and the brutal ferocity of their manners, render them decidedly the most frightful and disgusting of all the *Quadrumana.* Their home is Africa, where they frequent rocky ridges more than the forests. They live mostly on scorpions, which they find under stones and deprive of their stings by a skillful application of the thumb and finger. In the Baboon, the facial angle is reduced to 30°. The name is from the Italian Babbaino, from which comes the Latin word *Papio*, applied to these animals especially in the fifteenth and sixteenth centuries. In brilliancy of color, they vie with the gorgeous plumage of the tropical birds.

"They are distinguished from the *Apes*, by the equality of their members, their cheek pouches and ischial callosities; from the *Monkeys*, by the short robust make of their bodies and extremities, their tubercular tails, too short to execute the functions usually assigned to that organ, and the mountain rather than silvan *habitat* which this conformation necessarily induces."

Cynocephalus, (Gr. *Κύων*, *Kuōn*, a dog; *Κεφαλή*, *Kephale*, a head;) *Dog-headed.* *C. Mormon*, (Gr. *Μορμών*, *Mormōn*, a bogie.) This is the MANDRIL, or GREAT VARIEGATED BABOON.

The Mormon resembles the dog and bear. It is a native of Guinea and West Africa, has a short, erect and stumpy tail, by which, and the enormous protuberances of its cheeks, it is readily distinguished from the other species. This is not only the largest of all the Baboons, but the most brilliant in its colors. When upright, its height reaches five feet. The muzzle is of a bright scarlet color; a stripe of vermilion runs along the center of the nose, and spreads over the lip; the cheeks are also of a rich violet hue, and elevated on each side by a singular development of the bone, which forms a socket for the roots of the immense canine teeth. The hair is of a greenish brown color, caused by alternate layers of yellow and black present in each hair. On the temples it is directed upwards, so as to meet in a point on the crown of the head. The brilliancy of the colors is connected with the skin, and disappears when the animal dies or is sick. The Mandril frequents forests filled with brushwood, whence it sallies forth to plunder the nearest villages. Its bulk is great in proportion to its height and strength, and its ferocity great, so that it is a terror to the natives. Cuvier says he has seen it expire from the violence of its fury.

Semnopithecus, (Gr. σεμνὸς, *Semnos*, to be reverenced; πίθηκος, *pithēcos*, Ape.)

This genus includes animals resembling, in many points, the Gibbons. As in the latter, their extremities are of great length as compared with the size of the body, which in its form is long and slender. But they differ from the Gibbons in having the hinder extremities longer than the front ones, which is the reverse of what occurs in the Gibbons. They are distinguished by having a very long, slender and muscular tail, terminated by a close tuft of long hairs. The color of the adult animal is intensely black, except the breast, the abdomen, and the root of the tail, which are gray. The black hairs on the top of the head are tipped with gray, and as age advances, the latter color is extended to the upper parts of the body. The hair is long, soft, and silky. The eye-brows consist of long stiff hairs, pointing forward. The stomach is three fold, one of the divisions being puckered into a number of distinct sacs; and its teeth resemble, in some degree, those of a ruminating animal. It evinces less restlessness, petulance and curiosity, but has more of real intelligence than the common monkeys. The animals of this genus are found in Cochin China, the East Indies and the neighboring islands.

S. Maurus. (Gr. μαῦρος, *mauros*, a fool?) The BUDENG.

This species abounds in the extensive forests of Java, and forms its dwelling on trees. Troops of more than fifty individuals are found together. When approached, they scream loudly, and by their movements branches of decaying trees are often thrown down upon the spectators. The natives chase them on account of their fur; attended by their chiefs, attacking them with stones and cudgels, and often destroying them in great numbers. The furs of these animals are used both by the natives and Europeans, in preparing riding equipages and military ornaments.

S. Entellus. (Lat. the proper name of a Roman athlete.) THE ENTELLUS, OR COCHIN CHINA MONKEY. The HOONUMAN of the Hindoos.

This species is one of the most common in Hindoostan and the Indian Archipelago, and in India is the object of a blind adoration. According to the popular superstition, he who puts to death an Entellus Monkey, will surely die within the year. Its form is slight, the limbs long and slender, the length of the body from the muzzle to the tail is, in the full grown animal, four and a half feet, and the tail is even longer than the body. When young, they seem gentle and free from malice; but their characters do not improve by *age*. This animal is very active in the capture of serpents, stealing upon the poisonous reptile when asleep, and grinding down the reptile's head until the poisonous fangs are destroyed.

QUESTIONS ON THE QUADRUMANA.

What is the second order of animals? What three families does this order include? Give the derivation of the order and the families. Where is their location or *habitat?* Why was the name *Quadrumana* given to these animals? Have they any weapons for defence? With what are they endowed? For what does their peculiar structure adapt them? Of what benefit is their muscular strength? What is said of their leaping powers? What of the Barbary apes? Where are these found? Are any other of the Quadrumana found in Europe? How does the hand of the most perfect Quadrumana compare with man's? What is said of the thumb? Of the fingers and of the palm? What of the teeth and the hair? What is their principal food? What change occurs from domestication?

Spell and define the following words, giving examples of each as you proceed: *Carnivorous*, flesh-eating; (Lat. *caro*, flesh, and *voro*, to devour.) *Frugivorous*, eating fruits, seeds or corn; (Lat. *fruges*, corn.) *Granivorous*, eating grain, or feeding on seeds; (Lat. *granum*, grain.) *Herbivorous*, eating herbs, feeding on vegetables; (Lat. *herba*, herb.) *Insectivorous*, eating insects; Lat. *insecta*, insect, and *voro*, to devour.) *Apivorous*, bee eating; (Lat. *apis*, a bee.) Apiary, a place where bees are kept. *Piscivorous*, fish eating, living on fish; (Lat. *piscis*, a fish.) *Reptilivorous*, eating snakes, toads, and other reptiles; (Lat. *reptilis*, from *repo*, to creep.) *Omnivorous*, eating everything, devouring all kinds of food; (Lat. *omnis*, all.)

To what regions are quadrumanous animals peculiar? In what respects are they useful? What divisions do the SIMIADAE include? How are these divisions readily distinguished from one another? What is said of the monkeys of this family? Which is the genus first named, and from what is the name derived? Where is it found, and what is said of its resemblance to man? Describe its appearance, habits, &c. Why is it difficult to obtain it alive? Have attempts been made to import these animals, and with what success? What is said of Tommy? Describe him particularly. What is the difference between a biped and a quadruped? Ans. One is two-footed, (Lat. *bis*, two, *pes*, a foot;) the other four-footed, (Lat. *quatuor*, four, *pes*, foot.) What between a bimanous and a quadrumanous animal? From what language is the Orang-Outang derived? From what are the generic and specific names derived? Where is it found? What is said of it? Where does it live? For what kind of a residence is it fitted by its long arms and hooked hands? What does it construct among the branches of trees? Does it often leave them, and for what? How many species are found in Borneo? Which is the largest and most powerful? What is it justly named? Is the Orang of Sumatra of the same species? What is said of their size, and what account do persons give who have seen them in their native woods? Why do the Borneans dread them? What is the generic term for the long armed ape, or Gibbon? From what derived? What does this name import? Give the derivation of *syndactulus.* Why was it given to this species? Where does it live and in what country is it found? Describe its habits, size, gait, food, sounds, &c. For what is it most celebrated? What is said of the silvery or agile Gibbon, *H. agilis?*

What is the most striking peculiarity of BABOONS? What do they resemble? Where are they found? On what do they live? From what is the name derived? What is said of their colors? How are they distinguished from the apes? Describe the Variegated Baboon, or Mandril. Give the derivation of the generic and specific terms. What is said of its size, color, habitat, &c. How do the natives regard it? From what is *Semnopithecus* derived? In what respect does this genus resemble the Gibbons? How do they differ from the Gibbons? By what are they distinguished? What is said of their hair, eye-brows, stomach, disposition, intelligence, &c.? What is said of the Budeng? What of the Cochin China monkey?

SECTION VI.

AMERICAN MONKEYS.

These are a very numerous division found in South America, and arranged into two leading groups, viz.: the SAPAJOUS and SAGOINS; the former having muscular, grasping tails; the latter feeble ones, unfit for grasping. They are sometimes called the four-fingered monkeys, as the thumb is reduced to a mere rudiment, and in some species is entirely wanting. They are without cheek-pouches and callosities.

I. SAPAJOUS.

These may be regarded as representing the GUENONS, (*Cercopithecus*, Gr. *kerkos*, a tail,) of the Eastern Continent. The whole of them are very active, climb well, and are well formed for living and moving among the trees. The fore-hands show a less perfect organization than is seen in the monkeys of the Eastern Continent. The palms of both extremities are endowed with exquisite sensibility. These monkeys are of small size and playful disposition. Gathered in herds, they lead a merry life, feeding mostly on insects and fruits. The facial angle is about 60°. Among them we include the HOWLERS, (*Mycetes*,) as has been done by other naturalists. The Howlers differ, however, from the other Sapajous in some respects, particularly in having a facial angle of but 30°, but agree with them in having prehensile tails. Of the numerous species of these and other South American monkeys, we can notice only the most interesting and prominent.

Mycetes, (Gr. *μυκήτης*, *mukêtes*, a Howler.)

These are the largest monkeys of America, and remarkable for the development of the vocal organs. The bone at the root of the tongue, (the hyoid bone,) is, in these animals, very large, swelling into a capacious drum which communicates with the larynx, and gives a tremendous power and volume to the voice. They howl in concert, especially at the rising and setting of the sun; but the night is often made dismal with their frightful yells. One monkey begins the cry, and is immediately followed by the others; and their distressing, unearthly sounds have been heard at two miles distance. The canine teeth are, according to Swainson, six times as large as the incisors or cutting teeth. The part of the prehensile tail with which these animals lay hold of the branch of a tree, is naked below, and of course has a higher sensibility of touch. Their size is rather larger than that of the fox. In their dispositions they are ferocious and intractable; in habits social, and most of them have a thick beard. Their deep sonorous yells are supposed to be a call to their mates; in other words, a hideous love-song.

M. ursinus, (Lat. *ursus*, a bear.) The Ursine Howler, or Arguato.

This animal is, exclusive of the tail, nearly three feet long. The hair is of a golden color, and the thick beard is of a deeper color than the rest. Humboldt counted above forty of these animals in a single tree, and says, "their eye, voice, and gait denote melancholy." They feed upon fruit and the leaves of plants, and in traveling follow an old monkey as their file leader. This Howler has a membranous sack in the throat, connected with the wind-pipe and capable of being inflated, giving the power to utter terrific sounds.

Atelès, (Gr. *ατελης*, *ateles*, imperfect.)

This and the preceding genus are "Ordinary Sapajous;" (the term Sapajous also including the genus *Cebus*, or the Sajous.) This genus includes what are called the Spider Monkeys, so called from their long slender tails, and sprawling movements, which give them a spider-like appearance. It is termed *ateles*, or imperfect, because in most of the species the thumbs on the fore-arms are rudimental, or else entirely wanting; (they are, however, found on the hinder extremities, and large and opposable to the fingers.) They have four molar teeth more than man, making the number of teeth thirty-six, and are distinguished for their round heads and thick or corpulent bodies. The eyes are far apart; the nostrils open laterally, (or sidewise;) the hair is generally long, coarse, and of a glossy appearance. *Trees* are their home; on the ground they drag themselves along with their fore-arms, using them as crutches, and resting upon their half closed fists. Sometimes they crouch along on their hind legs.

Troops of them are found together, and they are said to "exercise a perfect tyranny over all the other arboreal mammals in their neighborhood." Though living chiefly upon leaves and fruit, they also hunt after insects and the eggs and young of birds, and are even said to fish for crabs with their long tails. They are uncommonly intelligent, easily domesticated, and evince a strong attachment for those who treat them kindly; and they have less of curiosity, mischief, and violent passion than the common monkeys. They use their prehensile tails as a fifth hand, even crossing streams by mounting to the topmost branches of some over-hanging tree, and forming themselves into a long chain. The last monkey keeps a good hold on the tree, while the living chain swings to and fro, until by the impetus thus gained, the foremost can reach a branch upon the opposite side, when the rear animal lets go his hold, and the whole are rapidly drawn up. The Indians esteem their flesh as an article of food, and it is said to be "white, juicy, and agreeable." It is related that the Spider Monkey, when shot, fastens its tail so closely to the branches that it remains suspended even after death. Among the most noted species are *A. Paniscus*, (*Πανισκος*, *Paniscos*, dim. of *Παν*, *Pan*, a little Pan.) This is the QUATA, or as the French write it, the COAITA, found in large companies in Guiana and Brazil.

A. Belzebub. The MARIMONDA.

The monkeys of this, like those of the preceding species, unite in large companies and form the most grotesque groups. All their attitudes evince the extreme of sloth. They will bend their long arms over their backs, and remain motionless in this position for hours together, under the heat of a tropical sun.

CEBIDAE.

(From *Cebus*, Gr. *κῆβος*, *kebos*, monkey. The SAGOU, or SAJOU.)

The animals of this genus are grouped among the SAPAJOUS, but denominated more distinctively the SAJOUS. They are also called CAPUCHIN MONKEYS, from the hood-like formation of the hair of the head.

C. Appella. The WEEPER.

Why this very common species received so dolorous a name is not apparent, as in confinement it is "good tempered, playful and hardy." It has a rather rich fur of a color inclining to olive, with a golden tinge on the lighter parts, and is distinguished by its yellow, flesh-colored face.

C. albifrons. (Lat. *albus*, white, and *frons*, forehead.) The OUAVAPAVI, or WHITE-FACED CAPUCHIN.

This animal has a grayish blue face, except the pure white orbits and forehead. The color of the body is grayish olive. Troops of these monkeys are found in the forests of Oronoco. The Indians often keep them as playthings, and derive from them much entertainment. Humboldt saw a domesticated one that caught a pig every morning, and rode him about the whole day, while he was feeding in the savanna. Another, in the house of a missionary, bestrode a cat which had been brought up with it, and patiently submitted to its rider.

C. fatuellus. (Lat., the same as *Faunus*, or *Pan*, a Roman divinity.) The Sagou Cornu, or HORNED MONKEY.

This species takes its name from the bushes of hair which elevate themselves on the base of the forehead, producing a resemblance to horns. The color in some of these animals is a deep brown, or purplish black; in others, reddish brown. It is a native of French Guiana.

II. SAGOINS.

These include several groups, which, though differing from each other in some particulars, agree in having tails that are feeble and *not* prehensile, but which they use for protecting themselves against the cold, of which they are very sensible. They are light and graceful in their movements; of a lively, timid, and irritable disposition. Their food consists of fruit, birds' eggs, and insects. Of the genera belonging to this division we name the

Callithrix sciureus. (Gr. καλός, *kalos*, beautiful, θρὶξ, *thrix*, hair.) *Sciureus*, the specific term, is from the Gr. σκιούρεος, (*skiureus*,) squirrel-like.

This is the SAIMIRI of Buffon, otherwise called the SQUIRREL MONKEY, and is a very beautiful little animal not quite a foot long, and with a tail three or four inches longer than the body. It is native to Brazil and Guiana. The head is rounded in form; the muzzle is short and dark colored; the ears very large, and it has a large bushy tail. Around the eyes are two circles of flesh. The general color is olive gray; but the fore-arms and legs are of a fine orange red. Its cry is a hissing sort of whistle repeated three or four times, and expressive of impatience or anger. The tail, though not properly prehensile, it sometimes winds around objects as a sort of feeler or support, so that this animal may be regarded as a link between this division and the Ordinary SAPAJOUS.

The SAKIS,

(Or those SAKIS which have long bushy tails, and hence have been denominated FOX TAILED MONKEYS; the term *Saki*, in its more general application, denoting any American Monkey which has *not* a prehensile tail.)

Pithecia. These are the largest of the SAGOINS. Of this genus, which has a facial angle of 60°, the most remarkable is the *Pithecia lugens*, (Gr. *πιθηκεία*, *pithekeia*, ape-like; *lugens*, Lat. mourning,)—the WIDOW MONKEY, so named from the contrast of black and white displayed in its natural dress. The general color is black, but the face and hands are white. The Creoles of South America say, "it wears the veil, kerchief and gloves of widowhood," according to the custom in South America.

Pithecia cheiropotes. (Gr. *χειρ*, *cheir*, hand; *πότης*, *potes*, drinker.)

THE HAND-DRINKER, so named because with its hands it conveys water to its mouth, from a vessel or running stream. This animal is the Capuchin of the Oronoko. It is distinguished by two distinct bushy tufts formed by the parting of the hair above the large, sunken eyes, and by its long crisped black beard. The fur is of a reddish chestnut color. It lives in pairs only, and is very shy.

But a more interesting species of these animals is the *Iacchus vulgaris*, (Gr. *Ἴακχος*, *Iakchos*, Bacchus.) The MARMOSET, OUISTITIS, or STRIATED MONKEY. This small species has a body about eight inches long, and a tail eleven or twelve inches. Upon its head are two tufts of white standing hair; the facial angle is 50°; the fur very soft. Some are black with yellow feet; others brown, striped with yellow, hence called *striated.* When removed from its native region to a colder climate, the Marmoset nestles itself among the materials of its bed, out of which it seldom emerges. It is very fond of insects: in captivity it will eat scores of the largest cockroaches, with many smaller ones, (rejecting the wing-cases and legs,) three or four times a day. Its chief and favorite food in the wild state, is the banana, though in that state it is almost omnivorous.

I. argentatus, (Latin, silvered.) This is the least and most beautiful of the SAGOINS, having silvery colored hair, which pleasantly contrasts with a tail of deep brown, inclining to blackness. In general habits, it is like the preceding.

What are the two leading groups of the numerous monkeys found in South America? What is a marked distinction of the SAPAJOUS? What of the SAGOINS? Why are they sometimes called four-fingered monkeys? What monkeys of the Eastern continent do the SAPAJOUS represent? What is said of their habits, manner of climbing, living, &c.? What is said of their fore hands? What of the palms of both extremities? What sort of a life are they said to lead? In what respect do the Howlers differ from the SAPAJOUS? In what particulars do they agree with them? Which are the largest American Monkeys? For what are they remarkable? Describe their howling, size, disposition, &c.? What is said of the URSINE HOWLER, or ARGUATO? How many did Humboldt count in a single tree? From what is *Ateles* the generic term for spider monkey derived, and what does it mean? Why are they called Spider Monkeys? For what are they distinguished? How do they move on the ground? Where and upon what do they live? How do they use their tails? How cross streams? What is said of their flesh? What is said of the Marimonda (*Ateles Belzebub?*) What is said of the SAJOUS, or CAPUCHIN MONKEYS? To what genus do they belong? With what are they grouped? What is said of the Weeper? What of the White-faced Capuchin? What does Humboldt relate of this monkey? From what does the horned monkey derive its name? Where is it found? What do the SAGOINS include, and in what do they all agree? For what do they use their tails? What is said of their movements, food, &c.? What is said of the Squirrel Monkey, and from what is the term derived? What does the term SAKI generally denote? Which of them are called Fox-Tailed Monkeys? Which genus of SAGOINS is the largest? Of this genus *Pithecia* which is the most remarkable? What do the Creoles of S. A. say of it? How is the Hand-Drinker distinguished? Why is it so named? Where found? What is said of the Marmoset, Ouistitis or Striated Monkey? What is said of its food in its wild state? What in captivity? Among what class of animals on the chart would you look for cockroaches? Which is the least and most beautiful of the SAGOINS?

SECTION VII.

LEMURIDÆ, (Lat. *Lemures*, ghosts.)

The Lemurs were so named by Linnæus, on account of their nocturnal habits and noiseless movements. The larger part of this family are natives of Madagascar; but some inhabit the African continent, and a few of them the East Indies. They resemble the monkeys in having opposable thumbs on both pairs of extremities; those of the hinder limbs are large, and much expanded at the tips; the nails are flat, except those of the first finger of each hinder limb, which are long, raised and pointed. They do not show either the mischievousness and petulance, or the sprightliness and curiosity of the monkey tribe. From them they also differ in size and form, and in respect to their teeth. The chief difference among the Lemurs themselves relates to *color;* the habits, manner and general figure being the same in all. The muzzle is very pointed, the tail very long; the fur woolly and soft. They are generally not larger than a fox, and some are smaller. The Lemurs of Madagascar and two or three adjacent islands appear to take the place of the Monkeys, none of which are found in those islands. Their habits, in a state of nature, have not been much observed. When in captivity, they are quite tame, and good natured; fond of attention, and leap about with surprising agility. They are evidently nocturnal. When undisturbed, they spend the greatest part of the day in sleep. If alone, they roll themselves up in the form of a ball, and wind their long tails in a very curious manner about their bodies, seemingly for the purpose of keeping themselves warm, for they are naturally quite sensitive to cold, and delight in basking in the rays of the sun, or in keeping themselves as close as possible to the fire.

At twilight they show more alertness, springing from perch to perch, and uttering a peculiar grunt of pleasure and satisfaction. At this time, they seem most desirous of food, which in confinement is usually bread and fruits. They are naturally climbing animals and exceedingly active, twisting their tails about objects, but not using it as a fifth hand.

They endure changes of air and climate better than the Monkeys; but "dust and wet not only annoy them, but produce disease and death." It is said that "one of their favorite situations is the edge of the fender, on which they will rest, spreading out their hands before the fire, half closing their eyes, and luxuriating in the genial glow."

The noise which the Lemur makes when alarmed, or suddenly startled, is a singular "braying, or roar of interrupted hoarse sounds, ending with abruptness." Their native food is not certainly known, but it is believed to be fruits and eggs, birds and insects. When in captivity, they refuse cooked meat. They live together in troops, clinging to the branches of trees, or when confined, to the bars of their cages, like the *sloth*, which in many respects they resemble. The eyes are full and of hazel color; in confinement, blindness is a common occurrence.

The whole are sometimes called MADAGASCAR CATS.

Cuvier arranges the Lemurs into five groups, viz.

I. The Makis, or Macaeos, the TRUE LEMURS.

II. The Indris, *Lichanotus*, (Gr. *lichanos*, index-finger; *ous*, an ear.)

III. The Lori group, Slow Lemurs, *Stenops*, (Gr. *Stenos*, narrow; *ops*, face or muzzle.)

IV. The Galagos, *Otilicnus*, (Gr. *ous*, an ear; *liknon*, a fan.)

V. The Tarsiers, *Tarsius*.

Among the most beautiful species of the first group, is the RED LEMUR, *L. ruber*, (Lat. *red*.) This is also one of the largest, and apparently suffers less than others by a removal from its native abode. Its fur is of a deep rich chestnut; but the face and fore-hands, as also the under parts and tail, are black. It is easily tamed, and very gentle.

A still more beautiful species is the *L. Macaco.* The RUFFLED LEMUR, the largest of the family. Its fur is varied with pure white and black, in nearly equal proportions; the hands, however, are black, and a white *ruff* surrounds the face. In habits and disposition, it is like the rest. All the species of the Lemurs are handsome, and worthy of attention; but it is sufficient for our purpose to name the above.

The *Indris*, (*Lichanotus*, Illiger.) These are found in Madagascar, and present two species, the long and the short tailed.

The BLACK OR TAILLESS INDRI, *I. brevicaudatus*, (Lat. with short or rudimentary tail,) is described as "a large animal three and a half feet high, entirely black except on the face and abdomen, which are of a grayish cast, and the rump which is white." The face is dog-like; the ears are short and much tufted; the hair is silky and thick, but in some places, curly; the nails are flat, but pointed. When young it is trained to the chase like a dog. Its note is spoken of as like a young child's crying; hence it probably derived its name INDRI, *man of the wood.*

The FLOCKY INDRI, *I. laniger*, (Lat. wool-bearing,) has a black face, and large and greenish gray eyes; five-fingered feet with long claws, except the thumbs which have rounded nails. It is said to be one foot nine inches long from the nose to the end of the tail, the tail being nine inches. The color above is a pale yellow ferruginous, or iron color, and white beneath. The fur is very soft and curly.

The LORIS. *Stenops*, (Illiger.) The animals of this genus have narrow, pointed muzzles, and are without tails. Their eyes are close together, and they have a grasp that is quite tenacious. Their movements are sometimes very slow; their habits nocturnal. "The base of the arteries of the limbs has the division into small branches which is found in the true *Sloths.*" The number of their teeth is thirty-six. The thumbs are widely separated from the fingers on both extremities. Two species are found in India and Ceylon, viz.

L. gracilis. (Lat. slender.) The SLENDER LORIS. This is a very small animal, being only eight inches in length. It has a long, dog-like visage, a thin and weak body, and long slender limbs. On each foot, the thumb is very distinct and separate from the toes. The color above is tawny; beneath whitish. According to Pennant, it is very active, and many of its actions are like those of an ape.

L. tardigradus. (Lat. slow-paced.) The SLOW-PACED LEMUR is "an animal of small size, scarcely equal to that of a cat." The largest yet noticed is but sixteen inches long. The apparent clumsiness of its form is much increased by the manner in which it usually contracts itself into a kind of ball. The large eyes have transverse pupils capable of being closed during the day, and very largely dilated at night. The hair is long, close and woolly, and of a deep ashy gray with a brownish tinge. A brown or chestnut band runs along the middle of the back. Under the true tongue is a second tongue, narrow and sharp pointed, which the animal projects in connection with the other when he drinks, and also when he eats, especially when eating flies, of which he is very fond; but he is able to retain the second within his mouth at pleasure. One of this species was a pet of Sir William Jones, during his residence in India.

Galago. The GALAGOS, found in Africa and India. These animals have round heads, short muzzles, and very large eyes and ears. The feet are five-fingered, with the exception of the first finger of the hind feet, which has a sharp awl-shaped claw. The tail is very long and hairy. Their large ears close when they sleep, but open upon their hearing any noise. They make their nests, squirrel-like, in the branches of trees, and cover it with a bed of leaves or grass for their young. Their food consists of soft fruits and insects. They are found in great numbers among the gum-trees of the desert of Sahara, and are particularly fond of the gum yielded by these trees. Thence they are taken by the Moors, and carried to the coast for sale, where they are named "animals of the *Gum.*" These animals are gentle and pretty, but small, the length of the body being only seven inches, and that of the tail, nine.

Of the several species, the one most worthy of notice is the *G. Moholi.* The MOHOLI. This singular but beautiful animal, peculiar to Africa, has a long glossy tail, very long hinder legs, large, bare and spreading ears. The color of the tail is a medium between a yellowish brown and cochineal red; the fur is throughout of the same color; that of the other parts is a dark slate color, except at and near the surface; the eyes are a deep topaz yellow. In its grimaces and active movements, it resembles the monkey. It is rarely seen during the day, which it spends in the nest it forms in the forks of branches, or in the cavities of decayed trees. Its length from the nose to the tip of the tail is sixteen inches.

The TARSIERS are found in the Molucca islands. These have *tarsi,* which are very long, and this gives to their hinder limbs a disproportionate extent. They have a rounded head, large eyes and a long tufted tail. The hands are small and delicate; externally covered with a soft down, but within they are naked. The nails of all the fingers of the hand as well as of the third and fourth finger of the feet are triangular in shape; on the index and middle finger of the feet they resemble the thorns of a rose bush. The fur is woolly and soft, the general color brown, inclining to gray. Two species are known. *Tarsius Bancanus.* The BANCA TARSIER, and *T. fuscomanus.* (Lat. *fuscus*, dark or swarthy; *manus*, hand.) These animals feed chiefly on lizards. Averse to light, they retire by day under the roots of trees. Dr. Horsefield obtained the BANCA TARSIER in Banca, near Iaboos, one of the mining districts, where, he says, it inhabits the extensive forests in the vicinity.

Cheiromys, (G. *cheir*, hand; *mus*, mouse.) The AYE-AYE. This quadruped, whose name signifies *hand-mouse*, resembles the *äi*, or sloth in its habits, but should not be confounded with that animal. Cuvier places it with the *Rodentia*, but it may properly be classed, as it has been by some naturalists, among the monkeys. Its specific name *Madagascariensis*, points it out as a native of Madagascar. It burrows under ground, and is slothful and nocturnal in its habits; has large flat ears, like those of a bat, and a tail like a squirrel's; but its most distinguishing peculiarity is the middle finger of the fore foot, the last two joints of which are very long, slender and without hair. This peculiarity aids the animal in drawing worms out of the holes in the trees, and in holding on to branches. Its length is eighteen inches, exclusive of the tail, and its general color ferruginous (iron) brown, mixed with gray.

Galeopithecus, (Gr. *galeòs*, a weasel; *pithecos*, an ape.) FLYING LEMUR.

This genus of animals is the connecting link between the Lemurs and the Bats. There are two species; some enumerate *three.* *G. volans*, the Flying Lemur,—is found in the most eastern islands of the Indian Archipelago. The chief peculiarity of this animal is the extension of its skin between the front and hind limbs, including also the tail, by which it receives a parachute-like support in the air, and is able to take long sweeping leaps from tree to tree, somewhat like flying; but it has not, like the bats, the power of continued flight. The general structure is like that of the Lemurs. During the day it sleeps suspended on the branches, with the head downward. At night it goes forth in quest of its food, which in addition to insects, consists of fruits, eggs and birds.

By whom were the Lemurs so named, and why? Where are they found, and of what do they there take the place? In what do they resemble Monkeys, and how differ from them? In what do Lemurs chiefly differ among themselves? What is said of their size, and is much known of their habits in a wild state? What are their habits in a state of captivity? In what do they delight? What food do they prefer, and what refuse? How do changes of climate affect them? What influences have wet and dust upon them? What is to them a favorite position? Do they live alone, or in troops? What animal do they strongly resemble? What general name is sometimes given them? Name the groups into which Cuvier arranged them? What is said of the Red Lemur? Which of the Lemurs is the largest and most beautiful? Give some account of the other groups, the Indris, the Slow Lemurs, the Galagos, and the Tarsiers. Describe the Aye-Aye, and the Flying Lemurs. Why is the name Aye-Aye given to the Chairomys? Ans. Because this name, as pronounced, is supposed to resemble the cry of the animal.

SECTION VIII.

Order 3. CARNIVORA.

(Lat. *caro*, flesh; *voro*, to devour.)

The two preceding orders, we have found specially characterized by the number and properties of their hands. In the animals we are now about to consider, the hands are modified into feet. At the head of the four-footed animals are the Carnivora, or flesh eating animals, which have the strongest thirst for blood, and with it the power and instruments for its gratification. These, in the structure of their teeth, their digestive organs, and general conformation, are adapted for preying upon other animals. In common with the first two orders, they have three kinds of teeth, and nails or claws on their feet; but unlike them, never have the front toe opposable to the other fingers. Their molar teeth, or grinders, are adapted for cutting and tearing rather than bruising or grinding. The greater or less development of the molar teeth as cutting or tearing instruments, indicates the kind of animal food suited for their support. Those Carnivora which have their molars, in whole or part, tuberculated, (covered with small knobs,) use vegetables, to a greater or less extent; those which have them serrated, or notched with points, live chiefly on insects. Other modifications of the molar teeth, fit them for crushing bones, or dividing flesh, as occasion may require. As a general rule, the jaws open and shut like a pair of shears, upwards and downwards, but do not admit of a side-wise movement. The *Carnivora* have no third lobe in the brain. The senses of sight, hearing and smell are exceedingly acute. Their feet are of a peculiarly soft structure, to enable them to steal silently upon their prey; and their supply being uncertain, they can endure long abstinence from food. The intestines of this order are suited to their flesh-eating habits, being shorter, and less voluminous than those of herbivorous animals. A kind providence has so arranged things that the larger and more formidable of the carnivorous tribes are but thinly scattered and more or less remote from the abodes of civilization.

The CARNIVORA may be divided into I. the CHEIROPTERA, Bats; II. the DIGITIGRADA, including the Cat, Dog and Weasel families; III. the PLANTIGRADA, the Bear family or tribe; IV. the AMPHIBIA, the PHOCIDÆ, or Seal family; V. the TRUE INSECTIVORA, including Shrews, Moles, &c.

By what were the first two orders characterized? How are the hands modified or changed in the CARNIVORA, and other orders of MAMMALS? What order stands at the head of four footed animals? What are their propensities, and have they the power to gratify them? For what are their teeth and digestive organs adapted? How many kinds of *teeth* have they? Describe them, and spell their names. Ans. *Incisors.* The fore teeth with sharp cutting edges for cutting or separating the food. *Canine* teeth are on each side of the incisors. These are very long and prominent in the Carnivora. (See plate IV. fig. 3.) Those in the upper jaw are called eye teeth in the human family. *Molars*, or grinders are of three kinds; *false molars* are more or less pointed, and stand next the canine teeth; next come the carnivorous teeth, especially adapted for dividing and lacerating muscle, and last the *Tuberculated* teeth, full of rounded knobs or pimples. *Serrated* teeth are notched with points like a saw, and show that the animal lives on insects. *Trenchant* teeth are very sharp and cutting. *Granulated* teeth are covered with small elevations, or grains. What have Carnivorous animals in common with the first two orders? In what are they unlike them? For what are their molar teeth adapted, and what is indicated by their variations? How can you distinguish by the teeth what food an animal lives upon? How do carnivorous animals generally open and shut their jaws? Have their jaws any side-wise movement? How many lobes has the brain in animals of this order? What is said of their senses? For what are their feet peculiarly adapted? Can they sustain long fasts? Are the intestines shorter in Carnivorous than in Herbiverous animals? How are the wisdom and goodness of God shown in the distribution of carnivorous and blood-thirsty animals? How are the CARNIVORA divided?

SECTION IX.

I. DIVISION OF THE CARNIVORA.

SUB-ORDER CHEIROPTERA, (Gr. *χειρ*, *cheir*, hand; *πτερον*, *pteron*, wing.)

These singular animals combine so much of the character of birds with that of quadrupeds, that it was long thought difficult to assign them a separate arrangement in the system of nature. It is now, however, settled that the structure of their bodies, their viviparous nature, their hair, etc., entitle them to a place among

PL. IV.

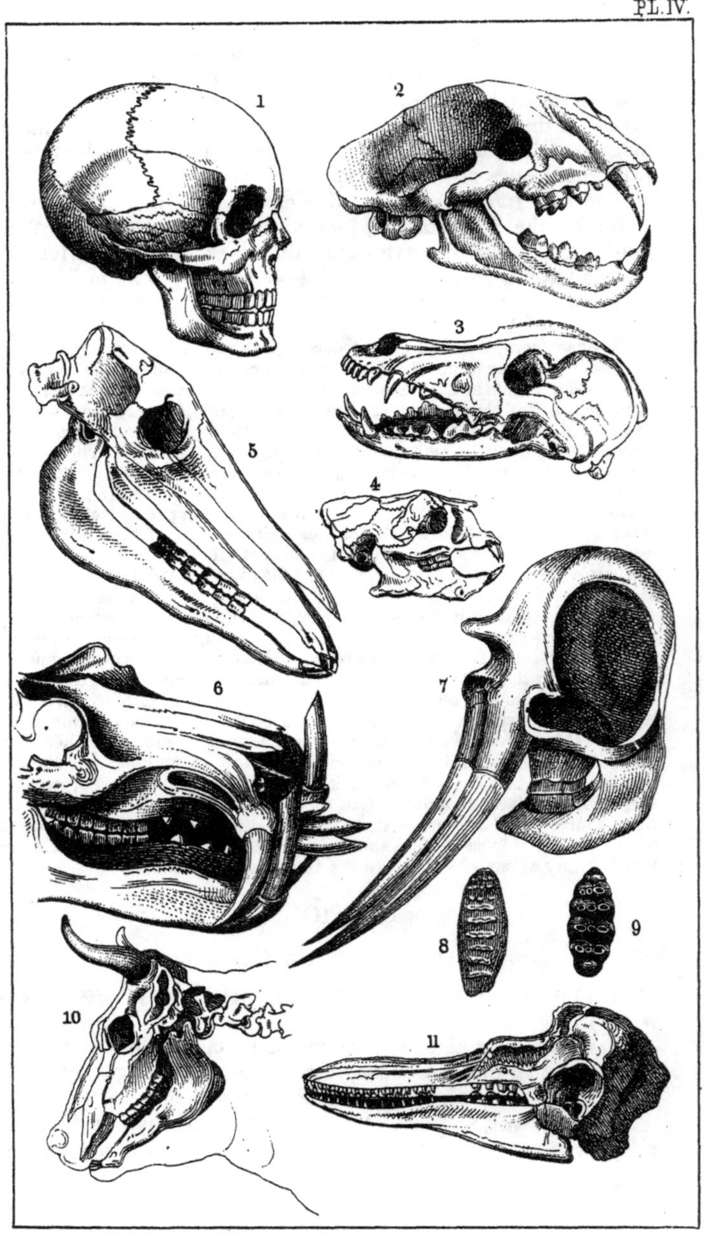

EXPLANATION OF PLATE IV.

DENTITION.

1. Skull of man, showing the omnivorous teeth of the order Bimana.
2. Tiger's head, showing the carnivorous teeth of the Cat family, (Felidæ.)
3. Dog's head, showing the carnivorous teeth of the Dog family, (Canidæ.)
4. Skull of a porcupine, showing the teeth of a gnawing animal, order Rodentia.
5. Horse's head, showing the vacancy for the bit between the front and back teeth, which space corresponds with the angle of the lips.
6. Hippopotamus' head, showing the canine teeth, (eye teeth,) developed into enormous tusks, with a chisel like edge.
7. Elephant's skull, showing the long, round, arched, pointed tusks or incisors projecting from the upper jaw.
8. A molar, grinding or back tooth of the elephant, of which there are never more than two on each side of the upper and lower jaws of the African elephant, and only one in a similar position in the Asiatic elephant.
9. Mastodon's tooth, showing the conical points whence the animal derives its name. For the tusks of the mammoth, see the Chart.
10. Skull of a cow, showing the dentition of a cud chewing animal, order Ruminantia.
11. Porpoise skull, showing how the numerous teeth interlock with one another when the jaws are closed.

the quadrupeds. Some of them are fruit eaters; but as a whole, we arrange them as Cuvier has done, with carnivorous animals. They are found both in the Eastern and Western Continents, and also in Australia. A climate tolerably temperate seems best suited to these animals; but they are largest in warm countries. Their most distinguishing character consists of a fold of the skin, which rising at the neck, extends over the lengthened limbs, as the silk over the whalebone of a parasol or umbrella, and gives them a winged appearance. Those genera which have the bones of the hand so developed as to spread a sufficient extent of this membranous skin, have power to perform all the evolutions which are required for flight. The *hand-wings* present a much greater extent of surface than those of birds, and the strong muscles attached as in the birds, to the sternum or breast bone, assist them to fly with great rapidity, and turn with astonishing swiftness. A lengthened bone proceeding from the heel, assists the tail in expanding that part of the membranous skin which is between the thighs, and where the tail is absent, performs that office alone; and thus gives the power of governing the direction of the flight, like the spread tail of a bird. By the extension of the upward curving of the tail and the hind feet, the *interfemoral* (between the thighs) part forms a hollow cradle into which the new born young is received. The thumb is free, short and armed with a strong hooked claw, by which they crawl along on the ground. The feeble hind feet have five toes, armed with sharp edged, curved and pointed claws, by which these animals suspend themselves, head downwards, in hollow trees, caves, or deserted buildings, where they are found during the day, going forth only at night. Their eyes are extremely small, but the external ears are often large, and with the wings, form an extensive surface endued with the most singular and exquisite sensibility, and enabling them, even when their eyes are sealed up, or removed, to pursue their rapid and wheeling flight, avoiding every obstacle, not even hitting threads stretched in various directions across their way, and passing through the narrowest passages without touching the sides. All are exquisitely susceptible of cold, and pass the winter in a state of lethargy, retiring to old ruins, caverns, or hollow trees, where they continue suspended by their claws until the genial spring warms them into activity. They are most active in the calm evenings of summer. Some of them are supposed to be migratory in their habits. In the *Cheiroptera* the teats are pectoral; in all the rest of the Carnivora, they are ventral. They perform a very useful part in the economy of nature in the destruction of insects.

The CHEIROPTERA, or VESPERTILIONIDÆ, are divided into five sub-families, each including many genera, viz. 1. Phyllostomatina. 2. Rhinolophina. 3. Vespertilionina. 4. Noctilionina. 5. Pteropina. They may also be arranged into 1st. the Frugiverous group, and 2d. the true or Insectivorous Bats. *Omnivorous* is, however, a term that more accurately describes the former group. Their teeth are, some of them, more trenchant than fruit eating habits would alone require. Cuvier says of these animals, "they know how to pursue birds and small quadrupeds," and it is quite probable they sometimes prey on the large insects found in the regions of their abode. Beside the variations in the teeth of the Fruit Eating or Omnivorous, and the Insectivorous Bats, there are other differences which relate to the stomach and intestines. The stomach of the former is very complicated, and the intestines very long, (in the *Pteropus*, seven times as long as the body,) whereas in the latter, the stomach is very simple, having but two divisions or portions, and the intestines are not more than twice the length of the body. Another difference respects the tail, which in the insectivorous bats is generally powerful; in the fruit eaters wanting, rudimental or comparatively inefficient.

From what is the name of the sub-order CHEIROPTERA derived? What do these singular animals combine? To what difficulty did this give rise? Is it now a settled question? What particularly entitles them to a place among quadrupeds? Are any of them fruit-eaters? How did Cuvier arrange them as a whole? Where are they found, and what climate suits them best? Where are the largest found? What is their most distinguishing characteristic? How do their wings compare with those of birds? How are the muscles attached? How is the cradle for the young formed? What is said of the thumb, and of what use is the hooked claw? By what do they suspend themselves, in what position, and in what places? What is said of their eyes, ears, wings, &c.? What of their exquisite sensibility, and what does it enable them to avoid? Are they affected by cold, and how do they pass the winter? At what season do they leave their retreats? When are they most active? Are any of them migratory? In what do they differ from all other CARNIVORA? Are bats useful? In what way? Into how many sub-families are they divided? How may they also be arranged? What does Cuvier say of them? What is said of their teeth, intestines, &c.? Name any further differences between the Insectivorous and Frugiverous Bats.

Spell, give the derivations and examples of each of these five sub-families.

1. PHYLLOSTOMATINA, (Gr. *Phúllon*, a leaf, *Stóma*, a mouth,) named from the leaf-like crest upon the nose. The Vampire, (*Vampirus Spectrum*,) of South America, is one of this blood-sucking family, acquaintance with which would divest it of half its terrors.

2. RHINOLOPHINA, (Gr. *Rhìn*, a nose, *Lóphos*, a crest.) These are the Horse Shoe Bats, of Java, which derive their name from the shape of the leafy membrane upon the nose. The genus *Nycteris*, (Gr. *Nucteris*, a bat,) inflate their bodies, and appear like small balloons.

3. VESPERTILIONINA, (Lat. *Vespertilio*, a bat.) These are found in all parts of the world, including Australia. The Flitter mouse of England, *V. murinus*, has the ears inclining backwards. The New York Bat, *V. Noveboracensis*, the Little Brown Bat, *V. subulatus*, (Lat. awl-shaped,) the Silver Haired Bat, *V. noctivagans*, (Lat. *nox*, night; *vagans*, wandering,) the Carolina Bat, *V. Carolinensis*, are all found in the United States and Canadas.

4. NOCTILIONINA, (Lat. *Noctilio*, from *nox*, night, and *eo*, to go.) These South American bats have side pouches for receiving their young.

5. PTEROPINA, (Gr. *Pteron*, a wing; *pous*, a foot.) These are the ROUSSETTES of the French, and the fruit-eating bats of Java. The Kalong, or Fox Bat, *Pteropus Javanicus*, is the largest, measuring five feet in the spread of its wings. They are found in large companies, suspended from trees.

SECTION X.

THE CARNIVORA PROPER.

The CARNIVORA proper are sometimes arranged into three divisions—the DIGITIGRADA, the PLANTIGRADA, and the PHOCIDAE or AMPHIBIA.

II. DIVISION OF THE CARNIVORA.

I. DIGITIGRADA, (Lat. *digitus*, a finger or toe; *gradior*, I walk;) walking on the toes.

This division of the CARNIVORA derive their name from their application of the toes to the ground in walking. It includes the Cat, Dog, and Weasel families. They are distinguished by their free, light and active step, their elasticity of motion, beauty of fur, and elegance of form. Many of them are nocturnal, slumbering by day in some dark den or deep recess, but prowling stealthily and noiselessly about during the night. Having satisfied their blood-thirsty dispositions and voracious appetites, when "the sun ariseth, they gather themselves together, and lay them down in their dens." Some animals of this division, as the wolf, are, however, more open in their movements, and in bands hunt their prey during the day.

1. FELIDAE, (Lat. *felis*, a cat.) The Cat family.

These include Cats, Lions, Tigers, Leopards, and Lynxes. Among them are the most eminently carnivorous and formidable of the mammalia, and they include a large number of animals that closely resemble each other in structure and appearance. They are among quadrupeds what birds of prey are among the feathered tribes. The size and strength of the Lion, Tiger and Leopard, combined with their thirst for blood, render them most fearfully dangerous.

The jaws and teeth of the FELIDAE are quite different from

those of the preceding orders; the jaws are much more powerful, the teeth longer and sharper. On their fore feet are five toes, and on the hind ones four, all armed with strong hooked and sharp claws. To prevent the claws from injury by coming in contact with the ground, they are, when not in use, drawn back. They are also elevated above the ground by the soft pad underneath, into sheathes, so that the point only just peeps out beneath the fur, and thus are not liable to be worn or blunted. (See Plate VI, fig. 7.) The tongue is very rough, as may be known by feeling that of the domestic cat. This roughness is occasioned by the innumerable *papillæ* which are turned backwards, and are like so many little hooks to assist the animal in tearing off any remnants of flesh that may adhere to the bones of their prey. Their sight is acute, and suited for vision both by night and by day. The expansive power of the pupil of the eye is so great that it takes in every ray of light. In the larger cats the pupil is circular; in those that roam at night and also see well by day, as our domestic cat, it is oval. Their long whiskers are delicate organs for the sense of smelling. These whiskers are each connected with a large nerve, and they are useful in indicating objects when the animal is prowling at night.

Felis Leo, the LION. This is the strongest and most courageous of the feline tribes, called the "King of Beasts," and "Monarch of the forest." He is regarded as the emblem of majesty and strength combined with generosity. His form supports the royal arms of England, and surmounts them as a crest. Many allusions are made in the Sacred Scriptures to his energy, power and majesty, (Rev. v., 5,) and his ferocious and sanguinary disposition. There are two kinds of Lions, *Leo Africus and L. Asiaticus.* The brown Lions of the Cape of Good Hope are more ferocious than the yellow variety found in that vicinity, and will carry off a heifer as easily as a cat would a rat. The Lion of Senegal has a thinner mane, and is of a deeper yellow than the Lion of Barbary. The Bengal Lion, the Persian Lion, and the Maneless Lion, are only varieties of the Asiatic Lion, *Leo Asiaticus.*

The Lioness is smaller than her mate, has two and sometimes three blind whelps at a litter, which she guards with great care. They are easily tamed when young, and live from twenty-five to thirty years, sometimes much longer. The great lion Pompey, which was in the Tower of London in 1760, had been there seventy years. One from the river Gambia died in the Tower at the age of sixty-three. Anderson, the African traveler, does not represent lions as so ferocious and formidable as we have been accustomed to consider them. They have a small horny prickle, or hook, fastened to the skin and concealed in the tassel at the end of the tail. It is easily detached, and its use is still unknown. Lions belong exclusively to the Eastern Continent, but the Puma is sometimes called the American Lion, and as it is the largest of the Cat family on the Western Continent, we shall give it a more particular notice.

Felis Concolor. The PUMA, COUGAR, PANTHER, PAINTER, CATAMOUNT.

This formidable animal is known under all these names in North and South America. Washington Irving, (see his "Astoria,") mentions it as seen at the mouth of the Columbia river. Dr. Goodman gives an account of a sportsman killed by one of these animals in the Catskill mountains. One of them, within the recollection of Dr. Dekay, was even seen a few miles from the city of New York. This animal was, no doubt, formerly found in all the Northern and Eastern States, west of the Rocky Mountains, and along the borders of the Pacific. A few yet remain in the less cultivated portions of the Atlantic States. In Florida and Texas it is quite abundant. It is also found within the tropics in Mexico and Yucatan, and has made its way through Panama into Guiana and South America, where it is called the Puma, and reaches its greatest size. From its likeness in other respects, to the lion of the old world, it is, though maneless, sometimes named the American lion. The courage of the Cougar is, however, not great, and unless very hungry or wounded and at bay, he seldom attacks man. The body is long and slender, (five feet in length and including the tail, eight;) the legs are short and stout. The general color of the Puma, when the animal is mature, is silvery grey, and hence it is sometimes called the silvery lion. In the United States the general color is tawny or fulvous; the under part is reddish white. The name "*concolor*," it obtains from its uniformity of color. The tail of the male is longer than that of the female, and without a tuft. The Puma lives much on trees, which it climbs with great ease; and its uniform dusky fur makes it so like the bark that it is not readily distinguished from the branches on which it rests. From trees, it falls suddenly upon monkeys, deer, and cattle as they pass by; or it lurks among reeds and thickets by the side of rivers and marshes, where it seizes the alligator as he raises his head above water, or crawls out upon the bank. In Florida, the animal inhabits the miry swamps and the watery everglades; in Texas he is sometimes seen in the open prairies, and his tracks are found in every crossing place of creeks and bayous where perhaps he may find some calf, cow or bullock that has been sunk and suffocated in the mire. The Cougar sometimes attacks young cattle, but is generally compelled to subsist on small animals, such as young deer, skunks, racoons, &c., or birds, and even will eat carrion when hard pressed by hunger. (Audubon.)

The Panther is nocturnal in its habits; not, however, from necessity, as it can see well in day light. It makes its way through tangled forests in searching for prey at night—perhaps arousing and affrighting some benighted traveler or wearied hunter, who has bivouacked at the foot of a large tree; and fortunate indeed is he if his rifle fail him not, or if by a burning fire-brand he can frighten away the hungry animal. At the sight of a Panther, horses are thrown into such fright that they "break all fastenings and fly in every direction." Audubon says, "a respectable gentleman of the state of Mississippi gave us the following account. A friend of his, a cotton planter, one evening while at tea, was startled by a tremendous out-cry among his dogs, and ran out to quiet them, thinking some person, perhaps a neighbor had called to see him. The dogs could not be driven back, but rushed into the house. He seized his horsewhip which hung inside the hall door, and whipped them all out, as he thought, except one, which ran under the table. He then took a candle, and looking down, to his surprise and alarm, discovered the supposed refractory dog to be a Cougar. He retreated instanter; the females and children of the family fled, frightened half out of their senses. The Cougar sprang at him—he parried the blow with the candle-stick, but the animal flew at him again, leaping forward perpendicularly, striking at his face with the fore feet, and at his body with the hind feet. These attacks he repelled by dealing the Cougar straight-forward blows on its belly with his fists, lightly turning aside and evading its claws as best he could. The Cougar had nearly over-powered him, when luckily, he backed towards the fire-place, and as the animal sprang again at him, dodged him, and the panther almost fell into the fire, at which he was so terrified that he endeavored to escape, and darting out of the door, was immediately attacked again by the dogs, and with their help and a club, was killed."

The female has three, four, and even five at a litter, but the usual number is two. She shows great affection for her young, never leaving them except to obtain food to support her strength.

Felis Tigris, (regalis.) The ROYAL TIGER. (Pl. IV. fig. 2.)

This animal infests Hindostan, and the parts of Asia between

Bengal and China. It is nearly equal to the lion in size, and though inferior to him in strength, surpasses him in activity and rapidity. Whole villages are sometimes depopulated by this most dangerous animal. The tigress has five cubs at a time, which are easily tamed but not to be trusted. Among American Tiger Cats may be enumerated the Ocelot, *F. pardalis*, of Tropical America; the Chati, *F. mitis*, (mild,) of South America, about one third larger than a cat, and the Pampas, or Jungle Cat, *F. Pajeros*, which lives on Guinea pigs. The Nepaul Tiger Cat, *F. nepalensis*, is two and a half feet long, including the tail. The Serval, *F. Serval*, an African Tiger Cat, plays like a kitten, and looks very cat-like.

The LEOPARDS, or SPOTTED Cats, are numerous, and found on both continents. They are distinguished for beauty and elegance. Their color, in the East, is a pale yellow, covered with rosettes of black, which contract into spots about the head, neck and limbs. The general length is about four feet and the height about two. The Leopard preys upon antelopes, deer and monkeys. So great is the flexibility of its body that it can make surprising leaps, swim, climb trees or crawl like a snake, with nearly equal facility. These animals are fierce and rapacious, and it is remarked that "though they are ever devouring, they always appear lean and emaciated."

The JAGUAR, *F. onca*, is the Leopard of this Continent. This formidable animal inhabits Mexico, and is met with in almost every part of Central America. In common with many of this family, he is often called the Panther. The Cheetah, *F. jubata*, is the Hunting Leopard of the Cape of Good Hope, and combines in some degree, the habits of both the cat and the dog. Its specific name *jubata*, (Lat. crested,) is derived from the thin mane running down the neck.

The LYNXES are distinguished by their tufted or tasseled ears, and shorter bodies and tails. Eight species are described. The Wild Cat, or Bay Lynx, *Lynx rufus*, looks most ferocious, but flies from its pursuers, moving by bounds or leaps, and raising all the feet at the same time from the ground. The Canada Lynx, (*L. Canadensis*.) is more retired in its habits, and its fur furnishes the most beautiful materials for muffs, collars, &c. The Caracal, *F. Caracal*, takes its specific name from the black tips of its ears, the word in Turkish meaning black. Its body is longer and more slender than in the true LYNXES. It is called the "Lion's provider." The domestic and the wild cat are supposed by many to be of distinct species. A marked difference is shown in the tails of the two; that of the wild cat is bushy and short, while that of our tame cats is long and slender. The varieties are numerous; among the most noted are the Tabby, or Brindled; the Maltese, of a bluish hue; the Tortoise-shelled or spotted; the Angora; the Egyptian; and the Manx Cats, of the Chartreuse, a species that have no tails. Another variety are said to have the fore paws divided into two parts. The cat is more attached to places than persons; is sly and suspicious; loves her ease and seeks the softest places for her bed; is fond of catnip and valerian; and is a great favorite, particularly with children. She is fond of rats, mice, squirrels and birds, and notorious for thievish propensities; dislikes cold water and bad smells. Her hair is electric, and always dry and glossy; average age, 14 years.

What three divisions compose the CARNIVORA PROPER? From what is *digitigrada* derived? What does it include? By what are they distinguished? What are their usual habits? At what time do they seek their prey? Which hunt in bands? At what time?

FELIDÆ.

What does *felis* signify? What does this family include? What is their character, and how do they resemble each other? To what are they compared? What renders them particularly dangerous? In what way do the teeth and jaws of the FELIDAE differ from those of the preceding orders? What is said of their feet and claws? How are the claws protected? By what are they elevated above the ground? What is said of the tongue? What causes the roughness? What do these hooks assist them to do? What is said of their sight and of the shape of their eyes? Of what use are their whiskers, and with what is each connected?

Which is the strongest and most courageous of the feline tribe? What is he called? How regarded? Where referred to? What species are here mentioned? What is said of the lioness and her young? Are they long lived? How does Anderson, the African traveler who was recently trodden to death in that country by elephants, speak of them? What is concealed in the tuft of hair at the end of the tail?

Where do lions belong? Which is the largest of the American Cats? Under what names are they known? Where have they been found? Where is it still found? Why is it called the American Lion? Why the Silvery? What is said of its courage? What of its general color in the United States? How do the male and female differ? On what does the Puma live? What is said of the appearance of its fur? Of what advantage is this? How does it secure its prey? Where are its haunts in Florida? Where in Texas? What does the Cougar attack? On what does it usually subsist? What are the habits of the Panther? Define and spell, Noc-tur-nal, (Lat. *nocturnus*, by night, from *nox*, night.) Di-ur-nal, (Lat. *diurnus*, by day, from *dies*, day.) Crepuscular, (Lat. *crepusculum*, twilight.) Are Panthers attached to their young, and what is their usual number?

How does the Royal Tiger compare with the Lion? What countries does he infest? What is said of his ravages? How many cubs has the tigress? When tamed, are they trustworthy? Where is the Ocelot found? Where the Chati? What is said of its size? Where is the Jungle Cat found? On what does it live? What other Tiger Cats can you mention?

Are Leopards or Spotted Cats numerous? Where are they found? For what are they distinguished? What is their color in the East? What their usual length and height? Upon what do they prey? What is said of the flexibility of their bodies? What of their disposition? Which is the Leopard of this Continent, and where found? What is he often called? Where is the Cheetah found? What called? What habits are united in him? What is the meaning of the specific name *jubata?* Why given?

How are the Lynxes distinguished? How many species are described? What is said of the Wild Cat, or Bay Lynx? What of the Canada Lynx? From what does the Caracal take its specific name? From what language is the name derived? What does it mean? Is the Caracal larger or smaller than the true lynxes? What is it called?

Are Domestic and Wild Cats of the same or different species? What is a plain difference? Mention the most noted varieties. Give the character of the cat. What is her average age?

SECTION XI.

SUB-FAMILY HYAENINA.

HYAENA, (Gr. *ὕαινα*, Huaina.)

The Hyaena has the head and feet of a fox, and the intestines of a civet. Linnæus placed it between the wolf and fox. It is one of the most ferocious, malignant and carnivorous of animals. There are three species, the Striped, (*H. striata,*) the Villose, (*H. villosa,*) and the Spotted, (*H. maculata.*) The Striped is the *H. vulgaris*, or Common Hyaena, (see Chart.) It often deceives its pursuers by feigning lameness at the commencement of a chase. It dwells in caverns and rocky places; prowling about at night to feed on dead animals, or such living prey as it can seize, seldom, however, assailing man unless in self defence. Hyaenas are useful as feeders on carrion, in cleansing the region where they dwell of the decaying remains of larger animals, and preventing the increase of poisonous effluvia. They are found in the train of armies, whose slain they feed upon, and sometimes even tear newly buried corpses out of their graves.

VIVERRIDAE, (Lat. *viverra*, a Ferret.) The CIVETS.

This entire group are noted for their perfume, which is secreted in a glandular pouch near the tail, and is of some importance as an article of commerce. It is called *Civetta*, (Arabic, *Zibetta,*) meaning scent or perfume, and gives name to the animal. They are nocturnal and predatory; inhabit Africa, Asia and the adjacent islands, and are particularly numerous in Abyssinia. Their general appearance is like that of the fox.

GENETTA, (Fr. *Genette.*) The GÉNETS or WILD CATS.

The Genets are similar to the Civets, but in contour of body, are most like the Weasels, having long and slender forms, short limbs and sharp pointed muzzles. They give out the same odor as the Civets, though the odoriferous pouches are much reduced in size.

HERPESTES ICHNEUMON, Pharaoh's Rat, or Mangouste.

This beautiful little animal is appropriately called *Herpestes*, (Gr. a creeper,) and *Ichneumon*, (Gr. a tracker.) It was anciently ranked among the sacred animals of Egypt; destroys reptiles and young crocodiles, and thousands of crocodile's eggs. It is kept tame in the houses of the east, to destroy unpleasant intruders.

What is said of the mixed form and nature of the Hyaena? What of its disposition and habits? How many species are there? Describe the one figured on the chart. Give its zoological gradations. Ans. The Common or Striped Hyaena is of the VARIETY, striata; SPECIES, vulgaris; GENUS, Hyaena; sub-family, Hyenina, FAMILY OR TRIBE, Canidae; SUB-ORDER, Digitigrada; ORDER, Carnivora; CLASS, Mammalia; WARM BLOODED division of the SUB-KINGDOM, Vertebrates, the highest branch of the ANIMAL KINGDOM. Give the meaning of these several gradations. Trace out every genus studied by the class in this way. To what deceptive expedient do the Hyaenas resort? Do they often assail man? Where do they live? Upon what do they feed? What do they sometimes do?

For what are the Civets noted? From what is their name derived? Where are they found? What is their general appearance?

What are the Genets most like? In what do they resemble the Civets?

How was the Ichneumon ranked, and for what is it useful in Egypt?

SECTION XII.

2. DIVISION OF THE DIGITIGRADES.

The CANIDAE, (Lat. *canis*,) a dog. (Pl. IV., fig. 2.)

This includes a large number of animals, some of which, in particular respects, resemble the Cats; others, the Weasels and Bears. The dog has, from olden time, been the friend and companion of man; yet some uncertainty still exists as to its original stock. It is quite like both the Wolf and the Jackal. Some naturalists incline to assign it a common origin with the former; others have identified it with the latter. The balance of the argument, however, seems in favor of the wolf as the original source from which the domestic dogs have sprung. Their skulls and skeletons are similar. The period of gestation is sixty-three days in both. Both open their eyes the tenth or twelfth day, and live fifteen or twenty years. We, however, prefer the position that when man first went forth to till the ground whence he was taken, the dog was given him by the Creator as his assistant and ally. The relation which he sustains to man differs much from that sustained to him by other animals. The dog is alone identified with his master's interests and occupations. Other animals may *endure* his rule; to the dog it seems a pleasure. He knows his looks, his voice, his walk, rejoices at his approach, and shows himself his willing defender. The classic scholar will remember that Homer, in the true spirit of nature and of poetry, represents Ulysses as recognized on his return to Ithaca by his old and faithful dog alone, which died with joy at his feet. The value of the dog's services, in the early stages of society, and in preparing the way for civilization, affords confirmation of our idea concerning its origin. In wild and uncultivated regions, and especially in northern latitudes, the very existence of man is often dependent upon the fidelity and ever ready aid of the dog. "He is the only animal which has followed man through every region of the earth." The intimacy of relation implied in this remark of Cuvier should be qualified in respect to its extent, as it is well known the Jews, Mohammedans, and Hindoos, regard the dog as impure and abominable, and will not touch it without ablution. The teeth of the Canine family, (including dogs, wolves and jackals,) are forty-two in number. The muzzle of these animals is more or less lengthened; the tongue small, and the pupil of the eye circular. The fore-feet have five toes; the hind feet four, and sometimes a fifth; the toes are not retractile.

Domestic Dogs.

Their legs are long, and hence their stature is elevated. Though carnivorous, their ferocity is not generally equal to their strength. They obtain their prey, not by sudden bounds, but by hunting it down by the aid either of sight or smell, often associating in packs for that purpose. Martin makes seven divisions, containing fifty varieties of the Domestic Dog.

Facts almost innumerable illustrate the docility, sagacity, and memory; the courage, faithfulness and love of this animal. The Esquimaux dogs, included in the first division, are peculiarly valuable to the dwellers in Arctic regions. They are used by them in pursuing the seal, the bear and the reindeer. Yoked to heavily laden sledges, they often drag them with untiring patience, fifty or sixty miles in a day. Capt. Parry's "Journal of a Second Voyage for the Discovery of a North West Passage," and Dr. Kane's "Arctic Explorations," abound in graphic descriptions of the manners of the Esquimaux themselves, and in interesting particulars showing the utility of their dogs. With good sleighing, six or seven of these dogs will draw from eight to ten hundred weight, at the rate of seven or eight miles an hour, for several hours together.

The GREYHOUND, (one of the second division,) is the swiftest of all the dogs, and is used principally in the chase of the hare.

The NEWFOUNDLAND dog is so named from the place whence it originated. It is not to be confounded with the Labrador dog, which is a larger and stronger animal. Both are trained to draw sledges and light carriages. The Newfoundland dog is well known for his care in guarding the property of his owner. He is remarkably fond of the water, and will bring out any object which his master points out in the water, and place it at his feet. Many have been rescued by this dog from a watery grave. He evinces the greatest fidelity and affection towards those who take care of him.

The WATER SPANIEL, (of the fourth division,) delights in taking itself to the water, which it does in pursuit of game. It is useful to persons who are shooting wild ducks, or water hens, as these fowl conceal themselves so closely that without aid they cannot be discovered. It will dive to a considerable depth, and bring up any small object from the bottom.

The BLOODHOUNDS, (fifth division,) are noted for the acuteness of their smell, and can trace a man or an animal with unfailing certainty. Sometimes they have been used in the capture of thieves, especially sheep stealers. It is about two feet four inches in height, and has a voice peculiarly deep, and that may be heard a considerable distance.

The MASTIFFS, (sixth division,) are distinguished by the shortness of the nose, and the breadth of the head, which is caused by the large muscles that move the jaw. Its powerful frame and deep voice have led to its selection as a house guard against burglars.

The TERRIERS, (seventh division,) are used for destroying rats and other vermin, and will boldly invade the covert of the fox or the badger. They become strongly attached to their masters, and can be taught many tricks for their amusement.

The Shepherd's dog, (of the same division with the Esquimaux,) is a rough and shaggy animal, having sharp pointed ears and nose.

To the shepherd it is an invaluable assistant. In point of intelligence, thoughtfulness and promptitude, it is not probably excelled by any of the varieties of dogs. A story is told of a dog belonging to the "Ettrick Shepherd," who had 700 sheep under his care. On a certain occasion, they broke away in the middle of the night, and in spite of every effort of the shepherd and his assistants, roamed to a distance across the hills. "Sirrah," said the afflicted shepherd to his dog, "Sirrah, my man, they're a' awa." Away went the dog in the darkness, the shepherd and his companions meantime scouring the hills, but seeing nothing of the flock or the dog. The next morning they found them at the bottom of a deep ravine, not one lamb of the whole flock missing, and the dog standing in front of them, keeping watch.

On the Alpine summits of St. Bernard, remarkable for its hospital, and covered with the snows of a ceaseless winter, the resident monks have been often known to issue forth in the midst of tempests and snow storms, and by means of their large dogs, of peculiar breed, have discovered travelers unable to track their way, and saved them from the cold embrace of death. We subjoin the following as illustrating the powers of *imitation* and *memory* possessed by the dogs.

A few winters since, a gentleman in Lawrence, Mass., one morning when the snow was covered with a smooth icy crust, noticed a little dog seated on his haunches, sliding down the steep bank before his house. He supposed that the dog had slipped, but noticed as he reached the bottom of the hill, he ran up again. He continued his sport for some time, apparently with great delight. P. H. Gosse, in his article on the dog, relates that "Lord Combermere's mother, (Lady Cotton,) had a terrier named Viper, whose memory was so retentive that it was only necessary to repeat to him once the name of the numerous visitors at Combermere, and he never afterwards forgot it. Mrs. H. came on a visit there on a Saturday. Lady Combermere took the dog up in her arms, and going up to Mrs. H. said, "Viper, this is Mrs. H." She then took him to another newly arrived lady, and said, "Viper, this is Mrs. B.;" and no further notice was taken. Next morning, when they went to church, Viper was of the party. Lady Cotton put a prayer book in his mouth, and told him to take it to Mrs. H., which he did, and then carried one to Mrs. B., at his mistress's order."

A man in Windsor, Vt., owned a large and valuable Mastiff dog, which had the misfortune to break his leg. The owner, after trying in vain to set the bones himself, sent for a physician, who speedily put the bone in its place, and splintered up the leg.

For several days the doctor visited the dog, and dressed the wound, and then told the owner he should come no more, but if any thing seemed to be wanting, to bring the dog to his office. He did so two or three times, and when he ceased going, the dog would go alone to the doctor's office every morning, and lie down until the doctor looked at his leg, and then he would return, continuing this practice until he was fully cured. Some time after this, the great dog found in the street a little one, with a broken leg; and after smelling around him for some time, he got him up on his three legs, and managed to get him to the before mentioned doctor's office, where he waited with the little dog, until the doctor came and set the bone.

Canis vulpes, (Lat. *vulpes*, a fox.)

The Fox is about the size of a small dog. He is by nature suspicious, timid and cunning; his sight is keen; his smell and hearing so acute that it is difficult to take him in any kind of trap. Unmolested, the fox lives from twelve to fourteen years; the first year he is called a cub; the second, a fox; and the third, an old fox. Audubon enumerated twelve species, four of which exist in North America. The skin of the Silvery Fox, (*C. argentatus*, Lat. silvered,) of Labrador, has been sold in London for five hundred dollars. Its fur is copious, and of a beautiful, lustrous, black hue, with the longer hairs of a silvery white. It is found in Oregon, and the northern parts of this continent.

The Common Fox of Europe, *Vulpes vulgaris*, is there the favorite object of the chase. The American Red Fox, *C. fulvus*, (Lat. tawny,) is somewhat larger; its fur is finer, and of a brighter color, and it has a larger brush tail. It eats fish as well as rats, rabbits, &c. The Swift-Fox, *C. velox*, (Lat. swift,) is the smallest of the fox tribe. The Cross-Fox derives its name, *C. decussatus*, (Lat. divided cross-wise,) from its markings, not from its nature. The Gray Fox, *C. cinereus*, (Lat. ash-colored,) is the annoyance of the southern planter, as the Red-Fox is of the northern farmer. The Arctic Fox, *C. Lagopus*, (Gr. *Lagos*, hare, *pous*, foot, Hare's-foot,) is covered with white woolly fur. The Antarctic Fox, *C. Antarcticus*, is called the Wolf-Fox, from its resemblance to that animal. It is tame, and barks like a dog. The Caama, *C. Caama*, is the smallest African fox. The Fennec, or Zerda, *C. Zerda*, whose place has been so often discussed by naturalists, has the skeleton and teeth of the dog family. Its fur is short and silky.

C. Lupus, (Lat. a wolf.) The Wolf.

The Wolf, in its habits and physical development, we have already intimated, is closely related to the dog. His proportions are larger, and his frame more muscular than those of that animal, and between the two there exists a most inveterate hatred. The well known traits of the Wolf are ferocity, cunning and cowardice. In the earlier periods of English history, it is often adverted to as a common and dreaded pest. In consequence of its ravages, many of the early British kings and chieftains, as if to render themselves more formidable, adopted its name with certain adjuncts. This is seen in such names as Athlewolf, (noble wolf;) Berthwolf, (illustrious wolf;) Eadwolf, (prosperous wolf,) etc. It was finally extirpated in England, about 1350, in Scotland, about 1600, and in Ireland, about 1700. It is still abundant in the northern countries of Europe, and in France and Western Asia. Wolves always hunt in packs, and evince great craftiness in waylaying and pursuing their prey. Sometimes they form a semicircle and advance upon the animal which they would reach, in this way forcing it over a precipice, or gradually hemming it in so as to prevent its escape. Winter is the time when they are most dreaded by those living in the regions which they inhabit. Then as hunger renders them peculiarly ferocious and daring, they, with the greatest obstinacy, follow after their prey, whether it be man or animal. Under the gnawings of famine, they will devour every sort of offal, and even disinter the dead. It is related that in the reign of Louis XIV. a large party of dragoons were, in the depth of winter, attacked at the foot of the Jural mountains, by a numerous band of wolves. The dragoons fought bravely, and killed many hundreds of them; but at last, overpowered by numbers, they and their horses were all devoured.

Of the Wolf, many varieties are found in both continents. The *C. lupus*, Common Wolf, is of a yellowish or fulvous gray color; covered with harsh and strong hair, and from twenty-seven to thirty inches high at the shoulders. Of this there is a variety, white, either as an albino, or as the effect of a northern or cold climate, also found in both continents, viz. *C. lupus albus*. The wolves of Lapland and Siberia are almost all of a whitish gray color; those of the Alps in Europe, and the Rocky Mountains of North America, become white or nearly so. The length of the American White Wolf (*albus*) is about four and a half feet, it being the largest of all the varieties of this animal. The Black American Wolf, *C. Lupus*, (*Niger*,) is of the same shape as the Common American Wolf, and rising three feet in length.

Packs of this animal, showing various shades approaching black, have been found occasionally in every part of the United States. In Florida the prevailing color is black. This is the most numerous variety among the Pyrenees of Europe, and also south of those mountains, where it is of larger size than the common wolf. Several varieties of wolves are met with in Asia. Those of Asia Minor are deeply fulvous, and show more of red than the wolves of Italy.

Numbers of such as the *C. lupus* (*nubilus*,) the Dusky Wolf, the Black Wolf, *C. lupus* (*niger*,) are found on the sandy plains east of the Rocky Mountains. They go in droves, and hunt deer by night, with dismal, yelling cries, and woe to the foxes if they find them on a plain at any distance from their hiding places! In the same districts, and associating in greater numbers than other wolves, are found the *C. lupus*(*latrans*,)the Prairie or Barking Wolf, intermediate in size between the large American Wolf and the Virginia Fox, and in many respects like the fox. In its bark or howl it greatly resembles the latter animal, as well as the domestic dog of the Indians. Their general color is ashy gray; their length two feet, ten inches. They are well known to the inhabitants of the western parts of Arkansas and Missouri, and to those who live on the borders of the Upper Missouri and Mississippi rivers.

Their skins are of some value, the fur being soft and warm, and constitute a part of the exportations of the Hudson Bay Company. The Prairie Wolf is found in California and Texas, and on the eastern side of the mountains of New Mexico, as well as on the western prairies.

C. lupus, (*rufus*.) The Red Texan Wolf resembles the common gray variety, is more slender and light than the White Wolf of the North-West part of this continent, and has a more fox-like aspect. The hair is not woolly like that of the White Wolf, but lies smooth and flat. The length is two feet, eleven inches. In habits, it is nearly like the Black and White Wolf. It is said that "when visiting the battle fields of Mexico, the wolves preferred the slain Texans or Americans, to the Mexicans, and only ate the bodies of the latter from necessity, as owing to the quantity of pepper used by the Mexicans in their food, their flesh is impregnated with that powerful stimulant." Audubon, in referring to the geographical distribution of this animal, remarks of quadrupeds generally, that toward the north they are more subject to become white; toward the east, or Atlantic side, gray; to the south, black; and toward the west, red.

C. aureus, (Lat. golden.) The JACKAL. This animal is found throughout the Levant, in Persia, India and Africa. It is called "*aureus*" on account of the yellow tint of its skin. The Jackal is supposed to be the fox of the sacred writers, (Judges xv. 4, 5.) Like the wolf, it hunts in packs, pursuing the antelope and other animals for prey, and making away with carrion in every state of putrefaction. It has been called the "lion's provider," for when the cry of the Jackal is heard, the Lion, aware of the cause, makes his appearance, and without ceremony seizes upon the booty. The Jackals, however, retaliate by aiding in the consumption of the larger prey which the lion destroys. They are useful in the east as scavengers, consuming the offal which in oriental cities is thrown into the streets, and might otherwise breed pestilence. Grapes are the special delight of the Jackal, and it often makes great havoc in vineyards. When hunting, these animals utter most piercing shrieks, which produce, it is said, a very terrific effect, "as resounding through the stilly darkness of night, and answered from a thousand throats."

The Jackal is rather larger than the fox, but its tail is shorter and less bushy. It is easily tamed, and is dog-like in disposition and habits. One species of the Jackal, *Canis Corsac*, the ADIVE, is not larger than a pole-cat, has a long tail, and is found in troops amidst the deserts of Tartary. Other species are the Cape Jackal, *C. mesomela*, (Gr. *mesos*, middle, *melas*, black,) and the *C. anthus*, (Gr. *anthos*,) of Senegal. All agree in manners and general disposition, and in exhaling a strong and offensive odor, which, however, is "scarcely perceptible" in a state of domestication.

Proteles Lalandii. The AARD-WOLF, or EARTH-WOLF, of South Africa. This animal has interest as connecting together the Civets, Dogs and Hyaenas. It has the bones and external appearance of a hyaena, the head and feet of a fox, and the intestines of a civet. The fore legs are considerably longer than the hind ones, and in this respect it is also like the hyaenas. It is about the size of a full grown fox, yet stands higher on its legs; but for its more pointed head, and the additional fifth toe of the fore feet, it might, at first sight, be easily mistaken for a young hyaena. The color is a pale ash, with a slight shade of yellowish brown. The fur is woolly, except the mane, which is coarse, stiff hair, and bristles up when the animal is provoked. One of these animals was brought from Africa, by the traveler, Lalande, from whom it received its specific name. The generic term is from the Greek *protelès*, and relates to the superior length of the fore legs.

Spell CANIDAE and give its derivation. What does the second division of DIGITIGRADES include? What other animals do they resemble? What is said of the origin and antiquity of the dog? What of their resemblance to Wolves and Jackals, and in what respects do they agree? How do his fidelity and attachment compare with those of other animals? In what state of society and in what regions is he particularly valuable? What does Cuvier remark respecting the dog? What qualification does this remark require? What is said of their teeth, muzzle, tongue, eyes, feet, claws, &c.? How many varieties of domestic dogs does Martin make? Does their ferocity equal or surpass their strength? How do they obtain their prey? To what people are the Esquimaux dogs of great value? What use is made of them? What works give interesting particulars respecting them? What is said of the Greyhound? Why is the Newfoundland dog so named? What is said of his fidelity and affection? What use is made of him? In what does the Water Spaniel delight? For what is it useful? For what is the Bloodhound noted? How are Mastiffs distinguished? Of what use are Terriers? In what does the Shepherd's dog excel? Relate the story of the Ettrick Shepherd and dog. What is said of the dogs of St. Bernard? For what are they trained? What anecdotes can you give showing the imitative power and memory of dogs? What is the size of the common fox? Describe him. How long does he live? How many species did Audubon enumerate? How many are found in North America? What fox furnishes the most valuable fur? Where is it found? Name and characterize the other principal species.

How do the dog and wolf compare with each other? What are the traits of the wolf? What is said of it in English history? From what places has it been extirpated? Where is it still abundant? How do wolves hunt? When are they most dreaded? Why at that time? What occurred near the Jural mountains? Are there many varieties of the common Wolf? Describe it. What is said of the Red Texan Wolf, and its preferences? What does Audubon say of the changes of color in quadrupeds?

Where is the Jackal found? What is it supposed to be? What is it called? Why? Of what is it particularly fond? What is said of Jackals' hunts? Are they of any use? What is their size? Why called *aureus?* Are they easily tamed? What is said of the Adive? What of the Aard, or Earth Wolf? What does it connect? Describe its habitat, size, color, fur, &c. &c.

SECTION XIII.

3. Division of the Digitigrades.

Mustelidae, (Lat. *Mustela*, a weasel.) The Weasel Tribe.

The weasels are readily distinguished by their long snake-like bodies, short muzzle, sharp teeth and predatory habits. Their relish for blood is strong. In pursuing their prey, they are bold, cautious and resolute, creeping toward their unsuspecting victim, usually a rabbit, rat or bird, and on a sudden, darting at it, and piercing its neck with its sharp teeth. Fixing themselves where some large vein invites them, they hang on until their prey expires, devouring its brain, and sucking its blood; but almost always leaving the flesh untouched. Their head is small, oval and flattened, and their bodies so pliable as to be capable of being insinuated into holes and crevices which it would seem they could not possibly enter; and their short strong limbs and sharp claws, enable them to climb with the greatest celerity and adroitness. In their habits they are more or less nocturnal. According to Audubon, about twelve species of the true Martens are included in this family, four of which inhabit North America.

Mustela vulgaris, (or *Putorius Vulgaris.*) The COMMON WEASEL.

This is the smallest of the tribe, and well known, especially by farmers, as they often have occasion to lament its onsets upon their young broods of poultry. For this, however, they have some compensation in the destruction, by this animal, of numerous rats and mice that infest their barns and out-houses; so that it is sometimes said, weasels "ought to be fostered as destroyers of vermin, rather than extirpated as noxious depredators." This active little creature is sometimes tamed, and by its playfulness and *unexpected* display of affection, has awakened much interest.

M. Erminea, or *Putorius Ermineus.* The STOAT, or ERMINE.

This species closely resembles the Weasel, but is a third larger, being about the size of a cat. In the summer, its general color is a yellowish brown, when it is called a STOAT; but it changes to a pure white in winter, when its fur is extremely beautiful, and it is called ERMINE. It is abundant in the northern parts of this continent, and in Europe and Asia. The fur of the Ermine is closest and most purely white in the most northern latitudes, and constitutes a valuable article of commerce. The white skins of this animal usually bring from ten to fifteen dollars per hundred. The tail remains black at the extremity, during all the changes of the color. Formerly, the official robes of judges and magistrates were lined with this fur. In predatory habits, it is like the kindred species. Hares and rabbits fall easy victims to this animal, which kills them with a single bite, penetrating to the brain. It frequents stony places and thickets, and in a short race will outstrip a dog.

Mephitis, (Lat. *a noxious odor or exhalation.*) MEPHITIC WEASELS.

The animals of this genus are so named from the intolerable odour which, when irritated, or for self-protection, they give forth. They have on their fore feet nails, strong and well suited for digging. The distinguishing color of the genus is black, striped with white, lengthwise along the back, and the tail is long and bushy. The Mephitic weasels all move slowly; seldom flee from man, unless when they are near their burrows. Though feeble and insignificant in some respects, yet they seem conscious of a power to "annoy beyond the point of endurance." Large numbers of them are sometimes found in the same hole. They feed on poultry, birds, eggs, small quadrupeds and insects. The head is short; the nose rather projecting; the snout generally blunt; the hairs on the tail are very long. Seventeen or eighteen species have been enumerated; one in South Africa, two or three in the United States, and the rest in Mexico and South America; but of these species there are almost endless varieties in respect to color and markings.

Mephitis Americana or *M. Chinga*. The COMMON AMERICAN SKUNK.

This animal is about as large as a cat, and generally is of a blackish brown, with white stripes running lengthwise on the back. In the markings of white, it shows many diversities, and it has a long bushy tail. All the varieties of this animal have a broad fleshy body, not unlike that of the wolverine. Its legs are short; the fur is rather long and coarse, intermingled with much longer smooth and glossy hairs. Its length from the point of the nose to the root of the tail is seventeen inches. No quadruped found on this continent is more universally detested than the skunk. The offensive fluid is contained in two small sacs situated near the root of the tail. By day it is so thin and transparent as to be scarcely perceptible; but at night has a yellow luminous appearance. He is himself a very cleanly animal, never suffering a drop of the fluid to touch his fur, nor does his burrow give forth any offensive smell. In the northern states, this animal retires to his burrow about December, and is not seen again till the following February. In the southern states he does not go into winter quarters, but continues to prowl at night during the winter. It is said his flesh is "well tasted and savory," and cooked and eaten by the Indians.

The LONG or LARGE TAILED SKUNK, *M. macroura*, (Gr. μακρος, *makros*, long, ουρα, *oura*, tail,) common in Mexico and Texas, is of the size of a common cat, and has five or six young at a time.

The *M. Zorilla*, or CALIFORNIA SKUNK has white spots on the forehead and on each temple, and four white stripes on the sides and back, with a broad tip of white on the tail; in form is a small image of the common skunk, and like it, so offensive as seldom to be approached. The African Zorilla, found at the Cape of Good Hope, has the tail spread out in the form of a plume, and does not give out the overpowering odor of other species.

The *M. mesoleuca*, (Gr. μέσος, *mesos*, middle, λευκος, *leukos*, white.) The MEXICAN SKUNK has the long and under fur of the whole back and the tail, white. The long tail of this animal is

4

often first seen in the high grass and bushes, and makes a beautiful appearance.

The Teledu, or Skunk of Java and Sumatra, *Mydaüs meliceps*, (Lat. *melis*, a badger, *caput*, a head,) has a short tail covered with a mere pencil of hairs. In some things, it reminds one of a hog.

M. Martes. The MARTEN. Of this there are three varieties, the Common Beech, or Stone Marten, the Pine Marten, and the Sable, of which the furs are exquisitely soft and beautiful. Their agile and graceful motions are not excelled by any of the Weasel tribe. They reside in woods, and prey chiefly on birds, and small animals. They also feed on rats, mice, and moles, and will sometimes eat seeds and grain. The general length is about a foot and a half; the tail is ten inches long, bushy, and of a darker color than the other parts. The Marten is of a dark, tawny color, with a white throat, and the under part is of a dusky brown; the muzzle is pointed, and the eyes bright and lively. The fur is of two sorts; the outer is long and brown, with varying shades, in different parts of the body; the inner, very soft, short and of light yellowish gray color.

M. fagorum, (Lat. of beech-trees.) The BEECH MARTEN is a variety with a white throat, found in Northern and temperate Europe, and Western Asia. It approaches the habitations of men oftener than the Pine Marten, resorting for prey to the vicinity of farm yards and homesteads. Its fur, which is much inferior to that of the Pine Marten, is called in trade, the Stone Marten. Many skins of this animal are obtained from the north of Europe, and the fur is dyed to represent Sable; though the practised eye easily distinguishes it from the latter. The richest furs of this Marten come from the most northern latitudes.

M. Abietum, (Lat. of fir-trees.) The PINE MARTEN. This variety with a yellow throat, varies much in color, so that it is difficult to find two specimens alike, but generally is yellowish, blended into a blackish hue in other parts. It is found in Mount Caucasus, and the northern parts of Europe; and is very numerous in the wooded districts of the northern latitudes of this continent. It is particularly abundant where the trees have been killed by fire, but are still standing. Specimens have been obtained from near Albany and the Catskill mountains, and the northern parts of Pennsylvania. Its southern range is about lat. 40°, and the northern about 68°. The length is one foot five inches. This Marten is, in its disposition, shy, cruel, cunning and active; does not approach the residences of men, but keeps rather in dense woods. The fur of this animal is valuable, next to the Sable; and when in fashion, Marten skins bring good prices. It is sometimes dyed, and efforts are made to palm it upon buyers as fur of a more costly kind. The Hudson's Bay Company have sold as many as 14,000 skins in a year, and upwards of 30,000 have, in the same time, been exported from Canada by the French. According to Sir John Richardson, Martens of the finest and darkest fur "appear to inhabit certain rocky districts." The flesh of the Pine Marten is rank and coarse, but is eaten by the Indians. In confinement, it appears tolerably gentle, and loses much of its "snappish character." They are trapped only in autumn and winter.

Mustela Zibellina. The SABLE. This is the most celebrated of all the Weasel tribe, not only on account of the richness of its fur, but from the perils connected with the chase of it, carried on in the depth of winter, and in regions the coldest and most desolate traversed by human footsteps. It has long whiskers, rounded ears, large feet, (the soles of which are covered with fur,) white claws, and a long bushy tail. The general color of the fur, of which the hair lies each way, is brown, with the lower part of the neck and throat grayish. These animals inhabit the northern parts of Europe and Asia. Vast numbers of them are killed in Siberia, and their skins form a very considerable article of commerce among the Russians. Sables' skins are in the highest perfection between November and January; and within that time they are sought after by large numbers of hunters. They are taken in snares, or traps, which are usually pit-falls, with loose boards placed over them, baited with flesh. Sometimes fire-arms and cross-bows are used in taking them.

Putorius Vison. The MINK. This animal is of a brown color, with a white chin and short ears. The feet and palms are covered with hair to the extremity of the nails, and the feet are semi-palmated. It is smaller than the Pine Marten, being thirteen inches long, and the tail is half the length of the body. It presents varieties which are striking and permanent, both in respect to size and color. Next to the Ermine, it is the worst depredator that prowls about the poultry yards of the farmer. The Mink catches rats like the weasel or ferret, holding them by the neck like a cat, and it has no *aversion* to fish; trout and salmon seem to be special favorites. It will steal them when it can, or dive after them in brooks and shallow water, swimming with considerable facility, and like the muskrat, diving at the flash of a gun. It resides of preference on the borders of ponds, and along the banks of small streams. This species is very numerous in salt marshes of the southern states, where it subsists principally on the marsh-hen, the sea-side finch, and sharp-tailed finch. It has not much cunning, and is easily taken in any kind of trap. When taken young, it becomes very gentle, and much attached to those who fondle it. It does not emit its unpleasant odor except when it is hurt. The skins of the Mink have been used for making muffs, tippets, &c., and sold for about fifty cents each. Some skins are of a beautiful silver gray color, the fur being quite unlike that ordinarily obtained. Such skins are rare; six of them suffice to make a muff worth at least a hundred dollars. (Audubon.) The Mink is constantly found in almost every part of North America.

Mustela furo, (Lat. *I rage.*) The FERRET. This useful but ferocious little animal is kept in Europe, in a domesticated state, and is employed for rabbit-hunting, and for destroying rats. Its general form is like that of the Polecat, but it is smaller, being usually about thirteen inches in length. It has a very sharp nose, red and fiery eyes, and round ears. In the slenderness of its form, and the shortness of its legs, it resembles the Weasel.

The head of *M. Canadensis*, the FISHER, or PENNANT'S MARTEN, is more like that of the dog than that of the cat. It catches and eats fish.

The BLACK-FOOTED FERRET is about a foot and a half long; found in woody districts, as far as the Rocky Mountains. The

P. pusillus, (Lat. very small,) is the smallest of the Weasels. It is one-third smaller than the Stoat, the Polecat, or Fitchet Weasel.

M. Putorious, (Lat. *Putor*, stink,) is stouter than the common weasel. The under coat of fur is short, silky and pale yellow; the outer is of a dark chocolate brown, and long and coarse. The fur is inferior to that of the Sable and Marten, but esteemed as an article of commerce under the name of Fitch.

Lutra, (Lat. Otter,) (Gr. λούω, louô, to wash.)

This genus includes a species known as the common or river Otters, whose habits are aquatic, and whose food is fish, and also the Sea Otters. In their skulls and muzzles, there are points of resemblance to the Seal, (*phoca vitulina.*) The limbs are short and strong, and so articulated as to allow of free motion; the animal being able to turn them easily in almost any direction, and bring them on a line with the body, so as to act like fins. The teeth are sharp and strong, and the tubercles of the molars very pointed; which aids them in taking and destroying their slippery prey. Their intestines are very long. The body is covered outwardly with long and glossy hair, with a softer, shorter, downy fur, intermixed. The Otter is fierce, wild, and shy, and its habits principally nocturnal. The hunt of this animal has been a favorite, but a cruel sport. Pursued, he betakes himself to the water, where he is more than a match for the strongest dog. His determined courage holds out to the last, and pierced with spears, he dies without uttering a cry. Eleven species are enumerated.

Lutra vulgaris. The COMMON OTTER. This species is about two feet long, and its tail fifteen inches in addition. The tail is flat and broad, and the toes of the feet are connected by a complete web. In its entire structure, the animal is well adapted for an aquatic life ; diving and swimming with great readiness, and with much ease and elegance of movement. It has a black nose, and long whiskers. The ears are small and erect, the eyes very small, and nearer the nose than in most animals. The color is brown except small patches of white on the lips and nose. The size varies from two to three and a half feet. When it has seized a small fish, it immediately leaves the water and eats it, beginning with the head, while the body is held in the fore paws. Larger fish are held down by the paws, and the head and tail often left uneaten. These animals destroy multitudes of fish, in ponds and rivers, eating but a small portion of the fish, when they have an abundance of prey. When fish are scarce, and they are pressed by hunger, it is said, they sometimes go far inland and attack lambs, sucking pigs and poultry, and even feed upon larvae and earth worms. The Otter's place of retreat is beneath roots of trees, or in holes near ponds and rivers. The female bears from three to five young at a time.

The Common Otter is capable of domestication, but most readily when taken young, and fed with small fish and water. Sometimes it shows attachment, but if offended, "bites grievously." In some instances, it has been trained to catch fish, or to assist in fishing. When tamed, "they will allow themselves to be gently lifted by the tail;" though they "object to any interference with the snout, which is probably with them the seat of honor." Usually they resort to fresh waters, but in some regions frequent the sea, and hunt far out from land. Few animals show more attachment for their young than the Otter. When these are taken from them, they express their sorrow in tones resembling the crying of children.

Lutra Canadensis. The CANADA OTTER.

This is larger than the Common or European Otter, having dark, glossy brown hair, with the chin and throat dusky white, and is five feet in length. The longer and outer hairs are glossy and stiff, but the inner fur is soft, dense, and nearly as fine as that of the Beaver. The ears are closer together than in the Common Otter, and the tail flattened horizontally for half its length. The American Otter frequents running streams and large ponds, and sometimes is found on the shores of some of our great lakes. It prefers those waters which are clear, and a burrow in the banks, the entrance to which is under water. Their favorite sport is said to be sliding down steep banks, head foremost, sometimes for the distance of twenty yards. When shot and killed in water, they sink from the weight of their bones, which are solid and heavy, so that in deep water, the hunter may lose his game. The American Otter, like the European, when taken young, is easily tamed, will follow its owner, and sometimes is playful. Audubon had one which was very familiar, and much attached to him. And he relates that a landlord in the interior of Ohio, had four Otters alive which were so gentle that they would come when he whistled for them, and approach him with much apparent humility. This species ranges almost the whole of North America, but is now obtained most readily in Maryland, and the western parts of the United States. The British provinces of North America annually furnish a considerable number. Their furs are much esteemed.

Enhydra marina. (*Mustela Lutris*, Linnæus.) SEA OTTER. The generic name *enhydra*, is from the Greek *ενυδρος*, (*enudros*,) *εν*, (*en*, in,) ‘*υδωρ*, (*hudor*, water.) The palmated feet, and the teeth of this genus are so modified as to connect this Otter with the Seal, (*Otia*,) which have ears. The color is chestnut brown or black; the fur exceedingly fine and velvety; the size about twice that of the Common Otter. In length it is from four to five feet. The hind legs and thighs are short, and better adapted for swimming than in other mammalia, seals excepted; the hind feet are flat and webbed, and clothed with glossy hairs. The hair, both on the body and tail, is of two kinds; the longer hairs are silky and glossy, but not very numerous; the fur is shorter hair, exceedingly fine and soft. This Otter runs very swiftly, and swims with great rapidity, either on its back or sides, and sometimes as if upright in the water. It has very long intestines, they being twelve times as long as the animal, while those of the Common Otter are but three and one-fourth times its length. It seems to have more the manners of a seal than a land otter; haunts sea-washed rocks, and lives mostly in the water. The female brings forth its young on land, and though the animal is marine, it is found occasionally, very far from the sea.

The Kamtschatdales, on whose coasts the greatest numbers of these animals are killed, exchange the skins with the Russians, for those of the fox and sable; and the Russian merchants formerly sold them to the Chinese, at a very high price, even as high as from eighty to one hundred dollars each. The fur is not prized so high as formerly. The Sea Otter is caught by placing a net among the sea weeds, or by chasing it in boats. It inhabits the waters that bound the northern parts of America and Asia, and the seas and bays from Kamtschatka to the Yellow Sea, on the Asiatic side. and from Alaska to California on the American.

How are Weasels readily distinguished? What is their character? What their habits? How many species of true Martens does Audubon include in this family? Who was Audubon? Answer. One of the most enthusiastic, industrious and observing American naturalists. Died near New York city four or five years since, aged 76. How many of these inhabit North America? What is said of the Common Weasel? Describe the Ermine or Stoat. Describe its winter and summer dress. Which is the Ermine dress? For what was this fur particularly used? What places does it frequent? What is its pace? Give the meaning of the generic term MEPHITES. Why is this genus so called? What is said of their name? What is said of the nails of these animals? What of the tail? What is their distinguishing color? How are they striped? What is said of their movements? What gives them their power? Upon what do they feed? How many species have been enumerated? How many in the United States? How many in Africa? Where are the rest found? What is said of their varieties? To what do these varieties refer? Give some account of the Skunk. What places does it frequent? Describe the Common American Skunk, and give its peculiarities. When in the Northern States, does it retire to its burrow, and when reappear? How is it in the Southern States? What is said of its flesh? What is said of the Large Tailed Skunk? Where found? What is said of the California Skunk? What of the African Zorilla? What of the Mexican? What of the Teluda of Java, and what does it resemble?

How many varieties of the Marten? What is said of their motion? Where do they reside? What is said of the fur? Where is the Beech, or Stone Marten found? What distinguishes it? What is said of its fur? What is it called in trade? Whence are many skins obtained, and what is said of their fur? What distinguishes the Pine Marten? What is the general color? Where is it found? In what places is it particularly abundant? In what part of the United States has it been found? What is said of its fur and flesh? Which is the most celebrated of the Weasel tribe? What countries does it inhabit? At what time are the skins of the Sable in the highest perfection? How are they taken? How does the fur differ from the Marten? What others are mentioned, either on the chart, or in the text? From what animal is the fur called Fitch obtained? Mention the varieties and habits of the Mink? Where is it numerous? What use is made of its skin? What is said of the Ferret? Which is the smallest Weasel? Give the derivation of LUTRA? Describe the Otters? Repeat the description given of the Common Otter? How does the Canada Otter compare with the European Otter? Give some account of it. How extensive is its range? What is said of its fur? What is the meaning of ENHYDRA? In what respect does the Otter resemble the Seal? What is said of its size, speed, fur, &c.? For what do the Russians exchange its fur? In what waters is it found?

SECTION XIV.

III. Division of the Carnivora.

II. Plantigrada. (Lat. *planta*, sole of the foot, *gradior*, to walk.)

This name is given to those carnivorous animals which apply the whole, or part of the sole of the foot to the ground in walking. They are able to raise themselves on their hinder limbs or haunches, and easily keep an upright position. There is a slowness and heaviness in their motions; their habits are generally nocturnal, and in northern latitudes, they are in a lethargic condition during the winter.

First in order are the Ursidae, (Lat. *ursus*, a bear,) the Bears forming a connecting link between this family and the herbivorous animals. These lay the whole of the foot upon the ground in walking, which occasions their well known heavy, shuffling gait, but allows them to raise themselves with facility, and to maintain an erect position. When in this position they frequently use the fore paws in self defence, or else to strike or hug an assailant to death, by muscular pressure. The entire sole of the foot is naked. The feet have five toes each, fortified with strong, curved, and somewhat obtuse claws, adapted for digging; their grinding teeth are more or less tuberculated, and the food is either animal or vegetable. In form they are generally robust. The genera of this family inhabit both continents.

Ursus. The Bear. Of this animal, according to Audubon, eight species have been described, "three existing in Europe, one of which, the Polar Bear, is common also to America; one in the mountainous districts of India; one in Java; one in Thibet; and three in North America." The head of the Bear, is large, the body stout, and thickly covered with coarse, shaggy hair; the ears are large and slightly pointed; the limbs are stout and massive; the five toes have strong curved claws, fitted for digging rather than for taking prey; the tail is short, and usually hidden in the hair of this animal; the teeth are forty-two in number; the grinders have flattened crowns, surmounted with tubercles, and are fitted for bruising vegetables, rather than cutting flesh, and the incisor teeth give these animals but a limited power of cutting it, so that they are ranked as the most omnivorous of all the Carnivora. Some of them subsist on vegetable food alone, and nearly all are capable of supporting themselves upon it. They are nocturnal, but often seen wandering about during the day. Their habits are unsocial, most of them frequenting the recesses of mountains and caverns, and the depths of forests. In winter, they dwell in caves and hollow trees, almost without food, and comparatively dormant. In that season the female produces her young. Though widely diffused throughout both continents, they are seldom met with in Africa. Bears are said to be very fond of honey, and will climb trees in order to get at the nests of wild bees, for though clumsy animals, they are expert climbers. In Russia and other northern regions, the skins of bears are among the most useful as well as most comfortable articles of winter apparel. They are made into beds, coverlids, caps and gloves, and used also for the hammer cloths of carriages, for pistol holsters, etc.; and the *leather* prepared from them is used in harness, and for other purposes where strength is requisite.

Ursus Arctus, (Gr. αρκτος, *arktos*, a bear.) This bear is found in mountainous districts of Europe, from very high latitudes to the Alps and Pyrenees.

It was once common in Great Britain; but centuries ago was there extirpated.

This bear of Northern Europe seems to be the only one with which Linnæus was acquainted. To the people of Kamtschatka it gives the necessaries, and even the comforts of life; its skins forming their beds and coverlids, bonnets for their heads and collars for their dogs; overalls are also made of the skins, and drawn over the soles of their shoes, to prevent them from slipping on the ice; the intestines yield them material for masks or covers for their faces, to protect them from the glare of the sun in spring, and as substitutes for glass, cover their windows. The flesh is much esteemed as food, and the hams and paws considered great delicacies. So great are the benefits which it yields, that the Laplanders, it is said, call it "the dog of God;" while the Norwegians say, "it has the strength of ten men and the sense of twelve." If this bear is unable to find a hollow tree or cavern for its wintry home, it constructs a habitation for itself, out of branches of trees, lined with moss, where it continues dormant and without sustenance until spring. The female produces two cubs at a birth, which at the first are about the size of puppies. The brown bear is long lived. One in the menagerie at Paris, France, is spoken of as forty-seven years old. This animal is four feet in length, and about two and a half feet in height.

Ursus ferox, (Lat. *ferox*, fierce.) The GRIZZLY BEAR is the most ferocious and powerful of the family, frequently attacking man. It sometimes weighs more than 1,000 pounds. The Indians fear it so much that a necklace of its claws, which may only be worn by one who has destroyed this bear, is an ornament that entitles the wearer to distinguished honor. In California it keeps among the oaks and pines, on the acorns and seeds of which it feeds. It is strong enough to overcome and carry off a Buffalo.

U. Americanus. The AMERICAN BLACK BEAR is smaller than the Grizzly bear, and of a more clumsy appearance. It feeds upon berries, succulent roots, and juicy plants. When in swamps, it wallows in the mud like a hog, living on cray fish, roots, and nettles; sometimes it seizes on a pig, or sheep, or calf, or even a full grown cow. In robbing bee trees it is peculiarly expert. The young are at first not larger than kittens. The Cinnamon Bear, which is a permanent variety of this species, is quite a northern animal, and its fur is more valuable than that of the black bear.

Ursus maritimus, (Lat. belonging to the sea,) or *thalarctos*, (Gr. θάλασσα, *thalassa*, the sea, αρκτος, *arktos*, a bear.) The POLAR BEAR.

This formidable species of bear has a long and narrow head, prolonged in a straight line with the forehead, which is flattened; a long neck, and long, soft hair or fur, of considerable value. Its average length, when full grown, is from six to seven feet. Capt. Ross brought back a specimen measuring seven feet ten inches, and the weight of which, after losing thirty pounds of blood, was 1131 lbs. Another specimen, described by Capt. Lyon, measured eight feet seven and a half inches, and weighed 1600 lbs. The Polar Bear is entirely white, except the tip of the nose and claws, which are jet black. Dr. Kane, in his "Arctic Explorations," remarks that this animal is, "next to the Walrus, the staple diet to the North; and excepting the Fox, supplies the most important element of the wardrobe." "The liver of the animal," he says, "is, for some reason, poisonous, though eaten with impunity by the dogs."

The chief diet of the Polar Bear is obtained from the floating carcasses of whales and fishes, which often carry him, as a swimmer, far away from the shore. He also makes unceasing war upon the seals and walruses, and neither refuses the animal exuviæ which the waters cast upon the land, nor the few berries afforded by the shrubs of an arctic climate. On the land, these animals prey upon hares, young birds, etc. Their lodges are dens formed in layers of ice which are piled up so as to make stupendous masses. The males are said not to hybernate, but to brave the severity of the winter upon the ice of the open sea, wandering along the margin and swimming from floe to floe in search of prey. The females, however, do not appear until the approach of milder weather, when they sally forth from their retreats, accompanied by two cubs. At this period, gaunt, lean and famished, they are peculiarly formidable, hunger and the presence of their young adding to their natural ferocity. This bear is, however, formidable at all times, strong and active as it is, running with great swiftness either on the ground or on the ice, and with its claws, easily ascending the slippery sides of icebergs. The affection of this animal for its young is much celebrated, and its sagacity is great.

U. ornatus, (Lat., furnished or adorned.) The SPECTACLED BEAR, in the Cordilleras of the Andes, in Chili, has two semi-circular marks of a buff color above the eyes, appearing somewhat like a pair of spectacles.

U. collaris, (Lat. *collare*, a collar.) The BEAR OF SIBERIA has a large white collar passing over the neck and shoulders, on to the breast.

U. Syriacus. The SYRIAN BEAR, mentioned in 2 Kings, ii, 23, is probably the first of which there is any record.

U. labiatus, (Lat. *labia*, a lip.) The LABIATED or SLOTH BEAR, was, sixty years since, called the Five-fingered Ursine Sloth. The cartilage of the nose is capable of extension, and the lips of considerable protrusion.

U. Malayanus. The MALAYAN BEAR.

The long tongue of this Bear aids it in feeding upon the honey of bees, of which, as of other delicacies, it is extremely fond. It has also a taste for the young shoots of the Cocoa trees.

The existence of bears in Africa was doubted by Cuvier, but there is now good reason to believe the animal is found in Abyssinia, and the mountains of Arabia Felix.

Procyon lotor, (Gr. *προκυων*, *prokuon*, *προ*, *pro*, before, *κυων*, a dog.) The RACOON.

The remaining animals of this group form a sort of connecting link between the plantigrade and digitigrade carnivorous tribes. The Racoon, which with one or two other species, was formerly included in the genus *Ursus*, is now separated from it, and included in the new genus *Procyon*. It is a native of this continent, and numerously found in its northern territories, also in the Eastern, Northern and Middle States of the American Union, and yet more abundantly in some of the Southern States. The average length of the animal is about two feet, from the nose to the tail. The head is somewhat like that of a fox, the forehead being broad and the nose sharp; the ears are short, and slightly rounded; the body is broad and stout; the back arched;

the limbs rather short, and the fore legs shorter than the hinder. The upper part of the body is of a grayish color mixed with black. The ears and under part whitish, with a black patch across the eye. Varieties, however, are seen, some of which are black, others, yellowish white. The tail is bushy, and rather long, with rings of black and gray. Albinos are sometimes found, with red eyes and only faint traces of rings on the tail. In its feet the Racoon is only partially plantigrade, and when it sits, it often rests the whole hind sole of the foot on the ground, in the manner of a bear. The nails are strong, hooked, sharp and without hair. The outer hair is long and coarse; the inner, softer and more like wool.

The Racoon is a cunning, and when mature and in good case, quite a handsome animal. It mounts trees with facility, and frequently invades the woodpecker's nest; and it digs up and devours the eggs of the soft-shelled turtle.

This animal sometimes makes great havoc among wild as well as domesticated birds, eating only the head, or the blood which flows from their wounds. Occasionally it ravages plantations of sugar cane and Indian corn, especially when the latter is young. Oysters are also a favorite article of food with the racoon. These it is very expert in opening, biting off the hinge, and dexterously hooking out the contents of the shells. Audubon remarks that "the habits of the muscles, (*unios*) which are found in our fresh water rivers, are better known to the Racoon than to most conchologists, and their flavor is as highly relished by this animal as is that of the best bowl of clam soup by the epicure in that condiment." Swampy or marshy lands, abounding in trees and coursed by small streams, are the Racoon's favorite resorts; it traverses the margins of creeks and other waters, looking after frogs and muscles, which are found along their banks. It feeds chiefly by night, keeping by day in its nest or lair, which is usually made in the hollow of some broken branch of a tree. It rolls itself up, with the head between the hind legs, and sleeps away the time until the approach of darkness, when it goes forth in search of food. Sometimes, however, it is seen in corn fields; occasionally it will make an onset upon poultry during the day. The universal testimony is that it shows great slyness and cunning in its tricks and devices for procuring food. When in captivity, kind treatment soon renders it docile; it learns to be active during the day and to remain quiet at night. It shows an insatiable curiosity, prying into every corner and crevice with the greatest assiduity. In its habits it then becomes omnivorous, eating any thing, "vegetable or animal, cooked or uncooked," with equal avidity. The Racoon exhibits a peculiar fondness for sweets of every kind, and a great dislike for acids. It is fond of water, and before eating its food usually washes it; hence its name *lotor*, or washer. When hard pursued by the hunter, the animal takes to a tree, but unless the tree is very large, the pursuer is still after the "coon." If he cannot be taken otherwise, the axe levels the tree to the ground, when he is soon dispatched. The more common method of taking him is by box traps, baited with an ear of corn, a fish or a squirrel. For several months during winter, this animal hibernates in the hollow of some large tree, leaving its retreat only occasionally and when the weather is warm. The flesh is eatable, and the fur considered by hatters next in value to that of the beaver.

Proycyon cancrivorus, (lat. *cancer*, a crab; *voro*, to devour.) CRAB EATING RACOON. This species has a longer and more slender body than the common racoon. As observed in California, it conceals itself during the day, in the holes of decayed oak trees, which exist in the branches, not in the trunk itself, (Aud.) Besides *crabs*, frogs and fish, it feeds on birds, eggs, fruits, etc., and is said to be specially fond of the sugar cane.

Nasua, (lat. from *nasus*, a nose.) The COATI-MONDI, found in Brazil, Guiana, and Paraguay,—is like the Racoon, characterized by nocturnal habits, a semi-plantigrade mode of progress, and facility of climbing, but is readily distinguished from the racoons by its snout, which is quite long and extremely flexible; also by its longer and more slender body, and by its feet, which are stronger and well fitted for digging. The animal uses its snout in routing the worms and insects, which it digs up. The size is about that of a large cat, and in addition to insects and worms, it eats birds and eggs, and sometimes roots. Like the cat, it descends a tree with the head downwards, and it is even more active than that animal. The smell of the Coati seems to be more highly developed than any other sense. It is easily tamed, but is irritable and not to be touched without caution.

Cercoleptes, (Gr. κέρκος, *kerkos*, a tail, λεπτὸς, *leptos*, thin.) *caudivolvulus*, (Lat. *cauda*, tail, *volvulus*, twisted.) The POTTO KINKAJOU, or MEXICAN WEASEL,—is found in Mexico, and the warmer parts of South America, resembling the Coati in its habits, but showing more activity, and having a long tail, which is prehensile, and used after the manner in which the spider monkeys use theirs. Its size is that of a cat, but its limbs are shorter, thicker and more muscular. The tongue is long, slender, and very extensible, and used for drawing out of crevices, insects which are beyond the reach of its paws. This animal is a great destroyer of the nests of wild bees, for the sake of obtaining the honey, of which it is very fond, and has, therefore, been called the "Honey-bee."

Meles, (Lat. *a badger.*) *M. vulgaris*, (Lat. *common.*) The BADGER. The Badger has teeth which are best suited for masticating and bruising vegetable substances, and is less carnivorous than any of the PLANTIGRADES, except perhaps the bears. It is about as large as a dog of medium size, being about two feet three inches in length, but stands much lower on the legs, and has a broader and flatter body. The hairs taken separately are yellowish white at the bottom, black in the middle, and ashy gray at the point; the last color alone appears externally, and gives a sandy gray shade to the upper parts of the body. The face is white, and a long band of black runs along each side of the head, to the upper parts of the body. It is a quiet and inoffensive animal, but is often subjected to such ill-treatment, that "badgering" a person is a phrase used to express irritating him in every variety of manner. This animal inhabits most parts of Europe and Asia, but in some places is less common now than it once was. It is rather solitary and stupid, seeking refuge in retired places, where it excavates deep burrows, and shuns the light of day. The cruel sport of "baiting the badger," which consists in putting him in a kennel,and setting dogs to bite him through his thick hair and tough skin, is in some parts still continued. The Badger defends itself with great resolution, and sometimes to the destruction of its assailants. The flesh is esteemed a delicacy in Italy, France and China, and may be made into hams and bacon. The skin, when dressed with the hair, is impervious to the rain, and makes excellent pistol furniture and covers for traveling trunks, while the hairs or bristles are made into paint brushes.

M. Labradorius. The AMERICAN BADGER. The general characteristics of the American are the same as those of the European Badger. There is, however, a difference in the teeth of the American animal, and it has one tooth less than the Common Badger, on each side of the lower jaw. The length of this species is about two and a half feet. The body is very thick and fleshy, the nose thinner than that of the European species, and the claws of the fore feet much larger in proportion, while the tail is comparatively shorter; its fur is also of a quite different quality, and its appetites more carnivorous. The hair of the head and extremities is short and coarse; that of the other parts is fine and silky. At the roots it is dark gray, then light yellow, then black tipped with white, so that in winter it has an aspect of hoary gray; but in summer is more nearly a yellowish brown. It abounds in the plains watered by the Missouri, and has been traced as far north as the banks of the Peace River. It is known to inhabit Mexico, but its exact southern range is perhaps not accurately determined. The sandy plains on the borders of Lake Winnipeg, are perforated with innumerable badger holes, which greatly annoy horsemen, particularly when covered with snow. Its burrows are sometimes six or seven feet deep, and run beneath the ground to the distance of thirty feet. It enlarges and penetrates the burrows of marmots, ground squirrels, etc., and feeds upon these animals, which it cannot obtain when the ground is frozen. During the snowy season, or from November to April, it remains in a half torpid state. The badger is a slow and timid animal, taking to the ground when pursued, and to escape from danger, burrowing in the sandy soil with the rapidity of a mole. "The strength of its fore feet and claws is so great that one which had insinuated only its head and shoulders into a hole, resisted the utmost efforts of two stout young men, who endeavored to drag it out by the hind legs and tail, until one of them fired the contents of his fowling piece into its body." Early in the spring, badgers come abroad, at first fat, but soon become lean. At that time, they may be easily caught by pouring water into their holes, for the water not penetrating the frozen ground, soon fills the hole, and the animal is forced to come out. In this as in the *Ovis montana*, the Rocky mountain sheep, the fur, during the winter, changes from a furry texture to a woolly covering. In confinement, the American Badger appears gentle, and "allows himself to be played with, and fondled by his keeper, but does not appear to be well pleased with strangers." It produces from three to five young at a litter.

M. collaris. The INDIAN BADGER, or BEAR PIG of the Hindoos,—is about the size of the common badger. It has the body and limbs of a bear; the snout, eyes and tail are those of a hog.

Gulo, (Lat. a glutton.) This genus includes the GLUTTON, or WOLVERINE, and the GRISON. These animals are semiplantigrade in their walk, but resemble the weasel tribe in their teeth, and their thoroughly carnivorous propensity, as well as in the lengthened form of their bodies. Four species of this genus have been described.

G. Arcticus. This species is found in the Arctic, or northern regions of both continents; in size is about equal to the badger, but is more slender in body, and much more active. It seems to be intermediate between the badger and the polecat; in its general figure and aspect resembling the former; in its teeth the latter. The hair is of a chestnut color, verging, in some instances, towards black; its head is something like that of the polecat, but broader, and indicates greater strength of jaw. The nature of the Glutton is indicated by its name; and its laniary teeth evince its voracious and blood thirsty appetite.

It is sometimes called the "Quadruped Vulture," from the fact that it preys occasionally upon dead bodies of quadrupeds, chiefly those which have been killed by accident. It is said, these animals "do more damage to the fur trade than all other animals conjointly. They follow the Marten hunter's path round a line of traps, extending forty, fifty or sixty miles, and render the whole unserviceable, merely to come at the baits, which are generally the head of a partridge, or a bit of dried venison. They are not fond of the Martens themselves; but they never fail to tear them in pieces, and bury them in snow at a considerable distance from the trap. Drifts of snow often conceal the repositories thus made of the Martens, at the expense of the hunter, in which case, they furnish a *regale* for the hungry fox, whose sagacious nostril guides him unerringly to the spot, and two or three foxes are often seen following the Wolverine for this purpose." Perhaps these attendant foxes have given rise to the remark that the Arctic Fox is the "Jackal or provider" of the Glutton.

The Glutton feeds upon meadow mice, marmots and other *rodentia*, and occasionally upon disabled quadrupeds of a larger size. It resembles the bear, but is not as fleet; is industrious, feeds well, and is generally fat. It goes abroad much in the winter, and the track of its journey in a single night, may often be traced for miles. From the shortness of its legs, it moves with difficulty through the loose snow. Sir John Richardson says "the Wolverine is a great destroyer of beavers." It must, however, be only in summer, when these animals are at work, that it can surprise them, for an attempt to break through their frozen mud-walled house, would drive the beavers into the water, to seek shelter in their vaults, on the borders of their dam. Whatever the boldness of the Wolverine, in defending itself against other quadrupeds, "it makes but a poor fight with a hunter, who requires no other arms than a stick to kill it."

This animal has two secretory organs, from which he, on occasion, discharges a yellowish brown fluid that gives forth an offensive odor. The female brings forth yearly from two to four cubs, covered with a downy fur, of a pale cream color.

The Wolverine remains through the winter, as far north as 70° 11′ latitude, but does not change its color on account of the intense cold. According to Lesson, it inhabits a complete circle around the North Pole, in Europe and Asia, as well as America. The skins furnished by Wolverines, do not compensate for their destructive habits. The fur resembles that of a bear, and is much used for muffs, and when several skins are sewed together, makes a beautiful sleigh robe. In Kamtschatka, the women dress their hair with the white paws of this animal, which they esteem a great ornament.

G. vittatus, (Lat. from *vitta*, a band or fillet,) the GRISON. A white line or band passes on each side of the front to the shoulders. They are most numerous in Guiana and Paraguay.

G. or Ratellus mellivorus, (Lat. *mel*, honey, *voro*, to devour.) The RATEL of the Cape of Good Hope, in general characters, corresponds with the glutton; in size is about equal to the badger. The color is of a dull ash gray, but whitest towards the head. It is said to feed principally upon the honey of bees, which inhabit the deserted lairs and burrows of the Ethiopian boar, the porcupine, etc.

Ailurus fulgens, (Lat. shining,) the PANDA, or WAH,—is found in the Himalaya chain of mountains, between Nepaul and the Snowy mountains. Cuvier declared this to be one of the most beautiful of quadrupeds, and included it in the Bear tribe. In the arrangement and form of the teeth, it shows some resemblance to the *Nasua* and *Procyon*. It is about the size of a large cat; the soft and thickly set fur is above, of the richest cinnamon red, behind more fulvous, and beneath, deep black, while the head is whitish, and the tail whitish, annulated with brown. Its loud cry resembles the word *wah*, whence its name. "This elegant animal frequents the vicinity of rivers and mountain torrents, passes much of its time on trees, and feeds upon birds and the smaller quadrupeds." The generic name is from the Gr. *ailouros*, a cat.

To what animals is the name PLANTIGRADES given? What is the derivation of the word? What their movements and habits? When and where are they in a lethargic state? From what is the family name URSIDAE derived? To what animals are the bears a connecting link? What is said of their gait? What use do they make of their fore paws? Describe their claws. To what kind of food are their teeth adapted? Where are the genera of this family found? According to Audubon, how many species of the genus URSUS have been described? Give their locations. Describe the bear. What is said of the number and kind of their teeth? Which of the carnivorous animals is most omnivorous? Do any bears subsist on vegetable food alone? What are their habits? What is their condition in the winter? In what part of the world are they seldom met? What use is made of their skins? Where is the common bear found? Is it now met with in Great Britain? What was the only species known to Linnæus? What does it furnish the people of Kamtschatka? What do the Laplanders call it? Why? What do the Norwegians say of it? What is said respecting its winter home? Is it long lived? What is said of the age of one in the Menagerie at Paris? What is its size? What is said of the ferocity of the Grizzly Bear? What use is made of its claws? How much does it weigh? Upon what does it feed? What is said of its strength? What is said of the size, appearance and food of the American Bear? What is the size of the young at first? What is said of the Cinnamon Bear? Describe the Polar Bear. What is its average length? What is said of its weight? What does Dr. Kane remark respecting this animal? What is its chief diet? On what else does he feed? What do these animals eat when on the land? What is said of their dens? How do the males spend the winter? Define and spell hybernate, migrate and emigrate. When do the female bears sally forth from their winter retreats, and what is their appearance, and the degree of their ferocity? What is further said of the Polar Bear? What is said of the Spectacled Bear? What of the Siberian Bear? What of the Syrian Bear? What of the Sloth Bear? What of the Malayan Bear? Are bears found in Africa? Were they known to exist there during Cuvier's life?

What is said of the remaining animals of this group? What name is given to the Racoon? Give the meaning of the generic and specific terms? How were the racoon and other species formerly arranged? To what continent does it pertain? On what part is it numerously found? Describe the animal in his appearance and habits? What is a favorite kind of food with the racoon? How does it get at the contents of the shells? What does Audubon say as to the racoon's knowledge of the habits of fresh water muscles? What are its favorite resorts? How does it appear in captivity? What is said of its curiosity? Why is it called *lotor?* How does it spend the winter? What is said of its flesh and fur? How does the Crab-eating Racoon differ from the Common Racoon? Where does it conceal itself in the day time? On what does it live? From what is the generic term *nasua* derived? Where is the Coati mondi found? What are its characters? How is it distinguished from the Racoon? How does it use its snout? What is its food? In what respects does it resemble the cat? What is further said of it? Give the derivation and meaning of CERCOLEPTES. Where is the Mexican Weasel found? What other names has it? What animal does it resemble? What is said of its tail and size? What use does it make of its tongue? What name has been given it? Why?

What is said of the Badger's teeth? What of its food? What of its size and hair? What does "badgering" a person mean? Where is the animal found? What are its habits? What is "baiting the Badger?" What is said of the flesh, and what use is made of the skin? Wherein does the American Badger differ from that of Europe? Where does it abound? What is said of its Northern and of its Southern range? What is said of its burrows? How does it annoy huntsmen? How long and at what season is it torpid? What is said of the strength of its fore feet and claws? How are these animals easily caught in the spring? How do they appear in confinement? What changes does the fur undergo? What is said of the Indian Badger? What does the genus Gulo include? Give the characteristics of these animals. How many species? To what is the species *Gulo arcticus* intermediate? What is it sometimes called? Why? How do these animals injure the fur trade? Upon what does the Glutton feed? What animal does it resemble? What more is said of it? How is the Grison marked, and where most numerous? Describe the Rattel? Where is the Panda or Wah found? What animals does it resemble? What are its resorts?

SECTION XV.

SUB-ORDER AMPHIBIA, (Gr. 'αμφίβιος, *amphibios*, having a double life.)

The term Amphibia, is, strictly speaking, applicable only to such animals as have double sets of lungs, or gills, giving them the power of living, indifferently, at the same time, either upon land or water; but it is commonly given to seal, otters, beavers, etc., and to many reptiles whose habits are at once terrestrial and aquatic. (Pl. VI. fig. 11.)

PHOCIDAE, (Gr. φώκη, *Phokê*, a sea-calf or seal.) This tribe of animals, belonging to the carnivorous order, show a peculiar adaptation to the sphere assigned them by the All-wise Creator. None of the four-limbed mammalia display such complete adaptation to residence in the water. Seals resemble quadrupeds in some respects, and fishes in others. They have round heads, and broad noses, not unlike those of dogs, with the same mild and expressive physiognomy; large whiskers; oblong nostrils, and large, sparkling black eyes. In the seal there is no external ear; but a valve exists in the orifices which he can close at pleasure, in order to keep out the water; a valve is also found in the nostrils, which is useful for the same purpose. The body is covered with stiff, glossy hairs, which are closely set against the skin; it is elongated and conical in form, gradually tapering from the shoulders to the tail. The feet of the seal differ from those of all other quadrupeds. They have the same number of bones, but are covered with a membrane which would make them resemble fins more than feet, but for the sharp, strong claws with which they are pointed. The limbs may be viewed as a sort of oars, or paddles. In the front pair, the arm and forearm are very short, so that but little more than the forearm advances from the body; the hind limbs are directed backwards so as to almost seem like a continuation of the body; the thighs and legs very short; the tail is short and thick; the foot is formed on the same plan as the forepaw; but the toes are in contact; the web is folded when not in use as a paddle; but spread out when the animal is swimming. The seal moves in the water with great ease and rapidity, but on the land, or on masses of ice, with extreme awkwardness. It is gregarious, living in herds more or less numerous, along the shores of the sea. The cellular tissue, situated between the skin and muscles, is very loose and fibrous, and seems to be a receptacle for the blood, during the suspension of breathing under water. It can remain in that element a long time without injury; when it is submerged, the blood not freely circulating, and thus accumulating in the larger veins. Its tissue appears designed in part to relieve the animal from the pressure of the superincumbent water. The blood is abundant and dark in appearance, showing that it has less oxygen than that of strictly terrestrial animals.

Seals are found in almost every quarter of the globe, but they are most numerous in frozen and temperate regions. They exist in vast numbers in the seas around Spitzbergen, and on the coasts of Labrador, and Newfoundland. About thirteen species are included in the genus *Phoca*. In their wide range, seals are sometimes found within the waters of the state of New York. About the middle of the Spring of the year 1857, one was taken in the Hudson river, and another on the borders of Long Island. Dr. Dekay (N. Y. State Nat. Hist.) describes a female seal caught in Long Island Sound, near Sand's Point. At a former period, these animals were abundant in our waters. "A certain reef of rocks in the harbor of New York, is called Robin's Reef, from the numerous seals which were accustomed to resort thither; *robin*, or *robyn*, being the name in Dutch for seal."* In the Kingston (U. C.) Chronicle, of February, 1823 or 1824, there was a notice of a seal taken on the ice of Lake Ontario, near Cape Vincent, (Jefferson county,) N. Y. In August, 1824, a seal was exhibited alive in New York, which had been taken in a seine in the Chesapeake, near Elkton, Maryland. A seal, said to have been beautifully spotted on the under side, was taken some years since near Lynn, Mass.

The length of the common seal, *Phoca concolor*, or *P. vitulina*, (Lat. calf-like,) (see Plate VIII. fig. 1.) is, on an average, about five feet; the color, yellowish gray, clouded with brown or yellow. The female produces her young during the winter, taking care of them at the place of birth for a few weeks, until they become sufficiently strong to be taken to the water, to which they are then removed by the parent, not without solicitude for their safety. By her they are taught to swim, and seek for fish, and when they are fatigued, she carries them on her back. As might be expected from the nature of its food, the seal has a fishy smell. It is reported that when assembled in numbers on shore, the odor is perceivable at some distance. In pursuing their watery prey, seals display much cunning and power of swimming.

* Nat. Hist. of State of New York.

The voice of the animal when old, is a hoarse, gruff bark; when young, a peculiarly plaintive whine. "With a good glass," says Dr. Kane,* "you may study these animals in their natural habitudes, undisturbed by suspicion. As thus seen, in the centre of a large floe, and within retreating distance of his hole, the seal is a perfect picture of solitary enjoyment, rolling not unlike a horse, stretching his hide, awkwardly spreading out his flippers, and twisting his rump towards his head. Again he will wriggle about in the most grotesque manner; the sailors call it 'squirming,' every now and then rubbing his head against the snow. The shapes of a seal, or rather his aspects, are full of strange variety. At a side view, with his caudal end *slued* round to the side from you, and his head lifted suspiciously in the air, he is the exact image of a dog, *chien de mer*. During his wriggles, he resembles a great snail; a little while after, he turns his back to you, and rises up on his side flippers, like a couching hunter, preparing for a shot, the very image of an Esquimaux." The seals are proverbially shy. The Esquimaux and Greenlanders, to whom these animals are of inestimable importance, as furnishing them with the chief means of subsistence, are from earliest youth, trained to the pursuit of them, and look upon the most successful hunters of them as their *great men*. "No one can pass for a right Greenlander who cannot catch seals." This is not strange, considering the manifold benefits furnished the northern tribes by these animals. The boat, or kajah in which they brave the violence of a northern sea, and the perils of the chase, consists of the skin of the seal placed over a light frame work of wood. The same skin furnishes the material for his dress; the flesh of the animal supplies him with his "most palatable and substantial food; the fat gives him oil for lamp-light, chamber and kitchen fire. He can *sew* better with fibres of seal's sinews than with thread or silk. Of the skins of the entrails, he makes the windows of his house, curtains for his tents, his shirts; and part of the bladders they use at their harpoons, and he makes train bottles of the maw or stomach." Seal skins and oil are to him also important articles of commerce. The fishing commences in autumn, and is practised by means of nets stretched across narrow sounds where the seals are in the habit of swimming. Only the young ones can be taken in these nets; the old ones are shot, or else the boatmen enter the recesses of the animals at night, with torches and bludgeons, and despatch them, which they do easily with a slight blow on the forehead or muzzle.

* Grinnell Arctic Expedition.

"To shoot seal," says Dr. Kane, "one must practise the Esquimaux tactics, of much patience and complete immobility. It is no fun to sit motionless and noiseless as a statue, with a cold iron musket in your hands, and the thermometer 10° below zero. Very strange are these seal! a countenance between the dog and the ape; an expression so like that of humanity, that it makes gun-murderers hesitate. At last, at long shot, I hit one. The ball did not kill outright; it struck too low. He did drown finally and sunk, and so I lost him. Curiosity, contentment, pain, reproach, despair, and even resignation, I thought I saw on this seal's face." . . . "A Danish boy who had joined us by stealth at Disco, told us that the animal's sinking was a proof that he had no blubber, and he was probably right." Though the orifice of the ear, as we have said, contains a valve which closes, yet the seal has a most delicate sense of hearing, and delights in musical sounds, a fact not unknown to the ancients. Laing, in his account of a voyage to Spitzbergen, states that when the violin was played, "a numerous audience of seals" would generally collect around the vessel, following her course for miles. In allusion to this peculiarity of the seal, Sir Walter Scott says,

"Rude Heiskar's seals, through surges dark,
Will long pursue the minstrel's bark."

The seal has often been domesticated, and it is said, made use of in fishing. The following is among the anecdotes illustrating this remark. "In January, 1819, a gentleman residing in the county of Fife, Scotland, completely succeeded in taming a seal. Its singularities attracted the curiosity of strangers daily. It appeared to possess all the sagacity of a dog, lived in its master's house, and ate from his hand. In his fishing excursions, this gentleman generally took it with him, when it afforded no small entertainment. If thrown into the water, it would follow for miles, the track of the boat, and though thrust back by the oars, it never relinquished its purpose. Indeed it struggled so hard to regain its seat, that one would imagine its fondness for its master had entirely overcome the natural predilection for its native element."

When companies of seals are seen at some distance "walking the water," their heads peering above it, they assume sometimes such appearances as have given rise to the stories of TRITONS, SIRENS and MERMAIDS, concerning which many marvelous things have been written.

The *Phoca Groenlandica*, or HARP SEAL, is about six feet in length, and noted for the variations of its color, as it advances towards maturity.

The *Phoca barbata*, (Lat. bearded,) is larger, and has thicker and stronger moustaches than the others. Its length varies from seven to ten feet. Dr. Kane speaks of one which was shot by Capt. Haven, of the Grinnell Arctic Expedition, measuring "eight feet from tip to tip; five feet eleven inches in his greatest circumference, and five feet six inches in girth behind the fore-flippers." "His carcass," says the Dr. "was a shapeless cylinder, terminating in an awkward knob, to represent the head."

P. cristatus, (Lat. crested,) or *Stemmatopus cristatus*, (Gr. *stemma*, a wreath; *ōps*, face,) or Hooded Seal, is distinguished for having a globular sac, which can be swelled upon the top of the head, in the male animal. This species reach the size of seven or eight feet, and live in the seas about Greenland and Newfoundland.

The Elephant Seal, or Sea Elephant, *P. Macrorhinus*, (Gr. *makros*, long, *rhin*, nose,) *proboscideus*, (Gr. *proboskis*, a trunk,) is the largest known species, being from twenty to thirty feet long, and having a girth at the largest part of the body, of eighteen feet. A full grown male of this species will yield seventy gallons of oil. This kind of seal is found on the southern coasts of Australia, Juan Fernandez, and the neighboring parts of South America. Its voice is like the lowing of cattle, and it is inert in its habits. The name "Elephant Seal," is given to the animals of this species, partly on account of the large size of their tusk-like canines, and partly from their power of lengthening the upper lip into a kind of proboscis. They are much sought after on account of the quantity of oil which they yield, and also of their strong skins, which are valuable for harness making.

The Sea Lion, *Platyrhyncus leoninus*, found on the north and south coasts of the Pacific, is from six to ten feet in length, and of a yellowish brown color. The males have a large mane upon their necks, partly covering the head and shoulders, and a very powerful voice, whence their *name*.

The Sea Bear, *Arctocephalus ursinus*, is so called from the fur and shape of the head. It grows to the length of five or six feet, and has small external ears. The membrane of the hind feet is prolonged into as many lobes as there are toes, and the fore feet are placed very far back. The color of the fur is brown, but when it is old, assumes a grayish tint. This species inhabits the coasts of the South Pacific, and is also said to be found in the northern hemisphere.

Trichĕcus Rosmărus, the Walrus, Morse, or Sea Cow.

This animal resembles the seal in its general conformation, but is much larger, and more thick and clumsy in its proportions. Its distinguishing peculiarity is the construction of the skull. The lower jaw is without incisor and canine teeth, and is compressed laterally to fit in between two enormous canine teeth, or tusks, which arise out of the upper jaw, and are inclined downwards with a gentle curve. The length of the tusks is sometimes two feet. The *alveoli*, or sockets of these tusks, occupy the whole of the front portion of the upper jaw, and give a roundness to the form of the muzzle; the nostrils do not end in a snout, but are far above the mouth, or what seems the middle of the face. The development of the brain is less in the Walrus, than in the seal, and it shows less intelligence. The ears are merely two small orifices; the head is small in proportion to the bulk of the body; the neck short; the lips are thick, the upper one divided by a longitudinal furrow, and studded with strong bristles; the skin is very thick and impenetrable, and covered with smooth, yellowish hair. This huge animal is often eighteen or twenty feet in length, and ten or twelve in circumference, around the chest.

The Walrus is found in the icy seas of the north. Like the seal it is gregarious. It is not a ferocious animal, but on account of its great strength, and formidable tusks, is dangerous when attacked; and the more dangerous because many hasten to the help of a companion when in trouble. They are said to be monogamous. The females defend their young with great resolution and perseverance. These animals resort to islands of ice, or the ice-bound shore. The tusks furnished them by the Creator, assist them to mount the slippery acclivities, or ledges of ice, they striking the points of the tusks into the glassy surface in order to secure themselves firmly, and drawing up their unwieldy bodies. It is said their hind feet are furnished with suckers, which act on the principle of cupping glasses, exhausted of air, so that the feet adhere to the ice, and thus help the animals to propel themselves forward. Thus the Walrus can climb the iceberg with security, pass over its surface and betake itself at pleasure to the waters of the ocean.

Captain Cook, in his Journal of his Voyages, speaks of meeting with Walruses off the northern coast of America. "They lie," says he, "in herds of many hundreds, upon the ice, huddling over one another like swine, and roar and bray so very loud that in the night, or in foggy weather, they gave us notice of the vicinity of ice before we could see it. We never found the whole herd asleep, some being always on the watch. These, on the approach of the boat, would awaken those next to them, and the alarm being thus gradually communicated, the whole herd would be awake presently; but they were seldom in a hurry to get away till after they had been once fired at; they would then tumble over one another into the sea, in the utmost confusion, and if we did not at the first discharge, kill those we fired at, we generally lost them, though mortally wounded. The dam, when in the water, holds the young one between her fore arms." The chief use of the walrus to man, is in its tusks, which yield the finest ivory, and in its abundant blubber, or fat, which yields oil. They, and indeed all the marine mammalia which are found in the Arctic seas, have abundant fat, as their defence against the cold. A beautiful and striking evidence of kind and intelligent design, of which numberless instances are presented to the student of Natural History, is seen in the fact that immediately beneath the skin, a thick layer envelopes the body, and being a bad conductor of caloric, besides other advantages already referred to, prevents the vital heat from passing off. With the Polar Bear, *U. Maritimus*, the Walruses have frequent and desperate conflicts. They feed upon shell fish, and marine vegetables, and perhaps a further use of their tusks is to root up their food from the spot to which it adheres. Their flesh, like that of the seal, is highly valued by the inhabitants of Arctic regions, and northern voyagers have often found it a most acceptable repast.

Give the derivation and meaning of AMPHIBIA. To what animals alone does it strictly apply? To what others is it commonly given? From what is PHOCIDAE derived? What is said of their adaptation to a watery residence? Describe the Seal. What is said of its habits? How is it enabled to remain in water a long time without injury? Where are Seals most numerous? How many species does the genus include? Where have they been found in this country? What is their size? What does Dr. Kane say of these animals? To what people are they of inestimable importance? Relate the particulars which are given respecting them. What has occasioned the stories respecting Tritons, Syrens and Mermaids? What is said of the Harp Seal? Give some particulars of the Bearded, Hooded and Elephant Seals. What is said of the SEA LION? Why is it so called? Give some account of the SEA BEAR. What animal does the Walrus resemble? What other names has it? What is its distinguishing peculiarity? How long are the tusks? Give its general characteristics. What is said of its intelligence? What is its length? Where is it found? What are its habits and disposition? With what are its hind feet furnished? What does Captain Cook relate respecting Walruses? Who was Captain Cook? Ans. A celebrated English circumnavigator, who was killed by the natives at Owyhee, Sandwich Islands, in 1779. What is their chief use to man? What evidence do they give of kind and intelligent design on the part of the

Creator? With what animal does the Walrus have severe conflicts? What is its food? What is said of its flesh?

SECTION XVI.

SUB-ORDER INSECTIVORA. (Lat. *insecta*, insect, *voro*, to eat.) The INSECTIVORA, as the term denotes, comprehends those animals whose food is especially insects, but not exclusively, as sometimes they feed on other, and even vegetable substances. They walk on the sole of the foot, (*plantigrada.*) The sub-order includes three families. Their motions are feeble, feet short and slender, snout lengthened. In cold climates they pass the winter in a dormant state.

HEDGE-HOGS, (*Erinaceadæ*, from *erinaceus*, Lat. for hedge-hog.) The true hedge-hogs are found in Europe, Asia and Africa, while others are found in Madagascar and the Oriental Islands. They are slow and inoffensive, but are self-defended by a coat of stiff, tough spines or prickles. They roll themselves up into a round ball, and thus the spines project from every part of the surface; and are a defence and safeguard. They lie concealed in some crevice between the moss-grown roots of a tree, among a mass of withered leaves, or in a hole which they have excavated; and in this condition, the animal remains during the day, protected from injury in the way before described, should its retreat be discovered. As the dusk of evening comes on, it issues from its lurking place and prowls about for food. If pursued it makes no defence, but rolls itself up and trusts to its spines for safety. These are, indeed, the only means of defence bestowed upon this little, weak and timid animal. It feeds upon insects, frogs, snails, fruits, and esculent roots. It is useful in gardens, and often kept in large kitchens for the destruction of beetles and cockroaches.

The TENREC, (*Centetes*, Gr. *κεντέω*, *kenteo*, to sting or prick,) called also the Asiatic or striped hedge-hog, of Madagascar, has no tail, but is covered with a spiny coat of mail. It rolls itself up in the way of the hedge-hog already mentioned, though not so easily, is nocturnal, and passes three months of the year in sleep. Some are not larger than a mole.

The species are Tenrec *Centetes acaudatus*, Lat. *a*, without, *cauda*, a tail.)

C. setosus, (Lat. bristly.) Its spines are short and rigid.

Varied Tenrec, *C. semi-spinosus*, (Lat. *semi*, half, *spina*, spine.) Its body is clothed with a mixture of spines and bristles.

SOREX, (Lat. *shrew.*) The SHREWS have usually been considered a kind of mice and of the order *Rodentia*. They are, however, distinguished from the latter by their teeth, and the conical form of the head, and nose tapering to a long point. They place the entire sole of the foot upon the ground, which makes their legs appear short. They have glands along the side of the body, which secrete a humor of an unpleasant and peculiar odor. Their shrill, piercing cry may often be heard in spring and summer. Water shrews, which are twice the size of the others, are found upon the banks of rivers, ponds, and marshes, and appear to collect their food, consisting of the larvæ of the ephemeral flies, from the loose mud. Stationing themselves at the mouths of their holes, they look intently on the water, and if a shoal of minnows pass by, they plunge in among them, diving with much adroitness. Their fur repels the water, and while submerged they appear almost white. The Common Shrew, *S. araneus*, (Lat. Spiders,) is covered with soft velvety fur, is easily distinguished from the mouse by its long, tapering and cartilaginous snout; the eyes, too, are very minute, almost hidden in the surrounding hairs, and the ears are round and close. It is usually of a reddish mouse color above, grayish beneath, and sometimes tinged with yellow. Its entire structure is well adapted to burrow under the earth, but it can also move rapidly upon the surface. Its length, from the snout to the tail, is about five inches; its tail is one inch long; it feeds upon insects, worms and grubs.

Sorex fodiens, (Lat. digging.) The WATER SHREW closely resembles the common shrew in its conformation. Its feet are rather broad and formed for swimming, having a lock of stiff hairs on the end of the toes; its tail is rather slender and fringed with stiff hairs. Its swimming is principally effected by the alternate action of the hind feet. The appearance of these animals, and their motions in water are quite amusing. A sort of musk is expressed from the region about the tail, and the skins are put into chests and wardrobes, among clothes, to preserve them from moths.

The DESMAN or MUSK RAT, *Mygale* (Gr. *spider-mouse*,) *moschata*. This is known as the Russian Musk Rat, is about the size of a hedge-hog and distinguished from the shrews by its long scaly tail, flattened at the sides. Under the tail of the Desman are two small follicles, containing a kind of unctuous substance of a strong musk odor, from which the name of musk rat is given to it.

The SCALOP, to which Linnæus gave the name of *Sorex aquaticus*, is a native of Canada and is now separated from the true shrews.

We come now to notice the MOLE (*Talpa*)—Family, *Talpidæ*. This animal is five or six inches in length and formed for an underground life. Its body is thick and cylindrical; the head is prolonged, especially the muzzle, which projects far beyond the jaws, and is very flexibile and strong, serving to convey the food to the mouth; it has no external ears, but the auricular apparatus is highly developed, and the sense is very acute; its eyes are very small and concealed by its fur, so that it is a vulgar opinion that it is deficient in these important organs. The head is not distinguished from the body by any appearance of neck; the legs do not project perceptibly from the body.

The mole is accustomed to burrow for its food, forming its abode or "encampment" under ground, and raising a larger hillock than the rest for the reception of its young. Its subterranean excavations are most distinctly and determinately made, having passages or "high roads" from one part of its domain to another. Into these roads open the excavations in which it daily searches for food. In this home, which is separated from that in which its nest is formed, it dwells from autumn to spring. The mole is *essentially* an accomplished miner, and unlike most of the mammalia, finds his happiness and his home in the subterranean (underground) galleries which he excavates with admirable skill and industry. Its fore feet, which are broad and muscular, are constructed like hands and form complete paddles for throwing the soil behind the animal. (See Plate VI, fig. 4 of Mole's foot.)

It has been mentioned that there is no external conch to the ears, as the auditory opening concealed by the fur is small. "A valve, capable of being raised or lowered like an eye-lid, the mechanism of which is visible if the fur be shaved away, closes this aperture at the will of the animal, so as to exclude any particle of earth or sand." The eyes, too, which are exceedingly small and buried in the fur for protection, may be uncovered at pleasure, when it emerges to the light. The Creator has given it the power of vision, but in a very limited degree; in fact it is in the very lowest stage of development, but it has all in this respect that is needed. Its keen sense of smell is its chief guide in searching for food, and dwelling as it does, in darkness, this sense is remarkably perfect.

The structure of the mole is such as to concentrate the whole force and energy of the animal in the anterior portion, and thus is adapted to its habits and mode of life; the hands are large, broad, and thick; the bones knit firmly and solidly together; the claws are enormous—these are the organs by which it throws up the earth; the head is an organ for boring or digging, very long and flat, with the cartilages of the nose ossified; the ligament of the neck, which in other animals is elastic, is here bone also, so that the strain in digging is better borne; the pelvis is very small; the bones of the hind limbs are small and slender and the hind feet, though having claws, are feeble in comparison with the spade-like hands, thus hindering not its course through its under-ground roads, but yet having sufficient strength, and not in the way.

In short, were we called upon for striking evidence of the design and attentive care of God, we would point to the habits and manners of the Mole, and the fitness and adaptation of the means and instruments with which it is provided. The mole does not, of its own accord, emerge from its subterranean abode, except to seek for some more favorable soil in which to construct its halls and winding galleries. Rich and cultivated meadows, abounding in worms and other insects, are its favorite localities in which it makes its burrows.

Unlike the dormouse or marmot, it is not less active in winter than in summer; the twilight hours of morning and evening are its period of labor.

The nest where the female mole nurses her helpless young, (of which she has one brood yearly, generally four or five, sometimes as few as three, rarely six,) is formed in a vault, carefully constructed at the center of diverging passages, made soft with leaves, grass, and scales of bulbous roots. "The parents afford a pattern of mutual affection and assistance."

The food consists of worms, insects, and when it can obtain them, small birds or quadrupeds, to which roots are also added. It is impatient of hunger, and cannot endure a fast of more than six hours' duration; an abstinence of twelve hours is said to produce death.

Agriculturists complain that they suffer injury from the young corn which moles carry off for constructing their nests; but its turning up and lightening the soil, and its destruction of insects, earth worms and noxious creatures found near the surface of the ground and so hurtful to grass, corn and other plants, furnish advantages to the farmer which probably more than counterbalance any injuries which he suffers from the doings of the mole; at the same time, we should guard the undue increase of these mining animals.

Condylura, (Gr. κονδύλη, *kondŭlē*, a knob, οὐρά, *oura*, a tail; knobbed tail.) Crested or Star-Nosed Mole. This name was given to this animal, by Illiger, under an *erroneous* impression that the tail is "knobbed." There is but one species well known, *cristata*, (crested,) found in various parts of the United States. The nostrils are surrounded by movable cartilaginous points that radiate like a star when expanded. The color is brownish black above, a shade lighter beneath. The head is remarkably large; the body thick and short, growing narrower towards the tail, which is smaller at the root, large in the middle, and tapering to a fine point at the tip; the fur on the body is very fine, soft and shining. The shape of the body resembles that of the common shrew mole, and it is similar in its habits.

The BANXRINGS, (*Tupaidæ*,) of Sumatra and Java, are remarkable insectivorous animals. They are nocturnal, and squirrel-like in their appearance and habits.

QUESTIONS ON THE INSECTIVORA.

How many families does the INSECTIVORA include? On what do they feed? What is said of their motions and habits? Where are the true Hedge-hogs found? Where others? How are they self-defended? How do they conceal themselves? How is the day spent? When does it seek its food? How act when pursued? For what is it useful? Where is the Tenrec found? What is it called? How covered? What are its habits? What its size? How many species? Give their names and derivation. To what order have Shrews commonly been referred? How are they distinguished from mice? How do they tread? What have they upon the side of the body? What is said of their cry? To what places do Water Shrews resort? What do they use for food? What is said of their watching for minnows? What effect has their fur upon the water? How is the shrew distinguished from the mouse? What is its color? For what is it well adapted? What is said of the Water Shrew? What of the Russian Musk Rat? What of the Scalop? For what kind of life is the mole formed? Describe the animal. How does it obtain its food? What is said of its excavations? How are its fore feet constructed? What is remarkable about the ear? What is said of the sight and smell? In what part of the body is the strength concentrated? Give particulars as to its structure. Wherein does it give proof of divine care? Why does it leave its subterranean abode? In what respect is it unlike the dormouse or marmot? What is said of its nest? What of its abilities to fast? Why do agriculturists complain of the mole? What benefits does it confer upon the farmer? From what is the term *Condylura* derived? Was it rightly given? Why is this animal called Crested or Star-nosed? Describe it. What is said of its shape and habits? What is said of the Banxrings?

OBS. Here, at the close of the order CARNIVORA, and every other order, let the teacher have a general review, naming the sub-orders, tracing out the genera, families, &c., giving the specific name to each as he describes the animal, pointing them out when on the chart, telling all he can remember about them, either from the book or chart. If he omits anything, let it be mentioned by other members of the class. No pupil should ever be permitted to pass the name of a person, or place, or even a word, without knowing who the person was, where the place is, and what the word means.

SECTION XVII.

FOURTH ORDER. MARSUPIALIA, OR MARSUPIATA.

(Lat. *marsupium*, a purse or bag.)

This order is arranged into two sections,—Marsupials and Monotremata. These are not unfrequently regarded as *separate* orders, constituting a sub-class termed *Ovo-vivipara*, (Lat. *ovum*, an egg; *vivo*, to live, and *pario*, to produce,) and intermediate between the truly viviparous mammals and the oviparous birds and reptiles. The animals of this order are numerous and quite different in their organs from all other mammals. So peculiar is their internal structure that Cuvier remarks they may be looked upon as containing several orders running parallel with the orders of ordinary quadrupeds. Their rank is low in the scale of intelligence. Of the two sections the marsupials show the least departure from the general type of the *Mammalia*. The most striking peculiarity, common to them all, is the immature state of the young at birth, they being much like the half formed chick in an egg which has been but a few days incubated; and their reception into a pouch or fold of a skin in the female, in which they are nourished, remaining there five or six weeks, until they increase in size and are able to take care of themselves. Even for some time after the young one can procure its own living, and runs and plays by its mother's side, it instinctively flies to the maternal pouch for protection from threatening danger. The pouch is supported by two bones placed amidst the abdominal muscles and called the marsupial bones. They are found in the male as well as in the female, and even in species where the pouch-formed fold of the skin is scarcely perceptible. It is remarkable that these mammals are confined almost entirely to Australia, including New Guinea and the islands immediately adjacent, excepting the Opossums, whose home is South America, but which are also found abundantly in the United States, residing in woods and thickets near hamlets and villages. Appearances of secondary rocks seem, however, to indicate that at former periods they were more widely spread over the earth's surface than they are at present.

The Marsupials include between seventy and eighty known species, arranged by Prof. Owen into sixteen genera. The whole are divided into five families, named from the more usual character of their food. I. The SARCOPHAGA, (Gr. σαρξ, *sarx;* φαγω, *phagō*, to eat.) FLESH-EATERS.

These are found in New Holland and Van Diemen's Land

alone; though remains of them have been found in the Stonefield slate, (England,) and in the gypsum quarries of Paris, (France.) They show great varieties of size, from that of a small wolf to a mouse, the larger ones being considerably fierce, destroying sheep, and even making their way into houses; others attack poultry and suck their blood. Those of the smallest size show a likeness to the Insectivora, and live on trees. Prof. Owen enumerates three genera of the *Sarcophaga*, viz.: *Thylacinus*, *Dasyurus* and *Phascogale*. These, with others of the order, show a tendency to the multiplication of teeth, and peculiarities of the arterial system and bodily organs. The *Thylacinus*, (Gr. *θύλακος*, *thulacos*, a sac; *ἶνις*, *inis*, offspring,) has incisors, $\frac{8}{6}$; Canines, $\frac{1\cdot1}{1\cdot1}$; Molars, $\frac{7\cdot7}{7\cdot7}=46$. The species *T. cynocephalus*, (Gr. *κυων*, *kuōn*, a dog; *κεφαλή*, *kephale*, head,) Dog-headed Thylacinus, Tasminian or Zebra Wolf, is an extremely active animal, of the size of a young wolf; has short smooth hair, of a dusky brown above, but barred or *zebraed* on the lower part of the back with about sixteen jet-black transverse stripes. This has to the other animals of the group, relations similar to those which the lion and tiger have to the larger quadrupeds of Africa and Asia. Formerly it preyed chiefly upon Phalangers and Kangaroos, rejecting the flesh of the Wombat, an animal common in the district which it inhabits. Since sheep have been introduced, its favorite food is mutton, which puts shepherds on the alert to destroy these animals by every possible means. The *Dasyurus*, (Gr. *δασύς*, *dasus*, thick; *οὐρά*, *oura*, tail,) has a conical shaped head, and on the hind feet the great toe is reduced to a tubercle, or entirely absent. It has four less molar teeth than the *Thylacinus*, making the number forty-two. One species is named *D. ursinus*, (Lat. *ursus*, a bear,)—*Ursine Dasyurus*—having very strong muscular jaws, and in its movements resembling the bear. Its vulgar name is "Native Devil." The *Dasyurus* is very destructive to poultry, eats raw flesh of all kinds and probably dead fish and blubber, as its tracks are found on the sea shore. In confinement it appears untamably savage, biting severely, and uttering at the same time a low, yelling growl. The *Phascogalē*, (Gr. *φασκώλιον*, *phaskolion*, a bag; *γαλε̄*, *galē* a weasel,) has seven molars instead of six, on each side, above and below, making the whole number forty-six. The species *P. penicillata*, (Lat. *penicillus*, a little tail,) lives on trees, has fur short, woolly and thick, and is rather larger than the brown rat.

II. FAMILY, the ENTOMOPHAGA, (Gr. *ἔντομα*, *entoma*, insect; *φαγω*, to eat.) INSECT EATERS.

These have three kinds of teeth in both jaws and a simple

stomach, like the preceding family, but more complicated intestines. This family includes three branches, or sub-families; *Ambulatoria,* (walking;) *Saltatoria,* (leaping;) *Scansoria,* (climbing.) The only genus of the Ambulatoria, or Walking section, is *Myrmecobius,* (Gr. μύρμηξ, *múrmēx,* an ant; βιόω, *bioō,* to live. The only species is *M. fasciatus,* (Lat. swathed,) which feeds on ants and has the reddish black of the body adorned with nine white bands, whence the specific name. Its length is ten inches. The *Perameles,* (BANDICOOTS,) is of the Leaping section, including animals which, in their general structure, form a link between the Opossums and the Kangaroos, evidently approaching the latter in their form, and particularly in the development of their hind quarters; with the Opossums they agree in having a simple stomach and ten incisors in the upper jaw. Some species, as *P. lagotis,* (Gr. λαγως, *lagōs,* a hare,) make large and almost exclusive use of vegetable food. In most of this family the pouch opens backwards, the reverse of what occurs in the other Marsupialia, though in *P. lagotis* it opens anteriorly. The species are found in Van Diemens' Land and in New Guinea. The *Scansoria,* or Climbing section, include the *Didelphidæ,* or OPOSSUMS, in their geographical distribution confined to this continent. These animals are all small, the largest being about the same size as the domestic cat, while some of them are no larger than mice. They number about thirty species, ranging from Brazil to Virginia, under one genus *Didelphis,* (Gr. δὶς, *dis,* double; δελφὶς, *delphis,* a pouch,) with the exception of a single species, found in Surinam, in size larger than a rat, and from its aquatic habits, as shown by its broad webbed feet, ranked as a sub-genus, under the name *Cheironectes,* (Gr. χεὶρ, *cheir,* hand; νηκτὴς, *nēktēs,* a swimmer.) The true OPOSSUMS, (*Didelphis,*) have fifty teeth, viz.: ten incisors above and eight below, four canines, twelve false molars, sixteen molars. The incisors are small and disposed in the form of a semi-circle; the canines are large and strong; the molars are crowned with sharp tubercles. The feet have each five toes, armed with strong curved claws; the inner toe of the hind feet, however, is destitute of a claw, and is so placed as to be opposable to the others, thus constituting a true thumb. The tail is more or less prehensile at the tip, and hence they are arboreal. The soles of their feet are covered with a naked skin of great sensibility; the ears and tip of the muzzle are likewise naked. In some species, as *D. dorsigerus,* (Lat. *dorsum,* a back; *gero,* to carry,) the pouch exists only in a rudimentary state, or slight folds of the skin. The young of these species, when of sufficient size, leave the pouch of the parent and are carried on her back, where they hold themselves by entwining their prehensile tails around that of the parent. (See Plate V. fig. 7.) The species best known is the common Opossum, *D. Virginiana*, of the United States, as early as 1649 thus described: "This beast hath a bagge under her bellie, into which she taketh her young ones, if at any time they be affryghted, and carryeth them away." The food of the Opossum is roots, poultry, and wild fruits. Like the spider monkeys, this animal uses the tail for climbing and swinging from branch to branch; it crawls slowly on the earth. When attacked it will feign itself dead, and no beating will induce it to show any signs of life. Even dogs are deceived, and turning it over, pass it by. The initiated determine whether it be alive or not "by the appearance of the last joint of the tail, which is never relaxed." From its assuming a feigned character, any adroit cheat, or sly deceitful acting, is said to be "possuming," or "playing possum." It has been said, "if a cat has nine lives, this creature surely has nineteen; for if you break every bone in their skin and mash their skull, leaving them for dead, you may come an hour after and they will be gone quite away, or perhaps you may meet them creeping away."—(Lawson.) The color of the Opossum is greyish white, darker along the sides; the flesh is very white and well tasted; for this it is hunted, but not for its fur. When disturbed or alarmed it gives out a very unpleasant odor.

The Virginia Opossum is about the size of a domestic cat. Its hair is of two kinds; the lowest a long woolly down, brownish at the tip, through which pass the long hairs of a pure white on the head, neck, and upper parts of the body. The tail is not so long as the body, covered at the base by long hairs, but only scantily furnished with bristles which come out from between the whitish scales that protect it for the greater part of its length.

III. Family, the CARPOPHAGA, (Gr. καρπος, *karpos*, fruit; φαγω, *phago*, to eat.) FRUIT EATERS have large and long incisors in both jaws; the canines sometimes wanting, and a still longer intestinal canal. They resemble the squirrel tribe, but are more closely related to the Kangaroos, the Kangaroo-rats, (*Hypsiprimnus*, Gr. ῾υψιπρυμνος, *hupsiprimnos*, high extremity or stern,) affording the connecting link.

Of this family are the PHALANGERS, *Phalangista*, (Gr. φαλαγξ, *phalanx*, plu. φαλαγγες, *phalanges*, small bones of the hands or toes, (see Plate III. figs. 3 and 4.) These are so named because they have the second and third toes of the hind feet united as far as the last *phalanx*, (or small bone,) in a common skinny sheath. They have short, woolly fur, and a long prehensile tail. Among these are the COESCOES, (sub-genus *Cuscus*,) of the Molucca Islands, said to suspend themselves by the tail at the sight of a man.

The *Petaurus*, (Gr. *petaō*, to fly; *aura*, air,) has thirty-eight teeth; no canines; the skin expands between the fore and hind limbs, enabling it to take very long leaps, supported in the air as by a parachute. In leaping, it is aided by its flattened and bushy tail.

P. sciureus, the NORFOLK ISLAND SUGAR SQUIRREL, or FLYING SQUIRREL, rests by day, but at night skims through the air, half leaping, half flying from branch to branch, feeding upon leaves and insects.

The IV. Family is the *Poephaga*, (Gr. *πόη*, *poē*, grass, *φαγω*, *phago*, to eat,) GRASS EATERS.

Sub-family *Macropidæ*, (genus *Macropus*, Gr. long-footed.) The KANGAROOS. The aspect of these animals is singularly striking—the front parts are light and graceful, while the hinder parts of the body, limbs and tail are very stout and muscular; the head is lengthened; the ears very large; the upper lip cleft; the whiskers very short and few; the hind limbs have very long tarsi, like those of the Kangaroo-rat, but are much longer and more robust; the tail is long, triangular and very muscular. The teeth are comparatively few, viz.: incisors, $\frac{6}{2}$; canines, 0; molars, $\frac{4-4}{4-4}=24$. The species are numerous. The one best known is the *Macropus major*, the GREAT KANGAROO. The natural position of these animals is sitting upon their hind legs, in which attitude they are supported by the strong, muscular, and tapering tail. Their movement on all fours is awkward and constrained, but they bound or hop along on their hind limbs with great facility, each leap being about fifteen feet. They easily clear obstacles seven or eight feet high. *M. Brunii*, Le Brun's Kangaroo, is the first of the Marsupials with which naturalists became acquainted. It is an inhabitant of New Guinea, and was described by Le Brun as early as 1711. The Kangaroo was discovered by Capt. Cook in his first voyage. Since that period, (1770,) it has been brought over in abundance to Europe and this country; has bred freely and might become an associate of deer in parks and forests. The conical and tapering form of the body at once suggests to the beholder the idea of great muscular power in the loins and lower limbs, just the opposite to the mole. Its fore limbs are of little use in its forward movements. The defensive weapon of these animals consists of the large claw of the hind foot, which is lengthened, strong, and armed with a hoof-like nail. With this they can inflict a severe blow; their eyes are full and bright; the mouth small; the ears large and pointed; the fore paws are divided into five fingers, armed with nails for scratching or digging; the hind feet have five toes, but the two inner ones are very small, and so united in their whole length under the skin as to appear but one. The Great Kangaroo inhabits New Holland and Van Diemen's Land, and is about five feet without the tail, the length of which is about three feet. The female, like the Opossum, carries the young about in its pouch, from which they emerge when they desire exercise, and leap back again on the least alarm. The largest weigh 140 to 150 pounds. The Kangaroo's flesh is much esteemed; it is hunted in Australia with a breed of dogs between the mastiff and greyhound.

The V. family is the *Rhizophaga*, or ROOT EATERS, (Gr. ῥιζα, *rhiza*, root; φάγω, *phagō*, to eat.) In this we find the WOMBAT, *Phascolomus*, (Gr. φασκώλιον, *phaskōliŏn*, a pouch; μῦς, *mus*, a mouse,) Sub-family Phascolomyidæ.

This animal has been described as follows: "The Wombat, or as it is called by the natives of Port Jackson, the Womback, is a squat, short, thick, short-legged, and rather inactive quadruped, with great appearance of stumpy strength, and somewhat bigger than a large turnspit dog. Its figure and movements, if they do not exactly resemble those of the bear, at least strongly remind one of that animal. Its length from the tip of the tail to the tip of the nose is thirty-one inches. The hair is coarse and about an inch and a half in length, thinly scattered; thinly set upon the belly, thicker upon the back and head, and thicker upon the loins and rump; the color is of a light and sandy brown of various shades, but darkest along the back." The Wombat will not compare with the Kangaroo in swiftness of foot, as most men could run it down. Its pace is a hobbling or shuffling, something like the awkward gait of a bear. The flesh is said to be excellent meat, and as it is nearly three feet in length, it is suggested that it might be worth naturalizing in other climates, specimens which have been taken to Europe having lived for years.

The whole of the Marsupialia, though some are active and sprightly in their manners, present but little appearance of real docility and intelligence; and this fact, connected with the low degree of development of their brain, points to their inferior rank among the placental Mammalia. To denote this inferiority the boundary lines of this Order are, on the chart, bent round towards the CETACEA. The earliest mammiferous animals whose remains are found in the secondary and tertiary formations, are those of this order.

Sub-order MONOTREMATA, (Gr. μόνος, *monos*, one; τρῆμα, *trēma*, perforation.) The animals of this sub-order have given occasion to naturalists for much discussion concerning their proper affinities and their appropriate position among the MAMMALIA. They are truly unique, both in their external form and their anatomical and internal arrangements, the details of which cannot be given in this work. We will only say that "in the form of the skull, the construction of the shoulder and the breast-bone, but particularly in the whole reproductive system of organs, the *Monotremata* present a manifest departure from a mammalian type, and a corresponding approach to that of the oviparous Vertebræ, tending to the reptiles more than to the birds." But however anomalous, it is evident they should have a place among the mammals; and also, though without any external pouch, that the marsupial bones in the skeleton require that they be placed next in order to the Marsupialia, "of which they constitute the lowest and most aberrant type."

These singular animals have no true teeth, but those of one genus have horny substances in the jaw which represent those organs. The muzzle is prolonged into a flat beak, more or less like that of a duck; the eyes are small; the ears are merely minute orifices and without any external *conch;* the limbs are short and strong, suited for digging; the feet have each five toes, furnished with stout claws, and on the hind foot is a kind of sharp spur.

The order includes but two genera, viz.; *Echidna* and *Ornithorhyncus.* Both are found exclusively in New Holland and Van Diemen's Land.

Echidna, (Gr. ἔχιδνα, *echidna*, a fabulous monster or viper.) Of this there is but one species, changing its name with the variations of its clothing at different seasons, viz.: *E. histrix*, (Gr. ὕστριξ, a porcupine,) to *E. setosa*, (Lat. bristly.) The muzzle of this animal is elongated and slender, terminated by a small mouth, having a long extensile tongue, similar to that of ant-eaters and pangolins; it is, however, more beak-shaped. The skin of this beak is thick and without hair. The animal has no teeth, but the palate is armed with many rows of small spines, directed backwards. The feet are very large, robust, and armed with claws, being formed for opening ants' nests. The upper surface of the body and of the short tuberculous tail is covered with stout and strong spines, intermingled with stiff, bristly hairs, and when alarmed, the animal can roll itself up like a hedge-hog or porcupine, with which latter it well compares in point of size. The chestnut colored, soft and silky hair is so abundant at a certain season as to half cover the spines, whilst at another, the hair entirely disappears. It lives on ants, with their larvæ and pupæ. It takes them with its extensile tongue, which it can protrude to a great distance, and which is always covered with an adhesive secretion. The Echidna digs for itself burrows in which it remains during the dry season, coming out of the earth only during the rains. It is supposed capable of enduring a long abstinence, and it has intervals of suspended animation which continue for more than three days at a time, and recur frequently when the animal is kept in confinement. Its strength has been thought to exceed, considering its size, "that of any other quadruped in existence."

Ornithorhyncus, (Gr. ὄρνις, *ornis*, a bird, and ῥύγχος, *rhunchos*, a beak, so named from its bird-like bill.)

Two species have been described, *O. fuscus*, (Lat. dusky,) and *O. rufus*, (Lat. red,) but the latter differs from the former only in having the fur softer and of a redder tint. It is said that in looking at this animal one would imagine that the beak of a shoveller-duck had been artificially fastened on the front of the head of a small otter. The beak, which is broader at the tip than at the base, is covered by a thick leathery skin. This skin projects in the form of a loose flap from each mandible, and protects the eyes from the mud in which the animal is perpetually dabbling for food. There are no true teeth, yet back of each mandible are two horny appendages resembling teeth, but without roots, which are of a form verging to a square, with a broad uneven surface, fitted rather for crushing than grinding. Beneath the skin of the face are capacious cheek pouches for the carrying of food. The eyes are bright, but very small and high set; the ears mere orifices which are opened and closed at the will of the animal; the feet have five well developed toes, all armed with long, curved, and pointed claws, connected by a leathery web, which in the fore feet extends considerably beyond the tips of the claws, presenting a broad and powerful oar when in the water, but folded back when the animal is digging in the earth. On the hind feet the web reaches only to the termination of the toes. In the male the feet are also armed with a stout, sharp, movable spur, formerly regarded as highly poisonous. The tail is broad and depressed; the fur combines the properties of an aquatic and also of a burrowing animal, readily expelling both water and dust. A full grown ornithorhynchus is about two feet long, measuring beak and tail. The general color is deep brown, with a white spot in front of each eye. These animals are called *Water Moles* by the colonists. Their favorite resorts are the borders of some stream covered with aquatic plants, where the banks are steep, shaded, and convenient for burrowing. They burrow in a serpentine direction, sometimes to the distance of fifty feet, and ending in a small chamber. In this chamber they place their nest made of dry grass.

QUESTIONS ON THE MARSUPIALIA.

From what is the term MARSUPIALIA or MARSUPIATA derived? Into what two sections is this order arranged? How are these sometimes regarded? What does Cuvier remark respecting the animals of this order? Which of the two sections deviates least from the general type of mammalia? What is their most striking peculiarity? To what part of the globe are these mammals confined? What is the ground of their division into families? What is said of the first family, SARCOPHAGA? How many genera of this family does Prof. Owen enumerate? What peculiarities do they show? Describe the Dog-headed *Thylacinus* or Zebra Wolf? What relation does it bear to the other animals of the group? What is said of the *Dasyurus?* What species of this animal is mentioned? On what does it feed? How does it appear in confinement? What is said of the *Phascogale?* What species is named? Give the names and characters of the second family. What three branches or sub-families does this include? To which of these does *Myrmecobius* belong? How many teeth has it? What species of this genus is named? What genus of the leaping section is mentioned? What link do the animals of this genus form? In what respects do they agree with the opossums? What species is named and what is said of it? What animals do the *Scansoria* include? How many species of them? To what region are they confined? How is the term *Didelphis* compounded? What sub-genus is named? How are the Opossums characterized and described? Describe the best known and only species found in the United States. Give the general character of Fruit-eaters, or the third family. How are they linked to the Kangaroo? What genera and species are mentioned? Describe and characterize the Kangaroos, or grass-eaters. Which is best known? Where is it found? Which of the Root-eaters is mentioned? How is it described? How do the MARSUPIALS rank among mammals? Why are the boundary lines of this order carried round next *Cetacea?* What is peculiar in the MONOTREMATA, and what is their general rank? How many genera do they include? Where found? What is said of the Spiny Ant-eater? Particularly describe the *Ornithorhyncus.*

What is said of this order along the branches of the Chart? Mention the animals of this order named or figured on it, tracing each.

SECTION XVIII.

Fifth Order, EDENTATA. (Lat. *toothless.*)

This name was originally given by Cuvier, to the animals of this order, from their agreement in the absence of incisive teeth from their jaws, and in the length of their claws. Apart from this agreement, they appear to have among themselves but little natural affinity. To several of the ant-eating tribe, which this order includes, the name *Edentata* is literally applicable; but in other genera it is limited to the front, or incisor teeth. In this order Cuvier included the Monotremata, but their most natural place seems to be with the *Marsupials.*

I. Family, Tardigrada, (Lat. *tardus,* slow, *gradior,* to step;) also named Bradypodidae, (Gr. βραδὺς, *bradus,* slow, πους, *poūs,* a foot.) This includes two genera, *Bradypus tridactylus,* (Lat. three-toed,) the Ăi, or Sloth, and *Choloepus,* (Gr. χωλὸς, *cholos,* lame, πούς, *pous,* a foot,) *didactylus,* (Lat. two-fingered;) the Unau.

These animals have no incisor teeth, four canines, two in each jaw, fourteen molars, eight in the upper and six in the lower jaw. The molar teeth consist each of a cylinder of bone, covered with enamel; hence their surfaces are always concave, the enamel wearing less rapidly than the soft interior. No *laminae,* or folds of the enameled substance penetrate the body of the teeth, as in most other animals; the canines are somewhat longer than the molars, and in form pyramidal. When these animals stand erect upon their hind legs, their fingers can reach to the ground; and when moving upon all fours, they trail themselves slowly and painfully along upon their elbows. Their claws surpass the whole foot in length, and are very sharp and crooked. (See Plate VI. fig. 5.) In a state of rest, they are drawn down upon the palm and wrist, and can be extended only by the will and muscular effort of the animal. Sharp, and bent in form, they are so many effective hooks for holding on; while the rigidity of the limbs gives a firm hold; the feet and thighs are jointed obliquely, which adapts them for embracing a branch; and the great length of the arms aids these animals in seizing a fresh hold, and drawing twigs and leaves, their usual food, to their mouths. They are born and live on the trees, and never leave them, unless from the operation of force, or accident, resting not *upon* the branches, like the squirrel, or monkey, but *under* them, and moving and even sleeping suspended from them. It is remarked of some which were in a state of captivity, that they assumed, during sleep, "a position of perfect ease and safety, on the fork of a tree," the head being supported between the arms and chest, and the face buried in the long wool which covers those parts, and thus protected during sleep, from the myriads of insects which would otherwise assail it.

The animals of the other genus, the Unau, or *Choloepus didactylus*, the two-fingered sloth, have essentially the same singular conformation and habits as the three-fingered sloth, and are with those of the other genus, found among the tropical forests of South America.

II. Edentata Proper. Myrmecophagadae, Ant-Eaters. *Myrmecophaga*, (Gr. μύρμηξ, *murmex*, an ant, φάγω, *phago*, I eat.)

The Ant-eaters are distinguished by being *entirely* without teeth, and also by their hairy covering. The latter peculiarity separates them from the Pangolins, (*Manis*,) or Scaly Ant-eaters, of Asia and Africa, which animals, in other respects, they closely resemble. In this family, the jaws are produced into a very long and slender muzzle, which has a mouth of very diminished size. (See Chart.) The phalanges, or small joints of the toes (particularly the last,) which bear the claws, are so formed as to allow them to be bent inwards only as in the Sloths; and to this end, have very powerful ligaments, which keep them in a state of repose, bent in along the sole of the foot, and do not allow the hand to be opened entirely, but only half extended, as seen in gouty or rheumatic people. (Plate VI. fig. 5.) The toes are of unequal size, and vary in number, in different species; as in the Sloths, they are united closely together as far as the claws, and are not capable of separate or individual motion; but this disability is more than compensated by the increased strength which it produces. The claws are all large and powerful, especially that of the middle toe, which is enormous. In walking, these animals tread upon the outer edge of the foot, which is provided with a large callous pad for that purpose; whilst their toes being bent inwards, along the palms, the sharp claws are preserved from being injured by the friction of the hard ground.

The Ant-eaters are remarkable for their very long and rounded tongues. With these, they take the ants which are their principal food. On approaching an ant-hill, the animal scratches it up with his claws, and then protrudes his slender tongue, which has the appearance of an exceedingly long tape worm. The tongue is covered with a glutinous saliva; it is nearly twice the length of the whole head and snout together, and when not extended, is kept doubled up in the mouth, with the point directed backwards. The ants adhere to his tongue when it is thrust into their hills, and by retracting it, he swallows thousands of them. The eyes of the *Myrmecophaga* are exceedingly small; their ears short and round; the legs robust and amazingly powerful, but so unfavorable for locomotion, that these animals are almost as tardy in their movements as the Sloths themselves, except when put to their speed, at which time, their motion is pretty rapid. Of the Ant-Eaters proper, we name three species. 1. *M. jubata*, (Lat. maned or crested,) the GREAT ANT-EATER. This animal is about four and a half feet in length, from the snout to the tail, which is three and one-quarter feet long, so that the entire length of the animal is seven and three-quarters feet; the height at the shoulders is three and three-twelfths feet, and but two and ten-twelfths feet at the croup, in consequence of which, being perfectly plantigrade, it necessarily stands lower behind than before, as is seen in the bear and badger; the toes are four on the front, and five on the hind extremities. It is sometimes called Ant-Bear, from its mode of defence, which resembles that of the bear. When assailed by a dog, he seizes him between his strong fore legs, and squeezes him to death, or else deals out severe blows with his sharp prehensile claws. The clothing of the Great Ant-Eater consists of long, coarse hair, forming a *mane* down the neck and back, and enveloping the tail in a thick brush, which trails upon the ground. On the head the fur is close and spare. The color is generally a grizzled black; a dark black stripe, bordered with white, passing obliquely from the side of the neck, to the upper part of the back. This singular animal has but a single young one at a birth, which for a whole year is carried about with the mother wherever she goes. Its digestive organs seem adapted for extracting nutriment from ants alone. In its habits, it is solitary as well as slothful. Like all other animals living upon insects, it can exist a long time without food. Its flesh, though black, and of a musky flavor, is sometimes found on the tables of Europeans, and by the Indians is highly esteemed.

M. Tamandua. (Cuvier.) The TAMANDUA.

This Ant-Eater is much smaller than the one just described, being not so large as a fox, or even a good sized cat; whereas the Maned Ant-Eater exceeds in length the largest greyhound, though much inferior to that animal in height, owing to the shortness of its legs. In the conformation of its extremities, and the number of its toes before and behind, the Tamandua is like the Ant-Bear; but it differs from that animal in the prehensile power of its tail, which makes it essentially an arboreal quadruped.

The hair differs also, being short and shining, and of a consistence which makes it a medium as to its qualities, between silk and wool. The colors, are, likewise, more variable than those of the Great Ant-Eater. The Tamandua is found in the thick primeval forests of tropical America, living on trees, upon termites, honey, and according to D'Azara, upon *stingless* bees, which have their hives among the loftiest branches of the forest. The female has but a single cub at a birth, which she carries about with her on her shoulders, for the first three or four months.

M. didactyla, (Lat. two-fingered, or toed.) The LITTLE or TWO-TOED ANT-EATER.

This is easily distinguished from the other two species, by its size, which does not exceed that of a large rat or squirrel; also by the number of its toes, four on the hinder, and only two on the front extremities. The length from the snout to the tail is but six inches; that of the tail is seven and one-quarter inches; towards the point the tail tapers, and becomes naked, and it is strongly prehensile. The snout is not so long in proportion to the body, as in the other two species; the legs are stout and short; the hair very fine and soft to the touch. Like the other species, the Little Ant-Eater has but one young at a birth, which it conceals in the hollow of some decayed tree.

Orycteropus Capensis, (Gr. ὀρυκτηρ, *oruktēr*, a digger; πους, *pous*, a foot.) The AARD-VARK, or EARTH-HOG.

This animal, of Southern Africa, is also to be numbered with the Ant-Eaters, though there has been some difference of opinion as to its proper location. It resembles both the Ant-Eater and the Armadillo, agreeing with the former in its general habits; but though without any scaly armor, more like the latter in its anatomical structure. Like the Armadillo, it has large and powerful claws, adapted for digging up roots and insects, and for making burrows in the earth. When full grown, it is five feet long, from the snout to the end of the tail, which is about half the size of the body. Its tongue is not cylindrical like that of the Ant-Eaters proper, but flat and slender, and cannot be protruded so far. The flesh, particularly of the hind-quarters, is dried for hams, and much esteemed as food.

Manis. (Linnæus.) The PANGOLIN, or SCALY ANT-EATER.

The name Pangolin, which is given to the animals of this genus, is said to be derived from the word Pangoeling, signifying, in the Javanese language, "an animal which rolls itself in the form of a ball." The Pangolins are limited to the warmest parts of Asia and Africa. In common with the Hairy Ant-Eaters, they are without teeth, and have a very long extensile tongue, covered with a glutinous mucus, for securing their insect nutriment, but they differ from them in their body, limbs and tails, which have as a panoply, their scales large, imbricated, (*i. e.* hollowed like a roof, or gutter-tile,) and overlapping each other; they differ also in being able to roll themselves up when in danger, by which means their trenchant or sharp cutting scales become erect, and present a defensive armor against their enemies. These animals are particularly remarkable for the strength and number of the vertebrae of the tail, (forty-seven in the large species.) By some they are regarded as a kind of link between viviparous quadrupeds and the Lizards.

M. macroura, (Gr. long-tailed,) or *M. tetradactyla*, (Lat. four-fingered.) (Linnæus.)

This species, found in Africa, is more than two feet in length, and the tail is more than twice as long as the body. The broad, striated and pointed scales, cover the whole body, except the under part; the legs are very short, and also scaled; on each of the feet are four claws, those on the fore feet being stronger than the others. The scales are of a uniformly deep brown color, with a tinge of yellow, and a glossy surface.

M. brachyura, (Gr. *βραχὺς*, *brachus*, short; 'ουρά, *oura*, tail,) or *M. pentadactyla*, (five-fingered.) The Short-Tailed Manis.

This Scaly Ant-Eater is a native of East India, where it receives different names, Tiled-Cat, Land-Carp, Caballe, &c. It has a much thicker and shorter tail than is found in the preceding species; the body is stout, and shorter than the tail. Each of the feet, as the specific term *pentadactyla* denotes, has five toes; those on the fore feet, except the outer one, which is small, being very strong. The scales differ in shape from those of the Long-Tailed Manis, and are much larger and wider in proportion to the body and the tail; they are so impenetrable that when the animal rolls itself up, the tiger, panther, or hyaena attempts to force it in vain. The middle claw of the fore paws, far exceeds the others in its proportions, and is admirably adapted for the destruction of the nests of termites, or white ants, which are a great part of its food. It is said the natives "have a method of making a hole in its skin with a knife, and thus of guiding and governing the animal at their pleasure, the point of the knife, which is kept in the hole, goading and irritating him." It is numerous in Ceylon.

III. *Dasypodidae*, (Gr. *δασυς*, *dasus*, hairy; *πους*, *pous*, a foot.) The Armadillos.

This remaining family are arranged by Cuvier into five groups. They are distinguished by having molar teeth alone, and appear to have a place between the Sloths and Ant-Eaters, the latter being without teeth, and the Sloths, in addition to the molars, having large and powerful canines. Ant-Eaters differ from the Sloths and Armadillos, not only by being without teeth, but also by the want of clavicles, or collar bones. The most prominent distinction of the Armadillos, is the peculiar nature of their external covering. This consists of a bony, tessellated crust, in which their bodies are enveloped; the hips and shoulders being covered by large, broad bucklers, while the intermediate back is shielded by transverse movable bands, similar in form and appearance to the plate armor of the middle ages. Hence the name Armadillos, (from *Armada*, armed, and of Spanish origin,) has been given to these animals. The transverse bands which are separated by narrow strings of membrane, overlap each other, as in the ancient coats of mail, so as to give greater freedom, and some degree of lateral motion. The tail, with the exception of one species, is covered with a series of rings; the limbs are incased in a hardened, tuberculous sort of skin, and are very short and strong; the toes have strong claws, adapted for digging or burrowing, a process, which, in the light sandy soil traversed by them, they accomplish with surprising celerity. The molar teeth with which they are furnished, are never less than twenty-six in the whole; and in one species amount to ninety-eight! those of one jaw fitting into interstices of the other as in the Dolphins. (See Plate IV. fig. 11.) The eyes are very small; the ears large; the long and slender tongue, like that of the Ant-Eaters proper, is lubricated with a viscid saliva, by means of which it readily takes up ants and similar insects, upon which it chiefly subsists. It however, also feeds on farinaceous roots, and on carrion, so that in Paraguay, deceased persons who are "interred at a distance from the usual place of sepulture, are obliged to be protected by a lining of strong boards." In searching for food, it is guided chiefly by the sense of smell; its sight is poor, but this is compensated by the acuteness of its hearing. The Armadillos burrow with such rapidity that they soon disappear in the earth, when suddenly surprised. Their movement is a sort of waddling run, but rather rapid, most of them easily outstripping a man. In captivity, this is kept up by the hour together, and without any apparent motive. The greater portion of them are nocturnal, never moving abroad while the sun is above the horizon, but remaining concealed in their burrows. The female bears annually, and frequently six, eight, or even ten at a birth. The Armadillos are able, more or less perfectly, to roll themselves up into a ball. These hardy animals thrive and breed rapidly, with a moderate portion of care, in most temperate countries, but their proper habitat is the tropical and temperate portions of South America. Of the nine or ten species, we particularize

1. *Dasypus Peba*, or *D. novemcinctus*, (Lat. nine-banded.) The PEBA OR BLACK TATU. Pl. VI.

This species, found in Paraguay, Guiana and Brazil, varies in the numbers of its bands, so that it is sometimes called *D. octocinctus*, (Lat. eight-banded,) and *D. septemcinctus*, (Lat. seven-banded.) Its length, from the snout to the tail, is sixteen inches; that of the tail is fourteen inches, and its circumference at the base, six inches. It is much hunted on account of the delicacy of its flesh, which when roasted in the shell, is fat and well tasted; said to resemble that of a sucking-pig. Of individuals of this species, found in the Zoological Gardens of England, it is remarked, "they are fed on vegetable diet, and appear to be in excellent health. During the summer, they are allowed the liberty of a little paddock, where, by the singularity of their actions, they attract a crowd of spectators, and come in for a share of the interest excited by the gambols of their fellow countrymen, the Spider Monkeys." (Martin's Quadrupeds.)

D. Apar. The MATACO.

The animals of this species are distinguishable from all others of the genus, by "the faculty which they possess of rolling themselves up like a hedgehog, into a round ball, in which situation they may be tumbled about, or even, it is said, thrown over precipices, without receiving any material injury." They are, however, less common than some of the other species.

D. gigas, (Lat. a giant.) The GREAT ARMADILLO.

This species have unequal toes and enormous claws, but what most distinguishes the animals of this group, is their possession of from eighty-eight to ninety-six teeth, a number greater than is found in any other mammal. (Pl. VI. fig. 6.)

The Great Armadillo is about three and one-quarter feet long, from the nose to the tail, which is one foot, five inches. It is separated from the other species of this genus, not only by its superior size, but by various remarkable characteristics. Its head is proportionably smaller; the forehead more protuberant; the face rather cylindrical in form, like that of the Peba; the ears are not very large, pointed, and crouched backwards; the bucklers of the shoulders and croup have nine and eighteen rows of plates respectively, and are separated by movable bands to the number of twelve or thirteen, formed of rectangular scales, about half an inch square. At the root, the tail is as much as ten inches in circumference, and covered with ring plates, at the base, and with crescent-shaped lines throughout the rest of its length.

The claws are very large and powerful. This animal confines itself to the great forests, and burrows with surprising facility, being assisted in this by the strength of its claws. "Those who are employed in collecting the Jesuit's bark, frequently meet with it in the woods, and report that when any of their companions happen to die at a distance from the settlements, they are obliged to surround the body with a double row of stout planks, to prevent it from being scratched up and devoured by the Great Armadillo."

Chlamyphorus, (φορεω, *phoreo*, I bear; κλαμὺς, *chlamus*, a cloak.) The PICHIAGO.

This edentate animal seems to blend in itself the characteristics of several distinct tribes. Like the Armadillos, it has a tessellated shield; this, however, is not, as in them, attached by integuments, to the entire under surface, but is connected with the back only, by a ridge of skin along the spine, and with the skull by two bony prominences from the forehead, the margins of which are beautifully fringed with silky hair. Its feet, eyes and snout, exhibit resemblances to the mole. From the appearance of the hind part of the tesselated shield, this animal has the specific name *truncatus*, (Lat. truncated, or cut off.) Naturalists have designated resemblances in it to the Sloth, the Aard-Vark, the Great or Maned Ant-Eater, the *Echidna*, and the *Ornithorhyncus*; and to the Ruminants and Pachyderms. Dr. Buckland regards it as "one of the nearest approximations to the *Megatherium*, particularly in regard to its coat of mail, and in the adaptation of the animal for digging." Dr. Harlan says, "taken collectively, it furnishes us with an example of organic structure, if not unparalleled, not surpassed in the history of animals." The Pichiago is quite small, the total length of the animal being only five inches and a quarter. "It is a native of Chili, but is so rare even there, as to be regarded by the natives as a curiosity."

IV. *Megatheridæ*, (Gr. μέγας, *megas*, great; θηρίον, *therion*, wild beast.) FOSSIL SLOTHS.

This is a group of animals of such gigantic size, and massive proportions, that even their fossil remains strike the beholder with wonder and astonishment. Of such a character are these remains, that we are constrained to bestow more space upon them than can be given to other fossil tribes. These are the MEGATHEROIDS of Professor Owen, whose descriptions of them are exceedingly elaborate and interesting. Of these fossils, the following genera have been enumerated by him, viz., *Megatherium*, *Megalonyx*, *Glossotherium*, *Mylodon*, and *Scelidotherium*, all of which are found in South America alone. Of the *Megatherium*, nearly the whole skeleton has been considered, by comparing different imperfect specimens, found after three unusually dry seasons, in the river Salado, running through alluvial plains, to the south of Buenos Ayres. This has given rise to the not improbable "suggestion," that the long continued drought brought these extinct gigantic animals to a slender stream, running between mud banks, and that they may have been "engulphed in their efforts to reach the water."

The *Megatherium* gives evidence in its remains, that it was more nearly allied to the Sloths and Ant-Eaters, than to the Armadillos. The skull is thought to resemble the former two; the rest of the body was adapted partly to the former and partly to the latter. When full grown, it is judged this enormous animal must have been not far from eighteen feet in length, and nine feet in height. (See fig. on the chart.) The thigh bone twice the thickness of the largest elephant's; the fore foot more than three feet in length, and more than one in width, and terminated by an enormous claw. The width of the upper part of the tail, could not have been less than two feet. The entire structure of this extinct animal, must have been admirably adapted for digging in the earth, so as to enable it to obtain the succulent roots which probably constituted the principal part of its food. Dr. Buckland, in his "Bridgewater Treatise," says, "The size of the Megatherium exceeds that of the existing *Edentata*, to which it is most nearly allied, in a *greater degree than any other fossil animal* exceeds its living congeners. The entire frame must have been an apparatus of colossal mechanism, adapted exactly to the work it had to do; strong and ponderous in its proportions, as its work was heavy, and calculated to be the vehicle of life and enjoyment to a gigantic race of quadrupeds, which, though they have ceased to be counted among the living inhabitants of our planet, have, in their fossil bones, left behind them imperishable monuments of the consummate skill with which they were constructed."

Megalonyx, (Gr. μέγας, *megas*, great; ὄνυξ, *onux*, nail or claw.) To the remains of this animal, this name was given on account of the size of its claws. Mr. Jefferson described it from some bones found in caverns in Western Virginia, and considered it to be carnivorous. He supposed it the largest of unguiculated animals, and probably the enemy of the *Mastodon*. Dr. Wistar, of Philadelphia, afterwards saw in the bones of the fossil foot, resemblances to those of the Sloth. Cuvier showed that it belonged to the *Edentata*. Professor Owen reviewed the whole subject, and arranged the animal as a distinct genus.

Glossotherium, (Gr. γλωσσα, *glōssa*, a tongue; θηρίον, *therion*, a wild beast.)

This genus is based "on a fragment of a cranium found in Mr. Damin's collection, discovered in the bed of the same river, in Banda Oriental, with the skull of the Toxodon." Reasoning from this fragment, Professor Owen found decisive evidence that the cranium was that of an extinct Edentate, and related to the genera *Myrmecophaga* and *Manis*.

Mylodon, (Gr. μυλή, *mulē*, a grinding mill; οδοὺς, *odous*, a tooth.)

This fossil Edentate, according to Professor Owen, "holds an intermediate place between the Ai and the Great Armadillo." It must have had the size and proportions of a Rhinoceros, but with the limbs still more massive. So great was probably its muscular strength, it could overthrow trees; and as it was a leaf-eater, and too bulky and ponderous for climbing, it could thus feast at its ease, on the abundant foliage.

Scelidotherium, (Gr. σκελὶς, *skelis*, a haunch, or thigh; *thērion*, a wild beast.)

This large extinct Edentate was allied to the *Megatherium*, and the *Orycteropus*, Cape Ant-Eater.

What is the fifth Order of the MAMMALS? Why did Cuvier give this name to the animals of this order? Is it strictly applicable to all the genera? To which is it applicable? Name the first family. What is its meaning? Give the other name of this family and its significations. How many genera does it include? What are the leading characters and habits of the *Sloth or Ai?* What gives them a firm hold on the branches of trees? Do they ever leave them? On what part do they rest? What is said of their sleep during captivity? What is said of the habits of the Unau and other tropical Sloths? Where are they all found? Give the name of the second family. What is its derivation? How are the ANT-EATERS distinguished? What peculiarly separates them from Pangolins? What is said of their jaws, phalanges (small bones of the fingers and toes,) and toes? Are the toes capable of separate motion? How do they walk? Describe the tongue and its uses. What is said of the other parts of the animal? How many species of the ANT-EATERS PROPER are named? What is said of the size of the animal? How many toes has it? Why is it sometimes called *Ant-Bear?* Why *Jubata* or crested? What more is said of it? What is said of the size of the *Tamandua?* In what respect does it differ from the Great Ant-Eater? How is the *Little or Two-Toed Ant-Eater* distinguished, and in what particular respect or feature does it differ from the other two species? Where is the *Aard-Vark* or *Earth-Hog* found? What animals does it resemble? What is its size? Does its tongue differ from that of the A. E. Proper? Why was the name *Pangolin* given to the *Scaly Ant-Eaters?* To what region is it confined? How does it resemble the *Hairy Ant-Eaters*, and how differ from them? For what are these animals particularly remarkable? Where is the Long-tailed species found? What is said of the scales? Where is the Short-tailed species found? What names has it received?

How does it differ from the Long-tailed species? What more is said of it? What is the name of the *Third Family?* How does Cuvier arrange it? Has it teeth? What is the chief distinction of the *Armadillos?* Describe them. What is the origin of the name? What is said of the tail? How does the number of the teeth vary? How do they resemble those of the Dolphin? What further is said of these animals? Where is the Peba found? How long is it? On what account is it hunted? What is said of animals of this species in the Zoological Gardens of England? What distinguishes the Mataco from all others of the genus? What is said of the toes and claws of the *Great Armadillo?* What is its size? How is it separated from the other species? What is reported by the collectors of the Jesuits' bark? Give some account of the *Chlamyphorus?* What does Dr. Harlan remark? What is its size?

What is the Fourth Family? What is said of the size and proportions of these animals? What of their fossil remains? How many species does Prof. Owen name? How has the structure of this animal been made out? What suggestion has been made respecting it? To what animals were they most nearly allied? What is said of their size? What of its fore feet and tail? For what was it adapted? Give the quotation from Dr. Buckland. Define the term *Megalonyx.* Why was this name given? State Jefferson's views of this animal. What did Cuvier show? Who arranged it as a distinct genus? What is the import of the term *Glossotherium?* Upon what was this genus based? How did Prof. Owen determine their relation to Ant-Eaters? Explain the term *Mylodon.* What place does Prof. Owen assign it? What is said of its size? Define the term *Scelidotherium.* To what does it relate?

What is said of the MEGATHERIUM on the chart? Give its dimensions and trace it from its position among the *Sloth Family*, BRADYPIDAE, through all its grades. Trace the Armadillo in the same way.

SECTION XIX.

SIXTH ORDER.—RODENTIA. (Lat. *rodo*, to gnaw.)

RODENTS or GNAWERS. The GLIRES of Linnæus.

The animals of this order may be at once known by their having, for the most part, two incisors or front teeth in each jaw, remote from the back teeth or grinders; (the *Hare family* have two, four, and sometimes six in the upper jaw.) There are no canine teeth, but a vacant space appears between the front and back teeth. The greatest number of cheek teeth is twenty-two. The incisors have no roots, but are deeply inserted in their sockets. The enamel of the front side being more durable than the other bony matter of the teeth, always preserves their chisel-like edge. The jaws are so articulated that the lower jaw, (besides opening and shutting,) simply moves backwards and forwards, or horizontally; so that the front teeth serve to file down, or reduce to fine particles, the hard substances which are brought under their action. To meet the wear of the enamel and other parts, the teeth constantly grow in a ratio corresponding with the decrease or wear. Should one tooth be lost by accident, or displaced, the counter one of the opposite jaw becomes enormously long, so as to impede its feeding, as is seen in rabbits. The molar teeth have flat surfaces, with ridges of enamel running transversely across, so as to be opposed to the horizontal movement of the jaw, and thus more readily grind their food. The entire dental arrangement evinces admirable beauty and simplicity of design.

The Rodentia, according to De Kay, include not far from 300 species, spread over the globe, (except Australia,) of which seventy are found in North America. They are generally inoffensive, being of a gentle and timid disposition, and trusting for protection to flight or concealment; seldom more than of a moderate size, while a portion of them are the smallest of the mammals. Of these last the Harvest Mouse is an example; the largest Rodents are the Beaver, Capybara, and Porcupine. The Rodents feed upon the harder sort of vegetable matter, as nuts, grain, roots, twigs, etc., (except rats and mice, which are omnivorous, eating anything that comes in their way, as most house-keepers know to their sorrow.) The Rodents have generally six or eight young at a birth, and this two, three, and even four times in a year. They are, however, kept from overrunning the earth by the rapacity of beasts and birds that live upon them.

Many are remarkable for their soft and beautiful fur. The Beaver, Chinchilla and Grey Squirrel are valuable in commerce. Some of them, as the squirrel and dormouse, use the fore paws to convey food to the mouth, to hold an object, and to climb. The form of the body is usually more or less conical, the chest and shoulders being small, whilst the loins and haunches are robust and muscular; the hinder limbs are longer than the fore ones, whence their movement is that of leaping or hopping along. "Most of them are nocturnal or crepuscular in their habits; many dwell in burrows; some conceal themselves amidst herbage, some among the foliage of trees, and some build for themselves habitations which have excited the interest and admiration of men." (Pict. Museum.)

We arrange the numerous animals of this order into eight families, viz.: 1. SCIURIDÆ, (Squirrels,) 2. ARCTOMYDÆ, (Marmots;) 3. GERBILLIDÆ, (Jerboas;) 4. CASTORIDÆ, (Beavers;) 5. HYSTRICIDÆ, (Porcupines;) 6. MURIDÆ, (Rats and Mice;) 7. CAVIADÆ, (Cavies;) 8. LEPORIDÆ, (Hares) Our limits will not allow us to do more than to give brief accounts of some of the principal genera and species.

I. Family Sciuridæ, (Lat. *Sciurus*, a squirrel,) SQUIRRELS.

This includes between sixty and seventy species. Audubon says about twenty well determined species are found in North America. They are arranged into two groups, viz.: I. Squirrels with free limbs; II. Squirrels with their limbs invested in the skin at the sides. These are not only the most elegant and sprightly, but the most numerous and widely scattered of the Rodents. They are distinguished by their simple grinders, having tuberculous summits, and the lower front teeth paired and much compressed at the sides. The toes are long and accompanied with sharp and hooked claws, and the rudiments of a thumb. There are four claws on each fore foot and five on the hind. The full development of the collar bones, (clavicles,) gives them much facility in using their paws as hands. In eating, the squirrels usually sit upon their haunches, and holding the food between the rudimentary thumbs of both paws, nibble it away until the whole is consumed. The head is proportionably rather large; the eyes full and prominent; the tail long, with the fur disposed on its sides like a feather; the ears in many species are tipped with a pencil of hairs. These animals are easily tamed, and from their playful and graceful manners, often become great pets. Most of the species resort to trees, but the Ground Squirrel, (*Tamias* or *S. striatus*,) burrows in the ground. The generic name, *Sciurus*, or Shadow-tail, is derived from Gr. σκια, (*skia*,) a shade, and ουρα, (*oura*,) a tail. Of this name the English term *squirrel* is a corruption; it refers to the fact that when the animal is at rest, its long and bushy tail is turned over the back and *shades* it.

I. GROUP.

S. vulgaris, COMMON RED SQUIRREL. This graceful and active little animal is generally about fifteen inches long from the nose to the tip of the tail, having the ears terminated by long tufts of hair; the color of the head, body, tail and legs of a bright reddish brown; the belly and the breast white; the eyes large, black and sparkling; the fore feet strong, sharp and well adapted to hold its food; the legs short and muscular; the toes long and the nails sharp and strong; the lip is cleft; the fur short and silky. It lives in pairs, constructing in the hollow of a tree, or in the fork between two branches, a water-proof nest of curiously interwoven moss, twigs and dry leaves. In May or June it commences to rear a young family, usually four or five in number. In the fall of the year it carefully hoards up its winter stores, which are concealed in holes and crevices of trees not far from its retreat. In Sweden and Lapland, the color of the Common Squirrel becomes gray in the winter season; in Siberia it is often seen entirely white; in other regions slight variations of color are also noticed.

In the varieties found on this continent, the pencil of hairs which tufts the Common Red Squirrel is wanting. The Gray Squirrel, (*S. Carolinensis*, or *S. migratorius*, Lat. migratory,) is one of the most common American species, found along the Atlantic, from Hudson's Bay to Carolina. Of this De Kay enumerates five varieties. It is about the same length as the Common Squirrel, (15 inches.) One of the most remarkable peculiarities of this species is a propensity to distant emigration in large numbers.

The Northern Migratory or Gray Squirrels are as much dreaded by the farmers of the West, as the devouring locust by the Eastern nations. Everything suited to their taste vanishes before them, and no obstacle can withstand their progress. It is believed by many that they pass rivers seated on a piece of bark brought by them for the purpose, and their tails hoisted for a sail. Audubon saw troops of squirrels cross the Hudson river at different places between Waterford and Saratoga, in the autumn of 1808 or 1809, but said they appeared to him unskillful sailors and clumsy swimmers.

S. vulpinus, (Lat. *vulpes*, fox.) The Fox-Squirrel abounds in the pine forests of the Southern States, feeding upon the seeds of the cones of the long-leaved pitch pine, (*pinus palustris*, Lat. marshy,) acorns and other nuts. It makes long journeys to visit corn fields when the corn is in the milky state, and often erects a temporary summer house in their vicinity.

S. Palmarum, (Lat. of palms.) The Palm-Squirrel, usually seen frisking about palm trees, is said to be remarkably fond of palm wine. They are often taken to England alive.

S. bilineatus, (Lat. marked with double lines.) The Plantain-Squirrel, kept by the Javanese as a pet; the tail trails gracefully upon the ground; when angry it bristles up like an irritated cat; when asleep, rolls itself up like a dormouse, with its tail encircling its body.

S. niger, (Lat. black.) The Black Squirrel of a glossy black with a lighter shade beneath; claws covered with hair; the hind legs have a few scattering hairs; the fur is softer and finer than that of the little Gray Squirrel, before which this species is said to be disappearing. The flesh of both these species tastes like that of a rabbit, but is more juicy; it is nice broiled, and makes excellent meat pies.

S. macrourus, (Gr. *makros*, long, *oura*, tail.) The Long-tailed Squirrel, of Missouri, is 22 inches long, the tail equaling in length both the body and head.

S. quadrivittatus, (Lat. four-striped.) The Four-Striped Ground Squirrel is a very beautiful species found in the Rocky Mountains.

S. striatus, (Lat. streaked. Tamias of Illiger.) The *Striped or Ground Squirrel* is characterized by its reddish brown color, a black stripe upon the back, and a shorter light colored stripe bordered with black upon the sides; by having the body shorter and more robust for its size than that of the Red Squirrel, and eight instead of ten molars in the upper jaw. It is also known under the names *Chipping Squirrel*, and *Chipmuck*. Usually it is seen running along fences; it is particularly fond of stone walls, which afford this animal a ready retreat. Under these it burrows and stores its winter food. Sometimes it makes its home in the center of a decayed stump. It does not ascend trees except when its retreat is cut off from its hiding place. The range of this squirrel on this continent is from 33° to 50° N. L.

II. Group.

Pteromys, (Gr. πτερον, *pteron*, wing; μυς, *mus*, winged-mouse.) Flying Squirrel. This genus comprehends ten or more species found on this continent, in Northern Europe and in Java. Some of them are nocturnal. These squirrels are distinguished by a membrane adhering to the sides, extending from limb to limb, so as to form a parachute, by the agency of which they can throw themselves from tree to tree to a great distance, and sustain a short flight. In the sailing movement, they are aided, and perhaps in part guided by their broadly expanded tail. The species of Northern Europe, (*P. volans*, Lat. flying,) is about the size of a large rat, and of a gray color. *P. volucella*, (Lat. dim. of *volucer*, flying.) The Small American Flying Squirrel has only a rudimentary membrane. The loose skin stretches forward by his fore legs, and backward by his hind legs as he springs, so as to buoy him up and enable him to leap a long distance at one bound. This squirrel is about ten inches long, including the tail; the fur of a brownish ash, tinged with cream-color, very fine, soft and silky. It is found in all the Atlantic States and in Canada West. In Canada East it is replaced by a species, (*P. sabrinus*,) one third larger than *P. volucella.* *P. alpinus*, of the Rocky Mountains, is still larger.

II. *Arctomydae*, (Gr. ἀρκτος, *arktos*, bear; μυς, *mus*, mouse.) MARMOTS.

This family is nearly allied to the Squirrels, among which Swainson places them. Sometimes they have been arranged with the RATS, They have a large and somewhat flattened head, ten molars above and eight below, heavy body, short bushy tail and short limbs. Some of the species have cheek pouches. They live in communities and all burrow and hybernate.

Spermophilus, MARMOT SQUIRRELS, sometimes ranked as a sub-genus, (Gr. σπερμα, *sperma*, seed; and φιλος, *philos*, lover.) This includes animals so named by Cuvier from their fondness for *seed*, and furnished with cheek pouches. Of them there are several species. One of these is *S. ludovicianus*, (Lat. *ludo*, to sport or frisk; *vicinia*, vicinity or neighborhood,) so named because living and sporting together in large communities. This is the *Prairie Dog* of Missouri and California, an appellation which Audubon says was probably given to the animal from its yelp,—"*chip, chip, chip;*" it is not like a dog in its form. The numbers of these animals are very great; they sit on their little mounds at the entrances of their burrows, "chirping and chattering to one another like two neighboring gossips in a village."—(Hon. C. A. Murray.)

Arctomys, MARMOT. Of this genus there are also several species, which have the form, teeth, and habits of the preceding, but only rudimentary cheek pouches. *A. alpinus*, the ALPINE MARMOT is about the size of a rabbit, of a grayish yellow color, approaching to a brown towards the head; inhabits the mountains of Europe, particularly the Alps and Pyrenees, just below the region of perpetual snow, and feeds on insects, roots and vegetables. Living in societies, these animals post a sentinel that gives a shrill whistle if danger approaches, when they retire for safety into their ingeniously contrived burrows.

A. monax. WOODCHUCK, GROUND HOG, OR MARYLAND MARMOT. This, when full grown, is of a reddish gray color and about as large as a rabbit; the young are reddish or of a uniform black; its wool is intermixed with long coarse hair; it has short ears and cheek pouches; the length is a little more than two feet, though in this respect as well as color, it greatly varies. The range of this marmot extends from Maine to California. It dwells in subterranean abodes, which are partitioned into chambers, feeds on clover and esculents, is easily tamed, and very neat and cleanly in its habits. In some places it selects forests of pine, in others cleared lands and old pastures for its residence. The Woodchuck is awkward and slow in its movements; its safety is found in its extreme watchfulness and sharpness of hearing. When at all alarmed, it flies to its deep and long burrows, "thirty or forty of which have been seen in a field of five acres." To these its dilated cheeks carry its winter stores. Its fondness for clover often renders it an annoyance to the farmer.

3d. Gerbillidae, or Dipodidae, Jerboas.

This family of the Rodents, sometimes called Jumping Mice, are apparently formed to live on prairies and sandy deserts. They have very short fore feet, and the hind ones very long, being Kangaroos in miniature; the tail is generally longer than the body, and used in leaping or walking; (Plate V. fig. 2.) the forefeet are employed in conveying food to the mouth, and seem of little or no use as organs of progression; the fur is soft; there are two cutting teeth in each jaw, the grinders simple or compounded, six or eight beneath; parts of the internal structure are bird-like. As far back as the time of Herodotus, these Rodents are alluded to as inhabiting Africa.

Dipus, (Gr. Δις, *dis,* two; πους, *pous,* foot.)

The animals of this genus have compound molars, and may be regarded as an intermediate link between the Squirrel and the Rat; but are more like the former than the latter. The fore legs are very short, and scarcely used in walking; the enormous hind legs and tail at once remind the beholder of the Kangaroo. When first seen, the animal seems supported in its rapid bounds by only two long legs; whence the name *Dipus*, two-footed. At a single bound, it moves four or five yards, and sometimes more; it feeds, sitting upon its haunches, like the Squirrel; is found abundantly in Egypt, Syria, and the north of Africa. The most common species is *D. sagitta,* (Lat. arrow,) the Gerbo, or Egyptian Jerboa, about the size of a large rat, living in large societies, and constructing burrows under ground.

Meriones. (Gr. μηριον, *mēriŏn,* a thigh.) The animals of this genus are small, with long, slender, and nearly naked tails; they have six composite molars beneath; their fore feet have a rudimentary thumb, with a small nail. They hybernate, and are nocturnal.

M. Americanus, or *Gerbillus Canadensis.* The Deer Mouse; Jumping Mouse. This is about the size of a common mouse, and of a reddish brown color; has very short fore legs, long hind ones, and a scaly, rat-like tail. It leaps ten or twelve feet at a time; is found in Canada and farther north, and as far south as Pennsylvania, in fields of grass and grain. (Pl. V. fig. 2.)

PL.V.

EXPLANATION OF PLATE V.

ORDER RODENTIA, FAMILY MURIDÆ, (Mice.)

Jerboas, or Jumping Mice.

1. Jumping Hare, Cape Jerboa, or Grand Jerboa, *Pedetes Capensis*, in the position in which it eats, using its small fore feet to bring the food to its mouth. With these feet it digs its burrow so expeditiously as quickly to hide itself; the hind legs are proportionally longer than in any other known quadruped. The tail is a most efficient organ; if deprived of it they can neither leap nor sit upright.
2. Labrador Jumping Mouse, *Meriones Labradorius*. a, animal sitting; b, jumping.
3. Pouched-Rat, or Sand-Rat, *Geomys*, or *Pseudostoma* (false mouth,) *bursarius*, (of skin.) The cheek pouches much resemble the thumb of a lady's glove in form and size, and hang down by the sides of the head. In the Canada Pouched Rat, or Missouri Gauffre, or Gopher, they are like pockets, extending from the sides of the mouth to the shoulders, lined with short, soft hairs, and opening on the outside of the mouth.
4. Woodchuck, Ground-Hog, or Maryland Marmot, *Arctomys monax*.

FAMILY LEPORIDÆ, (Hares.)

5. Common Hare, *Lepus timidus*, showing its long ears for collecting and conveying sounds, like an ear trumpet.

FAMILY CHINCHILLIDÆ, (Chinchillas.)

6. Chinchilla, *Chinchilla lanigera*, a woolly field mouse of S. America. It feeds in a sitting posture, conveying its food with its fore paws.

ORDER MARSUPIALIA, FAMILY DIDELPHIDÆ, (Opossums.)

7. Virginia Opossum, *Didelphis Virginiana*, showing the retreat of the young when threatened with danger, and the use they make of their prehensile tails.

Hélamys, (Gr. *Ἀλλομαι*, *allomai*, to leap; *μυς*, *mus*, mouse;) or *Pedestes*, (Illiger,) (Lat. *Pes*, a foot; *sto*, to stand.)

The animals of this genus have eight molar teeth beneath; the front legs are quite short; the hind ones very long, and both armed with exceeding long claws; the tail is long and very bushy. This includes the *P. Capensis*, of the Cape of Good Hope, the largest of the Jerboas; (length from nose to tail, about fourteen inches; of the tail, nearly fifteen inches;) which leaps from twenty to thirty feet at a bound, and sleeps in a sitting posture, placing the head between the legs, and holding its ears over its eyes, with its fore legs. It is a very strong and rapidly burrowing animal. (Plate V. fig. 1.)

Myoxus, (Gr. *μυοξος*, *muoxos*, a Dormouse.)

The Dormouse is intermediate between the Squirrels and Mice; is found in temperate and warm countries, and lives entirely on vegetable food. It has the two cutting teeth of the family, in each jaw, and the grinders simple, with divided roots; four toes before, and five behind, (the reverse of the preceding genus;) and naked ears. When in its winter retreat, this animal rolls itself up, and becomes torpid, occasionally rousing itself and partaking of its stores of food. Of this genus there are several species. *M. avellanarius*, (Lat. *avellana*, a filbert,) is the *Common Dormouse*, about as large as a common Mouse, but more plump, with a less sharp nose, and large black eyes; its color is a tawny red; the fur remarkably soft.

Chinchilla. This genus is regarded as a connecting link between the Hares and Jerboas. *C. lanigera*, (wool-bearing,) is found in the valleys along the line of the Andes; inhabiting regions where the temperature is below a moderate degree. It lives in companies, making burrows in the earth. Its food is entirely vegetable, and principally consists of bulbous roots. The Chinchilla has an exquisitely fine downy fur. The Creator has thus protected it against severe frosts. The length of the fur well adapts it for spinning; and the ancient Peruvians manufactured it into stuffs as articles of clothing. Numbers of these animals are annually destroyed for the sake of their skins. In size and appearance, they are like young rabbits; but the tail, like that of the squirrel, is usually held turned up over the back, and the ears, though long, are naked, broad, round and open. The color of the fur varies in depth, in different individuals; is of a dark, clear gray, lighter beneath. The Chinchilla is mild and inoffensive, but does not, in captivity, exhibit much sprightliness, or intelligence. Its length is about nine inches, exclusive of the tail, which measures about five. (Plate V. fig. 6.)

4th. The Beaver Family.

Castoridae, (Gr. *κάστωρ*, *kastor*, a Beaver.)

The animals of this family have bodies covered with two sets of hair, viz., fine and soft down, and long and rather rigid hairs. The tail is flattened and covered with rounded or hexagonal scales. The hind feet are the longest; the ears short. In habits, these animals are aquatic and social. Some species have webbed feet, and all a musky smell. The range of these animals on this continent is more limited than in former periods, when it extended from 68° to 30°. N. L. They are still common on the Euphrates, and along some of the larger European rivers, as the Rhone and the Danube. In England, they have not been seen since 1188.

Castor fiber, (Lat. Beaver.) The beaver is of a yellowish brown color, and from two to three feet long; it has four incisor teeth in both jaws; no canines, and sixteen molar teeth. The toes of the hind feet are webbed. It has also a glandulous follicle on the lower part of the body, producing an article called *castor*, (not castor oil,) and which is used in medicine. The flattened and scaly tail, it uses as a kind of paddle. By this, it is enabled, when loaded with a mass of timber, to stem a rapid current; and by making strokes up and down with its tail, it can dive or rise with great celerity: tradition says, but *untruly*, that the Beaver uses its tail in plastering its habitation. It moves more easily in water than on land; the eye is small, better suited to twilight than the glare of the sun. The external openings of the ear and of the nostrils are capable of being closed, which is a divine provision suited to its diving habits, and its continuance under water. The Beaver's great incisor teeth are his only tools; and most effective they are, for with them "he can divide a common sized walking stick at a bite, as cleanly as if severed with a knife." In doing his work, he *goes up the stream* from the site which he has chosen for his dwelling, so as to have the advantage of the current. Summer is the season; night the time of his labors. The skill, perseverance and toil which he exhibits in constructing his habitation, and storing it with food, have given to this animal great celebrity. In this, its instinct begins and ends; in other respects, it is very stupid, not comparing well with the Dog, Elephant and other quadrupeds. The fur of the Beaver is highly valued, especially for the manufacture of hats; and is an article of extended commerce. In one year, (1808,) Quebec alone exported nearly 127,000 furs, worth eighteen shillings sterling, each. *C. fiber*, (*Americanus*,) is a variety of this animal.

Fiber, (Illiger.) The animals of this genus have long, narrow, and somewhat flattened tails; twelve molar teeth; and the toes of the hind feet partially webbed. The species *F. Zibethicus*, is the MUSK-RAT, called in Canada, MUSQUASH; about the size of a small rabbit, and of a reddish brown color; sometimes black, or black and white. This animal has four strong cutting teeth, of which those in the under jaw are nearly an inch long; in instincts and disposition, it is similar to the Beaver. It receives its name from its strong musky odor, deposited in glands, near the origin of the tail. Its length varies considerably, but is generally from eighteen to twenty inches, while the tail alone is from seven to ten inches. The Muskrat frequents swamps and low, marshy grounds; and is specially fond of the *calamus* root, and of fresh water muscles, or clams. Its utility consists in its fur, which is soft and glossy, and used in hat making. The territorial range of this animal is similar to that of the Beaver.

5th FAMILY, PORCUPINES.

Hystricidae, (Gr. ῾υστριξ, *hustrix*, a porcupine; ῾υστριξ, from ῾υς, *hus*, a hog; θριξ, *thrix*, a bristle.)

We have already contemplated in the *Insectivora*, a group of animals (Hedgehogs,) protected by a coat of spines. In the present family, this spinous defence is more strongly and decidedly exhibited. The hollow tubes of the Porcupines are somewhat like the quills of feathers. They usually terminate in a fine point of hard enamel, but sometimes open at the end, as if cut off at their greatest thickness. These quills seem to be a smooth, glossy envelope of horn, with an inner pith or marrow of soft texture, and pure white. "They grow from a bulbous root, formed within a cell below the true skin, or *cutis*, and containing also a portion of fat, in which the vessels supplying its pulp and capsule, are imbedded. The capsules consist of three membranes, of which the innermost secretes the horny envelope, while the pulp supplies the pith of the spine." The spines vary in size; some are very long, slender and weak; generally, they are from four to eight inches in length, and very strong; thick in the middle, and tapering to a point at the extremity. (See fig. on the Chart.) They are less thickly set in the tail, which is short; their place there is supplied by numerous, open, hollow quills, raised on slender stalks, so as to vibrate with every movement. When angry, the porcupine clashes these hollow quills together, making a rustling noise, resembling that of a rattlesnake. In his undisturbed state, the spines lie down in regular order, with the points all directed backwards; but when he is angry, they are raised up by means of a peculiar muscular expansion under the skin, and joined to it, which by its action, influences their elevation and depression. When clashed violently together, one or two more loose than the rest, may be disengaged and fall; but the story that they dart out their spines like javelins, is pure fable; however, by pushing backwards or sidewise, quickly and with violence, the Porcupine can both defend itself, and inflict wounds on its enemies.

The head of this animal is thick; his eyes small; his face very round or convex; and his muzzle blunt. His cutting teeth are very large and strong, so that he can gnaw through the thickest and hardest boards. He is unsocial in his habits; when taken captive, is "neither familiar nor intelligent; in his native state, digs burrows in dry and barren situations, as far removed as possible from the haunts of men. These burrows have several entrances leading to a chamber in which it passes the day in silence and in solitude." As the light recedes, it cautiously ventures out in search of food, such as birds, roots, fruit and other vegetables. In winter, it goes out only occasionally for food.

The CREATOR has given to the Porcupine special endowments for his course of life. The animal burrows in hard and stony soil, and for that purpose is provided with digging implements; his limbs are short, strong and thick; and his toes, four before and five behind, on each foot, have thick and powerful nails or claws; the tongue is roughened with scaly prickles, directed backwards. The length of the Porcupine is about two feet; his general color a grizzled black, the spines being elegantly ringed with alternate black and white, and the limbs entirely black. This family of animals was originally introduced from Africa into Europe and America. The description above given is that of the COMMON PORCUPINE, viz., *Hystrix cristata*, (Lat. crested.) The *Hystrix dorsata*, (Lat. ridged,) (or *Hystrix Hudsonius*, of De-Kay,) otherwise called the Canada and North American Porcupine, ranges as far north as 67° N. L., and in New York, Pennsylvania, the northern parts of Virginia and Kentucky, and west as far as the Rocky Mountains. It is said to be increasing in the western parts of New York; in this species, which is from two to two and a half feet long, the spines are almost concealed by the hair with which they are intermingled; the fur of a soft and dusky brown color, is remarkable for its length and fullness; that of the Canadian animal is almost black. The incisors are of a deep orange color. This Porcupine is inoffensive, and of gentle manners; in size well comparing with that of a fox; it feeds on the leaves and bark of hemlock, bass-wood and ash trees; is fond of fruit and maize; and when confined, eats almost every kind of vegetable. The spines or quills vary in length from one to four inches; by a strong muscle in the skin, those of the back, when the animal is irritated, are erected and extended in various directions; the tail is also erected, and by a quite sudden movement, he is enabled to strike, leaving the loosened spines in the body of his assailant. The flesh of the Porcupine resembles young pork, and is by the Indians very highly esteemed. Spines dyed of various colors, form ornaments for their dresses. (De Kay.)

6th. The Mice Family.

Muridæ, (Gr. *μυς*, *mus*, a mouse.)

This numerous family have in each jaw, besides the two cutting teeth common to the *Rodentia*, six molars (usually) in each jaw, surmounted by blunt tubercles. The teeth of the upper jaw shelve backwards; those of the lower, forwards; the feet are neither webbed nor fringed with stiff hairs, but several species swim with much ease. The tail is round, usually naked or thinly haired. Most of this family are small burrowing animals; some genera are furnished with cheek pouches. Dr. De Kay, (N. H. S. N. Y.,) arranges all into two groups, I. those having, II. those not having cheek pouches. The ordinary food of these animals is grain, seeds, and other farinaceous matter, for bruising which their teeth are well fitted; but they are really omnivorous.

Mus decumanus, (Lat. tenth.) The Norway or Brown Rat is of a grayish brown color above and white beneath; in length, from the head to the end of the tail, about twenty inches, having the tail quite as long as the body. It was originally introduced into Europe from the southern parts of Asia; from its superior strength and ferocity, has in some places almost entirely expelled the Black Rat, (*M. rattus*.) It came to the United States with the foreign mercenaries during the war of the Revolution, and is now spread over the United States and Canadas. It infests wharves and has been called the Wharf Rat, or Dock Rat; the name *decumanus* alludes to the tithe or tenth of everything taken by this voracious creature.

M. musculus, (Lat. dim.) The Common Mouse is of a dusky gray color, has ears about half the length of the head, a long, bare and scaly tail, and in constitution and disposition is similar to the rat. It breeds at various seasons of the year, from six to ten at a litter; is omnivorous, but prefers vegetable food. The young are, in about a fortnight, strong enough to collect their own food. The mouse is said to be very susceptible to the power of music. An anecdote is related of a gentleman who was playing a violin, seeing a mouse run along the floor and jump about as if distracted. He continued the strain, and after some time the mouse, apparently exhausted with its exertions, dropped dead on the floor.

M. leucopus, (Gr. *λευκος*, *leukos*, white; *πους*, *pous*, foot.)

This little animal is of a brownish color above; the feet and all beneath, white; the ears large; the tail hairy and as long as the body. The whole length is six inches. The colors and proportions give this mouse a delicate and beautiful appearance. Like the Deer mouse, it is, from its agile, jumping movement, called the "Jumping Mouse." It feeds on grains and grasses.

M. messorius, (Lat. *messis*, a harvest.) This is the smallest and one of the most beautiful of the mammalia, called the HARVEST MOUSE. It is scarcely half the size of the common mouse. The color is of a reddish brown or squirrel-like aspect above; the under parts white; its eyes are dark; its action lively. In winter it lives under ground in burrows, but it breeds in grassy compact nests of the size of a cricket-ball, like those of a bird, made among the stalks of the standing corn, and supported on two or three straws. The principal food of the harvest mouse is corn; but it is also fond of insects.

Arvicola, (Lat. *arva*, corn-fields; *colo*, I inhabit.) This genus includes many species known under the names of FIELD MICE and FIELD RATS, differing from the mice proper in the structure of their teeth, and the length and hairy covering of the tail.

A. amphibius is the Water Rat common on the banks of rivers, brooks, &c.

Geomys, (Gr. *γη*, *ge*, earth; *μυς*, *mus*, mouse.) POUCHED RAT, SAND RAT. (See Plate V., fig 3.)

Of this genus there are several species, having eyes small and far apart; small ears; ten molars above and ten below; large and pendulous cheek pouches, opening, (according to Audubon,) outside of the mouth, and extending in some species along the neck to near the shoulders. These pouches are cold to the touch and of an oblong form when distended. This animal has been seen "when in the act of emptying its pouches into its paws like a Marmot Squirrel, and squeezing its sacs against the breast with its fore paws."

7th FAMILY.

Caviadæ of Tropical America.

The Cavies seem to hold a middle rank between the Mouse and the Rabbit.

Hydrochoerus, (Gr. *ὑδωρ*, *hudor*, water; *χοιρος*, *choiros*, a hog.) CAPYBARA, or WATER-HOG, of South America.

This animal attains to the size of a hog of two years old; lives on vegetables, sugar-cane and fish. To procure the latter, it betakes itself to rivers, swimming as readily as the otter. Like the Peccary, it is without a tail, and instead of a cloven hoof, its feet are partly webbed, thus fitting it for its aquatic life. In its manner of walking and its color, this animal resembles the pigs; but when seated on its haunches and watching any object with one eye, it has the appearance of the Cavies. The Jaguar destroys it in great numbers.

Cavia Cobaya. The COMMON CAVY or GUINEA PIG. This animal, which is about as large as a rat, is distinguished for the beauty and variety of its colors, and the neatness of its appearance and habits. It is a native of South America, but has been introduced into the Eastern Continent. It is the most prolific of all the *mammalia*, producing every two months from six to twelve young. A single pair soon multiply to the number of 1,000. It has no tail; its flesh is tasteless and its skin of little value.

C. Patachonica, the PATAGONIAN CAVY, or HARE-LIKE CAVY, is a burrowing animal, producing two or three young at a time; in its essential details of structure is a Cavy; but its long legs, long erect ears, combined with the general form of the head, lead casual observers to mistake it for the *Hare*, which in size it surpasses, sometimes weighing twenty or thirty pounds.

Dasyprocta, (Gr. δασυς, *dasus*, hairy or bristly, πρωκτος, *prōctos*, anus.) The yellow-nosed Cavy or AGOUTI of Brazil. This animal is found in great numbers in South America. It exhibits some resemblances in form and mode of living to the Hare and Rabbit, and indeed is called the Rabbit of that region. The toes have large and powerful claws. The *Agouti*, when eating, sits upon its hind quarters, using the fore paws, like squirrels, to hold its food. This consists ordinarily of yams, potatoes and other roots, though it is almost omnivorous. It does not burrow, but takes shelter in hollow trees. These animals are quite prolific, and very destructive to sugar cane, and therefore are caught and killed by the planters.

Leporidæ, (Lat. *lepus*, a hare.) HARE FAMILY. (Pl. V. fig. 5.) The most remarkable difference between this family and the other Rodents, consists in their having behind the two incisor teeth common to the group, two additional ones of smaller size in the upper jaw, making four, which in young ones are increased to six; the inside of the mouth and the soles of the feet are hairy; the tail is very short or wanting; the eyes are large and prominent; the hind legs are usually more developed than the fore ones; the clavicles wanting; the fur soft and copious.

Lepus. This genus includes, among other species, *Lepus timidus*, (Lat. timid.) The Common Hare. *L. cuniculus*, (Lat.

Little Rabbit.) *L. Totai*, the Totai, of Siberia; *L. Capensis*, of North and South Africa; on the American Continent, *L. nanus*, (Lat. dwarf,) or *Americanus*, the American Gray Rabbit. *L. Americanus*, the Northern Hare of America; *L. variabilis*, the Alpine Hare; *L. Hibernicus*, the Irish Hare. Upwards of thirty species are known, of which half belong to this continent, all agreeing in having a short erected tail, and the hind larger and more muscular than the fore limbs.

The Hare, *L. timidus*, is as large as a fox, of a grayish brown color, has long pointed ears, and is a native of Europe. The fur of this animal is an important article in the hat manufacture, and its excellent flesh often found in the market. This animal is remarkable for its extreme vigilance. Its senses are most acute and its fleetness, in proportion to its size, unrivaled. These are its means of defence. The general length of the animal is about two feet; the color verges towards an iron gray, with the chin and belly white. The eyes are placed laterally, and they are said to be constantly open, even during sleep. The usual and favorite residence of the Hare is in rich and rather dry and flat grounds. It feeds principally by night and remains concealed during the day; its food consists of herbage of various kinds. Of parsley it is especially fond. Sometimes it does great injury to wheat and other grains. So *timid* is it that it flees on the slighest alarm if disturbed while feeding. Acting like tubes applied to the ears of deaf persons, its long ears carry to it remote sounds. In their flight, these animals are apt to exhaust their powers at their first efforts, and hence are more easily taken than the slower but more wily foxes. Its voice heard when in distress or wounded, is like the sharp cry of an infant. In addition to the persecutions of mankind, it is assailed instinctively by every kind of dogs, and by the cat and weasel tribes: even birds of prey, snakes, adders, etc., drive it from its summer lodging place. This lessens the increase of these animals, which from their prolific tendencies, would otherwise be greatly multiplied.

The *L. variabilis*, sometimes called the Scotch Hare, is found not only on the Alps, but on the mountains of Scotland. Its tawny gray color in summer, is in winter changed to white, except the tips of the ears, which are black. In portions of Russia there is a variety called the *Russak*, which in Siberia is always white, but sometimes is entirely coal black. The winter dress of the American Hare is white, or white tinged with brown; the summer dress is more reddish brown with white beneath; the ears are but little shorter than the head; the length is twenty to twenty-five inches. It is said to be found from Canada to the Gulf of Mexico, though its range is not well determined.

L. cuniculus, (Lat. a little rabbit.) The RABBIT resembles the Gray Squirrel in size and shape; but when domesticated varies in these respects. It is native to the warmer parts of the Eastern Continent; is well known as a burrowing animal, and everywhere domesticated. The most common kind of the Albinos are the white with red eyes. The flesh of the Rabbit is insipid and its skin of no value, but its fur is made into gloves, stockings and hats. It has a litter of five or six young ones every month, and its great fecundity is in some places nothing short of a calamity.

L. silvaticus, (Lat. woody, wild,) is the AMERICAN GRAY RABBIT, in its wild state, having a color similar to that of the European burrowing Rabbit, but it does not change to the different colors which the European Rabbit shows when domesticated.

L. callotis, (Gr. *καλος*, *kalos*, beautiful; *ους*, *ous*, an ear.)

The BLACK-TAILED HARE, of Mexico and adjacent parts, mottled with gray and yellowish brown above, and white beneath; it has very long ears and long and slender legs and body, fitting it for long and rapid leaps. This is a very interesting species, and on account of the length of its ears, called in Texas, the Jackass Rabbit.

Lagomys, (Gr. *λαγως*, *lagōs*, a hare; *μυς*, *mus*, a mouse.)

This is a genus of the Rodents which is separated from the Hares proper, and includes four species; "one, the *Pika*, in the northern mountains of the Old World, one in Mongolian Tartary, one in the south-eastern parts of Russia, and one, (*L. princeps*, the Little Chief Hare,) in the Rocky Mountains of North America."—(Audubon.) They lay up winter stores, which is never done by the true Hares.

QUESTIONS UPON THE RODENTIA.

Give the derivation of Rodentia. How may the animals of this order be easily known? What family are an exception to this? Give some further account of the teeth of this order. How is the wear of the enamel and other part of the teeth counteracted? What is the result if one tooth be lost? How many species does the order include? How many in North America? Which of these is the smallest? Which the largest? On what do the *Rodents* feed? How are they kept from overrunning the earth? Which are valuable in commerce? Give the remark of the Pictorial Museum. Into how many families is the order divided? Give their names.

How many species does the *Squirrel* Family include? Into what two groups is the family divided? Describe the family. Give the derivation of the generic name. Describe the Common Red S. What is one of the most common American species? How many varieties of this? What characteristic of the common Red S. is wanting in this? What is one of its most remarkable peculiarities? Describe the Striped or Ground S. By what other name is it known? What is the meaning of its name? Where is it usually seen? In what respect does it differ from other S.? What is its range?

Give the derivation of the term *Pteromys.* How are the Flying Squirrels distinguished? Give some account of the American species.

Give the character of the Marmot Family. What is the derivation of the term *Spermophilus?* What animals does this genus, (or sub-genus,) include? What is the meaning of the term *Ludovicianus?* What is the common name of this species? What is said of these animals? Derive the term *Arctomys.* In what respect does it differ from the preceding? What is said of the Alpine Marmot? Describe the Woodchuck. Give the character of the Jerboas. What ancient writer alludes to them? Derive the term *Dipus.* What is said of the animals of this genus? Which is the most common species? Describe the second genus. What is said of the Jumping Mouse? Give some account of the animals included in the third genus. Describe the Dormouse. What is said of the common D.? Of what animals is the *Chinchilla* the connecting link? Describe this animal. Describe the Beaver Family. Give the Greek and Latin terms for Beaver. Name the most striking peculiarities and habits of the B. Where is the Musk-rat found? Why is it so called? Give some account of it. Give the derivation of the term *Hystrix.* What is the family name? What is the distinguishing peculiarity of this family? Describe its spines. How does this animal defend itself? What is further said of it?

Give the derivation of the term *Muridæ.* What is said of this family? Into what groups does Dr. De Kay arrange them? What is said of the Norway or Brown Rat? Describe the Common Mouse. Derive the term *leucophus.* Describe this animal. Why is it called the Jumping Mouse? Give some account of the Harvest Mouse. What is said of the species included in the genus *Arvicola?* How do these differ from the Mice proper? What is said of the *Geomys*, or Pouched Rat? What is said of the Cavies? What species are particularly named? Describe the Water Hog of S. A. What is said of the Guinea Pig? What of the Patagonian or Hare-like Cavy? What of the Agouti, or yellow-nosed Cavy? What is the derivation of the term Leporidæ? What is the chief difference between this family and other Rodents? What species are particularly named? Give some account of the Common Hare. What is said of the Rabbit? What species of the genus *Lepus* is referred to? What is the derivation of the term *Lagomys?* How many species does it include? Does it differ from the true Hares?

How would you trace the Beaver on the Chart? Ans. *Fiber* is the species, *Castor* the genus, Rodentia, or Glires, the order, Unguiculata the sub-class, Mammals the class, Warm-blooded the first division of the sub-kingdom Vertebrates. Trace the Squirrel and Porcupine in the same way, giving the meaning of the terms at each step. What other gnawing animals are mentioned on the Chart?

SECTION XX.

SECOND SUB-CLASS. UNGULATA.

ORDER 7. PACHYDERMATA, (Gr. παχὺς, *pachùs*, thick; δέρμα, *dérma*, skin.)

The animals included in this order are for the most part of large size; some of them are of truly gigantic proportions, being the largest of all land animals. They are called PACHYDERMATA, on account of the massive thickness and solidity of the skin; a peculiarity which strikingly distinguishes the more prominent species. These animals are thinly covered with bristly hairs, or else almost entirely naked; and their external appearance is frequently rough and coarse. They inhabit the warm latitudes of Asia, Africa and America. One genus, (*Sus*,) the Wild Boar, is found wild in Europe; and two or three others, used for purposes of economy, are now almost universally distributed by domestication. The Pachydermata, for the larger part, live upon vegetable food, such as grasses and watery herbage, and the succulent plants of the tropics. Their molar teeth are compound, often triple, with flattened crowns; in many there is a peculiar development of the canines or the incisors into curved and projecting tusks. The muzzle is frequently produced into a proboscis, or trunk, as in the Elephants, Tapirs, and, in a less degree, in the Hogs.

I. FAMILY PROBOSCIDAE, (Gr. προβοσκὶς, *proboscis*, a trunk;) including the Elephant Mammoth, and Mastodon. These are *Multungulate*, (many-hoofed.)

Elephas, (Gr. ᾽ελέφας, *elephas*) The ELEPHANT.

Of this magnificent animal there are two species, *Elephas Indicus*, or *Asiaticus*, and *E. Africus*. Both species are distinguished by their enormous tusks, which project downwards from the upper jaw of the male Elephants, of India, and of both males and females, of the African Elephants; also by the absence of front teeth from the lower jaw, and by having five hoofs on each fore foot. The enormously large tusks are seated in the bones, from which the incisor teeth proceed in other quadrupeds, and continue to grow while the animal lives. The grinders or molar teeth strongly resemble those of many of the RODENTIA. These are made up of a certain number of vertical laminae, each formed of bone, covered with enamel, and held together by a third substance, called the *cortical*, (Lat. *cortix*, bark.) They are changed six or eight times in the course of the Elephant's life. The manner in which they succeed each other is quite peculiar. The old tooth is not pushed up by the new one, as is usually the case; but the new one appears *behind* the old one, urging it forward, so that the latter wears away, and its place is finally taken by the new comer. The teeth are of immense weight, and with the tusk, are the most valuable part of the animal. (Pl. IV. fig. 7. & 8.) The tusk is hollow for a great part of its length; the cavity containing a vascular pulp, which supplies successive layers within, as the tusk is worn down without. Blumenbach, (see his Comparative Anatomy,) says that some modern naturalists consider the tusks a species of horn; and that balls with which the animal has been shot when young, have been found on sawing through the tusks, imbedded in their substance, in a peculiar manner. These organs, especially in the African species, are extremely large. Cuvier has a table showing their great size. The largest recorded in the table was a tusk sold at Amsterdam, which weighed 350 lbs. One possessed by a merchant of Venice, was fourteen feet in length. The largest in the Paris Museum is nearly seven feet long, and about five and a half inches in diameter, at the largest end. Professor Silliman, during his last tour in Europe, measured one in the British Museum, which was ten feet in length. One described by Hartenfels, in his Elephantographia, (Gr. *Elephas*, Elephant; *grapho*, to write,) exceeded fourteen feet. Ordinarily, the tusk of the Indian Elephant does not weigh more than from fifty to seventy-five pounds. The first or milk tusks, never attain much size; but are shed between the first and second year; and the permanent tusks of the female are very small, in comparison with those of the male. The feet have five toes, "encrusted," as Cuvier says, in the callous skin which covers the foot, and appearing in the hoof by the nails alone. The foot is enclosed in a horny shoe or sock, which, when detached, presents a cavity that is quite tight, and used by orientalists as a vessel to contain their food Professor Silliman measured the shoe of an Elephant, in the British Museum, and found it five feet in circumference. (Plate VI. fig. 9.)

The immense weight of the head, renders indispensable a powerful muscular apparatus, and to that end a large surface for the insertion of muscles is necessary. The extended surface of the head gives full room for the attachment of the muscles of the neck. These muscles are most powerful, not only supporting the neck, but assisting the animal in digging or employing the tusks as means of defence. The vertebrae of the neck are more fully developed than in the Ruminantia, and the spinous processes in the vertebrae of the back, are lengthened and strong. The entire structure is well compared to the Cyclopean walls of some ancient city, huge, shapeless, and piled over against each other as if destined rather to sustain weight, than to permit motion. The internal organization, as a whole, is more simple than that of the Ruminants; but still Elephants feed on nearly the same sort of food. The stomach is of a very lengthened and narrow form, its greatest diameter being only about one-fourth of its length. There seems to be a receptacle, though less extensive than that of the Camel, to enable the Elephant to retain or secrete water that may be used for moistening its food, but at times is also used to disturb the insects, which during a march, or in hot weather, annoy or torment it.

But the trunk is unquestionably the most remarkable part of this animal's structure. This is properly a continuation of the nose, and becomes more valuable as an organ of prehension, from the unwieldy size of the head, and the shortness of the neck. It is an organ of respiration, as well as prehension; and it is also a delicate organ of touch and smell. The short neck, made necessary by the weight of the head and tusks, prevents the Elephant from putting its head to the ground, or from stooping to the water's edge; but for this disahility it is fully compensated in the advantage of the trunk. This extraordinary organ has, according to Cuvier, not much less than 40,000 muscles, which enable the Elephant to shorten, lengthen, coil up, or move it in any direction. Its structure is cartilaginous, and composed of numerous rings; a partition runs from one end of it to the other, so that although outwardly it appears like a single pipe, it is inwardly divided into two. "Endowed with exquisite sensibility; nearly eight feet in length, and stout in proportion to the massive size of the whole animal, this organ, at the volition of the Elephant, will uproot trees, or gather grass; raise a piece of artillery, or pick up a comfit; kill a man, or brush off a fly. It conveys the food to the mouth, and pumps up the enormous drafts of water, which by its recurvature, are turned into it, and driven down the capacious throat, or showered over the body." Through the trunk the Elephant uses his trumpet-like voice; the end has two openings or nostrils, like those of a hog, and also a finger-like appendage, with which he picks up small objects. His skin is usually of a brownish gray color, sometimes slightly mottled with flesh color; generally it is full of scratches and scars, which it receives in its passage through thick woods and thorny places. The form of the head varies with the animal's age; and it increases immensely in those of full growth. The tail is long, and has a tuft of hair reaching nearly to the ground. The strength of the Elephant in union with its capacity, renders it a most efficient aid, where extraordinary animal force is required, as in dragging ships, heavy stores and ordnance. Its ordinary pace is equal to that of the horse at an easy trot. A consideration of the velocity of its motion, as compared with the mass of its body, may help one to judge of its very great force. Many arduous and difficult military operations in the East have been much indebted to the sagacity, patience and strength of the Elephant. The height varies considerably. The East India Company's standard for serviceable Elephants is "seven feet and upwards, measured at the shoulders, in the same manner as horses are." It has been said, they reach the height of seventeen or twenty feet; but there is reason to believe they seldom exceed ten feet in height. Those from Pegu and Siam are much larger than those of Hindostan.

The Elephant has long been the companion of the Orientalist, in great hunting parties, (see border of the chart,) and from a very early period, has been made to minister to the wanton and cruel pleasures of Eastern princes, by being stimulated to combat, not only with other Elephants, but with various wild animals. The ivory of these animals, which is now sought for useful purposes, and also for minor ornaments, was in great request with the ancient Greeks and Romans, for various domestic uses, as well as for the Chrys-elephantine Statuary, (Gr. *Chrusos*, gold; *Elephas*, elephant,) of Phidias, such as the Minerva of the Parthenon, and the Olympian Jupiter.

The exports of the tusks from the East have been, and still continue to be, large. In 1831–2, those to Great Britain alone, amounted to 4,130 cwt.; a weight of ivory, taking the average of the tusks to be sixty lbs. weight, involving the destruction of from 4,500 to 5,000 Elephants. It is said 45,000 tusks are now annually consumed in Sheffield, (England,) alone. The Western and Eastern coasts of Africa; the Cape of Good Hope; Ceylon; India; and the countries to the east of the straits of Malacca, are the marts whence the supplies of ivory are obtained. The chief consumption of this article in England, is in the manufacture of handles for knives; but it is extensively used for other purposes. Ivory articles are manufactured to a greater extent at Dieppe, on the French sea-coast, than in any other place in Europe. In preparations of ivory, the Chinese excel. No European artist has, we believe, succeeded in cutting concentric balls after the manner of the Chinese; and their boxes and other ivory articles are decidedly superior to any that are to be met with elsewhere.

Though captured in India, and reduced to servitude, and extensively hunted in Africa, on account of his tusks, the Elephant is still found in great numbers in remote, secluded districts of the East, where large streams or rivers running through a wide and level region, are fringed by a luxuriant vegetation. A traveler who accompanied some Elephant hunters in South Africa, was told by an experienced hunter that he had seen as many as three thousand in a troop, ranging along the banks of the Fish river. "A herd of Elephants," says Pringle, "browsing in majestic tranquility amidst the wild magnificence of an African landscape, is a very noble sight, and one of which I shall never forget the impression." Sometimes they "tear up immense numbers of mimosa trees, sprinkled over grassy meadows, which border the river's margin." Of the soft and juicy roots of these and other trees, they are very fond. In overturning the trees, they sometimes employ their tusks as we do a crowbar, thrusting them under the roots to weaken their hold of the earth, and facilitate the work of tearing up the trees with their proboscis.

The Elephant is known to have a strong relish for sweetmeats and *arrack*, a spirituous liquor distilled from rice; and by these things is occasionally encouraged to perform tasks requiring great skill and labor. In plantations of sugar cane, he revels with great delight. Sometimes he adopts curious methods to gratify his love of sweet things. "It chanced that a Cooley, laden with jaggery, which is a coarse preparation of sugar, was surprised in a narrow pass in the kingdom of Candy, by a wild Elephant. The poor fellow, intent upon saving his life, threw down the burthen, which the Elephant devoured; and being well pleased with the repast, determined not to allow any person egress or ingress who did not provide him with a similar banquet. The pass formed one of the principal thoroughfares to the capital; and the Elephant taking up a formidable position at the entrance, obliged every passenger to pay tribute. It soon became known that a donation of jaggery would ensure a safe conduct through the guarded portal, and no one presumed to attempt the passage without the expected offering."

The Elephant possesses all the senses in great perfection; that of smelling is in him exquisite. He is not often bred in captivity, it being found more advantageous to take a well grown animal from a wild herd, and discipline it for service. In captivity, it is very docile and gentle, but when provoked will take full revenge. This, some who visit menageries have found out to their sorrow. All Elephants are fond of the water, and sometimes submerge themselves, so far that nothing but the end of their trunk remains above the surface. There are various modes of capturing these animals. One of these is by decoy Elephants, which are well trained to their work. With two of these decoys, the hunters proceed into the woods. The females advance quietly, and by their blandishments so occupy the attention of any unfortunate male that they meet, that the hunters are enabled to tie his legs together and fasten him to a tree. His treacherous companions then forsake him. At length he is subdued by hunger and the fatigue of efforts to free himself from his bonds, and then the hunters drive him home between their two tame Elephants. When once captured, he is easily trained.

When in captivity, maternal affection does not seem strong in the elephant; but in the wild state, the animal has given striking illustrations of such affection, as well as of marital and filial love. The young animal is exceedingly playful. It becomes mature when between 18 and 24 years of age, and usually lives to a great age; Aristotle says, "more than 200 years;" it has sometimes lived even more than 400 years.

The Elephant appears deeply susceptible of influence from kindness. The natives in the East are wont to address him with persuasive and endearing epithets, which he seems to comprehend and by which he is stimulated to exertion. Sometimes his actions and display of comprehension appear almost the result of a reasoning process. An officer who served in India remarks, "I have myself often seen the wife of a *mohont*, (for the officers often take their families with them to the camp,) give a baby in charge to an Elephant while she went on some business, and have been highly interested in observing the sagacity and care of the unwieldy nurse."

Memory, which, as well as instinct, is given to animals for their well being, seems to have great strength in the Elephant. An illustration of this remark is given by Mr. Corse, in the *Phil. Tran.*, and quoted by Swainson in his "Habits and Instincts of Animals." "An Elephant which had escaped, and which was subsequently captured in company with a herd of wild Elephants, after an interval of eighteen months, was recognized by one of the drivers. When any person approached the animal, he appeared wild and outrageous as the other animals, and attempted to strike the person approaching him with his trunk, until an old hunter, riding boldly up to him on a tame Elephant, ordered him to lie down, pulling him by the ear at the same time, upon which the animal seemed quite taken by surprise, and instantly obeyed the word of command with as much quickness as the ropes with which he was tied permitted, uttering, at the same time, a peculiar shrill squeak through his trunk, as he had formerly been known to do. By this circumstance, he was immediately recognized by every person who had been acquainted with his peculiarity."

When bogged in swamps, the elephant shows a sagacity which is remarkable. "The cylindrical form of his leg, which is nearly of equal thickness, causes the animal to sink very deep in heavy ground, especially in the muddy banks of small rivers. When thus situated, the animal will endeavor to lie on his side, so as to avoid sinking deeper; and for this purpose will avail himself of every means to obtain relief. The usual method of extricating him is by supplying him liberally with straw, boughs, grass, &c. These materials being thrown to the distressed animal, he forces them down with his trunk, till they are lodged under his fore feet in sufficient quantity to resist his pressure. Having thus formed a sufficient basis for exertion, the sagacious animal next proceeds to thrust other bundles under his belly and as far back under the flanks as he can reach; when such a basis is formed as may be to him proper to proceed upon, he throws his whole weight forward and gets his hind feet gradually upon the straw, &c. Being once confirmed on a solid footing, he will next place the succeeding bundles before him, pressing them well with his trunk, so as to form a causeway by which to reach the firm ground. The instinct of the animal, and probably the experience of his past danger, actuate him not to bear any weight definitely, until by trial both with his trunk and the next foot that is to be planted, he has completely satisfied himself of the firmness of the ground he is to tread upon."—(Swainson.)

The general characters and habits of the two species *E. Asiaticus* and *E. Africus*, are the same, and yet there are some points of difference. The Elephant of India has a head or skull almost pyramidal in form; that of the African species is more rounded in contour. The tusks and ears of the latter are the larger. So enormously large are the ears that they cover the animal's shoulders, and are often "used by the natives as a sort of truck, upon which to draw various loads." The teeth, too, are different in numbers, the African species having eight molars, whereas the Asiatic has but four, and they are in the former also differently marked; the Asiatic is the larger in its frame and its color is a paler brown, and it has four nails on each hind foot, while the African has only three; it is considered essential to the perfection of the Asiatic Elephant that it have eighteen nails, five on each fore foot, and four on each hind one. The

Asiatic species is also deemed the superior of the two in point of sagacity; though Cuvier was of the opinion that even this species does not in intelligence surpass the dog, an opinion that finds corroboration in the size of the Elephant's brain, which is estimated to be only $\frac{1}{500}$ part of his body, while in man the brain is $\frac{1}{33}$ part.

E. primigenius, (Lat. *primus*, first; *gigno*, to produce.) The MAMMOTH.

This is the name of an extinct species of *Proboscidæ*, remains of which have been discovered in the tertiary fresh water deposits of the Eastern and Western Continents. Abundant remains of this species have been found in the frozen mud of Russian America; they have also been traced in smaller quantities as far south in the United States as Ohio, Kentucky, Missouri and North and South Carolina. The Chart figures one of these animals dug up at Newburg, N. Y., twelve feet high and fourteen feet long. Mammoth bones and tusks occur throughout Russia, and particularly in Eastern Siberia. The skeleton of one sixteen and one-half feet in length, obtained in Siberia, and having the skin attached to the head and feet, is preserved in the museum at St. Petersburg. The hair of this specimen consists of two kinds, common hair and bristles; showing in the arctic character of its clothing that it was capable of living in high northern latitudes, like the Rein-Deer and Musk-Ox of the present day. It is inferred from the teeth of these animals that their food did not probably differ much from that now used by their survivors in tropical countries.

E. Americanus. AMERICAN ELEPHANT.

Dr. De Kay, (N. Y. S. N., 189,) designates a species under this name from specimens of teeth found in a diluvial formation near the Irondiquoit river, in Munroe County, ten miles east of Rochester.

Mastodon, (Gr. *μαστὸς*, *mastos*, a nipple or udder; *οδους*, *odous*, a tooth.) (Plate IV. fig. 9.)

This is the name of an extinct genus of gigantic Pachyderms whose remains are found abundantly in tertiary and sometimes in secondary deposits. The animal must have equaled or exceeded the elephant in bulk, and greatly resembled him in shape; the tusks, proboscis, and the general conformation of the body and the limbs were similar. The principal distinction between the two genera was formed by the molar teeth, the crown of which, unlike those of the elephant, exhibited, on cutting the gum, large conical points of a mammiform structure, whence the animal derived its name. The whole number of teeth was twenty-six. The Mastodon was probably less exclusively herbivorous than the elephant. "There is scarcely a state east and south of the Hudson River which has not afforded specimens of the *Mastodon.*" The genus embraces species which "have been found in almost every part of the world, and in all latitudes." The term *mammoth*, which was specially applied by the inhabitants of Siberia to a fossil elephant, has sometimes been *improperly* given to this animal.

M. giganteus, now one of the attractions of the British Museum, was found near the banks of the river *La Pomme de Terre*, a branch of the Osage River, in Burton Co., Missouri, imbedded in a brown sandy deposit, full of the remains of cypress, tropical cane, swamp moss, stems of palmetto, &c. Five arrow heads were found with the remains, which were 20 feet 2 inches long, and 9 feet 6¾ inches high. These remains were exhibited in London in 1842–3, under the name of the Missouri Leviathan. At the Big Bone Lick, in Kentucky, were discovered the remains of 100 Mastodons and 20 Mammoths, with bones of the *Megalonyx Stag*. The grinders have been dug up in the streets of London. Mr. Woodward, in his Geology of Norfolk, Eng., says that "upwards of 2,000 have been dredged up by the fishermen of Happisburgh in thirty years." On the Hudson River, remains of the Mastodon have been repeatedly discovered. About three years ago, a very fine specimen was discovered in the inclined side of a marshy declivity, a few miles from the city of Poughkeepsie.

II. SUIDÆ, (Lat. *sus*, a swine or hog.) (Four hoofs.)

The SWINE FAMILY.

This is a family of the PACHYDERMS, highly valuable to man as food. The animals of which it is composed are characterized by having on each foot two large principal toes, shod with stout hoofs, and two lateral toes which are shorter and hardly touch the earth. "The incisor teeth are variable in number, but the lower incisors are all leveled forwards; the canines are projected from the mouth and recurved upwards." The muzzle terminates in a truncated snout adapted for turning up the ground. Ten species are enumerated as belonging to this family.

Sus scrofa, (Lat. *scrofa*, an old sow.) The HOG, or WILD BOAR.

The well known Hog is the domesticated descendant of the Wild Boar, an animal still found in the larger forests of Europe, Asia, and the Northern parts of Africa. The wild race may, however, be distinguished from the domestic breed by the color, which is a dark grizzled brown, by the longer limbs, the small, erect ears, the greater development of the snout, and by a more bony appearance.

In his native retreats, the Wild Boar is a truly formidable animal, and the hunt of it exciting and dangerous; for this fierce and powerful animal is armed with long, curved and sharp tusks, capable of inflicting severe and fatal wounds. After he has passed his fifth year, he becomes less dangerous, on account of the increased size of his tusks, which so turn up as often to hinder rather than assist his design of wounding with them. In the wooded regions of Europe, the chase of the Wild Boar is still continued, and it is also one of the most exciting sports of oriental countries. It is not now found in its natural state in Great Britain, but formerly it there rioted in the dense forests; and in so high estimation was the chase of the animal held, that by a forest law of William, the Conqueror, any who were found guilty of killing a Wild Boar had their eyes put out! The lair is generally in some wild, retired spot, not far from water, and commanding, by some devious path, an entrance to the open country. The young, or Marcassins, as they are termed by the French are striped with longitudinal bands.

The Domestic Hog, (*Sus scrofa,*) differs from the wild animal principally in having smaller tusks and ears larger, somewhat pendant, and of a more pointed form. It is known that it varies considerably in color as well as in size, but the prevailing cast is a dull yellowish white, marked or spotted irregularly with black; sometimes it is perfectly plain or unspotted, sometimes rufous, and sometimes totally black.

The Hog is, of all quadrupeds, the most gross in his manners, and is generally esteemed the very image of impurity. The Jews were forbidden to eat his flesh, probably from the tendency to cutaneous disease growing out of its use in the East; and the Mahometans follow the same prohibition. Late experiments with this animal, as kept for domestic uses, go to show that some injustice has been done and losses incurred, by the opinions which have been long entertained as to its proclivity for dirt and mud. If *fairly treated,* it is by no means a dirty animal, and most judicious farmers find their account in *reforming* the pens of hogs so that they be kept in a cleanly condition, instead of being saturated with filth. However fond of wallowing in the mire, yet, " with plenty of dry litter, space and water, the hog will keep himself scrupulously clean and will thrive all the better. Even the trouble of washing and currying him frequently will be well repaid. Wood's Natural History contains an imputation against the Hog, which we do not remember to have seen or heard of before. "I have," says he, "seen pigs suck the cows in a farm yard, while they were lying down and chewing the cud, nor did the cows attempt to repel them."

The Hog is remarkably prolific, bringing forth two litters in the year, of from eight to twelve, and sometimes fifteen or twenty each. It is plain, therefore, that they would soon become annoyingly numerous were they not diminished by being used for the support of man. Though perfectly useless during life, they are among the most important of the animals which are reared for the value of their flesh as human food. None convert a given quantity of corn or other nutriment so soon into fat, or can be made fat on so great a variety of food. "Their flesh," says Linnæus, "is wholesome food for persons of athletic constitutions, or those who habituate themselves to much exercise, but improper for such as lead sedentary lives." It is an article of great importance to us as a naval and commercial nation, as it takes salt better than any other kind of flesh, and hence can be longer preserved. "The largest animals are not the best. Fertility, a capacity of fattening with rapidity and with the least expense, the smallness of the bones, and the firmness and sweetness of the flesh, with its readiness to receive salt, are objects of higher importance than mere bulk." The introduction of the Chinese Hog, it is said, has in some places made great changes in the native breeds. This breed is remarkable for productiveness. Cuvier believed it to be specifically different from the wild Boar. It is of "small size, short and thick; the belly deep, and when fat nearly reaching the ground; the legs short and fine; the head very short; the neck thick." The pork of the Chinese animal is particularly delicate; but the common breeds are said to "yield the best bacon and hams."

The senses of the Hog are acute, especially that of smelling. The broad snout ploughs up the herbage, and not a root, an insect or a worm, escapes the olfactory sense. The animal is not stupid as compared with many quadrupeds, and when treated with kindness, sometimes evinces strong attachment. That it is docile is proved by the number of "learned pigs," and by "the famous sporting sow that went regularly out with the gun, and stood her game as staunch as any pointer."

S. Babirussa, (native word, Hog deer.) The BABYROUSSA.

This animal is nearly the size of a common hog, but differs from it in some marked respects. The form is longer and lighter, the limbs more slender, and of greater length. The skin is black, naked and warty, and when closely examined found to be sparingly set with short bristly hairs. It is remarkable for possessing four tusks The two recurved tusks of the upper jaw, instead of passing out between the lips, pierce through the skin half way between the eyes and the end of the snout, turning upwards towards the forehead, like the horns of the *Ruminantia;* the tusks of the lower jaw are also very long, sharp and curved; but not of equal magnitude with those of the upper. The tusks are very fine ivory, but neither so hard nor so durable as that of the Elephant; the eyes are small; the ears erect and pointed; the tail rather long and slender, and tufted at the end with long hairs. The food of the Babyroussa consists chiefly of vegetables, and the leaves of trees. When hunted closely, and in apparent danger, this animal takes to the water, and by facility in swimming, alternately diving and rising, is frequently able to escape from its pursuers. It is said that it "crosses without difficulty the straits that intervene between neighboring islands." It is capable of being domesticated, and its flesh is palatable and well adapted for food. The Indians ascribe these animals to a union of the Hog and the Deer. The Babyroussas abound in the Molucca islands, and are also found in a few other islands of the Indian Archipelago.

Two of these animals, (one of each sex,) were brought to France, some years since, and kept in the Paris Menagerie. "The female was much younger, and more active than the male, which was aged and very fat, and spent his short life in eating, drinking and sleeping. The female bred once after her arrival in Europe. When the male retired to rest, she would cover him completely over with litter, and then creep in under the straw to him, so that both were concealed from sight. They died of diseased lungs, about three years after their arrival."

Phacochoerus, (Gr. φακὸς, *phakos,* lentil; χοιρος, *choiros,* hog.) The Warty Hog.

This genus of the Pachydermata, found in Africa, is allied to the Swine, and from the projecting appendages about the head, called Warty-hog. Its feet are formed like those of the True Hog. Some of them have but sixteen, others twenty-four teeth, while the common hog has forty-four, and the Babyroussa thirty-two. Of the two species, one has incisors, the other none; both have tusks, lateral and directed upwards. Their system of dentition points them out as more herbivorous than omnivorous.

P. Aeliani, Aelian's Wart-Hog, is a native of the north of Africa. This, at all ages, has incisor teeth in the upper and the lower jaw, which clearly distinguishes it from the Wart-Hog of the Cape Colony.

The skin of this animal is scantily bristled and of an earthy color. A mane, commencing between the ears, runs along the neck and back, made up of long bristly hairs, some of them ten inches long. These bristles and those found on the other parts of the body, are light brown. With the exception of the back, the body has a naked appearance. The head is broad along the brow, which is rather depressed; the eyes are small and very high up on the head; and two large warts appear, one on or near the cheek, called the larger wart, and the smaller one along side the cheek. These warts are formed of thickened skinny tissue; are smaller in this species than in the Wart-Hog of the Cape. The eyes are small; the tail thin, and nearly bare, with a tuft of hair at the end. On the fore feet is a piece of thick, hard, protuberant skin. These animals haunt low bushes, and forests, creeping on their bent fore feet, in search of food, and in this posture, digging up the roots of plants on which they feed with their enormous canine teeth; the hind legs pushing the body forward as it moves in this position.

P. Aethiopicus, or Aethiopian Wart-Hog, has larger warts than the preceding, and a more singularly formed head.

Dicotyles, (Gr. δὶς, *dis*, two; κοτύλη, *kotulē*, hollow, or cavity.) The PECCARIES.

These animals are native to South America; of a short, compact form, thickly covered on the upper parts of the body with large and strong dark colored bristles, and marked by yellowish white rings; and round the neck is usually a whitish band or collar. By their general appearance and propensities, they are closely allied to the True Swine; but they differ from them in respect to their teeth, having four instead of six incisors in the upper jaw, and six instead of seven molars on each side; their tusks also differ from those of the common hog, not turning up and projecting out of the mouth, but having the usual direction as in other animals; the hind feet have only three toes, the external toe on each foot being absent; and the limbs are more slender, the head shorter, and the snout longer than in the common hog. The tail is merely rudimentary, and not visible. The most decided characteristic is its having a glandular opening on the loins, which secretes a fetid and disgusting odor, infecting the flesh when the animal is killed, unless immediately cut away; in that case it is tolerable food. There are two species of the Peccary.

Dicotyles torquatus, (Lat. *torques*, a collar.) The COLLARED PECCARY. It has its name "Collared" from a line of white which passes from the fore part of the neck, obliquely upwards, to meet over the shoulders; is found in Mexico, as well as in the greater part of South America. The food of this species consists of acorns, roots, and earth worms, and similar creatures, bred in moist and marshy places. The Collared Peccary has been domesticated in South America, and some of the West India islands, and in the domestic state is fed upon the same esculents as the common hogs, but its flesh is far inferior to theirs, both in flavor and fatness. "The comparative infertility of the Peccary, which only produces two young at a birth, is a bar to its superseding the domestic pig, which is equally fertile in all climates where it has been introduced." The gland also, presents a strong objection to the Peccary, as a domestic animal, however "neat and trim" it may be in its general habits and appearance. It is said that D'Azara "revelled in its scent, as a perfume," and that others have considered it "agreeable enough," but to most persons it proves extremely offensive. These animals haunt the thickest and largest forests, dwelling in hollow trees, or holes in the earth made by other animals. They go in pairs or small families, laying waste the cultivated fields and plantations of maize or sugar cane, if not driven from them; but they are not common in the vicinity of villages.

D. labiatus, (Lat. lipped.) The White-lipped Peccary.

This is larger, stronger and heavier than the Collared Peccary, often measuring three and a half feet long, and sometimes weighing one hundred pounds, whereas the Collared species seldom exceed three feet in length, or weigh more than fifty pounds. The prevailing color is brown; the lips are white. The White-lipped Peccaries are found in numerous bands, sometimes, as is said, amounting to upwards of a thousand, spreading over a league of ground, and directed, the natives say, by a leader who takes his station in front of the troop. They cross rivers, and ravage plantations on their march. "If they meet with any thing unusual on their way, they make a terrific clattering with their teeth, and stop and examine the object of their alarm. When they have ascertained that there is no danger, they continue their route without further delay."

Rhinoceros, (Gr. ῥὶν or ῥὶς, *rin* or *ris*, a nose; κέρας, *keras*, a horn.)

The Rhinoceros.

This large uncouth looking creature is a native of the hotter regions of the Eastern Continent, and next to the Elephant, the most powerful of all quadrupeds. There are several species of this animal, of which the chief peculiarity is the horn, consisting of fibres matted together like those of whalebone.

R. unicornis, (one-horned,) or *R. Indicus*, the common E. Indian Rhinoceros, is usually about twelve feet long, and seven in height, and the circumference of the body is nearly equal to its length. This species, as the name *unicornis* imports, has but one horn, slightly curved, and sharpened to a point, not far from three feet in length, and used as a most powerful and effective weapon. The upper lip protrudes considerably, and from its extreme pliability, answers the purpose of a small proboscis. The skin is thick and coarse, and has a knotted or granulated surface; it is disposed in several folds, on the neck and shoulders. The legs are very short, strong and thick; and the feet divided into three large hoofs. (Plate VI. fig. 10.) The Rhinoceros of India leads a quiet, indolent life, in the shady forests, or wallowing in the marshy borders of lakes and rivers, in the waters of which it occasionally bathes. Its movements are usually slow, and it carries its head low like the hog, ploughing up the ground with its horn, and forcing its way through jungles. Pennant and others are of the opinion that this is the Unicorn of the Holy Scriptures. The female brings forth one young at a time. The ordinary food of the Indian Rhinoceros consists of herbage, and the branches of trees. The flesh is said to be not unpalatable. One of these animals, which was taken to England, in 1790, ate twenty-eight pounds of clover, the same quantity of ship biscuit, together with a great quantity of greens, each day; and twice or three times a day, five pails of water were given to him. The Asiatic specimen in the Zoological Gardens, London, is fed on clover, straw, rice and bran. The skeleton of this animal approximates to that of the Tapir and the Horse; the stomach is more like that of a man or a hog. It has thirty-six teeth, twenty-one of which are molar, but none of them canines.

R. Javanacus. The Rhinoceros of Java. This has less rough or prominent folds than those of the Indian Rhinoceros; its range extends from the level of the ocean to the summit of mountains which are considerably elevated. Marsden, the Missionary, in his "History of Sumatra," says, that "both the one and the two horned Rhinoceros are natives of the woods;" and he denies the stories which have been told "of the desperate encounters of these two enormous beasts."

R. Sumatrensis, or *Bicornis*, (Lat. two-horned.) The Rhinoceros of Sumatra, has a skin covered with stiff brown hairs, and almost altogether without folds, and it has a second horn behind the ordinary one, in this respect resembling the African animal.

The Rhinoceros is not uncommon in Sumatra, but is very shy, and therefore rarely seen.

R. Africanus. The animals of this species range over a large part of Africa. They were formerly common in the vicinity of Cape Town, but their present limit on the South West coast is the twenty-third degree of latitude. In the interior of the continent, the tribe is still very numerous, but less so than in Asia. This species differs from the Asiatic in having a comparatively smooth hide, while almost all the Asiatic species have a very coarse one, which is covered with large folds, not unlike a coat of mail. Four distinct varieties are said to exist in South Africa, two of a dark, and two of a whitish hue, called the "black" and the "white" Rhinoceros. The common Black Rhinoceros, *Rhinoceros bicornis*, is called by the natives the "Boiële;" the other, the "Keitloa," *Rhinoceros keitloa.*

The upper lip in both species of the black Rhinoceros is capable of extension, and is so pliable as to twist round a stick, collect its food, or seize any thing which it would carry to its mouth. These animals are very fierce, and except the Buffalo, perhaps, the most dangerous of all the beasts of South Africa.

Of the white Rhinoceros, the two varieties are *R. simus*, the common white Rhinoceros, called *Monoohoo*, by the natives, and *R. Oswellii*, the KOBAABA, or long horned White Rhinoceros. The chief difference between these two species, relates to the horns, the front horn of the Monoohoo averages about two feet in length; that of the Kobaaba frequently is more than four feet. The latter variety is least often found, and confined to the more interior portions of Southern Africa.

The White Rhinoceros is of larger size than the black. The head is so prolonged that it is nearly one-third part of the entire length of the body, which is from fourteen to sixteen feet; the nose is square; the anterior horn is longer; the disposition of the animal milder, and the flesh better tasted than that of the black species. Its food is grass. The black species are very sullen and savage in their disposition. Their flesh is lean, and of an acrid taste, given to it by the "Wait-a-bit" thorn bushes, on which they feed, ploughing them up with their short horn.

The body of the *R. Simus*, (between fourteen and sixteen feet long, and ten or twelve round;) is exceeded in size only by that of the Elephant; its belly is large and hangs near the ground; the legs are short, round and very strong; and the hoofs divided into three parts, each pointing forward. The head is large; the ears long and erect; the eyes small and sunken, or deep set in the head. The horns are not affixed to the skull, but attached to the skin, resting, however, in some degree, on a bony protuberance above the nostrils. They take a high polish, and are worth half as much as Elephant's ivory, being much used for sword-handles, drinking-cups, rifle-ramrods, etc. People of fashion at the Cape, have the cups set in silver and gold. The Turks believe these cups will split asunder and fly into pieces, if poison be put into them! Even the chips and turnings of the horns are carefully preserved, being esteemed of great benefit in convulsions, faintings, and many other illnesses. The Rhinoceros is nocturnal in his habits, commencing his rambles at dusk, and visiting the pools or fountains between the hours of nine and twelve o'clock at night. Having wandered until sunrise, he spends the day in sleep, under the shelter of some rock or tree. All the beasts dread him—the lion avoids him—even the elephant, should they meet, retreats, if possible, without hazarding a combat, and he will also fight his own species. His hearing and smell are acute, but his sight is not good. The Rhinoceros is not gregarious, but yet of a social turn, and usually goes in pairs; sometimes browses and pastures in droves of a dozen.

The best time to shoot these animals is when they go to the pools to quench their thirst and wallow in the mire, which they always do once in twenty-four hours. Occasionally the Rhinoceros, like the Elephant, is taken in pitfalls. The mother is affectionate and guards her offspring with tender care. The young also show strong attachment to the mother, clinging to her for days after she has been killed. The general appearance of the African Rhinoceros is that of an immense hog, with the bristles off, excepting a tuft at the extremity of the ears and tail; it has no hair whatever, and is the "very image of ugliness." The full grown male of the common white species, weighs not less than four or five thousand pounds, or as much as three good sized oxen. The Rhinoceros is long lived, attaining, as is thought, the age of one hundred years. Unwieldly as he appears, he is still swift of foot, at least this is true of the black species. Gordon Cumming, in his "Adventures in Africa," says, "that a horse with a rider, can rarely manage to overtake it." The food of this animal consists of vegetables, grasses, shoots of trees, and all kinds of grain; but it is not a voracious feeder. The statement that the hide of the rhinoceros is "impenetrable to a bullet," is now regarded as pure myth; for "a common leaden ball will find its way through the hide with the greatest facility." In consequence of the solid structure of the head, and the great thickness of the hide in that part, and the smallness of the brain, a shot aimed at the head rarely proves fatal.

The cavity of the brain in this animal holds but one quart, while that of the human skull will contain nearly three pints. However severely wounded, the Rhinoceros seldom bleeds externally; the hide being so thick and elastic and not firmly attached to the body, but constantly moving, the hole made by a bullet almost immediately closes up. Very many of these animals are annually destroyed in South Africa. Anderson, from whom we gathered many of the particulars here given, states, in his "Lake Ngami," that Messrs. Oswell and Vardon killed, in one year, eighty-nine of these animals, and that he himself, "single handed," killed in the same time nearly two-thirds of that number. Cumming, in his "Adventures," states that these animals are attended by what are called "Rhinoceros-birds," which stick their bills in the ear of the Rhinoceros, and uttering a harsh, grating cry, warn him of impending danger. These birds feed upon the ticks and other parasitic insects which swarm upon these animals.

Hippopotamus, (Gr. ἵππος, *hippos*, a horse; ποταμὸς, *potamos*, a river.) The RIVER HORSE. (Four-hoofed.)

This gigantic inhabitant of the African rivers is formidable in his strength, and in bulk inferior only to the Elephant. The ancients named him River-Horse, on account of the similarity of his voice to that of a horse. The form of this animal is in the highest degree uncouth; the body being extremely large, fat and round; the legs very short and clumsy. So low, indeed, at times is the animal in the body, that the belly almost brushes the ground. The head is exceedingly large, the mouth of enormous width, and the teeth of vast size and strength. (Plate IV. fig. 6.) The canines or tusks of these animals, of which there are two in each jaw, sometimes measure more than two feet in length, and weigh upwards of six pounds each; so hard and strong are they that they strike fire with steel, which gave rise to the fable of the ancients that the River-Horse vomits fire from his mouth. The tusks of the lower jaw are always the hardest. The hoofs are divided into four parts, unconnected by membranes. The skin, nearly an inch thick, is destitute of covering, except a few scattered hairs on the muzzle, edges of the ears and tail. The color, when on land, is of a purple brown; but when seen at the bottom of a pool, it appears to be of a dark blue, or as described by Dr. Burchell, "of a light hue of Indian ink." As in the Crocodile, the upper mandible is said to be movable. The inside of the mouth has been described by a recent writer, as resembling "a mass of butcher's meat." The eyes, (which have been compared to the garret-windows of a Dutch Cottage,) the nostrils, and the ears are all on nearly the same plane. This gives the use of three senses, and allows of respiration, with a very small portion of the animal exposed, when it rises to the surface of the water.

Two species are found in Africa, viz.: *H. amphibius*, (Gr. αμφι, *amphi*, double; βιοω, *bioo*, to live,) and *H. Liberiensis*, (some consider these, however, one and the same species;) the latter is much the smaller of the two. They range from the Cape Colony to 22° or 23° N. Lat., being found in the lakes and rivers, but in no rivers which empty into the Mediteranean, except the Nile, and that part of it which flows through Upper Egypt, or in the fens and lakes of Ethiopia. They inhabit both fresh and salt water, but are retiring before the advance of civilization. This animal is believed to have once existed in Asia, but on that continent has now become extinct.

The adult male of this species, *H. amphibius*, attains a length of 11 or 12 feet, and is nearly the same number of feet in circumference. The height is seldom more than 4½ feet; the female is considerably smaller than the male. The water seems the native element of the Hippopotamus, in which it swims and dives like a duck, and taking into account its unwieldly bulk, in a manner truly astonishing. When on land, with its dumpy legs supporting so enormous a weight, its progress is anything but rapid. Seldom does it wander far from water, to which it immediately betakes itself when disturbed or alarmed.

It is nocturnal, rarely feeding except during the night, for that purpose taking to the shore, it being an herbaceous animal. It appears rather nice in the selection of its food, which consists of grass, young reeds and bulbous succulent roots. When near cultivated districts, it ravages plantations of rice and grain, destroying as much by the treading of its enormous feet as by its voraciousness.

These animals are gregarious, being found in troops of from six to thirty. It is said to be "amusing to watch them when congregated; to see them alternately rising and sinking, as if impelled by some invisible agency, in the while snorting most tremendously and blowing the water in every direction." Sometimes they continue perfectly motionless near the surface, with the whole or part of the head out of the water. When in this position, they are described as appearing, at a little distance, "like so many rocks." By some zoologists, they are represented as naturally mild and inoffensive. This may be true of them in regions rarely visited by the foreign hunter and the firelock; but it is certain they have at times shown themselves to be most ferocious and hurtful. Their memory seems good, and seldom do they expose themselves to a second attack in the same place.

That they naturally are fond of the aquatic element, is shown by the fact that if the mother be shot dead just after calving, the young one will immediately make for the water. The natives harpoon them in a manner similar to that adopted with the whale; if killed outright the animal sinks, but in half a day reappears. The flesh is highly esteemed; the tongue is regarded as a great delicacy, and the fat of the animal forms a capital substitute for butter.

The hides of these animals form no mean article of commerce in the Cape Colony; in Northern Africa they are made into whips for the dromedary, and also for punishing refractory servants. The ancient Egyptians used the hide largely in the manufacture of shields, helmets, javelins, etc. But the teeth, (canine and incisor,) are the most valuable part of this animal. They are considered much superior to Elephant's ivory, and when perfect and heavy, (say from five tc eight pounds each,) have been known to bring about five dollars a pound. They do not readily turn yellow, as is frequently the case with Elephant's ivory, and on that account are more valuable for artificial teeth.

Medicinal qualities are attributed to certain portions of the body of the River Horse. These animals are easily domesticated but are very voracious. One of them now in the Regent's Park Gardens, (London,) when first shipped at Alexandria, Egypt, and yet comparatively a "baby," consumed daily the milk of two cows and three goats. This portion, until supplemented with Indian corn, did not, however, suffice to satisfy his enormous appetite. "On his arrival at the gardens, oat-meal was substituted for Indian corn; and the change, with an extra supply of milk, seemed to give the gigantic infant great satisfaction." Vegetable diet was by degrees administered in place of milk; at the present time the animal is fed on clover, hay, corn, chaff, bran, mangel wurtzel, carrots and white cabbage. The three last named vegetables constitute his favorite food. On a daily allowance of 100 lbs. of this kind of food, he thrives astonishingly well, as is proved by the fact that weighing 1,000 lbs. when he first arrived, he now weighs more than 3,000 lbs. Not less than six bushels of chewed grass was found in the stomach of one examined by Mr. Burchell.

Tapirus. The TAPIR.

Of this genus of Pachydermatous quadrupeds there are three species; two of them found in South America, the other in Sumatra and Malacca. The general characters of this genus are the following; the molar teeth, which are seven on each side above and six below, "have their crowns crossed by two transverse and straight ridges, at least, until worn down by attrition; the incisors in each jaw are six; the canines two, separated from the molars by a wide interval; the nose is elongated into a short flexible trunk; the feet have four toes before and three behind; the skin is dense and thinly covered with short, close hair." In its general form and contour, this animal reminds the beholder of a hog; but it is distinguished from the hog by its flexible trunk, which answers partly the same purpose as that of the elephant. (See Chart.) The trunk has not, however, any finger-like appendage like that of the latter animal. The eyes are small and lateral, and the ears long and pointed.

The AMERICAN TAPIR, *T. Americanus*, is the largest animal of South America. It is of a deep brown color throughout, approaching to black; between three and four feet in height, and from five to six in length. The hair is short and very scanty, so that it is scarcely discernible at a short distance. The back of the neck is bristled with a thin mane of stiff blackish hairs. The inmost recesses of deep forests are the chosen resorts of this species, which is not gregarious, and avoids the society of man. For the most part, it is nocturnal in its habits, sleeping, or remaining quiet during the day, and at night seeking its food, which, in the natural state of the animal, consists of shoots of trees, birds, wild fruits, etc. It is, however, when in confinement, an indiscriminate swallower of every thing, filthy or clean. Its enormous muscular power and the tough, thick hide which defends its body, enable it to tear its way through the underwood in whatever direction it pleases. Its ordinary pace is a sort of trot; but it sometimes gallops, though awkwardly, and with the head down. It is very fond of the water, and often resorts to it, sometimes remaining below the surface for a considerable time. Its disposition is peaceful and quiet; it will, however, defend itself vigorously, and inflict severe wounds with its teeth, though it never attempts to attack either man or beast, unless hard pressed. The Jaguar often springs upon it, but is frequently dislodged by the activity of the Tapir, who rushes through the bushes as soon as he feels the claws of his enemy, and endeavors to brush him off against the thick branches. In some parts of South America the Tapir is domesticated. M. Sonnini saw several of them "walking at liberty about the streets of Cayenne, whence they were accustomed to stroll into the neighboring woods, returning at night to their home; nor were they by any means destitute of intelligence, but seemed fond of their masters, whom they not only recognized, but acknowledged by various tokens of attachment." In his opinion, the Tapir might, from its great strength and docility, be advantageously used as a beast of burden. The sight, hearing and smell of this animal are very acute. It is in much request by the natives for its flesh, which, though coarse and dry, they deem excellent food. The skin is also valuable, from its toughness and density.

T. Malayanus, (or *T. Indicus.*) The MALAY or ASIATIC TAPIR is larger than the American, which it resembles in form and general habits. Its back and sides are of a grayish white, abruptly edging the brown of the other parts, which gives the animal an appearance as if a white horse-cloth had been spread over it; the neck is destitute of a mane. Its flesh is eaten by the natives of Sumatra. In captivity, like the South American animal, it is gentle and inoffensive, "becoming as tame and familiar as a dog, feeding indiscriminately on all kinds of vegetables, and sometimes fond of attending at table to receive bread, cakes, or the like." The young, as is the case with the American species, differs in color from the adult, being, at the age of four or five months, black, beautifully marked with spots and stripes of a fawn color above, and white below.

A third species has been discovered in the Cordilleras of South America, covered with thick black hair, and with a more elongated trunk.

Hyrax, (Gr. ῞υραξ, *hurax*, from ῞υς, *hus*, a swine.) The animals of this genus are small, and aptly described in the Holy Scriptures, as "a feeble folk," but of great interest on account of the peculiarity of their organization. "They are rhinoceroses in miniature." Cuvier, by a recourse to the anatomy of the *Hyrax*, proved it to be a true Pachyderm; and "notwithstanding the smallness of its proportions," intermediate between the Rhinoceros and the Tapir. This animal has twenty-one ribs on each side, a number greater than that possessed by any other quadruped, the Unau excepted, which has twenty-three. Its molar teeth resemble those of the Rhinoceros, as it does also in the characters of its stomach and alimentary canal. The body is covered with thick hair, and "beset here and there with erinaceous bristles." It has a simple tubercle in lieu of a tail; four toes on each fore foot, and three on the hind ones.

H. Syriacus. The DAMAN of Syria, the CONEY of the Bible, is of a brownish gray color above, and has the lower parts white; these two colors being separated by a yellowish tint, and the head and feet being more gray than the body. The skin, where it is exposed, is of a blackish violet. The length is about two feet; the height eleven inches. It stands rather low on the legs, being partially plantigrade; its body is stout for its size, which is hardly equal that of a hare. It skips about with great agility, and its actions evince a sportive and playful disposition; in captivity it becomes docile and affectionate. The Conies make their houses among the rocks.. Proverbs xxx; 26. "The nature of their retreats renders the capture of these animals very difficult. To behold this creature among the craggy and broken mountain scenery of the land of the Psalmist, where he noticed it himself, and recorded the goodness of God in providing a refuge for a defenceless animal, surrounded with numerous enemies, the jackall, the hyaena, and the eagle, cannot but raise in the mind that train of reflections which led David to exclaim, "O Lord, how manifold are thy works! in wisdom hast thou made them all!"

The Conies or Damans, associate in considerable numbers, coming forth from their retreats during the day, but flee on any alarm, to their holes in the rocks, or in the steep declivities of mountains, there resting themselves in their nests prepared of leaves and grasses. Their food consists of the roots and vegetables of mountain districts. In walking they steal along as if frightened, with the belly almost on the ground, advancing a few steps at a time, and then pausing. Their whole manner is mild, feeble and timid; they are easily tamed, though if roughly handled at first, they will bite. It was classed among the unclean animals of the Jews. Lev. xi: 5. In Abyssinia, its flesh is considered unclean, both by Christians and Mohammedans. The Arabs, it is said, eat it and call it, perhaps in jest, "the sheep of the children of Israel."

H. Capensis. The KLIPDAS, closely resembles the Syrian animal. In winter, it is fond of coming out of its hole and "sunning itself on the lee-side of a rock," and in summer, of enjoying a breeze on the top; but in both instances, as well as when it feeds, a sentinel, which is generally an old male, is on the look out, and usually gives notice by a prolonged, shrill cry, of the approach of danger.

Palaeotherium, (Gr. παλαιος, *palaios*, ancient, and θηϱιον, *thērion*, a wild beast.)

This is the name given by Cuvier to an extinct genus of Pachydermatous animals, discovered in the gypsum beds of Paris, in company with *Anoplotherium*. Of this fossil genus, nearly fifty species have been discovered. Remains of *Palaeotheria* have been found in the tertiary formation near Rome, the department of the Gironde, Provence, etc., and in the lower and marly beds of Binstead, in the Isle of Wight. The zoological position of the genus appears to be intermediate between the Rhinoceros, Horse and Tapir. The habits of the animals which it included were probably like those of the Tapirs. Dr. Buckland supposes that they lived and died upon the margins of lakes and rivers. The species varied greatly in size, some having been as large as a Rhinoceros, and others having ranged from the size of a horse to that of a dog. In relation to the discovery of this fossil genus, Cuvier thus writes: "I found myself as if placed in a charnel house, surrounded by mutilated fragments of many hundred skeletons of more than twenty kinds of animals, piled confusedly around me; the task assigned me was to restore them all to their original position. At the voice of Comparative Anatomy, every bone and fragment of bone resumed its place. I cannot find words to express the pleasure I experienced in seeing, when I discovered one character, how all the consequences which I predicted from it, were successively confirmed. The feet accorded with the characters announced by the teeth; the teeth were in harmony with those previously indicated by the feet. The bones of the legs and thighs, and every connecting portion of the extremities were found to be joined together precisely as I had arranged them before my conjectures were verified by the discovery of the parts entire. Each species was, in fact, reconstructed from a single unit of its composing elements." The *Palaeotheria* were characterized by having twenty-eight complex molar teeth, four canines, and twelve incisors; six in each jaw.

Anoplotherium, (Gr. from *αγ*, *an*, neg; *οπλον*, *hoplon*, a weapon; *θηριον therion*, a wild beast.)

This name, signifying a beast without offensive arms or tusks, was given to a genus of extinct Pachyderms, found by Cuvier in company with the Palaeotheria, and named by him, ranging between the Pachydermata and Ruminantia.

III. Family of the Pachydermata. (Pl. VI. fig. 8.)

Equidae, (Lat. *Equus*, a horse.) Solid-ungulous, or one-hoofed Mammalia. These Pachyderms have but a single finger or toe, terminating each extremity; and this finger or toe is enclosed in a horny hoof or shoe. They include Horses, Asses, and the Zebra. Of these the Horses are far the most valuable and most widely distributed. Instead of the massive form and heavy tread of the Elephant, these animals approach to slender forms, and they (especially the horses,) resemble in their graceful proportions, and their fleet movements, the Deer and the Antelope. They are distinguished from other animals, not only by the undivided hoof, but by their stomach, which is simple and incapable of rumination. Their intestines are much lengthened, and thus adapted to their food, which consists entirely of herbage. "The Asses and the Zebras," says Dr. Gray, "are always whitish and more or less banded with blackish brown, and always have a distinct dorsal line, the tail only bristly at the end, and they have warts only on the arms, (fore legs,) and none on the hind legs; and have long hair on the tail, from its insertion to its extremity."

The number of teeth in this family is forty-two, viz. twelve incisors or nippers, four canines and twenty-six molars. While the animal is young, the incisors have their crowns furnished with a groove; the molars have square crowns marked or edged with four crescents of enamel. Between the canines (which are developed in the male alone,) and the first molar, is a broad, open space, which is capable of receiving the bit, (see Plate IV. fig. 5,) with which these animals are governed when in a state of domestication. The female ordinarily produces one young at a time, which is called a foal, and suckled during six or seven months. Africa and Asia are the native regions of the EQUIDÆ. They range in large herds over the extended plains and table lands which are uncovered with forests. Two species, the Horse and the Ass, have been domesticated and widely dispersed over the earth.

E. caballus, (Lat. a Keffel or Saddle Horse,) The HORSE.

It has been well remarked that if custom had not dignified the Lion with the title of "King of beasts," reason could no where confer that honor more deservedly than on the horse. His courage, strength, fleetness, his symmetrical form and grandeur of deportment, are unalloyed by any quality injurious to other creatures, or adapted to create the aversion of man, whose orders he implicitly obeys, whose severest tasks he undertakes with a cheerful alacrity, and to whose pleasures he contributes with animation and delight. On the battle-field, he shows the most resolute fierceness and courageous ardor. In the poetical language of the sacred writings, "His neck is clothed with thunder. The glory of his nostrils is terrible. He paweth the valley and rejoiceth in his strength. He goeth on to meet armed men. He mocketh at danger and is not affrighted; neither turneth he back from the sword." (Job, xxxix.)

Much discussion has existed concerning what region the Horse originally inhabited, and to what nation we are indebted for his first subjugation. It is well known this animal is found wild in the Western as well as the Eastern Continent; but there is no doubt it was at first confined to the latter, where the wild species, such as the Zebra, the Quagga, etc., still range in freedom. The testimony of the sacred writings is decidedly in favor of the eastern origin of the horse, making it quite evident that the Egyptians first subdued it to obedience and servitude. The first mention of the horse occurs during Joseph's wise administration in Egypt, who, it is said, gave bread to the famishing in exchange for horses; and when the remains of the patriarch, Jacob, were removed from Egypt to Canaan for burial, we read "there went up with him both chariots and horsemen." This shows that horses were used both for draught and burden 1650 years before the birth of Christ, which is earlier than the date of any profane history in relation to this subject. The Egyptian and Nubian horses are still among the handsomest, but Arabia bears the palm in producing the most beautiful breed of horses, and also the most generous, swift, courageous and persevering. "The Arab treats his horse as one of his family; it lives in the same tent with him, eats from his hand, and sleeps among his children, who tumble about on it without the least fear. Few Arabs can be induced to part with a favorite horse." It is related that "an Arab, the net value of whose dress and accouterments might be calculated at under seven pence half penny, refused all offers made to purchase a beautiful mare on which he rode, and declared that he loved the animal better than his own life." The Arabian horses are in height not often more than fourteen hands two inohes. They are found, though not in great numbers, in the deserts of that country, and the natives make use of every stratagem to take them, reserving the most promising for breeding, and instead of crossing, as is done in other countries, taking every pains to keep the breed pure. Some have supposed that to Arabia we are indebted for the primitive breed of this noble animal and its subjugation to the use of man. Mr. Bell, however, who is high authority in this subject, says in his "History of British Quadrupeds," "there is great reason to conclude that it was only at a comparatively late period that they were employed by that people. While Solomon was receiving from Arabia treasures of various kinds, it was from Egypt only that his horses were brought. There appears great probability in the opinion that Egypt, or its neighborhood, is the original country, and still more that this extraordinary people first rendered it subservient to man, and subsequently distributed it to other countries."

In Brande's Dictionary of Science, &c., it is remarked that "wild horses appear to be free from nearly all those diseases to which the domestic breed are prone."

The wild horse is found in immense numbers in the vast plains of Great Tartary, and also in South America in the rich pampas extending from La Plata to Paraguay. The wild horses of South America are undoubtedly descended from those of Andalusia, originally carried thither by the Spanish conquerors.

When caught, they are easily subjugated and valuable for their speed, hardness and strength. The usual method of taking them is by the *lasso*, a running noose at the end of a long leathern thong, thrown with wonderful precision and capable of bearing a sudden and violent strain.

The horse is a herbivorous animal, as its teeth indicate, and is more scrupulous in the choice of its food than most other quadrupeds, rejecting in the meadow several plants which the ox devours with pleasure. By the peculiar structure of some of the bones of his face, he is enabled to so move his jaws as to comminute and grind down his corn. The best method of judging of a horse's age is from a careful investigation of its teeth. "Five days after birth, the four teeth in front, called *nippers*, begin to shoot; these are cast off at the age of two years and a half, but are soon renewed, and in the following year two above and two below, namely one on each side of the nippers, are also thrown off; at four years and a half, other four next those last placed fall out and are succeeded by other four, which grow much more slowly. From these last four corner teeth it is that the animal's age is distinguished, for they are somewhat hollowed in the middle and have a black mark in the cavities. At five years old these teeth scarcely rise above the gums; at six their hollow pits begin to fill up and turn to a brown spot, and before eight years the mark generally disappears. A horse's age is also indicated by the canine teeth or tusks, for those in the under jaw generally shoot at three years and a half, and the two in the upper at four; till six they continue sharp at the points, but at ten they appear long and blunted. There are, however, many circumstances which render a decision as to the age very difficult after the marks are defaced from the lower incisors; and it should be observed that horses which are always kept in the stable have the marks much sooner worn out than those that are at grass, to say nothing of the various artful tricks resorted to by dealers and jockeys to deceive the inexperienced and unwary." We must refer to other works for interesting particulars respecting the various breeds of horses and the different kinds used, as the Race Horses, supposed to have been originally of the Arabian breed, the Hunter, the Roadster, the Carriage Horse, the Dray Horse, the Cart Horse, &c., and the smaller varieties, such as the Shetland Pony, Galloway, &c.

The Horse is capable of strong attachment to man and to other animals. Among the numerous anecdotes illustrating this remark are the following: "A horse and a cat were great friends, and the latter generally slept in a manger. When the horse was going to have his oats, he always took up the cat gently by the skin of her neck and dropped her into the next stall, that she might not be in his way while he was feeding. At all other times he seemed pleased to have her near him."

Two Hanoverian horses had long served together during the Peninsular war, in the German brigade of artillery. They had assisted in drawing the same gun and had been inseparable companions in many battles. One of them was at last killed, and after the engagement the survivor was picqueted as usual and his food brought to him. He refused, however, to eat, and was constantly turning round his head to look for his companion, sometimes neighing as if to call him. All the care that was bestowed on him was of no avail. He was surrounded by other horses, but he did not notice them; and he shortly afterwards died, not having once tasted food from the time his former associate was killed. A gentleman who witnessed the circumstance assured me that nothing could be more affecting than the whole demeanor of this poor horse."*

Of the age to which the horse would naturally arrive, it is difficult to speak with certainty. The animal sometimes exceeds thirty or forty years, and it has even exceeded fifty years in age, but from ill usage and over exertion the majority come to their end before they have seen nine or ten years. The horse is now spread over every part of the Western Continent. The natives of Terra del Fuego are well stocked with horses, each man having six or seven, and all the women and even the children have their own horses. With the horses are found herds of wild oxen. The number of mustangs or wild horses found in South America may be judged of from the fact that from 1838 to 1842, 90,000,000 lbs. of oxen and horses' hides, and 9,500,000 lbs. of horse hair were obtained within the limits of Monte Video and Buenos Ayres. In his "Fauna Boreali Americana," Sir John Richardson says that the horse is found amongst the wandering Indians who frequent the prairies of Saskatchewan and the Missouri, and who use it for chasing the Buffalo as well as a beast of burden. Among the Indians as well as the Guachos, the horse is eaten. It is also eaten by the Calmuck Tartars; and in many parts of Asia, mare's milk is taken as an article of diet. It is converted into butter and cheese, and a favorite beverage amongst the Tartars is made by fermenting it.

Equus Asinus, (Lat. Ass.) THE ASS.

The ass was anciently employed by the Orientals for common

* "Gleanings in Nat. Hist."

purposes more than the horse. They seem to have looked upon the horse as rather fitted for war and scenes of pomp and state. The ancients were ignorant of the art of shoeing the horse's hoof with iron, and found it more easily injured by travel on hard roads and by long journeys than the harder hoof of the Ass. The hoofs of horses were usually protected by sandals of thick close felt. From time immemorial, the ass has, in Oriental countries, been the servant of man, and its introduction into Europe and this country must be regarded as comparatively recent. Though out of its native climes, it seems strong, patient and hardy, still it is much different from the animal of the East. Here it is dull and slow, small and clumsy; there it is larger, well made, light, footed with a sprightly pace, and carries its head high. White asses were in ancient times highly valued. Upon them those rode who sat in judgment, Judges, v. 10; and our Saviour, too, appears to have honored this animal by riding "upon the foal of an ass upon which never man sat." Mark, xi. 3. Asses made no inconsiderable part of the wealth of the Hebrew patriarchs.

The MULE is a mixed breed between the horse and the ass, an animal not much sought after among us, but extensively used in the East for riding, and in Spain is the chief beast of burden. It is very strong and sure footed, and is for that reason employed in the Andes of South America instead of the Llama, and also used in other mountainous countries.

E. asinus onager, or *onagga.* The KOULAN or WILD Ass is of a pale reddish color in the warmer season, in the winter grayish. It is found abundantly in Mesopotamia and Persia, and on the shores of the Indus. Bishop Heber says no attempt has ever been made to break in the wild Ass for riding. The ears are two inches longer in the male than in the female. In the tame species the ears are elongated and acute, but in this and the other wild species are moderately short and rounded.

E. asinus hemionus. The DIZIGGUETAI.

The name Dizigguetai, given to this species, has been spelled in seven or eight different ways. It is said to refer, in the Mongol language, to the large ears of this animal, which exceed in size those of a horse. This species are found in Oriental Tartary, and the wild regions of China and Persia. This is the wild ass of Scripture. "Who hath sent out the wild ass free? or who hath loosed the bands of the wild ass? Whose house I have made the wilderness and the barren land his dwellings. The range of the mountains is his pasture, and he searcheth after every green thing." Job, xxxix chapt.

The Dizigguetai, or Dziggtai, is in size intermediate between the horse and the ass, and in disposition extremely wild and untamable. Its limbs are finely proportioned, showing the lightness and elegance of the stag, combined with the power of great muscular exertion. It runs with great rapidity, carrying its head erect, snuffing up the wind and defying the speed of the fleetest coursers. Sir R. Ker Porter gives an amusing account of an unsuccessful chase of one of these animals, which he was unable to overtake, though mounted on a very swift Arabian horse. This roamer in the "wilderness" and "barren land," lives in troops, like the wild horse, being guided by a leader or chief, on which the movements of the rest depend. The flesh is esteemed a delicacy by the Mongols, who occasionally manage to shoot the leader and thus throw the troop into confusion, so that several fall before they take their flight. The fur is short, smooth, and of a bright red bay, the legs of a straw color. The males are the more deeply colored and larger, often standing 14 hands high. These animals live partly on the plains and partly on the mountains, and from this fact perhaps, the lower surface of the hoof shows considerable variation in form and concavity. They are found in a climate where the temperature is below the freezing point in the middle of summer, yet they throw off their pale woolly coat during that season and become bright bay.

Equus montanus, (or *Asinus Zebra.*) The ZEBRA.

This species is native to Africa, confined to the mountains, for traversing which their hoofs are expressly formed. The ground color is white, with black bands on the head, body and legs to the hoofs; the nose is reddish; the belly and inside of the thighs not banded; the end of the tail is blackish; the hoofs narrow and deeply concave beneath. The Zebra has scarcely ever been brought under the bit. Sometimes, however, in spite of its vicious habits, it has been trained to draw a carriage. It is fierce and strong and universally admired for its fleetness and beauty. The voice of the Zebra is very peculiar and cannot be described. We have said above that the ground color is white; this is true of the female, but in the male the ground color is yellowish fawn. It is often a prey to the lion, and also to the untutored natives, by whom its flesh is regarded as a great delicacy.

The *Equus Burchelli*, or the PEECHI, is a native of the plains, inhabiting, in small companies, the flat country lying north of the Cape of Good Hope and stretching into the interior. This is a strong, muscular animal, and might be used as a beast of burden. The head, neck, shoulders and back are covered with alternate stripes of white and black; the nose is white, with faint

8

black markings; the black stripes of the body become fainter on the haunches and disappear on the under parts.

E. Quagga, or *A. Quagga*. The name, (Quagga,) of this animal, expresses the sound of its voice, which, in some degree, is like the barking of a dog. It is now sometimes tamed by the natives and used for purposes of draught, and occasionally a half domesticated specimen is offered for sale at Cape Town with a rider on its back; but in the most tractable state to which it has hitherto been reduced, it is regarded as vicious, obstinate and fickle.

ORDER PACHYDERMATA.

Give the derivation of the term PACHYDERMATA. Why are the animals of this order so called? Where are they found? What genus appears wild in Europe? What is the nature of their food? What peculiarity have the canine or incisor teeth? How is the muzzle prolonged? What is said of the size and strength of these animals?

From what is the family name PROBOSCIDEA derived? What does it include? How many species of the Elephant? By what are both species distinguished? How do they compare as to general characters and habits? In what respects do they differ? What is said of the molar teeth? What peculiarity in the manner in which they succeed each other? In what bones are the tusks seated. Describe the tusks. What is said respecting their length and weight? How is the foot enclosed? What is required by the immense weight of the head? What is said of the muscles of the neck? To what is the entire structure compared? What is said of the internal organization? Which is the most remarkable part of the Elephant? What are its uses? How many muscles has it? Describe it. What is found at its extremity? What is said of the skin, head, tail and general appearance of the Elephant? What indicates its very great force? What is the East India Company's standard as to height for a serviceable Elephant? Where are the largest found? What use has been made of them by Eastern princes? What use was made of ivory in ancient times? What weight of it was exported to Great Britain in 1831,2. How great a destruction of Elephants did this involve? Mention the chief marts whence ivory is obtained. For what purpose is it most used in England? Where in Europe are ivory articles manufactured most extensively? What people excel in preparing ivory articles? Where are Elephants still found in great numbers? Of what roots are they especially fond? What ingenious use do they make of their tusks? How is this animal stimulated to extra effort? Repeat the story showing his love of sweet things. What is said of his senses? What of his fondness for certain flowers and plants? Why is he not often bred in captivity? For what is he trained in India? What is said of his fondness for water? How is he captured? What allusions to the Elephant are found in the Holy Scriptures? What is said of their affections? How illustrated? How are they influenced by kindness? Give examples showing their sagacity and power of memory. Derive the specific name of the *Mammoth*. What is said of their remains? Give the derivation of the term *Mastodon*. Is this a living or extinct genus? In what deposits are its remains found? How is it principally distinguished from the Elephant? What more is said of it? To what is the term *Mammoth* properly applied? Where have numerous remains of the *Mastodon* been discovered?

From what is the term SUIDÆ derived? How many species does it include? What is the origin of the common Hog? Where is the wild animal still found? How distinguished from the domestic Hog? What is next said of him? What other differences between the domestic and the wild animal? What is said of the variety of color in the common Hog? What of its habits? What people were forbidden to eat its flesh? What do late experiments show? From what may it be inferred that this animal was designed to be food for man? Give the opinion of Linnæus. Why is the flesh of great importance to commercial people? Does the value depend chiefly upon the size? What is said of the Chinese breed? What further is said of the Hog?

What is the meaning of the word *Babirussa?* How does this animal compare with the common Hog? What is said of its tusks? Upon what does it chiefly subsist? How does it elude pursuit? In what countries does it abound? What is said of a pair taken to France? What is the derivation of *Pacochoerus?* Where are the animals of this genus found? Why is it called Warty-Hog? What is said of its teeth? What is indicated by its structure?

Give the derivation of the term *Dicotyles.* Where are PECCARIES found? By what are they allied to the true swine? How do they differ? What is their most decided characteristic? How many species? Why is the COLLARED PECCARY so called and where found? What objection to it as a domestic animal? What is said of the WHITE-LIPPED PECCARY? Are they found in great numbers?

What is the derivation of the term *Rhinoceros?* In what regions is this animal found? What is its rank? How many species? What is their chief peculiarity? What is the meaning of the specific term *unicornis?* Why is this name given to the Common East Indian Rhinoceros? Describe this animal and give its habits. What did Pennant suppose it to be? What is its ordinary food? Illustrate its capacity for food? What more is said of it? How does the R. of India differ from that of Java? What is its range? What does Marsden say? What is said of the R. of Sumatra? How extensive is the range of the R. of Africa? How does the African species differ from the Asiatic? How many varieties or species are found in South Africa? Describe the Black variety? What is the chief difference between the two varieties? How do the white and black R. differ? Describe the *R. simus.* What use is made of its horn? What account is given of the Rhinoceros?

Give the derivation of the term *Hippopotamus.* Describe the characteristics of this animal. How many species and what is their range? Is it found in Asia? What is the size of the adult male of the *H. amphibius?* What seems its native element? What is said of it on land? When and what does it eat? How do troops of them appear when in water? What fact shows their fondness for that element? How are they harpooned? What is said of their flesh and hides? What is the most valuable part of the animal? In what respects is their ivory superior to that of the Elephant? Are they easily domesticated? What is said of one in the Regent's Park, London?

How many species of the TAPIR? Where is it found? Give the general character of the genus. What animal does it resemble? How does it differ from it? Describe the Tapir of South America. What animal often springs upon it? What is said respecting its domestication? What are its uses? How does the Asiatic compare with the AMERICAN TAPIR in size? What appearance has this animal? How is it in captivity? Where has the third species been discovered? What Scripture animal is the *Hyrax?* How is it described? Who proved it to be a true Pachyderm? How many ribs has it? What larger animal does it resemble? Describe the *Daman* or *Hyrax* of Syria. Is it easily captured? On what does it feed? How was it classed among the Jews? What is said of the CAPE HYRAX?

Give the derivation of the term *Palaeotherium.* To what animal is it applied? How many species have been discovered? Where have the remains been found? What is the Zoological position of this genus? What were the habits? What does Buckland suppose? What does Cuvier say respecting the discovery of this fossil genus? How is it characterized? Give the derivation of the term *anoplotherium.* What does it mean? In what connection is it found?

Give the derivation of EQUIDÆ. What animals belong to the ONE-HOOFED Mammalia, or solipedes? Mention the UNEVEN-HOOFED animals, and the MULTUNGULATE, or many hoofed? What does the HORSE FAMILY include? Which are most valuable and widely distributed? How are the EQUIDÆ distinguished from other animals? What is said of the Ass? What of the Zebra? What is said of the teeth of the Equidæ? What are their native regions? Which species have become domesticated? What does the term *caballus* mean? What is said of the qualities of the horse? What of his native regions? What of his first subjugation? What countries produce the most beautiful breed of horses? How does the Arab treat his horse? Give Mr. Bell's remarks respecting the native country of the horse. Where is it now found? How usually captured? What is said of its food? How is its age determined? What is said of its attachments? Give examples. How long does it live? How widely diffused? What is said concerning the use of the Ass in Oriental countries? What of its introduction into other countries? What is remarked respecting WHITE ASSES? What of the breed of the MULE? Give some account of the ONAGGER or ONAGGA. To what does the name DIZIGGUETAI refer? Where is this animal found? Describe it. Where is the *Asinus Zebra* found? Give some account of it. Of what country is the *Equus Burchellii* a native? What is said of it? What is said of the *Asinus Quagga?* Trace the horse from the species, (caballus,) through the higher divisions.

Name the three families of the order Pachydermata from the chart. Mention the animals named in each, giving both the common and scientific names. Give the characteristics and peculiarities of each family, genus and species. Which is the largest animal of this order mentioned? Which the smallest figured on the Chart?

SECTION XXI.

EIGHTH ORDER.—RUMINANTIA. (Lat. *rumen*, a stomach or paunch.)

RUMINATING, OR CUD-CHEWING ANIMALS.

This pre-eminently useful order includes the oxen, sheep and goats, deer, giraffes and camels. They were very anciently recognized as a separate group, and taken as a whole are extremely compact and well defined. The camels alone present some slight exceptions to the general character. Each foot ends in two toes, covered with two sharp pointed horny hoofs, fitting each other as though a single round hoof had been cleft in the middle. Behind these are two small spurs, or rudiments of lateral toes. Hence they are called animals with "divided, or bifurcate hoofs." The RUMINANTS are well known as herbivorous. Their name indicates the singular faculty which they have of masticating or chewing their food a second time, and by which they are specially distinguished from all other animals. For this purpose they are furnished with four stomachs, or one divided into four distinct chambers or cavities, each having a distinct office to perform. The first is the *rumen*, or paunch, in full grown animals the largest of all, and covered with *papillæ*, or flattened warts. Into this passes the hard and coarsely masticated food from the beginning of the muscular canal, which is at the end of the æsophagus or gullet. From the *rumen*, the rudely bruised herbage is transmitted into the second stomach, called the *reticulum*, or hood, which is beautifully divided into hexagonal cells, like a honeycomb. Water is received from the mouth into this second cavity. The food is here moistened and moulded into small balls or pellets, and by a rapid and inverted action of the muscles of the gullet is propelled into the mouth, where it is more perfectly masticated, mixed with fluid and again swallowed, passing now into the *psalterium, omasus*, or manyplies, the *third* stomach. The inner coat of this division is set with parallel longitudinal *laminæ*, or folds, resembling the leaves of a book. In the sheep it has forty, in the ox as many as a hundred of these folds. In these plates the superfluous fluid, which might otherwise have too much diluted the gastric juice, is absorbed; and the sub-divided cud passes gradually into the fourth and last, or red stomach, (*abomasus*,) which is large and pear shaped, and wrinkled and hairy, as to its inner surface. This is the true digesting stomach, and in the young, while sucking, is the largest of the four.

For the purpose of assisting the reader to form a correct idea of this wisely arranged internal mechanism, we give a section of a stomach, as in Plate VI. fig. 13, with the following

EXPLANATION.

In the stomach of the Ruminants, (fig. 13,) the gullet or æsophagus (A) which is opened, expands into the paunch or *Rumen* (B) which is divided by a muscular wall; the valve (C) allows the food only to pass gradually, by the action of the paunch, into the *Reticulum* or hood, (D,) which is opened to show the folds and cells called the honeycomb, and from which the food, moistened and compressed, is passed back to the mouth and chewed again. When swallowed the second time, it passes to the *omasus*, or third stomach, (E.) The gullet has a fold running down and walling in the orifice of the *omasus*, (F.) The fibres surrounding this orifice contract on the application of crude vegetable matter; but when this matter has been elaborated in the *reticulum*, and chewed the second time, the orifice expands, and by the action of the muscular fibres of the stomach, is brought higher up into the gullet to receive the then welcome mass. The laminated or leaf-like structure of the *omasus*, sometimes called the leaflet, is shown in the figure. From this the food passes to the fourth stomach, (*abomasus*, G,) which has digestive powers similar to that of the simpler stomach in other animals. The third stomach is the least essential to ruminants capable of enduring long thirst and of living upon dry shrubs, like the Camel and Llama. It cannot properly be said to exist in them, and the opening leads directly into the *abomasus*. It is remarkable that the milk upon which young animals of this kind are fed, requiring no process of rumination, passes directly from the gullet into the *fourth* stomach.

Another character of the Ruminants is the possession of incisor or cutting teeth in the lower jaw only. Cuvier makes them consist of two divisions; first, those without horns, and secondly, those with horns. The larger part have horns, particularly the males. The few species which want these organs have the tusks, or cutting teeth, which are deficient in the others. The vegetable nature of their food renders the flesh of these animals wholesome and agreeable. Their milk furnishes butter and cheese; their skins, leather; their horns, combs; their wool, cloth and yarn; their hair is used in the making of matresses, sofas, etc. The fat has the property of hardening as it cools, and is distinguished by the name of suet. Their tallow is made into candles and soap, and their bones are in great request for manure. Indeed, they seem to have been formed with the express design of ministering to man's comfort and welfare. They inhabit the known world, with the exception of Australia.

Define the term Ruminantia. What animals does the order include? Is the group well defined? What exception is made? Why are the hoofs of this order termed bifurcate or bisulcate? Upon what do these animals subsist, and what peculiarly distinguishes them from all other animals? Describe in full the stomach of a ruminant, pointing out the parts as you proceed, on plate VI. fig. 13. How does Cuvier divide the *Ruminants?* To what species are the larger incisors confined? What is the nature of their food? What their uses? Where are they found?

Name the seven families of Cud-chewing animals on the Chart. Which are without horns? Which have solid horns? Which hollow horns? Which is the tallest? Which most useful for food? Which for clothing? Which wild? Which domesticated?

SECTION XXII.

Camelidæ, (Gr. κάμηλος, *kamelos*, a camel.) The CAMEL TRIBE.

This family of the Ruminants differ, in some respects, from the others, forming a connecting link between them and the Pachyderms, or thick skinned animals. They are without horns; the hair inclines to be woolly; there are fleshy bosses, or humps on the back "These humps are of a firm, fatty consistence, seeming like reservoirs of nutriment, being observed to diminish from absorption, during long abstinence, but to increase again when food becomes abundant." The eyes are large and projecting; the ears small. The Camels have canine teeth in both jaws, and two incisors in the upper jaw, which are wanting in other Ruminants. The lower incisor teeth are six in number; there are six molars on each side in the upper jaw, and five in the lower. The anterior one takes the form of an additional canine. This, however, is wanting in the Llamas. The upper lip is swollen and cleft in the center, and has a power of motion. It is used for feeling or examining the dry shrubby food on which these animals mostly live, before it is conveyed to the mouth. When in the midst of abundant pasture, they usually browse as much in an hour as serves them for ruminating all night, and for supporting them during the next day. But such pasturage they do not often find, and they are even thought to prefer nettles, thistles, cassia and other prickly vegetables to the softest herbage. They have seven callosities, or firm pads,

EXPLANATION OF PLATE VI.

FEET AND STOMACHS.

1. **Foot of an ox.** Bisulcated foot, or bifurcated hoof; cloven-footed; two-hoofed.
2. Camel's foot, showing the pad or cushion which prevents its sinking in the sand.
3. Llama's foot, showing the sharp hoof for climbing rocky hills.
4. Mole's foot, formed for digging or scooping out the earth.
5. Sloth's sharp, strong, crooked claws, for clinging to the branches of trees, on the under side of which they live suspended.
6. Foot of the Armadillo, *Dasypus*, fitted for rapid burrowing.
7. Lion's toe. A, represents the toe with claw sheathed. B, shows the retractile apparatus, with claw in same position. C, claw unsheathed. D, claw in same position, with tendons exposed.
8. Horse's foot; solipedes, hoofs whole, not cloven or divided. Solidungulate, one-hoofed.
9. Elephant's foot, showing the horny shoe enclosing all the toes.
10. Rhinoceros' foot; three toes on each foot incased in hoofs.
11. The fore foot or hand of a Seal, used as a fin for swimming.
12. A Dolphin's fore fin, flipper or paddle for swimming.

---

13. **Stomach of a ruminant or cud-chewing animal.**

A. Æsophagus or Gullet, expanding into the rumen or paunch.
B. Rumen or Paunch. It is the first stomach and much the largest in the adult animal, but small in the young.
C. Valve allowing the food to pass from the rumen into the reticulum.
D. Hood, Honey-comb-bag, Bonnet, or Reticulum. The second stomach.
E. Omasus, Manyplies, or Psalterium. This third and smallest stomach does not properly exist in the Camel or Llama.
F. Orifice of the Omasus.
G. Abomasus, the fourth stomach, the true organ of digestion, is next in size to the rumen or paunch. In calves it is the largest stomach, the milk passing from the gullet immediately into it. When salted and cured, this stomach of the calf is called rennet, and used for making cheese.

14. Cells of the reticulum or second stomach of the camel. These cells can be dilated so as to contain an unusual quantity of water.

PL. VI.

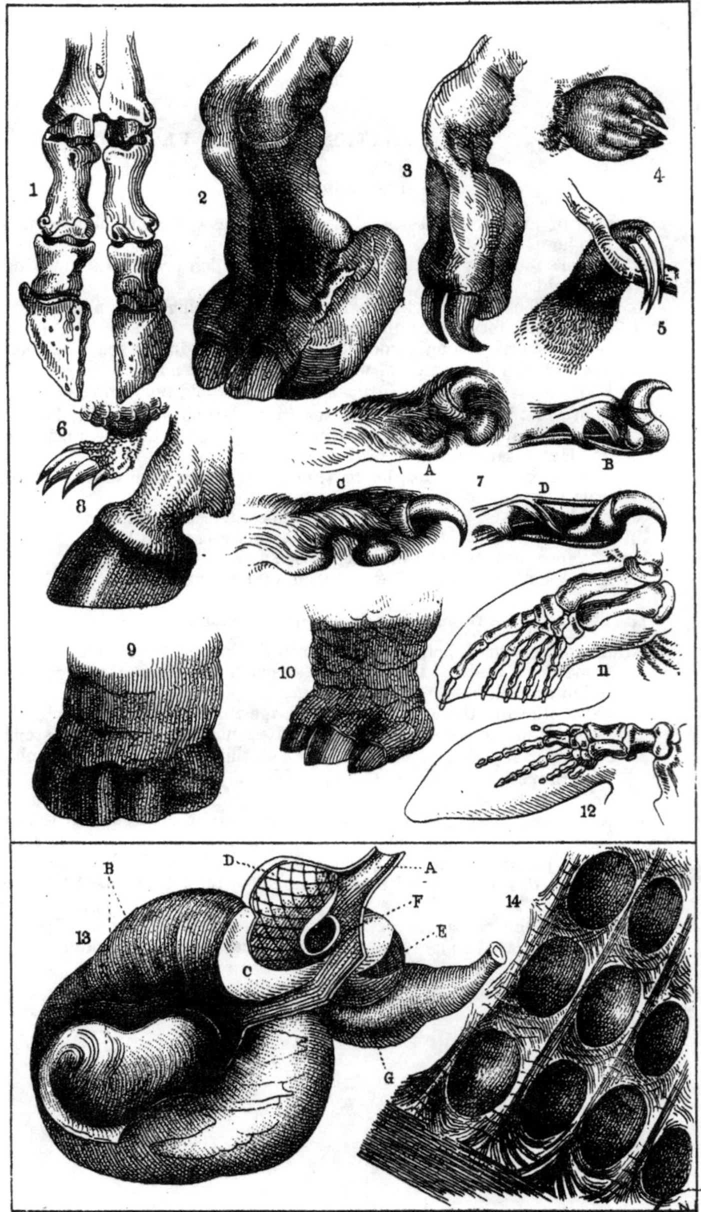

for the support of their burden; one on the breast, two on each of the fore legs and one on each of the hind ones. The toes do not present the true cloven figure, but are united underneath by an elastic pad, or cushion, connecting them together, but leaving the points free and separable, so that a larger surface thus comes into contact with the sandy earth. (Plate VI. fig. 2.) This, in connection with the elastic nature of the sole or cushion, enables the animal to tread with equal comfort over the yielding desert and the hard and arid plain.

The Camel has great difficulty in moving upon a soft and muddy soil, as it slips at every step. So great is its aversion to treading upon such soil, it is said, that its drivers "have been obliged to spread their tent coverings over the obnoxious ground in order to conceal its appearance and induce the animal to proceed." The step of the Camel is *noiseless*. "What always struck me," says the writer of a work on Constantinople, "as something extremely romantic and mysterious, was the *noiseless* tread of the Camel, from the spongy nature of his foot. Whatever be the nature of the ground, sand, or rock, or turf, or paved stones, you hear no foot-fall; you see an immense animal approaching you stilly, as a cloud floating in the air; and unless he wear a bell, your sense of hearing, acute as it may be, will give you no intimation of his presence."

The sense of hearing, in this animal, is very delicate. It seems greatly pleased with the sound of bells, and with the cheering song of its driver; its sense of smell, also, is remarkably acute. When the traveler across the desert is suffering from thirst, the camel, snuffing the gale, first indicates, by its dumb show, that the water is near of which the exhausted pilgrim must soon "drink or die."

The third stomach, or laminated *omasus*, of the Ruminants, is *wanting* in the Camels. The paunch or pannel is provided with a large number of cells, in order that water may be retained to serve the wants of the animal in case of extreme necessity. A longitudinal ridge of muscular fibres divides the paunch into two portions, the left containing a row of cells, which, (in the Arabian Camel,) holds four or five quarts of water; the right has a smaller series, holding about a quart. (See Plate VI. fig. 14.)

When the cells are filled, the fluid is kept from mixing with the food by the contraction of the orifice of each cell, and it can be forced out at pleasure by the action of a muscular expansion covering the bottom of the cellular apparatus. The deep cells of the *reticulum* are arranged in twelve rows, and are formed by muscular bands intersecting each other transversely. This compartment in the Camel appears destined exclusively as a reservoir of water, never receiving solid food, as in the ox and sheep. Sir E. Home is of the opinion that "Camels accustomed to journey for an unusual number of days without water, acquire the power of dilating their cells so as to make them contain a more than ordinary supply for their journey." When pressed with thirst, the Camel, by the contraction of the muscles, throws up water into the other stomachs, which serves to macerate its dry and simple food. As it drinks but seldom, it takes in a large quantity of water at a time; and travelers, when straitened for that article, have been often known to kill their camels for the water which they expected to find in them.

The large and prominent eye of the Camel enables it to take in a very extensive range; its vision is very keen, but the animal cannot look upward; in the horizontal position in which the head is carried, the brow overhangs the orb so as to shield it from the glare of the sun in a burning sky. The Camel has been called "the ship of the desert." Here the Simoon, or hot wind, blowing from the south-east, carries along with it dense yellow clouds of sand, which impede respiration, and are often suffocating to travelers. Even when the lighter winds blow, the fine particles of sand, driven along in volumes, and loading the atmosphere, would, to animals with wide and open nostrils, occasion the greatest suffering; but the nostrils of the camel being in the form of narrow oblique slits, which it can open or close *at pleasure*, it is, by breathing gently and gradually, enabled to exclude the suffocating mass. The Camel is full grown at the age of eight years. It generally lives forty years, sometimes much longer. It is said that instances have been known of Camels which have reached the age of one hundred years. The female has one young at a time which is suckled for a year. Her milk is described as rich, thick, and abundant, but rather strong in taste, though when mixed with water, it is a very nutritious diet.

The entire structure of this animal is wonderfully adapted to the region of its abode, and to the habits and uses of man. "The pad or sole cushion of the spreading foot dividing it into two toes, without being externally separated, which buoy up, as it were, the whole bulk, with their expansive elasticity, from sinking in the sand, on which it advances with silent step; the nostrils, so formed that the animal can close them at will, to exclude the drift-sand of the parching simoon; the powerful upper incisor teeth, for assisting in the division of the tough prickly shrubs and dry stunted herbage of the desert; and above all, the cellular structure of the stomach, which is capable of being converted into an assemblage of water tanks," must be included among the plainest and most striking evidences of the Creator's wise and benevolent care, as presented in the "Animal Kingdom."

The Camel combines within itself qualities, the possession of any one of which serves to render other quadrupeds absolutely necessary to human welfare. "Like the Elephant, it is manageable and tame; like the horse, it gives the rider security; it carries greater burdens than the ox and the mule; and its milk is furnished in as great abundance as that of the cow: the flesh of the young one is supposed to be as delicate as veal; the hair is more beautiful and in more request than wool; nay, there is scarcely a part of their frame of which it can be said, *it is useless.*"

C. Arabicus. The ARABIAN CAMEL.

Of the two species, this is the best known, and sometimes is called the Dromedary, or runner, (*C. dromedarius.*) Plate II. fig. 3. The term *dromedary* is, however, more strictly applicable to a lighter variety, *El-Heirie*, which is unfit for burdens, but employed when despatch is required. The Arabian Camel is more extensively used than the Bactrian; and from its constitution, appears able to endure, for a greater length of time, the fatigues and deprivations to which these animals are subjected. It is the wealth of the Arab, and nearly the only beast of burden in Turkey, Persia and the north of Africa. Having only a single hump, placed nearly in the center of the back, it is at once distinguished from the other species, which has two; it is also of a size and stature somewhat smaller, being from five to seven feet high at the shoulders. The muzzle is less swollen than that of the other species; the hair soft, woolly and very unequal, longest on the neck, the throat, and the hump. The color is always lighter than that of the Bactrian Camel, being, while the animal is young, of a dull, dirty white, but becoming, with age, of a reddish gray. The long woolly hair is woven into garments and tents, and the finer hair is imported into Europe for the manufacture of artists' pencils. The best is obtained from Persia. Of the varieties of this Camel, the Turkish and Arabian is the strongest and most hardy. In China there is a swift breed to which is given the poetical name of "the Camel with the feet of the wind."

The Arabian Camel is carefully trained, when young, to kneel and receive burdens. In temper, it is mild, submissive, docile and patient; but is said to be very obstinate when over-loaded, often refusing to rise if the burden is felt to be beyond its strength. Numerous caravans of these animals, each with a load of five or six hundred weight, and arranged in long rows, patiently pursue their toilsome way beneath a scorching sun, at the rate of about twenty-four miles a day; in some instances, fifty miles have been traversed in that time, but this could not be continued for successive days. Clapperton's Journal of Travels in the East, (continued by Lander,) after mentioning the arrival of five hundred Camels, with salt, from the borders of the Great Desert, says: "They were preceded by a party of twenty merchants, whose appearance was grand and imposing. They wore black cotton robes and trowsers, and white caps with black turbans, which hid every part of the face, except the nose and eyes. In their right hand they held a long and light polished spear, while with their left, they held their shields and retained the reins of the Camels. Their shields were made of white leather, with a piece of silver in the center. As they passed me, their spears glittering in the sun and their whole bearing bold and warlike, they had a novel and singular effect which delighted me. They stopped suddenly before the residence of the chief, and at the word, ('choir,') each of the Camels dropped on its knees, as if by instinct, while the riders dismounted to pay their respects." Sometimes, while attending caravans across the deserts, these animals of the swifter breeds perform the office of scouts, keeping a look-out for danger from wandering tribes and for the approach of the water stations. They will then travel from seventy to one hundred and twenty miles in twenty-four hours. The swift Dromedary has been known to perform a journey of six hundred and thirty miles in five days. It will continue at a long trot of eight or nine miles an hour for many hours together. A modern traveler, (see Morgan's Algeria,) says, it was often affirmed to him by the Arabs and the Moors that the express Dromedary "makes nothing of holding its rapid pace, which is a most violent hard trot, for four and twenty hours upon a stretch, without showing the least symptoms of weariness or inclination to bait; and that having swallowed a ball or two of paste made up of barley, and perhaps a little powder of dates among it, with a bowl of water or Camel's milk, if to be had, and which the courier seldom fails to be provided with, in skins, as well for the sustenance of himself as his Pegasus, the indefatigible animal will seem as fresh as at first setting out, and ready to continue at the same scarce credible rate for as many hours longer, and so on, from one extremity of the African desert to another." We know nothing of this animal in a wild condition, but in a domesticated state it has existed from the earliest times, in Egypt, Arabia, Palestine and the neighboring countries. Herds of Camels formed no small portion of the wealth of the scriptural patriarchs, (Job, xliii., 12,) and they are mentioned among the acquisitions of Abram on his first visit to Egypt, (Gen. xii., 16.)

The Camel was well known to Aristotle, and described by him in his "Natural History." Its native country extends from Mauritania to China, within a zone of one thousand miles in breadth. The Arabian Camel is found throughout the entire length of this zone, on its southern side, as far as Africa and India. It is numerous in the Canary Islands, to which it has been introduced, and found also in Pisa, Italy. After the conquest of Spanish America, an attempt was made to introduce Camels into that country; but the project was looked upon with disfavor by the "ruling Spaniards," and the animals gradually dwindled away.

Camelus Bractianus. The BACTRIAN CAMEL.

This species is found in the northern side of the zone above referred to, including the central portions of Asia and China and Thibet; occasionally it is seen in other countries. As already stated, it is easily known from the Arabian by its having two humps, one near the shoulders, the other near the croup. This is a stronger and heavier animal than the other species, and never used when dispatch is needed. It is larger than the Dromedary, being twelve feet in length and eight feet in height, between the humps. The hair is shaggy, particularly under the throat; the color generally dark brown, though variations occur in this respect, and also in respect to size, strength and fleetness, according to the breed and climate. The Bactrian Camel can carry a weight of twelve hundred pounds, but from five to eight hundred pounds is the usual burthen.

The Camels not long since purchased by the United States government, with a view of testing their utility in crossing the wide extended plains lying between the Mississippi valley and the Pacific ocean, are described by one of their superintendents "as very superior ones, presenting a far more sightly appearance than the miserable creatures which have been exhibited to crowds in the strolling menageries." He says, "their stride is about 3 feet in length, and with steady traveling they will average 3½ miles per hour. They do not kneel to receive their loads, as has been stated, at the word of command, but with a Kir-r-r, Kir-r-r, and a gentle pressure upon the neck, or a pull upon the halters, they assume the kneeling position." (Their not doing so, at the word of command, to receive their loads, may be owing to a defect in their early training.) The same writer remarks, "their cries are uttered to express their distress or dissatisfaction *at all times*. When half suppressed, they are the same as the lazy grunt of a hog whose repose is rudely disturbed; but when enraged, it is much more wild and greatly like that of a Bengal tiger when his keeper 'stirs him up with a long pole.' We had about 600 pounds of corn on each of them for the first day after leaving Howard's Ranch; but each day reduces it by feeding until we lay in another supply. They have worked admirably well so far, and promise to fulfill our most sanguine expectations in regard to the experiment."

Give the derivation of *Camelidæ?* What orders does this family link? What is said of their teeth? Which is wanting in the *Llama?* What is peculiar in the upper lip? For what used? What is said of their browsing? How many callosities? Where situated? What is said of the cushion on the foot, (see Plate VI. fig. 2.) and the aversion of the animal to mud? What results from the spongy nature of its foot? What is said of its senses of hearing and smell? Give some account of the cells of the stomach and their uses, (Plate VI. fig. 14.) In what direction does it look? How is the *Camel* protected against the effects of the *simoon?* How long does it live? In what respects does it show the wise and benevolent care of the Creator? What qualities does it combine? Which of the two species is best known? What is it called? What is said of its powers of endurance? How is it distinguished from the other species? Where is the *Bactrian Camel* found, and how does it differ from the Arabian C.? How large a burden does it usually carry? What is the habitat of the C.? What is said of the Camels purchased by the U. States?

Trace the genera of the Camel family upon the Chart, giving their specific names and a synopsis of each animal.

SECTION XXIII.

Auchenia. (Gr. 'αυχὴν, *auchen*, a neck.)

THE LLAMA.

We have in this genus the Camels of the Western Continent, inhabiting the Cordilleras of the Andes below the line of perpetual snow. They are found principally in Peru and Chili, though in much fewer numbers than formerly. Sometimes they have been taken into Mexico, but rather as curiosities than for any other purpose. The Llamas were first noticed at Rio Bamba, about ninety miles south-west of Quito, and not far from the snow-capped mountain of Chimborazo; and at this very spot, they are now seen in considerable numbers. Rio Bamba is 11,670 feet above the sea-level, and the temperature of the air corresponds to this elevation. But these animals, as many as five hundred in a herd, are found at elevations still higher,—say from 13 to 16,000 feet above the level of the sea, and where the mercury falls every night below the freezing point. They do not, however, advance so high as the line of perpetual snow, preferring rather a middle region affording congenial temperature and food. As a protection against the cold of their elevated abodes, they are clothed with a long and woolly fur. The name *Auchenia* refers to the long slender neck of these animals, in which they resemble the Camels proper. They are also like them in the great cellular development of the second stomach; the cellular apparatus of the paunch; the absence of the third or plicated stomach, and the concomitant power of enduring thirst, or rather abstaining from water altogether; in the large, full, over-hung eye; the division and mobility of the upper lip; and the fissured form of the nostrils, and the meagre limbs. Contrasting the location of these animals with that of the Camels, we naturally look for a deviation in the structure of the foot. The pad which connects the toes of the Camel beneath, would have afforded no very sure footing for an animal destined to climb the precipices of the Andes. We accordingly find in the Llamas toes which are armed with strong nail-like hoofs, (Plate VI. fig. 3,) completely separated from each other, and each defended with its own pad or cushion,—thus admirably adapted to firm progression either in ascent or descent; while there is nothing in the structure to hinder great rapidity upon comparatively smooth and level ground. The humps of the true Camels are not found on the backs of the Llamas, yet there is said to be in the latter a conformation resembling those excrescences, and "consisting of an excess of nutritious matter, in the shape of a thick coat of fat under the skin, which is absorbed as a compensation for want of occasional food."

The genus *Auchenia* is now generally considered to include three species, viz., the *Auchenia Llama*, or Guanaco, which is used for burdens, and the *Auchenia Alpaca* and the *Vicugna*, which are raised for their flesh and wool. Cuvier regards the Paco or Alpaca, as a variety of the Llama, with the wool more amply developed, but the Vicugna as a distinct species.

Llama is the common term with which the Peruvians designate their sheep. The wild Llama is usually of a deep rich fawn, verging to white on the under parts. The wool is long and shaggy, but shorter on the neck and limbs than on the body. The long slender neck is "held erect and swan-like;" the head is small; the lips are thick; the eye large and brilliant. On the breast, there is a bunch which constantly exudes a yellowish oily matter. The length of this animal is six feet; the height at the shoulders about four. The reclaimed Guanaco or domesticated Llama, is greater in size than the wild animal; the body is slender, and the limbs more muscular; the wool smoother and closer;—the physiognomy is no longer wild and independent, and its air betokens mildness and subjection. The color is white, brown, black, and sometimes mixed or piebald. Its step is slow and regular, and it has not the strength or energy of the wild Guanaco,—carrying at the most but one hundred and fifty pounds. Under this load, however, it will travel with firm and sure step fourteen or fifteen miles a day, along rugged mountain-passes and the narrow ledges of precipitous rocks; but if loaded too heavily, or urged beyond its wonted pace,—camel-like, it lies down and refuses to move another step. All that the conductor can do, in such a case, is to sit down by the animal and wait until "by his blandishments, he prevails on it to rise spontaneously." The difference in weight and speed between this animal and the Camel, the Peruvians make up in the great numbers which they use of these beasts of burden,—one drove sometimes including more than five hundred that subsist in traveling as they are able. Formerly these animals were used in bringing down the products of the mines, and 300,000, it is said, were once employed in the mines of Potosi alone. Mules, however, are now chiefly used for that purpose; though the Llama is still employed to some extent; its labor involving less expense. The white Llama is said to have been the presiding divinity of the natives of Callas before that province was annexed to the empire of the Incas; the "priests of the sun" sacrificed it, at stated seasons, to the orb of day. Frequently the Llama, but preferably the Alpaca, was a pet in the Indian's cabin. In intelligence these animals rank high among the ruminants. As regards patience and resignation, they are said to equal the ox, while in point of sensibility, they are unsurpassed by any other quadrupeds. The size and shape of the eye indicate a strong and quick sight as well as a peculiar capacity for bearing the reflection of the sun's rays in the same manner as the Camel resists that glare of the sands which in man so often produces *ophthalmia*. (Gr. blearedness of the eyes.)

Auchenia Llama. The GUANACO.

The animals of this species are rather larger than sheep, but smaller than heifers. Their compact bodies, their long legs, and their feet having toes armed with nail-like hoofs, fit them for dwelling in their wild state, among crags and precipices, where the hunter would be foiled if he dared to venture. Vast herds of wild Guanacos associate, during the summer, free as the air,—feeding upon the herbage of their elevated abode, and the grass or rush called *icho*, which covers the mountain slopes. As long as green and succulent vegetables can be procured, the animal never drinks. The cells of the stomach in this animal probably retain the moisture of the masticated vegetables for the necessities of the system, perhaps even adding to it by a liquid secretion of their own. It is a proof of Divine Providence that formed to dwell in such regions, the Guanacos are not only able to live without water, but if they can obtain their natural food, do not even require it.

Auchenia Alpaca. PACO, or ALPACA LLAMA.

The great peculiarity of this species is its long, fine and silky wool, covering the neck as well as the other parts of the body. The staple of our common wools is not more than six inches long; but that of the Alpaca averages from eight to twelve, and sometimes reaches twenty inches; acquiring strength without being accompanied by coarseness,—the reverse of which occurs in other woolly tribes. Each filament, or thread, appears straight, well formed, and free from crispness; and the quality is more uniform throughout the fleece. There is also a glittering brightness upon the surface, which gives it the glossiness of silk, especially, when it comes out of the dye-vats. It is distinguished by softness, an essential property in the manufacture of fine stuff; and being exempt from spiral, curly and shaggy portions,—when not too long, it spins easily, and yields an even and true thread. It is, besides, less liable than other wool to form *knots* difficult to unravel; it is not injured by keeping, nor does it lose in weight; and it is less subject to injury from moths, as the following fact will show. A small bundle of Alpaca wool, with a few locks of other wool mixed with it, was accidentally thrown into a closet and forgotten. At the end of twelve months, the closet was opened, when it appeared that the moths had nearly eaten up the common, without injuring, at all, the Alpaca wool. The fleeces of Alpaca wool range, in Peru, from ten to twelve pounds each, whereas, "those of our full sized sheep seldom go beyond eight pounds, and the small species four pounds." As far back as the days of Philip II., efforts were made to introduce these animals into Spain; but failed through the intervention of war. They were taken to France, in the days of Napoleon I., where they have found a congenial climate; and they have lived to their full period in the low lands of Spain. They have also been bred in Hamburg and in England, where the wool seems to improve. The staple of some Alpaca wool from the Earl of Derby's flock, was exhibited in England some ten or twelve years since. This appeared about a foot long, and it was estimated the animal had seventeen pounds of it on his back.

The meat of the Alpaca has been compared to "venison, and even heath-fed mutton." Its quality could hardly fail to be good as the animal eats nothing but the purest vegetable substance, and in habitual cleanliness, is said to *surpass every other animal.* The Alpaca is also far less subject to disease than sheep; and as it seldom perspires, the fleece does not require washing before it is taken from the back.

It has extraordinary foresight of storms, and power to contend with them, so that, in its native climes, seldom is one missing after a tempest.

The first marketable fabric made from the wool of this animal, was presented at Greetland, near Halifax, (Eng.,) about twenty-five years ago. It was there sold, at a very high price, in the form of ladies' carriage shawls and cloakings, as curiosities. The quantity manufactured and used since that time, has steadily and greatly increased. From Alpaca wool, plain and figured stuffs are produced, which have a beautiful luster. The difficulty which was at first found in dying it, being now overcome, the most delicate colors are obtained, such as royal blue, scarlet, green and orange, as seen in the *mousselines de laines*, and other ladies' dresses now in use. The blacks are superior, and the damask patterns very showy in their appearance. In some instances, Alpaca takes the place of Angola, or goat's hair wool; and in France, it has been used for cashmeres and merinoes. English capitalists have introduced the animal into the colony at the Cape of Good Hope, where it has succeeded well, the shearing yielding eight and a half pounds a sheep. Alpaca goods are, to some extent, manufactured, and largely imported, and used in the United States.

Alpaca Vicugna, or *Auchenia Vicuña*. THE VICUNA.

This is a much hardier animal than the Guanaco. It inhabits ranges nearer the line of perpetual snow, where the cold is intense, and is rather pleased than annoyed by snow or frost. In size, it is less than the Guanaco. The wool is of a pale yellowish fawn color, and exquisitely fine, having a texture which may be termed *silken*. It is used for manufacturing expensive shawls and other articles of dress. For the sake of it, eighty thousand of these animals, it is said, are killed every year. They are not unlike goats, except that they are larger, and have more horns. The Vicunas are found in flocks, appear timid, and flee at the sight of men and of wild beasts.

What is the habitat of the *Llama?* Where was it first noticed? To what does the name *Auchenia* refer? In what particulars do they resemble the Camels proper? What deviation is there in the structure of the foot? (Pl. VI, fig. 3.) Has it the humps of the Camel? How many species are included in the genus *Auchenia?* What are their respective uses? How did Cuvier regard the *Alpaca* and *Vicugna?* How do the Peruvians use the term Llama? Describe the wild Llama? What is the size? What is said of the reclaimed Guanaco as compared with the wild Llama? What of its uses as a beast of burden? What of the *White Llama?* How do these animals rank in intelligence? What is the size of the *Guanaco?* Upon what does it feed? What proof of divine providence is referred to? What is the great peculiarity of the *Alpaca Llama?* What is said of its wool? How early were attempts made to introduce it into Spain? When were they taken to France? In what other countries have they been bred? What is said of the cleanliness of the *Alpaca?* When and where was the first marketable fabric made from the *Alpaca* wool? Where is most of the spinning and weaving of this wool now performed? What is further said of its manufacture? How does the *Vicugna* compare with the *Guanaco* in hardiness and size? What is said of its wool? What animal does it resemble?

What Llamas are named upon the chart? What is said of them? Trace them?

SECTION XXIV.

CAMELOPARDAE. (Gr. κάμηλος, *kamelos*, a camel; πάρδαλις, *pardalis*, a leopard.)

THE CAMELOPARDS.

These singular and beautiful ruminants, in their general structure, most nearly approach the Deer, but have points of resemblance also to the Antelopes and Camels, besides striking peculiarities of their own. They have persistent horns, common to both sexes, and are the tallest of all known quadrupeds; frequenting the wooded plains and hills that skirt the arid deserts, or the verge of mighty forests where groves of mimosa trees beautify the scenery.

Camelopardalis Giraffe is the sole species, including two varieties,—the one native to Nubia, Abyssinia, and the regions adjacent, and ranked by Swainson as a distinct species,—the other, found in Southern Africa.* (See Chart.)

The general characters of the Giraffe are the following, viz., "Lip not grooved, entirely covered with hair, much produced before the nostril; tongue very extensile; neck very long, and having a short thick mane; body short; hind legs short; false hoof none; tail elongate, with a tuft of thick hair at the end." This animal at once impresses the beholder with its towering height, varying from fifteen to twenty feet. The males are generally fifteen or sixteen, and the females thirteen or fourteen feet in height, and their young at birth, six feet. Its thickness is not what might, perhaps, be expected from the height. In order to support its very long neck, (but having only the number of bones found in the human neck,) the withers are elevated; the spinal processes of the vertebrae are prolonged to meet the elastic ligament which runs along the neck, and assist to keep it in its natural position. It is said above,—"the hind legs are *short.*" This describes them as they *appear;* but in reality the front and hind legs are about the same length; the

---

* The Commentator on the "Pictorial Bible," where a good cut of the Giraffe is given, says, with reference to the word *Chamois*, used, Genesis iii., 21, "The Arabic version understood that the word Giraffe is meant here, which is very likely to have been the case, for the Chamois is not met with so far to the Southward as Egypt and Palestine." The Jews had, probably, many opportunities of becoming acquainted with the animal while in Egypt, as had also the seventy (translators of the Septuagint) who resided there, and who indicate their knowledge of it in their translation of the Hebrew name.'

thighs in front are so long in comparison with those behind, that the back of the animal seems inclined like the roof of a house; and this gives to it an appearance of unwieldiness and unfitness for active movements. But the seeming drawbacks related to its structure and condition, are balanced by marked and peculiar advantages. A man on horseback can, without stooping, ride under the body of the animal,—the height to the tip of the shoulder being ten feet. *Why* that neck of prodigious length? *Why* the disproportioned height of the fore and the hind parts of the body, giving to the animal its appearance of unwieldiness and clumsiness? The answer is,—the animal derives a large part of its food from the leaves of trees, particularly the *mimosa*,—a species of *acacia*, called *acacia giraffe*. The peculiarity of the Giraffe's form enables it to reach the high branches which are uncropped, because above the reach of ordinary animals; and a shorter neck, on the other hand, would not have allowed it to reach the earth in districts where woods are less common. In reaching the high branches, it is also aided by the *tongue*, which has the power of motion in such a degree, accompanied with the faculty of extension, that it performs "the office of the proboscis of an elephant in miniature." This organ may be extended seventeen inches after death, but in the living animal, can be so diminished in size as to be inclosed within its mouth. According to Sir Everard Home, its actions depend on the combined powers of muscular contraction and elasticity; its increase and diminution of size arising from the blood vessels being at one time loaded with blood, and at another empty. The Camelopard seizes the foliage with its long and narrow tongue, using it as a prehensile organ, and a beautiful accessary to the other parts of the structure,—rolling it around the object with considerable pliability.

The tongue is used as an organ of examination, for the power of prehension is so great, that when extended to the utmost, it can grasp an ordinary lump of sugar, of which the animal seems very fond. He retroverts the tongue for the purpose of cleansing the nostrils,—an office which its flexibility enables him to perform in the most perfect manner. The tongue, it is said, can be so tapered as to enter the ring of a very small key. The eyes are large and prominent, and soft and gentle in their expression; the ears large and spreading; the lips, especially the lower one, being movable; the head is small, but elegantly modeled, tapering to the singularly narrow muzzle, with a well-formed mouth.

Both the male and female Camelopard have horns,—*not* such as are periodically shed and renewed; nor yet true and prominent horns, like those of the Antelope, but consisting of two porous, bony substances, about three inches long, with which the top of the head is armed, placed just above the ears, and crowned with a thick tuft of stiff upright hairs; a considerable protuberance also rises in the middle of the forehead, between the eyes. By some, these horns muffled with skin and hair, are said to be "useless as instruments of defence,"—others say,—"We have seen them wielded by the males against each other with fearful and reckless force." The Giraffe does not butt by depressing and suddenly elevating the head; but strikes the callous obtuse extremity of the horns against the object of his attack with a sidelong sweep of the neck. The imperfection of the horns has been plausibly ascribed "to the state of the circulation of the blood in the arteries of the skull." The long neck is supposed to *impede* the circulation, so that the vital stream ascends with difficulty,—it rises slowly, in more moderate quantity, and is "inadequate for a supply of osseous matter, remarkable either for its abundance, or its rapid elaboration." Who does not see the wisdom of this ordering? What could the long-necked Camelopard do with the ponderous horns of the Moose, or the Wapite? "It is not for nothing that the neck is elongated, that the head is light, and the tongue made flexible;—it is not without *design* that the horns are rudimentary; for such modifications the instincts and the habits of the creature demand; the one part involves the other." Professor Owen has noticed a further beautiful provision in this animal, which is, that its nostrils are provided with cutaneous sphincter (Gr. σφιγγω, *sphingo*, to constrain) muscles, and can be shut at will, like the eyes. He supposes that the object of this mechanism, is to keep out the sand when the storms of the desert arise.

The hair of the Giraffe is short and close; the ground color of a light grayish fawn, marked with numerous triangular spots, with a darker hue, less regularly shaped on the sides than on the neck and shoulders. The Northern variety of the animal is of a paler color than the Southern.

The eyes of the Giraffe are so placed that he can see much of what is passing on all sides, and even behind, without turning the head. Hence it is difficult to approach him; and when surprised or run down, he directs most accurately the rapid storm of kicks with which his defence is made. Ordinarily, however, this animal seeks safety in flight. Its motion is extremely rapid, especially along rising ground; but cannot be maintained for a sufficient time to enable it to escape from the Arab mounted on his long-winded steed. The pace is an amble; the animal moves two legs on each side at the same time, but when put in motion, it can, for a while, keep a horse at a pretty smart gallop. The lamented Anderson says, in his "Lake Ngami,"—"It is a curious sight, a troop of Giraffes at full speed, balancing themselves to and fro in a manner not easily described; and whisking, at regular intervals, from side to side, their tails, tufted at the end, while their long and tapering necks, swaying backward and forward, follow the motion of their bodies. They are so long-winded, that a swift horse seldom overtakes them under less than two or three miles." The author of the "Menageries" remarks,—"Until the year 1827, when a Giraffe appeared in England, and one in France, the animal had not been seen in Europe since the 15th century, when the Soldan of Egypt sent one to Lorenzo De Medici, which was familiar to the inhabitants of Florence, where it was accustomed to walk at ease about the streets, stretching its long neck to the balconies, and first floors, for apples and other fruits, upon which it delighted to feed." In 1836, four Giraffes were introduced into England by the Zoological Society, at an expense of between eleven and twelve thousand dollars. One of them soon died; but the others lived, and one of the females had several young ones, which were sold and taken to different parts of the world. In our own country, the Camelopard is often exhibited. The animal, it is said, is often seen in a tame state, at Grand Cairo, in Egypt, and is found figured in the sculptured remains of that country. Pompey the Great exhibited in the theatre, ten of these animals, which he had brought from the scenes of his military enterprise. His rival, Julius Cæsar, also exhibited them. After him, several Roman Emperors showed them in the public games and processions. All these were probably obtained from the northern or north-eastern part of the African Continent, and by way of Egypt.

What is said of the structure of the *Camelopards?* How many varieties and where found? What are the general characters of this animal? What its size? How is its long neck supported? Are its hind legs really shorter than its fore legs? What compensation is referred to? What aids it to reach high branches? What is said of the tongue? Has this animal horns? Why are they imperfect? What provision is noticed by Professor Owen? How do the varieties differ in color? How do the eyes of the *Giraffe* assist him in self-defence? What is the remark of Anderson? What more is said?

What is said on the chart of its size?

SECTION XXV.

MOSCHIDAE, or MUSK DEER. (Gr. μόσχος, *moschos*, a Musk.)

These are so called, from the fact, that one species yields the well-known perfume, called *musk*. According to Cuvier, "they are much less anomalous than the Camels, and only differ from the other Ruminants in the absence of horns, in having a long canine tooth on each side of the upper jaw, which comes out of the mouth in the males, and, finally, in having in their skeleton, a slight *fibula*, (clasp, or connecting link,) which has no existence in the Camels."

The distinction of the other canine tooth noticed by Cuvier, is not, however, confined to the Musks,—as some of the males of other deer, the Muntjak, for example, show a similar formation; that of the *Mōschus moschiferus*, (Lat. musk-bearing,) is three inches long. In general form, the Musk deer differ only a little from other Deer; but the body is rounded and stouter, and the neck shorter,—the head is not carried erect, and the bearing not so bold; the limbs are more tapering, and the hind quarters considerably elevated; the face is narrow and lengthened, and they are destitute of horns. None of them have tear openings, or tufts of bushy hair on their legs, like the other deer. They have large, dark and brilliant eyes, rather small ears, and short tails; they have also front and hind hoofs,—the front hoofs being long, narrow and pointed, the hind ones high set, small and conical. In the *true musks*, however, the hoofs are broad and expanded; the hind ones large, almost touching the ground. Besides the true and celebrated Musk Deer, the family includes four other species, one found in Ceylon, and three in Java, including the *smallest*, and according to some, the most elegant of the Ruminants.

Moschus moschiferus. The THIBET MUSK. (Plate VII. fig. 1.)

This is a mountain animal,—timid, shy, and a lover of solitude, having somewhat the form of a roebuck, but thicker and more clumsy. It is six inches higher behind than at the shoulder, where it measures about two feet three inches. The ears are long, and rather narrow; in the inside, pale yellow, and dark brown, outside. The hair is long, coarse and harsh, and mixed with brown yellow, and whitish, which produces a dark red tinge on the back, fading off to whitish beneath,—the tail is nearly rudimentary, and covered by the hair; a tuft hangs on each side from the lower jaw. This animal being extremely cautious, and

PL. VII.

EXPLANATION OF PLATE VII.

DEER.

1. The Musk-Deer, *Moschus Moschiferus.*
2. The Common Stag, or European Red Deer, *Cervus Elephas.*
3. The Moose, Flat-Horned Elk, or Black Elk, *C. alces.*
4. The American Elk, Round-Horned Elk, or Wapiti Deer, *Elephas Canadensis.*
5. The Caribou, or American Reindeer, *C. rangifer* or *R. Tarandus.*
6. The Fallow Deer, *C. Dama* or *Dama vulgaris.*
7. The Roe-buck, *C. Capreolus* or *Capreolus Dorcas.*
8. The Muntjak, *C. vaginalis.*

OX.

9. The Musk Ox, *Ovibos Moschatus*, Little Bison of the Chipewyans and Copper Indians.

SHEEP.

10. The Mouflon, *Ovis musimon* or *Caprovis musimon*, Wild Sheep or Siberian Goat of Pennant.
11. The Argali, or Wild Sheep, *Caprovis Argalis.*
12. The Many-Horned Sheep, *Ovis polycerata.*

GOATS.

13. The Syrian Goat, *Capra Syriaca.* Its large pendulous ears are from one to two feet long, and at times so troublesome that the owners are obliged to trim them. Amos iii. 12.
14. The Ibex, *Capra Ibex.*

ANTELOPES.

15. The Kudoo, *Antilope strepsicoros.*
16. The Blessbok, *A. albifrons.*
17. The Prong-Horned Antelope, *A. Americana* or *Antilocapra A.*
18. The Common Antelope, or Sasin, *A. cervicapra.*
19. The Dorcas Gazelle, *A. Dorcas*, or *Gazella Dorcas.*
20. The Oryx, *A. Oryx*, or *Oryx Gazella.*
21. The Chamois, or Gems, *A. rupricapra*, or *R. Tragus.*
22. The Mhorr, *Gazella Mhorr*, or *A. Mhorr.*
23. The Gnu, or Gnoo, *A. Gnu*, or *Catoblepas Gnu.*
24. The Bekker-el-Wash, or Wild-Ox of the Arabs, *A. Bubalis*, or *Alcephalus Bubalis.*

residing among broken crags and precipices covered with pines, is yet eagerly, and often with peril of life, hunted for its perfume, peculiar to the male alone. Its habits are similar to those of the Chamois,—it climbs and bounds over the Alpine ridges of Central Asia with astonishing activity, assembling in herds, and sometimes in considerable numbers. Occasionally, it is killed with a cross-bow, a string having been set in the path of the animal. The bag containing the perfume, is kidney-shaped, and about the size of a hen's egg. It has two openings, the larger one oblong, the smaller round, and covered with hair. The musk, on the application of pressure, may be driven through the openings,—it is an unctuous, dusky red substance, and when dry, is more or less granulated. The hunters cut off the bag and tie it up for sale; but like many other articles of commerce, it is often adulterated by the addition of blood and other matter, and pieces of lead have sometimes been found enveloped in it for the purpose of increasing the weight. The quality and quantity of the musk in a given bag vary, according to the age of the animal. To the taste, it is bitter, and somewhat acrid. No substance is known to have a stronger, or more subtle and permanent smell. It strikingly illustrates the extreme divisibility of matter, for a single grain of it will perfume a whole room, and its odor continue for days without any diminution. When once introduced, it is exceedingly difficult to destroy its perfume. Vessels of silver do not for a long time part with the scent of musk that has been placed in them. When exposed in large quantity, its effect is really violent upon the nervous system; blood has been forced from the nose, eyes and ears of those who have imprudently inhaled a large amount of the vapor. Purchasers of the article sometimes secure themselves from the sudden effects of the smell by covering the face with a handkerchief several times folded. For nervous diseases and convulsions, it has been used as a medicine. Orientalists make warm winter dresses for themselves out of this animal's skin, with the fur preserved; they also prepare from it a soft and shining leather. The Romans and Tartars even eat the flesh, though that of the male is highly flavored with musk.

Moschus Meminna. The MEMINNA.

This beautiful little Musk, about seventeen inches in length, and weighing only five and a half pounds, is a native of Ceylon, frequenting woods and groves, but never found in the plains. It has large dark eyes, and smooth shining hair, of an olive color, clouded with reddish about the limbs. The sides are dappled with interrupted lines and irregular dots of white; the throat and chest are also white, and from the former, two lines of the same color on each side radiate backwards, the lower one extending to the shoulders. (This peculiar marking specifically varied in a slight degree, characterises the remaining species of this genus.)

Moschus Napu. The NAPU, or CHEVROTAIN.

This Musk Deer is a native of Java, and is about the size of a rabbit,—the legs are scarcely as thick as a common quill; the general color is a uniform ferruginous brown, clouded with black; and the animal has throat marks as above referred to.

To this species, Sir Stamford Raffles has given the specific name *Javanicus.* He remarks, that it "frequents thickets near the, sea-shore; and feeds principally upon the berries of a species of *Ardisia;* can be easily trained when taken young, and will become quite familiar."

Moschus Kanchil. KANCHIL MUSK DEER.

This is by some regarded as the most elegant, as it is one of the *smallest* of the *Ruminantia,*—and is also found in Java and Sumatra. Its height is about nine inches; its length, fourteen. The color is a deep yellow brown, approaching to black on the back, a bright bay on the sides, and on the under parts white. The markings of the throat have the upper line of white extending from the face to the shoulder, differing in this respect from those of the Napu. It has long canine teeth, and a tail tufted and white at the tip. Berries and wild fruit constitute its food. Among the Javanese, it is said to have a reputation for strategy similar to that of the fox. A Malay proverb describes a great rogue as being "as cunning as a Kanchil." "If taken in a noose laid for it, the Kanchil, when the hunter arrives, will stretch itself out motionless, and feign to be dead; and if, deceived by this manœuvre, he disengage the animal, it seizes the moment to start on its legs, and disappears in an instant." A still more singular expedient is mentioned, viz., "that when closely pursued by the dogs, the Kanchil will sometimes make a bound upwards, hook itself on the branch of a tree by means of its bent tusks, and there remain suspended, until the dogs have passed beneath."

Linnæus placed tbe Musk Deer between the Camels and Deer. Swainson places them between the Camelopards and Deer.

Why are the *MuskDeer* so called? What is Cuvier's remark respecting them? What is said of their general form, &c.? How many species does the family include? What is said of the size of these animals? Describe the *Thibet Musk?* What is its great peculiarity? What shows the powerful nature of the Musk? What property of matter does this illustrate? To what medicinal use has it been applied? Give some account of the *Meminna?* What peculiarity has it in common with the remaining species of this genus? Where is the *Napu* found? What does Sir Stamford Raffles say of it? Where is the *Kanchil* found? What is its size? What reputation has it among the Javanese? What Malay proverb is mentioned? How is its cunning illustrated? Where did Linnæus and Swainson place Musk Deer? Where are they placed on the Chart?

SECTION XXVI.

SOLID-HORNED RUMINANTS.

CERVIDAE. (Lat. *Cervus*, a stag.) The DEER FAMILY.

We come now to a group of animals which have been ever greatly admired. They seem, many of them, to have been formed to embellish the forest, and impart animation to the solitudes of nature. In their internal structure, they closely resemble the ox, but they are "without the gall-bladder; the kidneys are formed differently; and the spleen is larger in proportion to the size of the animals." Of the genus *Cervus*, the general characters are simple. Incisor teeth are found, eight in number, in the lower jaw alone; the grinders are six on each side above and below; the canine teeth are generally *wanting*. The pupils of the eye are elongated, and below the inner angle of the eye, there is a deep *fossa*, or opening, generally known as the *lachrymal sinus.* In some, this opening, called by the French, *larmiers*, (from Fr. *larme*, a tear,) is of considerable size. It has been supposed "to communicate with the nostrils, and assist them in maintaining respiration, during great exertion or swiftness;" but its use is not fully ascertained. The cavity secretes a wax-like substance, which sends forth a strong odor. The ears are large and pointed; the tail short; the legs slender; and the feet bisulcated. The horns, or antlers,—excepting in the case of the Rein Deer, found alone in the males,—are solid, and in a large part of these animals, annually shed and renewed. "The form of the horns is various. Sometimes they spread into broad palms, which send out sharp snags around their outer edges; sometimes they divide fantastically into branches, some of which project over the forehead, whilst others are reared upward in the air, or they may be so reclined backwards, that the animal seems almost forced to carry its head in a stiff, erect posture; yet, in whatever way they grow, they appear to give an air of grandeur to the animal." The geographical range of the Deer includes the entire globe, with the exception of Australia and Southern Africa. The species found in the colder regions, are generally marked by superior size, and a greater development of the horns; and by having a broad muzzle covered with hair.

The production, loss and renewal of the antlers of this family of quadrupeds, are among the most remarkable phenomena of animal physiology. The subject is treated with great ability and clearness in W. C. L. Martin's work on the *Mammalia*, from which we extract the following:

"The horns are seated upon an osseous peduncle, or footstalk, rising from each frontal bone at its central point of ossification,—these peduncles are enveloped in skin. It is not until in the spring, or beginning of the second year, that the first pair of horns begin to make their appearance. At this epoch, a new process commences, the skin enveloping the peduncle swells, its arteries enlarge, tides of blood rush to the head, and the whole system experiences a fresh stimulus. The antlers are now budding," for, on the top of their footstalks, the arteries are depositing layers of osseous matter, particle by particle, with great rapidity. As they increase, the skin increases in an equal ratio, still covering the budding antlers, and continues so to do until they have acquired their due development and solidity. This skin is a tissue of blood vessels, and the courses of the large arteries from the head to the end of the antlers are imprinted in the latter in long furrows, which are never obliterated. In ordinary language, the skin, investing the antlers, is termed *velvet*, being covered with a fine pile of close short hair. Suppose then, the antlers of the young deer, now duly grown, and still invested with this vascular tissue; but the process is not yet complete. While this tender velvet remains, the deer can make no use of his newly acquired weapons, which are destined to bear the brunt of many a conflict with his compeers; it must, therefore, be removed; but without giving a sudden check to the current of blood rolling through this extent of skin, lest, by directing the tide to the brain, or some internal organ, death be the result. The process then is this:—As soon as the antlers complete, (according to the age of the animal,) the footstalk, always covered with skin, they begin to deposit round it a bone, or rough *ring of bone*, with notches, through which the great arteries still pass. Gradually, however, the diameter of these openings is contracted by the deposition of additional matter; till, at length, the great arteries are compressed as by a ligature, and the circulation is effectually stopped. The velvet now dies for the want of the vital fluid; it shrivels, dries and peels off in shreds, the animal assisting in getting rid of it by rubbing his antlers against the trees. They are now firm, hard and white; and the stag bears them proudly, and brandishes them in defiance of his rivals. From the burr upwards, these antlers are no longer part and parcel of the system,—they are extraneous, and held only by their mechanical continuity with the footstalk on which they were placed; hence their deciduous character; for it is a vital law, that the system shall throw off all parts no longer intrinsically entering into the integrity of the whole,—an absorption process soon begins to take place just beneath the burr, removing particle after particle, till at length the antlers are separated and fall by their own weight, or by the slightest touch, leaving the living end of the footstalk exposed and slightly bleeding. This is immediately covered with a pellicle of skin which soon thickens, and all is well. The return of spring brings with it a renewal of the whole process, and a finer pair of antlers branch forth."

The rapidity with which this firm mass of bone is secreted, is worthy of particular notice. The budding horns of a male Wapite, are several inches high in ten days from their first appearance; a month afterwards there is an interval of two feet between them, measuring from branch to branch. When the process is ended that completes the horn, the deer seems conscious of his strength, and goes forth prepared to encounter any creature, even man himself, that may dare to invade his haunts. Thus he continues for a season,—but when he again sheds his horns, betakes himself to the recesses of the forest until they are replaced. The Common Stag sheds his horns about the end of February, or in the month of March; the Fallow Deer from the middle of April to the first week of May. In the Stag, the horns do not appear until the second year. The first shed, is straight, or single, like a small thrust sword or dagger,—whence the young male is termed *Daguet*, (Fr. *dague*, a dagger,) by the French; the next horn has commonly but one antler; the third has two, and sometimes three; the *fourth* has three or four, sometimes five or six. Up to this time, the animal is called a Young Stag,—the fifth horn has five or six antlers; the sixth is shed when the animal is about seven years of age. In addition to the growth of antlers, the horns become larger, have the furrows more marked, the burr more projecting; and the supports of the horns become, every year, shorter and wider. By these signs, the age of the animal, from eight years and upwards, is determined. After the seventh year, there is no fixed rule as to the antlers. They are multiplied towards the summit of the beam, where they are united into a sort of crown, and are said to be *palmated*. The oldest have not usually more than ten or twelve antlers; though it is said some have borne the enormous number of *thirty-three*. (See Plate VII. fig. 2.)

Deer are remarkable for the acuteness of their hearing and smelling, and it is therefore very difficult for the hunter to approach them when he follows the course of the wind. They are very nice in choosing their food, and will not eat that which has been handled or touched by any foreign substance. The flesh of many of these animals, as is well known, is used for food, and familiarly known under the name of *venison*. Strong and lasting leather is made from their skins. According to Dr. De Kay, (N. H. S. N. Y.,) this family "comprises forty-five real or nominal species, distributed, according to the ideas of systematic writers, into eight or ten genera. But six species are found within the United States, and of these three only exist in the State of New York."

Elaphus Canadensis, or *C. Canadensis*. The AMERICAN STAG, or WAPITI, or ROUND-HORNED ELK.

This animal, which is frequently called the Canada Stag, is of a much larger and stronger make than the Stags of Europe; and in fact is one of the most gigantic of the deer tribe, being from four to five feet in height and from seven to eight feet in length. Their horns are shed annually; they are round and very large, branching into serpentine curves, but never palmated, and measuring six feet from tip to tip. (Plate VII. fig. 4.) Under the throat of the male is a dewlap composed of black hair from four to six inches long; the tail, in both sexes is very short. Most of the upper parts of the Wapiti are of a lively yellowish brown color; the neck is mixed red and black; the rump yellowish, bounded by a dark, circular marginal line; the limbs on the front are deep brown; the tail yellowish. The Wapiti feeds on grass and young shoots of trees; is easily tamed and has been trained to the harness. It is said to make a shrill, quivering noise, "not very unlike the braying of an ass." The flesh is somewhat coarse, and not highly valued; but its hide, when made into leather is said not to turn hard in drying after having been wet, a quality which places it in high estimation. The Wapiti is found, not only in the northern parts of this continent, but on the western prairies, and in California, Oregon, and New Mexico.

C. axis. The AXIS. (So named by Pliny.)

Of this beautiful deer there are two varieties. The common Axis, in its size and general form, nearly resembles the fallow deer, being, at the shoulder, about two and a half feet in height.

It has a rich fawn-colored skin, spotted with white, and hence sometimes receives the specific name *maculosa*, (spotted.) Along the back the ground-color changes to nearly black; but the under parts are snow white. A broad dusky spot appears upon the forehead, and a line of the same color extends along the middle of the nose. The Axis is a native of India, and is particularly numerous on the banks of the Ganges. It roams among the thick jungles, near streams of water, and is hunted under the name of the Spotted Hog-Deer. This animal feeds in the night, is timid, mild and inactive, excepting when the females have young, at which time the male is bold and fierce. It has been kept with success in menageries and parks, to which, from its form and color, it is highly ornamental. The larger variety, *A. major*, (Lat. greater,) a native of Borneo and Ceylon, is about the height of a horse, and has horns which are three-forked, thick and rugged, and nearly three feet long.

Capreolus Dorcas, (Gr. δορκὰς, *dorkas*, a gazelle,) or *C. capreolus*, (Lat. Roebuck or Chamois.) The Roebuck.

This species of deer, once common in England, is now confined chiefly or entirely to the Highlands of Scotland. They are of less size than the fallow deer, being only two feet four inches in height, and three feet six inches in length. The color is reddish brown on the back, the chest and under parts of the body are yellowish, and the croup white; the horns are round, divided into three branches, and about nine inches long. (Plate VII. fig. 7.) The Roebuck does not live in herds, but singly and in pairs, amongst the shady thickets and rising slopes. This deer is very cunning, when pursued, sometimes baffling the dogs by making a few enormous leaps, waiting until the dogs have passed and then resuming its former track. It is said to be very fond of the *Rubus saxatus*, called in the Highlands, the Roebuck-berry. In winter, when the ground is covered with snow, these animals browse on the tender branches of the fir and birch. The flesh is delicate food, and the horns are used for carving-knives. By the old Welsh laws, a Roebuck was valued at the same price as a she-goat. It can be easily subdued, but never perfectly tamed, always retaining some portion of its natural wildness.

C. leucurus. (Gr. λευκὸς, *leukos*, white; οὐρα, *oura*, a tail.) The White-Tailed Deer.

This resembles the European roebuck. On the Columbia river it is the most common deer; the tip and under part of the tail are of a cream white.

C. macrotis. (Gr. μακρὸς, *makros*, long; οὖς, *ous*, ear.) The Mule Deer.

This takes its name from its long ears, which are half the length of the whole antler. The hair is waved or crimped like that of the elk; upon the thighs near the croup it looks like white thread cut off abruptly.

C. elaphus. (Gr. ελαφὸς, *elaphos*, a stag.) The RED DEER, or STAG.

This noble species is found native in the European forests and in those of Asia where the climate is temperate. It is the largest of the English Deer, associated with the *forest laws*, so oppressive that they affixed a less value to the life of a man than that of a stag; and it is blended with the legends of deadly feud, as in the celebrated ballad of "Chevy Chase." The Red Deer is distinguished by its brown color, and a pale spot on the rump, and sometimes attains a great size. Pennant speaks of one that weighed 314 lbs., exclusive of the entrails, head and skin. According to Buffon, the small size of some of these animals is owing to a deficiency of nourishment, as in rich pastures its size becomes greatly increased. The horns are round, having the antlers turned towards the front, the summit terminating in a fork, or snags from a common center. (Plate VII. fig. 2.) It is very common in France, and is supposed to have been originally introduced from that country into England. In the latter country it is now largely superseded by the common or Fallow Deer, which is of a more manageable and placid disposition and affords far superior venison. The Red Deer has a fine eye, an acute smell and a good ear; when listening, raises his head and erects his ears; when going into a coppice, or other half-covered place, stops to look around him on all sides, and scents the wind to discover if any object be near that might disturb him. He eats slowly, and after his stomach is full, lies down and leisurely ruminates.

The pursuit of this deer is a very favorite amusement in England, summoning into action all the energy of youth and manhood. The animal in stalking is generally shot; but when wounded and yet able to fly, the dogs are let loose in the chase. In olden times, the dogs were mainly relied on for taking and killing deer, so that fleet and courageous hounds became the pride of nobles and princes. It is said he is particularly delighted with the sound of the shepherd's pipe, and is by that instrument sometimes lured to his own destruction. In winter and spring, this animal rarely drinks, the dews and herbage being sufficient to satisfy his thirst; but during the parching heats of summer, he not only frequents the brooks and springs, but searches for deep water wherein to bathe and refresh himself. He swims with great ease and strength, particularly when he is in good condition, his fat contributing to his buoyancy. The female bears one young, seldom more, in or near the month of May. The fawn, or *calf*, as it is called, the first year, does not quit the dam during the entire summer. The female is most assiduous in concealing and tending the young one, which is needful to secure it against assaults, not only from the cat and dog tribes, but even from the stag himself, who is not *overstocked* with paternal affection.

C. Dama, (Lat. a Fallow Deer.) The FALLOW DEER.

This has the same general form, aspect and manners as the Stag, with a more gentle disposition. The size is smaller, but the chief difference between the Fallow Deer and the Stag relates to the horns, (Plate VII. figs. 2 and 6,) which, in the former, are broad and palmated, at their extremities pointing a little forward, and branched on their hinder sides. It is less delicate than the stag in its choice of food, and browses much closer; is at full maturity when three years old.

There are two varieties of this animal in England, where it adorns the modern parks. The beautiful dappled variety is supposed to have been brought from the south of Europe, or the western parts of Asia; the other very deep brown variety is said by Pennant to have been introduced by James I., from Norway. On the continent of Europe, as well as in England, they are confined in parks; but they are found wild in Moldavia as well as Lithuania. The venison of this Deer is of the richest and most delicate kind; the skins of the buck and doe are unrivaled for durability and softness; the horns, like those of the stag, are manufactured into knife handles and other articles, while from the refuse, ammonia or *hartshorn* is extracted. This species is represented in the sculptures of Nineveh.

C. Virginianus. The AMERICAN DEER.

This species resembles the English Fallow Deer, and is so named by Professor Emmons, (Mass. Report.) The color is bluish gray in the autumn and winter, dusky reddish in the spring, changing to bluish in the summer; the young animal is spotted with white. The horns are of moderate size, curving forward, having the concave part in front, "with from one to six points occasionally palmated." In the adult males the horns show a great variety, which is regulated by their age, the season of the year, and the abundance or scarcity of their food. These animals range from Canada to Mexico. In some places, the united attacks of men and *wolves* are largely diminishing their number. Their horns are usually cast in the winter. Dr. De Kay says the reason so few of the horns are found, is that as soon as they are shed they are eaten up by the *Rodents* or gnawing animals. In frontier countries these animals are exceedingly useful, not only for the food which they furnish, but for their skins, which form an important article of commerce. They live upon twigs of trees, shrubs, berries and grasses; for the buds and flowers of the pond-lily, they are said to show a peculiar fondness. The female has one, sometimes two fawns at a birth, in the latter part of spring or early in the summer.

C. alces. The Elk or Moose. Flat-Horned Elk, Black Moose or Elk. (See Plate VII. fig. 3.)

This animal, surpassing all the true deer in size and strength, is found in the northern parts of Europe and America. The name which it bears is of Celtic origin, coming from "Elch," whence is derived the latter word *alce* or *alces*, which is the Celtic transferred to the Roman language. In America, it is known under the various names of Flat-Horned Elk, Black Elk, or Moose. The latter, which is the more common term, is a corruption of the Indian appellation, *Moosoa* or *Musee*, wood eater.

The Elk is six or seven feet in length, and from four to five and a half feet high at the withers; the head is large and elongated, and is, including the upper lip, covered with short projecting and flexible hair, something like that of the Tapir; the eyes are moderately large, and placed near the base of the horns; the ears long and asinine; the neck very short and strong and furnished with a mane; the lachrymal pit is small; horns are found in the male only. The hair of the lips and throat, in connection with its very long and flexible tongue, serves to direct food to the mouth. The food consists of shoots and twigs of trees, particularly of striped maple; the Elk also feeds upon high coarse grasses, but when wishing to graze, reaches the ground with difficulty, and sometimes feeds leaning on its knees. It likewise peels old trees and feeds upon the bark. During the summer, Elks frequent the neighborhood of lakes and streams, often resorting to the water as a refuge from tormenting musquitoes, and feeding upon aquatic plants; like the *C. Virginianus*, they are said to be particularly fond of the roots of the pond-lily. In winter, they betake themselves to the wooded hills. The Elk can hardly be said to be gregarious, but two or three being seen together, except at particular seasons. Some naturalists consider the Moose of this country to be a different species from the Elk of Europe, asserting that in the heavy palmated horns of both, there is a difference which indicate a diversity of species; but according to DeKay, this difference is not uniform, and the animals should be considered of the same species. The horns, perfected in the fifth year, are from ten to twelve feet apart, and weigh from fifty to sixty pounds. The snags or branches sometimes amount to twenty-eight. The body of the Elk is round and compact, supported by legs of disproportionate length; the hair is full and coarse, longest upon the head and withers; it is black at the tips, gray in the middle, and white at the roots. The dress of summer is of a browner tint than that of the winter. (See the figure above the Camelopard on the Chart.)

In its ungainly form and awkward movements, this animal exhibits a strong contrast to the others of the same family. The shoulders being rather higher than the croup, it does not bound like the deer, nor gallop like the horse, but shuffles or ambles along, its joints or hoofs cracking at every step. Like those of the Rein Deer, the hoofs are broad and divided so that they diverge on pressing the ground, thus giving the animal a sort of *natural snow-shoes*. When each part is brought smartly together by the sudden raising of the limbs, the cracking noise above mentioned is produced and may be heard at a considerable distance. When increasing its speed, the animal straddles his hind legs to avoid treading on its fore heels, tossing about the head and shoulders when breaking from a trot into a gallop. In its progress, it holds up its nape so as to lay the horns horizontally back, and prevent their entanglement among trees. The Moose is a timorous and wary animal, and as its senses of hearing and smell are acute, must be approached with great caution. When it notices the coming of the hunter, it at once endeavors to escape, trotting off with great rapidity; at this gait, it soon leaves the hunter far in the rear, stepping with ease over fallen timber of the largest size. When hard pressed by the hunters wearing snow shoes, if it breaks into a gallop they soon overtake it; though in the winter it may sink at every step, it still keeps on its way, the sharp ice wounding its feet, and its lofty horns becoming entangled in the branches of the forest as it passes along. The trees are broken with ease, and wherever the Moose runs, the hunter perceives it by the snapping off of branches of trees as thick as a man's thigh with its horns. The chase may last in this manner for a whole day, sometimes for two or three days together; for the pursuers are often "not less excited by famine than the pursued by fear." The poor animal "at last quite tired and spent with loss of blood, sinks like a ruined building, and makes the earth shake beneath his fall." The flesh is highly esteemed; the nose and tongue in particular are thought to be great dainties. The Elk can be easily domesticated, and has been used for draught. The male sometimes becomes very large, attaining the weight of eleven hundred pounds. Elks were formerly used in Europe for conveying couriers, and could accomplish 36 Swedish, or 234 English miles in a day, when attached to a sledge. Dorelli, a Swedish gentleman, recommended that they should be used in time of war as flying artillery, to reconnoitre and carry dispatches. The skin is so tough that a regiment of soldiers was furnished with waistcoats made of Elk's hide, which could hardly be penetrated by a ball.

C. rangifer, or *Rangifer tarandus*. The REIN DEER.

The Deer of this species have received many names. They are found throughout the arctic regions of Europe, Asia and America; but those of Lapland and Spitzbergen are said to be the finest. Their general height is about four and a half feet; their horns are long and slender, having round, branched and recurved antlers, the summits of which are palmated; (Plate VII. fig. 5;) the body is of a thick and square form; the legs are stouter in proportion than those of the Stag; the size differs with the climate, those in regions farthest north being the largest; the color is brown above, varying, however, with the age of the animal and the season of the year. As the Rein Deer grows older, it often becomes of a grayish white beneath, and sometimes almost entirely white; the space about the eyes is always black. Both sexes have canine teeth; both also have horns, but those of the male are larger, longer, and more branched than those of the female.

The male sheds his horns about the last of November; the female retains hers until she brings forth; if barren, she drops them in the beginning of November. The horns, during the early part of their growth, are extremely sensitive, and the animal experiences much suffering from the gnats and musquitoes. The hoofs are long, large and black, as also are the false or secondary hoofs behind. While the animal is running, the latter hoofs, as in the Elk, make, by their striking together, a remarkable clattering noise, which may be heard at a considerable distance. Richardson, who has given many particulars respecting this Deer, thinks that in the fur countries of this continent, at least two varieties exist, called by him the "Barren Ground Caribou," and the "Woodland Caribou." The Woodland animal goes south in the spring, and is confined to wooded districts; the Barren Ground animal goes northward, retiring to the woods only in the winter, and passing the summer on barren grounds, or on the borders of the Arctic Seas. Bucks of this latter variety, when in good condition, weigh, according to Richardson, from 90 to 130 lbs., without the offal. Sir John Franklin states the weight of the Woodland Caribou to be from 200 to 240 lbs. It has been asserted that some Rein Deer have weighed as much as 400 lbs., though the correctness of this is questioned. The Rein Deer of Norway and Sweden are small when compared with those of Finland and Lapland, which, in their turn, yield to those of Spitzbergen, and these again fall short of the Polar races. The Barren Ground Caribous feed, in summer, upon the shoots of grasses growing in the valleys of the north, returning to the woods in September; they there feed upon the tree lichens and mosses found on the rocks and ground. They root for the lichen like swine in a pasture. The forehead, nose, and feet, are covered with a hard skin closely attached to those parts, and are thus guarded against injury by the icy crust which covers the surface of the snow. The Rein Deer of the Eastern continent are sustained by the same kind of food as the American animal. The Caribou is not less necessary for the support of our northern native tribes, than the Rein Deer of the Eastern Continent for that of the Laplander and other people of the north. Of the Caribou horns the Indians make their fish spears; the hide, dressed with close and compact fur and remarkably impervious to cold, forms their winter clothing, and from it is made a soft and pliable leather for moccasins and summer garments. When sixty or seventy skins are sewed together, they make a tent sufficient in size for the residence of a large family. By pouring one third part of melted fat over the pounded meat, and incorporating them well together, a composition called *pemmican* is made. This, if kept dry, may be preserved for three or four years, and containing much nourishment in small bulk, is well fitted for use in extensive journeys, as is abundantly proved by the experience of traders and others traversing the northern latitudes. Another mixture, called *thucchawgan*, made of pounded deer's meat and fish, is either eaten raw or made into soup.

The Caribous travel in herds varying in number from eight or ten to two or three hundred; their daily excursions being generally towards the quarter from which the wind blows. They are approached with more ease than any other deer found on this continent. A single family of Indians have sometimes destroyed two or three hundred in the course of a few weeks. To the Indians this animal is solely a beast of chase, not, as among the Laplanders, being used for purposes of draught. It is hunted or taken in traps or pounds, or lured to its fate by other artifice. Sometimes the hunter takes advantage of the animal's inquisitiveness, by creeping behind an object affording him partial concealment, where he imitates the bellowing of the animal, at the same time having his deer skin coat and hood drawn over his head. In this attempt he seldom fails to shoot down the animal before he comes within a distance of twelve paces. The rude inhabitants of the whole of northern Asia use the Rein Deer as a beast of burden; but in Lapland, where it is essential to meet the wants of a pastoral people, it is most highly appreciated. In that country the horse and ox could not exist; but the Rein Deer supplies their place, furnishing, as it does, food and clothing, and submissively and patiently yielding its labor. The movements of the Laplander and his habits of life are, in fact, *controlled* by his deer. He must go where they go in search of lichens and mosses, and is obliged to make periodical journeys involving much labor and fatigue, in order to keep them from being annoyed by the gadfly (*Oestrus Tarandi*,) which not only torments them with its sting, but even deposits its eggs in the wound which it makes in their hides. Often the hides are pierced in a hundred places, like a sieve, by this insect; and some deer die in the third year from this cause. The Laplander flees with his deer to the mountains in order to escape this insect, not only, but the scarcely less dreaded musquitoes, which are more ferocious in the cold climates than in the tropics. His deer are the Laplander's wealth. When in good circumstances he has three or four hundred of them, and can live in comfort. He who has only one hundred is thought to be in a condition somewhat precarious, while he who has but fifty commonly joins his animals with the herd of some richer man, and himself performs the necessary menial service. The civilization of Lapland, which is on the advance and promoted by intercourse with other nations, depends upon the Rein Deer as the only beast of burden and conveyance. When a traveler crosses the border line of Lapland, he must, for further progress, like Bayard Taylor, step into the sledge drawn by the rapid Rein Deer. The sledge is a light vehicle, running, not on wheels, but on its flat boards, which are covered with leather. The Rein Deer is yoked to it by a collar, and guided by reins attached to its horns.

> "Obsequious at their call, the docile tribe
> Yield to the sledge their necks, and whirl them swift
> O'er hill and dale, heap'd into one expanse
> Of marbled snow, far as the eye can sweep,
> With a blue crust of ice unbounded, glazed."

With the usual load of from two to three hundred pounds, they will trot over the glazed snow at the rate of ten miles an hour. Journeys, by these animals, of one hundred and fifty miles in nineteen hours are not uncommon. In truth, some stories of their swiftness would appear incredible, if not so fully attested. Pictet, with three deer, went in 1769 to the north of Lapland, in order to observe the transit of Venus. "The first performed 3089 feet, 8 inches and $\frac{96}{100}$ in two minutes, making a rate of nearly nineteen English miles an hour; the second went over the same ground in three minutes, and the last in three minutes and twenty-six seconds." One is recorded to have "drawn, in 1699, an officer, with important dispatches, eight hundred English miles in forty-eight hours; and the portrait of the poor deer, which fell dead at the end of its remarkable journey, is still preserved at the palace of Drottingholm, Sweden."

C. muntjac, or *Cervulus* (Lat. dim.) *vaginalis.* (Lat. sheathed.) The MUNTJAC or KIJANG, of India. (Pl. VII. fig. 8.)

This animal is a little larger than the Roebuck; has a pointed head and rather large ears; its eyes are large with lachrymal sinuses; the tail is short and flattened; the male has large canine teeth in the upper jaw; the female has none, and is without horns. The horns in the male are short and simple, "rising from a footstalk apparently beneath the skin, and running obliquely upwards, one on each side of the forehead, beginning as low down as the inner angle of the eye." On the face, two rough folds of the skin, following the direction of the prominent part of the forehead, unite so as to mark the face with the letter V. The general color is a reddish brown above; the under parts and front of the thighs, pure white. The Chinese Muntjak is of a grayish brown color, with pale ringed hair. The Muntjak is one of the most elegant and beautiful of the deer kind. It possesses "a great portion of craftiness, combined with much indolence." As it gives forth a strong scent, dogs easily follow its path. In its flight, it is at first very swift; but it soon slackens its speed, and taking a circular course, returns to the spot from which it started. After making several such circuits, if still followed, it thrusts its head into a thicket, and thus remains fixed, as in a secure place, unmindful of the approach of the sportsman. The male animal has a great share of courage, and when the dogs are at bay with him, he makes, with his tusks, a most vigorous defence, and many dogs are wounded in the attack. Dr. Horsefield, whose account of this animal is the most satisfactory, states that the Muntjak "selects for its retreat certain districts which it never voluntarily deserts. Many of these districts are known as the favorite resort of the animal for several generations. They consist of moderately elevated grounds, diversified by ridges and valleys, tending towards the acclivities of the more considerable mountains, or approaching the confines of extensive forests." These districts, common in Java, are "covered with long grass, and shrubs, and trees of moderate size, growing in groups or small thickets." The long grass, *saccharum spicatum*, and a plant called *Phyllanthus Emblica*, constitute the principal food of the Muntjak. The flesh is said to afford excellent venison, and is often found on the tables of European residents. Among the Mahrattas, this animal is called *Baikar*. It uses its long *sinuses* apparently for the purpose of smelling, "dilating them to a great extent, and applying them to various objects."

The SOUTH AMERICAN DEER form a beautiful group. Of these we can notice only 1st, *C. nemorivagus*, (Lat. *nemus*, a wood; *vagus*, wandering,)—the GAUZU-VIVA, a delicate little deer, which is but twenty-six inches in length, approaching, in its aspect, that of the sheep. In this species, the lachrymal sinus, or tear pit, is scarcely perceptible. The lower part of the head and legs is whitish; about the eyes, on the inside of the fore legs and under part of the body, the color is a palish cinnamon; the neck and other parts brownish. The horns are very short. It is found in Brazil.

2. *C. rufus*, (Lat. red.) The PITA.

This is about twenty-nine inches in height; in its general color reddish brown, but in some parts whiter. It lives in the low marshy grounds of South America; is found in large herds, and "as ten females are seen for one male," and as the former are without horns, the existence of deer on this continent, without horns, has by some been incorrectly reported. The Pita shows little power of endurance when pursued, being soon run down by dogs; sometimes it is captured by the lasso and balls.

FOSSIL CERVIDÆ have been discovered, the most remarkable of which is the *Megaceros* (great horned) *Hibernicus*, the gigantic Irish Deer, larger in size than the Moose; the antlers over five feet in length, from the burr to the tip, in a straight line, and nearly eleven feet apart, reckoning from the extreme tip of the right to that of the left antler.

QUESTIONS.

What is said of the internal structure of the DEER FAMILY? Give their general characters. Are the horns found in both sexes? What is said of their form? How extensive is the range of the Deer? What is remarked of the loss and renewal of the antlers? Briefly describe the process. At what time does the *Common Stag* shed his horns? How soon, in the young animal, do the horns appear? What determines the age of the Stag?

How many antlers have the oldest? What is said of their hearing and smell? How many species does the family include? How does the *Wapiti* compare with the European Stag? What is said of his horns? What other characteristics are given? Upon what does it feed? Is its flesh highly valued? Where is it found? How many varieties of the *Axis?* Describe the Common *Axis*. Where found? What do the hunters call it? What is said of the larger variety? Where is the *Roebuck* now found? Give its size and other particulars. What is said of the *White-tailed Deer?* What of the *Mule* or *Long-eared Deer?* Where is the *Red Deer*, or Stag found? With what is this associated? How is this distinguished? What is said of its size or weight? What species has largely superseded this in England, and why? What is said of the chase of the Deer? Give other particulars. How does the *Fallow Deer* compare with the Stag? When is it mature? How many varieties in England? Where is it found wild? What is said of its venison? From what part of the animal is hartshorn obtained? Which English sp. does the American or Virginia D. resemble? Describe it. What is its range? Why are so few of its horns found? How is it useful in frontier countries? What is said of the size of the *Elk?* What is the origin of its name? What is the animal called in this country? Explain the term *Moose*. Name its characteristics. Of what roots and twigs is it particularly fond? Does the Am. differ from the Eur. sp.? What is said of its horns, hair, &c.? How does it contrast with other Deer? What is peculiar in its hoofs? What of its efforts to escape from hunters? For what purpose were Elks formerly used in Europe? Of what regions is the *Rein Deer* a native? Give its size and other characteristics. How many varieties, according to Richardson, are found on this continent? Give the weight of each? Which Rein D. are the largest? On what does the *Caribou* feed? What is said of its uses? What is pemmican? What is thucchawgan? How does the Caribou travel? For what do the Indians use it? How is it hunted? What are the uses of the Rein D. on the Eastern Continent? How does it affect the character and condition of the Laplander? What is said of the size of the *Muntjac?* Give some account of its disposition and habits. What S. American Deer are mentioned? Give some account of them. Which is the most remarkable of the fossil Deer?

Compare the description of the *Flat-horned Elk* with the figure above Camelopard, on the chart. Give the genera, species, &c., of the *Round-horned Elk* or *Wapiti*. What else is it called? Trace the *Rein D.* and compare the description in the book with the figure on the chart.

SECTION XXVII.

BOVIDAE. (Lat. *Bos*, an ox.) The OXEN.—Bisulcated. (Lat. *Bis*, two; *sulcus*, furrow, two hoofed or furrowed.)(Pl.VI. fig.1.)

The animals of this family have characteristics easily recognized and generally familiar. Both sexes have horns which are permanent, hollow and smooth, except at their base, where they are ringed; also rounded and tapering to a point, so as to form a crescent. The horns are supported by bony cores, having cavities, or cells communicating with the interior of the skull;

the muzzle is large; the neck thick, deep and compressed,—its skin forming a pendulous dewlap; the body is heavy and massive; the limbs stout; there is a distinct ridge upon the back, which is sometimes produced into a dorsal hump; the expression of the countenance is often, particularly in the males, malignant and threatening, betokening the ferocity that belongs to several of the species;—the Cow and Ox, however, exhibit a quiet, decided gentleness of physiognomy. The oxen are social in their habits; and some are gregarious, associating in immense herds, as the Bison or Buffalo. The organs of digestion in this family are after the same plan as those of the other ruminant, or cud-chewing animals, and need not be here particularly described. The main food of the Ox family is herbivorous; for although they do browse upon shrubs and trees, yet grass and herbage they prefer. (For the kind of teeth in this family, see Plate IV. fig. 10.) When hungry, they have been known to feed on plants not designed for their use, and by which they have been injured. Meadow-Saffron, (*colchicum autumnale*,) for instance, is deleterious to them if taken in any large quantity; and Hellebore, (*Helleborus*,) is said to be poisonous to them; Yew, (*taxus baccata*,) is fatal to them, as it is to herbivorous animals generally. In a state of "domesticated nature,"—that is, when not stall-fed, or at all using artificial grasses, but roaming at large, oxen are said to eat two hundred and seventy-six plants, and to reject two hundred and eighteen. Heifers waste away in enclosures where the Meadow-Sweet, (*spiraea ulmaria*,) grows in abundance, and covers the ground; but to the GOAT this is nourishing food. The present races of wild cattle are probably all descended from those which were, at some period, subservient to man. The ancient *Urus*, or Wild Ox, was a savage, untamable animal, with large spreading horns, and of great size.

Bos taurus. (Lat. a Bull.) This animal, with flat forehead, and the withers not humped, was properly regarded as the type of the entire tribe. This species includes the Common Ox which is so widely diffused, and of such extended and varied utility;—of which more than forty synonyms have been given. The horns differ much as to their form and direction, from the influence of domestication; the colors are various, as reddish, white, gray, brown and black. "The male is called a *bull;* the female, a *cow;* and the young, a *calf;* the name Ox is given to the gelded male; and he is called an ox-calf, or bull-calf, until he is twelve months old; a steer until he is four years old. and after that an ox or bullock."

The Ox is less used for farming purposes than formerly; the

horse and improved agricultural implements taking its place; it reaches its full vigor in three years, and its term of life is about fourteen. The breeds of the animal are numerous, and generally distinguished by the length or shape of the horns. The "Durham," or short-horned breed, is perhaps most valuable for the dairy, as well as for a tendency to fatten rapidly, and at an early age." The "long-horned," the "middle-horned," and the "polled," or hornless breeds, have each their particular values. The "Alderney Cow," with "crumpled horn," has long been celebrated for the richness of its milk. Within the last half century, many and successful efforts have been made to improve the breed of cattle both in England and in this country.* Considerable benefit has resulted from the labors of Agricultural Societies, and, in particular, from the stimulus which, by the offer of premiums, they have given to the raising of cattle for exhibition at the annual COUNTY and STATE FAIRS. The uses of the Ox are well known, and we need not describe them; every part of the animal is of value. Formerly, the cruel sport of bull-baiting was much practiced; and in some countries, particularly Spain, it is still a popular diversion.

Bos Indicus. The ZEBU, or BRAHMIN BULL, of India. (See Chart.)

This is distinguished for a more lengthened form of the head, with a decidedly concave line of profile; an arched neck; a lump of fatty substance rising from the withers; an arched back, sinking and rounded off on the hinder part; an enormous dewlap dangling down in folds; long, pendulous ears; a mild and sleepy eye; and long and tapering limbs. The size varies from that of a large mastiff to that of a full grown buffalo. Over the whole of Southern Asia, the islands of the Indian Archipelago, and the eastern coast of Africa, the Zebu supplies the place of the Ox. In some places, it is saddled and ridden, or harnessed in a carriage; traveling from twenty to thirty miles in a day. Its beef is inferior to that of the Ox. The hump is deemed the most delicate part. This sometimes becomes greatly increased in size, and has even been known to reach "the enormous weight of 50 lbs." Among the Hindoos, the Zebu has a "charmed life." They venerate this animal, and hold its slaughter to be a sin; though they do not object to work it. In the streets of Calcutta, "some particularly sanctified" Zebus may be seen wandering at their ease in the public streets, and taking their food where they list. The utmost a native does when he sees them honoring his goods too much, is to "urge them by the

* See "American Herd Book," and other Agricultural works..

gentlest hints, to taste some of the good things in his neighbor's stall." If lying down in some narrow way, a person must not disturb them ; but he must either proceed by another road, or wait until the sacred animals are pleased to rise !

B. Dante. The Dante. This is an Egyptian species, resembling the preceding, figures of which are found on ancient tombs of Egypt.

Bison. (Gr. *Βίσων*. named from the Thracian *Βίστονες*, Bistones.) The Bison.

This generic name first used by Pliny, applies to two living species,—one of them European, and now almost extinct; the other American, and still found in great numbers. Audubon enumerates five species, three of which, however, are more generally arranged either with the genus *Bòs*, or the genus *Bubabus*. The European Bison is now found living in the Moldavian and Wallachian districts, and in some parts of the Caucasus ; the other species at one time "ranged over nearly the whole of North America ;" it is now found in vast herds in some of the Western prairies, and is thinly scattered along the valleys which border upon the Rocky Mountains. The districts which these animals inhabit, are described very graphically in Washington Irving's "Tour in the Prairies." They delight in level prairies, covered with luxuriant vegetation, bordering the hills of limestone formation, where saline springs or marshes abundantly occur. The American species, *B. Americanus*, has fifteen pairs of ribs ; the European has fourteen, (one more than the common ox.) This points out the main difference between the two species.

The Bison is marked by its broad and slightly arched forehead, and the long and wavy hair upon it, forming on the chin and breast a kind of beard ; by the elevation of the withers, arising from the lengthened spinous processes for the attachment of the ligament and enormous muscles of the neck, serving to support the large and ponderous head ; and by a continuous fatty deposition, or sort of hunch,—from which the back gradually declines, the hind quarters appearing disproportionably weak and small ; and by its short but amazingly powerful limbs. The horns are short, tapering and erect ; the general color dark umber brown, becoming in winter tinged with a grayish white. The aspect of this animal is fierce, wild and malicious; the eyes being small, fiery, and half hid in the shaggy hair intermingled with wool, which copiously overspreads its head and shoulders. The height at the shoulders is upwards of six feet ; the length (exclusive of the tail, which is twenty inches) is eight and a half feet; the weight of a fat bull is generally near two thousand pounds; that of a fat cow, nearly twelve hundred, which is considered a good weight in the fur countries. The Indians have long been hunters of this animal, which they call the Buffalo; using bows and arrows, which, wielded by their skillful hands, strike the huge creature to the ground. The female is beyond all comparison swifter than the male, and is the constant object of the hunter, from the superior quality of her flesh. The Bison is a shy and wary animal; usually it flies before its pursuers; but sometimes, led by an infuriated individual, the whole herd will turn, and rushing towards the hunters, trample them down in their headlong course. Next to man, the enemies which these animals most greatly dread, are the grizzly bear and the wolf, by which many of them are destroyed; the wolves assailing them in packs and making great havoc, especially among the smaller animals.

While feeding, they are frequently scattered over a vast surface; but when they move onwards in a mass, they form a dense, impenetrable column, which once fairly in motion, is scarcely to be turned. They swim large rivers in nearly the same order in which they traverse the plains; and when flying from pursuit, it is vain for those in front to halt suddenly, as the rearward throng rush madly forward and force their leaders on. The Indians sometimes avail themselves of this habit. Driving a herd of these animals to the vicinity of a precipice, and setting the whole in rapid motion, they, by shouting and other artifices, impel the affrighted animals onward to their own destruction. The herds of these animals found together, sometimes number "countless thousands." Lewis and Clark say, that "20,000 would be no exaggerated number" for a herd which they saw, and which "darkened the whole plain." To Catlin's account of his travels among the North American Indians, reference may be had for many interesting accounts of "buffalo hunts." The risk of this chase is considerable, but its rewards are great; few animals minister more largely to the wants, and even to the comforts of man, than the Bison. The flesh is said to be juicy, bearing "the same relation to common beef that venison does to mutton." The tongue, well cured, is thought to surpass, as a relish, that of the common ox,—the hump also is esteemed peculiarly rich and delicate. Much of the *pemmican* used by Northern voyagers, or by those attached to the fur companies, is made of bison meat,—one bison furnishing meat and fat enough to make 90 lbs. of the article. The Indian tribes make every part of the animal subservient to their necessities and comfort,—the "Buffalo robes,"—the skin dressed with the hair on,—defending them against the cold; the horns are converted into powder-flasks; and the ribs of the animal, strengthened by some of the stronger fibres, are made to furnish the bow, by which others of the species are to be destroyed. Catlin says, that "there are, by a fair calculation, more than 300,000 Indians who are now subsisting on the flesh of the buffaloes, and by these animals supplied with all the luxuries of life which they desire, as they know no others." The advance of white population over the regions of the West, bearing with them the institutions of civilization is, however, modifying this statement, and gradually contracting the range of the Bison.

Bubalus Buffalus, or *Bos Bubalus*. The BUFFALO of Asia.

This animal, in its general aspect and carriage, resembles the Bison, or perhaps the Domestic Ox, though larger and stronger,—but differs from the Bison in its horns, which are enormously large, bent down and recurved at the tip; in its ears, which are half the length of the head, and slightly covered with hair; and in the fur, which is rough, irregular and bristly. Of this species, there are two varieties, the *B. Arnee*, (Shaw,) and the *B. Rhainsa*. The Arnee is the wild Buffalo of India, found on the margins of old and thick forests; and, like the Rhinoceros, confining itself to the most swampy parts of the region where it dwells. Its horns are often five feet in length, and so inclined together at the points, as to form a figure somewhat lyre-shaped. It is also remarkable for the shortness of its tail, which reaches no lower than the hock. It is one third larger than the Rhainsa, or tame Buffalo, being ten and a half feet long, and six to six and a half feet high at the shoulders. Its strength is so great, it is a formidable enemy even to the tiger, who shuns an encounter with him; and such is the power of his charge, that he frequently prostrates a well-sized elephant. The Rhainsa is universal in India and adjacent countries, and was formerly, as now, used as a beast of burden in Egypt, Greece and Italy. In the latter country, it is, on account of its great strength, very useful for carrying purposes, especially in marshy and swampy districts, where the roads are two or three feet deep with mud. The hide of the Asiatic Buffalo is peculiarly thick and strong, and in great request for making harness.

Bos Gaurus, the *Gour*, or *Gaur*, of mountainous parts of Central India.

This has the hind hoof only half the size of the fore one,—the general color is brown, but the legs are white; the horns are bent downwards at the front; "the limbs have more of the form of the deer than any other of the bovine genus." It is asserted that the tiger has no chance in a combat with a full grown Gour. This animal does not, like the Buffalo, wallow in swamp and mire. The large quantity of milk given by the cow, is said to be occasionally so rich as to cause the calf's death.

Poephagus. (Gr. πόη, *poe*, grass; φάγω, *phago*, I eat,) or *Bos grunniens*, (of Linnæus.) The YAK.

Of this genus, there is but one species, *P. grunniens*, found in the woods and recesses of the Thibet mountains. It has fourteen or fifteen pair of ribs, and resembles the Buffalo in its form, but is smaller. Both sexes grunt like a pig, whence the specific name, *grunniens*, (Lat. grunting.) The tail has full flowing hair like that of a horse, and is used in India as a fan or whisk to keep off the musquitoes,—when fixed into an ivory or metal handle, it is called a *chowrie*. Elephants are sometimes taught to carry a chowrie, and waive it about in the air. The neck and back are surmounted by a sort of mane; the hair of the body is black,—smooth and short in summer, but thick and harsh in winter; the back and tail are often white. The Yaks dislike the heat of summer, and hide themselves in the shade and water. The hair is applied to various purposes by the Tartars. They weave it into cloth, of which they not only make articles of dress, but also tents and the ropes which sustain them. There are two varieties,—those used for the plough, and those used for riding. The former are ugly and short-legged, and guided by the nose, carry their heads very low; the latter much handsomer, having twisted horns, a noble bearing, and an erect head; also a stately hump, and a rich silky tail reaching nearly to the ground.

Bos moschatus. (Lat. musky,) or *Ovibos moschatus.* The MUSK OX. (Plate VII. fig. 9.)

This animal has sometimes been removed from thė genus *Bos*, in consequence of the absence of the naked muzzle which is possessed by others of the bovine groups, and ranked as a connecting, or intermediate link between the ox and the sheep; hence the generic term *ovibos*, (Lat. *ovis*, a sheep; and *bos*, ox.) It may be doubted, however, whether, on this account, it should be separated from the bovines. The full-grown male is about the size of a small two year old cow; the female is considerably smaller; the horns are united at the top of the head,—flat, broad, and bent down against the cheeks, but become round and tapering, and turning up, end in a sharp point about the level of the eyes. The animal is covered with long bushy hair, which reaches almost to the ground. The general color of the hair is brown, or brownish black, except a portion in the middle of the back, which is dirty gray; in the female, the general color is black; the head is large and square; the eyes moderately large; the ears short, and scarcely visible through the surrounding long hair. Under the hair of the body, is an admirable second coat, consisting of brown, or ash-colored wool; the legs are short and thick, covered with close hair, unmixed with wool; the tail very short; the hoofs are small compared with the size of the animal,—resembling those of the Rein Deer. It is said "none but an experienced hunter can distinguish the difference of the impressions made by the toes on the snow." Its food is also like that of the Rein Deer,—lichens in winter;—grass in summer. The length of the Musk Ox from the nose to the root of the tail, is about five and a half feet; and its weight, according to Parry, about 700 lbs. It is gregarious, being found in herds, twenty or thirty in number. The home of these animals is in the barren lands of North America, in regions above the 60th degree of latitude. They are hunted by the Esquimaux, but not without danger, as when provoked or wounded, they are apt to turn upon the pursuer. The poor creatures seem to fancy that the report of guns is thunder, and crowd together in a mass, so that they afford a good mark. If, however, they get sight of one of their assailants, they instantly charge at him, and then they are very dangerous enemies. Sometimes the Esquimaux turn the animals' irritation to good account;—for, after the adroit hunter has provoked the animal, and induced it to attack him, he wheels around it more quickly than it can turn; and by repeated stabs, puts an end to its life. The speed of the Musk Ox in running, is great, and it climbs rocky paths and broken and uneven sides of hills, with great agility. Sir John Richardson says, the wool of this animal "resembles that of the Bison, but is perhaps finer, and would be highly useful in the arts, if it could be procured in sufficient quantity." The same author informs us, that "when the animal is fat, its flesh is well tasted, and resembles that of the Caribou, but has a coarser grain." When lean, these animals "smell strongly of musk, their flesh, at the same time, being very dark and tough, and certainly far inferior to that of any other ruminant animal in North America."

QUESTIONS UPON THE BOVIDAE, (OX FAMILY.)

How is Bovidae derived? What is said of the general character of this family? What of the horns in particular? What of the appearance and habits of these animals? What kind of food do they use? What plants are hurtful to them? How many plants do oxen eat? How many do they reject? Which species furnishes the type of the entire tribe? How many synonyms have been given? Give the different names appropriated to this animal? How are the breeds of this animal usually distinguished? What is said of the Durham breed? What other breeds are mentioned? By what means has the breed of cattle been improved? How is the *Zebu* distinguished? How extensively does it supply the place of the ox? What uses are made of it? How is it regarded by the Hindoos? What is said of the Egyptian species? How many species of Bison are there? Where is the European species now found? What has probably prevented its entire extinction? In what part of North America is the other species found? What is the main difference between the European and the American species? How many ribs has the common ox? Give the distinctive marks of the Bison? Describe its disposition and habits. What enemies does it most dread? How do the Indians avail themselves of the habits of this animal? What is said of the largeness of the herds? Mention the uses made of the different parts of the Bison. How many Indians does Catlin estimate are daily supported by its flesh? How does the Asiatic Buffalo differ from the Bison? How many varieties of this species? What is said of them? In what countries is the animal used? Where is it especially useful? What is said of the Gour? How does it differ from the Buffalo? Give the derivation of the term *Poephagus?* How many species of the Yak? What renders the specific name appropriate? What use is made of its tail? What of its hair? How many varieties of this animal? Why is the generic *Ovibos* applied to the Musk Ox? What is the composition of that term? What characteristics are given? Where is the home of this animal? What more is said of it?

Name the genera and species of the Ox Family found upon the chart, tracing and giving some account of each as you proceed.

SECTION XXVIII.

Ovidae. (Lat. *ovis*, a sheep.) The SHEEP.

These differ so slightly from the Goat in anatomical structure that both genera are by some naturalists united.

The chief distinctive characters consist "in the sheep having no beard; in the horns being directed backwards, and then inclining spirally more or less forwards; in having a convex forehead; and in the existence of a sac, or fossa, situated at the base of the toes, lined with hair, and furnished with sebaceous follicles." The males also differ from the goat in being inodorous. The age of sheep is reckoned from the first shearing. Their value, both for food and clothing, is well known, and is incalculably great, while they are reared upon soils where other animals could not obtain sufficient for their support. The filaments of wool taken from a healthy sheep, present a polished, glittering appearance; those of a sickly, or half-starved animal, exhibit a paler hue. The dressed skin is largely used for the binding of books, and for different kinds of apparel. The bones, when calcined, are employed as tests in refining processes; from the entrails are prepared strings for musical instruments. Sheep furnish milk which is thicker than that of cows, and yields a greater quantity of butter and cheese. In some cases, water must be added in order to produce whey. The history of these animals is intermingled with poetical descriptions and national customs and enactments. They are mentioned in the earliest scripture records, and formed the chief wealth of the ancient patriarchs. Among the Jews, under the economy of Moses, the lamb was offered in sacrifice,—pointing to "Christ, the Heavenly Lamb;" and in the New Testament these animals are the subjects of many beautiful and touching parables.

Ovis aries. The Common Sheep.

This exhibits numerous varieties, and many of its form have been raised to the rank of species. The *Ovis Hispanicus*, the Spanish, or Merino Sheep, is among the most celebrated. These sheep, it is said, are the regenerated stock of the sheep of Boeotia, and survived the conquest of Spain by the Goths and Vandals. They have been transferred to Great Britain, Germany and the United States; and are remarkable for the fineness of their wool. In Germany, the wool has been brought to the highest perfection. Merino Sheep were introduced into Great Britain in 1787. The original stock in this State, (N. Y.,*) was derived from Holland; the Merino variety was first introduced in 1801; though their importance was not fully appreciated until seven or eight years after that period; when the excitement respecting them became very great, and they were sold at enormous prices.

Of the Merino Sheep, there are three varieties, viz., the Paular, the Negretti, and the Gaudaloupe breeds. The quality of the wool has been improved by the introduction of Saxony Sheep, (originally of the same Merino race.)

The breeds of sheep are distinguished by the comparative length of the fibres, which compose their fleece. They are designated as short-wooled, middle-wooled, and long-wooled sheep. To the short-wooled division belong the "Merino, Saxony and Australian breeds, whose short, fine and silky wool is used in the manufacture of broadcloths. The middle-wooled breeds, such as the English South-down, Suffolk and Cheviot, furnish material for the coarser cloths, flannels and similar fabrics. The Leicester breed, and some others, are long-wooled. The fibre of the wool in these sheep is strong and transparent, but is deficient in the power of *felting*, on which the compactness of cloth depends. This wool is used for merinoes, *mousselines de laine*, hosiery, etc. Welsh sheep are noted for the superior flavor of their flesh, and "in the London market Welsh mutton is always in demand."

---

* DeKay.

O. Ammon Argalis, or Siberian Sheep. The ARGALI.

This is one of the varieties of wild sheep, native to Siberia, and ranging over the mountains of Asia,—a strong, muscular, and active animal, about as large as a small fallow deer, and having thick, roughly ringed horns. (Plate VII. fig. 12.) In summer, its hair is smooth, and of yellowish gray color; but in winter, it becomes thick, harsh and reddish; the muzzle, throat and under parts, continuing white at all seasons. The whole form of this animal appears better adapted for agility than that of the common sheep.

O. Canadensis. The TAYE, or BIG HORN SHEEP, of Canada.

This is identical with the *O. Montanus*, of Geoffrey, and a variety is the *O. California*, of Douglas, which Dr. Gray says is probably the same as the *Ammon*, of Siberia.

O. Musimon, or *Musmon*. The MOUFFLON, of Cyprus, Candia, and Corsica. (Plate VII. fig. 10.)

This differs from the Argali, only in being rather smaller, and in the horns being very small, or altogether absent in the female. Like the Argali, it makes its home upon the mountains. It has been supposed that the primitive stock may be traced either to this, or the preceding species,—the hair of both species possessing the essential character of wool,—an imbricating scaly surface,—which gives to the covering of the domestic breeds the remarkable *felting* property upon which its utility so much depends.

O. polycerata. (Gr. πολύς, *polus*, many; κέρας, *keras*, horn.) The MANY-HORNED SHEEP.

This species found in Iceland and the most northern parts of the Russian dominions, resembles the common sheep in its body and tail, but has three, four, five or more horns. (See Plate VII. fig. 13.) The wool is long, smooth, hairy, and of a dark brown color. Under its outer coat, is a fine, short and soft kind of wool, or fur.

O. laticauda. (Lat. *latus*, broad; *cauda*, tail.) The BROAD-TAILED SHEEP,—is common in Tartary, Arabia, Persia, Barbary, Syria and Egypt. This sheep is chiefly noted for its large, heavy tail, often so loaded with fat as to weigh from ten to twelve pounds, and according to some, double that weight, and a foot broad; sometimes it is necessary to support it artificially. The upper part is covered with wool, but it is bare underneath, and the fat, of which it consists, is regarded as a great delicacy.

O. strepsiceros. (Gr. στρέφω, *strepho*, to twist; κερας, *keras*, horn.) THE CRETAN SHEEP.

This is chiefly found in the Island of Crete, but is kept in several parts of Europe on account of its singular appearance;

the horns being very large, long and spiral, those of the male upright,—of the female, at right angles with the head.

O. Guineensis. The AFRICAN, or GUINEA SHEEP,—found in all the tropical climates of Africa. It is large, with rough, hairy skin, short horns and pendulous ears, a kind of dewlap under the chin, and a long mane reaching below the neck. It is stronger, larger and more fleet than other sheep, and better suited to a forest life; but the flesh is quite indifferent food.

SHEEP.

What are the distinctive characters of the *Sheep?* How does the wool of the healthy sheep appear? What are the uses of this animal? With what is it associated? What Scripture references are given? What is said of the varieties of the *Common Sheep?* Which is the most celebrated? What is said of their origin? Where is their wool brought to the highest perfection? When was this variety first introduced into the State of New York? How has the quality of the wool been improved? How are the breeds of sheep distinguished, and how designated? What breeds are included in the *Short-Wooled* division? What in the *Middle-Wooled,* and what in the *Long-Wooled?* In what respect is the fibre of the *Long-Wooled Sheep* deficient? For what is the wool much used? What is said of the Welsh variety? Where is the *Argali* found? What is said of it? How does the *Moufflon* differ from the *Argali?* In what respects does it resemble it? What is said of the *Many-Horned Sheep?* Where is the *Broad-Tailed Sheep?* What is said of its tail? Where is the *Cretan Sheep* found? What is said of its horns? Where is the *Guinea Sheep* found, and what is said of it?

Trace the varieties mentioned on the chart,—tell where they are found, and their peculiarities.

SECTION XXIX.

Capridae, (Lat. *capra,* a goat.) The GOAT FAMILY.

The distinguishing characteristics of the Goat family are that they have hollow horns turned upwards and ringed; that they have eight cutting teeth on the lower jaw and none in the upper; and that the male has a beard. The muzzle is comparatively narrow, with no naked space about the nostrils; the tail is short; there are no fissures, or tear-pits, beneath the eyes, nor tufts of hair upon the knees. Either "native or naturalized," this animal appears in almost every part of the world. It is capable of enduring all kinds of weather, being found in high northern latitudes, and also thriving in the hottest parts of Africa and India. The internal organization of the animal is almost entirely similar to that of the sheep, (*Ovidæ.*) "He is, however, stronger, lighter, and more agile, and less timid than the sheep. The suppleness of his organs, and the strength and nervousness of his frame, are hardly sufficient to support the petulance and rapidity of his natural movements." (Buffon.)

The milk of the Goat is sweet, nutritious, and medicinal, owing to the character of its food, which consists chiefly of what is obtained from high hills, or from pastures where aromatic shrubs abound. Anciently the skin was deemed valuable for clothing; the best Turkey or Morocco leather is made from it, and from the skin of the kid is prepared the softest and handsomest leather for gloves. The strong odor of the Goat is well known, and it is said to be "refreshing" to horses. The female bears, generally in the last of February, usually two, sometimes three and even four young. Among the Greeks and Romans the Goat, because an enemy to the vine, was sacrificed to Bacchus. This animal is remarkably sure footed. Pennant says, "two yoked together, as they often are, as if by consent, take large and hazardous leaps, and yet so time their mutual efforts as rarely to miscarry in the attempt." The Goat butts, raising himself on the hind legs, and then coming down sidewise against his enemies. The varieties are numerous, and some of them have been exalted to the rank of species.

Hircus (or *Capra*) *Aegagrus.* The WILD GOAT. This is regarded by Cuvier and others, as the parent stock of the Domestic Goat in all its varieties.

It is found in herds, freely ranging in the great mountain chains of Asia. In Persia it is called the *Paseng.* The size is rather larger than that of the domestic breed; the horns also usually exceed those of the common Goat; the color is a brownish gray above and white beneath. The male has a large brownish beard; the female neither beard nor horns.

Capra hircus. The DOMESTIC GOAT. (Lat. *hircus*, a he-goat.)

This animal, like others reclaimed and subject to man, exhibits great varieties in respect to size, color, the quality of the hair, and even the largeness and number of the horns.

C. Angorensis. The ANGORA GOAT. (See Chart.)

This is a native of Angora, in Asia Minor; generally is of a milk-white color, short legged, with black, spreading, and spirally twisted horns and pendulous ears; its silk-like wool, which is its chief excellence, covers the entire body in long, hanging and spiral ringlets, and from it the finest camlets are made.

The CASHMERE GOAT, which is found in Thibet and roams the pastures of the Himalaya mountains, has an undercoat of wool, exquisitely delicate and fine. From this are manufactured the Cashmere shawls so highly valued by the fashionables of both hemispheres. It is remarked that the lower the temperature where the animal pastures, the heavier and finer is its wool. The Goats which feed in the highest vales of Thibet are of a bright ocre color; in lower ground the color changes to a yellowish white, and still lower down to entirely white. The highest parts of the Himalaya mountains inhabitable by man have a kind of black Goats, which yield wool from which are made shawls that in India command the highest price. The fine curled wool of these Goats lies close to the skin, just as the under hair of the common Goat lies below the coarse upper hair. The flesh of the Himalaya Goats is said to taste as well and its milk to be as rich as that of the common Goat.

The Angora Goat loses the delicacy of its hairy covering when exposed to a change of climate and pasture. It is said the people of Cashmere constantly work 16,000 looms, each loom giving employment to three men, the annual sale being calculated at 30,000 shawls.

The "NATURALISTS' LIBRARY," (RUMINANTIA, part II. by Sir William Jardine,) says that "a fine shawl, with a pattern all over it, takes nearly a year in making. The persons employed sit on a bench at the frame, sometimes four people at each, but if the shawl is a plain one, only two. The borders are marked with wooden needles, there being a separate needle for each color, and the rough part of the shawl is uppermost while it is in a process of manufacture. The Cashmeres which are obtained from the kingdom of that name are most sought after. India, however, produces several Goats besides the true Cashmere breed which yield wool from which shawls are made. Twenty-four pounds weight of the best wool of Thibet, sells at Cashmere for twenty rupees."

C. Jaela, or *C. Nubiana*. The ABYSSINIAN GOATS, found in the mountains of Abyssinia and Upper Egypt, and also on Mount Sinai, differ from the Goats of Thibet, in having close smooth hair, a convex forehead, and a projecting lower jaw.

The SYRIAN GOAT, (*Capra Syriaca*,) is distinguished by its large pendulous ears, (see Plate VII. fig. 14,) which are usually from one to two feet in length, and sometimes so annoying to the animal that the owners are obliged to trim them to enable it to feed with more ease. It has black horns which bend a little forwards, and are only about two inches long. The hair is colored like that of a fox, and it has two fleshy protuberances under its throat. It is very numerous in Syria, where it finds pastures specially adapted to its wants. Pennant says that "it supplies Aleppo with milk." It is no unimportant part of the wealth of a pastoral people, its flesh being used for food and its hair wrought into cloth. This was one of the animals offered in sacrifice by the ancient Hebrews; it was this Goat over which the Jewish High Priest, putting his hands on the Goat's head, "confessed the iniquities of the children of Israel," and then "sent him away by the hand of a fit man into the wilderness." The long ears of this animal illustrate those words of scripture, Amos iii. 12, "As the shepherd taketh out of the mouth of the lion . . . . a piece of an ear." So large and thick are the ears of this Goat that they make a considerable *mouthful* even for a *lion.*

C. Ibex. The IBEX.

Of this species there are several varieties in the mountain ranges of Europe, Asia and Africa, but more especially those of Asia and the bordering parts of Europe, all, however, resembling each other in their structure and general habits. This animal is much larger and stronger than the common domestic Goat. "The color is a deep hoary brown, the under parts of the body and insides of the limbs are of a much paler and whitish hue; the body is thick, short and strong; it has a small head, large eyes, and strong legs; very short hoofs; a short tail; and extremely large and long arched, brown colored horns, with knobs on the upper surface." (Plate VII. fig. 15.) The fore legs are considerably shorter than the hind, which enables the animal to ascend more easily than he can descend lofty mountain heights. In manners and voice the Ibex is much like the *Chamois.* It is found in small flocks consisting of ten or fifteen individuals. When hard pressed, these animals sometimes turn upon the hunter, hurling him down the most frightful declivity. It is a native of the Carpathian and Pyrenean mountains and of the Alps.

C. Americana. ROCKY MOUNTAIN GOAT.

These animals inhabit the lofty chain of mountains whence they derive their name, ranging from 40° to 65° North Latitude. They resort to grassy knolls begirt with craggy rocks as affording them places of refuge against the onsets of dogs and wolves, visiting, daily, caves in the mountains said to be encrusted with an effervescence of salt, of which they are fond; they are of larger size than the common Goat, have black horns, which are smooth and polished at the tips, and curved backwards, and obscurely ringed at the base, where they are sometimes a foot in circumference. On account of the great size of the horns, this animal is called by the hunters, the "Big-horn." The muzzle is extremely small; the color white; the hair long and straight; the skin very thick and spongy, and principally used in making moccasins. The flesh, when it is in season, is said to exceed in flavor the venison obtained in the same region, and the fleece is also highly valued, being next to that of the Cashmere Goat in fineness.

Give the chief characteristics of the *Goat Family.* What is Buffon's remark? What are the uses of the Goat? What remark is quoted from Pennant? Has this family many varieties? Which species is regarded as the purest stock? Where is it found? How does it compare in size, &c., with the domestic breed? In what respects does the *Domestic Goat* vary? Describe the *Angora Goat.* Where is the *Cashmere Goat* found? What articles are manufactured from its wool? What shawls command the highest price in India? What is said of the wool of which they are made? Has change of climate any effect upon the wool of the Angora Goat? What is said of the manufacture and sale of the Cashmere shawls? How do the *Abyssinian Goats* differ from those of *Thibet?* For what is the *Syrian Goat* distinguished? What is said of its hair, &c.? What use was made of it by the *Ancient Hebrews?* What words of Scripture do the ears of this animal illustrate? What is the habitat of the *Ibex?* What is said of its varieties? How do they compare with the Domestic Goat? Give the character of this animal. Does it ascend or descend most easily? What animal is it much like? What is the range of the Rocky Mountain Goat? What are their particular resorts? What do the hunters call this animal? For what are its skin and flesh used?

Name, trace and characterize the species on the chart.

SECTION XXX.

Antelopidæ. "Bright eyed." (Gr. ἄνθος, *anthos*, a flower or beautiful ornament; ωψ, *ōps*, eye.)

ANTELOPES. (Bisulcated or Cloven-footed.)

This beautiful family of Ruminants is by some considered a connecting link between the Goat and Deer families. Like the Goats, they never shed their horns; in size and general structure, the nature and color of their hair, and their swiftness of foot, they resemble the Deer. The hind limbs, like those of the hare, are much longer than the fore ones. This not only helps them to be more fleet, but increases their security in climbing precipices, which they are delighted in doing. The larger part of the species are brown on the back, and white on the under part of the body, with a black stripe between the brown and white. The tail is of various lengths, but always covered with pretty long hair; the ears, which are beautiful and well placed, terminate in a point. The hoof is cloven like that of a sheep; the perennial horns are conical, bent back, and ringed at the base, never showing the angles and ridges which distinguish those of the sheep and goats. This last is, perhaps, the *most general character* of the family. The case of the horns is thin, and as a group, the Antelopes are numbered among the HOLLOW-HORNED ANIMALS. A large part of them have lachrymal sinuses or "tear-pits," as seen in the Deer, and which can be opened at the will of the animal. These are furnished at the bottom with a gland that secretes an oily, viscous substance of the color and consistency of ear-wax, and turning black upon exposure to the air. The common Indian Antelope, and the Gazelle, according to observations of them away from their native climes, use this organ when any strange substance is brought to their notice, particularly if it be odoriferous; and they appear to derive great pleasure from protruding the sinus and rubbing it against the odorous body. The possession of sinuses distinguishes the Antelopes from the Goats and the Sheep; and this, connected with the absence of horns in the females of many species, also makes this family an intermediate link between the rest of the Hollow-Horned Ruminants and the Cervine, or Solid-Horned Animals. A few species of Antelopes have an additional gland running lengthwise between the sub-orbital sinus and the mouth, but having no internal opening, and secreting an oily substance. Another and more general character of this family than even the lachrymal sinuses, is the inguinal pores or folds opening inwards and secreting a substance similar to that of the other glands to which we have now referred.

The form of the upper lip is quite various. In some species it forms a broad naked muzzle, as in the ox; in others it is hairy and attenuated, as in the goat, and in still others it shows a modification of both these characters. The hair of the Antelope is usually short and smooth, and of an equal length on every part of the body; some, however, have bristly manes along the neck and shoulders, and a very few species, like the Gnu, have a beard on the chin and throat. Generally these animals are found in large herds, but some species reside in pairs or families. Africa may be regarded as the "head quarters" of the Antelopes. The nature of their *habitat* varies in different species.

This family has been arranged into two grand divisions, the ANTELOPES OF THE FIELDS, and the ANTELOPES OF THE DESERT, between which the most obvious distinction is that in the Antelopes of the Fields "the nostrils are free from hairs, whilst in the Antelopes of the Desert, the nostrils are beaded within, or covered with bristles." (English Cyclopedia.)

I. ANTELOPES OF THE FIELDS. These are arranged into three groups.

1st. *True Antelopes,* "which have a light, elegant body; slender limbs; small hoofs; a short or moderate tail, covered with elongated hairs at the base; lyrate or conical horns, placed over the eye brows."

2d. *Cervine Antelopes*, "approaching the Deer in character. They have a rather heavy, large body; strong, slender limbs; a long tail, cylindrical at the base, with the hair longer at the end, often forming a compressed ridge." The muffle resembles that of the Deer.

3d. *Goat-like Antelopes*, having "a heavy body; strong legs; large hoofs and false hoofs; very short tail, flat and hairy above; recurved, conical horns."

The species in each of these groups are quite numerous; but though all are handsome creatures, we must content ourselves with noticing the more prominent.

True Antelopes.

A. Dorcas. (Gr. δέρκομαι, *derkomai*, to see.) The Gazelle, or the Corinne. (Plate VII. fig. 19.)

This is perhaps the most beautiful of all the Antelopes. Its large, mild, and black eyes beam with lustre, and its light and graceful figure has made it a favorite with Oriental poets. In the sacred writings it is alluded to under the name of the *Roe*, "swift upon the mountains." The Gazelle is common in the northern parts of Africa, where large troops of them bound along with such amazing fleetness that they seem bird-like. The Ariel (*A. Arabica*) a variety of this species, abounds in Arabia and Syria.

"The wild Gazelle o'er Judah's hills
Exulting still may bound;
And drink from all the living rills
That gush on holy ground."

So swift is this animal that the greyhound is generally unable to overtake it, unless aided by falcons which fly at its head, and thus check its speed until the dogs regain their lost distance. In some parts of Syria, the gazelle is taken by driving a herd into an extended enclosure surrounded by a deep ditch. A few openings are made through which the affrighted animals leap and fall into the ditch, when they are easily taken. If pursued in the open field, it flies to some distance, then stops to gaze a moment at the hunters, and again renews its flight. A flock when attacked in a body, disperse in all directions, but soon come together again, and when brought to bay, defend themselves with courage and obstinacy, uniting in a close circle, with the females and fawns in the center, and presenting their horns at all points to their enemies; yet notwithstanding their courage, they are "the common prey of the lion and panther, and are hunted with great courage by the Arabs and Bedouins of the desert." When taken young, the Gazelle is easily domesticated; and it is frequently seen at large in the court-yards of the houses in Syria, the exquisiteness of its form, and its great beauty and playfulness rendering it a special favorite.

The size of the Gazelle's body, (3½ feet long,) about equals that of the Roebuck, but the legs are considerably longer, and the entire form is lighter and more elegant; the fur is short and close pressed; the color a dark fawn above, and white beneath, the upper parts being divided from the lower by a deep dark band along the flanks. The horns are black, lyre-shaped, and have twelve or fourteen rings. Upon the monuments of Egypt and Nubia, this animal is frequently found sculptured. A circumstance of this creature's extreme affection, and which ended fatally, occurred not very long since in the island of Malta. A female gazelle having suddenly died from something it had eaten, the male stood over the dead body of his mate, butting every one who attempted to touch it; then suddenly making a spring, struck his head against the wall, and fell dead by the side of his companion.

A. (*or G.*) *mhorr.* The MOHR. (Plate VII. fig. 22.)

This Gazelle is 4 feet 2 inches long, and 2½ feet high at the shoulder, (8 inches taller than the preceding,) found in Western Africa, and much sought after by the Arabs on account of producing the bezoar stones, called Mohr's eggs in Morocco, and valued in eastern medicine. The Mohr is said to live in pairs, not in flocks like the other species.

A. euchore. (Gr. εὖ, *eu*, well; χορὸς, *choros*, dance.) The SPRINGBOK, or the SPRING-BUCK.

This animal of Southern Africa, in the gracefulness of its proportions and the beautiful variety of its colors, is scarcely surpassed by any other of the Antelope tribe. It is nearly a third larger than the Gazelle; its horns are black, irregularly lyrated, and of moderate length. The most marked peculiarity of this species is a line of long white hairs arising from two longitudinal foldings of the skin, commencing about the middle of the back and extending to the tail. In their ordinary state, the edges of these foldings approach each other, and are so near together, as to conceal, in a great measure, the stripe of white. But when the animal leaps, as it sometimes does, perpendicularly to the height of six or seven feet, the folds are expanded and form a broad circular mark of the purest white extending over the whole croup and hips, producing a very remarkable and pleasing effect. Immensely large herds of these animals are found on arid plains of the interior of South Africa; but when the pools and pastures to which it has been wont to resort, are dried and burnt up by the excessive heat, it migrates to the cultivated districts of the Cape. Travelers who have witnessed these marches estimate the numbers that unite in their migrations at from 10,000 to 50,000. "Cumming's Adventures" give some graphic views of these "grand migrations." Before the migration is closed, it is said, those which happen to get in the rear of the troop are lean and half starved, being left nearly destitute of food in consequence of the cropping of the scanty pastures almost bare by the preceding ranks; but when the troop begin to retrace their steps northward, those which formed the van during the advance, are necessarily in the rear returning; hence they soon lose their plump condition, and, in their turn, are subjected to want and starvation. In their approaches to the settlements of men, thousands of these animals are killed for food. Great numbers of them are also destroyed by panthers, hyænas and wild dogs. On the return of the rainy season, they retrace their steps to the plains of the interior, and in a brief period not a Spring-Buck is to be seen. So fearful is this animal of man, it is said, that "if it has to cross a path over which a man has passed before, it does not walk over, but takes a leap ten or twelve feet high and about fifteen feet long, at the same time curving its back in the most extraordinary manner." It is from this habit of leaping, the dwellers at the Cape have given it the name of *Spring-Buck.*

A. cervicapra. (Lat. stag-goat.) The COMMON ANTELOPE, or SASIN, of India.

This species is spread in large families, over every part of India's rocky and open plains. It is remarkable for the form and beauty of its horns, which are ringed and spirally convoluted, (Plate VII. fig. 18,) having two or more turns, according to the age of the animal. When full grown, it is four feet long and two and a half feet high; almost black above and white beneath; on the knees are tufts of long bristles, forming small knee brushes; the other parts have the hair short and close. The Sasins are so swift that except when taken by surprise, greyhounds are slipped after them in vain; the dogs are more likely to be injured than the game. Capt. Williamson, in his "Wild Sports of the East," says he has seen an old buck-Antelope lead a herd of females over a net at least eleven feet high; and that these animals frequently vault to the height of twelve or thirteen feet, passing over ten or twelve *yards* at a single bound. They are usually hunted by the Cheetah, which "creeps cat-like towards the herd and bounding upon a selected victim, dashes it to the ground with a blow." In size they equal the fallow deer. They are bold and familiar in captivity, and would be graceful ornaments to public parks. The fakirs and dervishes of the East polish their horns and wear them at their girdles instead of swords and daggers, which their religious vocation prevents them from using.

A. tragulus. (Lat. dim. goat.) The STEIN-BOCK, or STONE-BUCK. (3 ft. 4 in. long, 1 ft. 7 in. high.)

This ranks as one of the most elegant and graceful of the Antelope tribe. The legs are longer and smaller in proportion to its bulk than in any other species. A remarkable distinction in this species, (existing also in the Spring or Prong Buck,) is the total absence of spurious hoofs, both on the fore and hind feet, a character which "no other ruminating animals of the hollow-horned family possess." The Stein-Buck resides in pairs on the stony plains and mountain valleys of South Africa. When closely pressed, and without power to escape, it will hide its head in the first hole or corner it meets with, and thus patiently resign itself to its fate.

A. oreotragus. (Gr. ὄρος, *oros*, mountain; τράγος, *tragos*, goat. Mountain-goat.) The KAINSI, or KLIPPSPRINGER. (3 ft. 2 in. long.)

This is an antelope which inhabits the most barren and inaccessible mountains of the Cape, and appears to supply, in South Africa, the place of the Chamois and Ibex; the general color of the hair above is a lively mixture of yellow and green, and light sandy yellow tinged with red beneath; the texture of the hair in this, as in the Spring or Prong Bock, is so fragile that it breaks with the slightest touch, crushing like straw between the fingers, and it is so wanting in elasticity that it never regains its original form. The legs are more robust than in most other species; and the hoofs, instead of being pointed and flat beneath, are entirely round and cylindrical, being worn only at the tips, upon which alone the animal treads. This, with other peculiarities of structure enables the Klippspringer to bound with very surprising agility among the most dangerous rocks and precipices.

A. saltiana. (Lat. leaping or bounding.) The MADOQUA. (2 ft. long, 14 in. high.)

This antelope is found in all parts of Abyssinia, where it was first discovered by Bruce, and lives in pairs in mountainous districts. It is well nigh the smallest of all horned animals, being "scarcely the size of a good English hare;" the color is like that of the American Gray Squirrel, intermixed with deep reddish brown above, and pure unmixed white beneath; the tail is a mere stump; the legs very long in proportion to the weight of the body, and so small that they scarcely equal the little finger in thickness.

A. perpusilla. (Lat. very small.) KLEENE-BOC.

This is an exceedingly small species, about a foot high with horns only an inch and a half long; found at the Cape of South Africa, and called by the Dutch Colonists, Kleene-Boc, (Little Goat.) When domesticated, it soon becomes familiar, and learns to answer to its name.

CERVINE ANTELOPES.

A. oryx. (Gr. ὄρυξ, *orux*, a gazelle.) The GEMS-BOC or ORYX.)

This strong cervine animal is about five feet long, and from three to four feet high, found in the southern and central parts of Southern Africa, and once common but now rare in the Cape Colony. It possesses many of the beautiful peculiarities of the antelopes, but in form it is somewhat anomalous. The horns are black and almost perfectly straight, and situated in the plane of the forehead, about 2½ feet long, blunt in the male, but very sharp-pointed in the female; (Plate VII. fig. 20,) the general color of the body is dark rusty iron gray above, but the head and under parts are white. There are beautifully black bands on the head and flanks, producing a contrast of colors which has a singular effect upon the animal's appearance. In coloring and height, the Gems-Boc resembles the Ass; but in its erect mane and its long sweeping tail it is like the horse, while its head and hoofs are those of the antelope. It always keeps to the open field, living in small families.

Anderson says "it is the swiftest quadruped he met in South Africa, and lives on grass, succulent plants, (often of a very acrid taste,) shrubs, &c. It rarely if ever attacks man, but can defend itself with its formidable horns, even against the lion." Others say that even "the lion himself is afraid to attack this powerful and courageous animal, and that sometimes when pressed by famine, he has ventured to do so, he has been beaten off with disgrace, or even paid for his temerity with his life." The Oryx has been said to live without water, but Anderson remarks that "troops" of this animal "have been found dead or dying near pools purposely poisoned by the natives to capture wild animals."

A. leucoryx. (Gr. λευκὸς, *leukos* ; ὄρυξ, *orux*, the gazelle.) The WHITE ORYX.

This species, called by some the "Milk-White Antelope," is perhaps the most celebrated of all the Antelope genus, it being the one that gave rise to the fabulous unicorn of the ancients. The horns are more distinctly ringed for about half their length than in the preceding species, gradually curved throughout the whole course, and in a side view appearing to be one and the same. The neck, throat, and some portions of the face are brown; but the other parts are milk-white. This species is found represented on the monuments of Egypt and Nubia; "in the inner chamber of the great pyramid at Memphis, a whole group may be seen, (with one exception,) shown in profile, so that but one horn appears." The White Oryx is gregarious; its range is more northern than that of the Gems-Boc, including Nubia and Senegal; its food consists of different species of acacias.

GOAT-LIKE ANTELOPES.

A rupicapra. (Lat. *Rock-Goat*,) The CHAMOIS, or GEMS.

This interesting animal is the only Antelope of Europe, being found in all the high mountain-chains of that region, and also those of Western Asia. The horns of the Chamois are usually but six or seven inches long, nearly parallel in their whole extent,—and bent backwards like hooks at their tip. (Plate VII. fig. 21.) Its length is about three feet three inches; and its height at the shoulders, a little more than two feet; the face is straight and goat-like; the ears are small, erect, and pointed; the long hair of the body hangs down over the sides, and is of a deep brown color in winter, a brownish fawn in summer, and in spring, slightly mixed with gray; the pale yellow of the head is banded with dark brown on each side. Beneath the external covering, is a short thick coat of fine wool, which lies close to the skin, and protects the animal from the severe weather of cold mountainous regions, and the bruises to which, from its *habitat*, it is liable. The hoofs are admirably adapted for security, enabling it to avail itself of every little roughness and projection, either from the naked granite, or from the icy glaciers.

In its elevated home, the Chamois displays all the vivacity, restlessness and agility of the Common Goat. It does not bear heat, and is, therefore, in summer found on the tops of the highest mountains, or in deep glens where the snow lies during the year; in winter, it descends to lower ridges, and then only is it hunted with any prospect of success. All its senses are exceedingly acute; and these, combined with its agility, are its means of security. Its sense of smell, it is said, will enable it to perceive an aggressor at the distance of one and a half miles. It is restless, and very much alarmed until it gets a sight of the object of its terror, leaping upon the highest rocks at hand, in order to obtain a more extensive prospect. When undisturbed, its voice is a low kind of bleating; if excited by the approach of a hunter, it utters a suppressed whistle, or hissing sound, and all the while, shows much agitation; but when the hunter comes near, it flies with its utmost speed,—bounding from ledge to ledge, where the eye can mark no footing,—and from crag to crag, and point to point —sweeping over the glacier,—throwing itself down precipices of fearful depth, and pitching, almost by miracle, upon the slightest projection. "It does not descend at a single bound, nor in a vertical direction, but by projecting itself obliquely or diagonally forwards, striking the face of the rock three or four times with its feet for the purpose of renewing its force, or directing it more steadily to the point it aims at; and in this manner, it will descend a rock almost perpendicular, of twenty or thirty feet in height, without the smallest projection upon which to rest its feet."

The hunting of the Chamois, is among the most perilous of human undertakings, and involves "a perversion of mental energies capable of better things." It has been remarked; "no Chamois hunter ever dreams of any other death than that of falling from the brink of a precipice, or being buried in some chasm beneath the treacherous snow;" yet urged on by a sort of fascination, "he pursues his course of life with feelings allied to those of the gambler, alternating with hopes and fears."

The Chamois seldom drinks. Its food consists of mountain herbs, flowers, and the tender roots of trees and shrubs. This gives a richness and a fine flavor to the flesh, which is much esteemed as a venison. For this and the skins, the Chamois hunters jeopard their lives. The animal can seldom be captured alive, and rarely thrives in captivity. "Like the Swiss, its congenial home is among its native mountains, and in its native liberty."

A. furcifer. (Lat. Prong-bearer;) or *Antilocapra Americana.* The Prong-Horned Antelope.

The absence in this animal of inguinal and lachrymal openings, and of accessary hoofs, together with the fact that it has branching horns, (Plate VII. fig. 17,) of which no instance occurs among the other species of Antelopes, led Audubon to refer it to the genus *Antilocapra*, derived from the two genera, *Antilope* and *Capra*, Goat-Antelope. This Antelope is confined to the Western portions of North America, and is never seen East of the Mississippi, but ranges as far South as California and New Mexico, feeding on moss, buds, &c.

It is shortly, but more compactly built than the Virginia Deer, but in its elegant and stately form, resembles more the Antelope than the Deer family. The horns of the male are curved upwards and backwards, with a short triangular prong about the centre. In winter a ridge of coarse hairs, resembling a short mane, appears on the back of the neck, of which, in summer, only a black stripe remains,—the color is a reddish dun, with the throat and the clink on the hinder parts white. The head, ears and legs are covered with short close hair of the common description, but that of the body is long and padded, and of a texture altogether different from that of other animals; it being hollow like the feather of a bird, brittle, and when bent, not returning to the original straight form. The animals are gregarious, sometimes several hundreds being found together, and they migrate from North to South according to the season. When the ground is clear, their speed surpasses that of most other animals, but a good horse easily outstrips them after a slight fall of snow. They are sly, but extremely curious; and the Indians, and even the wolves, it is said, know how to take advantage of their curiosity to get within reach of them, by crouching down and moving forwards, or stopping, alternately. These Antelopes will wheel round and round the object of their attention, decreasing the distance at every turn, till at last they approach sufficiently near to be shot or captured. Sometimes they are caught in pens, in nearly the same manner as the bison; but in the deep snow of winter, when they are suffering for want of food, they are generally dispatched with clubs,—Audubon says, "principally by the women." They are fattest in autumn. "Their liver is much prized as a delicacy, and we have heard that many of these animals are killed simply to procure this choice morsel." (Anderson.) Their flesh, however, is not highly esteemed by the Indians, who hunt them only in times of scarcity.

II. Group.—Antelopes of the Desert.

A. Gnu. The Gnoo, or Horned-Horse.

This equine Antelope is sometimes called *Catoblepas*, (Gr. καταβλέπων, *Katablepōn*, looking down,) a name well expressive of its sinister aspect, shaded as its face is by overgrown horns, bent down and outwards, on the sides, broad at the base, and bent up at the tip. (Plate VII. fig. 23.) It has a wide and bristly nose, with large covered nostrils; and in size about equals a well grown ass. The neck, body and tail, precisely resemble those of a small horse, and the pace also, which is a species of light gallop, is so perfectly similar, that a herd of Gnoos, when seen at a distance, flying over the plains of South Africa, "might be readily mistaken for a troop of the wild zebras, or quaggas, which inhabit the same locality, if their dark and uniform color did not distinguish them." They are naturally wild and difficult to approach, and when provoked very dangerous if wounded, turning upon the hunter and pursuing him, dropping on their knees before making an attack, and then darting forward with amazing force and velocity. "When the hunter approaches the old bulls, they commence whisking their long white tails in a most eccentric manner; then springing suddenly into the air, they begin prancing and capering, and pursue each other in circles at their utmost speed. Suddenly, they all pull up together, to overhaul the intruder, when two of the bulls will often commence fighting in the most violent manner, dropping on their knees at every shock; then quickly wheeling about, they kick up their heels, whisk their tails, with a fantastic flourish, and scour across the plain enveloped in a cloud of dust." (Cumming's South Africa.) They are said to be subject to a cutaneous eruption at particular seasons of the year, which they sometimes communicate to domestic cattle, and which invariably ends in death. Their flesh is in good repute both among the natives and colonists.

A. Caama. The LECAMA, or HARTE-BEEST.

This species of Bovine Antelopes inhabit the plains of South Africa, and are the *most common* of all the large Antelopes in that country. They are of a gray-brown color; reside in large herds; and are much hunted by the natives and colonists. Their pace resembles a heavy gallop, but yet is tolerably quick. In their manners, they are mild and tractable; but when put upon their defence, they make good use of their powerful lyrate horns, like the Gnoo, dropping upon their knees before charging, and after advancing some distance in this position, suddenly darting with great force against the hunter. The flesh is much esteemed, being more like ox-beef than that of any other Antelope, except, perhaps, the *Eland*.

The Strepsicerae (twisted horns) is another small group referred to in the "Penny Cyclopædia," under the name of Antelopes, and including some very interesting Ruminants. They are named from the subspiral, or twisted form of their horns; and distinguished among the "Hollow-Horned Bovine Ruminants," by being marked with white stripes and spots. Agassiz has remarked, that the horns of the *Strepsicerae* and the sheep are twisted in opposite directions.

A. Stepsiceros. The KUDOO.

This magnificent animal is found in South Africa. It is one of the largest of the Antelopes, being upwards of eight feet long, and four feet high at the shoulder. The horns of the Kudoo, for which it is most remarkable, are nearly four feet long, and beautifully twisted into a large spiral form, of about two turns and a half. A bold ridge runs over the horns and follows their curvature. (Plate VII. fig. 15.) The leading color is a bright fallow-brown, with a narrow white stripe along the spine. In its external aspect, the animal more nearly resembles the ox than the Antelope. Although large and heavy, it can leap with wonderful activity. The weight of the horns is considerable, and in part to relieve itself from that weight, and in part also to keep the spreading horns from entanglement in the bushes on which it lives and feeds, the Kudoo usually bends its head back and rests its horns upon its shoulders. When closely pursued, it takes to the water, and seeks to escape by its power of swimming.

A. oreas. (Gr. ὀρειάς, *oreias*, of the mountain.) The ELAND, or the BOSELAPHUS, (ox-stag,) of the ancients.

We have in this animal *the largest* of the Antelopes,—measuring eight feet two inches in length, and full five feet in height at the shoulder—being quite as large as a good sized horse. It has very thick, nearly straight horns, about a foot and a half long, and covered, for the most part, with a thick spiral wreath. The ears are large. A protuberance, of the size of a man's fist, appears on the larynx; from this organ, the animal probably derived the name of *Eland*, (as it is called at the Cape Colony.) When full grown, it weighs from seven to nine cwt.; and, contrary to the usual rule observed among Antelopes, is commonly extremely fat. The flesh is more highly prized than that of any other animal in South Africa. The Eland is mild and inoffensive in its disposition, so that a man may penetrate into the very midst of a herd without alarming them. Being quite heavy, the great object in hunting this animal, is to turn the game in such a direction as to drive it close to the residence of the hunter before it is killed; and the Cape farmers, it is said, "very frequently succeed in accomplishing this master piece of South African field sports."

A. picta, (painted.) The NYL-GHAU.

This large and magnificent Antelope is about the same size as the Gnoo, standing about four feet high at the shoulder. It is found in the forests of N. W. India, ranging thence as far as Persia. The face of this species is long and narrow, surmounted with short, smooth, and nearly parallel horns. The fore-quarters are considerably raised, and there is a slight elevation upon the withers; the neck is long and horse-like; from the throat and shoulders hangs a dense bunch of hair; the haunch is small and low, so that the hinder limbs are short. The Nyl-Ghau is less graceful in its proportions than the Stag, but more muscular and powerful. The color of this animal is a slaty blue; it has, however, several white spots which, contrasting with the slaty blue, or dark brown of the other parts, suggested the specific name of *picta.* It is extremely vicious, and cannot be approached without danger. In making an attack, it first falls upon its knees, like the Gnoo, and then springs violently forward. It is the common prey of the tiger; and hunters erect their platforms near the mangled remains of this animal, well knowing that the tiger will return to glut himself with the remainder of his prey. During the day, the Nyl-Ghau conceals itself in the forests, and at night leaves its coverts to feed, often doing harm to adjacent cultivated fields. It has been often taken to England, where it breeds, and is not an uncommon animal.

QUESTIONS ON ANTELOPES.

What is the derivation of the word Antelopidae? What families are the Antelopes thought to connect? In what respect are they like the goats? In what like the deer? What advantages do their hind limbs give them? State the color of the larger part of them. What is said of the tail, ears and hoofs? What is the most general character of the family? Are their horns solid or hollow? What is said of their tear-pits? From what do these distinguish them? What makes this family an intermediate link between the two kinds of horned animals? What additional glands are spoken of? What is said of the form of the upper lip? What of the hair? Are they gregarious? What two grand divisions do the Antelopes embrace? What is the most obvious distinction between the two? Name the groups of *Antelopes* of the *Field*, with their characters. Where is the Gazelle found? Give some account of its peculiarities and habits. Where is the *Mohr?* Why is it sought after by the Arabs? What is the locality of the *Springbok?* Is it larger or smaller than the Gazelle? What is its most marked peculiarity? What is said of its leaps? What of its migrations? How widely is the *Common Antelope* diffused? What is said of its horns? Illustrate its swiftness. What animals are used in hunting it? How large is it? What Antelope is next mentioned? What is said of it? Where is the *Klipspringer* found? What is peculiar in its hair? What enables it to bound with very great agility? Who first discovered the *Madoqua?* What is said of its size, color, &c.? Where is the *Kleenbok* found? What is said of it? To which division of the Antelopes of the Field do the preceding ones belong? Where is the *Oryx* found? What is its size? Give some description of it? What is said of it by Anderson and others? Which is perhaps the most celebrated of all the Antelopes? To what fabulous animal did it give rise? How? Where is it found sculptured? What is its range? To what division do the two preceding Antelopes belong? Which is the only Antelope of Europe? What characteristics are mentioned? Further describe it. What is said of the hunt of this animal? How is its flesh esteemed? To what genus is the *Prong-Horned Antelope* referred? Name its characteristics and habits. To what division do the two last named species belong? What are Antelopes of the second group called? Which of these is first mentioned? Name its distinctive traits and habits. Give some account of the *Lecama*. What other small group of Antelopes is mentioned? Why are they so named? What distinguishes them? What has Agassiz remarked? What is said of the *Kudoo?* For what is it most remarkable? What animal does it most resemble? Which is the largest of the Antelopes? What gave it the name *Eyland?* What is said of it? Where is the *Nyl-Ghau* found? Give a description of it.

Name the species on the *Chart*. Trace them. Give the most prominent characteristics of each as a general review.

SECTION XXXI.

NINTH ORDER. CETACEA. (Gr. κητος, a whale.) WHALES, DOLPHINS, ETC.

MARINE-MAMMALS.

This is an order of mammiferous animals inhabiting the sea; surpassing all others in size, though lower in organization than those living upon the land. Moving in the water by means of fin-flippers, or paddles, "the earlier naturalists placed them among the fishes; but all now unite in placing them among the mammals." Like them, they are viviparous, (born alive,) suckle their young, have warm blood, and breathe by means of lungs. The contour of the body, is fish-like, no neck being distinguishable, and the whole tapering down gradually from the head to the tail. The tail, however, terminates, not vertically as in fishes, but horizontally, in a cartilaginous fin, and is moved upwards and downwards by muscles of enormous force and volume. In length, it is only five or six feet; but in width, from eighteen to twenty-six. So powerful is it in the largest varieties, that they frequently force themselves out of water. The greatest velocity is given by the upward and downward strokes; a slower motion is obtained by cutting the water sidewise, and obliquely downwards, as a boat is forced along by a single oar in the operation of *skulling*. So rapid are the movements of the Cetacea, they have been called the "birds of the sea." The flippers, or paddles, the anterior limbs, are generally stretched out in a horizontal position. When dissected, the bones of the paddles are found to be short and flattened, yet distinct and handlike; but the whole of this osseous frame-work is enveloped in a cartilaginous covering, so as to form an undivided oar. The chief use of the paddles seems to be that of balancing the animal, for as soon as life is extinct, it falls over upon its back; they are also employed in turning and giving direction to the velocity produced by the tail.

The Cetacea regularly resort to the surface to take in a fresh supply of air. They also descend into the remotest depths of the ocean; in the case of the larger animals sometimes encountering a pressure which has been estimated at two hundred thousand tons, or one hundred and fifty times as great as that of the atmosphere, and sufficient to force water through the hardest wood, causing it to sink like so much lead. For sustaining so vast a pressure, their structure is most wisely adapted.

The body is covered with a coat of peculiar elasticity. The naked skin is itself much thickened; but by an open texture of its interwoven fibres, it is made to contain within itself, a thick layer of oil or blubber, and thus the animal can endure, without injury, the greatest weight of water. "A soft wrapper of fat, though double the thickness of that usually found in the Cetacea, could not have resisted the superincumbent pressure; whereas, by its being a modification of the skin, always firm and elastic, and in this case, being never less than several inches, and sometimes between one and two feet thick, it operates like so much india-rubber, possessing a density and resistance which, the more it is pressed, resists the more."* As the blubber is specifically lighter than water, it also makes the animal more buoyant. A dead whale floats; but the body, when stripped of its fat, sinks immediately. Another important use of the blubber, is to preserve the vital heat of the body in a cold medium, which has a constant tendency to abstract caloric. Without this layer of blubber, which is one of the worst conductors of heat, the whale would be unable to resist the low temperature of the Arctic Seas, and must perish from cold. The eyes of the Cetacea are admirably adapted to the dense medium in which the animals dwell. As compared with the size of the body, the eyes are small,—generally not larger than those of an ox; in the Beluga, or White Whale, they are smaller than the human eye; in the Porpoise, not so large as those of a sheep. In the Cetacea, "the humours of the eye are so adjusted in their form, density and refractive power, as to prevent any *dispersion*, or decomposition of the rays." The refractive power of the aqueous humor,

---

* Naturalist's Library, Mammalia, VII., 48.

which is great in respect to land animals, would, in *water*, be comparatively weak; this defect is, in the case of the Cetacea, supplied by the *spherical form* and *great refractive power* of the lens of the eye. The outer, or sclerotic coat is, in these animals, remarkably thick and tough, it being as dense as tanned leather, serving both to preserve its spherical form, and to defend the animal from injury. This coat increases in thickness towards the back part, and is full five times the thickness behind, that it is on the front part. To this, Dr. Paley (see Nat. Theol.) has well referred, as strikingly evincing Divine contrivance. The front part sustains the pressure from without, and needs no additional support; but were the back part to yield, the globe of the eye would be distended in that direction, and the whole interior of the eye suffer derangement. As a safe-guard, the sclerotic coat is, therefore, remarkably strengthened behind.

One of the most extraordinary things in the economy of the Cetacea, is the length of time during which they can suspend respiration. While, in most animals, it can be suspended only for a few minutes, in some of the larger whales it may be suspended from one to nearly two hours, they remaining under water for that time. This fact points to the peculiarity of their breathing apparatus. The whale has a reservoir wherein there is an overplus of oxygenized blood which, on occasion, is emptied into the general circulation; it is thus able to continue longer under water, and less frequently resorts to the surface in order to inhale oxygen from the atmosphere. Whales have no nostrils, properly so called, and their mouths are seldom opened in the free air. The process of breathing is therefore carried on through tubes, called blow-holes, or spiracles, opening on the top of the head, and allowing a free passage to and from the lungs. These openings are called *blow-holes*, because the expulsion of the long-confined and heated air, as the animal rises to the surface, is attended with considerable noise, and the casting forth of water or steam. The "spoutings" are heard as far as two miles, and sometimes reach the height of twenty or thirty feet. They are most conspicuous in the larger genera; quite marked in the intermediate dimensions; but in the smaller, seldom or never visible. After the "spoutings are out," as the whalers say, most of the Cetacea descend into the depths of the ocean. The lungs are guarded from injury that might hence arise, by the conical stopper which, like the cork of a bottle, fits itself to the blow-hole so perfectly, as to exclude every drop of water. Habitually, the whales take their sustenance under water; but, "by a slight alteration in a few cartilages at the top of the windpipe, and in the direction of the air tubes, they feed as safely in the deep ocean as others do in the most balmy atmosphere."

The external opening of the ear is minute, and in some species, hardly discoverable. This can be closed at pleasure. The hearing, as well as sight, is quick beneath water,—whales have the sense of smell in some degree; showing themselves sensible of the noxious smell of bilge-water, pumped from the hold of vessels. The senses of taste and touch they possess, but in a less degree than other animals.

The stomach of the whale is divided into five, and sometimes seven distinct sacs, or pouches; instead of a single spleen, they have several, which are small and globular. The teguments of the tongue are soft and smooth. Those of the Cetacea which are possessed of teeth, have them all of conical shape, and all alike.

The brain in these animals is small, though the size of it varies in different genera. In a young Greenland whale it was found to be $\frac{1}{3000}$ part of the whole animal. The proportion in the Dolphin is much greater, the brain being $\frac{1}{25}$ part of the whole, and approaching quite near to that of man.

The degrees of intelligence manifested by the Cetacea, are various, and so are their dispositions; but all agree in the mutual regard which they entertain,—the mother for her young, the cub for its parents; and members of the same family, or shoal, for one another. The female has but one young at a time, in the early spring, which is about ten or twelve feet long at birth; the mammae are two in number, and situated near the vent.

The Cetacea may be divided into four families, viz: I. The DELPHINIDAE, including Dolphins, Porpoises, etc.; II. CATADONTIDAE, or PHYSETERIDAE, Spermaceti Whales; III. BALAENIDAE, True Whales, Common, Right or Whale-Bone Whales; IV. HERBIVOROUS CETACEA, including the Manatees, Dugongs, etc.

I. DELPHINIDAE. (Gr. Δελφὶς, *Delphis*, a dolphin.)

This is the Dolphin tribe, characterized by the moderate size of the head, and usually by the presence of teeth in both jaws. They are voracious feeders; their flesh is, for the most part, rank, oily, and unwholesome. They include seventeen genera, and sixty-four species, and are the most numerous family of the Cetacea; are scattered in all seas, and frequently ascend rivers. One genus (Inia) is found in the mountain lakes of Peru,—the fountains of the Amazon, and a thousand miles from the ocean. "They are the *Carnassiers* of the waters, preying upon the fishing tribes, which they chase in all directions; and their teeth are modified accordingly."

Delphinus. (25 species.)

Delphinus Delphis. The COMMON DOLPHIN. (Plate VIII. fig. 4. and Pl. VI. fig. 12.)

The animals of the genus *Delphinus*, have more teeth than any other of the Mammalia, the number averaging ninety in each jaw; in form, simple and conical, but adapted for seizing only. The jaws of these animals project so as to be like a slender beak, separated from the forehead by a groove, or furrow, that resembles the bill of a goose. (Plate IV. fig. 11.) There is a fin upon the back. The Common Dolphin is usually six or seven feet long, sometimes nine or ten feet. Its form is admirably adapted for swimming. The tail is large and powerful. This animal is familiar to fishermen and mariners, and cannot but be regarded with interest, on account of its beautiful and graceful form, the fleetness with which it darts through the waters, its gambols and leaps, and its social habits.

So smooth are the bodies of Dolphins, that "their sportive gambols create surprisingly little disturbance of the water." To the ancients, the manners of the Dolphin were well known, and to them, its playful, social disposition, made it a great favorite. It is accurately figured on many of their coins. Among the Greeks, it was sacred to Apollo, who was worshiped at Delphi with Dolphins for his symbols. It early appeared on the shield of some of the princes of France, gave name to a province of that empire, and a title to the heir-apparent of the crown.

The brilliancy and variety of many of the Dolphins found in the Southern and Equatorial seas, cannot be adequately represented by pictures, or exhibited in words. The Dolphin, "with its many dying colors," of which poets have sung, is, however, not the *true animal*, but a scomberoid *fish*, *Coryphaena hippurus*, the Dorado, of the Portuguese; though, as Dr. J. E. Gray remarks, "to this fish, which changes color in dying, most maritime persons generally confine the name of Dolphin." But, however it may be celebrated in story and in song, the Dolphin appears quite wolfish in its habits,—in troops, hunting down its prey,—in its rapid course, forcing the flying fishes to take refuge in the air; but continuing the chase until the exhausted victims are taken. Of the many wonderful stories related by ancient naturalists respecting the Dolphin, we have room for only the following. Pliny says that, "in Barbary, near the town of Hippo, a Dolphin used to frequent the shore, and receive food from any hand that supplied it,—that it would mix with persons bathing, allow them to mount its back, and obey their direction, with all celerity and precision." The ancients speak of the

Dolphin as peculiarly partial to children; and here we quote a further incident from Pliny, which has pathos, whatever may be thought of its *truth.* A Dolphin, which he says had penetrated the Lake of Lucrinus, in Campania, every day received bread from the hand of a child, answering to his call, and transporting him on its back to school on the other side of the lake. This intimacy continued for several years, when the boy dying, the affectionate Dolphin, overwhelmed with grief, soon sunk under its bereavement.

Monodon. (Gr. μόνος, *monos,* solitary; οδοὺς, *odous,* a tooth.)

Monoceros, the specific name is from Gr. μόνος, (*monos,*) and κερας, (*keras,*)horn.

This is the Narwhal. (*Nar,* signifies in Icelandish, a *horn; whal* or *wale* is synonymous with our word, *whale,* and derived from the same Teutonic root.) It is also called the Sea Unicorn, or Unicorn Whale. This Whale has no teeth, properly so called; it has, however, two tusks, one on each side the head. Only the left tusk projects, (from the upper jaw of the male,) the other remaining within the head, whence the name *Monoceros,* or *Unicorn.* This horn, or tusk, is eight or ten feet long, tapering, with a rope-like twist, to a point, and harder and whiter than ivory. Formerly, the tusks brought a high price. Many medicinal virtues were attributed to them. They are still of value as an article of trade. The kings of Denmark are said to have a magnificent throne made of these tusks, which is preserved with great care, in the castle of Rosenburg. The length of the Narwhal is from twenty to thirty feet, and, including the tusk, between thirty and forty. It is thought the animal uses the tusk to pierce the ice for the purpose of breathing, and also in capturing the fishes on which it feeds. It usually precedes the *Mysticetus,* both using the same kind of food. Hence, when Greenlanders see unicorns, they prepare for fishing in *earnest.* It is on record that the thick oak timbers of a ship have been pierced by the horn. Sometimes the Narwhal drives it into the sides of the huge whale, and greedily receives the oily blubber which oozes from the wounds thus inflicted.

Dr. Scoresby describes the Narwhals as active and inoffensive,—often sporting about his ship, sometimes in bands of about twenty together, raising up their long tusks, and crossing them with each other as if fencing. Our own lamented Kane says, "the play of a group of Narwhals is graceful, striking and beautiful." The blubber yields a superior oil. This and the flesh also are highly valued by the Esquimaux and Greenlander. Their tusks afford them weapons of defence, and even the intestines they use for lines.

Beluga. The WHITE WHALE.

Of this genus there are two species, viz.: the Northern Beluga, (*B. catodon,*) the Australian Beluga, (*B. Kingii.*) The shape of the Beluga is that of a double cone, one end of which is considerably shorter than the other, and extremely well adapted to motion in the water. It is known by its white color. The length varies from twelve to twenty feet. The tail is powerful, bent under the body in swimming, and impels it forward with the velocity of an arrow. The eye is scarcely larger than that of a man; the iris is blue. It has no olfactory nerve, no external ear, and the mouth is small when compared with the bulk of the animal. Its food is codfish, haddocks, and other fish. The favorite resorts of the Northern Beluga are the higher latitudes of the Arctic regions, Hudson's Bay, Davis' Straits, and the northern coasts of Asia and of this continent, where they frequent large rivers. They are found in the Gulf of St. Lawrence, and go with the tide as far as Quebec; and there are fisheries for them as well as for porpoises in the river St. Lawrence. They yield a considerable quantity of oil which is said to be of the finest quality. Of their skins a sort of morocco leather is made, which, though thin, is strong enough to resist a musket ball. They are not shy, but often follow ships and tumble about the boats in herds of thirty or forty, bespangling the surface with their sparkling whiteness. The whaler seldom disturbs these beautiful creatures; they being very active, it is difficult to strike them; the harpoon often gives way, and they are of comparatively little value when killed. They are said to visit the west coast of Greenland about the end of November, and are then very useful to the natives as their provisions fall short. In taking them, harpoons and strong nets are employed. The internal membranes are used for windows and bed-curtains, and the sinews for thread. The flesh resembles beef, but is to some extent oily.

Phocæna. (Gr. φώκαινα, *phōkaina,* a porpoise.)

The characteristics of this genus are as follows: "Head rounded, not much elevated; mouth terminal; snout, short and rounded; a dorsal eminence, (as in the Globicephalus,) usually of a small size; gregarious; piscivorus." (N. H. S. N. Y.)

Phocæna communis. The COMMON PORPOISE, or PORPESSE. (French, *Porcpoisson.*) (Pl. IV. fig. 11.)

Of all the Cetacea, this and the allied varieties are most common, being found in almost all the seas of Europe, and in large numbers on the coast of the American continent. It is common in our rivers and bays. It was formerly "so abundant on the shores of Long Island as to have induced the inhabitants to form establishments for its capture." The Common Porpoise, like the Common Dolphin, is the smallest of the varieties. Between the two there is a general resemblance in color, shape and disposition. The scarf skin of the porpoise is very soft to the touch, and easily detached. The eye has the iris of a yellowish hue, and the pupil in the form of V reversed. The opening of the ear is not larger than the prick of a pin; that of the blow-hole is on the top of the head, between the eyes. The dorsal fin, or eminence, is not bony, but composed entirely of fat, and incapable of separate movement; and the tail is without any osseous part within. The fat, or blubber is white, from one to two inches thick, and when heated yields an oil that is fine and much valued. It is "cut through on the back and belly and is peeled off in halves; it is scraped off with an instrument resembling a currier's knife, and the skin is then sent to the tanner. The leather made from this skin is said to be the strongest known, and is used more particularly for the upper leather of boots and shoes." (N. H. S. N. Y.) The deep bluish color of the Porpoise fades away on the sides, till it acquires a silvery whiteness. It has ninety-two teeth, cutting and somewhat rounded at the edge. The brain is large and has deep convolutions lying over the cerebellum. The porpoise, the dolphin, and the monkey are the only animals that in this respect resemble man. The food of porpoises is chiefly fish, and they occasionally pursue shoals of herring and mackerel, which they drive into the bays in very great apparent terror. They are great enemies of salmon, which, when pursued by the porpoise, often spring several yards out of the water; but from the quickness of their foe, are unable escape. The flesh of the porpoise was once esteemed a voluptuous kind of food, and is said to have been found on the table of the old English nobility as late as the time of Queen Elizabeth. Later than this it was extensively used in some countries, especially during the time of Lent.

GRAMPUS. (*Phocæna orca.*) The term Grampus is a corruption of the French, *Grand-poisson*, great fish, pronounced by the Normans, *Grapois*, whence came the English word GRAMPUS. American sailors have given it the names of "Killer and Thrasher." By some, (see Cat. of British Museum in Eng. Cyc.,) a portion of the animals once included under the genus Grampus, has been formed into a new genus, "Orca," which includes the KILLERS proper, and has four species.

The body is thick in proportion to its length, and of oval shape. The snout short and roundish; the lower jaw somewhat bent upwards, broader, but not so long as the upper. The teeth are forty-four in number, eleven on each side above and below, varying in number with age, sometimes are as many as sixty, and interlocking when the jaws are shut. The dorsal elevation, improperly called a fin, is from four to six feet high; the pectoral or swimming fins are large and oval, and it has a strong tail. The color is glossy black above, white beneath; occasionally there is a large white patch behind the eye, resembling an eyelid. The length is from twenty to thirty feet; the circumference from ten to twelve. The favorite abode of the Grampus is the coast of Greenland, Spitzbergen, and Davis' straits; it is also found in the Atlantic and Mediterranean. It was formerly numerous on the coast of New York State. It is a very powerful and voracious animal, devouring great numbers of fishes, large ones especially, such as the cod, haddock, and turbot, and even seals and porpoises have been found in their stomachs. "They go in company by dozens, will attack a young whale, and bait him like so many bull-dogs." The oil which they yield is of excellent quality. Fishermen sometimes call them Finners, or Black-fish Whales. Stories are told of their attacking whales, joining in herds for that purpose; but these perhaps need confirmation. Sir Joseph Banks says of one that was captured in the Thames, (Eng.,) "It pulled the attached boat twice from Blackwall to Greenwich, and once as far as Deptford, at the rate of eight miles an hour, and it was for a long time unimpeded by the lance wounds which were inflicted on it when it came to the surface. So long as it was alive, no boat would venture to approach it; and the dying efforts of this formidable creature were terrible. It was finally killed opposite the Greenwich Hospital."

G. Cuvieri, or *Phocæna grisea*, (of Lesson,) is a handsome species inhabiting the North Sea; has been taken on the West coast of France; is ten or eleven feet long; has only eight teeth, and these in the lower jaw. It is famous for uttering loud cries like the *Deductor* (or Howling) whale, and associating in groups like that whale.

Delphinapterus. This genus includes two species, *D. Peronii*, Peron's Dolphin, and *D. Borealis*. The head is rather convex in front, nose depressed, forming a slender beak, and there is no dorsal fin. The form and proportions are elegant. The snout, as far as the eye, and the under parts of the body and the tail are of silvery whiteness; a bluish black covers the upper parts of the body, giving it the appearance of having on a black cloak. The iris is of an emerald green color. The *D. Peronii* is the Right-Whale Porpoise of the Whalers, found in the higher southern latitudes. The D. Borealis inhabits the North Pacific. (See Peale's description in the U. S. Exploring Expedition.)

Globicephalus. (Globe-headed.) This contains five species, viz.: *G. Swineval*, Pilot-Whale, (North Sea;) *G. intermedius*, the Black Fish, (N. America;) *G. affinis*, Smaller Pilot-Whale, (locality unknown;) *G. Sieboldii*, Naiso Gota, (coast of Japan;) *G. macrorhynchus*, South Sea Black Fish. They are characterized by the absence of a snout, by having a globular head, an eminence resembling a fin on the back, and a single spiracle, situated near the back of the head. The length varies from sixteen to twenty-four feet; the pectoral fins are from six to eight feet, and the tail five feet in length. The second species resembles the Grampus in size, and is probably often confounded with it. The teeth are from twenty to twenty-eight in number in each jaw, and when the mouth is closed, they "shut together like a rat-trap." It is called the *Deductor*. With blind confidence, these animals follow one as a *leader*, the main body keeping close to him, "as sheep follow the wether." Efforts are therefore made to entrap the leaders, and then many others are taken. They are inoffensive, and so timid that men in boats, with ineffective weapons, and with shouts and noise in the water, drive them in great numbers to the shore, to their own destruction. When any one strikes the ground, it is said that it sets up a howling cry, and immediately others crowd to the spot as if for its relief. This circumstance has given it the name of the ca'ing (calling) whale. It is also called the Black Whale Fish, (species *G. intermedius*, or *melas*,) and Bottle-head. Of all the Cetacea it is the most sociable, vast numbers being found together, whence it is named the Social Whale. Large herds of these whales are frequently stranded and perish on the coast, particularly in high northern latitudes. "At Wellfleet, near Cape Cod, in 1822, a herd of one hundred, varying in length from ten to fifteen feet, were stranded and captured. In 1823, one was taken in Salem harbor, Mass.; in 1832, one at Fairfield Beach, Conn.; in 1834, two on the east end of Long Island." (Nat. His. S. N. Y.)

II. CATODONTIDÆ. (Gr. κατα, *kata*, under; οδους, *odous*, a tooth;) or PHYSETERIDÆ, (Gr. φυσητήρ, *Phuseter*, a blow-pipe, or bellows.) TOOTHED WHALES.

This family of the *Cetacea* are distinguished by the enormous size of the head, which occupies more than one-third of the whole bulk of the animal, and ends in a broad muzzle, appearing as though it had been abruptly cut off. The lower jaw is narrow, slender and pointed, and has numerous stout conical teeth, while the upper jaw contains either none or a few which do not perforate the gums. Hence the name *Catodontidæ.* The blow-holes have but one orifice, situated at the top of the muzzle. The three genera constituting the family, agree in their essential characteristics; we therefore omit a detailed and separate description and confine our remarks to the *Catodon*, or *Physeter macrocephalus*, (μακρος, long; κεφαλη, a head,) the Northern Sperm-Whale. It is sometimes called the *Cachalot*, a term derived from Cachon, a tooth, in the Basque (Spanish) language. The Sperm-Whale (Plate VIII. fig. 3,) is of enormous size, being between seventy and eighty feet in length, and from thirty to thirty-five in circumference. From its frequent paroxysms of fury it is one of the most dangerous monsters of the deep. It is found in all latitudes, but is a native of the Arctic and Antarctic Seas, where it is seen attended by its young. Sperm-Whales usually appear in parties of from two to five hundred, guarded by one or two males of the largest size. In the upper part of the head there is an immense cavity, divided into compartments and smaller cells, filled with oil which is fluid when the animal is alive, but hardens when cooled or after the animal is dead, and is known under the name of spermaceti. A hole is made in the head as soon as the whale is killed and the spermaceti is baled out with buckets. When the first process of squeezing and draining the oil is over, the yellow, unctuous and impure mass of cetine is put into bags made of hair or woolen, and further pressed between plates of iron in a screw press until it becomes hard and brittle; it is then broken into small pieces and thrown into boiling water, where it melts and the impurities are separated from it. After being cooled and taken from the first water, it is put into a boiler of clean water and a weak solution of potash is gradually added. This is thrice repeated, after which the whole is poured into coolers, where it crystalizes, and on being cut, exhibits the beautiful flaky appearance belonging to the spermaceti of commerce. An ordinary sized whale will yield from ten to twelve barrels of crude spermaceti. Ambergris, which is used as a perfume, and often found floating on the surface of the sea, is a fatty concretion formed by disease in the intestines. Upon the ivory teeth of the Sperm-Whales, sailors often show their taste in carving figures of various kinds. These whales produce but one young at a time, about fourteen feet in length, and having a skin much thicker than that of the old ones. The milk by which the young are nourished resembles that of quadrupeds. The throat of the Sperm-Whale is capacious enough to give passage to the body of a man, presenting a strong contrast to the contracted gullet of the *mysticetus*, or Greenland Whale. The mouth is lined with a pearly white membrane. The eyes are small in proportion to the size of the animal, and furnished with eye-lids; the skin is usually smooth, but in old whales sometimes wrinkled. At each breathing time, the Cachalot makes from sixty to seventy expirations, remaining at the surface of the water ten or eleven minutes. It continues below the surface for periods of from an hour to an hour and twenty minutes, consuming about one-seventh of its time in respiration. The Sperm-Whale feeds upon seal and fishes, which it pursues with great pertinacity; but a large species of cuttle fish, (*Octopus*,) is said to constitute its principal food. Its forty-eight huge teeth, which it sometimes employs in biting boats, make it formidable to whalers. Sometimes it swims off to a distance, and then rushes at the boat with its head, thereby knocking it to pieces. One of these whales sunk a ship by three or four blows from its head. The Sperm-Whale fishery is a principal branch of the industry of the United States, hundreds of ships being engaged in this important branch of the fisheries.

The names of the genera as given in the Catalogue of the British Museum, are Genus I. *Catodon*, 3 species; *C. macrocephalus*, Northern Sperm-Whale; *C. colueti*, Mexican Sperm-Whale; *C. polycyphus*, South Sea Sperm-Whale. Genus II. *Kogia*, one species; *K. breviceps*, Short-Headed Whale. Genus III. *Physeter*. *P. tursio*, the Black Fish.

III. Balaenidæ. (Gr. βάλαινα, *balaina*, a whale.) True or Whale-Bone Whales.

These include but a limited number of species, comprised in four, or according to Dr. J. E. Gray, three genera. They equal the Sperm-Whale in size. The head is very large, but does not, like theirs, terminate in a broad, abrupt muzzle. They have two nostrils, separate and longitudinal. The jaws are toothless; the blow-holes distinct, situated on the top of the head and each a foot long. The absence of teeth specially distinguishes these from other whales; their place in the upper jaw, which is extremely narrow, is supplied by *baleen*, or whalebone, consisting of pendent, horny plates, or *laminæ* (see Chart,) each fringed so closely as to fill up the cavity of the mouth and form a strainer retaining the *Clio Borealis*, minute crustaceans, and other small tenants of the sea. These are carried by thousands into the vast spoon-shaped lower jaw. The laminæ or plates are three or four hundred in number on each side, the longest often fifteen feet long; the Baleen of the *Balaena* alone is designated as Whalebone, or Whalefin, as it is called in commerce. That of the other genera, (*Balænoptera* and *Megaptera*,) is called Finner-Fin, or Hump-back-Fin; the tongue is very large, thick and fleshy, fat, soft and spongy, not unfrequently twenty feet long, and nine or ten wide. The blubber obtained from these whales is extremely abundant, a single whale often yielding forty tuns, or three hundred and twenty barrels of thirty-one and a half gallons each; much more than this is frequently yielded. The Arctic and Antarctic Seas are the principal, but not the exclusive resorts of the True Whales. See "Note" at the end of the "Cetacea."

Balaena mysticetus. (Gr. μύσταξ, *mústax*, a moustache; κῆτος, a whale.)

This is the Common Greenland Whale, sometimes called the Black Whale and Right Whale. Though not the largest of the tribe, it is, in a commercial point of view, most valuable for its oil and other products. It is without a fin on the back. The two pectoral fins are about two feet beyond the angle of the mouth, about nine feet long and five broad. It is thirty feet in height, and from sixty to eighty feet long; in weight, from sixty to one hundred tons, or as heavy as three hundred fat oxen. The enormously large and fat tongue is very soft and delicate, giving it the appearance of white satin; it is entirely incapable of protrusion, being fixed from the root to the tip. The front extremity of both jaws is surmounted by a few scattered hairs, to which the name *Mysticetus* has reference. The back, most of the upper jaw and part of the lower jaw, together with the fins, are black; the other parts gray and white, with a tinge of yellow. The older the animals the more they contain of white and gray, and some are all over piebald. When of the largest size they yield a ton of baleen. The blubber resembles the substance of salmon; in the younger whales is yellowish white, from eight to twenty inches thick, and when fresh, free from all unpleasant smell. A Greenland whale, sixty feet in length, will frequently yield more than twenty tuns of pure oil.

The flesh of a young *Mysticetus* is of a red color, and if cleared of fat, broiled and seasoned with pepper and salt, is said to have a relish not unlike that of coarse beef. That of the old whale becomes blackish and is exceedingly coarse. The tail is very fibrous and sinewy, and extensively used in the manufacture of glue. The bones are quite porous and contain large quantities of fine oil, and the jaw bones, from twenty to twenty-five feet in length, are often preserved, chiefly on account of the oil which drains out of them. The external surface, even of the most porous bones, is, however, compact and solid. The Greenland Whale remains at the surface to breathe for about two minutes, "blows" eight or nine times, then descends for five or ten, sometimes when feeding, for fifteen or twenty minutes. It blows most strongly and densely when alarmed, or when coming to the surface after having been a long time down. When harpooned, it has been drawn up by the attached line, and found to have broken its jaws, and sometimes the crown bone, by the blow which in its descent was struck against the bottom. Having no teeth, the *Mysticetus* cannot prey on its own kind, or on the larger fishes. Its throat is exceedingly straight and narrow, not more than an inch and a half in width. So very small is it that it could not dispose of a morsel which might be swallowed by an *ox!* In this respect it differs widely from some others of the Cetacea. Divine beneficence has, however, abundantly provided for its sustenance. A considerable proportion of the limits within which this whale is found, is occupied by what is called *green water*. This forms about one-fourth part of the Greenland Sea, between 74° and 80° N. Lat., equal to about 20,000 square miles. This body of water is colored by immense numbers of animalcules, for the most part invisible except with the aid of the microscope. These afford sustenance to multitudes of minute crabs, lobsters and sea snails by which the Mysticetus is nourished. This whale seems to attain its full growth at the age of twenty or twenty-five years. It is thought to attain a great age. Our limits do not allow us to enter into details of the perils and hardships connected with the chase and capture of the whale. We may say here, however, the instinctive attachment between the parent and its offspring, is a circumstance of which whalemen often avail themselves in order to secure their prize. The young cub, reckless of danger and easily harpooned, is often struck as a snare to the mother. Says the well known Capt. Scoresby, "at such a time, she joins her young one at the surface of the water whenever it has occasion to rise for respiration; encourages it to swim off; assists its flight by taking it under her fin, and seldom deserts it while life remains. One of my harpooners struck a sucker with the hope of its leading to the capture of the mother. Presently she arose close by the 'fast boat,' and seizing the young one, dragged about a hundred fathoms of line out of the boat with remarkable force and velocity. Again she rose to the surface; darted furiously to and fro; frequently stopped short or suddenly changed her direction, and gave every possible intimation of extreme agony. For a length of time she continued thus to act, though closely pursued by the boats; and inspired with courage and resolution by her concern for her offspring, seemed regardless of the danger which surrounded her. At length one of the boats approached so near that a harpoon was hove at her; it hit, but did not attach itself. A second harpoon was struck; this also failed to penetrate, but a third was more effectual and held. Still she did not attempt to escape, but allowed other boats to approach, so that, in a few minutes, three more harpoons were fastened. and in the course of an hour afterwards, she was killed."

The Right Whale was formerly found in great numbers along our own coast. The whale fishery, including this and the Sperm-Whale, is prosecuted largely and with great success by individuals and companies of men, subject however to great fluctuations. "The first vessel constructed expressly for this fishery was built at Nantucket in 1690."

Of this genus, the other species are the *B. marginata*, Western Australian Whale, Cape Whale; *B. Japanica*, Japan Whale; *B. antarctica*, New Zealand Whale; *B. gibbosa*, Scrag Whale; and the *B. australis*.

Balaena Australis is the Cape or Southern Whale, inhabiting the South Seas and of a uniform black color, measuring from thirty-five to fifty feet. Its baleen, owing to the great curve of the upper jaw, appears relatively longer than in the Northern *Balaena*, usually reaching to about nine feet in a whale of forty feet. The head is frequently covered with barnacles, layer above layer, which, concealing its true color, give it a whitish appearance quite unlike that of its northern relative. The pectoral fins are longer and more pointed, while the lobes of the tail are less marked than in the former species.

II. Genus. Megaptera. (Large-finned.)

This genus includes the Hump-Backed Whales, easily known from the "Finners" by "being shorter and more robust, in having the skull nearly one-fourth the entire length, the head wide between the eyes, the mouth larger, the lip warty, and the nose large and rounded. The plaits of the belly and throat are broad. The skull is intermediate between that of the Balaena and the Balaenoptera. Four species are enumerated: *M. longimana*, found in the North Sea, described by Dr. Johnson from a specimen cast ashore at Newcastle, Eng., and called Johnson's Hump-Backed Whale; *M. Poeskop*, the Poeskop, or Cape Hump-Back; it is the *Roqual du Cap*, of Cuvier, the Hump-Backed Whale of Ross'

"Antarctic Voyage," and an inhabitant of the Seas of the Cape of Good Hope; *M. Kuzira*, the Kuzira, inhabiting the Japanese Seas; *M. Americana*, the Bermuda Hump-Back, is of a black color with a white belly, and has its head covered with tubercles. It is found at Bermuda from March to the end of May, when it departs. The baleen of this whale is extensively imported from Bermuda.

Balaenoptera Rorqualus. The RORQUALS.

These include several species closely allied to those of the genus *Balaena*, but which have been separated from it, and formed into a distinct genus. Among them are the largest of the Whale tribe, and probably the largest and most powerful animals found on our globe. They are often from a hundred to one hundred and twenty feet in length; the head is about one-fourth part of the length. These whales differ from the *Mysticetus*, in having bodies which are longer and in their form more slender and cylindrical; in possessing a dorsal fin; in having blubber which is thinner, being generally not more than six inches thick, and yielding an oil of inferior quality and less in quantity; in their greater speed, quicker action and bolder conduct; in their more violent blowing; and in having shorter and less valuable baleen. Hence they are avoided by whalers as not repaying for the hazard of their capture. The upper jaw of the Mysticetus is relatively longer and more curved; consequently, the plates of baleen are long in the Mysticetus and short in the Rorquals. In the latter, the longest laminæ measure only three or four feet; the smallest are reduced to mere bristles, so that the animal has not fewer than four or five thousand distinct plates of whalebone. The posterior arch of the palate is so large that it could easily admit some modern Jonah, forming a great vestibule to the wind-pipe and gullet. This last is somewhat larger than a man's fist. The Rorquals feed not only upon the small medusæ, shrimps, etc., which form the food of the *Mysticetus*, but upon medusæ of a larger size, and such fish as herring, haddock, salmon, etc. This could not be unless the baleen were coarser and the *swallow* larger than in the *Mysticetus*. The Rorquals are sure to be in the track of the fish just referred to, and they devour them in quantities almost beyond imagination. M. Desmoulins states that six hundred great cod, and immense quantities of pilchards have been found in the stomach of one of these whales. Unlike the Common Greenland Whale, the animals of this genus often leave their native seas and stray far away to other waters and shores.

N. B. In the Catalogue of the British Museum, the genus *Balaenoptera* has but one species,—the *B. rostrata*, Pike Whale—

the *Rorqualus rostratus*, of DeKay. It is of a black color, underneath of a reddish white; inhabiting the North Sea, and has been found at Volognes, in France, in the Thames, at Deptford, Eng., and in the bay of New York. The other species, eight in number, are included in a fourth genus, *Physalus*. The names of the species, as given in the catalogue above referred to, are *P. antiquorum*, the Razor Back, or Great Northern Rorqual; *P. Boops*, of which a specimen, thirty-eight feet in length, is in the British Museum. This is probably one of the smaller Rorquals, and was taken in 1846. *P. Sibbaldi*, another Rorqual, of which a specimen is found in the Museum at Hull, Eng., forty feet long; *P. fasciatus*, the Peruvian Finner, found on the coasts of Peru; *P. Iwasi*, the Japan Finner. It is very rare. A specimen, taken nearly a century ago, was twenty-five feet long;—*P. antarcticus*, so named by Dr. I. E. Gray, from the baleen of a New Zealand species; *P. Brasiliensis*, the Bahia Finner, named from baleen, brought from Bahia. *P. australis*, the Southern Finner, found in the seas of the Falkland Islands.

The genus *Balaenoptera*, is divided into two sections,—one distinguished by the smoothness of the skin, of the throat and under parts, of which there is one species, *Balaenoptera physalus*, called the *Finfish*. The other section is characterised by the deep longitudinal regular folds into which the skin of the throat and under parts is thrown, and which are supposed to be capable of great dilatation. Of this there are several species. The name *Rorqual*, which they bear, is of Norwegian origin, meaning "whale with folds." The Rorquals have sometimes been arranged into greater and lesser Rorquals, (*majores et minores.*) Twenty-five feet is said to be the limit, as to length, of the smaller division. Their baleen is white and short; the folds are of a rosy tint. They frequent the rocky bays of Greenland, (especially during summer,) and the coasts of Iceland and Norway, rarely descending into lower latitudes. They are very active in their habits, so that, although valued in northern climates for the extreme delicacy of their flesh, yet the natives do not attempt to harpoon them, but wound them with their darts and spears, and after a fortunate hunt, hope to discover them dead and stranded. The smaller Rorquals yield an oil peculiarly delicate, and esteemed by the Icelanders as an article of their *materia medica*. The Rorqual of the Southern Seas, *B. Australis*, resembles the Northern Rorqual. Its great power and velocity make it difficult of capture, and its products by no means repay the risk and labor of taking it. It is sometimes called the Black Whale, and has been found in considerable numbers on the shores of California.

These Southern Whales are fond of placing themselves in a perpendicular position, for the purpose of surveying more easily the expanse of waters; at a distance, resembling large black rocks in the midst of the ocean. Fossil Rorquals have been found in Britain and other parts of Europe.

IV. Herbivorous Cetacea, or Aquatic Pachyderms.

This family of the Cetacea have teeth with flattened surfaces, and adapted to the herbivorous nature of their food; the skin is thick, and more or less horny; the stomach divided into four cavities. They have stiff moustaches on their lips, and pectoral *mammae*,—peculiarities which, when their bodies are partly raised out of the water, give them a somewhat human look, and probably are connected with the fanciful stories which have been often told about "Mermaids." They are frequently called "Sea Cows, Sea Calves," etc. The favorite haunts of these animals, are the mouths of rivers or straits, where the water is only three or four fathoms deep. Here, where the sea-weeds grow luxuriantly, they feed in troops, rising frequently to the surface, in order to take breath. This group includes three genera, and about twice as many species.

Manatus or *Lamantine*. The Manatee, or Sea Cow. (Plate VIII. fig. 2.)

The animals of this genus, are confined to three or four species, having oblong bodies, which are from ten to fifteen, and sometimes twenty feet long; long, rounded tails, and eight grinders in each jaw. They are gregarious, and strongly attached to each other, as well as their young, which the female defends, regardless of her own danger. Their flesh resembles fatted pork, and when salted, makes excellent sea-store. They are much sought after, being captured with a harpoon attached to a stout cord. The skin is of a blackish color, very tough and hard, full of inequalities, and sprinkled with a few bristly hairs about an inch in length. The *Manatus Americanus* is found at the mouth of the Amazon, Oronoco, and other rivers of South America, and one species is still hunted among the lagoons and keys of Florida. Its exhibits rudimentary nails upon its flippers, and by their aid, sometimes drags its unwieldy body on shore to bask in the sun, or seek for herbage growing on and near the banks.

Halicore Dugong.

This genus is similar to the preceding one; has one species, (Dugong,) and is found in the waters of the East Indies, and

EXPLANATION OF PLATE VIII.

Fig. 1st. Common Seal, or Sea-Dog, *Phoca vitulina.* It has five or six rows of white whiskers, short fore feet, with webbed toes, serving as oars for swimming, but upon land only available for creeping or shuffling along as it comes out to bask in the sun. The hind feet have short flattened claws, of which the three middle ones are smallest, giving the feet a forked appearance.

Fig. 2d. Manatee, Sea Cow, Siren, or Mermaid of the ancients, *Manatus Americanus.* The nostrils are in the skin, near the end of the muzzle. It has flippers or pectoral fins, or fin-like forearms, having their five fingers enveloped in a membrane or skin, with nails, or rudiments of nails, which terminate four of the fingers.

Fig. 3d. Sperm-Whale, *Physeter*, or *Cachalot macrocephalus.* The head forms one-third of its bulk ; the nostril, spiracle, or spout-hole, is a slit a foot long and shaped like the letter *f.* The case above the brain contains the sperm oil. In a large whale the cavity will contain a tun, or more than ten barrels. The portion just above the mouth, called the Junk, is formed of elastic, strong fibres, permeated with fine sperm oil and spermaceti. The eyes are small, and the two are said to be unequal in size. The ear-openings are behind the eyes, and only large enough to admit a small quill. On the neck is the Bunch, (Bunch of the neck,) and on the back the Hump. Although this animal is of enormous size, the Small, near the Flukes or Tail is not thicker than a man's body.

Fig. 4th. The True Dolphin of the ancients, *Delphinus delphis*, has a spiracle or blow-hole on the summit of the head, above the eyes, which are small and low down, near the angle of the mouth. The beak is of the same length as the head, with from forty to forty-eight teeth on each side, above and below, interlocking with each other. The swimming paws are placed low and half way between the end of the beak and the dorsal eminence.

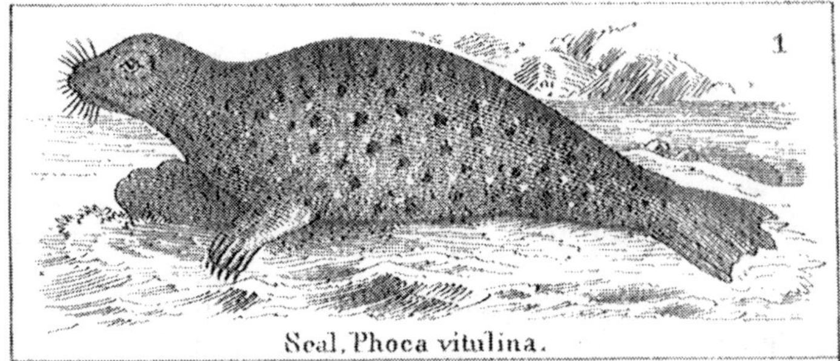

Seal, Phoca vitulina.

Manatee, Manatus, Americanus.

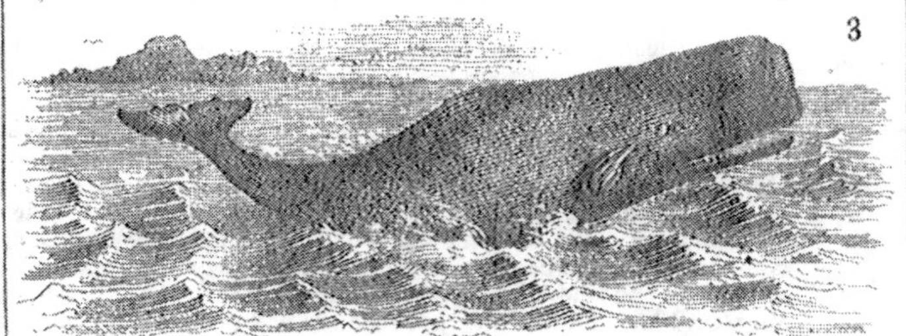

Sperm Whale, Physeter macrocephalus.

Dolphin, Delphinus Delphis.

those of the Northern line of Australia. The most conspicuous difference between this and the Manatus is, that it has no traces of nails. The tail is truncated, or two-lobed, and there are five grinders in each jaw; the body is round and tapering. To aid it in browsing upon *fuci* and submarine vegetables, the front part of the jaw is bent downwards, so as to bring the mouth in nearly a vertical position. A great peculiarity of the animal is, that the ventricles of the heart are detached from each other, being connected at their base only. The body is seven or eight feet long; the flesh tender and not unlike beef.

Halicore Tabernaculi, the *Dugong* of the Red Sea, is considered a distinct species by Rupel, who gave it the specific name, supposing, that with its skin, the Jews were required to veil the tabernacle.

Stellerus. This third genus is found in the Polar regions, and especially in the vicinity of Behring's Straits, where, in great numbers, the animal frequents the shallow parts of the shore, and the estuaries of rivers. The skin is remarkably thick and hard. The head and mouth are small; the lips appear to be double, *i. e.*, inner and outer. The space between them is filled with strong bristles, (one and a half inches long, and as thick as pigeons' quills,)—which are, to this animal, what the whale-bone is to the largest Cetacea. It has no teeth; but in place of them, has two horny substances adhering, the one to the palate, the other to the lower jaw. The length of the body is twenty-eight feet; the weight of a large one, eight thousand pounds. The skin is used by the natives for covering their boats. They esteem the blubber of the animal as good as "May butter." The flesh of an old *Stellerus*, when well boiled, resembles beef; that of the young one is like veal.

Zeuglodon. This is an American fossil, whose name was suggested by the back molar teeth, which resemble *two teeth yoked* together,—as the generic name signifies. Remains of this extinct animal have been found in Maryland, Alabama and Arkansas, (U. S.) Professor Owen supposes it to be allied to the Dugong and Manatee.

NOTE.

According to Lieut. Maury, (see Phys. Geog. of the Sea,) "the tropical regions of the ocean are to the Right or True Whale, as a sea of fire, through which he cannot pass, and into which he never enters;" but "whales, with harpoons in them bearing the stamp of ships known to cruise on the Baffin's Bay side of the American Continent, have repeatedly been taken near the Behring's Strait side;—and as, in one or two instances a very short time had elapsed between the date of the capture in the Pacific and the date when the fish must have been struck on the Atlantic side, it was therefore agreed that there is a north-west passage by which whales passed from one side to the other, since the stricken animal could not have had the harpoon in him long enough to admit of a passage around either Cape Horn, or the Cape of Good Hope." He adds, "It is found also that the Right Whale, of the Northern Hemisphere, is a different animal from that of the Southern."

QUESTIONS UPON MARINE MAMMALS.

What is said of the size, organization and movements of Marine Mammals? What have they been called? How do the paddles appear when dissected? What is their chief use? Why must the Whale resort to the surface? How great a pressure does it sometimes encounter? How is it guarded from injury which might thence arise? Describe the uses of the blubber. Give particulars respecting the eyes of the *Cetacea*. What author has referred to them as showing Divine contrivance? What is one of the most remarkable things in the economy of the Whale? How are they able to remain so long under water? In what manner do they breathe? Why are these called blow-holes? What is said of their sportings? How do the Whales usually take their food? What is said of their hearing? Have they the sense of smell? What is said of their stomach and other organs? What more is said of them? Into what families may the Cetacea be divided? Give a general view of the Dolphin tribe? How many species does this family include? What is said of the number of the teeth in the Common Dolphin? What of their form and use? How long is it? What gives interest to this animal? How was it regarded by the ancients? What is said of the colors of the Dolphin? Does its color change while dying? What are its habits? How are these illustrated by Ancient naturalists? What is the meaning of the term *Monodon?* What of Narwhal? What other names are given to it? Has it teeth? What is said of its tusks? What virtues have been attributed to them? What use does the animal make of them? What other species of Whales does this precede? Does Dr. Scoresby describe the Narwhal? What is said of them by Dr. Kane? Name the sp. of the *Beluga*, or White Whale? Give some account of this W. What are their favorite resorts? Why are they not more often pursued by whalers? At what season are they very useful to the Greenlander? Give the characteristics of the Porpoise. What is said respecting the wide diffusion of the Common P.? What of its resemblance to the Dolphin? What is said of the leather made from its skin? What is said of the brain and teeth of the P.? What of the word Grampus? Upon what do they feed? How was its flesh formerly regarded? What is the meaning of the term *Grampus?* What is it called by American sailors? Describe it. Where is its favorite abode? What is said of its voracity? What incident is related by Sir Joseph Banks? Name the sp. found on the coast of France. For what is it famous? Give the sp. included in the gen. *Delphinapterus*, and describe it? What do whalers call it? Where is it found? How many sp. do the gen. *Globicephalus* include? How are they characterized? How do the teeth shut into one another? What is this W. called? By what method are they taken? What additional names are given to this W.? How are the second fam. of Cetacea distinguished? Why is the name CATODONTIDAE given to it? What names are applied to the *Sperm W.?* Describe it. What peculiar substance does it yield? Describe the process for obtaining the pure Spermaceti of commerce. What is Ambergris? What is said of the throat of the *Sperm Whale?* What other peculiarities of structure are mentioned? On what does it feed How does it sometimes use its huge teeth? What is said of the Sperm W. Fishery? What is said of the size of the True W.? How are they specially distinguished from other W.? What supplies the place of teeth in the upper jaw? Describe the Baleen and its uses. How much blubber is yielded by one of these W.? What are their principal resorts? Which of the W. is most valuable in a commercial point of view? Give the derivation of its specific name. What renders it appropriate? What is said respecting the quantity of baleen and oil yielded by a Greenland W.? Of what use are the bones? What is said of their blowing? In what respects do they differ from other W.? How extensive is the area of the *green water?* What occasions its peculiar color? What is said of the age of these W.? How are their instinctive attachments illustrated? When was the first American whale-ship built? What is said of the *Humped-backed W.?* How many sp. are enumerated? To what are the *Rorquals* allied? What is said of their size and strength? How do they differ from the *Mysticetus?* Why are they avoided by sailors? What is said of the arch of the palate? On what does the animal feed? Illustrate its voracity. Name the two sections of the genus *Balaenoptera?* What is the meaning of the term Rorqual? How large are they? What waters do they frequent? What is said of the oil which they yield? Are they easily captured? In what position are the Southern Rorquals found?

Give the peculiarities of the *Herbivorous Cetacea.* What are they frequently called? Where are their favorite resorts? How many sp. does this group include? Describe the *Manatee*, or *Sea Cow.* Where is the *M. Americana* found? Where is the *Dugong* found? What is the chief difference betwen this and the Manatus? In what respect is it very peculiar? Where is the *Stellerus* found? Describe it. Mention the name and localities of the American fossil sp. What species of Whale is figured on the Chart? Trace and describe it. What other Cetaceans are named upon the Chart? Refer to the figures and species of this order, tracing each, giving their characteristics and habits, size, &c., &c.

ORNITHOLOGY.

SECOND BRANCH OF ZOOLOGY.

SECTION I.

Ornithology. (Gr. ὄρνις, *ornis*, a bird; λόγος, *logos*, a discourse.) Birds, (*Aves*,)—2d Division of the Warm Blooded Animals.

The numerous class of vertebrated animals which this term includes, are prominently distinguished from the Mammalia by their general form and feathery covering, and by producing their young from eggs. They fall below quadrupeds in the scale of nature,* but they far surpass fishes and insects in point of sagacity, and in the structure of their bodies. Though called bipeds, they nearly approach to quadrupeds, as may be seen by looking at a plucked pigeon, and observing how, in respect to limbs, it resembles a skinned rabbit, except that the forelegs have no feet or toes at their tips.

Birds are formed for flight; they have been, not unaptly, styled, "the Insects of the vertebrated series." The organization of the greater part is wisely adapted both for suspension in the air, and for motion through it. Their activity in so subtle a medium, required various conditions and adaptations of structure which an attentive examination clearly presents, as combined in their organization by the all-wise Creator.

The shape of the body is sharp before, to pierce and make way for itself through the air; it rises by gentle swelling, and falls off by an expansive tail, that helps to keep it buoyant, while the foreparts are cleaving the air by their sharpness. Hence, they may be compared to a bark making its way through the water;—the trunk of the body answering to the hold; the head to the prow; the tail to the rudder; and the wings to the oars.

---

* The following scale is given, showing the size of the brain as compared with that of the body:—Eagle, 1-260th of the body; Sparrow, 1-25th; Chaffinch, 1-27th; Redbreast, 1-32d; Blackbird, 1-68th; Canary-bird, 1-14th; Cock, 1-25th; Duck, 1-257th; Goose, 1-360th.

The rapidity and variety of motion of which Birds show themselves capable, may well excite admiration. The Swallow and the Eagle can dart through the air at the rate of sixty miles an hour; the Falcon at the rate of forty or fifty miles in the same time; the Passenger Pigeon outstrips the wind, which, when most violent, traverses only sixty miles an hour; the Curlew traverses three elements with ease,—running rapidly upon the ground; perfectly at home on the ocean wave, and borne in the air as it flies swiftly from one continent to another.

For this fleetness of motion, the skeleton of a bird is strikingly adapted. (Plate XII. fig. 2.) It unites lightness with firmness; the great bones of the limbs, and many of those of the body, are hollow reservoirs of air, communicating with the lungs. Sacs, or bladder-like receptacles, which can be filled with air, are distributed about the body; some of them internally; others between the muscles and the skin, down the throat and chest, or along the tendons of the shoulder; and these communicate with each other and with the lungs. The last named organs adhere closely to the ribs, occupying (Plate XII. fig. 2, E.) the hollows between them and on each side of the spine; the lungs are very large, but can be very little expanded or contracted. To compensate for this incapacity, which might impede their breathing, the ends of the branches of the wind pipe open into them; but these again communicate with the membranous sacs, or buoyant air cells, that run along the whole length of the body,—so that a probe thrust into the lungs of a fowl, easily finds a passage into the internal parts of the frame; and air blown into the wind-pipe, will be seen to distend the bird's body like air blown into a bladder. By the great development of the breathing apparatus, the blood is more rapidly and effectually oxygenized, and muscular energy accumulated for the action of flight; while, by the animal heat which is thus given out, the air contained in the complex respiratory apparatus is rarified; and thus the body is increased in bulk, but rendered specifically lighter.

The wings of a bird correspond to the arms and hands of man; but the hand in the bird consists of only two fingers, (Plate XII. fig. 2d, N.,) and a thumb, all of which are rudimentary. From the bones of the bird's hand (M) arise the *primaries*, or great quill-feathers of the wings,—ten in number, and by their form, stiffness, and relative strength, indicating the character and the power of the flight. The *secondaries* spring from the principal bone of the forearm, (K. L.;) the number of these varies in different species; they are generally stouter, longer, and more flexible than the primaries, and differ less in form from the general covering of the body. The bone of the upper arm (*humerus*, I.) gives rise to another series of feathers, called *tertiaries*, which, in such birds as the Plovers, Curlew, etc., are greatly lengthened; but in their structure are yet weaker than the secondaries. Fastened to the little bone which represents the thumb, are two or three short and stiff feathers, called the *winglet*, (or bastard-wing;) they lie upon the base of the first primaries, at the edge of the wing. Corresponding with the series of feathers, there are both on the outer and inner surface of the wing, several rows of smaller ones, called *coverts*, from their office of protecting the basal part of the quills; the feathers covering the shoulder-blade, or *scapula*, are called *scapulars*.

If we examine each feather separately, we find it composed of two parts; (1) a light but firm shaft, hollowed below for strength and lightness, into a horny tube, containing the blood-vessels by which it is sustained; (2) the vane, or beard of the feather, composed of a double series of layers, or thin parallel plates on each side of the shaft, and set at an angle to it. Towards the shaft of the feather, these layers are broad, and of a semi-circular form, to serve for strength, and for the closer grafting them one against the other, when in action. Towards the outer part of the vane, the layers grow slender and taper, to be more light. On their wider side, they are thin and smooth; but their upper outer edge is parted into two hairy edges, each side having a different set of hairs, broad at the bottom, and slender and bearded above. By this means, the hooked beards of one layer always lie next the straight beards of the next, and lock and hold each other. No resistance is offered to the flight of birds by this arrangement; while beneath these there is a layer of soft down, which preserves them from cold, to the effects of which, but for this provision, they would have been much exposed.

The wings are usually placed at that part of the body which serves to poise the whole and to support it. The *feathers of the wing* overlap each other, and present a continuous surface of great breadth, by repeated strokes of which upon the air, the bird performs its flight. (Plate X. fig. 2.)

"Each feather is concave, whether we regard it transversely or longitudinally; its stem, or middle, is remarkably strong, though very light, and the beards which present their edges in the direction of the stroke, are linked to each other by a series of minute hooks." These arrangements add to the power of the wings in their downward strokes upon the air.

Every part of a bird, except the beak and the lower, or posterior extremities, is, in general, clothed with feathers. The feet are protected by a naked, scaly skin which, in some cases, extends above the tarsus, and partly up the leg. The soles of the toes are covered with a granulated and callous modification of this skin. (Plates IX. and X.)

In Birds, particularly those of extended and powerful flight, the greatest part of the muscular force centers in the wings. The muscles which produce the downward stroke of the wing, are enormous; for their attachment, the breast-bone is greatly enlarged, and its surface is also increased by having its middle portion raised into a perpendicular ridge, the two faces of which, from their direction, afford an advantageous point of resistance, or *purchase.* A bird can move its wings with a degree of strength which, when compared with the animal's size, is almost incredible. The flap of a Swan's wing has power sufficient to break a man's leg; and a similar blow from an eagle has been known to produce instant death.

The powerful muscular action involved in flight, would naturally tend to draw the shoulders together; but this tendency is resisted by the insertion between the two bones (coracoids) to which the shoulder bones (H) are joined,—of a singular arched bone, called the wish-bone or *merrythought.* (G.) In the domestic fowl, the bone is feeble; but in birds of powerful flight, as the Hawks, the Swallows, and the Humming-Birds, it is very strong and elastic. On the other hand, when the bird never rises upon the wing, as in the case of the Ostrich and Emu, this bone is reduced to a mere rudiment. The bones of the lower, or posterior extremities also differ materially in structure from those of quadrupeds. These consist of (1) a thigh-bone, (or *femur,*) R.; leg-bones, (*tibia* and *fibula,*) S.; (these leg-bones are really two, but the *fibula* is very small, and becomes anchylosed to the *tibia;* i. e., immovably fixed by a continuation of bony secretion;) (2) the metatarsal, or shank-bones, U. U., at the lower end of which there are as many processes as there are toes, each process being furnished with a pulley for moving its corresponding toe; (3) the toes, of which the usual number is four,—a number never exceeded; while a few birds have only three; and the Ostrich only two. The three toes are directed forwards, and one, answering to the great toe, backwards. This, at least, is the general rule. The back, or great toe, is wanting in some birds. In the Swallows it is directed forwards; but in the Climbing Birds, as the Parrots and Woodpeckers, the outer toe and back toe are both directed backwards; while the Swifts have all the four toes directed forwards.

As the upper limbs, or anterior extremities, are exclusively for

EXPLANATION OF PLATE IX.

BEAKS AND HEADS OF BIRDS.

1. Owl's head, showing the egrets or tufts of feathers on each side, close to the ears.
2. Head of an Eagle, showing the strong curved beak of a raptorial bird. Order Raveners, (birds of prey.)
3. Falcon's beak, showing the notches and teeth near the tip of each mandible, for holding its prey.
4. Recurved beak of the Avoset, Stilt Plover, &c.
5. Merganser's bill, with serrated margins and abruptly hooked tip.
6. Long, slender, curved bill of the Curlew, formed for penetrating the mud on the sea-shore, in search of insects, slugs and small testaceans.
7. Bill of the Humming Bird, formed for searching out insects in flowers and sipping the honey dew. Order Tenuirostres, (slender bills.)
8. Beak of the Crossbill, the crossed points of which it inserts to open the cones of pines and fir trees, upon the seeds of which it feeds.
9. Beak of Parrots, Mackaws, Lories, &c., having the upper mandible greatly curved over the lower, which is considerably shorter.
10. Beak of the Petrel, so furrowed as to appear of distinct pieces.
11. Puffin's beak, transversely furrowed on both sides, appears as though a sheath had been slipped over both mandibles.
12. Duck's bill, broad, long and soft, with plaits on each side for straining insects, worms, &c., out of the mud.
13. Beak of the Hornbill, of a light honeycomb structure; the horn or helmet is hollow, and by some supposed to act as a sounding board.
14. Beak of the Whip-poor-Will, deeply cleft; the mouth fringed with strong, stiff hairs, (vibrissæ.) Order Fissirostres, (split bills.)

TAILS OF BIRDS.

15. Fan shaped or rounded tail.
16. A forked tail, indicating a swift flyer, as in fly-catchers, (Muscicapidæ.)

17 and 18. Lyre shaped tails, as in the Chatterers, (Ampelidæ.)

19. Doubly forked tail, as in *Psalurus bifurcatus*, a species of night-jar found in Brazil; very rare.

FEET OF BIRDS.

20. Foot of the Golden Eagle, showing its feathered tarsi, and the powerful talons of a rapacious bird.
21. Foot of a scratching bird, showing the three front toes united by a membrane up to the first joint, the hind toe articulated upon the tarsus, and the horny spur in the male bird. Order Rasores, (scratchers.)
22. Foot of the Ptarmagin, covered with hairlike feathers as far as the claws.
23. Woodpecker's foot, with toes in pairs, (yoke-footed, or Zygodactylus.) Order Scansores, (climbers.)
24. Webbed foot of a swimming bird. Order Natatores, (swimmers.)
25. Foot of a Phalarope, with lobate membranes or festoons on the toes sufficiently broad to assist them in swimming.
26. Grebes' foot, not webbed as in most water birds, but each toe flattened so as to serve as a separate paddle.
27. Foot of the Sacred Ibis, a wading, shore, or stilt bird. Order Gralliatores, (waders.)
28. Claw of Heron, showing the comb-like divisions of its inner edge.

BEAKS AND HEADS OF BIRDS.

TAILS OF BIRDS.

FEET OF BIRDS.

flight, and the bird depends principally on its bill for gathering its food, it became necessary, as the bones of the back have scarcely any motion, that the neck should be as it is, long and flexible. Hence, while in the mammals, the vertebræ of the neck are seven,—there being no more even in the Camelopard,—the deficiency of motion in the back is made up in birds by an increase in the vertebræ of the neck, (B.) proportioned to their wants Thus the Raven has twelve neck bones, the Domestic Cock thirteen, the Ostrich eighteen, the Stork nineteen, and the Swan twenty-three, which is the largest number yet ascertained, while the smallest is ten. These are so joined together, that the head can be turned completely around, (the position which the bird takes when at rest,) or moved in any direction, so that the bird can touch every point of its body with its bill.

The trunk is sustained on the thighs by very powerful muscles ; another set of which passes from the lower part of the thighs to the toes, turning over the knee and heel, in such a way that the flexion or bending of these joints shall shorten them. Hence, the simple weight of the body flexes the toes, so that birds are enabled to sleep perched on one foot. But the pectoral muscles, as a general rule, show the greatest development. The breast bone, or *sternum*, is made to project forwards with an elevated ridge or keel. To this, the powerful muscles which depress the wings, are attached, The depth of the keel is a partial criterion as to the power of flight ; in the Harrier it is deep ; in the Ostrich, where the wings are not sufficiently developed to raise the bird from the ground, it is quite flat.

The jaws of the bird are not furnished with teeth, but the place of these organs is supplied by a casing of horn, terminating in a point at the tip, and brought to an edge on the side of the jaw. This horny casing is known as the beak or bill ; the name mandibles is given to the upper and lower divisions. In Birds of Prey, (see Plate of Birds' Beaks and Heads,) the beak is like a carving or dissecting knife ; in the Woodpeckers it is an effective chisel ; in the Snipe and Woodcock, it is a long and slender probe, furnished at the tip with copious nerves of sensation for feeling in the deep earth of bogs and marshes ; in the Parrots, it is a climbing hook, or a fruit knife ; in the Swallows and Goat Suckers, it is a fly-trap; in the Swans, Geese and Ducks, it is a flattened strainer, with nerves on the inside for the detection of the food remaining after that particular operation which almost every one must have observed a duck perform in muddy water ; in the Storks and Herons, it is like a fish-spear ; in the Cross-Bills, or Seed-eating Birds, it forms a pair of Seedcrackers for extricating the kernel from the husk which covers it.*

Varying as the beak does in different kinds of birds, it in no instance performs a proper masticating function; though it may divide flesh, crack a nut, and with the assistance of the tongue, shell it; and though it may separate the grain from the husk, as is constantly seen in the Goldfinch and Canary. A nearer approach to mastication, is the bruising down of hard seeds by means of a knob in the middle of the palate, as is seen in the Buntings.

The stomach in Birds, consists of three parts, (not always, however, distinctly developed,) viz., the crop or craw, the membranous stomach, and the gizzard. From the want of masticating power in the bird, it, of course, swallows its food entire. When the food is flesh, the process of digestion is sufficiently simple, and so rapid as to need no preparation. To prepare for the digestion of hard grains and seeds, which are the food of so large a number of species, a sort of internal grinding mill is furnished by the gizzard.

This organ, which is seen to most advantage in grain-eating birds, is made up almost entirely of two semi-globular masses of dense muscle, whose flat faces, covered with a thick leathery skin, work over each other like a pair of millstones, and by the aid of small angular stones, sand, etc., swallowed for the purpose, very quickly grind down the hardest substances. In the Museum of the College of Surgeons, (London,) is a large glass bottle entirely filled with pebbles, &c., taken from the stomach of an ostrich. The experiment has been made, without injury, of conveying bullets beset with needles, and even lancets into the stomachs of granivorous birds, with the effect of the total destruction of those sharp instruments in a short period.

The organs of the voice in birds bear a striking resemblance to certain musical wind instruments. The larynx is made up of two parts; the second part, or lower larynx, contains a second *rima glottidis*, (cleft or opening of the throat,) furnished with tense membranes which perform, in many birds, the same office that a reed does in a clarionet, or hautboy, while the first or upper *rima*, (cleft or opening,) of the throat, like the ventage or hole of the instrument, gives utterance to the note. None of the endowments of this interesting class more minister to the pleasure and delight of man than their powers of song.

The development of the senses of birds varies in the different

* See Penny Cyclopedia; Art. Birds.

tribes, according to the mode in which they are adapted to take their prey. Of the five senses, sight, smell and hearing are most acute in birds. The crystalline humor of the eye is flat in birds; the vitreous humor is very small. The color of the iris varies in different species, and in many cases is very brilliant. Birds have three eyelids, two of which, the upper and lower, are closed in most of the race by the elevation of the lower one, as may be seen in our domestic poultry. But the third eyelid, or nictitating membrane, forms the most curious apparatus. When at rest it lies in the corner of the eye; but by the combined action of two muscles which are attached to the back of the sclerotic coat, (the white,) of the eye, it can be drawn out so as to cover the whole front of the eye-ball, like a curtain, and its own elasticity restores it to its resting place. This, it is said, enables the eagle to look at the sun.

The rapacious birds seem most remarkable for their *length* of sight; others, as the swallow tribe, which fly with extraordinary swiftness, have an almost inconceivable *quickness* of sight.

The sense of hearing in birds appears to be in general tolerably acute, especially in the nocturnal birds of prey, which have what other birds are without, an external cartilaginous ear.

The sense of smell does not seem to be very highly developed in the birds as a class, but is strong in the vultures.

Few of them have a tongue which serves as an organ of taste, but some of the swimmers and the parrots generally have one that is soft, thick and covered with papillæ, and there can be no doubt that these taste food of a soft or fluid nature, and select that which they like best; for the most part, however, the tongue is an organ for taking food rather than of taste. The sense of touch is in birds generally very obtuse.

The dress or plumage of birds is admirable for its fitness to the ends for which it was designed; for its softness, smoothness, compactness, and various hues. The most brilliant colors are lavished upon the "winged denizens of the air." This is particularly true of birds of the torrid zones. Those of the temperate zones are not so remarkable for the elegance of their plumage; but the smaller kinds make up for this defect by the melody of their voices. While the birds of the warmer regions are very bright and gorgeous in their colors, they have screaming voices, or are totally silent. The frigid zones, where the seas abound in fish, are stocked with birds of the aquatic kind far more than any other regions. These usually have a warmer coat of feathers, or they have large quantities of fat lying underneath the skin, which serves to defend them from the rigors of the climate.

Birds are oviparous; in other words, are produced from eggs consisting of a living point attached to a globular sac of nutriment called the yelk, surrounded by a layer of *albumen*, the glair or white part, and enclosed in two series of membrane and a hard calcareous shell. The egg is developed into a living, active chick, by the warmth which the parent supplies while sitting upon the nest wherein the eggs are deposited; or it may be hatched in a *breeding machine*, by means of artificial heat of about 96° Fahrenheit. The instinct of birds is wonderfully exhibited in constructing and locating their nests as places of comfort and concealment, and in the exquisite workmanship displayed in some of them.

The process of incubation lasts a few weeks, when the young is ready for exclusion. By means of a horny pointed scale attached temporarily to the tip of its beak, it succeeds in breaking the egg-shell and forcing its way to freedom. "At the end of the second day, the first moving of the heart is perceptible, and on the fifth, the whole frame of the little creature can be distinctly seen in motion. The feathers make their appearance in a fortnight. At the commencement of the fifteenth day, the chicken begins to breathe, and on the nineteenth it is able to peep." The gallinaceous and swimming birds can run about and pick up their own food as soon as they escape from the egg; but more commonly the young are, for several days, unable to quit the nest, and as is well known, are anxiously fed and cared for by the parent birds.

Few things have attracted more attention than the migration of birds. The immediate cause of this is doubtless to be traced to temperature and to food, particularly that which is adapted to the sustenance of the young; and the instinct of the bird accordingly leads it from one climate to another.

The change of plumage, termed *moulting*, takes place in all birds at least once a year, and sometimes oftener. Apart from the ailment connected with this change, birds are subject to very few diseases. In all countries they are said to be more long lived than the quadrupeds of the same climate.

Fowls sometimes live	20 years.	Linnets and other little birds shut up in cages often live 15 years.
Pigeons,	longer.	
Canaries,	25 years.	Robins, 17 do.
Parrots,	30 do.	Eagles, over 100 do.
Pelicans,	80 do.	Cockatoos, reach 120 do.
Geese,	80 or 90 do.	Swans, from 300 to 360 do.

The fossil remains of birds, though not numerous, are entitled to attention. Considerable interest was awakened by the discovery, by Prof. Hitchcock, of Amherst College, of many large bird tracks in the new red argillaceous sand-stones of the valley of the Connecticut river. Remains of colossal birds of the Ostrich type have also been obtained from Australia, which possess peculiar value from their relation to such birds as the DODO, known to have existed at a former period, but now no longer to be found.

The classification of birds into orders is founded upon characters derived from the beak and feet. The subordinate divisions take their rise chiefly from the form of the beak, and run into each other by almost imperceptible gradations. The number of species has been estimated at six thousand. These have been arranged into two grand divisions, viz.: LAND and WATER, or TERRESTRIAL and AQUATIC BIRDS, the former being divided into five, the latter into two orders, making the whole number seven.

ORDERS.	RAPTORES, (Raveners,)	LAND BIRDS.
	INSESSORES, (Perchers,)	
	SCANSORES, (Climbers,)	
	RASORES, (Scratchers,)	
	CURSORES, (Runners,)	
	GRALLATORES, (Stilts or Waders,)	WATER BIRDS.
	NATATORES, (Swimmers,)	

What is the second branch of Zoology called? Give the derivation of the word. To what division of animals do birds belong? How are they distinguished from the Mammalia or first division of warm blooded animals? Are they Bipeds or Quadrupeds? What have they been styled? What does their formation for flight involve? Explain or show how their structure is adapted to it. What is said of the rapidity and variety of their motions? Explain the skeleton of the bird as given on Plate XII. Name the different kinds of feathers and their situation, as illustrated in Plate X. Name the parts of which each feather is composed. In what does the greatest part of the birds' muscular force center? Illustrate its power. Show the use of the wish-bone or merrythought. Give some account of the bones of the lower extremities as illustrated in Plate XII. State the variations as to the number and direction of the toes in different birds. Show the benefit accruing from the numerous joints or vertebræ in a bird's neck. State the number found in different birds. Give some account of the muscles of a bird and their action. Strictly or properly speaking, have birds any teeth? What supplies their place? What are its upper and lower divisions called? Describe the different kinds of beaks. Do birds really masticate their food? What approach to mastication is seen in the Bunting? Of how many parts does the stomach consist? What are their uses? What facts show the power of the gizzard? What do the birds' organs of voice resemble? Illustrate this. What senses are most acute in birds? What is said of the eye and its appendages? What birds are most remarkable for their length of sight? What for quickness? What is said of their senses? What of their plumage? How are their young produced? What is said of their nests? Describe the different stages of the incubated egg. What is said of the migration of birds? What does the term *moulting* mean? How often does it occur? Are birds long lived? What facts show it? What is said of their fossil remains? Upon what is the classification of birds based? What is the estimated number of species? What are the Grand Divisions of birds? In how many orders are they respectively included? Name them.

SECTION II.

LAND-BIRDS. (First Division.) FIVE ORDERS.

I. Order.—RAPTORES. (Lat. *rapio*, to seize.) Birds of Prey, or Raveners.

This order is sometimes named Accipitres, (Lat. Hawks,) and divided into two sub-orders: 1. *Accipitres diurni*, (Lat. Hawks of the Day,) 2. *Accipitres nocturni*, (Hawks of the Night.)

The Birds of Prey are among the largest and the most muscular and powerful of all the feathered tribes. They are easily known by their strong hooked beak, and large acute talons or claws. (Plate IX. fig. 20.) By the claws the first blow is given to the prey, which, when grasped by the feet, is torn open by the bill. For aid in this operation, the typical Raptores, (the Falcons proper,) have a strong and sharp tooth-like projection from one or both of the mandibles; (Plate IX. fig. 3;) but in those of this order that feed upon carrion or small animals, this projection is nearly or entirely deficient. The base of the beak is covered with a naked skin called the *cere*, in which the nostrils are pierced; the stomach consists of a membranous sac, without a muscular gizzard.

The flight of these birds is lofty, vigorous and long-sustained. Their increase is slow, and they are comparatively few in number. They annually produce not more than two or three eggs, generally pure white. The females of the order, contrary to the general rule in birds, are one-third larger than the males. The Birds of Prey are found in all parts of the world. They include three distinctly marked families, viz.: 1. *Falconidæ*, the Falcons; 2. *Vulturinidæ*, the Vultures; 3. *Strigidæ*, the Owls.

First Family.—The Falcons.

Falconidæ, (Lat. *falco*, a falcon.)

This family is arranged by Swainson into five sub-families, viz.; (1) *Aquilinæ*, Eagles; (2) *Milvinæ*, Kites; (3) *Buteoninæ*, Buzzards; (4) *Falconinæ*, Falcons; (5) *Accipitrinæ*, Hawks.

The Falcons, as a family, exhibit, in their structure and habits, the highest development of the destructive faculty. In these the head is wholly covered with feathers, except the cere at the base of the beak; and the leading genera have, in the beak, the sharp projecting tooth referred to above. (See Plate IX. fig. 3.) The eyebrows usually overhang the eye, giving a stern expression to the countenance. The points of the strong and highly curved talons are kept from injury by a mechanism for raising them from the surface on which the bird rests; a process analogous to the sheathing of the claws in the Cat Family of the Mammalia.

The falcons are widely diffused; some species have been reclaimed and trained for the pursuit of game.

NOTE.—It would be interesting and might be instructive to give minutely the characteristics and habits of *all* the genera and species included in this and in the other orders and families of the Birds; but from the restricted limits of the present volume, and the great number of species, (6000,) in the feathered tribe, our notices, when given, will necessarily be brief, while very many species must be passed by altogether.

1st SUB-FAMILY. *Aquilinæ.* (Lat. *aquila*, an eagle.)

The first place is given to the EAGLE, (in treating of the Falcon tribe,) not because it presents most distinctly the family traits, but on account of its great size and strength, the grandeur of its aspect, and the dignity of its movements. This bird was honored by being in the Holy of Holies of the ancient Jewish temple, and every tyro in classical study knows that the old Romans regarded it as the "Bird of Jove."

Eagles are birds of high and powerful, but not of rapid flight. Usually they prefer to strike their prey upon the ground. They breed in solitude on the inaccessible crags of lofty mountains. In these birds, the notch or tooth of the upper mandible is almost obliterated; the claws are remarkably strong and curved; the under surface is grooved; the hind and outer claws are the longest. (See Plate IX. figs 2. of Beaks, and 20 of Feet.)

The GOLDEN EAGLE, *Aquila chrysaëtus*, (Gr. *chrusaëtus*, golden eagle,) is a truly magnificent bird, about three feet in length, having plumage of a deep and rich umber brown, glossed on the back and wings with purple reflections; the feathers of the head and neck are of an orange-brown hue, and when under the rays of the sun, have an almost *golden* appearance; the tail is striped with gray and obscure brown, but in the young bird is, in the under part, white. (This variation in the plumage of the young bird has led some to describe it as a distinct species, by the name of the Ring-tailed Eagle.)

The longevity of the Golden Eagle is almost proverbial. One that died in Vienna, (Austria,) is said to have lived in confinement one hundred and four years. This species is found throughout the middle and north of Europe, and in North America.

The BALD OR WHITE-HEADED EAGLE, *Haliaëtos*, (Gr. *hals*, the sea; *aëtos*, an eagle;) *leucocephalus*, (Gr. white-headed,) is, when full grown, three feet long, and more than seven feet across at the expanse of wings. The general color of the upper parts is a deep, dark brown; the head, chief part of the neck, and the lower parts are white. The wing is admirably adapted for the support of so large a bird, measuring two feet in breadth on the greater quills, and sixteen inches on the smaller.

The Bald Eagle is an occasional visitant to the Northern Hemisphere of the old world; on this continent it is common, breeding, according to Audubon, as far south as Virginia, though its nests are most frequent in the fur countries.

The ardor and energy of this bird might awaken deep interest were they not associated with so much of robbery and wanton exercise of power; for it habitually spoils the Osprey or Fish-Hawk, (Plate X. fig. 3, *a*,) of his watery prey, and will even, in "hard times," steal from the vultures the carrion on which they are feeding. The falls of Niagara are one of its favorite haunts on account of the fish caught there, and the attraction presented by the numerous remains of squirrels, deer and other animals which perish in attempting to cross the river above the cataract. The nest of the Bald Eagle is usually placed on a very tall tree that is destitute of branches to a considerable height; it is never seen on rocks. Dr. Franklin thus speaks of this eagle, the emblem of our national union:

"For my part, I wish the Bald Eagle had not been chosen as the representative of our country. He is a bird of a bad moral character; he does not get his living honestly. You may have seen him perched upon some dead tree, where, too lazy to fish for himself, he watches for the labors of the fishing-hawk; and when that diligent bird has at length taken a fish, and is bearing it to its nest for the support of his mate and young ones, the Bald Eagle pursues him, and takes it from him. With all this injustice he is never in good case, but like those among men who live by sharping and robbing, he is generally poor, and very often lousy. Besides, he is a rank coward; the little King-bird, not bigger than a sparrow, attacks him boldly and drives him out of the district. He is therefore by no means a proper emblem for the brave and honest Cincinnati of America, who have driven out all the ***King-Birds*** from our country, though exactly fitted for that order of knights which the French call ***Chevaliers d' Industrie.***"

WASHINGTON SEA-EAGLE, *H. Washingtonii*, is a new species first observed by Audubon in 1814, who named this "magnificent bird," the largest of the eagles, after Washington, the "father of his country."

The HARPY EAGLE, *Harpuia*, (Gr. *harpuia*, from *harpazo*, to seize,) *destructor*, (Lat. destroyer,) is one of the "Short-Winged" Eagles found in New Mexico, New Granada, and Guiana. In size and powers of body it equals the Golden Eagle. (See Chart.) The beak above is convex; the upper mandible is slightly notched, somewhat like the True Falcons; the tarsi are lengthened, very strong and feathered at the base. When full grown, this bird's head has a thick downy plumage, of a light slaty-gray color. On the back part of the head, there is a crest of dull black feathers, which ordinarily is slightly raised above the level of the feathers on the back of the neck, but on any sudden excitement is elevated at right angles to them; the back and wings are black; the under surface is pure white; the tail has four transverse black bands, alternated with whitish, or ash-colored spaces. (See fig. on Chart.) In the nakedness of its legs, it approaches the Sea-Eagles.

The Harpy is a solitary bird, frequenting the thickest forests, where it feeds upon the sloths; it also preys upon fawns and young quadrupeds. One of these birds, taken near the mouth of the river Amazon, while on its passage to England, is said to have destroyed and eaten a King of the Vultures. After its arrival, a cat was put into its cage, and the eagle, with one blow of its immense foot, broke its back.

The BRAZILIAN CARACARA EAGLE, *Polyborus* or *Aquila Braziliensis*, differs from the Harpy Eagle in having more slender and lengthened tarsi, and in the comparative weakness of its toes. It is of the size of the common Kite, and has a tail nine inches long. It is all over covered with dusky and blackish feathers; hence is called by some *morphnus*, (Gr. *morphnos*, obscure, or dark.) For its food it seems content with any animal substance; it is by no means shy, but is seldom attacked, as it rarely molests domestic poultry.

The FISH HAWK, or BALD BUZZARD, *Pandion*, (Gr. proper name,) *halietus*, has very long and curved talons, the outermost *versatile*, or capable of being revolved. These are well adapted for holding this bird's slippery fishy food. As already said, it has, in the Bald Eagle, a persecutor that often snatches from it its hard-earned prey. (Plate X, fig. 3, *b*.) The plumage of

Fish-Hawk is much like that of water fowl, white below, with a few brown streaks or speckles on the throat. This bird is spread over Europe and part of Asia; it is found in North America from Labrador to Florida.

The HARRIER, CIRCUS, (Gr. *kirkos*, a circle,) has the sides of the head furnished with a *circle* of feathers much like the disk of the Owl's head.

The COMMON HARRIER, *C. cyaneus*, (Gr. *kuaneos*, dark blue,) ranges from Labrador to Texas, and sometimes is seen in the Western prairies in flocks of thirty or even forty in number. The notes of this bird, while on the wing, "sound like the syllables, *pee*, PEE, PEE, the first slightly pronounced, the last louder, much prolonged, and ending plaintively." The Common Harrier feeds on insects, small lizards, frogs, &c., but occasionally will attack partridges and plovers.

SECOND SUB-FAMILY. KITES.

Milvinæ. (Lat. *milvus*, a kite.)

The length of the wings and the forked tail are the characters which most separate the Kites from the rest of the Birds of Prey.

The COMMON KITE, or GLEDE, *M. ictinus*, (Gr. *iktinos*, a kite,) or *M. regalis*, (Lat. royal,) is found in various parts of Europe. According to Charles Lucien Buonaparte, it is very common near Rome, (Italy,) especially about herds of cattle. Formerly, more numerous than now, it was a great scourge to the poultry yard. In falconry, the Glede was very docile, being used both as pursuer and pursued.

THE SWALLOW-TAILED HAWK, or FORK-TAIL, *Nauclerus*, (Gr. *naukleros*, a ship master,) *furcatus*, (Lat. forked,) differs from the true kite, (*milvus*,) in having a more largely forked tail. The Fork-tail, in steering its course through the air, reminds one of the helmsman who is guiding some noble bark amidst the waves. Its flight is graceful and long protracted. It has been seen as far north as Pennsylvania; in Mississippi and Louisiana it is abundant. The Swallow-tailed Hawks always feed upon the wing. In calm and warm weather they are seen soaring very high, and pursuing the large insects called "Musquito Hawks." The upper plumage is black with reflections of purple; the head and under parts white.

The genus *Elanus* has several species, among which are *E. melanopterus*, (Gr. black-winged,) the BLACK-WINGED SWALLOW-HAWK, of Africa; *E. dispar*, (Lat. dissimilar,) the BLACK-SHOULDERED HAWK, a beautiful bird found from Texas to South Carolina.

The genus *Ictinia*, (Gr. *iktin*, a kite,) is characterized by its strong and short bill, the upper mandible of which is "somewhat angularly festooned," and the lower distinctly notched.

The Mississippi Kite, *I. plumbeus*, (Lat. leaden or dull,) is by some naturalists ranked among the Buzzards. Twenty or more of these birds are sometimes seen together, sweeping around some tree, and catching the locusts which are numerous early in the season. The back and wings of this bird are of a slate blue; the head and under parts whitish, spotted with brown. This *Kite* is said to fly to a great height, where it remains for a while poised; it cleaves the air rapidly, in order to seize the insects which are its prey, added to reptiles and birds.

Third Sub-Family. The Buzzards.

Buteoninæ. (Lat. *buteo*, a buzzard.) *Buteo*, sub-genus of *Falco*.

The Buzzards are, as a group, distinguished by their short beaks, expanded wings, and squared tails. They are common in most of the wooded districts of Europe, and the adjacent parts of Asia, and have been met with in the fur countries of North America. They are indolent, sluggish birds, often remaining perched on the same bough the greatest part of the day; and generally feed upon small quadrupeds, reptiles, and various species of insects. The skins of the Buzzards are covered with fine down. In Cairo, (Egypt,) and in some other places in the East, the skins, after the feathers are removed, are tanned with the down upon them; as thus prepared, they are used by wealthy Turks and Persians for lining their silk robes.

The Common Buzzard, *B. vulgaris*, has a general plumage of chocolate brown; the primary feathers are black with the inner webs white, barred with brownish black; the tail has ten dusky bars on a reddish brown ground; the under parts are yellowish white. This bird builds its nest on high trees, though it has been known to construct it upon rocks. It often seizes upon the nest of a crow, which it enlarges and lines with wool and other soft materials. The female lays from three to five eggs of a whitish cast, spotted with pale brown, and almost without any of the tinge of red which is peculiar to diurnal birds of prey. The length of the Common Buzzard is about twenty-nine inches; the expanse of the wings about fifty inches.

The Red-Tailed Buzzard, *B. Borealis*, (Lat. northern,) peculiar to the American Continent, is found throughout the United States. This is a very wary bird, and hard to be approached by any one bearing a gun. In common with some other Falcons, it is much annoyed by parasitic fly-ticks. Unlike the eagles, pairs of these birds, after rearing their young, "become as shy to each other as if they had never met, and will "chase and rob each other of their prey on all occasions." Farmers, to whom this bird is known as the *Hen-hawk*, usually regard it with anything but complacency.

Other species of the genus *Buteo* found in the United States, are the RED-SHOULDERED BUZZARD, *B. lineatus*, (Lat. marked with lines,) of the Western and Southern sections of the Union; the Rough-legged Buzzard, *B. lagopus*, (Gr. hare-footed;) found east of the Alleghanies, also in the north of Europe, and noted as a great destroyer of meadow mice.

The HONEY BUZZARD, *Pernis*, (Norman, *perner*, to take,) *apivŏrous*, (Lat. bee-eating,) is distinguished by having a feathered band about the eyes. This bird is found in the warmer parts of Europe and of Asia. Its food does *not* consist of honey, as its specific name seems to indicate, but of bees, wasps, and their larvæ. It is a bird of passage, leaving Europe in the beginning of winter. Its length is about two feet; expanse of the wings fifty-two inches.

FOURTH SUB-FAMILY. The FALCONS, Proper.

Falconinæ. (Lat. *falco*, a falcon.)

(*Falco* is the typical genus of the *Falconidæ*, and includes the greater portion of this sub-family.)

The PEREGRINE OR SULTAN FALCON, *F. peregrinus*, (Lat. wandering,) is one of the most remarkable members of the Falcon family. It ranges over Europe, the north of Asia, America and New Holland. When full grown it is a foot and a half in length. We have already referred to the Falcon's strongly notched beak. The beak is of a blue color, approaching to black at the point; the back and upper surface of the bird is of a bluish slate color; the breast reddish white, with dark brown transverse bars. On account of the large size of this bird's feet, it is called the Great-footed Hawk; from its successful chase of ducks, it is sometimes named the Duck-Hawk. The flight of the Peregrine Falcon is amazingly rapid. It does not merely dash at its prey and grasp it with his claws, but strikes its victim with its breast, and actually stuns it with the violence of the blow before seizing it with its claws.

Peculiar interest attaches to the "Sultan" bird from its connection with falconry, an art of great antiquity and extensively practiced by English nobles from the period of the Heptarchy to the days of Charles II. Indeed, a person of rank in England would at one period of her history scarcely be seen out of doors unless he had upon his hand a hawking bird. This in old illuminations and ancient seals is the criterion of nobility. At the present day hawking is still practised in some Oriental countries.

So bold is the Falcon, it was generally employed to take the formidable Heron. When used anciently for hawking, the falcons were taken into the field with hoods over their eyes, and with little bells on their legs; the sportsman carried a lure to which the bird had been trained to fly by being fed regularly upon or near it with fresh killed meat. When the falcon closed with its prey, they both came to the ground together, and it was the sportsman's business to reach the place of conflict as soon as possible, and assist the falcon in vanquishing its prey.

This bird constructs its nest on ledges of rocks, laying four eggs of a reddish brown color.

The GYRFALCON, or JERFALCON, *F. gyrfalcon*. The name Jerfalcon is a corruption of *Hierofalcon*, Sacred Falcon. This bird is by some regarded as the boldest and most beautiful of the tribe, approaching in size nearly to the Osprey. It is a native of Iceland. In the days of falconry, the Jerfalcon was highly esteemed, and used for the larger game, such as cranes and herons. Its plumage is white with dusky lines.

The MERLIN or STONE FALCON, *F. æsalon*, (Gr. *aisalōn*,) is the smallest of the European species, being not much larger than a black-bird; in olden times it was considered as the "lady-bird," and used for taking partridges, which it would kill by a single stroke of the neck. It is not uncommon both in Europe and America, and is a migratory bird.

The KESTRAL, or WINDHOVER, *F. tinnunculus*, (Lat. a kestral,) inhabits Asia and Africa, as well as Europe. Its length is from fourteen to fifteen inches. Farmers often mistake it for the Sparrow-Hawk, and take every opportunity to destroy it; but as its natural food is field-mice, they ought to look upon it as a benefactor, and protect, instead of remorselessly killing it. Its nest is usually built in some deserted one of a crow or magpie.

FIFTH SUB-FAMILY. HAWKS.

Accipitrinæ. (Lat. *accipiter*, a hawk.)

The Hawks have short beaks, hooked from the base; and short wings, reaching no farther than two-thirds of the extent of the tail. The upper mandible has a *festoon*, or prominence in place of the notches of the true Falcons.

Of the genera, we name *Astur*, (proper name,) characterized by its short beak, its somewhat oval nostrils, and the scutellated acrotarsia, or highest parts of the tarsi.

The GOSHAWK, (or Goosehawk,) *A. palumbarius*, (Lat. from *Palumbes*, a wood-pigeon,) receives its name *palumbarius*, from its preying upon pigeons. These, together with pheasants, partridges and grouse, constitute its food. Hares and rabbits, also, it sometimes devours. A full grown female is about twenty-four inches in length; the male bird is one-fourth, and sometimes one-third less. The upper surface of the wings and tail feathers is black; the throat and under parts nearly white, with spots and bars of black. The Goshawk flies low, pursuing its prey in a line after it, or in a manner which falconers call "raking." It abounds in the forests of Continental Europe, and is found in the temperate regions of Asia and America. It has been seen in the neighborhood of Philadelphia, (Penn.)

The Goshawk was also one of the falconry-birds; the female generally "flown" at the large-winged bird; the male at partridges.

Birds of the genus *accipiter* have smooth and elongated tarsi.

The SPARROW-HAWK, *A. fringillarius*, (Lat. *fringilla*, a chaffinch,) is the type-bird of this Sub-family. It is widely spread throughout Europe; found also in Japan and Southern Africa. The adult male is about twelve inches, the female fifteen inches long. The individuals of this species show considerable diversity of color.

The Sparrow-Hawk is a great enemy of quadrupeds and small birds, and is often very destructive to poultry. When taken young, it is easily tamed, and then will "associate with quite incongruous companions." "A gentleman had a Sparrow-Hawk which used to live in his dove-cote among his pigeons; would accompany them in their flights, and be uneasy if separated from its strange friends."

It builds its nests upon high trees, laying four or five eggs of bluish white, marked with dark brown.

The CHANTING-HAWK, *Melierax*, (Gr. *meli*, honey; *hierax*, a hawk,) is the only known bird of prey whose voice has any sweetness, or is at all pleasant to the ear. It is a native of Africa. This Hawk chants every morning and evening; sometimes it continues its notes the whole night long. In size it equals a Goshawk.

Second Family. The Vultures.

Vulturidae. (Lat. *vultur*, a vulture.)

This Family of the Raptorial Birds, feed on the flesh of animals already dead. Decomposition is not, however, a necessary condition of their food, as is shown by the fact that they may frequently be seen regaling themselves on the flesh of an animal within half an hour after it has been killed. Their geographical distribution is confined to warm climates, where they act as scavengers to purify the earth from the putrid substances with which it otherwise would be encumbered. They are generally protected by the natives of the countries which they inhabit, on account of their utility in disposing of decayed animal remains.

It has been a disputed point, whether Vultures are directed to their fetid food by the sight, or the smell. Audubon was in favor of the former, as the directing power; and their lofty flight and telescopic eye, are extremely well adapted to assist these birds in detecting any dying or dead animal; but many facts might be adduced, tending to show that these birds are guided to their food by the action of *both* sight and smell.

The Vulture tribe are, on the whole, considerably larger than the Falcon birds, but they are much less courageous. The beak is lengthened, curved downward at the point, and not in the least notched; the talons are comparatively weak, by no means corresponding with the stature of these birds, and used by them far less than the beak. In order that the parts of the bird which come in contact with its offensive food might not become soiled or matted, as feathers, of course, would be by such contact, the head, and sometimes the neck, in a greater or less degree, are naked, or else covered only with a thin down; the legs, also, at the lowest part, are covered with scales, and not with feathers, as in the Eagle. The wings are strong and large, and the general plumage uncommonly thick and coarse.

The Griffon Vulture, *Vultur fulvus*, (Lat. tawny,) is found throughout a large extent of the Eastern Continent. This Vulture has its head and neck covered with close set, short and white downy feathers. The general color is yellow brown; the length more than four feet.

Cathartes. (Gr. *Kathartes*, a purifier.) The Vultures of this genus have a stout beak, but not the fleshy crest which these birds sometimes exhibit; and the head and neck are plumeless.

The Turkey Vulture, or Turkey Buzzard, *C. Aura*, (Gr. *aura*, air,) is a species that inhabits a vast range of territory in the warmer parts of this continent. In the Northern and Middle States of the American Union, it is partially migratory, the greater part returning to the South on the approach of cold weather. When full grown, it is not far from three feet long; and the wings expand six and one-third feet. The Turkey Buzzards live upon all sorts of food; "they suck the eggs and devour the young of many species of birds, and even eat birds of their own species when they find them dead. They are daily seen in the streets of the Southern cities, along with their relatives, and often roost with them on the same trees." (And.) In Jamaica, this bird is protected by a fine of five pounds sterling inflicted upon any one who destroys it within a prescribed distance from the principal towns.

The Black Vulture, or Carrion Crow, *C. atratus*, (Lat. clothed in black,)—is smaller than the preceding;—less elegant in form, and less graceful in flight. It is a constant resident of all the Southern States, and is seen, during the whole day, in the principal Southern cities, flying or walking about the streets. Those of Charleston resort at night to a swampy wood across the Ashley river. Audubon, and "his friend John Bachman, visited this roosting place together." They estimated the number of these vultures which they saw, at several thousands, spread over an extent of two acres.

The California Vulture, *C. Californianus*, is found in the valleys and plains of the Western Slope of this Continent, and in size bears the same proportion to the other species as a Golden Eagle to a Goshawk. The length of this Vulture is fifty-five inches. It builds its nest upon the loftiest trees; the eggs are two, nearly spherical, and *jet black.*

The Egyptian Vulture, *Neophron percnopterus*, (Gr. *perknos*, black; *pteron*, wing;) sometimes called Pharaoh's Chicken, or Hen, has a pure white plumage, except the great quill-feathers, which are black; the length is a little more than two feet.

Sarcoramphus. (Gr. *sarx*, flesh; *rhamphos*, beak.) The Vultures of this genus, have a fleshy tuft growing on their beaks, somewhat like the wattles, or fleshy excrescences of the Turkey; they have also the Turkey's naked neck and long and oval nostrils.

The Condor, *S. gryphus*, (Gr. *grups*, a griffon,)—is a bird respecting whose magnitude exaggerated statements have been often made. It may, indeed, be ranked among the largest birds which have the power of flight; but the greatest authentic measurement makes its length not more than five feet, and its expanse of wings not more than fourteen.

The beak of the Condor is four inches long, and straight at the base, but the upper mandible becomes arched to a point, and terminates in a strong, well covered hook,—so strong as to be able to pierce the body of a bullock. Around the lower part of the neck in the male and female birds, there is a broad white ruff of downy feathers, which forms the line of separation between the naked skin above, and the true feathers covering the body below it; on the head is a species of comb. (See figure of Condor on the Chart.)

The Condor is found in the Andes from one end of South America to the other; but is most numerous in Peru and Chili, and is frequently met with at an elevation of from 10,000 to 15,000 feet above the level of the sea. Here, amidst perpetual snow, Condors may be seen in groups of three or four, but never in larger companies, like the true Vultures. The Condor descends to the plains only when driven by the demands of appetite; but soon leaves them again for a lighter atmosphere. "The peculiarities of structure," says Dr. Roget, "have probably a relation to the capability we see them possess, of bearing with impunity, very quick and violent changes of atmospheric pressure. The Condor of the Andes is often seen to descend rapidly, from a height of above twenty thousand feet, to the edge of the sea, where the air is more than twice the density of that which the bird has been breathing, thus encountering, in its descent, variations of barometrical pressure extending from twelve to twenty-nine inches."

The general color of the Condor is brownish; the feathers on the back, however, are sometimes perfectly black. This bird does not build any nest, but, after the manner of many sea-birds, lays two white eggs, somewhat larger than those of a Turkey, on the bare lofty rock. It is very strong, and highly tenacious of life. Two Condors will attack and kill the Llama, or even the Puma; by their repeated buffeting and pecking, wearying it so completely that it finally yields to their power. So destructive is the Condor, that various methods are employed by the natives in S. A. to capture it. Sometimes a person clothed in the skin of a newly killed animal, goes out, and entices the Condor to attack him; while companions, who have secreted themselves, rush out from their hiding places, and seize it.

The King Vulture, *S. papa*, is much smaller than the Condor, but of a brighter plumage, and among the handsomest of the Vultures. When pressed with hunger, he will, in the absence of his favorite carrion, feed upon snakes and lizards. He is said to be called the King Vulture, because he keeps some smaller

Vultures under subjection, and "does not suffer them to approach a dead body until he has completely satisfied his own appetite, which is certainly none of the smallest." This Vulture is a native of South America; near the central portions of which it is abundant, and it is occasionally seen in Florida, U. S.

Gypaëtus. (Gr. *gúps*, a vulture; *aëtos*, an eagle.) The birds of this genus, are included by Gmelin, under the genus *Falco.* On the Chart, these are accordingly arranged with the *Falconidæ*, but they perhaps approach most nearly to the Vultures in habits and conformation, and we have thought best to place them with the Vulture Family, as is done by Prince Buonaparte and Mr. Gray. They have the weaker talons of the True or Griffon Vultures; the head and neck, however, are feathered, like the Eagles, and they reject putrid matter unless hard pressed by hunger.

THE LAMMERGEYER, (German, *Lamb's Eagle,*) or BEARDED VULTURE, *G. barbatus*, (Lat. bearded,) is a celebrated bird,—not, however, strictly a true Vulture, as its head and neck are feathered, and it rejects putrid flesh, except when pressed with hunger.

The term *bearded* is applied to this bird on account of the long tuft of hairs with which each nostril is clothed. It destroys hares, and young or sickly sheep and goats; when emboldened by hunger, it does not fear even to attack man himself. The young Chamois, the Mountain Hare, and various kinds of birds fall victims to its appetite. The head and neck of this Vulture, are a dirty white; the lower parts of the neck, breast, and belly, orange red; the back, and wing-coverts, deep gray brown. The Bearded Vulture, the largest bird of Europe, is a little more than four and a half feet in length; the expansion of its wings is from nine to ten feet. This Vulture inhabits the highest mountains of Europe and Asia, and is also found in the lofty mountains of Central Africa.

THE SECRETARY-VULTURE, or SERPENT-EATER, *Gypogeranus.* (Gr. *gups*, a vulture; *geranos*, a bird,) has, in regard to its true position, been a puzzle to naturalists. This very remarkable bird has long legs like a wading bird, but, in other respects, seems to rank between the Vulture and the Eagle. It feeds exclusively upon reptiles and serpents. A pendent crest appears on the back of the head, reminding the beholder of the pen stuck behind the ear by writing clerks,—hence, the name "Secretary." It chiefly inhabits the arid plains in the neighborhood of the Cape of Good Hope. Attempts have been made to introduce this bird into the Antilles with a view to diminish the Yellow Serpent, *Trigonocephalus*, (triangular-head,) which is six or seven feet long, poisonous, and in those islands, very abundant.

THIRD FAMILY. OWLS.

Strigidæ. (Gr. στρίξ, *strix*, a screech-owl.)

The family of *Strigidæ* comprise the Hawks of the night, (*Accipitres nocturni.*) They have the head very large, with great, dilated and projecting eyes, looking forwards, and capable of taking in every ray of light. The power of vision is increased by the manner in which the eye is fixed in a bony socket, just like the watch-makers glass. The pupil is so long that the bird is dazzled in full day, and hence in part arises the stupid appearance which Owls exhibit in the sun-light. Each eye is encircled by a concave disk, or circular fringe, formed of singularly diverging feathers, and assisting these birds to concentrate their whole faculty of sight upon the object directly before them, just as we use a tube in looking at a painting, or some object which we wish to see more distinctly. In those Owls which are partly diurnal in their habits, this circular fringe is scarcely discernible. When the feathers which form the hinder part of the disk are separated, the great ear is seen, enclosed between two valves of thin skin, from whose edges these feathers grow, and which are capable of being widely opened, at the bird's will, so as to catch every sound that may give notice of its prey amidst the silence and darkness. The plumage is loose and downy,—a character which reaches even to the wing-quills; hence the flight of the Owl is almost, or entirely *noiseless.* The downy feathers present various tints of dull yellow, and brown and white; often they are spotted minutely, and very delicately penciled. The Owls have the strong hooked beaks and acute claws of the raptorial birds.

In some species of this family, there is a series of feathers more or less lengthened, on each side of the top of the head, and which can be erected at pleasure; when raised, they have some little resemblance to horns, or to the erect ears of a cat. These are called HORNED, or EARED OWLS. (Pl. IX. fig. 1.)

The geographical distribution of the Owl family is very wide, species being found in Europe, Asia, Africa, America and Australia. They feed on birds and quadrupeds, and some species on fish. The large-horned Owls, of Europe and America, attack hares, partridges, grouse, and even the turkey; but mice, shrews, small birds, snakes and crabs, suffice for the inferior strength of the smaller Owls.

THE OWL FAMILY may be arranged into three divisions: (1.) the BARN OWLS; (2.) the TUFTLESS OWLS; and (3.) the HORNED

OWLS. The species are exceedingly numerous, and our notices of them must be few and general.

(1.) BARN OWLS.

Strix. (Gr. from *strizo*, to screech.) This genus includes the Barn, or typical group of Owls, having great ears, covered with a large operculum, or ear-flap. The beak is lengthened, and covered only towards the point; the tarsi are rather long and feathered, and the toes clothed with hair. The Barn Owls are eminently nocturnal; they are without egrets, or tufts of feathers upon the head; their colors are generally white and pale buff, marked and speckled with bluish gray; their voices loud and discordant.

The Barn, or Screech Owl, *S. flammea*, (Lat. flaming,) is common in the temperate and warmer regions of Europe. It hides during the day "in deep recesses, among ivy-clad ruins, in antique church towers, in the hollows of old trees, in barn-lofts, and in similar places of seclusion." At night it sallies forth for prey, which consists of mice, rats, moles, and shrews. The length of this Owl is about thirteen inches.

THE AMERICAN BARN OWL, *S. Americana*, is closely allied to the European Barn Owl; the color is of a darker brown, with the ruff red, and the length from seventeen to eighteen inches. It is much more abundant in the Southern section of the Union than in the other parts, and is also found quite plentifully in Cuba.

(2.) TUFTLESS OWLS.

These differ from the rest of the family, (excepting the Barn Owl,) in the extraordinary extension of the fringes of feathers about the eye; and also differ among each other in their adaptation to diurnal or nocturnal habits. Like the Barn Owls, they are widely diffused over the globe.

Surnium. (Gr. owl, or inauspicious bird.) THE HOOTING OWLS. The Owls of this genus hoot, and are of very large size. The legs are rather short, with the toes feathered. The plumage is very soft and downy; the facial disks are complete; the wings, very large and much rounded.

THE GREAT CINEREOUS OWL, *S. cinereum*, (Lat. ash-colored,) is very large, the female being about two and a half feet long. They range from the North-East coast of the United States to the Columbia river. The comparatively small size of their eyes seems to indicate that they hunt by day, as Audubon suggests;

the unusually small feet and claws also indicate that they do not prey on large animals. They are not found in any great numbers.

Surnia. (Gr. owls.) This is a genus of Owls having small heads, feathered claws, and wedged tails. Unlike the Barn Owl, they seek their food during the day. Among the species is the

BURROWING OWL, *S. cunicularia.* (Lat. from *cuniculus*, a rabbit,) found on the plains near Columbia river, and throughout the whole extent of the Rocky Mountains. It resides in the forsaken burrows of the Badgers and Marmots, (see Prairie Dogs in our account of the *Rodentia;*) it does not, however, appear to live on terms of intimacy with those animals. The burrow selected by this bird, is usually found at the foot of the worm-wood bush, (*artemisia absinthium,*) upon the summit of which, this Owl often perches. The plumage, as Mr. Townsend states, swarms with fleas, probably left in their burrows by the Marmots and Badger. "I know," says Mr. T., "of no other bird infested by that kind of vermin." The eggs of this Owl are about as large as those of the common House-Pigeon. The length of the male is ten inches, of the female, eleven inches.

THE GREAT SNOWY OWL, *S. nyctea,* (Gr. from *nux*,) is nearly as large as the Eagle Owl, and on account of its snowy whiteness, one of the most beautiful of the tribe. It is found in the high mountain latitudes of both continents.

(3.) HORNED OWLS.

These are so called, from having the head furnished with a pair of tufts of feathers longer than the rest, which are placed above the ears. The tufts are termed *egrets*, and in many species, can be raised or lowered at will. Of this division, is the genus *Bubo*, (Lat. Horned Owl,) in which the ear-opening is small.

THE GREAT HORNED OWL, EAGLE OWL, *B. maximus*, (Lat. greatest,) is one of the largest of the nocturnal birds, being not much inferior in size to the Golden Eagle. It is very destructive to grouse, hares, and even fawns. Formerly, this bird was sometimes used by falconers to entrap the Kite. It inhabits the great forests of Europe; but is seldom seen in England. Pliny refers to it as an ill-omened bird, on account of whose visits ancient Rome twice underwent lustration.

THE VIRGINIA HORNED OWL, *B. Virginianus*, is a native of North America, being found in almost every part of the United States, and in the fur countries where the timber is of large size. Audubon represents it to be quite equal to the Eagle Owl in size. "It sails," says Wilson, "with apparent ease, in large circles, and rises and descends without the least difficulty, by merely inclining the wings or its tail as it passes through the air." Even when "not more than fifty yards distant, it utters its mournful *hoo, hoo, hoo-e,* in so peculiar an under tone, that to those not acquainted with the bird, it might seem they were produced by an Owl more than a mile distant." This Owl is very powerful and daring, attacking half-grown Turkeys with success, and making large havoc among other fowls.

THE LITTLE SCREECH OWL, *B. Asio,* sometimes called the MOTTLED OWL, and when young, the RED OWL of Wilson,—is only about ten inches long; and usually found about farm-houses, orchards, and gardens. Audubon "carried one of the young birds in his coat-pocket from Philadelphia to New York, traveling alternately by water and by land. It remained generally quiet, fed from the hand, and never attempted to escape." The Little Screech Owl is found in the Eastern States, and in Virginia and Maryland.

Otus. (Gr. *ous,* an ear.) In this genus, the conch of the ear is of enormous size.

THE EARED OWL is common to the Eastern and Western Continents; in Pennsylvania, it is much more numerous than the White or Barn Owl. (Wilson.) Of this genus, there are two species,—the *O. vulgaris,* the Long-Eared Owl, and *O. brachyotus,* (Gr. *brachus,* short,) the Short-Eared Owl. In the latter, the head tufts are inconspicuous. The size excepted, these Owls resemble the Great Horned Owl.

What is the first order of LAND BIRDS? What other name is sometimes given to it? Into what sub-orders is it divided? Give the general characters of the Birds of Prey. What is the comparative size of the females of this order? How many families do they include? Into what SUB-FAMILIES does Swainson divide the FALCONS? What is said of the Falcons as a family? Why in treating of the Falcon Tribe is the first place given to the Eagle? Have Eagles a distinct notch or tooth in the upper mandible? What is said of the Golden Eagle? Why is it called Golden? Is the Ring-tailed Eagle a distinct species? What fact is given showing the longevity of the Golden E.? What is said of the size, plumage, &c. of the Bald or White-Headed E.? What is one of its favorite haunts? What does Dr. Franklin say of it? When was the Washington Sea E. first observed? What is said of the Harpy Eagle? What of the Caracara Eagle? What is said of the Fish-Hawk or Bald-Buzzard? What of the Harrier?

What characters separate the Kites or Second Family from the other Birds of Prey? What is said of the Common Kite or Glede? What of the Swallow or Forked-tailed Hawk? What of the genus *Elanus* and its species? What of the genus *Ictinea?* What species is mentioned, and what is said of it?

Give some account of the Third Sub-Family. What is said of the Common Buzzard? What of the Red-tailed B.? What other species of the genus *Buteo* is mentioned? What is said of the Honey Buzzard?

What is the Fourth Sub-Family? Which is the typical genus of the FALCONIDÆ? What is said of the plumage and flight of the Peregrine Falcon? What gives peculiar interest to this bird? What is meant by Falconry? State some particulars respecting it? What is said of the Jer-Falcon? What of the Merlin and Kestrel?

What is the Fifth Sub-Family? Mention their characters. Give an account of the Gos-hawk? Which is the type bird of this Sub-Family? What is said of it? Where is the Chanting Hawk found? Why is it so named?

Which is the SECOND FAMILY of the Birds of Prey? On what do they feed? To what climates are they confined? How are they treated by the inhabitants? What has been a disputed point? How do they compare with the Falcons? What characteristics are given? What is said of the Griffon Vulture? What of the Turkey V. or Buzzard? What of the Black V. or Carrion Crow? What of the Egyptian V.? To what genus does the Condor belong? Give particulars respecting it. What is said of the King V.? To what genus does the Lammergeyer or Bearded V. belong? What is said of it? What of the Secretary Vulture or Serpent Eater?

Which is the THIRD and last FAMILY of the Birds of Prey? Give the characteristics and habits of the Owl Family. What is their Geographical distribution? Into what divisions are they arranged? What species of the Barn Owls are mentioned? What is said of them? How do the Tuftless O. differ from the Barn O.? What is said of the Great Cinerous O.? What of the Burrowing Owl? What of the Great Snowy O.? Why are the Horned Owls so called? What is said of the Great Horned O.? What of the Virginia O.? What of the Little Screech O.? What Eared Owls are mentioned of the genus *Otus?* Trace thus every bird of this Family mentioned on the Chart. The HARPY EAGLE is of the species *destructor*, genus *Harpyia*, sub-family AQUILINÆ, family FALCONIDÆ, order RAPTORES, sub-class, LAND Birds, class BIRDS, division of WARM-BLOODED ANIMALS, sub-kingdom, VERTEBRATES. Give the derivation of these several terms.

SECTION II.

ORDER II.—INSESSORES, OR PASSERES. PERCHING BIRDS.

These birds are of smaller size than those of the other orders. Naturalists regard them as exhibiting, in the highest degree, those properties by which, as a class, birds are distinguished.

So many are the variations of form and structure which are found in this group, (about equaling in number that of all the other orders taken together,) that but few positive characters can be assigned, which are common alike to the whole group and to a particular division. Its distinctions are mostly negative; for the group includes neither swimmers, waders, nor climbers, neither rapacious nor gallinaceous birds; and yet, by comparing the various tribes which it includes, a general resemblance of structure becomes apparent.

These birds have the power of grasping the branches and twigs of trees with their feet, and are accustomed to rest upon them; hence they are called *Perchers;* (Plate X. fig. 4.) the hind toe is always present and placed on the same level or plane as those in front; and the claws are incapable of being raised as in the Birds of Prey. The larger portion of the species usually dwell in woods and thickets. All have the faculty of flight in full perfection, and in the Swifts and Humming Birds it may be regarded as at its highest development. The beak in the Perchers differs greatly in form, but its common shape is that of a cone, more or less lengthened. In some of the genera a notch appears near the tip of the upper mandible, indicating some affinity for the habits of the Falcon tribe; but this gradually disappears in the others.

The food of these birds is various in its kinds; but by far the larger part feed either upon insects or the seeds of vegetables, which they almost always procure by the beak alone.

This order has peculiar interest as including the sweet songsters whose soothing influence is so widely felt and acknowledged.

The *larynx*, or organ of voice, is in these birds always of complex structure, so that there are few of them that do not, during the pairing season, either sing or utter some peculiar note or chatter analogous to song.

The instinct of birds in building their nests, is in those of this order most strikingly displayed. Admirable indeed are the compact felted nests of the Humming-bird, of the Goldfinch or Yel-

PL. X.

Raptores. LANDBIRDS. Insessores.

Scansores. Rasores. Cursores.

Grallatores. AQUATIC BIRDS. Natatores.

EXPLANATION OF PLATE X.

Fig. 1. The Dunlin or Purre, (*Tringa variabilis*,) showing the principal parts of the plumage, particularly those most conducive to flight; a, the front; b, the throat; c, the occiput; d, the cheek; e, auricle, or auditory conch; f, breast; g, back; h, scapularies; i, i. lesser coverts; k, k. winglet or spurious wing; l, l. greater coverts; m, m. the primaries or greater quill feathers of wing, which are succeeded by n, n. the secondaries, and these by o, o. the tertials; p, the upper tail coverts; q, the under tail coverts; r, the tail feathers, (rectrices.)

Fig. 2. The wing of common Buzzard, (*Buteo vulgaris*,) stripped of all its feathers excepting those which give it power and expansibility, and which are those arising from the hand and ulna, termed quill feathers. They form two sets; the first set, (m,) consists of those arising from the hand, (metacarpus and phalanges,) constituting the most important of the series, being mainly instrumental, by their length and shape, their stiffness or flexibility, in determining the character or the power of flight. They are termed the pinions or primary quill-feathers, and are ten in number, but they differ in form, as in relative length. The second set arises exclusively from the ulna, and are termed the secondaries or secondary quill-feathers, (n.) They are usually shorter, broader, and less rigid than the former. Their number varies. From the small bone which represents the thumb, arise certain stiff feathers, lying close upon the quills of the primaries, and constituting the spurious wing or winglet, (k.) Besides these, there is a group of feathers termed tertiaries, arising from the humeral joint of the fore-arm, and which in many birds, as the curlews, plovers, lapwings, &c., are very long, forming a sort of pointed appendage, very apparent during flight: in most birds, however, they are very short, or not to be discriminated from the rest of the greater coverts, of which, in fact, they are a continuation; hence they cannot be strictly reckoned among the quill-feathers. The same observation applies to the feathers (o,) attached to the upper part of the humerus and termed scapularies; these lie along the sides of the back, and in many birds are of great length.

LAND BIRDS.

Order 1st. Birds of Prey, Raptores or Accipitres.

Fig. 3. a, White-headed, Sea, or Bald Eagle, seizing the fish just obtained by b, the Fish-hawk or common Osprey.

Order 2nd. Perching Birds, Insessores or Passers.

Fig. 4. a, Long-eared Podargus; b, Ruby-throated Humming-bird; c, Blue-jay; d, Green Tody; e, Black-cap Titmouse; f, Wagtail.

Order 3rd. Climbers, Scansores.

Fig. 5. a, Cockatoo; b, Green Woodpeckers.

Order 4th. Scratchers, Rasores.

Fig. 6. a, Ruffed Grouse; b, California Partridge or Quail.

Order 5th. Runners or Travelers, Cursores.

Fig. 7. Cassowary or Asiatic Ostrich.

WATER BIRDS.

Order 6th. Waders, Stilts, or Shore Birds, Grallatores.

Fig. 8. a, Crowned Crane; b, Virginia Rail; c, Little Sand-piper.

Order 7th. Swimmers, Natatores or Palmipedes.

Fig. 9. a, Northern Diver; b, Patagonian Penguin.

low-bird, and of the Bottle-tit or Penduline Tit-mouse, and the woven, purse-like nests of the Oriole and the Starlings. (Plate XI.)

The Perchers always live in pairs; in general, the female is smaller and less brilliant in her plumage than the male. The young leave the egg in a blind and naked state, and for a while are entirely dependent upon parental care for their subsistence.

For convenience, this large order has been arranged into four tribes or sub-divisions, founded on the varying form of the beak, viz.: (1) *Fissirostres*, (Split-bills;) (2) *Dentirostres*, (Toothed-bills;) (3) *Conirostres*, (Cone-shaped bills;) and (4) *Tenuirostres*, (Slender-bills.)

First Division of the Perchers. Split-billed Birds

Fissiròstres, (Lat. *fissura*, a slit; *rostrum* a beak.)

This division of the Insessores is a comparatively small one, but is readily distinguished from all the others by the beak. This is short but broad, and more or less flattened horizontally, often hooked at the tip, and very deeply cleft, so that the opening of the mouth, (or gape,) is extremely wide. (Plate IX. fig. 14.) Most of the species feed upon insects, which they take when on the wing, receiving them in full flight into their open mouths. One genus, *Alcedo*, the King-Fisher, subsists on fishes. The *Fissirostres*, like the birds of prey, may be divided into diurnal and nocturnal. Their principal home is in tropical countries. Some species are found in the temperate zone, but rather as migratory visitors than as permanent residents, and, on the approach of winter, they depart to more congenial climes. Many of the species are celebrated for the brilliant hues which adorn their plumage.

The Split-bills are divided into six families, viz.: (1) *Caprimulgidæ*, (Night-jars;) (2) *Hirundinidæ*, (Swallows;) (3) *Meropidæ*, (Bee-eaters;) (4) *Trogonidæ*, (Trogons;) (5) *Todidæ*, (Todies;) (6) *Alcedinidæ*, (King-fishers.)

First Family. The Night-jars.

Caprimulgidæ, (Lat. *Caprimulgus*, Goat-Sucker.)

These birds have the beak exceedingly small, but the gape enormous; (Plate IX. fig. 14;) its sides are, for the most part fringed with long stiff bristles called *vibrissæ;* and the interior of the mouth is moistened with a glutinous secretion, both which aid them to secure their insect prey. The wings are long and formed for powerful flight; the feet very small and feathered to the toes, which are connected at the base by a membrane. The claw of the middle toe, in most of the genera, is extended on one side, the edge being cut into regularly formed teeth, like those of a comb, and used, as is thought, for cleaning their plumage.

The Night-jars are nocturnal or crepuscular in their habits, chasing their insect food by night, or at dusk, when the beetles and large moths are on the wing, for the capture of which the formation of the mouth is admirably fitted.

In their nocturnal movements, their feathered feet, their large ears and eyes, and in other additional respects, an analogy is discoverable between these birds and the Owls, and one which is recognized in the common names, Fern Owl, Churn Owl, &c., applied to some of the species. Indeed, the Night-jars are evidently to be regarded as a connecting link between the Perchers and Birds of Prey.

The species of these birds are widely spread. Their colors are usually various shades of black, brown, gray and white, beautifully intermingled with minute waves, lines and spots.

Instead of being noxious and mischievous, they are the most harmless and useful of birds, destroying the scavenger beetles and moths, those great enemies of vegetation.

The term Goat Suckers, also applied to these birds as far back as the days of Aristotle, is derived from a silly notion that they suck goats, an idea about as credible as the one sometimes entertained that hedgehogs suck cows, or cats the breath of children. The voices of the Night-jars, like those of the Owls, are often harsh and strange; and sometimes they show a peculiar vibratory or quivering character. Some of these nocturnal birds, (*Podargus*, Gr. *pous*, a foot; *argos*, inactive,) have a beak nearly as strong as an Owl's; others of them, (*Psalurus*, Gr. *psalis*, scissors; *oura*, tail,) have forked tails of excessive length; and one species, (*C. diurnis*, Lat. diurnal,) is "seen in cloudy days in troops of fifteen or twenty, skimming over the surface of ponds precisely in the manner of swallows."

The Common Goat-Sucker of Europe, *C. Europæus*, (see fig. on Chart.)

This is a beautiful Night-jar, in its migrations reaching England about the middle of May, and departing near the end of September. Its length is about ten inches. It builds no nest, but lays two mottled eggs on the ground. Frequently this bird sits on a branch or a fence-rail, and with the head as low as the feet, utters, with swollen, quivering throat, its singular *jarring* note, for a long space at a time, and without seeming to draw breath.

The CHUCK-WILL'S-WIDOW, *C. Carolinensis*, is an interesting American Night-jar, but rarely found beyond Mississippi or the Carolinas on the sea-board. It is the southern species of the United States. In sound and articulation, it seems to express the words of its name, putting the chief emphasis upon the last word. Its head and back are of a dark brown color, mottled with red, and streaked sidewise with black; the lower parts are of a dull reddish yellow. The length of the male is twelve and three-fourths inches; of the female thirteen and one-fourth inches. The notes of this bird are seldom heard in cloudy weather, and never when it rains. It forms no nest; its eggs are oval, of a dull olive speckled with brown, and are placed in a little space carelessly scratched amongst the dead leaves. If the eggs are touched, both parents remove them to some other place of deposit in the woods, where they cannot easily be again discovered.

The WHIP-POOR-WILL, *C. vociferus*, (Lat. *vox*, voice; *fero*, to bear or give forth,) is seen at the approach of spring in most parts of the Western and Southern States, and in small tracts, thinly covered with timber, in the Middle States also. Like its near relative, the Chuck-will's-widow, it is not often seen during the day, except when discovered casually in a state of repose; and it is much distressed by being forced to face a brilliant light. In the dusk of the evening, however, this bird becomes active and diligent in securing its insect prey. Its flight is light and noiseless, the motion of its wings only causing a gentle undulation in the air, scarcely noticed by a person a few feet distant. An imagined resemblance of its notes to the syllables *whip-poor-will*, has given this bird that common name. Its song is prolonged for several hours after sunset. The male bird is nine and one-half inches long. (See Chart.)

Steatornis, (Gr. *stéar*, fat; *ornis*, a bird.)

The birds which this term includes were ranked by Cuvier in the genus *Podargus*, but on account of their peculiar food and habits, were erected by Humboldt into a separate genus.

The GUACHARO-BIRD, *S. Caripensis*, (belonging to Caripe,) takes its name from the mountain of Guacharo, near the valley of Caripe, South America. It has a wedge-shaped tail, is about the size of a common fowl, and covered with plumage of a brownish gray color, mixed with small furrowed lines and black dots. The Guacharo mountain is noted for its large cave, pierced in the vertical profile of the rock, eighty feet broad and seventy-two feet high, which was entered by Humboldt and his companions. After penetrating not far from four hundred and fifty feet, they heard from afar the hoarse cries of the Guacharo Birds. These birds quit the cavern only at nightfall, especially when there is moonlight. Humboldt remarks that they are the only frugivorous birds of the night yet known. They feed on very hard fruits, and reject the insect food of which other Goat-Suckers are fond. Once a year, near midsummer, this cave is entered by the Indians. Armed with poles, they ransack the greater part of the nests, while the old birds hover over the heads of the robbers, as if to defend their brood, uttering, at the same time, horrible cries. The young which fall down are killed upon the spot. The inner parts of these birds are laden with fat; darkness and repose, as Humboldt suggests, favoring its formation, as in the case of *geese* and *oxen.* The fat of these birds, when melted, is called the butter or oil of the Guacharo; it is half liquid, transparent and inodorous, and so pure that it will keep a year or more without becoming rancid. The crops and gizzards of the young birds, when opened in the cavern, are found to contain all sorts of hard and dry fruits, which are conveyed to them by their parents. These are preserved, and, under the name of Guacharo-seed, are considered a remedy against intermittent fevers. The cave of the Guacharo is situated in South Lat. 10° 10′.

Second Family. The Swallows or Martins.

Hirundinidæ, (Lat. *Hirundo*, a swallow.)

The family of Swallows resemble the Night-jars in the smallness of the beak, and the great width of the gape, as they do also in the weakness and greatly reduced size of the feet. They, however, differ from the Night-jars in being active *during the day*, and hence are included in the sub-tribe, *Fissirostres diurni*, or Diurnal Split-billed Birds. The Swallows are also of far more powerful wing than the Night-jars, nor have their feathers the lax softness, or the mottled style of coloration common to birds of the night; but on the other hand, the plumage is close, smooth, and often burnished with a metallic gloss; while the prevailing shades are black, (more or less changing into blue or green,) above, and white, often varying with dull red, beneath.

The smallness of their feet is in correspondence with their almost perpetual flight; they even drink on the wing; and their feet, being small and weak, are little used, yet as these birds often cling to rocks and walls when they do rest, their toes are furnished with sharp crooked claws, and the hind toe can either wholly, as in the Swifts, or partially, as in the common Chimney Swallow, be brought to point forward.

The Swallows are widely scattered over the globe, but still are eminently fond of warm latitudes; they roam, indeed, over the temperate zones, and even advance as far north as the Arctic circle, but only in the summer season; when cold weather approaches, they hasten to equatorial climes. Everywhere they are known as birds of great speed, for which they are fitted by the firm and close plumage of their bodies, their long, stiff, and pointed wing-feathers, and their long and forked tails.

Cypselus, (Gr. *kupselos*, a martin.) SWIFTS or MARTINS.

The birds of this genus have the toes thickly feathered, and all the four toes directed forwards. The species *C. apous*, (Gr. without feet,) is the SWIFT, or BLACK MARTIN. This specific name is given to this bird on account of the exceeding *smallness* of its feet. It is spread over Europe in the summer season, and is popularly known as "Jack Screamer." This is one of the swiftest of the Swallow family, appearing to spend the whole day on the wing, and occasionally soaring almost out of sight, but *screaming* so shrilly that the sound is plainly heard. The Black Martins destroy a very great number of insects, retaining them in a kind of pouch under the tongue for the use of their young, and constantly renewing the supply.

Chaetura, (Gr. *chaitē*, bristle; *oura*, tail.) SPINE-TAILS.

The Spine-tails have the tarsus bare and longer than the middle toe; the tail short and even; the shafts prolonged into sharp points.

C. pelasgia, (Gr. *pelazo*, to come near,) is the AMERICAN SWIFT, or CHIMNEY SWALLOW, a bird which seems to show its appreciation of the progress of civilization by leaving its old abodes in the hollows of trees, and taking possession of chimneys free from smoke in the summer season. This bird builds its nest in a semi-circular form. The nest is glued together with the saliva or unctuous matter secreted in glands provided for that purpose, and with the same saliva it is fastened to trees or to a chimney wall. When the nest is in a chimney, it is usually placed on the east side, from five to eight feet from the entrance; when in the hollow of a tree, it is placed high or low, according to convenience. Audubon counted more than a thousand that "entered one chimney before dark," and he estimated that nine thousand roosted in a single tree which he watched near Louisville, Kentucky. This Swallow rears two broods in a season. It does not migrate farther east than Nova Scotia. In the State of New York it appears about the last of April.

The ESCULENT SWALLOW, *H. esculenta*, is the maker of the edible bird-nests, (see Chart,) esteemed such a delicacy among the Chinese, and a considerable article of their commerce. These nests are made of a species of sea-weed, (*fucus.*) The bird macerates and bruises it before forming the material in layers so as to construct the whitish gelatinous cup-shaped nests. The finest are those obtained before the nest has been contaminated by the young. These are pure white, scarce and valuable. The inferior ones are dark, discolored, or mixed with feathers. These are generally converted into glue. The only preparation for sale which these bird-nests undergo, is that of simple drying, without exposure to the sun, after which they are packed in small boxes. They are assorted into three kinds for the Chinese market, according to their relative values, and distinguished into first or best, second and third qualities.

These nests are found in Java, and they are particularly abundant in Sumatra. They are regarded as an article of expensive luxury, and sold at most extraordinary prices; they are consequently consumed by persons of rank alone. The sensual Chinese use them under an impression that they are powerfully stimulating and tonic; but probably their most valuable quality is their perfect harmlessness.

The BARN SWALLOW, *H. rustica*, (Lat. of the country,) ranges in the spring from New Orleans to Newfoundland. The same name is given to the Chimney Swallow of England, which in its song this bird entirely resembles. The nest of this bird is something like a section of an inverted cone, and is attached to the side of a beam or rafter in a barn or shed. The Barn Swallow surpasses in speed *every other species*, except the Humming Bird. The tail is deeply forked, the side feathers of which much exceed the wings in length.

The PURPLE MARTIN, *H. purpurea*, (Lat. of a purple color,) is seen early in April, and for its reception in our cities habitations are sometimes furnished; occasionally its nests are seen in the corners of houses. Its flight is easy and graceful, but not so swift as that of the Barn Swallow. Audubon was of the opinion that this Swallow goes farther south than any other of our migratory birds.

Other interesting species are *H. riparia*, (Lat. *ripa*, a bank,) the BANK SWALLOW, or Sand Martin, which perforates sand banks and makes its nest in the holes; (Plate XI. fig. 10;) *H. thalassina*, (Gr. *thalassa*, the sea,) the VIOLET GREEN SWALLOW; *H.*

fulva, (Lat. tawny,) the REPUBLICAN, or CLIFF SWALLOW. This latter species was described by Gov. De Witt Clinton, in 1824. Its winter retreat is in Mexico. In summer, it is found in different parts of the United States, taking the course of the valleys of the Mississippi and Ohio. Its nest is composed of mud or clay, with a narrow tubular neck, and resembles a coarse retort. (See Plate XI. fig. 11.)

THIRD FAMILY BEE-EATERS.

Meropidæ, (Gr. μέροψ, *merops.*)

We place the Bee-eaters next the Swallows, following in that arrangement the classification of Swainson, though sometimes, on account of the lengthened form of the beak in this family, they have been placed immediately before the tenuirostral, or thin-billed birds.

In addition to the long, slender and tapering beak, the Bee-eaters are distinguished by their long pointed wings; the first quill, for the most part, being nearly or quite as long as any other.

These birds are generally of a green color varied with blue. They associate in flocks, and in their appearance and rapid flight are much like the Swallows. The food of the *Meropidæ,* consists of large insects, which they capture and eat during flight. One species is said to perch and watch for prey on the horn of the Rhinoceros, giving notice to that animal of the approach of the hunter; but usually they take their food on the wing. These birds are entirely confined to the continents and islands of the eastern hemisphere. They do not construct nests, but lay their eggs in holes.

The EUROPEAN BEE-EATER, *M. apiaster,* (Lat. a bee-eater,) in its coloring and shape, is not unlike the King-fisher. It annually visits the countries bordering on the Mediterranean, appearing in flocks of twenty or thirty, and skimming over the vineyards and olive plantations in pursuit especially of bees and wasps. It is remarkable that these birds are never stung; they seize the insect, and with their strong beak crush it at once. The flesh of the Bee-eater is sufficiently esteemed to be sold in the markets both of Italy and Egypt. The boys of Candia, it is said, take it after this manner: they bend a pin like a hook, and tying it by the head to the end of a thread, they thrust it through a *Cicada,* (as boys bait hooks with a fly,) holding the other end of the thread in their hand. The *Cicada* so fastened, nevertheless continues its flight, which the *merops* perceiving, pursues and catches it, swallowing pin and all, whereby she is captured.

The ROLLERS, which are in some respects intermediate between the Swallows and Bee-eaters, have been variously arranged by naturalists. With Swainson, we place them among the *Meropidæ*. These are represented by the genus *Coracias*, (Gr. *korakias*, raven-like,) the birds of which have a straight and moderate sized bill, and very short perching or insessorial feet. One species, *C. Abyssinica*, has two long, loose processes, terminating the two external quills.

The COMMON, or GARRULOUS ROLLER, *C. garrula*, (Lat. talkative,) is plentifully found in most parts of Europe. The mouth is slightly furnished with bristles like those of the Night-jar; the voice is loud and chattering, whence its specific name. To the species *C. orientalis*, (Lat. eastern,) Linn, the name of DOLLAR BIRD is given. Swainson refers it to a sub-genus *Eurystomus*, (Gr. *eurus*, broad; *stoma*, mouth.) It resembles the common Roller, but has a shorter and wider bill, and longer wings; the sides of the gape are smooth. It is a native of South Australia. The natives near Sydney call it the NATAY-KIN; the Colonists name it Dollar Bird. (See Chart.)

THIRD FAMILY. TODIDÆ, (Lat. *todus*, a small bird.) TODIES.

The Todies are a small family, resembling the King-fishers in their general form, and found chiefly within the tropics of both hemispheres. Their legs are rather long; their wings short and rounded, and incapable of any but the most feeble flight. The beak is broad and much flattened, usually blunt or rounded at the tip. In their habits they resemble the Fly-catchers, hopping about among the slender branches of the trees, and occasionally making a short flight to capture insects; these form their principal food, to which, in the case of some species, berries are added. "They have scarcely any voice except at pairing time, and their color closely resembles that of the trees in which they dwell." The species included in the genus *Todus* are confined to Tropical America.

The GREEN TODY, *T. viridis*, is very common in the greater West India Islands. This is a very familiar and beautiful bird; (Plate X. fig. 4d.) while sitting upon some twig or low bush, watching for flying insects, "it will often let a man come within a few feet and look at it for minutes together, before it moves." It is interesting to note the various means Divine Wisdom has ordained for the attainment of a given end. The Swallow and Tody live upon the same food, (insects;) the Tody's short, hollow and feeble wings, are to him as effectual for securing his prey, as are, for the same purpose, the long and powerful pinions of the Swallow to him.

Eurylaimus, (Gr. *eurus*, broad; *laimos*, throat,) is a genus of Todies peculiar to the Eastern Continent. One species is *E. Javanicus*, in which the beak is, at the base, nearly as broad as its length. Sir Stamford Raffles says: "It frequents the banks of rivers and lakes, feeding on insects and worms. It builds nests pendent from the branch of a tree or bush which overhangs the water."

Fourth Family. Trogonidæ. Trogons.

This is not a large family of birds, but one pre-eminent in beauty and brilliancy of plumage. The color is usually a metallic golden green, strongly contrasted with scarlet, black and brown.

The Trogons have two toes behind and two before, as in the Woodpecker; still they have not the habit or power of climbing. The wings are very short but pointed; the quill-feathers stiff; the general plumage soft and thick. The beak is short, triangular-shaped and strong; the tip, and generally the edges are notched; the gape is wide. The head is rather large, and the form full and plump; the tail remarkably long and ample; the feathers regularly decrease in length outward; and in one genus, *Calurus*, (Gr. *kalos*, beautiful; *oura*, tail,) the tail-coverts are so greatly developed as to conceal the tail, and hang down in narrow flowing plumes of great length. (See Chart.)

The food of the Trogons consists principally of insects, "which," says Mr. Gould, "they seize upon the wing, as their wide gape enables them to do with facility; while their feeble tarsi and feet are such as to qualify them merely for resting on the branches as a post of observation, whence to mark their prey as it passes, and to which, having given chase, to return. Dazzled by the brightness of the meridional sun, morning and evening twilight is the season of their activity." The recesses of the thickest forests form their chosen abode for the entire year. The Trogons of the most exquisite plumage are found in South America. According to Mr. Gould, twenty-three species are inhabitants of America and its Islands, ten of the Indian Islands and India, and one of Africa.

The remarkable plumage and shy habits of the Trogons were closely observed by the ancient Mexicans. According to Cortes, three hundred men were employed in taking care of the Royal

Menagerie in which large numbers of these birds were kept; physicians were also appointed to watch their diseases and apply timely remedies. This was by order of the King, who not only delighted in the sight of so many species, but was very careful of their feathers for the sake of the famous mosaic images and pictures, as well as other works which were made of them.

The *Trogon (Calurus) resplendens*, (Lat. shining brightly,) is the QUESAL of Guatimala. Unlike some others of the family, its beak is not serrated; the head is surmounted with a compressed and elevated crest; the upper tail-feathers are so enormously developed as to hide the tail. (See Chart.) "It is scarcely possible," says Mr. Gould, "for imagination to conceive anything more rich and gorgeous than the golden-green color which adorns the principal part of the plumage, or more elegant and graceful than the flowing plumes which sweep pendent from the lower part cf the back, forming a long train of metallic brilliancy." From the feathers of this, "the most beautiful of a beautiful tribe,"—not excluding, however, those of some other species—the Mexicans made Mosaic pictures, together with ornaments for their head-dresses. A picture in mosaic, made from the feathers of this bird, is preserved in the Ashmolean Museum, Oxford, Eng. The subject is "Christ fainting upon the cross." The entire picture is about as large as the size of the palm of the hand, and the figures are only half an inch in height; yet it is said, the very expression of the features is preserved.

Prionites, (Gr. serrated) The MOTMOT.

The name of this genus is derived from the *serrated* margins of both the mandibles, in which particular it differs from the other Trogons. The Motmot, *P. Braziliensis*, is a very curious and handsome bird, inhabiting many parts of South America.

What is said concerning the size and number of the PERCHING BIRDS? What of their variations in form and structure? Why are they called PERCHERS? What is said of their power of flight? What of their beak, food, and musical powers? What of their instinct in building their nests? How do they always live? Into what TRIBES are they arranged, and upon what is the arrangement based? How are the Split-Bills distinguished from all the others? How and upon what do they feed? Is there any exception to this? Where is the home of the Fissirostral birds? Are any found in temperate zones? Name the FAMILIES into which this tribe is divided. Give the characteristics of the Night-jars. In what respects do they resemble the Owls? Why are these birds called Goat-Suckers? What is said of their voices? What sp. are mentioned? Which is diurnal and what is said of it? Describe the Common G. S. of E. What is said of the

Chuck-wills-widow? What of the Whip-poor-will? Repeat what is said of the Guacharo B.

What is the second FAMILY of the Split-billed birds? Are they nocturnal or diurnal? In what respects do they resemble the Night-jars? How do they differ from them? What is said of their flight? What other characteristics are noted of the SWALLOW FAMILY? What popular name has the Swift or Martin? What is said of this bird? What characteristic of the Spine-tails are mentioned? What is the scientific name of the AMERICAN SWIFT or CHIMNEY SWALLOW? What is its signification? Repeat what is said of this bird. What is said of the construction, uses, &c. of the edible bird-nests? Where are they found and how regarded? What is the range of the Barn Swallow? What English bird does it resemble? What other sp. are mentioned? How are the BEE-EATERS distinguished? What is the general color of their plumage? What birds do they resemble? To what birds are the ROLLERS intermediate? What habit is peculiar to one sp.? To what hemisphere are the Rollers confined? To which sp. of R. is the name Dollar Bird given? Describe the E. Bee-Eater? How do the boys of Candia take this bird?

Give the characteristics of the third FAMILY or TODIES. To what region are the birds of the gen. Todies confined? What is said of the Green Tody? What. gen. is peculiar to the Eastern Continent? What is said of the number, plumage and size of the TROGONS, or the fourth FAMILY? What are their characteristics? Where found? How many sp. according to Mr. Gould? How did the ancient Mexicans regard these birds? What use did they make of their feathers? What does Mr. G. remark of the QUESAL of Guatimala? Which is the most beautiful of the Trogons? What mosaic picture was made of its feathers? Where is the MOTMOT found?

SECTION IV.

SECOND DIVISION OF THE PERCHERS. TOOTH-BILLED BIRDS.

Dentirostres. (Lat. *dens*, a tooth; *rostrum*, a beak.)

The upper mandible in this division is notched on each side near the tip, whence the name *Dentirostres*, or Tooth-billed. In the Shrikes, or Butcher Birds, the indentation is very decided, and attended with a projecting tooth, so as to show a connecting link with the Birds of Prey; the beak being also very strong, hooked, and sharp pointed, and the habits of the birds ferocious and carnivorous. Even the Shrikes, however, differ from the Falcons in having the notch confined to the horny surface of the beak, whereas, in the Falcon, it is a true process, extending into the bone itself.

The favorite food of the Tooth-billed birds consists of insects, though some of them join with this food, berries and other soft fruits. Excepting the Finches, belonging to the Cone-billed birds—all the musical birds, including the Nightingale of the Old World, and the Mocking-bird of the New, belong to the division which we are now to consider.

The *Dentirostres* are spread over the globe. They are comprised in five families, viz: (1.) *Silviadæ*, or Warblers; (2.) *Merulidæ*, or Thrushes; (3.) *Muscicapidæ*, or Fly-catchers; (4.) *Ampelidæ*, or Chatterers; (5.) *Laniadæ*, or Shrikes.

First Family. The Warblers.

Silviadæ. (Lat. *Sylvia*, or *Silvia*, a wood.)

The small singing birds comprised under the general name of Warblers, form a very numerous, as well as interesting group. The bill in these birds is slender, straight, awl-shaped, higher than it is wide at the base, and furnished with bristles, the lower mandible being straight. Audubon enumerates no less than forty-four species of these birds found on the American Continent. The habits of the different species vary considerably; but in general, the Warblers frequent groves and woods, and search for the small insects, which are their food, among the leaves and twigs, and the crevices in the barks of trees, rather than on the wing, like the Swallows.

Excepting the Humming Birds, we find among this group, the smallest birds of the creation. The diminutive Golden Crests, *Regulus;* the Nightingale, *Philomela;* the White-throat, or Petty-chaps, *Silvia*, or *Curruca*, (Lat. caterpillar;) the Wood-warbler, or Wood-wren Sylvicola, (Lat. wood inhabitant,) are examples of genuine warblers.

Diffused over all parts of the habitable world, it seems to be the office of these birds to prevent an undue multiplication of the innumerable insects which lurk within the buds, the foliage or the flowers of plants. The smallness of these insects, causes them to elude the notice of the Thrushes and the larger insectivorous birds, whilst their habits secure them against capture by the Swallows and other birds that take their prey only when on the wing.

The Warblers are, for the most part, migratory birds. When the increasing warmth of spring is ushering the insect tribe into renewed life and activity, the return of these birds is providentially and wisely ordered, to prevent its troublesome increase. In autumn, when the hosts of insects begin to diminish, and no longer require to be kept in check, these useful little creatures take their flight to other climes.

The Warblers may be conveniently arranged into five groups,

having different tribes of insects allotted to them respectively, and showing a correspondent diversity in their favorite haunts.

(1.) THE GOLDEN CRESTS, *Sylviadæ*, and WOOD-WARBLERS, *Sylvicolidæ*, are the true warblers, confining themselves mostly to the taller trees, where they search for winged insects among the leaves, or capture them, like the Fly-catchers, when attempting to escape. Of these, the Gold-crested Wren, or Kinglet, *Regulus*, (Lat. *dim.* of *rex.* king;) *cristatus*, (see Chart,) is one of the most attractive species, and the smallest of the European birds, three and a half inches long.

The Golden-crested Kinglet, of America, *Regulus satraps*, is half an inch longer than the European species,yet agrees with it in its general appearance. The color is olive green; beneath whitish, but the crown is orange, or gold colored. It is an active and restless bird, generally found in groups, on the extremities of twigs and bunches of leaves. The Blue Bird, *Erythaca*, or *Sialia Wilsonii*, or Blue Robin, as it is called in some districts of the Union, bears considerable resemblance to the Robin Redbreast, of Europe. It is a lovely warbler, found in all parts of our country, appearing in New York early in the spring, (March,) but leaving in November for the South, as far as Mexico. It is very useful as a destroyer of multitudes of noxious insects. The Myrtle Bird, so called from its feeding in autumn and winter on Myrtle-wax berries, (*Myrica cerifera,*) or the Yellow-crowned Wood-warbler, *Sylvicola coronata*, is perhaps the best representation of the Wood-warblers,—it is very common in the State of New York, and ranges from Mexico to 65o N. Lat.

The *summer* YELLOW BIRD, *Sylvicola æstiva*, (Lat. of summer,)—so called to distinguish it from the Common Yellow Bird, (*Carduelis tristis,*) is "remarkable for its instinctive sagacity in getting rid of the eggs of the Cow Black Bird, (*Molothrus pecoris.*) As the egg is too large to be thrust out, this Yellow Bird commences a new nest above it; thus almost horizontally closing it up, and then proceeds to deposit her own eggs."

The TAILOR BIRD, *Silvia*, (Lat. a tit-lark,) constructs a nest of a curious kind,—by sewing leaves together. (See Plate XI. fig. 8;)

(2.) THE REED-WARBLERS and NIGHTINGALES, of Europe, *Philomelinæ*, which haunt the vicinity of waters, or the more dense foliage of hedges, for insects found in such situations. These are larger than the true Warblers, and live partly upon fruits as well as insects.

(3.) THE STONECHATS, *Saxicolinæ*, (Lat. *saxum*, rock; *colo*,

I inhabit,) which prefer dry commons, or wide extended plains, and feed on the insects peculiar to such localities. The Robin Red-breast, *Erithacus*,* (Gr. *erithacos*, Red-breast;) *rubecula*, (Lat. a Red-breast,)—of ballad and song celebrity, and a bird that sings throughout the whole year,—belongs to this third group. It is smaller, and more familiar in its habits, than the American Robin, *Turdus migratorius*.

(4.) THE WAG-TAILS and TIT-LARKS, *Motacillidæ*, (from *motacilla*, Lat. for wag-tail,) in some respects like the Wading birds, and which have for their food the insects that frequent humid and wet places.

The American Pipit, or Tit-lark, *Anthus*, (Lat. a tit-lark,) *ludovicianus*, is of this group,—a little bird about six and a half inches long, varying in its plumage with age and sex; in the male, of grayish brown on the upper parts, and dusky white beneath. It feeds on minute shells, shrimps, and aquatic insects found on rocky shores and the banks of streams, or on insects and various seeds which it finds in meadows and ploughed grounds; when feeding in the latter places, these Tit-larks are seen in small flocks; to this the specific name probably refers. These birds appear in New York about the first of May,—but range far North and West; wintering in Louisiana and still farther South.

(5.) THE TIT-MICE, *Paridæ*, (from *Parus*, Lat. for tit-mouse, or tom-tit,)—birds which search assiduously for insects among the buds and tender shoots of trees. At the same time, they are quite omnivorous, sometimes laying up stores of grain, and even eating small and sickly birds, when they are able to destroy them. Of this group, the Black-cap Tit-mouse, or Chickadee, *P. atricapillus*, (Lat. black-haired,) is a familiar example, (Plate X. fig. 4e,) a truly Northern species, and so abundant in the fur countries, that companies of them may be found in almost every thicket. The penduline Tit-mouse, or Bottle-tit, *Parus pendulinus*, derives its name from its purse-like, or bottle-shaped nest, suspended on the branch of a willow or some other aquatic tree, with an opening on the side for the ingress and egress of the bird and its young. (Plate XI. fig. 5.)

THE CHESTNUT-CROWNED TIT-MOUSE, *P. minimus*, (Lat. least,) of the Wahlamet, (near the Pacific coast,) constructs a curious nest, resembling a long purse, and hanging from a low bush. (Plate XI. fig. 1.) It is made chiefly of moss, down, and lint of plants, and lined with feathers; the female lays six white eggs.

---

* This is spelled *Erythaca*, on the Chart, after the manner of Swainson and others, but the true orthography is that given in the text.

The length of this pretty bird is only four and a half inches. Linnæus included the entire group of Warblers under the one genus *Motacilla*.

SECOND FAMILY. THRUSHES.

Merulidæ, (Lat. *merula*, a black bird,) or *Turdidæ*, (Lat. *Turdus*, a thrush.)

The Thrushes are the most numerous and diversified of the tooth-billed division of birds. The average size is considerably greater than that of the Warblers. The beak of these birds is as long as the head, and compressed at the sides; the upper mandible arched to the tip; the notch is well marked; the gape is furnished with bristles; the feet are long, with curved claws adapted for walking as well as perching, for exercising on the ground as well as moving among the trees. The food on which the Thrushes subsist, is less restricted than that of the Warblers; for, besides insects and their caterpillars, snails, slugs, earthworms, etc., they feed largely on pulpy and farinaceous berries. Many of the species are gregarious in the winter, and some, as the common Fieldfare, *T. pilaris*, (Lat. like a ball,) are so during the year.

The colors of the Thrushes are, for the most part, sombre, but often elegantly arranged; various shades of olive are the prevailing hues, and these often take the form of spots running in chains upon the breast and under parts. The Orioles are distinguished for their fine contrasts of rich black and golden yellow; the Breves, *Pitta*, (Gr. pitch,) with remarkably short tails, and found in India and Australia,—are distinguished for their dazzling blue and green; while some of the African Thrushes shine like the metallic lustre of burnished steel.

The Thrushes are common in all parts of the world, and many of them are eminently birds of song. As illustrating the general character and habits of the family, we may refer to the Song Thrush, Mavis, or Throstle, (*T. musicus*,) which sings with sweet and varied note from the commencement of spring, and even earlier, to the close of summer; to the Blackbird, (*T. merula*,) whose song is less varied, but still richer and mellower; and to the Fieldfare, or Gray Thrush, (*T. pilaris*.)

THE AFRICAN SHORT-LEGGED THRUSHES, *Brachypodidæ*, (Gr. *brachus*, short; *pous*, a foot,) are a sub-family; from having four long bristles on the back of the neck, sometimes called Bristly-necked Thrushes.

Another sub-family, is the BABBLERS, or LONG-LEGGED THRUSHES,

Craterpodidæ, (Gr. *crater*, a wine-cup, or opening; *pous*, a foot,) of Australia, which have large and strong feet, and send forth loud and disagreeable notes. Among the true Thrushes, *Merulinæ*, is the AMERICAN ROBIN, *T. migratorius*, (Lat. migratory,) whose cheerful note is always most welcome in the opening spring, and whose large nest, in which may be seen five beautiful sea-green eggs, appears to be regarded, even by boys, as more sacred than others.

THE AMERICAN MOCKING BIRD, *Orpheus*, (Gr. proper name of a famous musician;) *polyglottus*, (Gr. many-tongued,) is another true Thrush,—having a voice capable of every variety of modulation, surpassing, in this respect, even the European Nightingale; but not noted either for its gay, or its brilliant plumage. The Mocking Birds are much sought after on account of their extraordinary vocal and imitative powers, which remain undiminished even in confinement. They are easily raised; a single bird sells for from seven to fifteen or twenty dollars; the sum of fifty dollars has been paid, and that of one hundred dollars refused for an exquisite singer. The first brood of the Mocking Bird are always largest and stoutest. This bird is usually about ten inches in length. The CAT BIRD, *T. lividus*, (Lat. livid,) is an aberrant form, of *Orpheus*, and, in some parts of the United States, very numerous. It is nearly as large as the Mocking Bird, (see Chart,) and has a note which closely imitates the cry of a young kitten. The MISLETOE THRUSH, (*T. viscivorus*, (Lat. *viscus*, a misletoe; *voro*, I devour,) is said to "surpass all other Thrushes in size, and is decidedly the largest songster of the European birds." It is particularly fond of the berries of the misletoe; next to these, it prefers the berries of the mountain ash. Its length is eleven inches. Some persons call it the STORMCOCK, as it "pours forth its melody when the bleak winds of winter roar through the leafless trees." The WATER-OUZEL, or DIPPER, *Cinclus*, (Gr. *Kinklos*, the name of a bird,) *aquaticus*,—is an interesting bird, found principally in hilly places, where there are clear and rapid brooks and rivulets. It dives for considerable distances with apparent ease, and may be seen perched on the top of a stone in the midst of a torrent, in a continual dipping motion, while watching for its food, which consists of small fishes and insects. It has been said to possess the extraordinary power of "walking, in quest of its prey, on the pebbly bottom of a river, and with the same ease as on dry land."!! Respecting this alleged power, it has been well remarked, "If the Water Ouzel, which is specifically lighter than water, can manage, by some inherent power, to walk on the ground at the bottom of a rivulet, then there is great reason to hope that we, who are heavier than air, may, any day, rise up into it, unassisted by artificial apparatus, such as wings, gas, steam, or broom staff."

Third Family. Fly Catchers.

Muscicapidæ, (Lat. *musca*, a fly; *capio*, I catch.)

This very musical family, which receives its popular name from the expertness of the birds which it comprises, in catching the flying insects upon which they feed, is found widely diffused throughout both the Eastern and Western Continents; and includes many of the most beautiful of the feathered tribes. They appear to be a connecting link between the Split-billed and the Tooth-billed birds. Like the former, they have a beak which is broad at the base, and hooked at the tip, while the gape is surrounded with bristles; like them, also, their feet are unusually feeble, or less developed than the wings, and they feed upon insects which they take in their flight. Indeed, they are the most insectivorous of all the tooth-billed birds. In their generic details, the Fly-Catchers widely differ; all, however, are united by common peculiarities of structure; and, particularly, by having the beak strong, broad, flat, angular on the summit, or *culmen*, and notched at the tip; and by having the side of the mouth defended by thick bristles.

The European birds of this family, are sometimes called "Restricted Fly-Catchers," having shorter bristles around the mouth, and much more slender bills than the others. Of these there are but two species, viz., *M. grisŏla*, the Gray, or Spotted Fly-Catcher, known by several provincial names, all derived from its habits,—as the "Beam Bird," from a favorite site of its nest, and the "Cherry-chopper," from its supposed taste for the fruit of that tree. In some portions of England, it is called the "Post-bird," and in other parts, the "Bee-bird." It is about six inches in length; its breast is of a dullish white, slightly tinged with a dull orange, and the upper part of the body is brown. It is a very tame bird, often building its nest in places where persons are constantly passing and repassing,—seeming particularly partial to the vine and sweet-briar, as the support of its nest. The note of this Fly-Catcher is a weak chirp, and even that is not often heard.

The other European species, *M. luctuosa*, (Lat. *sorrowful*,) is the Pied Fly-Catcher, about as large as a Linnet, found in Prussia, Sweden, and sometimes in England. It has been called "a Magpie in miniature." The bill, and the crown of the head, are black; the other parts of the body present various shades of black, brown and white; there is a white spot on its forehead, from which its name is derived. The female lays five very pale blue eggs.

Of the numerous American Fly-Catchers, we name first, the King Bird, or Tyrant Fly-Catcher; *M. tyrannus*, (Linn.,) or *Tyrannus intrepidus*, (Lat. undaunted,) ranging during summer from the temperate part of Mexico to the remote interior.

It receives its name from the authority which it assumes over other birds during the time of breeding; the eggs are five in number, of a pale green color, or dullish white. At the breeding season, the King Bird's extreme affection for his mate and for his nest and young, makes him suspicious of every bird that happens to pass near his residence, so that he attacks, without discrimination, every intruder. In the months of May, June and part of July, his life is one continued scene of broils and battles; in which, however, he generally comes off conqueror. Hawks, and Crows, the Bald Eagle, and the Great Black Eagle, all equally dread an encounter with this dauntless little champion, which, mounting to a considerable height above these birds, darts down upon their backs, sometimes fixing himself there, to the no small annoyance of his powerful antagonists. In teasing the Eagle, he constantly keeps up a shrill and rapid twittering; this, in fact, is his only song. The Purple Martin, however, from its more rapid flight, is more than a match for the King Bird, eluding all his attacks, and teasing him as he pleases. "I have," says Wilson, "also seen the Red-headed Woodpecker, while clinging on a rail of the fence, amuse himself with the violence of the King Bird, and play *bo-peep* with him around the rail, while the latter, highly irritated, made every attempt, as he swept from side to side, to strike him,—but in vain."

In fields of pasture, the King Bird often perches upon the tops of the mullein, and other rank weeds, near the cattle, and makes occasional sweeps after passing insects, particularly the large gad-fly, so annoying to horses and cattle. This bird preys upon bees, but in his watchings of the bee-hive, it is said, he picks out only the drones, and never injures the working bees.

He must, however, be regarded as the farmer's friend, in destroying great multitudes of insects, whose larvæ prey upon the productions of his fields and gardens. Like all Fly-Catchers, the King Bird disgorges the harder parts of insects. This bird reaches New York the last of April or the first of May; it leaves the Middle States earlier than most other species. The King

Bird is eight inches long, and fourteen in the expanse of its wings.

THE PHEBE BIRD, or PEWIT FLY-CATCHER, *M. fusca*, (Lat. dark or dusky.) The notes of this bird are pleasing, not for any melody which they possess, but from their association with the returning verdure of spring. The favorite resort of the Phebe Bird is by streams of water, under or near bridges, in caves, &c. Near such places, he sits on a projecting twig, calling out, *pe-weé*, *pewittitee pe-weé*, for a whole morning,—occasionally sallying after insects, and returning to the same perch. The Pewit appears in New York State the last of March or the beginning of April. It lays four to five white eggs, with a few reddish spots near the larger end, and it sometimes rears three broods in a season. Insects are its summer food; berries and seeds, its winter fare. Whenever the Pewit appears, Mr. Bartram says, it is safe to plant almost all kinds of esculent garden seeds, as, after the arrival of this bird, there are rarely frosts severe enough to injure them. The plumage is a dark olive brown, the bill entirely black; the tail emarginate, the feathers whitish on the outer web. This familiar and favorite little bird, winters from South Carolina to Mexico.

THE WOOD PEWEE, *M. virens*, (Lat. green, or lively,) is generally found in the interior of forests; it is considerably more abundant than the Phebe Bird; is rather late in entering the Middle States, seldom reaching Pennsylvania and New York until from the 10th to the 15th of May, but it advances as far North as Labrador, and is seen on the Rocky Mountains.

THE AMERICAN RED-START, *M. ruticilla*, (Lat. red, inclining to golden yellow,) is found, during the summer, throughout the United States, but winters between the tropics; it is shy and solitary, and varies much in the brilliancy of its colors.

Genus *Culicivora*, (Lat. gnat-eaters,) includes the Blue-grey Fly-Catcher, *C. cœrulea*, (Lat. dark blue,)—a lively little bird, four and a half inches in length, noted for its being frequently the foster parent of the young Cow-bunting, the real mother of which drops her egg in its nest. It ranges from Texas northward.

The GREENLETS, which by some naturalists are included among the Fly-catchers, are by Audubon and Dr. Dekay erected into a separate family, *Vireonidæ*.

They include about eight species, and are peculiar to America. The bills of these birds are of moderate size, but strong, and broader than high at the base, which is furnished with bristles. The upper mandible is notched, and the tip bent; the tarsus of moderate length, as is also the tail, which is more or less emarginate. We can only glance at the principal species. They feed upon insects and berries.

The YELLOW-THROATED GREENLET, *V. flavifrons*, (Lat. yellow-front,) is of an olive green color; it winters in Texas and Mexico, but in summer advances as far north as Nova Scotia. This Greenlet is said to construct a pendulous nest. The length is from five to six inches.

The SOLITARY GREENLET, *V. solitarius*, has a bill that is very short and nearly as broad as the Fly-catcher's. The color of this Greenlet is dusky olive; the length five inches. It winters in Mexico, but is seen as far north as Nova Scotia and as far west as the Columbia river.

The WHITE-EYED GREENLET, *V. novoboracensis*, (of New York,) has a short, straight bill, abruptly curved at the tip and flattened at the base. Its general color is a dark olive, with white underneath; the third quill is the longest; the length five inches. The notes of this bird are uncommonly sweet. It uses bits of newspaper in making its nest, and Wilson says is, therefore, sometimes called *politician*. This Greenlet is seen in New York early in April, and is common; it leaves for the South in October.

The WARBLING GREENLET, *V. gilvus*, (Lat. pale yellow,) is of a pale green color above, but whitish beneath; the first and fifth primaries are equal; the bill short; the length five inches. This is a musical little bird, wintering in tropical America but reaching New York early in May, and advancing to 46° N. L.; it is seen westward as far as Columbia river. Its nest is pendent, containing from four to six white eggs, marked with brown spots and lines. The length is five inches.

The RED-EYED GREENLET, *V. olivaceus*, (from Lat. *oliva*, an olive,) is of a light olive green above; beneath whitish, with a yellowish tinge on the sides. It has a long and strong bill. This is a common species, ranging from Mexico to 55° N. L.

The LONG-BEAKED GREENLET, *V. longirostris*, (Lat. long-beaked,) has wings not reaching to half the length of the tail; the first quill shorter than the fourth. It is found in the Antilles; the length is five and one-half inches.

Another genus of the Greenlets is ICTERIA, with but one species, the YELLOW-BREASTED CHAT, *I. viridis*, (Lat. green,) connecting the Fly-catchers with the Greenlets, and also forming the passage between the *Merula* and the *Vireo*. This bird has a strong and lengthened beak, curved and with small divergent bristles at the base; the wings are rounded; the third and fourth primaries the longest; the first scarcely longer than the sixth. The color is a greenish olive; the length seven inches. This bird attracts attention by its singular notes, and the oddity of its motions. It comes from the tropical regions of America early in May; along the Atlantic does not advance farther than the southern part of New York; it is, however, not uncommon in the Western States. It leaves us among the earliest, going South about the middle of August.

FOURTH FAMILY. CHATTERERS, OR WAX-WINGS.

Ampelidæ, (Gr. *αμπελις*, *ampelis*, a vine or singing bird.)

The beak in the Chatterers is stouter in proportion to its length than in the Fly-catchers, the form of the lower mandible approaching that of the cone-billed birds; the upper mandible is, however, rather broad at the base, flat, with the upper edge more or less angular and ridged, and the tip distinctly notched. The feet are, for the most part, stout, with the outer toe united to the middle one as far as, or beyond the first joint. In many, the wide gape extends beyond the eye, and in some it is nearly as wide as in the Night-jars. The absence of bristles from the gape indicates that the wide opening is not to catch insects on the wing, as in the Swallow family. The Chatterers feed chiefly on berries and other soft fruit, which they swallow whole; and this food naturally requires a wide passage; occasionally they feed on insects. Their home seems to be in fruit-bearing trees, and they very seldom come to the ground.

The species in this family are not very numerous; but they are of varying forms, and widely scattered. Many of them are distinguished for their soft and silky plumage and the brilliant colors which adorn it. The plumage of the head forms a long and pointed crest, which is capable of being erected, and is common to both sexes. Some of these birds are distinguished by having singular appendages to the secondaries of the wing, and sometimes to the feathers of the tail; the shaft of the feather being extended beyond the vane, and its tip dilated into a flat oval appendage of a brilliant scarlet hue, and exactly resembling the appearance of red sealing-wax. Hence they are sometimes called *Wax-wings;* from the silky softness and smoothness of their plumage, and particularly that of the tail, they are also named SILK-TAILS.

The BOHEMIAN CHATTERER, or SILK-TAIL, *A. garrulus*, or *Bombycilla*, (Gr. *Bombux*, silk-worm,) *garrulus*, is the only species known in Europe; south-east of Germany it is quite abundant. Its general plumage is of a purplish red hue; the crown and crest are of a chestnut brown; some five or six of the secondary feathers, and, in very old males, some of the tail feathers also, have the extended scarlet appendages which are referred to above. Prince Bonaparte gives a very amiable character of the European Wax-wing in a state of nature. In the spring it eats all sorts of flies and other insects; in autumn and winter different kinds of berries. It is fond of the berries of the mountain ash; of grapes it is exceeding greedy, and is, therefore, with reason, called *ampelis*. When taking wing it utters a note resembling the syllables *zi, zi, zi*, but it is generally silent, though it bears the name of Chatterer. In captivity it eats almost any vegetable substance, losing at the same time, all its vivacity and its amiable social habits. Its length is nine or ten inches. This bird was seen by Dr. Richardson in N. Lat. 50°, in flocks, near the Great Bear Lake; it has also been procured in the vicinity of Philadelphia.

The CEDAR WAX-WING, or CEDAR BIRD, *B. Carolinensis*, ranges from Texas as far north as the Fur countries, and westward to the Columbia river. Its nest is built in the fork of a cedar or apple tree, and is composed of stalks of grass, coarse without and fine within. In this it lays three or four eggs of a bluish white, marked with dots of black and purple. It devours every fruit or berry that comes in its way. Dr. Brewer says it remains all the year round at Boston, and confers great benefit on the farmer by destroying thousands of the destructive canker-worm. Audubon thinks the name of *Fruit devourers* would be more appropriate for these birds than that of *Chatterers*. "By way of dessert," however, they eat largely of winged insects, being troubled with most voracious appetites.

The *A. cotinga*, (Gr. from *kōtillō*, to chatter,) (see Chart,) an inhabitant of Brazil, is sometimes called the *Pompadour* Chatterer, from having been introduced into Europe by the thoughtless and extravagant mistress of Louis XV.

The RED, or JAPANESE WAX-WING, *B. phœnicoptera*, (Gr. *phoinicos*, red; *pteron*, wing,) bears great resemblance to the Cedar Bird. It was discovered by means of the scientific mission to Japan, instituted by the government of the Netherlands.

The BELL-BIRD, *Procnias*, (Gr. *proknē*, a proper name or Swallow,) *carunculata*, (Lat. from *caruncula*, a small piece of flesh,) is a species of the Chatterers distinguished by the soft carbuncle or fleshy excrescence at the base of the beak. It is the celebrated Campanero of South America, whose voice, during the stillness of mid-day, it is said, exactly resembles the tolling of a bell.

At uncertain intervals, the Chatterers appear in particular districts in immense flocks, and so remarkable have such visitations appeared, that they have been recorded as events of history, and regarded as ominous, in some way, of great public calamities.

Fifth Family. Shrikes, or Butcher Birds.

Laniadæ, (Lat. *lanius*, a butcher.)

The structure of these birds closely resembles that of the Perchers, but their beak is very similar to that of the falcons, in its strength, its arched form, its strongly hooked point, (see Chart,) and in the distinct tooth which precedes the usual notch of the tooth-billed tribe. This peculiarity of beak is accompanied by a carnivorous appetite, a rapacious cruelty, and a courage altogether raptorial, and which, as indicating a kindred nature, have induced naturalists to associate them with birds of prey.

The Shrikes not only devour the larger insects, especially grasshoppers, but even attack and overpower small birds and quadrupeds, seizing them with their beak or claws, and bearing them to some station near to tear them in pieces with their toothed and crooked beak. These birds live in families for a few weeks after the breeding season; they fly irregularly and precipitately, uttering shrill cries; nestle on trees or in bushes; lay five or six eggs and take great care of their young. Many of them have the curious habit of impaling their prey upon a large thorn, and then pulling it to pieces and devouring it at their leisure. Hence they have derived the name of ***Butcher-birds***. Mr. Bell, when traveling in Russia, had one of these birds given to him, which he kept in a room, having fixed up a sharpened stick for him in the wall; and on turning small birds loose in the room, the Butcher-bird instantly caught them by the throat in such a manner as soon to suffocate them, and then stuck them on the stick, pulling them on with bill and claws; and so served as many as were turned loose, one after another, on the same stick.

The power which the Shrikes have of clutching with their toes is remarkably great. They always hold their prey in one foot, resting on the tarsal joint of that foot, unless when they have fastened the prey upon a thorn, when they pull it to pieces in a contrary direction. They show great boldness in defending themselves and their nests from their more powerful enemies; and the parents evince great attachment to each other as well as to their young. This family comprises a large number of species, distributed through all quarters of the globe. Some of them have a remarkably melodious song.

The GREAT GRAY SHRIKE, *L. excubitor*, (Lat. a sentinel,) receives its specific name from its habit of watching for birds of prey, and chattering loudly as soon as it perceives them. Bird-catchers sometimes avail themselves of this peculiarity in taking hawks. A pigeon is fastened to a net by way of bait. A string is attached and brought within the turf hut where the bird-catcher sits. Close to the hut a shrike is tied to the ground, and two pieces of turf are set up as a shelter for the bird from the weather, and as a refuge from the hawk. As soon as the hawk appears in the distance, the shrike becomes agitated; as it draws nearer, he begins to scream with fright; and just as the hawk pounces on the pigeon, he runs under his turf, which is the signal to the bird-catcher to pull the string, thereby enclosing the hawk within the folds of the net. The nest of this bird is built on trees, and contains about six grayish-white eggs, ash-colored on the larger end. The length of the Great Gray Shrike is from nine to ten inches.

The GREAT AMERICAN SHRIKE, *L. borealis*, (Lat. northern,) is larger in size, but in other respects, does not differ much from the preceding European Shrike.

The RED-BACKED SHRIKE, L. *collurio*, (Gr. *kolluriōn*,) has derived its English name from having the back, scapulars, and wing-coverts of a rusty red color. (See Chart.)

What is the 2d DIVISION of the PERCHERS? Why are they so called? In what birds is the notch most remarkable? What are their habits? How does their beak differ from the Falcons'? What is said of the TOOTH-BILLED BIRDS? Do they include all the musical birds? What is the exception? Into how many FAMILIES are these birds divided? What is the 1st FAMILY? What is said of their numbers? How many American sp. does Audubon enumerate? What is said of the habits and size of these birds? What office do they perform? What is said of their migrations? Into how many GROUPS may they be arranged? Which are the TRUE WARBLERS? What sp. are particularly mentioned? Which is the smallest of European birds? What birds are included in the 2d GROUP? What in the 3d GROUP? What celebrated bird is found in this group? How does it differ from the A. Robin? What is the 4th GROUP? What A. sp. is mentioned? What is the 5th GROUP? What is a familiar example? Why is the Penduline Tit or Bottle Tit so called?

Name the 2d FAMILY. What is said of their numbers, favorite haunts and average size? What of their beak, food and plumage? For what are the ORIOLES distinguished? For what the BREVES and AFRICAN THRUSHES? What sp. illustrate the general character and habits of this family? To what sub-family does the AMERICAN ROBIN belong? What is said of the Mocking-bird? What of the Cat-bird? Of the Misletoe Thrush? Of the Water Ouzel or Dipper?

What is the 3d FAMILY? Why are they so called? How do they appear to connect the Split-billed and Tooth-billed birds? Into how many genera does Audubon arrange them? What are the EUROPEAN FLY-CATCHERS sometimes called and why? Mention the sp. What is said of the King-bird? What of the Phebe B.? What of the Wood Pewee and the American Redstart? How have the Greenlets been arranged? What is said of them?

What is the 4th FAMILY? Give the characteristics of these birds. What is their food? What is said of their plumage? Why are they called WAX-WINGS? Name and describe the only E. sp.? What is said of the Cedar B.? What of the Pompadour Chatterer? Of the Asiatic Wax-wing? Of the Bell B.? What has been inferred from the appearance of immense flocks of Chatterers in certain districts?

What is the name of the 5th and last FAMILY? What is said of the structure of these birds? What accompanies their peculiarity of beak? Is their food confined to the larger insects? What curious habit have they? What name is hence given to them? What is related by Mr. Bell? What is said of the power of their toes? What of their boldness? Does this family include many sp.? What is said of the G. G. SHRIKE? What of the Great American S.? What of the Red-backed S.?

Trace those mentioned on the Chart.

SECTION V.

THIRD DIVISION OF THE PERCHERS. CONE-BILLED BIRDS.

Conirostres. (Lat. *conus*, a cone ; *rostrum*, a beak.)

This division is less numerous than the Dentirostres, but still includes a great number of birds of varying size, structure and habits. Naturalists regard this tribe as typical, not only in the Order of the Perchers. but in the whole Class of Birds.

The chief character by which they are associated together, is found in the beak, which, though differing in shape and comparative size, is generally short; at the same time, it is thick, and very strong, more or less *conical* in form, and usually without a notch at the tip. In one pretty large group, however, the TANAGERS, of Louisiana, and South America,—gay, fire-colored birds,—the beak, while partaking of the conical form of this division, is distinctly notched; constituting them one link of connexion between this and the preceding tribe, (Dentirostres.)

The feet in the Cone-billed Birds are, upon the whole, formed more for perching than for walking, though many birds of this division, walk habitually upon the ground.

Seeds and grain are the principal food of these, the "Hard-billed" Birds; and for opening the different capsules and seed vessels, as well as for crushing hard seeds themselves, their stout and horny beaks are peculiarly fitted. Some of these birds, however, join insects to vegetable food; and a part of them are nearly or quite omnivorous. As the form of the beak varies from that of a short and broad cone, so does the appetite proportionably vary from an exclusive seed diet. The Cone-billed Birds, particularly the FINCHES, seem to prefer the temperate and colder to the warmer regions; but they are represented in all the countries of the globe. The families of this tribe are the following, viz: (1.) *Corvidæ*, or Crows; (2.) *Sturnidæ*, or Starlings; (3.) *Fringillidæ*, or Finches; (4.) *Loxiadæ*, or Cross-bills; (5.) *Bucerotidæ*, or Horn-bills; (6.) *Musophagidæ*, or Plantain Eaters.

FIRST FAMILY. THE CROWS.

Corvidæ, (Lat. *corvus*, a raven.)

THE CROWS are among the largest of the Passerine, or Perching Birds. They are widely spread, but yet comparatively few in number. Their beak is powerful, more or less compressed at the sides, conical, but long, having the upper mandible usually arched, the gape nearly straight, and the nostrils concealed by stiff bristles, pointing forwards. The plumage is dark and sombre, often black, more or less glossed, and sometimes varied with gray or white. To this sombre coloration, the Jays, however, are an exception, being usually arrayed in the richest azure and purple. They are also more exclusively arboreal than others of the family which walk a great deal on the ground.

The *Corvidæ* are birds of firm and compact structure; their wings are long, pointed and strong; their feet and claws robust. Their disposition is bold and daring; they are very sagacious; easily tamed, and rendered familiar. Most of them have the faculty of imitating with much accuracy the sounds which they hear, and even words of human language. They show a strange propensity for thieving, and for hiding substances that can be of no use to them whatever, particularly if they display metallic or polished surfaces, or brilliant colors. They may be ranked as omnivorous; insects and their larvæ, grain, fruits, bread, flesh, both when fresh and when putrid, they can, by turns, devour with avidity.

"THE CROW, (*Corvus*,)" as Swainson strikingly remarks, "is the type of types, or the preeminent type of all birds, uniting a greater number of properties than are to be found in any other genus of birds. Like the Hawk, it soars in the air, and seizes living birds; like the Vulture, it devours putrid substances, and picks out the eyes of young animals; like the Climbers, it discovers its food when hidden from the eye, by pecking; like the Parrot family, it has a taste for vegetable food; has great cunning, sagacity, and powers of imitation, even to counterfeiting the human voice; like the Waders, it walks with facility, and has great powers of flight; like the Aquatic birds, it can both catch and feed upon fish. Thus it unites some of the properties of all other birds, and stands the preeminent type of the Perchers."

The largest and most powerful species of the genus *Corvus*, is the well known *Raven, C. corax*, (Lat. a raven,)—the CORBIE, of Scotland, celebrated even from the time of the universal deluge, and ever looked upon as a bird of dark omen. It is twenty-five inches in length, and fifty inches in the spread of the wings,—ranging from the Arctic seas to the Cape of Good Hope, in the Eastern Continent, and from the same seas to Mexico, on the Western; unchanged in character, amidst all the variations and extremes of heat and cold; traveling in pairs, and flying so high that it would escape notice but for its frequent crying; in all times and places, showing itself possessed of acute and powerful sight and smell; and at perpetual variance with all other feathered tribes.

THE COMMON CROW, *C. Americanus*, is seventeen inches in length, being somewhat smaller than the Common Carrion Crow of Europe, from which it differs in its voice, its gregarious habits, and the shape of its tongue. Both are regarded and treated as nuisances. Tens of thousands of them are shot every season. They may be of some use to farmers in ploughing time, by picking up worms and the larvæ of insects; but of other good deeds of the Crow, we are ignorant. No sooner are the seeds in the ground, than he begins to search after and devour them; for Indian Corn and eggs he seems to have a wonderful inclination; and even relishes young chickens, turkeys and goslings; at the same time, he is very cunning in avoiding the snares which are devised to entrap and destroy him. The FISH CROW, *C. ossifragus*, (bone-breaker,)—found on the sea-coast as far North as New York, like the Raven and Common Crow, robs other birds of their eggs and their young; but, being regarded as inoffensive, it is usually unmolested. It takes the liberty, however, to feed with great freedom on the best garden fruits. The MAGPIES, *Pica*, (Lat. magpie,) and the JAYS, *Garrulus*, (Lat. chatting, or talkative,) are near relatives of the Crows. The well known BLUE JAY, *G. cristatus*, (crested,) is capable of living in cold as well as warm climates, and is found in all parts of the United States. It is truly omnivorous, and, in times of scarcity, has been known to feed even on carrion. Though extremely beautiful in its appearance and graceful in its movements, (see Plate X. fig. 4c,) it is a deceitful, and often a very mischievous bird. The NUT-CRACKERS, *Nucifraga*, (Lat. *nux*, a nut; *frango*, to break,) all belong to this family. In their habits, they resemble both the Jays and the Woodpeckers,—climbing trees and perforating their bark, and devouring all sorts of fruits and insects, as well as small birds. The FRUIT CROWS, *Coracinæ*, are a sub-family of South American Birds, about whose proper place there has been some question among naturalists, but which are placed by Swainson with the *Corvidæ*. The most remarkable of these are the Capuchin Baldhead, *Coracina gymnocephala*, (Gr. baldheaded,) a bird about as large as the Common Crow, of Spanish-snuff color, or, as some say, capuchin color. Its large beak and ample forehead, bare of feathers, to which the specific name refers, give it a very singular appearance. The Crested Crow, *C. cephaloptera*, (Gr. head-winged,) is also a singular looking bird, of a uniform blue-black hue, having the head and base of the bill ornamented with a crest, forming a sort of parasol, to shade the face, and reaching to the end of the bill, compressed in the same manner as in the *Rupicola*, or Cock of the Rock. "The sides of the neck are naked, but long feathers forming a loose pelerine, and hanging down lower than the breast, spring from beneath the throat and from the sides of the neck. This crest and feathers of the pelerine give metallic reflections." (Lesson.)

In the family of the *Corvidæ* are included the Birds of Paradise, which some naturalists have, with reason, erected into a separate family, called *Paradiseadæ*, including some of the most singular and magnificent of the feathered tribes. They are natives of New Guinea, to which they are almost confined. Of these birds, splendid as they are, fiction has presented many strange and exaggerated descriptions. For a long time, it was asserted that some of them are without legs! They considerably resemble the Crows in their general structure, and they also approach them in size. In these birds, the wings are long and round, the tail varying in length at the extremity, or else rounded. The tarsi are robust, long, and covered by a single feathered scale; the toes long and strong, especially the hind toe; the claws large, curved, and powerful. The sides of the body, the neck, the breast, the tail, and sometimes the head, are ornamented with lengthened and peculiarly developed showy feathers; the plumage of the face and throat, is commonly of a scaly or velvety texture, and most richly glossed with metallic hues, while other parts of the body are frequently arrayed in rich and brilliant colors.

There are several species of these birds, but the EMERALD BIRD OF PARADISE, which is figured on the Chart, is the one best known. It is impossible adequately to describe its beauty of form, and the vivid and changing tints of its plumage. The generic part of the scientific name, *Paradisea apoda*, is from the Greek *Paradeisos*, a pleasure-ground; the specific name, which means *footless*, was given it by Linnæus, "because the older naturalists called it footless." The truth is, the natives of New Guinea were accustomed to dry birds of this species, (having first cut off the legs,) and to offer them for sale. They were taken to other countries in this "footless" condition; and hence, conjectures arose that they lived in the air, buoyed up by the lightness of their feathery covering; that the shoulders were used for a nest; that the only rest which they took, was by suspending themselves from a branch by the filamentary feathers of the tail; that their food was the morning dew,—and other things of a like character,—amusing enough, but entirely without foundation, in fact. So far from living wholly on dew, this bird eats no small amount of insects, such as grasshoppers, etc., which, however, it will not touch when dead; it also feeds largely on the seeds of the teak tree, and on figs and aromatics; when alive, it is about the size of a Common Jay, or Pigeon; its note is like that of the Starling. The body, breast, and lower parts, are of a deep rich brown; the forehead is clothed with close-set feathers of a velvety black shot with green; the throat of a rich golden green; the head yellow; the sides of the tail are clothed with a splendid plume of downy feathers of a soft yellow color. By these are placed two long filaments, or thread-like shafts, which extend nearly two feet in length. (See figure on the Chart.) "Of these beautiful feathers, the bird is so proud, that it will not suffer the least speck of dirt to remain upon them, and it is constantly examining its plumage, to see that there are no spots on it. When in its wild state, it always flies and sits with its face to the wind, lest its elegant flying plumes should be disarranged." The female is without these floating plumes of the male, and her colors are less lustrous. The Emerald, in its motions is lively and agile, and, in general, it perches only upon the tops of the most lofty trees. These birds are killed by the natives with blunt arrows, and sold to the Europeans; this forms a gainful traffic; and hence, the Chinese, it is said, fabricate imitations of these "celestial fowls," of the feathers of Parrots and Paroquets, which they sell at high prices to strangers.

Second Family. The Starlings.

Sturnidæ. (Lat. *Sturnus*, a starling, or stare.)

The Starlings are a numerous and widely distributed family; larger, for the most part, than the average of the perching birds; but of less size than the Crows, which, in structure and manners, they much resemble. The beak in these birds, is of a form well adapted for penetrating the earth in search of the worms and underground larvæ upon which they feed.

The plumage is commonly of dark colors, but has a peculiar richness; black, glossed with lustrous hues of steel blue, purple, or green, of the prevailing color, but occasionally it is relieved by broad masses of crimson or yellow, (and, in a few instances, of white,) as in the *Icterus*, or Baltimore Oriole.

The Starlings live in societies, sometimes immensely numerous, and seem universally to prefer the locality of plains frequented by cattle; in this particular, resembling the Maize Birds.

The Common Starling, (*S. vulgaris,*) it has been observed, becomes wonderfully familiar in the house; is very docile; always gay and wakeful; soon knows all the inhabitants of the house, remarks their motions and air, and adapts himself to their humors; he repeats correctly the airs which he is taught, imitates the cries of men and animals, and the songs of all the birds in the same room with himself; but his acquirements are of little value, for he forgets as fast as he learns. The Starling lays, twice in a year, from four to six eggs of a delicate pale blue, or of an ashy green color. It is about the size of the Blackbird.

The Meadow Starling, or Meadow Lark, (*Sturnella ludoviciana.*) is a beautiful bird, found abundantly throughout the United States, and as far North as the Fur countries, wintering in the Carolinas, or Florida. It builds its nest at the foot of some tall, strong grass. This bird, though useful in destroying thousands of larvæ in meadows, is a little too fond of scratching up the seeds of grain, and of plucking up young corn; it has been known even to kill and eat small birds. The male is about eleven inches in length.

THE RED-WINGED STARLING, *S. prædatorius*, (Lat. plundering,) ranges from Labrador to Mexico,—north of Maryland being migratory. From its strong predilection for corn or maize, and its extensive depredations upon the young ears, it has acquired a bad reputation, having among other names, that of CORN or MAIZE THIEF. A remarkable characteristic of this bird is, that the male is nearly two inches longer than the female, and of proportionate magnitude.

THE BOAT-TAILS are American Birds, and the largest of the Starling family, and might easily be mistaken for Crows. Their tails are so concave on their upper sides as to resemble a boat, whence the sub-family name, *Scaphidurinæ*, (Gr. *skaphis*, a boat; *oura*, a tail.) The typical birds of this group, (*Scaphidura*,) are found in South America.

Another genus, sometimes included in the Boat-tail Birds, is *Quiscalus*, which has several representatives in the United States. Among these are (1) the Boat-tailed Grackle, or Great Crow Black Bird, *Q. major*, (Lat. greater,) about sixteen inches in length, and found in the Southern States, particularly on the sea-coast. The food of this species consists, principally, of the small crabs, called "fiddlers;" (2,) the Purple *Grackle*, or Common Crow Blackbird, is a constant resident in the Southern States, but migrating very far North. It appears in the State of New York about the middle of April, and is notorious, and dreaded for its attacks on Indian Corn; (3,) the Rusty Crow Blackbird, *Q. ferrugineus*, (Lat. iron-colored,) of similar character and habits with the preceding, but ranging still farther North.

The sub-family, *Lamprotorninæ*, (Gr. *lamprotes*, splendor; *ornis*, a bird,) includes Grackles found in Asia and Africa, in which they represent the Boat-tails of America.

THE ORIOLES, or Hang-Nests, *Icterus*, (Gr. Yellow Thrush,) —sub-family, *Icterinæ*,—are a numerous and beautiful group of American Birds, of which the Baltimore Oriole, *I. Baltimore*, is the most noted. This is sometimes called Golden Oriole, Golden Robin; and also Fire-Bird, Fire Hang-Bird, from the bright orange seen through the green leaves, and resembling a flash of fire; but more generally, the Baltimore Bird; its colors of black and orange, resembling those of the arms or livery of Lord Baltimore, formerly proprietary of Maryland. The materials which this bird uses for making its nest, vary with the temperature. In Louisiana, its nest (see Plate XI. fig. 4) is constructed of moss, woven throughout, so that the air can easily pass through it, and it is placed in the coolest position; so strongly is it secured, that no wind can carry it off without breaking the branch to which it is suspended. In Pennsylvania and New York, the nest is constructed of the warmest and softest materials, and so placed as to be exposed to the sun's rays. In summer, the Baltimore Orioles are dispersed over the United States, and as far North as Nova Scotia. The song of this bird, is a clear, mellow whistle, repeated at short intervals. The male, according to Audubon, does not receive its full plumage until the third spring. The principal food of the Oriole consists of caterpillars, beetles, and bugs, particularly one of a brilliant glossy green.

Dr. DeKay, in the Natural History of New York, includes the Crow Blackbirds, (*Quiscalus,*) the Orioles, (*Icterus,*) the Cow Bunting, (*Molothrus,*) and the Bob-o'link, (*Dolichonyx,*) in one family, *Quiscalidæ.* But Audubon arranges these together, with the Marsh Blackbird, *Agelaius,* in the family *Agelainæ,*

MINO BIRD.—Among the Starlings we also place the Mino Bird, *Eulabes,* (Gr. *eulabes,* circumspect, or religious,) *Javanacus,* or *Gracula religiosa ;* following Swainson in this arrangement, who deems it quite unreasonable to place this long-legged Grackle close to the short-legged ROLLERS, as M. Lesson has done. This bird has a short and stout beak, with the tip distinctly notched. Its plumage is of a deep velvety black, with a white space in the middle of the wing; behind the eye spring fleshy carbuncles of a bright orange color.

The Mino Bird feeds on insects and fruits. It is easily tamed; learns to whistle and talk with great facility, and is therefore a great favorite with the Javanese. Marsden says, it has the faculty of imitating human speech in greater perfection than any other of the feathered tribe. There is said to be a smaller variety of this bird in India.

THIRD FAMILY. FINCHES.

Fringillidæ. (Lat. *Fringilla,* a finch.)

The Finches are a large and interesting family, the smallest of the Perchers, and, for the most part, excellent songsters. They have short, thick, and powerful beaks; both mandibles are usually of equal thickness, and their length and breadth nearly alike, so that when the beak is closed, it generally appears like a very short cone divided in the middle by the gape. In some

EXPLANATION OF PLATE XI.

Fig. 1. Nest of the CHESTNUT CROWNED TITMOUSE, suspended from the fork of a twig, nine inches long, more than three in diameter, entrance at the top less than an inch wide, made of the softest materials.

Fig. 2a. Nests of the AFRICAN WEAVER-BIRDS, (REPUBLICAN GROSBEAKS of Swainson.) The numerous entrances to this BIRD TOWN lead to regular streets, having nests on each side, at about two inches distance from each other; the general roof or cover is built by the united labors of the birds, and sometimes shelters hundreds. That from which this figure was taken was thought to contain a society of eight hundred or a thousand.

Fig. 2b. Hive nests of the SOCIABLE WEAVER-BIRDS. The lower surface abounds with perforations admitting the birds to their nests, but excluding snakes and other intruders. They never occupy the old nests, but continue to add successive tiers until the branches yield to the accumulated weight.

Fig. 3. Nest of the WOOD SWALLOW.

Fig. 4. Nest of the BALTIMORE ORIOLE, closely interwoven with flax, hemp, tow, hair, and bits of thread, cord, &c., stitched through and through with horse hair, securely suspended from the branch of a tree.

Fig. 5. Nest of the PENDULINE TITMOUSE, or BOTTLE TIT, made of the down of the willow, poplar, and thistle, lined with feathers, containing from ten to fourteen eggs.

Fig. 6. Nest of the PENSILE WEAVER-BIRDS, or WEAVER FINCHES, shaped like a Chemist's retort; suspended over water from trees; entrance from beneath.

Fig. 7. Nest of the WREN; of hay, if against a hay-stack; of moss, if against a mossy tree.

Fig. 8. Nest of the TAILOR-BIRD, or TAILOR WARBLER, of Ceylon, curiously formed by stitching with plant fibres or threads of cotton a dead leaf to a living one; nest open at the top and filled with fine down. A species in Italy are said to sew their materials together with spiders' webs.

Fig. 9. BAR-TAILED HUMMING BIRD, of Peru; the nest of soft delicate materials, is often warped or woven together with spiders' webs.

Fig. 10. Nests of the BANK SWALLOWS or SAND MARTINS, numerous in sand banks or artificial excavations, such as gravel-pits. Audubon says, "the little creatures are so industrious he has known a hole dug to the depth of three feet four inches, the nest finished in four days, and the first egg deposited on the morning of the fifth."

Fig. 11. GOURD-SHAPED nests of the REPUBLICAN or CLIFF SWALLOW, built of muddy sand under the eaves or cornices of buildings, or attached to rocks overhanging rivers, where they are found grouped by hundreds.

NOTE.—The nests of RAPTORIAL birds are seldom met with, as they are usually built in lofty trees or inaccessible precipices. OWLS do not usually construct nests, but deposit their eggs in some hole, in a tree, an old building, or in the ground. INSECTIVOROUS birds are solitary builders; among the Shrikes, Thrushes, Warblers, Tit-mice, and Fly-catchers, there is not one instance of a species either living or building in societies. Pensile nests are altogether peculiar to perching birds, and are more common in tropical than temperate latitudes. Hundreds of hang-nests may be seen in Brazil attached to a single tree; some of them are said to measure between four and five feet. Other nests are said to have a portico or ante-room where the male bird often sits during the time of the female's incubation.

PL. XI.

genera, however, the conical form is less obvious, by the lateral and vertical swelling of its outline. The GROSBEAK, or HAW-FINCH, *Coccothraustes*, (Gr. *kokkos*, grain; *thrauō*, I break,) has a beak *enormously* thick in proportion to its length, and in comparison with the size of the head. In this bird, and, indeed, in all the Finches, the great strength of the beak well adapts it for the uses to which it is destined, as the food of this bird consists of seeds often enclosed in woody capsules of great hardness, or the kernels of stone fruits, which must either be opened by a forcible wrench, or crushed by a strong pressure. The Finches, besides seeds, also feed on grain, and occasionally on insects.

These birds frequent fields, groves, and woodlands; numbers of them are found in gardens, building their nests in bushes. Many of them, in a state of captivity, are rendered subservient to human improvement, and become favorite domestic pets. So numerous are the genera and species of this family, it is impossible, within the limits of this volume, to give any more than the briefest notices of some of the more prominent ones.

1. We notice the WEAVERS, sub-family, *Ploceinæ*, (Gr. *plokeus*, a weaver.)

These birds build their nest upon branches extending over a river or pool of water; it is shaped exactly like a chemist's retort, (Plate XI. fig 6;) and is suspended from the head; and the shank, of eight or nine inches length, at the bottom of which is the opening, almost touches the water. It is made of green grass, and curiously woven. The Weaver Birds also construct the celebrated hive-shaped nests. (See nests of the Social Weaver Birds, Plate XI. fig. 2.) The *Textor*, (Lat. Weaver,) *erythrorhyncus*, (Gr. *eruthros*, red; *rhunchos*, a beak.) The RED-BEAKED WEAVER, of South Africa, companies with Buffaloes, and obtains from their hides its supply of food. It serves these animals by ridding them of the insects with which their hides are infested, and by flying up on any alarm, it becomes to them as a sentinel, indicating the approach of danger, or of any thing unusual. This bird does not appear to attach itself to any quadruped but the Buffalo.

THE WIDOW BIRDS, or Whidah Finches, ranged by Swainson under the sub-genus *Vidua*, (Lat. a widow,)—have long boat-shaped tails, with the two middle feathers excessively lengthened, and generally broad and convex. In Senegal and South Africa, is found the Widow Bird of the "English salesmen and fanciers," *V. paradisea*, about the size of a Canary bird,—but the two feathers next to the middle tail-feathers are a foot in length from the base, and about three-fourths of an inch in width; the two middle feathers have very broad webs on their basal half, (or extending about three inches midway,) but the remainder of the shaft becomes like a plumeless, hair-like process of the same length. The term, "widowed," is applied to this group from the sombre hue which prevails in the plumage, "suggesting the idea of widow's weeds." Among the *Plōceinæ,* Swainson places the JAVA SPARROW, *Amadina,* represented by the *Tiaris,* or Creslet, in South America, in which the thickness of the beak is enormous in proportion to its length, and the middle feathers of the tail are the longest. This bird is frequently kept as a pet in cages, living on seeds.

II. THE BUNTINGS, sub-family, *Emberizidæ,* are an interesting group of Passerine birds, differing from the Finches proper, chiefly by having a knob on the "palate," or on the under mandible,—the sides of the under mandible bending inwards; their strong conical beak is well adapted for breaking the seeds which constitute their principal food.

THE LARK BUNTINGS, *Plectrophanes,* (Gr. *plectron*; *phaino,* to display,) have moderately long tarsi; the side toes of equal length; the hind toe strong, with a lengthened and nearly straight claw. There are several species of these birds, among which is the SNOW LARK BUNTING, *P. nivalis,* (Lat. snowy,) which appears in the Eastern part of the United States early in November, and in some parts, remaining until March. The summer plumage of the Snow Bunting, is pure white and black, but it is found in all varieties of plumage. In the Highlands of Scotland, it is called the Snow Flake; in Labrador, New Foundland, and elsewhere, the White Bird; and also the White Snow-Bird, to distinguish it from the COMMON SNOW BIRD, *Struthus,* (Gr. *strouthos,* a sparrow;) *hyemalis,* (Lat. of winter.) The Arctic Bird, the Lapland Snow-Bird, or Bunting, *P. Lapponicus,* breeds in moist meadows, on the shores of the Arctic seas; and in the State of New York, is seen during the extreme cold of winter. Audubon observed these birds in Kentucky and Missouri. They have been seen as far North as 74o Lat.

THE BUNTINGS, *Emberiza,* include a large number of species. The BLACK THROATED BUNTING, *E. Americana,* is abundant in the Middle and Atlantic districts of the Union, but exceedingly so in the vast prairies of the West. Its simple and unmusical notes, are said to resemble those of the CORN BUNTING, of Europe, *E. miliaria,* (Lat. of millet.) Its length is six inches. The YELLOW WINGED BUNTING, *E. passerina,* (Lat. sparrow-like,) is a small bird, only four and a half inches long, which "passes, unobserved, from Mexico to Connecticut. The individuals seem to move off in a sulky mood, and in so concealed a way, that their winter-quarters are yet unknown." The FIELD BUNTING, *E. pusilla*, (Lat. very small, or weak,) breeds from Maryland to Maine. It is social and peaceable, and trills its notes like a young Canary Bird. In length it is six inches.

THE CHIPPING BUNTING, or CHIP-BIRD, *E. socialis*, (Lat. social,) is known to all. It is confined to the United States and the adjacent Eastern provinces; associating with the Song Sparrow, or Finch, *Fringilla melodia*, and other birds of the same genus. The Chip-Bird builds its nest on some low bush and lines it with cow-hair; lays from four to five bright greenish blue eggs, spotted with brown chiefly at the larger end. It seems determined to make up in quantity any defect in the quality of its notes, for it sings all the day long. It migrates to the Southern States in the winter, and is among the earliest of the Spring birds. This bird may be noticed, gleaning up crumbs from our yards, and our very doors,—it will even approach the threshold to pick up the crumbs thrown to it,—in this social characteristic, it is singular; it is distinguished by its black bill and frontlet. Its length is five and a half inches. This bird seems to represent, in America, the Common, or House Sparrow, of Europe, *Pyrgita domestica;* but it is less bold and crafty than the latter bird, and probably less voracious also. Buffon estimated that a pair of Sparrows will destroy about 4,000 caterpillars weekly in feeding their young; this is some compensation for the birds' devastation in granaries and barns.

THE TREE SPARROW, or CANADA BUNTING, *E. Canadensis*, breeds in the Fur countries. Audubon thinks it also breeds in Maine. This bird may be seen in the magnificent elms that ornament Boston and its adjacent villages. It is a sweet songster.

The well known SNOW BIRD, *Struthus hyemalis*, or *Niphæa*, (snowy,) *hiemalis*, Aud., migrates from the North, at night, as far as 30° N. L. It is common to the northern parts of the continent of Europe. This is a shy, timorous bird, and is rarely seen except in snow-storms, when it appears in flocks around dwellings. At night, it resorts to stacks of corn or hay, making there a hole for its resort in cold weather. Its nest is built on the ground; the eggs are usually four in number, of a spherical form, yellowish white, and sprinkled with reddish brown dots. Length six and a quarter inches.

THE INDIGO BUNTING, or INDIGO BIRD, *Spiza*, (Gr. from *spizō*, to chirp;) *cyanea*, (sky-blue,) is one of our beautiful birds coming from the South, and appearing in New York late in May,—it is seen throughout the United States. Its note nearly equals that of the Canary, but is not so sonorous. This bird seems gradually to lose its brilliant tints when caught and caged, as does the PAINTED BUNTING, *S. ciris*, (Gr. *keiris*, name of a bird,) of Carolina, Louisiana, and South America. In certain lights, the plumage of the Indigo Bird appears of a rich sky-blue, and in others, of a vivid verdigris green ; so that "the same bird, in passing from one place to another, before your eyes, seems to undergo a total change of color." (Wilson.) Its length is five and a half inches. The Painted Bunting is found in the orange groves of the South. It is abundant in the vicinity of New Orleans, where it is caught in trap-cages.

THE SHORE FINCHES, *Ammodramus*, (Gr. *ammos*, sand ; *dramēin*, to run,) are found on the Atlantic coast from Texas to Massachusetts,—and in summer, in our salt marshes, where they breed. The Seaside Finch, *A. maritimus*, (of the sea,) feeds chiefly on marine crustacea, and such insects as are found on the seashore. Its builds its nest on the ground ; and lays from four to six grayish white eggs, speckled with brown. Length from seven to eight inches. The SWAMP SPARROW, *A. palustris*, forms the principal food of the Sparrow Hawks and Hen-Harriers. In New York, it is often called the Red Grass-bird. Swamp Sparrows have been found abundantly in the marshes of Cayuga Lake. Their note is a harder tone than that of other Sparrows. The length is about six inches. This bird ranges from Texas to Labrador. It is said to be abundant about Boston during the winter ; has a short, conical bill, higher than broad at the base, and very acute at the tip.

THE LINNETS, *Linaria*, include several species. The BROWN LINNET, *L. linota*, is a song-bird common in every part of Europe. Of this Linnet, it has been said, "it is the cleanliest of birds, delighting to dabble in the water, and to dress its plumage in every little rill that runs by. The extent of voice in a single bird is not remarkable, being more pleasing than powerful, yet a large field of furze, in a mild sunny April morning, animated with the actions and cheering music of these harmless little creatures, united with the bright glow and odor of this early blossom, it not without its gratification."

The Common Linnet frequents commons and neglected pastures, and builds its nest in the center of a large and dense brush.

THE LESSER RED-POLL, *L. minor*, in length about five inches, and the MEALY RED-POLL, *L. borealis*, in length, five inches and a half, are found within the United States. The Pine Linnet, *L. pinus*, (Lat. pine,) sings while on the wing, like the Goldfinch.

It feeds among the branches of the tallest Fir trees, as well as on the seeds of Thistles, much in the manner of the European SISKIN, on the *Fringilla Spinus*, (Lat. black-thorn.) Its length is a little less than five inches.

THE AMERICAN GOLDFINCH, OR YELLOW BIRD. *Cărduelis*, (Lat. a thistle-finch,) *tristis*, (Lat. sad,) (see Chart,) is a well known and handsome bird, similar in its song and flight to the Goldfinch. Its plumage and notes make it universally agreeable. The Yellow Bird is abundant in the middle districts of the Union, in summer, and so hardy is it, that it often remains there during the whole winter. It ranges from the tropics to the northern and southern regions. Its length is four and a half inches. This bird feeds principally on the seeds of hemp, the sun-flower, and various species of thistles. From its fondness for the thistle down, it has been called the THISTLEFINCH. It is sometimes kept in cages for song, and will live to a great age in a cage or room. Audubon says he has known instances in which birds of this species had been confined for ten years. They had been taken in trap-cages, as the writer has taken them, in the vicinity of New York city. This bird is not only beautiful, but seems to give evidence of unusual *sagacity*. It can be trained to draw water for its drink from a glass,—and when it alights on a twig covered with bird-lime, for the purpose of securing it, "it no sooner discovers the nature of the treacherous substance, than it throws itself backwards, with closed wings, and hangs in this posture until the bird-lime has run out in the form of a slender thread considerably below the twig, when, feeling a certain degree of security, it beats its wings, and flies off,"—and, says Audubon, from whom we now quote, "I have observed Goldfinches that had escaped from me in this manner, when about to alight on any twig, whether smeared with bird-lime or not, flutter over it, as if to assure themselves of its being safe for them to perch upon it." Its length is four and a half inches. Several species of Goldfinch are found in the United States.

THE FINCHES PROPER, *Fringilla*, include quite a number of species, among which are the SONG SPARROW, *F. melodia*, (Gr. song,) which presents two varieties; one having spots generally distributed over the breast; the other having fewer spots on the breast, but a large black one in the center,—appearing among us even before the Pewee and Blue Bird. The SONG SPARROW is the harbinger of spring; it is "the earliest, sweetest, and most lasting songster." The first named variety builds its nest in low shrubs a few feet from the ground; the other builds it upon the ground. It feeds chiefly upon insects. Its length is about six and a half inches.

THE FOX-COLORED FINCH, or SPARROW, *F. iliaca,* (Lat. from *ilia,* flanks,) is one of the largest of the genus, being seven and a half inches long, and breeds in countries North of the United States. It has been seen as far North as 68o Lat., and ranges South to within 30o of the equator.

THE BAY-WINGED SPARROW, or GRASS BIRD, GREY GRASS BIRD, *F. graminea,* is ranked by Audubon and Wilson with the Buntings. We follow Dr. DeKay in placing this familiar Sparrow with the Finches proper. It feeds on grass seeds and insects. Length five and a half inches.

THE WHITE-THROATED FINCH, *F. Pennsylvanica,* is an active Northern Sparrow, appearing in New York, more or less, during the whole year, and advancing as far as 66o North.

THE GROUND FINCHES, *Pipilo,* (Lat. to peep, or chirp,) scoop out the earth and build their nests on the ground. They live on grubs and earth and wire-worms.

THE TOWHEE GROUND-FINCH, *P. erythrophthalmus,* (Gr. *eruthros,* red; *ophthalmos,* eye,) is found in large numbers on the Pine Barrens of Kentucky.. It breeds in New York State, and is known "under the name of CHEWINK from its peculiar note, and of GROUND ROBIN, from its seldom attempting to fly high." In Louisiana it is called GRASSET, and esteemed by epicures.

There are several species of PURPLE FINCHES, *Erythrospiza,* (Gr. *eruthros,* red; *spiza,* a bird like a sparrow.) The CRESTED PURPLE FINCH. *E. purpurea,* (Lat. purple,) frequently associates with the Cross-bills, and feeds upon the same trees,—it ranges from Texas to Labrador. Length six inches. This bird is seen on the Atlantic coast of New York State as late as December and January.

THE PINE BULL-FINCH, or Common Pine Finch, *Corythus,* (Gr. *korus,* a crest;) *enucleator,* (Lat. kernel, or seed-sheller,) is a most beautiful bird, and a charming songster; of a red color, (the female olive-green,) with the wings and tail brown,—ranging from Pennsylvania to Newfoundland, and breeding from Maine northward. The length is eight and a half inches. It has been seen in large flocks in the vicinity of New York city. Nuttall, Bonaparte, and others, name this bird *Pyrrhula,* (Gr. *purrhoulas,* from *puros,* red,) *enucleator.*

THE CARDINAL GROSBEAK, or CRESTED RED BIRD, *Pitylus,* (Gr. *pitulos,* frequent agitation and movement,) *cardinalis,*—is a bird which no one can see without admiring. In richness of plumage, elegance of motion, and strength of song, this species surpasses all its kindred found within the United States. Length eight inches. It breeds abundantly from Texas to New York. In some parts, it is called the Virginia Nightingale.

THE BLUE GROSBEAK, *Coccoborus cœruleus*, and the ROSE-BREASTED GROSBEAK, *Coccoborus ludovicianus*, are also very beautiful species.

THE SUMMER RED BIRD, *Pyranga æstiva*, (Lat. of summer,) coming from Mexico and farther South, is seen among us in the hottest part of summer, rarely moving eastward of New York. It feeds on insects, particularly the largest beetles. This bird cannot bear cold, or even temperate weather, and its stay in the United States, (where it breeds,) scarcely exceeds four months. Length seven and a half inches. This bird is also called *Tanagra*, (Gr. a brazen-vessel,) in allusion to the color of the female bird. The BLACK-WINGED RED BIRD, or TANAGER, is seven inches in length; reaches New York about the middle of May, and goes as far as 49° N. L. It migrates by night in September.

LARKS.

Sub-family *Alaudinæ*, (Lat. *alauda*, a lark.)

Of these singing birds there are many species, characterized by a long and straight hind claw, a strong straight bill, and by being able to raise the feathers on the back part of the head in the form of a crest. The greater part of them are migratory; they build their nests on the ground and may be regarded as peculiarly birds of the fields and meadows. The Larks are every where distinguished for their vigilance and their song. The conformation of their feet does not adapt them for perching, but rather for walking on the earth. They accordingly always build on the ground, making usually a rather slight, though neat nest, and laying about five eggs, for the most part of a grayish white, with specks of a brown color. They frequently rear two broods of young during the summer.

These birds are famed for singing while in flight, and soaring to great heights in the air. From the situation of their nests, they are much exposed to the attacks of predaceous animals of the weasel kind, which destroy a great many of the eggs and young. During their migrations, immense numbers of these singing birds are, contrary to our sense of justice, taken in nets to increase the pleasures of the table, particularly on the continent of Europe. Swainson considers the genus *Alauda* to be of the Fissirostral type, but they are more commonly ranked with the Cone-billed birds, where we have placed them.

The SKY-LARK, *A. avensis*, the *Alouette* of the French, the *Feld Lerche* of the Germans, and the *Lodola* of the Italians, is widely celebrated for its inexpressibly beautiful song, chanted far up in the air, when the bird is at liberty and in its natural state. It commences to sing early in the spring, and continues its song during the entire summer. "When this Lark first rises from the earth, its notes are feeble and interrupted; as it ascends however, they gradually swell to their full tone, and long after the bird has reached a height where it is lost to the eye, it still continues to charm the ear with its melody." Its food consists of insects and their larvæ, with many sorts of seeds and grain. The Sky-lark is about seven inches in length. It is found throughout Europe; also in Asia and the northern parts of Africa.

The WOOD-LARK, *A. arborea*, is smaller and can perch on trees, a power denied to the Sky-lark.

The HORNED LARK, *A. cornuta*, (Lat. horned,) is an American species of a dusky brown color, seven and a half inches in length. Its head has erectile feathers. This Lark ranges from 68° N. Lat. to Texas. It is seen during the coldest weather.

Dr. Buckland figures a Lark, (*alauda*,) among the land mammals and birds of the third period of the Tertiary series, in the first plate of his illustrations of his "Bridgewater Treatise."

SUB-FAMILY. The COLIES.

Coliadæ, (Gr. κολιὸς, *koliòs*, the name of a bird.)

The Colies are ranked by Swainson among the *Muscophagadæ*, or Plantain-eaters. Others rank them among the Finches. Gosse, in his work on Birds, raises them to the rank of a family, and places them between the Finches and Plantain-eaters. They are few in number and confined to Africa and India. The two mandibles of the short, conical beak, are, in these birds, arched, the point of the upper slightly overhanging the lower. The feathers of the tail are exceedingly long and stiff; like the Humming Birds, they deviate from the general rule of twelve tail-feathers, having but ten, agreeing in this respect with the Swifts, and also in having the hind toe capable of being turned forwards, so that all the four toes point in one direction. In their general form and habits, they do not, however, show any likeness to the Swifts. The Colies live mostly on trees, climbing about much in the manner of Parrots. They go in large flocks and even breed in communities, constructing numerous large and round nests in the same bushes; in each nest five or six eggs are deposited. It is said these birds sleep suspended from a branch, with their heads downwards, many of them together; and that when the weather is cold, as it sometimes is in South Africa, they are found so benumbed in the morning that they may be readily taken, one after another, without their making an effort to escape. The plumage of the Coly, (*Colius,*) is short, thick, and smooth, with a silky appearance. The feathers of the head are lengthened, forming a long pointed crest, which can be erected at pleasure. The prevailing colors are gray or ashy, from which circumstance, and that of their crawling about trees, they are, at the Cape of Good Hope, called Muys-vögel, or Mouse-birds.

The Colies live chiefly on fruits, the buds of trees, and the tender sprouts of vegetables. On account of the mischief which they do in gardens, they are much disliked. They are bad walkers, but expert climbers, clinging to the branches in all sorts of attitudes. Their cry is monotonous, (the wind-pipe, (*trachea,*) being furnished with but a single pair of vocal muscles,) and that of the largest species is said to resemble the bleating of a lamb. The flesh of the Colies is of a delicate flavor and highly esteemed. It forms the common food of several species of the Birds of Prey. The *C. Senegalensis*, as its name imports, is a native of West Africa. It has a pearly-gray plumage, with greenish reflections; the forehead is yellow; the under part of the body ruddy; and a naked reddish skin surrounds the eye.

FOURTH FAMILY. CROSS-BILLS.

Loxiadæ, (Gr. λοξὸς, *loxos*, oblique.)

The beak of the CROSS-BILLS, (Plate IX. fig. 8,) is of unique form, the mandibles curving to the right and left, and always in opposite directions to each other. In some of these birds the upper mandible is turned to the right, the lower mandible curved to the left; in others the position of the mandibles is reversed as to their direction. The upper mandible has a limited degree of motion on the head or cranium, the upper jaw bones and the nasal ones being united to the frontal bone by flexible osseous laminæ. The lower jaw is remarkably strong, and the muscles by which this and the upper mandible are moved, are large, particularly in the lower jaw, and act with great power in a sidewise direction. By this extraordinary bill, these birds are enabled to extract the seeds from pine cones with remarkable facility; and they are confined to localities in which these cones can be obtained, such as the Hartz, or great pine forests of Germany. They first fix themselves across the cone, then bring the points of the mandibles from their crossed position to be immediately over each other. In this reduced compass, they insert their beaks, and then opening them, not in the usual manner, but by drawing the lower mandible sidewise, they force open the scales. In this process, they are aided by the beautiful and peculiar adaptation of the tongue, an additional portion, partly osseous, with a horny covering being articulated to the front end of the bone of the tongue, (*os hyoides.*) Underneath this grooved appendage is another small muscle which is attached at one end to the bone of the tongue; at the other, it is joined to the movable piece, and by its erection bends the point downwards and backwards; whilst, therefore, the points of the beak press the shell from the body of the cone, the tongue, brought forward by its own muscle, is enabled, by additional ones, to direct and insert its cutting scoop beneath the seed, and the food thus dislodged is transferred to the mouth.

While these birds are at work on the fir cones, they send forth a gentle twitter, and may be seen climbing among the branches like parrots; but they are also said to have a pleasant song, poured forth only in the winter months, or at the season of incubation. The Cross-bills are subject to considerable changes of color.

The male of the Common Cross-bill, *Loxia curvirostra*, (Lat. curve-beaked,) varies from a beautiful red to an orange color on the head, neck, breast and back; the female is generally of a dull olive green on those parts which are red in the male. It is sometimes called the German Parrot, and on account of its sweet and well tasted flesh, is in special request in the bird-market of Vienna, (Austria,) for the purposes of the table. This bird is five and three-fourths inches long. It is a regular inhabitant of all our pine forests (situated north of 40° N. Lat.,) from the beginning of September to the middle of April, building its nest on the highest part of the fir trees, and making use of the resinous matter which exudes from them for fixing it to the trees.

The American Cross-bill, *L. Americana*, is of a red color, with brownish tail and wings, from six and one-half to seven inches in length; feeds on the cones of the hemlock, and on apples and other fruits, which the bird breaks open for its seeds. Bonaparte and other naturalists consider this species as distinct from the European Cross-bill. It is a northern-bird, but breeds as far south as Pennsylvania.

Another species is the White Winged Cross-Bill, *L. leucop-*

tera, (Gr. white winged,) which is somewhat less than a Goldfinch, (according to DeKay, six and one-half inches long.) It ranges from 40° to 68° N. Lat., and is common on the shores of Lake Ontario. (N. B. These Cross-billed birds are included by Audubon in the family of Finches. In the N. Y. State Nat. Hist. they have a like arrangement.)

FIFTH FAMILY. The HORN BILLS.

Bucerotidæ, (Gr. βούκερως, *boukerōs*, ox-horned.)

The characteristics of the birds of this family which most arrest the attention, are the enormous extent, and singular protuberances of the beak. In many of the species this organ is considerably larger than the head; there is a large, uncouth looking projection, various in form, on its summit. This projection sometimes resembles a horn, or the crest of a helmet which often encroaches upon the skull towards the crown of the head. The mandibles in adult birds are both notched on the edges. The protuberance on the upper mandible is small when the bird is young, and does not attain its great size until the bird is fully grown. (Plate IX. fig. 13.) By a beautiful provision of the Creator, for birds supporting so large an organ, the horny case of the beak is very thin, thus diminishing the weight; and, at the same time, the bony core is hollowed into numerous cells of various sizes and forms, with very thin walls between them, so that the needed firmness is preserved in union with remarkable lightness. The bones of the body are also permeated with air more than those of any other bird. The tongue in the Horn-billed birds is fleshy, and like that of the Birds of Prey, short and deep in the throat; the tail is long, broad, and more or less rounded at the extremity, consisting of only ten feathers; the feet are short, strong and formed for walking or perching; the claws short and blunt.

The Bucerotidæ are large sized birds; they are gregarious and noisy, and live both on animal and vegetable food; few are smaller than a Crow, and some are much larger than a Raven. The plumage is usually of a sombre cast, but frequently relieved with masses of white; the beak and naked skin often exhibit bright colorings during life. The abode of these birds is limited to Africa, India, and the large islands adjacent.

The Horn-bills seem to be most nearly related to the Crows on the one hand, and to the Toucans on the other, thus connecting the Perchers and the Climbers. That they form a link between these two orders was proved from anatomical examinations of the bird made by Professor Owen. More than twenty species of the genus *Buceros* have been named.

The ITALIAN RAVEN, *B. hydrocorax*, (Gr. Water Raven,) found in the Moluccas and in Africa, has the walk of the Crow, but is unlike that bird in its food, rejecting carrion, and being particularly fond of nutmegs, devouring them so greedily as often to do serious damage. The flesh is very delicate, and when roasted, possesses an aromatic flavor derived from its food.

The RHINOCEROS HORN-BILL, *B. Rhinoceros*, differs from the Indian Raven in living upon carrion; it "casts forth a strong smell, and hath a foul look, and much exceeds the European Raven in bigness." This bird is about the size, though rather more slender than a hen-turkey; its color is black, except the lower part of the belly and the tip of the tail, which are white; the bill is usually about ten inches long and of a yellowish white; the upper mandible red at the base; the lower, black; the legs are short, strong, and of a pale yellow color. The cry consists of a short hoarse croak, but when the bird is excited, this is changed to a loud discordant noise. It breeds in the hollows of lofty trees. The flight of the Horn-bills is sailing and resembles that of the crow; on the ground, they advance by a leaping kind of movement, assisted by the wings. When making their leaps on the highest branches of trees, and in their loud call note to their mates, the hollow protuberance of their beaks seems to be to them like a sounding board, increasing the reverberations of the air. The beak, it is thought, "constitutes a necessary defence against monkeys and other animals which may seek to assail its nest;" or it may be used in "drawing snakes and lizards from their lurking places, and young birds and eggs from the recesses of old and decaying trees."

SIXTH FAMILY. PLANTAIN-EATERS.

Musophagadæ, (Gr. μοῦσα, *musa*, gen. term for plantain; φάγω, *phagō*, to eat.)

This family, though a small one, includes birds of uncommon elegance and richness of plumage. They have a short beak; the upper mandible is much arched, and has its edges cut into minute saw-like teeth; the lower mandible is thin and narrow. The feet are short and formed for climbing, the outer toe being capable of a partial reversion; it is, however, united to the middle toe by a short membrane. The tail, as in the Colies, consists of but ten feathers; the head is generally clothed with a long and elegant crest. In their habits they show affinity to the Toucans, among the Climbing Birds, with which they are probably connected by the Horn-bills, (*Bucerotidæ.*) Some of them, as the PLANT-CUTTERS, *Phytotoma*, (Gr. *phuton*, a plant; *temno*, to cut,) show an affinity to the Bull-Finches. The Plant-Cutters are small, the CHILIAN PLANT-CUTTER, *P. rara*, being about the size of a quail. They feed on plants and have the destructive habit of cutting them off close to the root; often they capriciously cut off a quantity without touching them any further. On this account the rustic inhabitants carry on a continual war against these birds, and children who destroy their eggs, are rewarded. The nest is built in obscure places and on lofty trees, and thus the Plant-Cutters escape, in some degree, the persecutions of their enemies.

The PLANTAIN EATERS PROPER, are confined to Africa, where they subsist almost entirely on fruits. Their movements are extremely light and elegant, and unlike the Colies, they pass with an easy gliding flight from tree to tree. The first and fourth toes are directed laterally. It is said they, therefore, usually perch *lengthwise* on the horizontal branches, along which they walk, clasping the bough with their two toes arranged side-wise, while the other two point forwards. These birds live either in pairs or in families, according to the season. They construct a nest like the Parrots, in which they lay four eggs, delicately white.

On the gold coast and in Senegal is found the *Musophaga violacea*, the Violet-colored Plantain Eater—a "magnificent bird."

The TOURACOS, belonging to this family, include seven species, arranged under the generic name *Corythaix*, (Gr. κορυθαιξ, *koruthaix.*) They have a brilliancy of plumage, elegance of form, and grace of motion. Their long and broad tail and their high pointed crest add much to their beauty. The color of these birds is almost always rich green, set off with gorgeous crimson or purple on the expanded wing. One of the most attractive species is the *C. erythrolophys*, (Gr. red crested.)* When under excitement, the crest of this bird is elevated into a somewhat conical form, compressed at the sides, so that the head appears as if covered with a warrior-like helmet. To this appearance the generic name refers, signifying a warrior, or one who moves the helmet. In a state of repose, the crest feathers fall down

* One of these Red or Fire Crested Touracos lived for some years in the garden of the Zoological Society, (London.) An engraving taken from this bird during life, may be found in the "Penny Cyclopedia."

upon the head and project behind. The Touracos are about twenty inches in length.

What is the 3d DIVISION OF PERCHERS? What is said of their number? Mention the chief characteristics by which they are united. What peculiarity of the beak is found in one genus of these birds? What link do they thus form? To what are the feet generally adapted? What is the principal food of these HARD-BILLED BIRDS? What other food do they use? In what proportion does their food vary from an exclusive seed diet? What regions do these birds frequent? Mention the families into which they are divided.

Is the CROW FAMILY a numerous one? Describe the beak and plumage. What group form an exception in respect to plumage? In what other respect do they differ? What is said of the structure and disposition of the Crow? What of their propensity for thieving? What of their food? Show how the Crow is a remarkable type of the Birds. Which is the largest and most powerful of this family? How large is it and how regarded? What more is said of it? What is the size of the Common Crow, and how does it compare with that of Europe? In what respects is it a nuisance? Does the FISH CROW differ from the raven and common crow in character? What Birds of Prey are near relatives of the Crow? What is said of the BLUE JAY? What group do the NUTCRACKERS resemble in their habits? What is said of the FRUIT CROWS? What species are mentioned? What remarkable birds are included in this family? Of what region are they natives? Give a general description of these birds. Which species is best known? Give the meaning of the generic and specific name. Why were the birds of this species considered footless? What strange conjecture respecting them arose? On what do they feed? Give further particulars.

What is the 2d FAMILY of the Cone-Billed Birds? What is said of their number and distribution? What of their beak and plumage? Are they solitary or social in their habits? What is said of the COMMON STARLING? Of the MEADOW STARLING? Of the RED WINGED STARLING? Which are the largest of the Starling family? Why are they so called? Where are the Typical Birds of this group found? What other genus is sometimes included in the Boat tails? What Grackles are found in Asia and Africa?

What other SUB-FAMILY is mentioned? Which is the most noted species? What is said of it? Describe its nest. What is said of the Mino Bird? Where is it found? What faculty has it in great perfection?

What is the 3d FAMILY of CONE-BILLED BIRDS? What is said of their number, size, and musical powers? Describe their beaks. What is peculiar in that of the Grosbeak or Hawfinch? What use do they make of it in obtaining their food? What Sub-Family is first noticed? Describe their nests. Whence does the Red Beaked Weaver obtain its food? What is said of the Whidah Finches? Why is the term Widow applied to this group? What is said of the Java Sparrow? What SUB-FAMILY is next mentioned? What is said of the LARK BUNTING? Describe the Snow Lark Bunting. The Lapland Snow Bird. What species of the Buntings are mentioned? Describe the Chipping Bird. What European bird does it represent in this country? What fact shows the usefulness of Sparrows? What Bunting breeds in the fur countries? Describe the SNOW

BIRD. What is said of the Indigo Bird? What of the PAINTED BUNTING? Where are the SHORE FINCHES found? What species are mentioned? What is the generic name of the Linnets? What has been said of the Brown Linnet? What other sp. are mentioned? To what group does the American Goldfinch or Yellow B. belong? What is said of it? Mention the species which are given of the FINCHES PROPER. What is said of the GROUND FINCHES? What. sp. of Purple Finches are mentioned? What is said of the Pine Grosbeak? What of the Cardinal Grosbeak? What other sp. of Grosbeak are mentioned? What is said of the Summer Red B.? Why is this bird called the Tanagra? What SUB-FAMILY is next mentioned? How characterized? For what are they famed? What is said of the Sky-Lark? What of the Wood-Lark? Of the Horned-Lark? What additional SUB-FAMILY is mentioned? How do others rank them? To what countries are they confined? Describe them. What is said of their social habits? What of their plumage? Why are they much disliked? What is said of their cry? Of their flesh? What sp. is named?

What is the 4th FAMILY? Describe the beak. What use do these birds make of it? What aids them in this process? What is remarkable in this organ? What sp. are mentioned?

What is the 5th FAMILY? What are their most noticeable characteristics? What renders their large beak supportable? What is said of the bones of the body and the tongue? On what do they live? Describe their plumage. To what regions are they limited? To what other birds are they nearly related? How many sp. of the gen. *Buceros* have been named? What is said of the Indian Raven? What of the Rhinoceros Horn-bill?

What is the name of the LAST FAMILY? Give their leading characteristics? To what other birds are they related? Why are they called PLANTAIN EATERS? What is said of the Plant-Cutters? To what region are the P. Eaters confined? Describe their movements, &c.? What species is found on the gold coast of Africa? What other birds belong to this family? What is their generic name? What is its signification? What is said of this group? Which is the most attractive? What is said of their crest feathers?

Name and trace those mentioned on the Chart.

SECTION VI.

FOURTH DIVISION OF THE PERCHERS. THIN BILLED BIRDS.

Tenuirostres, (Lat. *tenuis*, thin, or slender; *rostrum*, beak.) This group of birds, M. Vigors considers "the most interesting of the animal world." They are characterized by the length and the slenderness of the bills, which are frequently curved and notched at the tip. The tongue is often divided at the end into two or more filaments; sometimes the slender filaments are so numerous as to resemble a painter's brush. The peculiar conformation of the bill seems chiefly intended to protect the tongue, by which, and not by the bill, these birds suck, or lick up the nectar of flowers, drawing in with the honied liquid, multitudes of minute insects, which form the solid part of their food. The feet are very short and delicate.

The smallest birds, and those the most brilliantly adorned, are found in this group. Many of the genera are clothed with a plumage of metallic lustre; on particular parts of their bodies, especially the forehead and throat, they have feathers of a scale-like appearance, which reflect the varying hues of precious stones. The Thin-billed Birds are principally to be found in the tropical regions, but many species visit the temperate zones, and a few are permanent residents of high latitudes. They are arranged into five families: (1,) *Promeropidæ* or *Upupidæ*, Hoopoes; (2,) *Cinnyridæ* or *Nectarinidæ*, Sun Birds; (3,) *Trochilidæ*, Humming Birds; (4,) *Meliphagidæ*, Honey-Eaters; (5,) *Certhiadæ*, Creepers.

First Family. Hoopoes.

Promeropidæ, (Gr. προμεροψ, *promerops*,) or *Upupidæ*, (Lat. *upupa*, a hoopoe.)

The Hoopoes are a small family of birds confined to the Old World, and most of them found in Africa and India. They exhibit some relations to the Bee-Eaters of the Fissirostral division. One species, the Common Hoopoe, visits Europe in company with the Bee-Eaters and other Swallow-like birds, but unlike them, walks upon the moist ground and newly turned earth, in search of insects and their larvæ; but the species of the genus *Promerōps*, seek for minute insects in the corollas of flowers. The Common, or European Hoopoe, *U. epops*, (Gr. *epops*, a hoopoe,) receives its name from the cry of the male bird, which is "*hoop, hoop.*" It has a very long and slender beak, slightly curved throughout its length, and compressed at the sides; long and rounded wings, and a long and broad tail. The toes are three before and one behind; the hind toe is long, with a long and nearly straight claw. The head is furnished with an erectile crest, the feathers of which are of a ruddy buff-color, terminated with black; the plumage presents striking contrasts of color, black, gray, buff, yellowish-white and white. The length is twelve and a half inches. It has been known to breed in England; building its nest in hollow trees, and laying from four to seven eggs, of a pale bluish gray hue. The Red beaked Promerops, *P. erythrorhyncus*, (Gr. red-beaked,) has a very long wedge-like tail, but is without an erectile crest. The long, slender beak, is of a coral-red; the entire plumage varies with metallic blue and green. This brilliant bird is found in South Africa. It lives in small flocks. The Grand Promerops, *Epimachus*, (Gr. *epimachos*,) *magnus*, (see Chart,) has a graduated tail, three times as long as the body; the feathers of the sides are lengthened, raised and curled. They glitter on their edges with steel-blue, azure, and emerald-green, like precious stones,—those of the body are of a deep, or brownish black. It inhabits the coasts of New Guinea. Swainson says it is "a bird of such excessive rarity, that only two perfect specimens have been known to exist in Europe."

SECOND FAMILY. SUN BIRDS.

Cinnyridæ, (Gr. *κιννα*, *kinna*, a grass; ὕριον, *hurion*, honeycomb?) Genus *Cinnyris*, Cuv.—or *Nectarinidæ*, (Gr. *νεκταρ*, *nectar*, nectar.) Genus *Nectarinia*, Illig.

The Sun-Birds, so called from their splendid glossy plumage, are arranged into two groups, (1.) *Cinnyridæ*,—genus *Cinnyris*,—of Africa, India and the islands of the Eastern Archipelago, which have comparatively slender bills and feet, and the tongue retractile and simply forked; and, (2,) the *Nectarinidæ*, of South America and the Pacific islands, which have the beak and feet comparatively strong, and hold an intermediate rank between the Creepers, (*Certhiadæ*,) and the Sun Birds, (*Cinnyridæ*,) and the Humming Birds, (*Trochilidæ*.) The Nectarines are to the New World what the Sun Birds are to the Old; their tongue ends in a sort of pencil or brush; and they hop from flower to flower, seeking the nectar of each; while the Sun Birds and the Humming Birds make no use whatever of their feet as they extract their food, but in feeding, are poised upon the wing. The Sun Birds and Humming Birds, as M. Vigors remarks, approach each other in the slendernes of their bill, the vividness and changeable lustre of their plumage, and the habit of hovering on the wing when they feed, and being chiefly separated from each other by the comparatively stronger foot and bill of the Sun Birds. Both groups of Sun Birds are included by Mr. G. R. Gray in one family, *Nectarinidæ;* but Swainson arranges the several genera under the name *Cinnyridæ*.

Some of the Sun Birds add the charm of song to that of brilliancy of plumage; and the music of one has been compared to that of the Nightingale. Their nest is usually suspended, and of a globe-like form, having an opening on one side, generally near the bottom.

In the case of the *Nectarinia*, of South America, the nest is placed in the worm-eaten trunks of mimosa-trees, and contains four or five eggs, entirely white. The Sun Birds of the genus *Melithreptes*. (Gr. *meli*, honey, *trepho*, to nourish,) included in the Nectarine group, are found in the Hawaiian islands. The beautiful yellow feathers of these birds, interspersed with a few of a scarlet color, are worn as ornaments of the head. "These feathers are among the most celebrated productions" of the above named islands. "Each bird yields only a few, and some thousands are required to form a head-dress. The wreath, or tiara, is sometimes valued as high as two hundred and fifty dollars. The birds, (*Melithreptes Pacifica*) are taken by means of bird-lime, made from the *pisonia*, and the catching of them is practiced as a trade by the mountaineers. The wearing of these feathers is a symbol of high rank."*

Third Family. Humming Birds.

Trochilidæ, (Gr. τρόχιλος, *trochilos*, a trochil, or wren.)

The family of Humming Birds is one of great interest. Mr. Gould, in his recently published work, enumerates about sixty genera, and his collection contains more than three hundred species. They are, however, separated from each other by comparatively slight variations in the length and curvature of the beak, the form of the wings, and the greater or less development of the tail, and of other parts. These birds are all confined to this Continent and the West India Islands: some species penetrate, in summer, to high latitudes on each side of the equator. The gorgeous flashings and changing tint, and the lustre, as of burnished metal, which are, to some extent, seen in the Sun Birds, are in the birds of this family preeminently conspicuous.

They are the smallest of the feathered races; (See Chart;) some species are exceeded both in size and weight by several of the insect tribe, while a few species are as large as a Swallow.

The Humming Birds are not less remarkable for their structure than for the remarkable splendor of their plumage. The excessively long wings are moved by pectoral and other muscles, which form nearly the whole of the fleshy substance of the bird; those of the feet "are reduced to the least possible quantity consistent with the requisite stability,"—all this showing, that they were adapted by the Creator to spend, as they do, the most active

* Narrative of the United States Exploring Expedition.

part of a highly active life in the air. The humming noise made by these birds, is produced by the extremely rapid movement of their wings.

The tongue is their principal organ for obtaining their food in the honied juices of flowers and insects; and like that of the Wood-peckers, it is so framed, that it can be darted out of the bill, as a spring suddenly released from its restraint. It is of such a length, that it can be protruded some distance from the bill. The long and slender beak comes admirably in aid for inserting the tongue into the nectaria of flowers. The sight of the Humming-Birds is very acute within the range required for its exercise, as is also the sense of hearing.

The females are without the splendid plumage of the males, and are clothed in modest dress. Some species living, as they do, from ten to fifteen thousand feet above the level of the sea, have the tarsi warmly and largely protected with white plumelets, and look as if they had downy muffs on their legs.

Humboldt notices the religious belief of the Mexicans, that Toyamiqui, the spouse of the god of war, conducted the souls of those warriors who had died in defence of the gods, into the mansions of the sun, and transformed them into humming-birds; and it must be owned, they form an image of the soul, scarcely less spiritual than the butterfly of the Greeks.

The nests of these birds are as wonderful as any that are made. (Plate XI. fig. 9.) They vary greatly in form and structure; but in all, the soft and delicate materials are so put together as to furnish as much warmth as possible, that being an object of the highest importance when the body of the animal is generally so small, and the quantity of animal heat given out accordingly diminished. The eggs are two in number, of an elongated form, and in some species, extraordinarily small. These birds are very valiant in defence of their nests. When attending their young, they attack any bird, indiscriminately, which approaches the nest. This display of valor, it is suggested, probably fostered the Mexican belief, that the bodies of these diminutive creatures contained the souls of slain warriors.

Among the most beautiful species, are the Sickle-winged Humming-Bird, *T. falcatus*, (Lat. from *falx*, a sickle;) the Recurved-bill Humming-Bird, *T. recurvirostris*, (Lat. recurved-beak;) Gould's Humming-Bird, *Ornismus*, (Gr. Bird-mouse,) *Gouldii;* the Bar-tailed Humming-Bird, (Plate XI. fig. 9,) *T. sparganurus*, (Gr. band-tailed;) the Double-crested Humming-

Bird, *T. cornutus*, (Lat. horned,) in length a little more than four inches.

Four species are found within the limits of the United States. (1,) The Mango Humming-Bird, *T. Mango*, found on Florida Keys; four inches and three-quarters in length; (2,) the Anna Humming-Bird, *T. Anna*, found on the Rocky Mountains, towards California; three inches and three-quarters in length; (3,) the Red-throated Humming-Bird, or Red-throated Honey-Sucker, *T. colubris*, (Lat. serpentine;) three and a half inches in length; ranging from Mexico to 57° N. Lat. This is the species most commonly seen in the State of New York, (Plate X. fig. 4b,) and well known for its golden green color, and its ruby colored throat; and (4,) the Ruff-necked Humming-Bird, *T. rufus*, rather more than three inches and a half in length; discovered by Capt. Cook, who found it abundant at Nootka Sound; it is met with also in the vicinity of the Blue Mountains of the Columbia River.

Fourth Family. Honey-Eaters.

Meliphagidæ, (Gr. μέλι, *meli*, honey; φάγω, *phágō*, to eat.)

The birds of this family in some measure depart from the tenuirostral type in the increased stoutness of the beak. This organ is, in these birds, awl-shaped and arched, and has the tip distinctly notched; the hind toe is so strong and robust, that it serves as a support to the bird while taking its food; the tongue is still capable of protrusion, but in a subordinate degree, and is terminated by a brush of hairs.

These birds are chiefly confined to Australia, where they feed on the nectar and pollen of flowers. As in that country the fields are never without blossom, they have in the luxuriant vegetation, a support that never fails. They also live on insects and berries. Usually, they are of sombre colors, black or olive-brown, without any metallic lustre. Their nests are cup-shaped, constructed in the forks of small branches of shrubs, not far removed from the ground. The Honey-Eaters are larger than most of the Thin-billed Birds; several species equal a Thrush in size, and some are of considerably greater dimensions. The Warty-faced Honey-Eater, *Meliphaga phrygia*, is described as sometimes to be seen in great numbers, constantly flying from tree to tree, particularly among those known as the blue gum, from the blossoms of which they extract the honey with their tongues as they pass along. One species is said to pick holes in the bark of trees, and thence to extract insects, very much in the manner of the Woodpecker; indeed, these birds probably represent in Australia, the true Woodpeckers, which are not found in that region. The Tui, or Poe-Bird, ***Prosthemadera Cincinnata***, of New Zealand, is about the size of a Black-Bird; from its great imitative power, it has been called "the Mocking-Bird;" and from its peculiar plumage, the "Parson-Bird."

Fifth Family.. Creepers.

Certhiadæ, (Gr. κέρθιος, *kerthios*, Creeper kind.)

The birds of this family manifestly deviate from the tenuirostral type, and approach the order of the Climbers, (*Scansores.*) We therefore follow Cuvier and Charles Lucien Bonaparte, who, while including them among the Thin-Billed Birds, place them on the confines of the present order. (See Chart.) In these birds the tongue is still capable of protrusion, but is no longer divided into filaments; the tip, however, is sharp, horny, and fitted for transfixing insects, which are sought beneath the bark of trees, in crevices of walls, and similar concealed situations. To aid them in taking their insect prey, the beak also is generally slender, sharp-pointed and strong, curved in various degrees; sometimes, as in the Wall Creeper, *Tichodroma*, (Gr. wall-runner,) *muraria*, (of a wall,) a species of Southern Europe, the beak is almost straight; and at others, as in the Tree-Creeper, *Dendrocolaptes*, (Gr. tree-beater,) found in Brazil, the beak is bent almost to a semi-circle.

This family are Climbers, but still have not the feet of the Climbers proper, (Scansores,) with which M. Vigors arranges them. The outer toe is not reversible, but the back toe is considerably larger and stronger than it is in the greater part of the perching birds.

Some of these birds, as the Tree Creepers, have the shafts of the tail feathers strong and rigid, and their tips are lengthened beyond the barbs, as in the Woodpeckers, and to meet the same exigency, viz.: the wearing away of the more fragile parts by the constant friction of the tip of the tail against perpendicular surfaces; in the species now referred to that organ being thrown in and pressed against the tree or wall for support in climbing.

The Common Creeper, *Certhia familiaris*, is not more than five inches in length, of a yellowish brown color above, the under parts being white. It is generally distributed throughout Europe and the United States. Wilson says: "The Brown Creeper is an extremely active and restless little bird. In winter it associates with the small spotted woodpecker, nuthatch, titmouse,

&c., and often follows in their rear, gleaning up those insects which their more powerful bills had alarmed and exposed; for its own slender incurvated bill seem unequal to the task of penetrating into even the decayed wood; though it may enter into holes and behind scales of the bark." It builds its nest in some rent or cleft in a tree, where a branch has been broken off, or where a hole has been chiseled by a woodpecker, and deposits in it six or eight ash-colored eggs, marked with dusky reddish spots. The voice of the Creeper is a monotonous cry, not very loud, but often and suddenly repeated, especially in its flight from tree to tree. The food on which it lives consists principally of small beetles, bugs and flies, which it draws from their places of concealment. Wilson mentions having found in its stomach, the seeds of the pine tree and large quantities of gravel. Did our limits permit, we would give details of genera and species found in South America and Australia.

The Nuthatches, *Sitta*, are allied to the Titmice on the one hand, and the Woodpeckers on the other. They have a stronger bill than that of the Tree Creepers; and it is straight and pointed like that of the Woodpeckers, used rather to scale off the bark than to perforate it; and they do not support themselves upon the tail. They run about the trunk and branches of trees, seeking for insects and their larvæ, berries and nuts; they are noted for their instinct of fixing a nut in a chink while they pierce it with the bill, swinging the whole body as on a pivot, to make the stroke more effective. The name Nuthatches is given to these birds on account of the hatches or hammerings which they make on hard nuts in search of the larvæ within. From four to six species of these birds are found within the United States. The White-breasted Nuthatch, *S. Carolinensis*, is about five inches long, of a slate blue above and pure white beneath; it ranges from Mexico to Maine. The eggs are whitish, spotted with brown at the larger end, and from four to six in number. The Red-bellied Nuthatch, *S. Canadensis*, is four and a half inches long and lead-colored. This is a more northern bird than the preceding, ranging from Maryland to Nova Scotia. The Brown-headed Nuthatch, *S. pusilla*, ranges from Texas to Maryland. The Pygmy Nuthatch, *S. pygmea*, is found in California; it is less than four inches long.

The Wrens, *Troglodytes*, (Gr. *trōglodutes*, a creeper into caves.) are properly included in the present family, though they have been differently arranged by some authors. (See Chart.) The House Wren, *T. aedon*, is a familiar little bird which has become inviolable, like the robin, from the confidence which it shows in courting the neighborhood of man. This Wren is of a dark brown above with blackish bands; beneath it is whitish, with faint or obscure bands. It builds its nest in boxes or houses prepared by man, in which it lays six or eight flesh-colored eggs. It is said that it seldom or never builds a distinct nest, but always conceals it in things "placed for its convenience around houses, or in the hollow of trees." The nest is proportionably very large. Audubon figures one beautifully as built in an old hat. The House-Wren shows great antipathy to cats. "Although it does not attack puss, it follows and scolds her until she is out of sight." It ranges as far as the 57° N. L. Audubon thinks it spends the winter southward of the United States. Its length is four and a half inches.

The Winter Wren, *T. hyemalis*, closely resembles the European Wren, *T. Europæus;* its song is energetic and musical; it lays ten or twelve whitish eggs. This Wren is small, being only three inches and a half in length.

The Wood Wren, *T. Americanus*, is nearly the same as the House Wren, but spends the winter within the limits of the United States.

The Mocking Wren, *T. ludovicianus*, is noted for its mimicry and song; it is about five and a half inches long, and ranges from Texas to New York.

The Ox-peckers, or Ox-eaters, *Buphagidæ*, genus *Buphaga*, (Gr. *bous*, an ox; *phago*, to eat,) found in Southern Africa, are also included among the Creepers. These birds have a large obtuse and nearly quadrangular bill, the lower mandible being stronger than the other, and both swollen towards the point, it somewhat resembles a pair of pinchers or scissors. The Ox-peckers fasten themselves with their strong, hooked claws and elastic tails upon the backs of ruminant quadrupeds, such as oxen, buffaloes, antelopes and camels, and also, some travelers say, upon the backs of the Rhinoceros and Hippopotamus, and with their beak dig and squeeze out from their backs the larvæ (or maggots) which the gadflies have deposited. Wherever, by the presence of an elevation, the bird is aware of the existence of a maggot, he extracts it with strong blows of his bill. This treatment the animals willingly bear, seeming to look upon these birds as their benefactors, as really they are, especially in a region where such insects abound.

What is the 4th Division of the Perchers? How does Vigors regard them? What are their characteristics? For what does the bill seem chiefly designed? What is said of their size and plumage? What of their distribution? Name the families into which they are arranged. Where are the HOOPOES found? What sp. visits Europe? From what does it receive its name? What is said of it? What is the food of the gen. Promerops? What is said of the Red-Beaked P.? What of the Grand P.?

What is the 2d FAMILY? Why are they so called? Into what groups are they arranged? To what regions are the CINNYRIDÆ confined? What is said of their bills and feet? Where are the birds of the 2d GROUP found? To what birds are they intermediate? What is said of the NECTARINES? How do they differ from the Sun B. and Humming B. in their mode of procuring their food? Are any of the Sun B. musical? What is said of their nests? What use is made of the feathers of one sp. of this bird?

What is the 3d FAMILY? Are they numerous? What is said of the distinction between the sp.? What of their size, structure, and plumage? How and whence do they obtain their food? What is said of their tongue? What of their sight and hearing? How are some of them protected against the cold in elevated regions? What does Humboldt notice? What is said of the nests of these birds, &c.? What species are named?

Mention the 4th FAMILY? What is said of the beak of these birds? To what region are they chiefly confined? On what do they feed? What is said of their plumage? Of their size? What sp. are mentioned?

What is the 5th FAMILY? Are they strictly Tenuirostral birds? What is said of their tongue? On what do they feed? Are they strictly Climbers? What is said of the tails of the TREE CREEPERS? Describe the Brown Creeper. Describe the bill of the NUTHATCHES. What use do they make of it? Why are they called Nuthatches? What sp. are found in the U. S.? Which is the smallest? What is the generic name of the WRENS? What is its signification? Describe the HOUSE WREN. What other sp. are mentioned? Where are the Ox-peckers found? Repeat what is said of them.

Name and trace those figured on the chart.

SECTION VII

THIRD ORDER. CLIMBERS.

SCANSORES, (Lat. *scando*, to climb.)

The birds of this order are unlike in their food and in their general structure and habits; but as a distinguishing character common to them all, they have four toes rising nearly to the same level, the outer toes being turned backwards more or less permanently, like the thumb, so that these are opposable to the middle and inner toes, which point in the opposite direction. This peculiar disposition of their toes gives these birds great facility in climbing the branches of trees, but it renders walking more difficult to them. As they pass most of their lives in trees, their powers of flight are usually moderate. Their nests are ordinarily constructed with less skill than those of the Perchers, these birds often employing for this purpose the hollows of decayed trees, and one family depositing their eggs in the nest of other birds. They feed on insects and fruits, and the species feeding upon each may be known by the greater or less robustness of the beak.

This order is divided into four families, viz.: (1.) *Ramphastidæ*, (Toucans;) (2.) *Picidæ*, (Woodpeckers;) (3.) *Psittacidæ*, (Parrots;) (4.) *Cuculidæ*, (Cuckoos.)

(Swainson also includes in this order the *Certhiadæ*, (Creepers,) which have the rigid tail of the Woodpeckers, but the feet of the Perchers, among whom they were placed by Prince Bonaparte.)

First Family. Toucans.

Ramphastidæ, (Gr. ῥαμφαστής, *ramphastēs*, a pike.)

The Toucans are all natives of Tropical America. They are large birds, clothed with brilliant plumage, and found in the depths of magnificent forests. They associate together in small companies, which are said sometimes to include even distinct species.

These birds are easily recognized by the extraordinary size of the beak, which in the typical genus, *Ramphastos*, is nearly as large and as long as the body itself; it is rendered light in the same way as that of the Horn-Bills, being permeated by a very thin and fragile net work of bony fibres, of a honey-comb appearance, and is said to be borne with so much of ease and grace as entirely to remove the idea of uncouthness which its appearance suggests to those who look at it only in figures and stuffed specimens.

The edges of the mandibles are both regularly notched at wide intervals, and curved downwards to the tip; the tongue is narrow, lengthened, and barbed on the sides like a feather; the feet are formed more for grasping than flying, having two toes before and two behind, and accordingly these birds are seen on trees, hopping from branch to branch. Their general movements are light and elegant, but having short, rounded wings, their flight, though rapid, is labored and in straight lines.

Their powers of smell are exquisite; the nerves of that sense are so distributed in the beak as to enable them more readily to discover their food. This is both animal and vegetable; but they prefer the eggs and young of other birds; in obtaining these from the deep hanging nests found in the regions which they inhabit, they use their enormous beaks, the surface of which is endowed with sensibility, enabling them to explore the contents of these nests. It is said these birds are remarkably fond of bathing in cold weather. They nestle in the hollows of trees, laying two white and delicately rounded eggs.

The Toucan takes great care of his bill, packing it away and covering it carefully with the feathers of its back before sleeping, when it exhibits the appearance of a large round ball of feathers.

Mr. Gould arranges the Toucans into two sections; (1.) the TOUCANS proper, *Ramphastos*, (from Gr. *ramphos*, a beak,) including eleven species; (2.) the ARACARIS, *Pteroglossus*, (Gr. *pteron*, wing; *glōssa*, a tongue.)

In the former the beak is without grooves; but in the latter it is notched at wide intervals. The tail in the Aracaris is shorter than in the Toucans proper, and is graduated instead of squared.

The true Toucans are generally black on the upper parts, with vivid colors, chiefly red and yellow, on the throat and breast. The beak is often tinted with brilliant hues which vānish after death. The TOUCAN, *R. Toco*, (see Chart,) is one of the largest of this section, being twenty-seven inches in total length, of which the beak is seven inches and a half. It ranges from the River La Plata to Guiana.

The KEEL-BEAKED TOUCAN, *R. carinatus*, (Lat. *carina*, a keel,) is conspicuous for the number and brilliancy of the hues adorning its beak, which is keeled along the upper edge.

The ARACARI, *P. pluricinctus*, (Lat. many-girdled,) has the breast marked with two broad bands of black, the upper separated from the throat by an intervening space of yellow, dashed with red; a similar but broader space separates the two bands of black, the lower of which is bounded by scarlet, advancing as far as the thighs, which are brownish olive. The total length is twenty inches; the bill four inches and a half. It is a native of Brazil.

SECOND FAMILY. PARROTS.

Psittacidæ, (Gr. ψίττακος, *psittakos*, a parrot.)

These birds are remarkable for their beautiful colors, their powerful bill, their fleshy tongue, and their imitation of the human voice. The articulation of some of the species is so perfect, that when unseen, it is difficult to suppose that the words pronounced do not come from the mouth of man. The power of moving the upper mandible is much more highly developed in this family than in other birds, that organ not being connected into one piece with the skull, by elastic and yielding bony plates, as is the case with the birds in general, but constituting a particular bone, distinct from the rest of the skull, and joined to it. This mobility becomes more conspicuous, for the reason that their vigorous jaws are set in motion by a greater number of muscles than are found in other birds. The advantages of this peculiarity of structure are apparent, when we remember the use which a Parrot makes of the beak, as a third hand, to assist it in climbing from bough to bough, or about the bars of its cage when in confinement. The beak appears to be well supplied with nerves of sensation, as the bird not only seems to enjoy holding its food with the tip of its bill, but sometimes scratches that organ with its foot, plainly showing that there must be sensation. The thick and fleshy tongue of the Parrots, is a very delicate organ of taste; it is covered, like that of the *Mammalia*, with *papillæ*, and being moistened by a constant secretion of saliva, they are able to select and taste different kinds of food. In some of the Australian species which suck the nectar of flowers, the tongue, while retaining the thick form and fleshy structure common to the family, is distinguished by the peculiarity of terminating in a number of very delicate and close-set filaments, which can be protruded and expanded like a brush. One of these species, the Australian LORIKEET, is of a predominant azure color, and is sometimes called the Blue-mountain Parrot, *Trichoglossus*, (Gr. hair-tongued;) *hæmatodus*, (Gr. of blood-color,)—when shown, in confinement, a colored drawing of a flower, it applied the tip of its tongue to it, as if it would suck it, and on another occasion, made a similar attempt on seeing a piece of furniture calico.

The most prevalent hue of the Parrots, is a soft and lustrous green, varied, however, with scarlet, yellow and blue in profusion, usually arranged in broad and well defined masses.

The Parrot tribe have been arranged into several groups, founded, to a great extent, upon variations of plumage.

(1.) The TRUE PARROTS (*Psittacus*) are, for the most part, found in tropical America. Their prevailing color is green. The Ash-colored, or Gray Parrot, *P. erythacus*, is seen in Africa. This group excels all the others in powers of imitation. The species of the Green Parrots are numerous. The best known, are the FESTIVE PARROT, *P. festivus*, and the AMAZON'S PARROT, *P. Amazonicus*. The latter has superior mimic propensities, but the Festive Parrot is the larger in size. The Amazon Parrot can be easily taught to repeat many words and sentences. It lives on fruit, particularly that of the Mangrove-tree. The COMMON GRAY PARROT, *P. erythacus*, is thought superior to all others in docility and mimicry; when well taught, it completely imitates the human voice; and is clear in its articulation. A Roman cardinal, it is said, "gave a hundred gold pieces" for one of these birds which had learned to repeat distinctly the "Apostle's creed." La Vaillant mentions one which had lived in confinement ninety-three years. The Parrots of this group are square-tailed, and have no crests. (2.) The LONG-BILLED PARROTS, (Australian genus *Nestor*,) are the connecting link between these and the Cockatoos. (3.) The COCKATOOS, *Plyctolophinæ*, (Gr. with washed, or folding crests.) are natives of Australia and the Indian Islands. These are also square-tailed, but have crests upon their heads. (Plate X. fig. 5a.) They are white birds, with the crests and under parts of the tail-feathers yellow; quite gentle in disposition, and easily domesticated, with the exception of a large Black Cockatoo, found in Australia. Their imitative powers seldom go beyond a very few words added to their own cry of "Cockatoo." (4.) The LOVE-BIRDS, *Psittacula*, are a group of beautiful and diminutive birds, nearly allied to the True Parrots, and found on both Continents. They are distinguished by their slightly graduated tails; and they have no *furcula*, or wish-bone. (5.) The PARRAKEETS, or PARRAQUETS, *Palæornis*, (Gr. *palaios*, old; *ornis*, bird,) are natives of India and the adjacent islands. Some eleven or twelve species are enumerated, one of which is found in Australia; their color is green, with the under parts scarlet. One species is named *P. Alexandri*, after Alexander the Great, in whose time these birds were first introduced into Europe. They have ever been noted for their beauty of form and movement; their powers of imitation, and their show of affection when kindly treated. Amid the luxuries of Rome, the "Indian-Bird" was kept in cages of the most costly materials, nor was any price, however great deemed extravagant, or beyond its value. A species which Wilson calls the *Parrakeet*, but which is named by Audubon *Psittacus Carolinensis*, is found as far north as Cincinnati, Ohio. The Parrakeets have long pointed tails. (6.) The MACAWS, or MACCAWS, *Macroceros*, (Gr. long-horned.) are American Birds. Those of South America and the Antilles, are the largest and most highly colored. Their imitative powers are much less than those of the True Parrots, but when domesticated, they become greatly attached. These birds are long-tailed, and the largest of the family. The *M. Ararauna*, of Brazil, has a plumage of rich hue above; the under parts light saffron. It is thirty-nine inches long, including the tail, which measures twenty-four. (7) The LORIES, *Lorius*, are a group found in the Moluccas and the Eastern Islands,—remarkable for the very rich and mellow hues of their plumage; blending scarlet with green, violet-purple, violet-blue, and orange-yellow. They are lively and active, and of an affectionate disposition, and show great docility in the articulation of words and sentences. The beak of these birds is lengthened, and comparatively feeble; the tail rounded, or graduated. They feed upon the juice of flowers or the pulp of the softest fruits.

THIRD FAMILY. WOODPECKERS.

Picidæ, (Lat. *picus*, a woodpecker.)

These birds are, in their whole organization, adapted to climbing, and eminently entitled to be called *Scansores*.

The feet are short, but very strong; the toes are placed in pairs, two pointing forward and two backward, (Plate X. fig. 23;) the claws are large, much curved, and very hard and sharp, enabling the bird to cling firmly, and creep on trees in all directions. The tail-feathers terminate in points, and are uncommonly hard, so that, being pressed against the bark, they assist the bird in its progress, or in keeping its position. The bill, destined for the laborious operation of penetrating the wood, or stripping off the bark of forest-trees, is beautifully adapted for the purpose, being wedge-shaped, and in one species, (*Picus principalis*,) nearly of the color and consistency of ivory, whence it has been termed the IVORY-BILLED WOODPECKER. This bird obtains its food, consisting of the larvæ of wood-boring insects, by chiseling away the bark and surrounding wood, until the subtle grub is exposed. The head then acts as a hammer, of which the beak is the face or point, and the curved neck the handle, and being moved by muscles of great energy, the sharp and wedge-like beak-tip is propelled against the tree in a succession of strokes given with remarkable force and activity.

To help in this work of chiseling out its grub-worm food, the Woodpecker also has a worm-like tongue, barbed at the point, and capable of being protruded to a great length; for which purpose there is a peculiar structure and arrangement in the muscles at the base of the tongue. By means of its protruding tongue, this bird transfixes the insects which it dislodges from their hiding places with its powerful bill. Added to this, there is on each side of the head, a very large gland which secretes a glutinous substance; this gland being compressed by the muscular action which protrudes the tongue, the viscid matter is poured out upon the sides of the tongue as it is thrust forth, and this is sufficiently adhesive to attach to itself small insects, such as ants, small grubs, beetles, &c., which are rapidly drawn in and swallowed. "But as many of the boring larvæ are too heavy thus to adhere, and would hold on by their tuberculous feet, or by their strong jaws, the capture of such is effected by a horny tip of the tongue being set with numerous fine barbs on each side, pointing backwards; the fine point readily pierces the skin of the insect, the barbs yielding as it enters, but when once within, it cannot, without much force, be withdrawn, the barbs having expanded within the skin, and so the insidious grub, despite his efforts to maintain his tenancy, is dragged forth by the powerful contraction of the Woodpecker's elastic tongue." All this is to be placed among those beautiful contrivances of the Divine Mind, which are so conspicuous in the "Animal Kingdom," and which, in so interesting and striking a manner, exhibit the benevolent and fatherly care of Him, without whose notice not even "a Sparrow falleth to the ground."

The Woodpeckers are widely scattered over the Eastern and Western Continents. As yet, however, no representative of this family has been found in Australia. The prevailing hue of these birds is black, often handsomely spotted with white, and varied with brilliant red, the latter especially upon the head. They lay their eggs and bring up their young in capacious chambers, which are hollowed out of the trunks of trees. Among the birds of this family, is included the *Yunx*, more properly *Iünx*, (Gr. ἴυγξ, *iunx*,) or Wryneck, (*Y. torquilla*,) of old described by Aristotle, and known to classical scholars as referred to in the second Idyl, of Theocritus. Its general color is ash, spotted with brown or black; its beak is short, straight, and depressedly conical. The Wryneck is a companion of the Cuckoo, appearing and departing about the same time; and in captivity, is a great favorite.

The species of Woodpeckers are quite numerous. Audubon mentions twenty-one as found in the United States.

The Imperial Woodpecker, *P. imperialis*, of California and the Rocky Mountains, is the largest, being two feet in length. The Green Woodpecker, *P. viridis*, (Lat. green,) is found on the European Continent. *P. torquatus*, (Lat. collared,) is a species of Green, or blackish-green Woodpecker, (Plate X. fig. 5b,) found in California and the dense forests bordering on the Columbia River. It has a band of dull white running over the back of the neck, and joining a patch of a reddish color on the front and part of the breast. The European species is thirteen inches long; the American eleven. The HAIRY WOODPECKER, *P. villosus*, (Lat. hairy,) is a constant resident of New York during the whole year. Length eight and a half inches.

FOURTH FAMILY. CUCKOOS.

Cuculidæ, (Lat. *Cuculus*, a cuckoo.)

This family of birds have a beak of a medium length, rather deeply cleft; both mandibles compressed, and more or less curved downward; the nostrils exposed; the wings, for the most part, short, but the tail lengthened. Their skin is remarkably thin; the plumage thick and compact, generally of subdued, but chaste and pleasing hues, with more or less of reflected lustre; the long tail is often graduated, and handsomely barred with black and white.

"So faintly," says Swainson, "is the scansorial structure indicated in these birds, that but for their natural habits, joined to the position of their toes, we should not suspect they were so intimately connected with the more typical groups of the tribe, as they undoubtedly are. They decidedly climb, although in a manner peculiar to themselves. Having frequently seen different species of the Brazilian Cuckoos in their native forests, I may safely affirm, that they climb in all other directions than that of the perpendicular. Their flight is so feeble, from the extreme shortness of their wings, that it is evidently performed with difficulty, and it is never exercised but to convey them from one tree to another. All soft insects inhabiting such situations lying in their route, become their prey, and the quantities that are thus destroyed, must be very great."

The Brazilian hunters give to their Cuckoos the general name of Cat's-tail, their long hanging tails and mode of climbing presenting some resemblance to that quadruped. Swainson thinks the long tail is given to the Cuckoo as a sort of balance, just as a rope-dancer, with a pole in his hands, preserves his footing when otherwise he would fall. It is a peculiarity of the Cuckoo, that the outer hind-toe can be made to form a right angle with that which is next it in front, so that it is termed *versatile*,—a term not, however, strictly applicable, as the toe cannot be brought more than half-way forward, although it can be placed entirely backward. The Cuckoos are really half perching and half climbing birds, not only in their feet, but in their manners. They are divided into two sub-families; (1.) *Cuculinæ*, which include the genuine Cuckoos, having the bill broader at the base than it is high. These, with the exception of the birds included in the genus *Molothrus*, are the only known parasitic birds,—making no nests for their own use, but taking possession of those of small insectivorous birds, usually of the Dentirostral tribe. "The whole care of hatching and rearing the young, is now left to the foster parent; and as the wants of so large an intruder, additional to those of their own offspring, would be more than the efforts of the selected nurses could supply, an instinct is implanted in the young Cuckoo, by which, even from the very day of its birth, it is impelled to eject from the nest the rightful tenants of it. This is a well known habit of the Common Cuckoo, whose notes as harbinger of spring, are pleasing, but whose reputation is bad, on account of the ruthless murders which, in its early days, it is supposed to have committed." The Toucans, however, seem to act as avengers. The favorite nests of the Cuckoo, are those of the Hedge Sparrow, the Pied Wagtail, the Pipit and the Robin.

(2.) The sub-family, *Coccyzinæ*, from the generic name, *Coccyzus*, (Gr. *kokkuzo*, I sing as a cuckoo,)—have a bill of a lengthened and oval shape, and are not to be regarded as parastic. *Coccyzus Americanus*, (*Cuculus Carolinensis*—Wilson,)—is well known by its notes, which seem to represent the word *cow, cow*, repeated eight or ten times with increasing rapidity. This is the Yellow-billed Cuckoo, which honorably builds its own nest, and lays four or five eggs of a green color. Sometimes it is called the Cow-bird. There are two other American species, viz., the Black-billed Cuckoo, *C. erythrophthalmus*, (Gr. red-eyed,) and Mangrove Cuckoo, *C. seniculus*, (Lat. a little old man.)

What is the Third Order of Birds? What characteristic is common to them all? For what does this fit them? What is said of their powers of flight and their nests? Into what Families is the order divided?

What is the First Family? Of what region are they all natives? What is said of their plumage? How do they associate? How are they easily recognized? Describe the beak. What other characteristics are given? What is said of their flight? What of their powers of smell? What food do they use? What assists them in obtaining it? What more is said of these birds when sleeping? How does Mr. Gould arrange the Toucan? State the differences between the two groups? Which is the largest of the Toucans Proper?—What species of the Aracari is mentioned? What other species of Toucans is mentioned?

What is the Second Family? For what are the Parrots remarkable? What is peculiar in their upper mandible? Describe it. Has the beak nerves of sensation? What is said of the tongue? What peculiarity attaches to some of the Australian species? What fact is mentioned in relation to the Australian Lorikeet? What is the prevailing hue of the Parrot? Where are the True Parrots for the most part found? What is said of the imitative powers of this group? Which are the best known of the Green Parrots? What is said of them? Which is superior in docility and mimicry? Are the True Parrots crested, and what is the form of the tail? What Parrots connect them with the Cockatoos? Where are the Cockatoos found? How do they differ from True Parrots? What is said of the color of their plumage? What is said of the Love-Birds? What group is next mentioned? Where are these found? What is their color? When were they first introduced into Europe? For what are they celebrated? What Parrokeet is found in the U. S.? Where are the MACCAWS found? What is said of their size? What species are mentioned? Where are the LORIES found? What is said of them?

What is the THIRD FAMILY? For what are they eminently adapted? How is this shown? What use do they make of the bill? Why is the Ivory-billed W. so called? What organs help them to obtain their food? With what glands is the head furnished? What purpose do these serve? How are they enabled to secure the larger larvæ? What do these marks of adaptation illustrate? What is said of the diffusion of these birds? How are they colored? Which is the largest species? Where found? How large? What is said of the WRY-NECK? Are the sp. of Wry-necks numerous?

What is the FOURTH FAMILY? What characteristics are mentioned? What does Swainson remark of these birds? What purpose is served by their long tails? What is said of the outer hind toe? What are the Cuckoos really said to be? What SUB-FAMILY includes the GENUINE CUCKOOS? Why are they called PARASITIC-BIRDS? Are there any other Parasitic-birds? Which are the favorite nests of the Cuckoo? What is the other SUB-FAMILY? Are they Parasitic-birds? What American species are mentioned?

SECTION VIII.

FOURTH ORDER. SCRATCHERS.

RASORES, (Lat. *rado*, to scratch;) or GALLINÆ, (Lat. *gallus*, a cock.)

This order, which includes the *Gallinaceæ*, or Poultry tribes, consists of birds having bulky forms and strong legs, and especially adapted to live on the dry ground. The Poultry are chiefly confined to the continents, few, comparatively being found on the adjacent islands.

The wings of the Scratchers are muscular, but not proportionate in size to the bulk of their bodies, so that their power of flight is comparatively small. Most of them have strong, arched beaks, long necks, and large, ample tails; many have their heads adorned with elegant crests; the tail has more than the usual number of feathers, having from fourteen to eighteen. Their food is, with few exceptions, vegetable, being chiefly derived from the seeds and grains of plants. These birds multiply with great rapidity, are easily domesticated, and as furnishing man with a large quantity of wholesome and delicate food, deserve special regard. Some of them, as the Peacock and Pheasant, are also interesting for the beauty and stateliness of their forms, and the diversity of their plumage. In the few species of this family which associate in *pairs*, such as the Ptarmigan and Partridge, the male and female birds are nearly alike, both in size and color.

The Scratchers are arranged into seven families, viz.: (1) *Columbidæ*, Pigeons; (2) *Cracidæ*, Currassows; (3) *Megapodidæ*, Megapodes, or large-footed Birds; (4) *Phasianidæ*, Pheasants; (5) *Tetraonidæ*, Grouse; (6) *Chionidæ*, Sheath-bills; (7) *Tinamidæ*, Tinamous.

First Family. Pigeons.

Columbidæ, (Lat. *Columba*, a dove or pigeon.)

The food, habits, and internal economy of these birds, and the form of their bills entitle them, in the judgment of Cuvier and others, to a place among the Rasores. They, however, show resemblances to the Perchers, which have led some naturalists to place them in that order. The feet of the Pigeons, though following the type of the Perchers, allow them to spend most of their time on the ground, and many of them perch very little. They differ from the Gallinaceous birds, in pairing, which is contrary to the habits of the latter, also in having the hind toe on the same level with the others, whereas the Gallinaceous birds have the hind toe higher up. The variations of the Pigeons from both the Scratchers and Perchers, have induced yet other naturalists to erect them into a separate order, *Gyratores*, (Gr. *guros*, a circle.) or Circling Birds, a name referring to their mode of flying in circles.

The Pigeons include a large number of elegant and amiable birds, spread over every part of the world. One of their principal peculiarities is the *crop*, which ordinarily is thin, but which, when the young are about to be hatched, becomes expanded on each side of the gullet, and very irregular as to its internal surface. From this organ the parent bird supplies its young with food, previously rendered suitable by the action of a milky fluid that is secreted in the crop; this fluid, it is said, coagulates with acids and forms curd. This apparatus constitutes among the Birds the nearest approach to the Mammal tribes: hence the term "pigeon's milk."

The beak in the Pigeons is of moderate length, and swollen towards the tip, which is curved downwards; the wings vary in length and in adaptation to powerful flight; the feet have three divided toes in front, and a single one behind. The structure of the feet varies, however, in different genera.

In the WOOD-PIGEONS, (*Columba*,) of North America and the Eastern Continent, the outer and inner toes in front are equal.

In the GREEN PIGEONS, (*Ptilinopus*, Gr. feather-footed,) of Australia and the East Indian Islands; and the AROMATIC VINAGOS, (*Vinago*,) of inter-tropical Asia and Africa, a group which includes the Thick-billed species of those countries, the inner toe is much shorter than the outer, so that they are more fitted for grasping than walking; but this proportion is reversed in the PASSENGER PIGEON, genus *Ectopistes*, (Gr. *ektopizo*, to migrate.) In the genus *Peristera*, (Gr. for dove,) which comprises the beautiful BRONZE-WINGED PIGEONS, of Australia, and the GROUND PIGEONS of this continent, the tarsi are higher, the hind toe shorter, and the inner toe is the longest.

The Pigeons generally nestle in trees and in the holes of rocks, laying but few eggs at a time, but breeding very often, so that their increase is very rapid. The prevailing hues of the plumage in the typical genus, *Columba*, are various shades of blue and gray, merging, sometimes, into purple, and at others, into white. Many of this family exhibit metallic reflections of great beauty, mostly confined to particular parts, especially the neck. The countenance in these birds is meek and gentle in its expression; the eye, large, liquid and engaging. The voice has a soft and mournful character; it is known by the term *cooing*.

The ROCK-PIGEON, *C. livia*, (Lat. livid,) in its wild state widely distributed, is the original stock of the COMMON, or DOVE-COTE PIGEON, and most of the curious varieties which are fostered by "pigeon breeders." Among the varieties are the TUMBLERS, so called from their singular habit of falling backwards when on the wing; the POUTERS, or CROPPERS, so named from their inflated crops, of which they seem exceedingly vain, and which they are enabled to fill so full of air that the head is almost hidden behind it; the CARRIERS, or Messenger Pigeons, trained to carry letters fastened under their wings or to their feet, celebrated in the verse of Anacreon, (ODE, *eis peristeran*, to the pigeon.) Victors in the games of ancient Greece sometimes employed these birds to announce their success; the Crusaders used them; they figure in Tasso's "Jerusalem Delivered," who sings of one that was attacked by a falcon, and rescued by the hero, Godfrey; but, though they continued to be used down to modern times, and at last, for such ignoble purposes as heralding the felon's death, increasing the gains of stock-jobbers, or bearing messages from the race course and prize ring, (see Hogarth's print in the Penny Magazine,) yet since the invention and application of the electric telegraph, their "occupation is" almost "gone." A well trained carrier-pigeon, it is said, has "performed the distance of forty miles in half an hour;" and "one has been known to fly nearly one hundred and fifty miles in an hour!" Their more usual rate of flight probably does not exceed forty miles an hour.

Other "fancy varieties" might be mentioned, but those given must suffice.

The astonishing fecundity of the domesticated pigeon is shown by the fact, that hatching as they do, nine or ten times a year, a single pair may produce, in four years, 14,760 young!

The TURTLE-DOVE, *Turtur*, (Lat. turtle-dove,) *risorius*, (laughing,) or *Columba risorius*, is deemed a fitting emblem of constant and faithful connubial attachment; it expresses its affection by "billing and cooing in the gentlest and most soothing accents." This bird reaches England early in the Spring, and leaves late in August; its length is rather more than twelve inches. The specific name, (*risorius*,) is given to it from a "fancied resemblance to the human laugh in its cooings."

The CAROLINA TURTLE-DOVE, *Columba* (*Ectopistes*) *Carolinensis*, is twelve inches long, and ranges and breeds from Texas to Massachusetts. The plumage of the upper parts is light yellowish brown, with the crown of the head and upper part of the neck, bright greenish blue; the under parts are brownish yellow. Wilson says: "This is a favorite bird with all who love to wander among our woods in the spring, and listen to their varied harmony. They will hear there many a sprightly performer; but none so mournful as this. The hopeless woe of settled sorrow, swelling the heart of female innocence itself, could not assume tones more sad, more tender and affecting. They are generally heard in the deepest shaded part of the woods, frequently about noon, and towards the evening. There is, however, nothing of real distress in all this; it is the voice of LOVE, for which the whole family of doves are celebrated, and none more so than the species before us."

The PASSENGER PIGEON, or Wild Pigeon, *Columba* (*Ectopistes*) *migratoria*, is found in all parts of North America, and in particular districts is, at times, wonderfully abundant. It is usually of a bluish-slate color, with white underneath, though there are considerable variations of color. The Passenger Pigeons have great acuteness of vision; they are also noted for their rapid flight. These Pigeons have been killed in New York with Carolina rice still in their crops. As the digestion of these birds is extremely rapid, they must have flown between three and four hundred miles in six hours, giving an average speed of a mile in a minute. Wilson and Audubon have both felicitously described the arrivals and departures of the almost innumerable multitudes of Wild Pigeons which they saw. Wilson estimated one multitude seen by him to contain above two hundred thousand millions of pigeons! Audubon judged that a flock seen by him contained one billion one hundred and fifteen millions!! The breeding places of these birds are sometimes of very great extent. One of these near Shelbyville, Kentucky, Wilson judged to have been several miles in extent, and upwards of forty miles in length. These birds usually raise two broods in a year. Their nests are composed of a few dry twigs crossing each other, and are supported by forks in the branches of trees. On the same tree, it is said, from fifty to an hundred nests may often be seen.

The BRONZE-WINGED PIGEON, or Ground Dove, *Phaps* (Gr. a pigeon) *chalcoptera*, (Gr. brazen-winged,) group *Peristerinæ*, is an extremely beautiful species found in Australia. The predominant colors are gray tinged with purple, and brown tinged with green; the wing coverts are bluish gray, but the outer webs of every feather have a large egg-shaped spot, exhibiting various shades of metallic brilliancy according to the direction of the light. The length of this bird is eighteen inches. Its cooing is so loud that when heard at a distance it has been compared to the lowing of a cow.

The CROWNED PIGEON, *Lophyrus*, (Gr. having a remarkable crest,) *cristatus*, is a native to the East Indian islands. The size of this bird, (28 inches long,) compares with that of a turkey, and its flesh is of excellent flavor. The greater part of the plumage is of a fine purple or bluish ash; other portions are of a dark reddish-brick color. It coos and shows the manners of pigeons, but in structure seems to approach the Curassows.

The WATTLED GROUND PIGEON, *Geophilus*, (Gr. lover of the ground,) *caruncülata*, (Lat. wattled,) is a native of South Africa, in size about as large as a turtle dove, but with the body stouter and more rounded. In its bill and plumage it conforms to the Pigeons, but in the naked red wattles of the forehead and chin, and in some other respects, it appears to approach the Gallinaceous Birds.

The GROUND DOVE, *Columba passerina*, (Lat. Sparrow-like,) is an American species, only six and three-fourths inches in length, ranging from Louisiana to Cape Hatteras.

SECOND FAMILY. The CURASSOWS.

Cracidæ, (Gr. *krax*, from *krazo*, to cry out *like* a crow.)

The hind toe in these birds is articulated on the same plane as the others, touching the ground on its length in walking, so that the foot is constructed after the model of the Perchers; hence, they are much more arboreal than the Poultry-birds, forming their nests among the branches of trees and feeding upon their buds and fruit. The curved form of the claws, their compressed sides, and their sharp points indicate that these birds are not habitually occupied in walking and scratching upon the ground; the toes, unlike those of all other gallinaceous birds, are destitute of any connecting membrane; the tarsi are without spurs, but in other respects the Curassows conform to the distinctive characters of the order.

These birds are found in Central and South America.

The COMMON CRESTED CURASSOWS, *Crax elector*, (Gr. *alĕktōr*, a cock,) are natives of Mexico, Guiana and Brazil. They are very common and furnish excellent food; are about the size of a turkey, and have the head adorned with crests of long, narrow, erectile feathers, curled at the tips. They usually perch upon trees, are found in numerous flocks and easily domesticated. These birds build their nests upon trees, laying but once a year; the eggs are from five to eight in number, and nearly as large as a turkey's. The plumage is of a deep black, with slight glosses of green above; the under parts are dull white.

The GUANS, *Penelope cristata*, do not differ much from the Curassows in their habits. They are known in Brazil by the name of Jacu, (pronounced *Yacou*,) derived, it is said, from their note. The length is thirty inches.

The HOAZINS, *Opisthocomus*, (Gr. *opisthen*, behind; *komē*, hairs or bristles,) *cristatus*, live in pairs or small companies of six or eight, in the flooded savannahs of South America. They seek for their food the leaves of a species of arum which is found in such places. Unlike other gallinaceous birds, their toes are without, or have only rudimentary membranes. In stature and gait they resemble the peacock. The generic name refers to the bristles which diverge from the base of the bill. The name *Hoatzin*, or Hoazin, is given to these birds from its imagined resemblance when pronounced, to their shrieking cry. They are nearly as large as the Guans.

THIRD FAMILY. MEGAPODES, OR GREATFOOTS.

Megapodiidæ, (Gr. *μέγας*, *mégas*, great; *πούς*, a foot.)

This family are scattered over Australia and the islands of the Indian Archipelago.

Their characters may be given thus: the beak is vaulted, somewhat compressed; the wings short and rounded; the tail short, varying in the n. mber of its feathers from twelve to eighteen; the feet of disproportionate size and strength, the tarsi being stout, elevated, and strongly scaled; the toes long, robust, and armed with strong, flat, rasorial claws.

The flesh of these birds is white, and much valued for its tenderness and flavor. The eggs are enormously large, as compared with those of other birds.

The BRUSH TURKEY, *Talegalla Lathami*, (of Latham,) is so called from being found principally in the thick brushwood of New South Wales. Mr. Gould has given an account of the curious nests of these birds. In making them, the bird never uses its bill, but always grasping a quantity of material in its foot, throws it backward to the common centre; and thus clears the surface of the ground for a considerable distance so completely, that scarcely a leaf or blade of grass is left. After heat is engendered in the mound, the eggs are planted at the distance of nine or twelve inches from each other, and buried nearly at arm's depth, perfectly upright, with the large end upwards. They are covered up as laid, and allowed to remain until hatched. It is said nearly a bushel of eggs is not unusually obtained, at one time, from a single heap; and as they are delicious eating, they are eagerly sought.

The MOUND-MAKING MEGAPODE, *Megapodius tumulus*, (Lat. a mound) confines itself to thickets near the sea-shore, and is called the Jungle-fowl. It is of a bright red brown color, about as large as a common fowl, and lays its eggs in mounds, not at intervals, like the Brush-Turkey, but at the bottom of the mound, usually five or six feet in depth. Sometimes the mounds are excessively large. One is spoken of as fifteen feet in height, having a circumference of sixty feet at its base! From their small brain, and not sitting upon their eggs, but leaving them to the warmth of the sun's rays, or the fermentation of vegetable matter, the Megapodes are supposed to be the lowest representatives of their class.

FOURTH FAMILY. PHEASANTS.

Phasianidæ, (Gr. φασιανός, *phasianos*, a pheasant, *i. e.*, a bird from the river Phasis, in Colchis.)

Sub-family *Pavoninæ*, (Lat. *pavo*, a peacock,) sometimes ranked as a family, and so presented on the Chart.

This extensive family includes birds of a large size and magnificent plumage; the flesh of all of them is in good esteem. They have an arched beak, and the nostril is covered with a naked and horny scale. The wings are characteristic of the order, in being incapable of rapid or long-sustained flight. The feet are large and powerful; the tarsi naked, covered in front with large plates, or scales, and have one or more curved and pointed spurs; the claws are slightly curved, and obtuse at the point; the hind toe is placed higher up on the tarsus than the three front ones, so that, in walking, its tip alone reaches the surface. The tail consists of eighteen feathers, which, in all, are developed well, and sometimes in an extraordinary manner; the tail coverts are also, at times, greatly lengthened. The males generally are of superior size and magnificence to the females, shining with rich, but not, usually, showy hues, reflecting the refulgence of precious stones or polished metal. Many, particularly the males, are ornamented with wattles, combs, or feathery crests. The most gorgeous species are found in the warmer regions of Eastern and Southern Asia.

Europeans date back their possession of the Pheasant twelve centuries before the Christian Era. From the most ancient time the Peacock has been a domesticated bird, as the references to it made by the earliest Greek poets, very clearly show.

The COMMON PEACOCK, (*Pavo cristatus*,) was regularly imported from the East in the fleets of Solomon; and its remarkable beauty was referred to at a period still more ancient, (Job xxxix, 13.)

The feathers of this bird do not constitute its tail; they begin to grow far up on the back, so that, when erected and spread, scarcely more than the head and neck of the bird appear in front of them. The true tail is situated beneath, being concealed by these, and consists of eighteen brown feathers, about six inches long. Immense flocks of these birds, identical with the domestic races, are found in the forests of India,—seeming to cover them with their beautiful plumage. The flesh of the Peacock, when not old, is juicy and savory, but is not eaten now so much as in former times, when it formed an important addition to great banquets, being served up dressed in its own brilliant plumage.

The TURKEY, (*Meleagris gallopavo*) was so called from an erroneous impression, that it came originally from the country of the same name. It appears to have been introduced into Europe about the year 1600. The generic name is the Lat. for Guinea-fowl; the specific, is Lat. from *gallus* and *pavo*, combined.

The habits of the Turkey, in a domestic state, are too well known to need description, and its utility on the score of food, most people are capable of appreciating. A few continue in a wild condition in some of our Western States; they are partly migratory in their habits, moving in the latter part of October, towards the Ohio and the Mississippi, seldom, however, using their wings, except when attacked, or in order to pass over a river. The stronger ones can cross a river of a mile in breadth, but the weaker frequently fall into the river, and then paddle to shore with some rapidity. C. L. Bonaparte, in his "American Ornithology," speaks of an ingenious method in which the Turkey escapes the onsets of large Owls, by suddenly "dropping his head, squatting, and spreading the tail over his back, in which case the Owl glances over him without doing any injury.

This fowl lays in the spring, usually, from fourteen to eighteen eggs, which are white, mixed with yellow or reddish freckles. Dr. Franklin expressed a wish that the Turkey, rather than the Bald Eagle, had been selected as our national emblem. In point of character and usefulness, it certainly much transcends the latter bird.

The COMMON PHEASANT, *Phasianus Colchicus*, is now spread over the greater part of the Old World. Fable says, it was introduced into Europe "by Jason and his companions, who brought it from Colchis in the good ship Argo." In size, this bird is about equal to the domestic Cock. Its plumage presents the finest tints of beautiful yellow and green, united with the richest ruby and purple, set off with spots of glossy black. The long wedge-shaped tail, partakes of the beautiful coloring of the body, and the whole bird has an air of great elegance. Several varieties have been produced by climate and domestication, such as the White, the Pied, and the Ringed Pheasant.

The GOLDEN PHEASANT, *P. pictus*, is among the rare species. It is a native of China, and remarkably elegant in its plumage. The tail of this bird is longer and more richly tinted than that of the European species; it is distinguished by a crest, which can be raised at pleasure. Cuvier supposes this Pheasant to be the Phœnix of Pliny. But the most splendid of the tribe, is

The ARGUS PHEASANT, *Argus giganteus*, (Lat. gigantic,) as large as a Turkey, found in Sumatra, and the South-Eastern parts of Asia. The "wings, the secondaries of which are three times as long as the primary quills, are painted and ocellated (having little eyes) in a manner which defies description." This bird derives its name from the shepherd Argus, fabled to have an hundred eyes.

THE GUINEA FOWLS, or PINTADOS, *Numida meleagris*, were originally brought from Africa, and in the swamps and pestilential regions of the Western portion of that Continent, they are found in immense flocks.

Dr. Livingston says, "the woods were literally alive with them,"—that his "guides roasted them on skewers in the off-hand fashion which is common among these people. They think it is waste of time to strip the bird of its feathers before roasting it, as the fire itself performs that operation." The flesh of these birds is considered a great delicacy, as it is tender and well flavored. Even in their wild state, they are not good flyers; indeed, they make more use of their legs than their wings. Their speed on the ground is surprising; but when chased for a while, they become fatigued, and sit still until they are picked up. Guinea-Hens are easily domesticated, and have been widely distributed. They are frequently seen in the poultry-yard where they are noted for their peculiar cries and unusual gait. During night, they always perch in high situations, or on trees. In Jamaica where these Hens do much mischief to some of the crops, they have resumed their wild habits, and are shot like other game.

DOMESTIC POULTRY BIRDS.

Gallinaceæ, (Lat. *Gallus*, a cock; *Gallina*, a hen.)

The Domestic Fowls are too well known to require a lengthened description. Some of the varieties are the following, viz:

The GAME FOWL.—some years ago much sought after for use in the cruel sport of cock-fighting, which, in some places, is still continued.

The COCHIN CHINA FOWL, (a variety of the Java Fowl,)—enormously large, and by some regarded as the origin of the Barn-door Fowl; though others suppose the Jungle Fowl, of India, to be the parent stock. The principal advantage connected with raising the Cochin China breed, seems to be that the chickens, from their large size, are ready for market earlier than those of the ordinary fowl.

The BANTAMS,—small, but very courageous, sometimes even

venturing to attack a Turkey. Some of them are feathered down to the toes. The long neck-feathers of this and the preceding fowl, are used by anglers for making artificial flies.

The SHANGHAI FOWL,—introduced from Shanghai, China, in 1848, by Capt. Forbes. Their general plumage is of a gold color, variegated with dark brown and red; their movement appears proud and showy, but their legs are rather too long for beauty.

The POLISH FOWL,—a small but beautiful breed, having deep black plumage, with a white tuft on the crown of the head.

The DORKING FOWL,—a large and delicate variety. Its chief peculiarity is the double hind toe,—it thus having five instead of four toes.

The MALAY FOWL,—a long-legged and timorous bird, which, for the first six months, has scarcely a feather to cover its nakedness; its flesh, except in pure breeds, is coarse and stringy.

The JUNGLE FOWLS,—are large and spirited, with plumage of purple and deep golden green, which, in the sun, has a splendid appearance. The Chinese use these birds as Game Fowls.

FIFTH FAMILY. GROUSE.

Tetraonidæ, (Lat. *tetrao*, a heath-cock, or moor fowl.)

This family are distinguished from the Pheasants by the absence of naked crests and wattles, that are so common among those birds, as well as of the brilliant colors and the metallic lustre of their plumage. The only naked skin about the Grouse is the space which surrounds the eye; this, when present, is of a scarlet color. The tail is very short, and, in some species, rudimentary. In the larger Grouse, of Europe and America, and the Pintails, of Africa, this organ is, however, largely developed. The birds of this family differ from the Pheasants in having the hind toe small and weak, and in some genera, reduced to a mere rudiment. Some are found in the warmer regions, but the larger and most typical part of them, in the cold regions of the Northern Hemisphere, and on Alpine summits. As a protection against the cold, these have the feet more or less clothed with feathers.

The Grouse, unlike other birds of the order, for the most part, pair at the breeding season; though several species congregate in large flocks. They all lay their eggs upon the ground, usually in large numbers; in their general habits, they are terrestrial, running with much ease and swiftness. In cold climates, they sometimes perch on the low stunted trees. They feed on the unexpanded leaf-buds of trees, upon grains, grass, seed, and pulse. The flesh of all of these birds is much esteemed for its tenderness and flavor.

The largest birds of this family, are included in the genus *Tetrao.*

The CAPERCAILLIE, or COCK OF THE WOOD, *T. urogallus*, (Gr. *oura*, a tail; *gallus*, a cock,) is common in most parts of Northern Europe. The male is a large bird, almost equaling a Turkey in size, but the female is considerably smaller. In the early spring, the male bird is noted for his "play," in which his movements are "much like those of an angry Turkey-cock, and he utters a call somewhat resembling *peller, peller, peller;* these sounds he repeats at some little intervals, but as he proceeds, they become increasingly rapid, until after a minute or so, he makes a sort of gulp in his throat, and ends with sucking in, as it were, his breath." The nest is made on the ground, and contains from six to twelve eggs. Mr. Yarrell gives the length of a specimen of this bird, as three feet four inches. The general plumage is such a blending of black and white, as to give it a gray hue. This bird feeds upon berries and young shoots.

The COMMON PARTRIDGE, or RUFFED GROUSE, *T. umbellus*, (Lat. a small tuft,) is found only on this Continent,—ranging as far South as Mexico. Its form is bulky, ar d it has a slight crest. (Plate X. fig. 6a.) The plumage is mottled with reddish and dusky brown. The length is eighteen inches. The Partridge is remarkable for producing a drumming noise, chiefly in the spring, but occasionally at other seasons.

The COCK OF THE PLAINS, or PHEASANT-TAILED GROUSE, *T. urophasianus*, (Gr. *oura*, a tail; *phasianos*, a pheasant,) is found in the Rocky Mountains, and in size not much less than a Turkey, being thirty inches in length. On each side of the lower part of the neck in front, this bird has a large bare space, capable of being inflated into a hemispherical sac.

The PINNATED GROUSE, also known as the Prairie-Hen, or Heath-Hen, *T. cupido*, is another species, which in its voice, manners and peculiarity of plumage, is perhaps the most singular, and in its flesh, the most excellent of the tribe found in the United States. It is nineteen inches in length.

The QUAILS, *Ortyx*, (Gr. *ortux*, a quail,) in the Southern and Western parts of the Union, called Partridges, are also included in the present family.

The COMMON AMERICAN QUAIL, *O.* or *Perdix*, (Lat. partridge,) *Virginiana*, is found abundantly from Texas to Massachusetts.

In Texas it keeps principally on the prairies. This bird is nine or ten inches long; the bill is short and thick, with the upper mandible curved from the base; the color a reddish brown, varied with black and white. It makes its nest on the ground, and lays from eight to eighteen pure white eggs. As it is timorous and restless in its habits, it is hard to domesticate. Its whistle, in the spring, is thought to resemble the words, *Buckwheat, Bob White.* The Quail is caught in large numbers by traps, horse-hair nooses, and nets.

The California Quail, or Partridge, *O. California*, resembles the Common Quail, but has a crest, which it can erect or depress at pleasure. (Plate X. fig. 6b.)

The Ptarmigan, *Lagopus*, (Gr. Hare-footed,) *albus*, inhabits the Northern parts of Europe and America. It has the legs and feet thickly covered with hair-like feathers reaching as far as the claws. (Plate IX. fig. 22.) Like the fur of the Ermine, the plumage changes in winter from an almost tortoise-shell color to a pure white. The length is about fifteen inches. In Norway, the peasants take them in snares. The captured birds "are kept in a frozen state until the dealers come, and one of these will sometimes sell 50,000 Ptarmigans in a season."

Sixth Family. Sheath-Bills.

Chionidæ, (Gr. χιών, *chiōn*, snow.)

The birds of this small family inhabit the high mountains or dry plains of South America, or the remotest parts of the Southern Ocean. They resemble the grouse, but have the nostrils surrounded by a sort of sheath; hence are called *Sheath-bills.* The typical genus is *Chionis*, a term suggested by the snowy white plumage of these birds. They are often found far out at sea, but chiefly inhabit the rocks washed by the tide, feeding on sea-weeds and shells, and have, therefore, been placed by some naturalists among the Wading Birds. The species *C. necrophaga*, (Gr. *nekros*, a dead body; *phago*, to eat,) found in New Holland, is about the size of a large partridge. It frequents the sea-shore, and feeds on dead animal matter thrown up by the tide. The Small Sheath-Bill, *C. minor*, is found on the dreary and iron-bound shores around Cape Horn; it is about as large as a Lapwing. This bird feeds on limpets and sea-weeds, not rejecting animal substances thrown upon the shore by the waves.

SEVENTH FAMILY. TINAMOUS.

Tinamidæ, (genus *Tinamus.*)

These birds include a very small number of species. They inhabit the immense grassy plains of South America, and are intermediate in form between the Partridges and Bustards, having the long neck and legs of the latter, and the nostrils covered with a naked scale, like the Pheasants. The beak varies in length; the wings are short, and the tail and hind toe rudimentary or entirely wanting. In South America they appear to take the place of the Partridges and Quails. Their appearance is such that they have been said to represent "a Bustard in miniature." Swainson considers their flesh, "both in whiteness and flavor, infinitely superior to that of the Partridge and the Pheasant." The size of the Tinamous varies from that of a Pheasant down to that of a Quail.

The GREAT TINAMOU, *Tinamus Braziliensis*, is eighteen inches long; it inhabits extensive forests. The general plumage is grayish brown, inclining to olive, with a mixture of white underneath and on the sides, and greenish on the neck. The female lays twelve or fifteen eggs, the size of those of a hen, and of a beautiful green color, in a nest formed of moss and dried leaves, and placed on the ground among the thick herbage near the root of some large tree.

The RUFESCENT TINAMOU, *T. rufescens*, is the most beautiful of the genus. It is fifteen inches and a half in length. It resides among thick herbage, and feeds on it night and morning, when it regularly utters its melancholy and feeble cry. The female deposits seven eggs of a fine bright violet color, in a hollow situated beneath tufts of grass.

The ANDALUSIAN TURNIX, or HEMIPODE, *Turnix tachydromus*, (Gr. swift runner,) is found in Spain and the northern parts of Africa. It is scarcely larger than a lark, of a yellowish brown color above, spotted and barred with chestnut, black, and white; the under parts yellowish white. It has three toes before, entirely divided; no hind toe; hence its name HEMIPODE, (half-footed.)

What is the fourth order of Birds? What useful group does it include? What characteristics are given to birds of this order? To what limits are the POULTRY BIRDS chiefly confined? State further particulars respecting the birds of this order. Name the families which it embraces. In what respects do Pigeons differ from Gallinaceous birds? Have they been treated as a separate order, and under what name? What chief peculiarity is mentioned? What does this apparatus constitute them? What is said of the beak and feet? State the variations in the feet of the different groups. Where do they nestle? What are their prevailing hues? What sp. is the origin of the Common, or Dové-cote Pigeon? Mention the fancy varieties. What fact illustrates the remarkable fecundity of the domestic P.? What is said of the Turtle Dove? What of the Carolina Turtle D.? Relate the particulars given respecting the Passenger Pigeon? What is said of the Bronze-winged P.? What of the Crowned P.? What other sp. are mentioned?

Give the general character of the CURRASSOWS, the 2nd FAMILY. Where are they found? Name the different sp. and repeat what is said of them,

---

What is said of the 3rd FAMILY? Where are they found? Give their characters. What is said of the flesh and eggs of these birds? Recite what is said of the BRUSH TURKEY. What of the Mound-M. Megapode?

What is the 4th FAMILY? What is said of the size, &c.? Give the general character. Where are the most gorgeous sp. found? When were the PHEASANT TRIBE introduced into Europe? What is said of the COMMON PEACOCK? What of the TURKEY? What of the COMMON PHEASANT? Of the GOLDEN P.? Of the ARGUS P.? Of the GUINEA FOWL? Name the varieties of poultry birds. Give particulars respecting them.

What is the 5th FAMILY? How is it distinguished from the Pheasants? Where are the GROUSE found? What are their habits? What genus includes the largest? What is said of the Capercallie? What of the COMMON PARTRIDGE, or RUFFLED GROUSE? What of the Cock of the Plains? What of the PINNATED Grouse, or PRAIRIE Hen? By what name are Quails known in the S. and W. States? What is said of the American Quail? What of the California Q. or P.? What of the Ptarmigan?

What is the 6th FAMILY? Mention their habitat? What suggested the name of the typical genus? On what do these birds feed? What sp. are mentioned?

What is the 7th FAMILY? Is it numerous? Where are these birds found? To what birds are they intermediate? What is said of the beak, wings, &c.? What do they represent in S. A.? What does Swainson remark of their flesh? How does their size vary? What is said of the Great Tinamou? Of the RUFESCENT T.? Of the Andalusian TURNIX, or HEMIPODE?

SECTION IX.

Fifth Order. RUNNERS.

Cursores, (Lat. *cursor*, a runner, from *curso*, to run hither and thither.)

This order contains a small number of species arranged in one family, *Struthionidæ.* These species differ from each other considerably, yet they all agree in having wings which are remarkably short, while the hind limbs are increased in size and strength of muscle, proportioned to the decrease of those in front. The pectoral muscles are small and slender, and the breast bone exhibits a uniform convex surface, like that of a shield, but not keeled, as in the Swallows and Humming Birds.

The Runners are all birds of large size, most of them equaling, if not surpassing the average height and bulk of the Mammalia, to which class they approach nearer than any of the other feathered tribes. They are found in the immense plains of the Southern Hemisphere. Most of them are remarkable for the peculiarity of their incubation. Many females occupy one nest in which a great number of eggs are laid, to be incubated chiefly by the male; when disturbed, he feigns lameness, as is common with birds that nestle on the ground. The hind toe is wanting in all these, except that singular one, the *Apteryx*, or Kivi-Kivi, of New Zealand, where it is found in the form of a small rudiment.

The Ostrich Family.

Struthionidæ, (Gr. στρουθος, *strouthos*, an ostrich.) Genus *Struthio.*

This family includes the true Ostrich, the American Ostrich, (*Rhea*,) the Cassowary, the Australian Cassowary, or Emu, and the Kivi-Kivi, or Apteryx, (for which see chart.)

These birds are very large, and the neck and legs of great length. Their plumage is loose and flexible; the thighs short and muscular. The toes vary, the Ostrich having but two, (and only one of these furnished with a nail somewhat resembling a hoof;) the Cassowary and Emu, three; the Apteryx, (including the rudimentary hind toe,) has four. (See Chart.)

The Ostrich, *Struthio camelus*, or Camel Bird, is so called from its resemblance to the Camel, which is very striking. Both "are furnished with callous protuberances on the chest and abdomen, on which they support themselves when at rest; they both lie down in the same manner, and the feet and (in some respects) the stomachs of both are similarly constructed; both are capable of subsisting on a scanty vegetation, of enduring thirst, and of traversing arid sands and desert regions." Anderson says, "their cry resembles that of a lion, so as even to deceive the natives; they are so swift and strong they will outstrip an English horse in speed, with two men mounted on their back, and it takes a long time to exhaust them. Their food, in the wild state, consists of seeds, tops and buds of various shrubs and plants; in confinement, they swallow, with avidity, stones, pieces of wood, iron spoons, knives, leather, hair, cordage, glass, minerals, and all sorts of indigestible matter, so that this bird has been called the Iron-eating Ostrich." Although capable of enduring thirst for a long time, yet "they flock daily, about noon, to the pools, where they swallow the water by a succession of gulps. This is one of the most favorable times to shoot them. The Ostrich, like the Capercaillie of Europe, has a plurality of wives, from two to six, each laying from four to six eggs in the same nest, which is a simple cavity scooped out in the sand; both male and female assist in hatching them.* The bird sits astride over them with its legs pointed forward. Some eggs are always placed outside the nest to serve as food for the young; when hatched, the chicks are about the size of pullets, and of a pepper and salt color, covered with neither down nor feathers, but a kind of prickly external. They are scarcely to be distinguished from the gravel or sand of the plains, or the stunted vegetation among which they dwell. The flesh of the young is not unpalatable, but that of the old bird is anything but agreeable, tasting much like the meat of the Zebra." Under the Mosaic law the Ostrich was an unclean animal, and the Jews were forbidden to eat it. The Arabs of the present day still adhere to this prohibition. Some of the less fastidious tribes of Southern Africa partake of it with a relish, more especially when fat. The brains of hundreds of these animals often made a dish at the luxurious suppers of the ancient Romans. They were considered great delicacies, and the Emperor Heliogabalus, it is said, was served with six hundred of them at a single feast. The eggs of the Ostrich are

* There is no inconsistency in this statement with the passage Job, xxxix, 14, which refers to the Ostrich as found in the torrid zone, where the intense heat renders incubation unnecessary, and the bird hence, "leaves her eggs in the earth, and warmeth them in the dust," showing little of maternal care or solicitude. The remarks here given from Anderson apply to the bird as seen in the cooler regions of Southern Africa.

much prized by travelers as well as by natives. They weigh about three pounds, and contain as much as two dozen of the eggs of our common barn-door fowls. One might be considered a sufficient meal for any man, but the Damaras sometimes eat two at a meal. "The shells are valued as ornaments, as well as drinking vessels, or to hold liquids, for which purpose they are covered with a sort of net-work, and slung across the saddle; grass, wood, etc., serving as substitutes for corks. The Copts suspend them in their churches, passing the cords of their lamps through the shells to prevent the rats from coming down to drink the oil; they look upon the shells as emblems of watchfulness. Dissolved in vinegar, or reduced to powder, they are used medicinally."

Stones as large as a bean or pea, are said to be sometimes found in the eggs. Barrows speaks of nine found in one egg and twelve in another, of a pale yellow color, about as large as a marrowfat pea, and exceedingly hard. A full grown Ostrich is seven or eight, sometimes nine or even eleven feet high, and weighs two or three hundred pounds, some say thirty stone, (420 lbs.) This bird is supposed to live between twenty and thirty years.

The general color of the female is a grayish or ashy brown, slightly fringed with white. The lower part of the neck and body of a mature male is of a deep glossy black, mixed with whitish feathers. In both sexes, the large plumes of the wings and tail are perfectly white; the thinner the quill, the longer and more wavy the plume, the more highly it is prized. Seventy to ninety feathers go to the pound; but though half this number may be obtained from a single bird, only a small portion are of any value. The best plumes are obtained soon after the moulting season. The price varies, as the market is fluctuating at the Cape of Good Hope. From five to fifty dollars are paid for a pound of the finest feathers. Those obtained from living birds are less liable to be attacked and injured by insects or worms than such as are taken from dead ones. The Damaras and Bechuanas manufacture handsome parasols from the black feathers, which serve as a sign of mourning, and to protect the complexion! These Ostrich parasols are used in hunting wild animals, as a Spanish bull-fighter uses a red cloth; just as a wounded beast charges a man, "he thrusts the support of the nodding plumes into the ground, and slips off, while the infuriated animal vents his wrath upon the feathers." The skin is also held in great request for manufacturing defensive armor. Ostriches usually dwell far from the haunts of men, but occasionally approach the settlements, trampling down grain and eating it. Domesticated, they are quiet, dull and heavy looking; in their native haunts they are restless, wary and difficult of approach. The senses of touch, taste, smell, and hearing are in these, as also in the other birds of the family, strongly developed. The eye is well formed; the sight is piercing, so that the Ostrich has a wide range of vision, and can discover danger at a considerable distance.

THE AMERICAN OSTRICH, *Rhea Americana*, (the Nhandu Guacu of the Brazilians,) prominently differs from the Ostrich of the Old World, in having three toes, all furnished with claws, and in its smaller size, being only about half as large as the African bird; it is also thinly covered with feathers. It has the same propensity for swallowing iron, stones, &c., as the Ostrich of the East. Haunting the banks of rivers, it runs so swiftly and cunningly as not only to evade the pursuit of dogs, but the weapons of the natives. These birds, like other ostriches, lay their eggs in the sand. "The males," it is said, "sedulously perform the office of incubation." The natives pursue them on horseback, and kill them by throwing the "bolas," or leathern thong, loaded at the end with a heavy stone or leaden ball. The Rhea frequently swims across rivers several hundred feet in width, thus exceeding the powers of the Ostrich and Cassowary. It feeds upon flesh and fruits, and upon the small fishes which are washed upon the sand; its flesh is said to equal that of geese and swans, and it is easily tamed.

A second (smaller) species, *R. Darwinii*, has been discovered in Patagonia, but it is rare.

The CASSOWARY, *Casuarius Casoar*, (or Emeu,) is a native of the Eastern part of Asia. Its wings are shorter than those of the Ostrich, and quite useless in aiding progression; the head is surmounted with a bony prominence, covered with a horny substance; the skin of the head and upper part of the neck is naked, tinged with cerulean blue and flame color, and has wattles like those of a turkey; the feathers are composed of two long, thread-like ones, proceeding from the same root and having the appearance of hair; the wing feathers are round, black and strong, and resemble the quills of a porcupine. At the end of the last joint of the wing is a sort of spur. This bird, next in size to the Ostrich, when erect and five feet in height, resembles the latter bird in its general form and aspect, (Plate X. fig. 7,) but differs from it in its digestive organs. The Cassowary lays a small number of green eggs, which it leaves to be hatched by the heat of the climate. Its food "consists of vegetable substances, and it will frequently swallow a large apple entire, trusting to the pebbles, &c., in its stomach to bruise it." The name Emu, formerly given to this bird, is now restricted to the following.

The *Emu, Dromaius*, (Gr. *Dromaios*, running swiftly,) is a native of New Holland, and in size and other respects closely resembles the Cassowary; but its plumage is thicker as its feathers are more barbed; the wings are small and hardly to be distinguished; but as a runner, it outstrips the swiftest greyhound. The dogs are shy of this bird on account of its powerful kicks, so powerful that by means of them it can break a man's leg.

The Kivi-Kivi, *Apteryx*, (Gr. *a*, priv.; *pterux*, wing,) of New Zealand, is a remarkably odd bird, appearing to hold among the feathered tribes of Polynesia, a position parallel to the Ornithorhyncus, or New Holland Mole, among the quadrupeds. Its bones are not hollow like those of other birds, and it has no abdominal air cells. It has no wings and only the most simple rudiments, ending in a sharp hook, which seems to be an instrument of defence; it is also tailless. Upon its very long and slender beak it leans forward as an old man would upon a stick. It is a nocturnal bird, pursuing its prey on the ground by the smell rather than by the sight. The olfactory openings are near the point of the beak; and thus it scents the worms on which it feeds, far below the surface of the ground. In the Zoological gardens, London, (Eng.,) is the only one ever seen out of New Zealand. The native name, Kivi-Kivi, is given to it on account of its peculiar cry. The apteryx is becoming quite rare in its native clime, and it is thought will, in a few years, become extinct.

Dinornis. This word represents a genus of struthious or Ostrich like birds formerly existing in New Zealand, and known there by the name of Movie, or Moa; but now, however, extinct, having been exterminated by human agency within a recent period; or if any of the species whose bones are found in a fossil state are still living, they are probably of the smaller forms and related to the Apteryx, "the only living diminutive representative of the stupendous Ostrich-like birds which once trod the soil of New Zealand."

Mr. W. Mantel, son of Dr. Mantel, of Eng., while on a visit to New Zealand, collected between seven hundred and eight hundred bones belonging to birds of various sizes, which were submitted to the examination of Prof. Owen. The Professor referred these to the genera *Dinornis*, *Palapteryx*, *Notornis*, and *Aptornis*. A part of the bones were found on the banks of the river Waingougou, on the western shore of North Island. With these were mixed fragments of egg shells. The eggs to which the fragments belonged were supposed to be about the size of a tea cup. In connection with this fact, interest attaches to a discovery recently made in Madagascar. "In a report to the French Academy of Science, M. St. Hillaire describes three fossil eggs from Madagascar, and small bones belonging to the same bird. The Captain of a merchant vessel trading to Madagascar, one day observed a native using, for a domestic purpose, a vase which much resembled an egg, and upon an examination proved to be one. The native stated that many such were to be found in the interior of the island, and eventually procured the eggs and bones exhibited by M. St. Hillaire. The largest of these eggs is equal in bulk to 135 hen's eggs, and will hold two gallons of water. M. St. Hillaire proposes the name of *Epiornis*, for the monster biped of which these marvelous eggs and bones are the first evidence brought under the notice of naturalists." Casts of these eggs have been made and are to be seen in various museums.

Gnathodon, (Gr. *gnathon*, a jaw; *odous*, a tooth,) is a genus of birds in the South Sea Islands, described by Sir William Jardine, from a specimen which was presented to him.

The upper mandible of the beak is strongly hooked, as in the Dodo; the under one is deeply notched; hence the name. The only known species, *G. strigirostris*, (owl-beaked,) is rather larger than a partridge, having the upper parts of a deep chestnut red, and the under of a glossy green black. Mr. Gould supposes it to feed on fruit or grass.

Didunculus, is a name given to a genus of birds found by Com. Wilkes, in the South Sea Islands, and thought to be the same as the preceding.

The Dodo, *Didus*, about whose proper place much doubt has existed, should perhaps have a position in the present family. To this bird, as now extinct, reference has already been made, (see section on Birds,) but fossil remains of it have been discovered, and there is abundant historical and other evidence of its former existence. Clusius, in a work published in 1605, gives a figure of a Dodo copied from a rough sketch taken by a Dutch navigator, who had seen the bird while on a voyage to the Moluccas in 1598. Bontius, (1658,) translated by Willoughby, describes it as "for bigness of mean size between an ostrich and a turkey, from which it partly differs in shape and partly agrees with them, especially the African ostriches, if you consider the rump, quill and feathers, so that it was like a pigmy among them, if you regard the shortness of its legs. It has a great, ill-favored head, with a kind of membrane resembling a hood, ... great black eyes, an extraordinary long, strong, bluish-white bill."

In the British Museum are the head and foot of one of these birds; also a painting said to be a copy of a picture taken from a living bird brought from Mauritius or St. Maurice's island, East Indies.

The SOLITARY, *Le Solitaire*, has sometimes been confounded with the Dodo, or represented as a species of that bird. Leguat, who (1631) resided in the island Rodriquez, gives a somewhat fanciful description of this bird as existing in his time on that island. His account of it makes it resemble a turkey, though taller and almost without either wings or tail. Subsequently it appears to have become extinct. Bones were, in 1832, discovered in the island, believed to be those of the Solitary.

What is the 5th order of birds? Does it contain many sp.? In what family are they included? What is said of their differences and agreement? What of the muscles of the breast and the breast-bone? What of their size? To what class of animals do they approach? Where are they found? For what are most of them remarkable? Have they any hind toe? Name the birds included in the OSTRICH FAMILY. What is said of their plumage, &c.? Why is the O. called the CAMEL-BIRD? Point out the resemblances between them. Relate the particulars given by Anderson. Were the ancient Jews allowed to eat it? How is it with the Arabs? How with the tribes of Southern Africa? What use was made of the brains of these birds by the ancient Romans? What is said of their eggs? What of the egg-shells? Give the height and weight of a full grown Ostrich. What is the general color of its plumage? What further is said of them? What is said of the A. Ostrich? Where has the second sp. been discovered? What is said of the CASSOWARY? Is this bird properly called the Emeu or Emu? Repeat what is said of the EMU. Where is the APTERYX found? What is its native name? Why was it given? Describe this bird. Where is the only one ever seen out of N. Zealand? Is this a numerous sp.? What does DINORNIS represent? What collection of fossil bones is mentioned? To what genera did Prof. Owen refer them? What is said of the fossil eggs discovered in Madagascar? What of the GNATHODON? With what other genera is it supposed to be identical? Repeat what is said of the DODO and LE SOLITAIRE.

Mention and trace the birds of this order named upon the chart. Let each pupil give an account of one of these birds.

SECTION X.

AQUATIC BIRDS. (Second Division.) TWO ORDERS.

First Order. Grallatores, (Lat. *grallæ*, stilts.) Waders or Stilt Birds. *Grallæ*, (N. H. S. N. Y.)

The Waders or Shore-Birds appear to hold an intermediate rank between the Gallinaceous or Poultry Birds, and the Natatorial or Swimming groups, which are confined to the water. M. Vigors is of the opinion that they, of all birds, enjoy most equally the advantages of land and water. They are distinguished by the great length of the tarsi and legs, which raise up their bodies as upon stilts; (Plate IX. fig. 27;) thus elevated, they frequent the banks of rivers, lakes, marshes and the shores of estuaries. The tibia or lower portion of the leg (a) is bare, so that they can wade to a considerable depth without wetting their plumage, and thus seize fishes and other aquatic animals on which they feed. In this they are aided by the length of the beak and neck, (Plate IX. fig. 6.) Such as are more especially aquatic have webs to their toes. Their wings are long and powerful, their flight strong and well sustained, enabling them to migrate with the seasons, which most of them do; thus becoming widely distributed. In flying, they stretch out their long legs behind, (Plate X. fig. 1,) as a counterbalance to their long necks; and the tail being very short, its office as a rudder is transferred to the legs. These birds have the power of maintaining a motionless position upon one leg for a considerable time. The most aquatic of them place their nests among the reeds and herbage of marshy places, or as the Herons, (*Ardeidæ*,) they build in company on trees; those that frequent dry and stony places, often lay their eggs upon the bare ground. The eggs are usually colored and spotted, of a lengthened form, with one end much pointed. The young run about as soon as hatched, except in those species which live in pairs.

This order includes the following families, viz.: (1) *Charadriadæ*, the Plovers; (2) *Ardeidæ*, Herons; (3) *Rostridæ*, Spoonbills; (4) *Tantalidæ*, Ibises; (5) *Scolopacidæ*, Snipes; (6) *Otidæ*, Bustards; (7) *Rallidæ*, Rails.

First Family. Plovers. (French *pluvier*, from Lat. *pluvialis*, rainy.)

Charadriadæ, (Gr. χαραδριὸς, *charadrios*, a kind of bird, from χαράσσω, to excavate.)

The Plovers are distinguished by having long and slender feet, adapted for running; the toes are rather short; the hind one is either entirely wanting, or so short as not to reach the ground. The wings are large and these birds are swift and strong in flight, moving in circles somewhat after the manner of pigeons, and wheeling round at no great height, uttering piping cries. The head is thick, with large dark eyes placed far back; the beak short and often slightly notched; the nostrils are pierced in a long groove. Resorting to the sea-shore, with their beaks they penetrate the ground for worms, to obtain which they are said to stamp with their feet, causing the worms to rise. Those with feebler bills resort to meadows and newly ploughed land, where they can more readily obtain their food; such as have stronger bills also feed on grain, herbage, &c. The colors of the Plovers are chaste and beautiful, consisting of various shades of brown, mingled with yellow, white and black, and often disposed in bands. Many of them are active during the night. The Plovers are dispersed over the entire globe.

The Lapwing, or Pewit, *Vanellus cristatus*, is a beautiful species; in summer spread over Europe and particularly plentiful in Holland, but passing the winter in warm latitudes. The plumage of the upper parts is green, with brilliant reflections; of the under parts, pure white. In its winter dress, the male has the head feathers very long, loose, barbed and curved upwards, forming a sort of crest, which is glossy black. When flying, the black and white colors of this bird make it very conspicuous. Sometimes thousands may be seen at once, gleaming in the setting sun, or appearing like a dense, black, moving mass, between its light and the spectator.

The Lapwings are about as large as pigeons; their eggs are laid upon the bare ground, and esteemed a luxury for the table. The Lapwing takes its name from the device by which it lures away intruders from its nest, dropping its wings in flight, and appearing as if wounded, to entice them away, and thus often inducing them to follow to a considerable distance. This bird is sometimes kept in gardens, and is useful for the destruction of vermin. Another European species is

The Great Plover, or Stone Curlew (of Europe,) *Œdicne-*

mus, (Gr. *oidos,* a swelling; *knēmos,* knee or shank bone;) *crepitans,* (Lat. making a rattling noise,) is about sixteen inches in length, and esteemed a delicate bird for the table. The generic name is given to this bird on account of the dilated or *swollen* form of the upper part of the tarsus, and the size of the knee-joint in the young birds. Their shrill evening cry pierces the ear, and may be heard, in the night, for nearly a mile.

The Golden Plover, *Charadrius Virginianicus, C. marmoratus,* (marble-colored,) Aud., differs slightly in size from the European Golden Plover, *C. pluvialis,* and in having the long axillary feathers dull brown instead of pure white. It ranges from 23° to 75° N. L., breeding in the arctic regions. From the general greenish appearance of their plumage, these birds are called *Greenbacks.* They are highly prized by the epicures of the fur countries, and they figure largely in the bills of fare of the old English nobles. The Golden Plover is about ten inches in length, and found in every continent. The generic name denotes a bird found in cavities or hollows like those worn by a rapid stream or torrent.

The Oyster-Catcher or Sea Pie, *Hæmatopus,* (Gr. *haima,* blood; *ōps,* face,) *palliatus,* (mantled,) is named from the red appearance of the bill and feet, and from the black of the upper plumage, which, contrasting with the pure white under plumage, has the appearance of a *mantle.* It feeds on oysters and other sea bivalves. The bill in this genus is long and wedge-shaped, with the tip much compressed; the feet have three toes, all directed forward and bordered with a narrow membrane. The Oyster-catcher breeds from Texas to Labrador. Its flesh is tough and unsavory.

The Gray Plover, or Gray Lapwing, *Squatarola cinerea,* (ash-colored,) has a rather strong cylindrical bill, swollen half way from the tip; the feet are four-toed, (the hind toe quite small.) In the warmer parts of the United States, this bird often has two breeds in a season; it runs well; its whistle is like that of the Golden Plover, but not so shrill. If killed in good season it is fine for the table.

The Turnstone, *Strepsilas,* (Gr. *strepho,* to turn;) *interpres,* (Lat. interpreter.) is widely distributed in both continents. It has four toes; the bill is compressed at the base, swollen in the middle, and blunt at the tip. The name Turnstone is given to it from its habit of *turning over stones* to obtain the small crustaceans and molluscous animals which constitute its food. It is known to sportsmen under the names of Brant-bird and Beach-bird. From its fondness for the eggs of the Horsefoot, (*Limulus polyphemus*,) it is also called the Horsefoot Snipe. Length from nine to ten inches.

The DOTTEREL, *C. morinellus*, has had credit for possessing great powers of mimicry. He has also been charged with so great stupidity that

> "Acting every thing, he doth never mark the net,
> Till he be within the snare which men for him have set."

For this, however, there seems not sufficient reason. When first seen, it shows but little fear of man; but this might be ascribed to its freedom from persecution in its native wilds; after a short experience of human annoyance, it becomes more cautious. Its "mimicking the action of the fowler by stretching out its leg, wing, or head," may be little more than the actions of other birds when aroused from their repose. The Dotterel feeds by night on insects, slugs, and worms; in common with others of the Plovers, it rests and sleeps during the day, and on this account, may allow of a close approach, as is true of the Golden Plover. These things considered, it can hardly be deemed proper to call it a *stupid* bird. The upper parts of its winter plumage are of a blackish-ash color with a tinge of green; a portion of the breast and under parts white; the face is white, *dotted* with blue. In their winter migrations, these birds visit Italy and Spain; they are particularly abundant in the Eastern parts of Europe and Northern Asia, where the larger part of them breed.

The SWIFT-FOOT or COURSER, *Cursorius*, (from Lat. *curro*, to run,) is found in the hot regions of Asia and Africa. One species, the BLACK-BELLIED COURIER, *C. Temminckii*, inhabits Abyssinia. The other species, *C. Isabellinus*, or CREAM-COLORED COURSER, is a native of Africa, but has occasionally been seen in Europe.

Glareola is a genus of Plovers confined to the Old World, and including three species, one of which, the PRATINCOLE, *G. Pratincola*, (meadow inhabitant,) is spread through the warm and temperate regions of Asia, Africa and Europe. It has very long wings and a greatly forked tail, and is remarkable for its rapidity and power of flight. Mr. Gould speaks of it as "an elegant and graceful bird." Its length is nine inches.

Other species are the KILLDEER PLOVER, *Charadrius vociferus*, which has its name from an imagined resemblance of its two notes to the word *Killdeer*, and which breeds from Texas to Massachusetts.

The WHISTLING PLOVER, *S. Helvetica*, called the Bull and

Beetle-Head Plover, common to Europe and America, and resembling, in its autumnal dress, the Lapwing of Europe.

SECOND FAMILY. HERONS.

Ardeidæ, (Lat. *ardea*, a heron ;) *Gruidæ*, (Lat. *grus*, a crane.) *N. H. S. N. Y.*

These birds are decidedly carnivorous in their appetite, feeding on fishes, aquatic reptiles, small mammalia, worms, and insects. The CRANES, in their terrestrial habits and in their food, approach some of the Gallinaceous or Cursorial birds, joining with an animal diet, grains, seeds, and herbage. The legs and feet of the Cranes are long and slender, as is also the neck, which is very flexible; the toes are four in number, the hind toe usually long and resting on the ground ; the beak is long, straight, sharp pointed, firm in texture and very powerful. The wings in this family are, in general, well developed, and some of the birds are capable of high and powerful flight.

The HERONS are the type of the group. They rank as the most beautiful of all the Waders, not so much from the shades of their plumage, though these are chaste and agreeable, as on account of their tapering and graceful forms, the curves of their slender necks, their elegant hanging crests, and the long plumes that adorn various parts of their bodies. They may be seen watching on the margin of the water or within the shallows; on the appearance of their fishy prey, it is transfixed by a sudden stroke of the pointed beak and swallowed entire. In their decidedly carnivorous habits, they differ from the Cranes. They are distinguished by their larger and more pointed bill, and the greater length of their legs ; their stomach is a large undivided sac, only in a small degree muscular. All the Herons have comb-like divisions on the inner edge of the middle claw, (Plate IX. fig. 28,) probably designed to free the plumage from insect-parasites. They are generally solitary in their habits ; but they build in companies, usually in trees not far from the banks of rivers. We give some specimens in each division of the family.

CRANES.

The AMERICAN CRANE, *Grus Americana*, when mature, has white plumage with the quills and their shafts black; when young, bluish gray, with the quills and their shafts brownish white. This is called the *Whooping Crane* ; it migrates as far North as the 68° Lat. Its length is fifty-four inches.

The COMMON CRANE, *G. cinereus*, (ashy,) of the Eastern Continent, is three feet eight or ten inches in length; in its general plumage ashy-gray; migratory and gregarious in its habits; in its contour and gait somewhat like the Ostrich; in its strong and muscular stomach, it differs from the Herons. The flesh is well tasted, and was formerly highly prized.

The DEMOISELLE, *Anthropoides*, (Gr. of human-like form;) *virgo*, (Lat. a virgin,) or *Ardea Virgo*, of Linn, is an African bird, but occasionally seen in Europe; its general plumage is slaty-gray; length about three feet. One of these birds hatched in the menagerie at Versailles, (France,) and lived there twenty-four years. Great numbers are seen in Egypt during the inundations of the Nile. This bird exhibits much delicacy and elegance of attitude, and a graceful playfulness in all its movements. Its food consists, principally of grain and seeds, though it occasionally eats small fishes, mollusks, and insects.

The STANLEY CRANE, *A. Stanleyanus*, (*A. paradisæus*,) is a beautiful East India species, named after Lord Stanley, late President of the Zool. Soc., Lond.; it is three feet and a half in length, and in manners and gestures, like the Demoiselle; its general plumage is bluish gray.

The CROWNED CRANE, *A. pavonicus*, (Lat. of a peacock,) is supposed to be the Balearic Crane, (*Balearica*,) of the ancients. Its plumage is of a bluish slate color; when full grown it is about four feet in length. Under the throat is a wattle like that of a turkey. This stately bird is found in Northern and Western Africa. (Plate X. fig. 8a.)

HERONS PROPER.

Of these there are quite a number of species. The use of these birds in hawking, has already been noticed. The destruction of their eggs was in Europe formerly visited by a heavy penalty; they seem to have ranked as high at the tables of the great as they did for their exploits in the field; now, however, their flesh is in low estimation.

Audubon includes the EGRETS and BITTERNS with the Herons, under one genus, *Ardea*, and enumerates twelve species found in America.

The COMMON HERON, *A. cinerea*, is remarkably light in proportion to its bulk, weighing scarcely three pounds and a half, though its length is upwards of three feet, and its expanse of wings above five. This Heron is found in Europe, Asia, Africa, and America.

The GREAT BLUE HERON, *A. Herodias*, is a species allied to the Common Heron of Europe, and met with in every part of the United States. It is over four feet in length, and six feet in the expanse of its wings. The bill is seven or eight inches long and very sharp pointed. This bird is partly nocturnal in its habits; a portion of its breast is covered with a down which is said to be phosphorescent; it is to be found from Texas to South Carolina; feeds on crabs, eels, and various other fish.

The GREENISH BLUE HERON, *A. virescens*, (Lat. verging to green,) is more generally known than most other American species, being widely spread in spring, summer, and autumn. It has the popular names of *Chalk-line*, *Polk*, *Fly-up-the-Creek*, &c. The length is about seventeen inches.

The BLACK CROWNED NIGHT HERON, *A. discors*, (Lat. discordant,) is from twenty-six to twenty-eight inches in length. In many respects it resembles the Common Heron in its habits, breeding like that bird, in company with others, on the topmost branches of trees. During the day it roosts in the recesses of woods in the vicinity of swamps and rivers, which it visits at night in quest of prey. It feeds on fish, aquatic reptiles, sea-lettuce, (*ulva latissima*,) grasshoppers, and other large insects. The popular name of Quawk, or Qua-Bird, is given to it on account of its deep guttural cry. It closely resembles the *A. nycticorax*, (Gr. raven of the night,) the Night-Raven of Europe.

The GREAT AMERICAN WHITE EGRET, *A. leuce*, (white;) *A. egretta*, (Wilson, Aud. and Bonaparte,) is forty inches in length; of a snowy white plumage, sometimes tinged with yellow. Its food consists of frogs, salamanders, mice, moles, &c. This bird is found from the Equator to 43° N. L. It is closely allied to *A. alba*, or *Herodias alba*, the White Heron of Europe.

BITTERNS.

These are represented in the genus *Botaurus*, (Lat *boo*, to cry out; *taurus*, a bull.) They are widely diffused and solitary birds, haunting woody swamps and marshes; hid all day and feeding at night. As might be conjectured from their haunts, they feed mostly upon aquatic animals. They spread over both hemispheres, but are not found in Australia. Everywhere they are noted for their voracity. The names Mire-Drum and Bull of the Bog are sometimes given to these birds on account of the drumming or bellowing noise for which they are famous. The English name Bittern was formerly spelled *Bittour*, and like the generic term, is supposed to refer to its deep-toned, bull-like voice.

The COMMON BITTERN, *B. stellaris*, (Lat. starry,) was well known to the ancients. It is referred to by Aristotle under the name of *Asterias ;* in the palmy days of falconry it was much sought for. It is not daunted when wounded, and therefore it was the duty of the falconer to plunge the Bittern's bill into the ground to prevent injury to the hawk; both the falcon and the falconer were sometimes endangered by the sharp beak of their victim. In the time of Henry VIII., of England, its flesh was in high esteem; when the bird is well fed it resembles that of a hare, and is not rank or fishy like that of the associate birds. The long claw of this bird's hind toe is prized as a tooth-pick, and in the olden time it was thought to have the property of preserving the teeth. The Common Bittern is crested and about two feet and a half in length, being smaller than the Common Heron. The general color of the plumage is dull pale yellow, varied with spots and bars of black.

The AMERICAN BITTERN, *B. lentiginosus*, (Lat. freckled,) or *Ardea minor*, is not quite so large as the Common Bittern. It is familiarly known by the names *Poke*, *Indian-Hen*, *Indian-Pullet*, &c., and migrates over most parts of the U. S. The color is a rusty yellow, mottled and sprinkled with deep brown.

The SMALL BITTERN, *A. exilis*, (Lat. small or slender,) is sub-crested and only eleven inches in length; of a chestnut color above, but whitish beneath. It ranges from Mexico to 45° N. L.

The COMMON BOAT BILL, *Cancroma cochlearia*, (Lat. snail-shells or spoons,) approaches in form quite closely to the Heron, except in the bill, which is not unlike the bowls of two spoons placed one upon the other, with the rims in contact. It perches on trees by the side of rivers, where it lives on fish, and *not* on crabs, as the name *Cancroma* indicates, though Linnæus supposed it to feed on crabs and so named it. Latham says: "We are certain fish is its most common, if not only food." This bird is native to South America.

STORKS.

The Storks, *Ciconia*, are not so aquatic as the other birds of the family, but are among the largest. They build their nests on turrets, steeples, and chimneys,—each pair, after wintering in Africa, returning to the same place in the spring. Their bills are very long and straight, resembling a lengthened cone. They live in marshes, and feed principally on reptiles, frogs, and their spawn, as well as on fishes. Wherever found, the Storks are a privileged race, on account of the havoc which they make among noxious animals. They migrate in numerous flocks, and are easily tamed. All the species make a clattering noise with their bills.

The COMMON WHITE STORK, *C. alba*, (see Chart,) is about three feet in length; when well treated, it approaches, without fear, the habitations of men. In the towns of Continental Europe, domesticated Storks, taken when young, "may often be seen paddling about the markets, where they are kept as scavengers to clear the place of the entrails of fish and other offal, which they do to the satisfaction of their employers." In Holland, and especially in Germany, this bird is a welcome guest. Dr. Shaw witnessed the annual migration of flocks of these birds from Mount Carmel; each flock that he saw, "was half a mile in breadth, and occupied three hours in passing over."

Among the ancients, to kill a Stork, was regarded a crime, which, in some places, was punished with death; and, like the Ibis, this bird became an object of adoration. It is noted for its great affection for its young, but more particularly, for its care of its parents in old age.

The BLACK STORK, *C. nigra*, or *A. nigra*, is, like the White Stork, a migratory bird, spending the winter in Southern Europe, and passing on to high northern latitudes in summer.

The ADJUTANT, *Leptoptilus*, (Gr. *leptos*, thin; *ptilos*, down, or plumage,) *Argala*, is a remarkable bird, native to the warmer parts of India, and highly useful there in devouring noxious animals and carrion, which it does with avidity. It stands from five to seven feet in height, and measures from the tip of the bill to the claws, seven and a half, while the expanse of wings is not less than fourteen feet. The beak is extremely large, stout, and strong; under it hangs a downy pouch, or bag, like a dewlap, which is capable of being inflated. The upper part of this bird is of an ashy-gray color; the under part white. The voracity of the Adjutant is not more extraordinary than its capacity for swallowing; it makes but one mouthful of a rabbit, a fowl, or even a small leg of mutton, and when domesticated, its habit of purloining, makes it necessary to keep all kinds of provisions out of its reach. Dr. Latham says, "These birds, in their wild state, live in companies; and when seen at a distance, near the mouths of rivers, coming towards an observer, which they often do with their wings outspread,—may well be taken for canoes upon the surface of a smooth sea; when on the sand-banks, for men and women picking up shell-fish, or other things on the beach."

The AFRICAN GIGANTIC STORK, or CRANE, *C. marabou,* (see Chart,) resembles the Argala, but is not so large,—seldom exceeding five feet in length; its pouch is also much shorter. Another similar species is found in Java. These species furnish the beautiful plumes, esteemed superior to those of the Ostrich, known by the name of *marabou* feathers.

The JABIRU, *C. mycteria,* (Gr. *mukter,* a nostril, or proboscis,) is native to Senegal, in Africa. In the enormous size of the beak, as also in devouring carrion, the Jabiru resembles the Adjutant; the greatest part of the head and body of this bird is entirely bare; the plumage of the latter white; its size is somewhat larger than that of the Swan.

THIRD FAMILY. SPOONBILLS.

Rostridæ, (Lat. *rostrum,* a beak.)

The Spoonbills have many characters in common with the Herons, and are often included with them. The peculiar form of the bill has gained for them the name which they bear. It is very long, strong, and much flattened; the point is widened and rounded so as to present the form of a *spoon.* The face and head are partially, or entirely naked; the neck and feet are long; the nostrils basal and linear; the toes are four; in some, the hind toe is very small, and articulated high up; the feet are partially webbed; the wings are ample, moderate in length; the first quill nearly as long as the second, which is the longest of all. The Spoonbills live in companies, in wooded-marshes, generally not far from the mouths of rivers, and are rarely seen on the seashore. Their food consists of small fish, spawn, and minute fluviatile testaceous mollusks, reptiles, and aquatic insects. According to circumstances, they build their nests either in high trees, in bushes, or among rushes. The young bird does not take the confirmed plumage of the adult until the third year. The crest makes its appearance at the end of the second year. (Temminck.)

The COMMON WHITE SPOONBILL, *Platalea,* (Lat. Spoonbill,) *leucorodia,* (Gr. *leukorodon,* a white rose,) is generally distributed throughout Europe, but is most numerous in Holland. It has a very full, long crest of loose feathers on the back of the head. The length is two feet and a half. The old males have a fine white plumage, with a patch of reddish yellow on the breast. This bird winters in Africa, and is found as far South as the

Cape of Good Hope. Its flesh, when well fed and fat, is said nearly to resemble in flavor that of a goose.

The ROSEATE SPOONBILL, *P. ajaja*, is a beautiful, though singular bird, constantly found in Texas and South Florida; it is seen as far eastward as North Carolina. The beak and wings are of a delicate rose-color; the lower parts of a deeper tint; the head is yellowish green; the neck white. The length is about thirty-one inches. This Spoonbill is usually fond of being with the Herons, whose keen sight and vigilance apprise it of danger, and allow it to take flight in due time; it breeds in flocks on trees, low bushes, or cactuses. The feathers of the wings and tail are manufactured into fans by the Indians and Negroes of Florida; and at St. Augustine, form an article of trade. (Aud.) The flesh is oily, and undesirable for eating.

FOURTH FAMILY. IBISES.

Tantalidæ, (Gr. ταντaλος, *Tantalos*, a proper name.)

These are birds which, in their general habits and conformation, closely resemble the Storks; they chiefly inhabit warm countries, but except in very cold regions, they are to be found in all parts of the world. The bill is very long, robust at the base, and curved at the tip; the face is naked; the throat dilatable; the legs are long, and have four toes; the front toes are webbed at their base as far as the first joint; the hind toe is very long, and rests upon the ground. The Ibises frequent the borders of rivers and lakes, feeding on insects, worms, mollusks, and occasionally on vegetable matter. They perform powerful and elevated flights, extending their neck and legs, and uttering a hoarse croak. The family includes between twenty and thirty species, which are distributed over the globe. Four of these are found within the limits of the United States.

The GLOSSY IBIS, *Tantalus falcinellus*, (Lat. from *falx*, a sickle,) is about two feet in length. In the matured bird, the neck, breast, top of the back, and all the lower parts of the body, are of a bright red chestnut; the wing coverts, quills, tail-feathers, and the rest of the back, of a dusky green, glossed with bronze and purple; but the bird varies much in its plumage at different ages. This species nestles in Asia, and is found on the streams and lakes in flocks of thirty or forty. They migrate periodically to Egypt, and pass in considerable numbers into Europe. The Glossy Ibis is also found in the United States. Audubon saw flocks of it in Texas, but it is only a summer resident there, associating with the White Ibis. Vast numbers of it are seen in Mexico. Cuvier says, this, to all appearance, is the species which the ancients called Black Ibis.

The White or Sacred Ibis, *Ibis religiosa*, (see Chart,) is, perhaps, the most celebrated species. Arriving in Egypt about the time that the inundation of the Nile commences, its numbers increase or diminish with the increase or diminution of its waters. It migrates about the end of June, at which time, it is first noticed in Ethiopia. This species does not collect in large flocks, more than eight or ten seldom being seen together. They are about as large as a hen; the head and neck are bare; the body white; the primaries of the wings tipped with shining, ashy black, among which the white forms oblique notches; the secondaries are bright black, glossed with green or violet; the quill-feathers of the tail, white. This, and the preceding species, were venerated by the ancient Egyptians, who used to rear them in their temples, and after death, to embalm them. Their mummies are found, to this day, in the vast catacombs of ancient Memphis. Herodotus supposed that the Egyptians worshipped the Ibis for services which it rendered in freeing them from winged serpents. But this is contradicted by the bird's structure: its bill is not fitted either to divide or pierce serpents, but rather for dabbling in marshes and moist grounds. This species is found throughout the extent of Africa.

The White Ibis, of the American Continent, *Ibis alba*, is about two feet in length. It is a constant resident in South Florida, where it abounds, but also breeds along the coast to Texas,—sometimes inland as far as Natchez and Red river, and Eastward to New Jersey.

Audubon says, Sandy Island, near Cape Sable, in Florida, is remarkable for the number of these birds found there. He counted forty-seven nests in a single palm-tree. "The nests are fifteen inches in their largest diameter, formed of long twigs, intermixed with fibrous roots and green branches of the trees growing on the island; the interior of them is flat, being furnished with leaves of the cane and some other plants." The bird lays but three eggs, once a year. Its flight is described as "swift and long continued. Sometimes it rises to an immense height in the air, while it performs beautiful evolutions." It feeds on small crabs, slugs, and snails; showing great ingenuity in procuring cray-fish;—breaking up the upper part of the mud which the latter throws up in forming its hole, and dropping the fragments into the cavity. The cray-fish, burdened by the load of earth, makes its way to the entrance of the burrow, when the

Ibis immediately seizes it with its bill. This bird is known in Louisiana by the name of "Spanish Curlew."

The SCARLET IBIS, *Ibis rubra*, is a splendid bird, sometimes, though rarely, seen in Louisiana and the adjacent States, but in the hottest portions of this continent, is found in large flocks. This bird flies rapidly, but rarely, except at morning and evening, in search of food. The plumage is scarlet; beak naked; part of the cheeks, legs and feet, pale red. Its length is twenty-nine inches.

The WOOD IBIS, *Tantalus loculator*, is an extremely large species, being forty-four inches, with a bill that is nine inches in length. It is found in deep woody swamps, (where it breeds on trees;) also in fresh water lakes. These birds, after gorging themselves with their fishy or reptile-food, taken in shallow, muddy streams, walk to the nearest margin and arrange themselves in long rows, with all their breasts turned towards the sun, in the manner of Pelicans and Vultures, and thus remain for an hour or more. In flying, their long necks and legs are stretched out to their full extent, the pure white of their plumage contrasting beautifully with the jet black of the tips of the wings. Although generally fat, they are unfit for food, their flesh being tough and oily. They are resident from Texas to North Carolina. Other species of the Ibis are found in India, Madagascar, and the Cape of Good Hope.

The OPEN-BEAKED BIRDS, *Anastomus*, (Gr. *ana*, through; *stoma*, mouth,) are included by Swainson in this family. The generic name was given to these birds by Illiger; that of Lacépède, is *Hians*, (opening and gaping,) They have a straight head and heavy bill, marked with wrinkles running lengthwise. The upper mandible is very straight; the base thickened at the top, and as high as the crown; the tip notched; the margin dentated; the under mandible is greatly curved upwards, only touching the upper at the base, and at the tip, and thus leaving an *opening* through or between the two portions of the bill. Not much is known respecting the economy of these birds, and hence it is not easy to explain the purpose of the unique structure shown in the beak. Cuvier places the genus *Tantalus* between these birds and the Spoonbills, (*Platalea.*)

FIFTH FAMILY. SNIPES.

Scolopacidæ, (Gr. σκολώπαξ, *scolopax*, a snipe.)

The most prominent characteristic of this family, is the extreme length and slenderness of the beak. This is covered with a soft skin, which is extremely sensitive; and the organ is much used in probing the soft mud or earth for the capture of minute insects. The hind toe is pointed on the tarsus above the level of the fore toes, and so short as to be unable to reach the ground; in some of the family, it is wanting. The Snipes have moderately long feet and necks; the wings are long and pointed; and their flight swift and well sustained; the tail is short and even; the front toes are frequently united by a membrane, more or less large. The plumage is of subdued and varied shades; black, white, and red being intermingled and contrasted; sometimes, the prevalent hue is a grayish olive. The flesh of these birds is held in high esteem. They frequent marshes, the banks of lakes and rivers, or the sea-coast, on which they run with great swiftness. With considerable powers of flight, they have also the faculty, in part, both of swimming and diving. The females are usually larger than the males. They lay four eggs, with but little nest, on the ground, of inland moors or fens. The young, when they escape from the shell, are clothed with down, and immediately begin to run about. The Snipes are widely distributed, and more or less migratory in their habits.

Mr. G. R. Gray divides them into six sub-families, viz: (1) *Numeninæ*, of which *Numenius*, (Curlew,) is the typical genus; (2) *Totaninæ*, typ. gen. *Totanus*, Tatler, Sand Lark, or Willet; (3) *Recurvirostrinæ*, typ. gen. *Recurvirostra*, Avoset;) (4) *Tringinæ*, typ. gen. *Tringa*, Sand-piper; (5) *Scolopacinæ*, *Scolopax*, Common Snipe; (6) *Strepsilinæ*, *Strepsilas*, (included by others in the *Charadriadæ*, or Plovers, which see;) (7) *Phalaropodinæ*, *Phalaropus*, Phalaropes. Of these, he enumerates thirty-four British species. De Kay (1843) says, this family contains, at present, upwards of one hundred species, distributed over the globe; of these, about twenty-eight, (according to Audubon, thirty-two,) are in the United States.

CURLEWS.

The Long-billed Curlew, *Numenius longirostris*, is the largest of the genus found in North America,—known to sportsmen under the names of Big Curlew, and Sickle-bill. Its length is from twenty-five to twenty-seven inches; the color is blackish brown above, with spots of a red hue beneath. The great length of the bill, (Plate IX. fig. 6,) (seven to nine inches,) distinguishes it from every other species. The Curlew forms a small nest for its young on the ground. The day, the Curlew spends in the sea marshes, but resorts at night to the sandy beaches of the sea-shore. The number collected at their nightly retreat, it is said, sometimes amounts to several thousands. The food of these birds consists chiefly of the small crabs, called "fiddlers;" they are also fond of small salt-water shell fish; and thrust out the bill to its full length into the wet sand in search of sea-worms and insects. The Long-billed Curlew resides in Texas and on the Islands of South Carolina; but wanders North along the coast, and is occasionally seen in the interior. Dr. Kirtland observed it in Ohio; others have seen it in Kentucky and Missouri.

Smaller species are the JACK CURLEW, or Short-billed Curlew, *N. Hudsonicus*, closely allied to the Whimbrel, of Europe; it breeds in the Northern regions. This species, in addition to the food of the preceding species, makes use of berries. Its length is sixteen inches.

The SMALL ESQUIMAUX CURLEW, *N. borealis*, is known under the names of Little Curlew, and Dough-Bird, and much esteemed by epicures. Its length is fifteen inches. Nuttall says it ranges from Paraguay to the 70o N. L.

TATTLERS.

The TELL-TALE TATTLER, *Totanus vociferus*, receives its name from its frequent cries, uttered quite as much for its own sake, as tc give warning to others. It ranges widely over the United States, and is found at all seasons. In Maine and New Brunswick, it is called "Humility,"—a name that does not seem to agree with its "vociferous habits." These birds "congregate in great numbers in the inland marshes of Florida, and along its rivers, during the winter." Though found near both salt and fresh water, they seem to prefer the latter, selecting ponds of which the water is shallow, and the shores muddy, affording places where they can walk and wade with ease. In the Western country, it is called the "Great Yellow-Shank." The upper parts are generally black, glossed with green, each feather margined with white triangular spots; the throat, breast and abdomen, are pure white. Length about fourteen inches.

The SPOTTED SAND-LARK, *T. macularius*, is a familiar bird, of a glossy olive brown color, with blackish waves,—found throughout the Union in small families, along almost every stream, and the borders of ponds and lakes. In allusion to its notes, it has the common name of *Peet-weet;* from its repeated grotesque, jerking motions, it is called *Teeter*, and *Tiltup*. It feeds on insects and worms; breeds in New York and farther North. Occasionally it is found in Europe. The length is eight inches.

The GRAY PLOVER, *T. Bartramius*, is twelve inches in length, and is much esteemed for game, but shy, and not easily obtained by sportsmen. It is described as *Bartram's Tattler* and *Sand Piper;* among its common names, are GRASS-PLOVER and FIELD-PLOVER. This bird is not found on the coast,—its bill is very short, scarcely longer than the head. In July and August, it is seen in large flocks on its way South. It is fond of grass-hoppers.

AVOSETS.

The AVOSET, *Recurvirostra*, (Lat. up-turned bill,) *Americana*, (see Chart,) is, from its "perpetual clamor and flippancy of tongue, called by the inhabitants of Cape May, the Lawyer; the comparison, however, reaches no further; for our Lawyer is simple, timid, and perfectly inoffensive." Wilson. The back and under parts, are white; the wings brownish black, with a broad band of white. The bill is more than twice the length of the head, very slender, tapering to a point, and somewhat re-curved, or upturned, (Plate IX. fig. 4;) the legs are very long and slender. This bird builds its nest among the tallest grass. The eggs, like those of other Waders, are four in number, pear-shaped, of a dull olive color, with blotches of black. Like the Roseate Spoon-bill, it moves its head "to and fro sideways," while it is passing its bill through the soft mud in search of insects; in deeper water, it immerses the entire head and a part of the neck, after the manner of the Spoon-bill and Red-breasted Snipe. The notes of this bird resemble the syllable *click.* Length eighteen inches. The Avoset ranges from Texas northward, and is abundant in the Rocky Mountains and the Fur countries. In New Jersey, where it breeds, it is, from the color of its legs, called *Blue-Stocking.* The food varies with its place of resort, consisting of insects, crabs, fishes, marine worms, and small mollusks..

The BLACK-NECKED STILT, *Himantopus*, (Gr. *himas*, a thong; *pous*, a foot,) *nigricollis*, (Lat. black-necked,) has white plumage with the head, neck, back, and wings, above, black. To this bird is assigned the name of LAWYER, (N. H. S., N. Y.,) it is also called Tilt and Longshanks. (See Chart.) It is a rare species; but ranges from the Equator to the 41° N. L. Its length is about fourteen inches.

NOTE.—Dr. DeKay (see N. H. S. N. Y.) has arranged the species of the two preceding genera into a separate family, *Recurvirostridæ.*

The KNOT, or RED-BREASTED SANDPIPER, *Tringa cinerea*, (Nutt.,) *T. canutus*, (Linn.,) has a slender, straight bill, rather longer than the head; the toes have a narrow membrane. This bird varies much in its plumage, and has, therefore, received different names. It is common to Europe and America; ranging in the latter, from the tropics to Labrador, and breeding in the Fur countries to a high latitude. The Knot is seen on the shores of New York in May, and is called by sportsmen, the Robin Snipe. From August to October, it migrates Southward in large flocks, when, in place of the red feathers, it has a white plumage, spotted with dusky, ash-colored above; it is then called White Robin Snipe, and Gray-back. Its length is ten inches.

WILSON'S SANDPIPER, *T. pusilla*, or the Little Sandpiper, (Plate X. fig. 8c,) is about four inches long, with a slender, dusky green tapering bill, and short neck; the tail is doubly emarginate. In summer, it is blackish and rufous; beneath white; in winter, ash; beneath, whitish, spotted with dusky. It is known as the *Peep*, so named from its usual note; and as the *Ox-eye*, from the size and brilliancy of its eye. This species pervades North America from Mexico to 68° N. L., and is one of the most abundant of the group, being found in the interior as well as on the sea-coast.

The SANDERLING, *Calidris arenaria*, or *T. arenaria*, has a straight bill, shorter than the head; thin in the middle, and widened towards the tip; the tail is short, the middle and outer feathers the longest; the toes have a warty membrane on each side. They are three in number, while the preceding genus has four. The female is larger than the male, being about seven inches long. The plumage above is bluish in summer, but light ash in winter; in both sexes, it varies quite as much as in the Turnstones, (*Strepsilas.*) In flying, these birds have fewer evolutions than the Sandpipers. They afford good eating, especially when young. In autumn, they are very fat, and highly relished by epicures. They are said to occur all over the globe.

The RUFF, of the Old World, (female REEVE,) *machetes*, (Gr. a fighter,) *pugnax*, (Lat. combative,) the *Combattant* of the French, has a long and slender bill; legs very long, slender, and naked high above the tarsal joint; three toes before, and one (short) behind; the tail is rounded. The hues of the plumage are so variable, that it is very difficult to find any two that perfectly resemble each other; but the prevailing ground color is brown, inclining to ash, with lateral, and under covers, white; in the autumn or winter, the plumage is more spotted, particularly in the under part, and a bunch of feathers or ruff appears on each side of the head in the male. The females, which are called Reeves, are smaller than the males, and have no ruff. The food of these birds consists of worms and insects, which they pick up in marshy places; in captivity, they are fed with bread and milk, or boiled wheat. They have sometimes been caught in nets, being decoyed by stuffed birds of their species. The Ruff is a very pugnacious bird; it weighs seven ounces, and is a foot in length.

PHALAROPES.

The RED PHALAROPE, *Phalaropus*, (Gr. *phalaros*, bald or naked; *pous*, foot,) *fulicarius*, (Lat. from *fulica*, coot,) has a long, slender, weak, and strait bill, both mandibles furrowed to the point, and the end of the upper curved over the lower one; the front toes are united up to the first joint; the others with festooned or lobated membranes, (Plate IX. fig. 25,) toothed on the edges; the hind toe without a membrane. The Red Phalarope is found in flocks in Kentucky, on the Ohio, and during autumn, is often seen at sea, as far as Newfoundland. It breeds in high northern latitudes, as far as Melville Peninsula. The route of this species towards the warmer regions, is along the Pacific coast. The length is seven and a half inches. These birds are said to breed in great numbers far North; their flight is rapid, resembling that of the Red-backed Sandpiper; sometimes they skim over the water, when they increase their distance from each other. They feed chiefly on insects and crustaceans, which live on the surface of the water.

The HYPERBOREAN PHALAROPE, or LOBEFOOT, *Lobipes hyperboreus*, procures its food principally upon the water, on which "they alight like Ducks, float as light as Gulls, and move about in search of food with much nimbleness." Length six inches.

MARLINS OR GODWITS.

The MARLIN OR GREAT MARBLED GODWIT, *Limosa fedoa*, has a recurved and tapering bill of great length, and long and slender legs; the tibia is bare for about one-third of its length; the toes are four in number; the hind one small, and touching the ground at the tip,—the plumage above is dark brown, varied with red and gray; below, pale reddish brown or buff, with small dusky spots on the neck. The length is from about sixteen to nineteen inches; the female is considerably the larger.

This is a very shy and vigilant bird, moving in large flocks, with irregular and rather quick flight, though less rapid than the Curlews. The flesh is tender and much esteemed. It is sometimes called the Red Curlew, the Strait-billed Curlew, and Dough-bird. The Marlins move along the coast in immense flocks, as far as Massachusetts, (reaching New York in May,) and are supposed to cross the land to Saskatchewan, where they breed. (Aud.) They return from the North in August, remaining in New York until their removal, in November, to their wintering places South of the United States. They feed on aquatic insects, leeches, small marine mollusks, crabs and worms.

The RING-TAILED MARLIN, *Limosa Hudsonica*, called in Boston, the *Goose-bird*, is sixteen inches long, and breeds in high northern latitudes.

The SNIPE, or WOODCOCK, *Scolopax*,—common species, *S. Wilsonii*, Wilson's Snipe, is about eleven inches long, and much sought by the younger gunners, and sometimes, by the keenest sportsmen. Its summer range is considerably beyond the northern boundary of the United States. It resembles the Common Snipe, of Europe, *S. gallinago*, and is sometimes called the English Snipe, but is, in fact, a different species. It breeds from Virginia northward,—it does so abundantly in New York. It resides in Kentucky and the Southern States, during the winter. In flying early in the spring, it soars high in the air, making a remarkable booming sound; its notes are said to differ from those of the Common Snipe, of Europe. It is fond of marshy, swampy places, and selects such for breeding. On the back, the brownish black feathers are edged with cream color, and barred minutely with reddish brown; the throat and breast are buff, spotted with brown and gray.

The GREAT SNIPE, of Europe, *S. major*, has a tail composed of sixteen feathers,—(the normal number is fourteen.) Sir Humphrey Davy, in noticing the breeding of this species in the great royal decoy, near Hanover, says that they require solitude and perfect quiet, and their food being peculiar, they need a great extent of marshy meadow. They feed on the larvæ of *Tipulæ*, (Father Longlegs,) or kindred flies, and, according to the same author, their stomach is the thinnest among the tribe of Snipes. The nest of the Great Snipe, like that of the Common Snipe, is usually placed on the borders of a swamp, and on a tuft of grass, or a bunch of rushes,—often it is found near willow-bushes. The eggs are three or four, yellowish olive brown, with great spots of reddish brown. Two other Snipes, according to Mr. Gould, exceed this in size,—one found in the hilly districts of India, the other in Mexico.

The AMERICAN WOODCOCK, *Rusticola minor*, was separated by Nuttall from the genus *Scolopax*. It has a straight and knobbed bill, slightly drooping at the tip.

Sixth Family. Bustards.

Otidæ, (Gr. ωτίς, *otis*, a bustard.)

The proper position of these birds has been a disputed point among Ornithologists. Temminck ranks them with the Runners, (*Cursores*,) and includes them with the Ostrich family. M. Vigors also places them among the same birds; but in locating them on the Chart, we have followed Cuvier and others, who have included them among the Stilt-birds.

The Bustards are comprehended in one genus, *Otis*. The bill in these birds, is of the length of the head, or shorter, straight, conical, and slightly depressed at the base; the point of the upper mandible is a little arched; the feet are long, and naked above the knee, with three toes in front, short, united at their base, and bordered by membranes; the wings are of moderate size, the third quill longest in each wing. The chin feathers and moustaches, (seen in the male bird) are composed of long wiry-feathers, and the barbs disunited and short; the scapulars are of a buff orange color, barred, and spotted with black; the back and tail coverts, reddish orange, barred, and variegated with black; the greater coverts, and some of the secondaries, are bluish gray; the sides of the neck white, tinged with gray; the lower part of the neck is fine reddish orange; the under parts white. This description of plumage applies to the Great Bustard, *O. tarda*. The male bird of this species, is about four feet long, and nine feet in the expansion of the wings, being (except the Lammergeyer,) the largest of the European birds. Its weight is, on an average, twenty-five pounds. The female is not more than half the size of the male. This bird is noted for its shyness, and its power of running; the young birds have sometimes been run with greyhounds. And yet, in its wild state, unlike the Ostrich, the Great Bustard, upon being disturbed, rises easily upon the wing, and "flies with much strength and swiftness, usually to another haunt, sometimes to the distance of six or seven miles." It was formerly said, this bird "has a pouch in the fore part of the neck, capable of containing nearly two quarts,"—but Mr. Yarrell, in dissecting a male Bustard, "failed to detect this organ." This Bustard is common in some parts of Europe, but is becoming very rare in England. It feeds upon corn, seeds of herbs, colewort, dandelion leaves, &c., and also upon insects and worms. Turnip-tops are said to be peculiarly agreeable. The eggs of the Bustard are two in number, generally, sometimes three, laid upon the bare ground, a little hollowed out for the purpose, either among clover, or, more frequently, in cornfields. The flesh is highly esteemed for food; it is dark in color, and short in fibre, but sweet and well-flavored.

The LITTLE BUSTARD, *O. tetrax*, is another smaller species, found in Europe and Africa.

The BLACK-HEADED BUSTARD, *O. nigriceps*, is a native of Asia, verging towards five feet in length, and having a crested head. It is found in large flocks in the open country of the Mahrattas, as well as in the highlands of the Himalaya. Its flesh is considered a very great delicacy.

The AFRICAN BUSTARD, *O. Denhami*, is a magnificent species, which was discovered by Mr. Denham in Africa, near the larger towns. It frequented moist places where the herbage was pure and fresh, and almost always appeared singly. This bird was ever found in company with the Gazelles; "whenever a Bustard was observed, it was certain that the Gazelles were not far distant." The eye is said to be large and brilliant; the Arabs "are accustomed to compare the eyes of their most beautiful women to those of the OUBARA,"—the general name for the Bustards in Africa.

The KORI BUSTARD, *O. Kori*, is a species discovered by Mr. Burchell in South Africa,—the most gigantic of the family,—standing upwards of five feet high. Mr. B. says, "its body was so thickly protected by feathers, that our largest sized shot made no impression, and, taught by experience, the hunters never fire at it except with a bullet."

The AGAMI, or GOLD-BREASTED TRUMPETER, *Psophia*, (Gr. *psopheo*, to make a noise,) *crepitans*, is an interesting bird, deriving its name from the peculiar noise which it makes without opening its bill. It is about the size of a Pheasant or large Fowl, being twenty-two inches in length; has long legs, and a long neck, but a very short tail, consisting of twelve black feathers, over which the rump-plumes hang droopingly. It inhabits the forests of South America, where it is found in numerous flocks; it is a swift runner, and when pursued, trusts more to its legs than its wings. When domesticated, it shows great fondness and fidelity; and is so regardful of its owner's interests, that it attacks the dogs and other animals that venture near him. Sometimes it is used to protect domestic poultry from the onsets of birds of prey.

Seventh Family. Rails.

Rallidæ, (Genus, *rallus*, a rail.)

The Rails are separated from the other families of this order by the shape of the body, (Plate X. fig. 8b,) which is compressed and flattened at the sides, in consequence of the narrowness of the sternum. The compressed and keel-like form assists their motion in the water, and as M. Vigors remarks, "is intended to counterbalance the deficiency in the formation of the foot, which separates them from the truer and more perfectly formed water-birds." It is certain that the greater portion of these birds are excellent swimmers; and in such habits, as well as in the shortness of their *tarsi*, they are found to deviate from all the remaining groups of the present order.

The Rails have been designated by that name on account of their peculiarly harsh notes. They differ from the Sand-pipers and Plovers in the great size of the leg, and the length of the toes. Swainson speaks of the structure of their bodies as specially adapted to the tangled recesses in which they live, consisting of reeds and aquatic vegetables, which clothe the sides of rivers and morasses. Their flesh is delicate, and from living chiefly upon aquatic seeds and vegetable aliment, they may be regarded as aquatic *Gallinaceæ*. Many of them build nests of accumulated materials, and lay a great number of eggs. The length of the toes enables these birds to walk, without sinking, on aquatic herbage, or in the soft mud of morasses. Although their feet are not webbed, they swim and dive with a facility unsurpassed by that of any of the ducks. The sternum is narrow; wings short and sustained by feeble muscles; hence, the flight is but for short distances, and is slow and heavy; while on the ground, whether among the reeds or tall grass of the meadow, they thread their way with surprising ease and celerity.

Among the well known species is the Common Gallinule, or Water-Hen, *Gallinula chloropus*, (Gr. *chloros*, green; *pous*, a foot.) This bird swims in the open water of rivers and ponds, and with much grace and swiftness, constantly nodding its head; it also dives with great skill and rapidity. It is shy and easily alarmed, in which case it dives under floating herbage, and remains with its beak above water until the danger is over. On account of this habit, it is impossible to take it unless accompanied by a dog. The nest of the Water-hen is built among sedges and reeds near the water, and contains from five to nine eggs, of a cream color spotted with brown. These birds show great sagacity in protecting their young, as the latter do in obeying the monitory signals of their watchful parents. The young have their legs and feet of their full size even while the feathers are only beginning to appear, showing how the organs of flight are subordinate to those of walking and swimming. The female has, contrary to the usual rule, a richer plumage than the male. The pike is the chief enemy of the Water-hens, and destroys many by darting at them from under the cover of water-lilies or other plants.

Other species are (1) the SALT WATER MEADOW HEN, *Rallus crepitans*, fourteen inches long, sometimes called the *Clapper Rail*, or *Mud Hen*, which is seen in New York the last of April, leaves for the South in October, and during the season is very abundant. It lays from eight to fifteen whitish eggs with reddish spots, which are highly valued and much sought for; (2) the FRESH WATER MEADOW HEN, or Great Red-Breasted Rail, *R. elegans*, a rare species eighteen inches in length; (3) the MUD HEN, or Virginia Rail, *R. Virginianus*, (Plate X. fig. 8b,) length ten inches; (4) the SORA RAIL, *Ortygometra*, (Gr. migrating with the quails,) *Carolina*, the same as the English Rail, or Coot, *O. krex*, (Gr. *krex*, a name derived from its cry,) and the species of the Southern States; numerous in New Jersey, and ranging to the 62° N. L.; length nine inches; (5) the NEW YORK RAIL, *O. Noveboracensis*; length five and one-half inches; a shy bird, and not seen in flocks like the preceding species; feeding on seeds and aquatic insects; breeding extensively throughout the United States; (6) the FLORIDA GALLINULE, *Gallinula galeata*, (Lat. helmeted,) fourteen inches in length, closely allied to the European species, *G. chloropus*, (referred to above,) and ranging from Mexico to Massachusetts.

SUB-FAMILY. FLAMINGOES.

Phoenicoptinæ, (Gr. φοινικόπτερος, *phoinikopteros*, red-winged.)

These birds are included in one genus, *Phoenicopterus*. Their proper position has been a matter of considerable doubt. Swainson places them with the Ducks, among the Swimmers, though he remarks: "The Flamingo, which has the longest legs in the Natatorial order, is so good a walker that it only swims occasionally." We give them a place among the *Grallatores*, to which order they have more commonly been assigned, but immediately *before* the Swimmers. The genus *Phoenicopterus* has the bill strong, higher than it is large, toothed and conical towards the point; the upper mandible is suddenly bent, curved at its point on the lower mandible, which is larger than the upper; the legs are of excessive length; the feet also very long, three toes in front, hind one very short and articulated high up on the tarsus; the wings moderate; first and second quills longest.

The EUROPEAN FLAMINGO, *P. ruber*, is found in the warmer parts of Europe, but is common in Asia, and the coasts of Africa. The beak is evidently adapted to its long and flexible neck. When this bird wishes to feed, it merely stoops its head to the water; the upper mandible is then lowest, and as is the case with the Duck, the edges of the beak filter what is received. Pestilent marshy places, which urge man to a distance, are boldly and safely frequented by this bird. Its plumage is a deep brilliant scarlet, except the quill feathers, which are black. Arranged in a line, these birds appear like a file of soldiers; but the miasma of the regions in which they dwell, is more deadly than the rifle, and its breath more surely fatal than the ball of the cannon. The nest of the Flamingo is a conical structure of mud, with an opening on the summit, in which are placed two or three dusky white eggs, somewhat larger than those of a goose. The nest is so high as to permit the bird to sit, or rather stand, her long legs hanging down on each side at full length. The height of this bird is five or six feet, (see fig. on Chart.) The flesh is said to be pretty good meat; the young are thought by some equal to a partridge. Juvenal, in his Satires, notes the Flamingo, (*Phoenicopterus ingens*,) as among the luxuries of the table; the brains and the tongue formed one of the favorite dishes of Heliogabalus. By some, however, the flesh is thrown away as fishy, while the feathers are used to ornament other birds served up at special entertainments.

The American species, *P. ruber*, or *P*, *chilensis*, scarcely differs from the European. It is remarked that "the development of the gizzard in this genus makes it very probable that vegetable substances form part of the diet of the Flamingo; but it is not likely that large fish, or indeed water animals of any great size, are ordinarily devoured by these birds. The bill is a colander, admirably contrived for separating the nutritious portions whether animal or vegetable, from the mud and other useless parts." The Red Flamingo is found in the warmer regions of North America. C. L. Bonaparte says it is very rare and accidental in the neighborhood of Philadelphia. In South America and the West India Islands it is also found. It is particularly abundant in the Bahamas, where it breeds.

What is the 1st order of Aquatic Birds? To what birds are they intermediate? Mention their distinguishing characteristics. What is peculiar in their flying? What power do they possess? Where and how do they build their nests, &c.? What Family does the order include? How are the Plovers distinguished? What use do they make of their beaks? What is said of their plumage and diffusion? Repeat what is said of the Lapwing or Pee-wit. What is the generic name of the Great P. and why given? Is it an American sp.? What is said of the Golden P. of A.? How does it differ from the Golden P. of Europe? What is it sometimes called? What is the generic name of the Oyster-catcher and why given? What is said of it? What of the Gray P.? Of the Turnstone? Of the Dotterel? Of the Swift-foot? Of the Pratincole? What other sp. are mentioned?

What is the 2nd Family? What Groups of Birds does this include? What is said of the Cranes? Of the Spoon-bills? Of the Herons? Repeat what is said of the A. Crane. What of the Common C. of Europe? Where is the Demoiselle found? What is said of its plumage, size, &c.? How long did one of these birds live in Versailles, (Fr.?) What is said of the Crowned Crane? Are the True Herons numerous? What use was formerly made of them, and how were they esteemed for food? In what genus does Aud. include the Egrets and Bitterns? What is said of the Common Heron? What of the Great Blue Heron? Of the Greenish B. H.? Of the Black-Crowned Night H.? Of the Great American White Egret? What is said of the diffusion, &c. of Bitterns? For what are they noted? What popular names have been given them and why? To what does the English name Bittern refer? Was the Common B. formerly sought in falconry? In what estimation has its flesh been held? What use was made of its hind claw? How does it compare in size with the Common Heron? What is said of the American Bittern? Of the Small B.? Why is the Common Boat Bill so named? On what does it feed? How do Storks compare with the other birds of this family? Where do they build their nests? What is the shape of their bills? Why are they a privileged race? What is the length of the Common White S.? What is said of its familiarity and of its appearance in European towns? What did Dr. Shaw witness? For what has the S. ever been noted? How regarded among the ancients? What is said of the Black Stork? Where is the Adjutant found? Describe this bird. What is said of its voracity? What does Dr. Latham say of these birds? What is said of the Marabou Crane or Giant Stork of Africa? Where is a similar species found? What name is given to the beautiful plumes of these birds? What is said of the Jabiru?

What is the 3d Family? With what family are these birds often included? Give their characters and habits. What is said of the Common White Spoonbill? What of the Roseate S.? What is the generic name of the Open-Beaked Birds? Describe the beak. Where does Cuvier place them?

What is the 4th Family? To what birds are those of this family allied? What countries do they chiefly inhabit? What characters are given? On what do they feed? How many species does this family include? How many in the U. S.? Where is the Glossy Ibis found? With what sp. known to the ancients is it identical? At what time does the sacred Ibis appear in Egypt? What is said of its size and plumage? How was this and the preceding sp. regarded by the ancient Egyptians? What is the length of the American White Ibis? What Island is noted as a resort for these birds? How many of their nests did Aud. count in a single tree? What else is said of the White I.? Give an account of the Scarlet I. Of the Wood I. Where are other species found?

What is the 5th Family? What is the most prominent character of this family? What use is made of this organ? What other characters are mentioned? What is the color of their plumage? To what places do they resort? What is said of their distribution? How does Mr. G. R. Gray divide them? Which is the largest of the Curlews? What is the generic name? What other sp. are mentioned? What is the generic name of the Tattler? Why is the Tell-Tale Tattler so called? What is said of it? What of the Spotted Sand L.? Of the Gray Plover? Of the Avoset? Of the Blue-necked Stilt? Of the Knot? Of Wilson's Sand-Piper? Of the Sanderling? Of the Ruff?

Mention the 6th Family. In what genus are they comprehended? What is said of the bills, &c.? How large is the Great Bustard? Does any other European bird exceed it in size? For what is it noted? Has it a gular pouch? What else is said of it. What sp. of Bustards are mentioned? What is said of the Agama, or Gold-breasted Trumpeter?

What is the 7th Family? How are they separated from the other families of this order? How are they aided by their keel-like form? In what respects do they deviate from the other groups of wading birds? Why are they called Rails? What does Swainson say of their structure? What is said of their flesh, nests, toes, &c.? What of their motion in the air and on the ground? Which are the different sp. and what is said of them? What Sub-Family is mentioned? What is said respecting their proper position? Give the characters of the gen. Phoenicopterus. Where is the European Flamingo found? What places does it frequent? What is said of it? Does the American sp. differ much from the E.? Upon what does it feed? Do vegetable substances form any part of its diet? What may its bill be called, and why? Where is the Red Flamingo found?

SECTION XI.

Second Order. Web-Footed Birds.

Natatores, (Lat. Swimmers.) Anseres, (Lat. *anser*, a goose.) Linn.

We come now to the last order of birds, viz.: those which are web-footed. These are numerous and widely distributed. Moving in an element which is everywhere essentially the same, we find, as we might naturally expect, that these birds are represented, not only by peculiar genera in every part of the world, but that particular species, as of the Ducks, the Terns, and the Petrels, encircle the globe.

The foot of the Grebes is not webbed, but has each toe separate and flattened, (Plate IX. fig. 26,) somewhat like that of the Coot in the last order, with this exception, the Swimmers are all marked by having the toes united by a membrane, giving to the foot the form of a powerful oar, as in the common Duck or Goose, (Plate IX. fig. 24.) In those species which are eminently aquatic, the feet are placed far back on the body, (see Auk on the Chart,) which renders their gait clumsy and shuffling on land, but gives to the backward stroke of the foot in the water an impetus that helps them in swimming; the tarsus is also flattened sidewise, diminishing the resistance to progression in the water.

The form of the body is flattened horizontally, (not laterally, as in the Waders,) the better to float on the surface. The plumage is remarkably thick and close, particularly on the under parts of the most aquatic kinds; besides which the skin is covered with a dense coat of soft down. The outer surface is usually polished and satin-like, probably from the oily secretion which the bird frequently applies to it. The larger part of the Swimmers have a copious and peculiarly oily secretion of fat.

As Cuvier remarks, these are the only birds in which the neck is longer than the legs, which is sometimes the case to a considerable extent, for the purpose of enabling them to search for food in the depths below, while they swim on the surface. The tail is generally short, and so are the wings; hence, flight is in most feeble, and in some altogether denied: and yet it must be noted, that in the order Natatores are found examples of the longest wings, and the highest powers of flight of the entire class of Birds, as, for example, in the Frigate Pelican. The Petrels and Terns have also great length of wing.

The web-footed fowl resort to fens, morasses, broad rivers, inland lakes, rocky coves, &c., and they are found also on the ocean's wide expanse. The marine kinds are more numerous in the colder seas of the North, than in those of tropical regions.

This order includes the following families: (1) *Anatidæ*, Ducks; (2) *Colymbidæ*, Divers; (3) *Alcidæ*, Auks; (4) *Procellaridæ*, Petrels; (5) *Laridæ*, Gulls; (6) *Pelecanidæ*, Pelicans.

First Family. Ducks.

Anatidæ, (Lat. *anas*, a duck.)

This numerous family have the beak thick and broad; high at the base, and covered throughout almost its whole extent with a soft skin, the tip alone being horny; the edges are cut into thin parallel ridges, or small teeth; the tongue is large and fleshy, with its edges toothed ; the wings are of moderate length. The males have, for the most part, the wind-pipe enlarged into a bony chamber, varying in form and size ; sometimes this tube is much prolonged, and bent back in folds within the swollen keel of the breast bone, peculiarities of organization probably connected with the loudness of the voice. The gizzard, especially in the land species, is large and muscular.

The Ducks mostly build their nests upon the ground, but some on trees, and lay numerous unspotted eggs. The young are at first covered with soft down, and can run and swim as soon as they leave the shell. The laminated structure at the edges of the mandibles, (Plate IX. fig. 5,) has often been referred to as showing special adaptation to the habit of feeding in birds of this family, enabling them to take with facility minute animals which swarm in rivers, and those equally numerous found on the sides of rivers and inland streams. By means of their broad beak, they capture at one effort, considerable numbers, and as they are drawn forth, covered with mud, this offensive part is thrown out between the interstices, or tooth edges of the mandibles, (Plate IX. fig. 5,) which, however, are not sufficiently wide to allow of the passage of the insect food at the same time, so that the beak operates as a sifter, expelling the refuse, but retaining the food. It is probable that the large and fleshy tongue is an assistant in this separating process.

Geese seem to form the connecting link between the Swimming and Wading Birds, retaining as they do the manners of the Waders, but walking much more than they swim. Their food consists more of grains and insects than of fishes; their legs are long, and they have a considerable space above the tarsal joint. These birds, in common with the Swans, have rather long necks.

The True Ducks include a large variety of species and are found in almost every part of the world.

The Shoveler, or Spoon-bill, *A. clypeata*, (Lat. furnished with a shield.) is in length from seventeen to twenty inches; it is named from its broad, shovel-like bill. Usually it breeds far North, but to this there are exceptions. Some think its flesh exceeds that of the Wild Duck.

The Mallard, or Wild Duck, *A. boschas*, (Gr. *boskas*, a mallard, from *boskē*, a pasture,) is the parent of our domestic broods. The ordinary length is about two feet, but one variety is said to measure thirty inches. Its flesh is much esteemed. Richardson says "the Widgeon or Wild Duck is a strange eater of grass ;" to this the specific term refers.

The Soft-Billed Shoveler, *Malacorhyncus*, (Gr. *malakos*,

soft; *rhunkos*, a bill,) found in Australia, has a very peculiar bill, the edge of the upper mandible having on it a thin membrane or skin, which hangs down like a wattle on each side.

The TAME DUCK, (from *A. boschas*,) is nearly omnivorous in its indiscriminate appetite and its voracity. In the natural state the Duck is a little more particular in its diet.

The GREEN WINGED TEAL, *A. Carolinensis*, is during the autumn and winter, common in all our fresh water lakes and ponds; its flesh is very well tasted. Length fourteen inches.

The PIN-TAIL DUCK, *A. acuta*, (Lat. sharp,) affords similar food to the preceding. It is about two feet long. This Duck is particularly abundant on the shores of Lake Ontario; ranging, however, during winter and spring, across this Continent; and breeding in high northern latitudes.

The BLACK DUCK, *A. obscura*, (Lat. obscure or dark,) breeds from Texas to Labrador. Its length is about two feet. Few Ducks are more highly prized than this species.

The AMERICAN WIDGEON or BALD-PATE, *A. Americana*, is very generally distributed. It feeds chiefly on aquatic vegetables, and is esteemed for its delicate flavor. Length from eighteen to twenty inches. The Widgeon of the Eastern Continent, *A. Penelope*, is also found in this hemisphere.

SUB-FAMILY *Fuligilinæ*, (from Lat. *fuligo*, soot.) SEA-DUCKS.

The Sea-Ducks include four genera, with a variety of species. They principally frequent the sea; but many of them are to be found in the fresh water lakes and rivers, where the water is deep. Their plumage is very close and thick, in comparison with that of the True Ducks, (*Anas*,) and the covering of the female differs much in hue from that of the male. The Sea-Ducks are not good walkers, though they can run or shuffle along rapidly. They swim remarkably well, but low in the water, and excel in diving, on which they rely when in danger, more than on their power of wing. Usually, they fly low, laboriously, and with a whistling sound. They are mostly found at the north; but some species are spread over the entire globe. Large flocks migrate periodically, chiefly on the line of the sea-coast, flying and feeding generally by night. They often make their nests near fresh water; both parents, in several of the species, strip off their down as a covering for their numerous eggs.

The genus *Somateria*, (Gr. *soma*, body; *eria*, wool, or *teiro*, to wear away,) includes Ducks which are peculiarly marine; according to Sir John Richardson, never found in fresh water.

Their food consists principally of mollusks found in the Arctic Sea.

The EIDER DUCK, *S. mollissima*, (Lat. very soft,) is remarkable for its exquisite and elastic down, so valuable in commerce, and so essential in preserving the proper balance of animal heat in the icy regions in which it dwells. The beak is prolonged on the forehead into two narrow flat plates, which are separated by an angular projection of the frontal plumage. This species is, in severe winters, seen as far South as the Capes of the Delaware. Northern explorers have repeatedly attested its value. Dr. Kane writes thus of its appearance. "The Eider Duck is an awkward animal on the wing, and hardly graceful in the water. The position of the legs, set very far back, throws the body, Penguin-like, nearly upright; and they move about erect, but easily, and with animation." His party gathered two hundred eggs from a gleaned field, one morning before breakfast. A whaler which they met, had four hundred and fifty dozen eggs on board: formerly, from a quarter to half a million of eggs were, during a single season, taken from Melville Island.

The Duck and Drake build the nest in company, and line it with down. This is of two kinds,—the dead down and the live down; the former is taken from a dead bird, and is of inferior quality; the latter is that which the Duck strips from herself to cherish her eggs; its lightness and elasticity are such, that it is said, two or three pounds of it squeezed into a ball, will swell out to such an extent as to fill a case large enough for a foot covering of a bed. The skin of the Eider Duck, with the feathers on, forms an article of commerce, particularly among the Chinese. The length is twenty-five inches.

The KING DUCKS, *S. spectabilis*, (Lat. deserving notice,) are also found in the Arctic regions, but in their migration do not pass so far South as the Eider Duck. According to Sir John Ross, they afford a valuable and salutary supply of fresh provision to the crews of vessels employed in the Northern Seas, and their down is equal to that of the Eider Duck.

The SURF DUCK, *Oidemia*, (Gr. from *oideo*, to swell,) seek their food at sea chiefly, and have their name from frequenting its shores. The prevailing color is black in the male and brown in the female. The generic name was suggested by the *swollen* appearance of the beak. The species *O. fusca*, (Lat. tawny,) has a very thick and close plumage, and is called the VELVET DUCK. The down is similar to that of the Eider Duck. They are very numerous at Hudson's and Baffin's Bay, The length is twenty-four inches.

The CANVASS-BACK DUCKS, *Fuligula*, (Lat. *fuligula*, a fen-duck,) *valisneria*, (botanical name of the tape-Grass, of which this species are very fond,)—breeds from 50° N. L. to the extreme northern limit of the Fur countries. About the middle of October it arrives on the sea-coasts of the United States. This Duck is shy, but much esteemed, as few birds grace the table better. It haunts the sea, its bays and estuaries. The length is twenty-four inches. In swimming, the tail is erected, and from the shortness of the neck, is nearly as high as the bird's head, so that, at a little distance, the bird seems to have two heads.

The SPIRIT DUCK, *Clangula*, (Lat. *clango*, to clang;) *albeola*, (partly white;) is abundant during the summer, on the rivers and fresh water lakes of the Fur countries; in autumn and winter, common in the United States. It is a most expert diver; the artful way in which it conceals itself after it has vanished under water, has given it the name of *Spirit Duck* or *Conjurer*. Its flesh is not in high repute. In Pennsylvania and New Jersey, it becomes so fat, it is called "Butter-Box" or "Butter-Ball." Length fourteen inches.

The LONG-TAILED DUCK, *Heralda glacialis*, (icy,)—the Old Wife and Swallow-tailed Duck, of Hudson Bay residents,—is noted for its very long tail of fourteen feathers. It swims and dives with all the expertness of the Spirit Ducks. The young Ducks are juicy and tender; the old ones not much valued for the table. This species is found in the Arctic regions of both Continents. Length twenty to twenty-one inches.

GEESE.

The SNOW GOOSE or WHITE BRANT, *Anser Hyperboreus*, breeds in high northern latitudes. It is from twenty-seven to thirty-one inches in length. Its feathers are valuable, and Richardson says, its flesh is far superior to that of the Canada Goose in juiciness and flavor.

The BRANT, *A. bernicla*, is deemed one of the most savory birds; its length is about two feet; it breeds near the Arctic Ocean; is found on both Continents.

The BERNICLE, or BARNACLE GOOSE, *Bernicla leucopsis*, (Gr. white-faced,) is found in the northern regions of both hemispheres.

The GRAY-LAG GOOSE, *A. ferus*, (Lat. wild,) in length two feet and nine inches, is the origin of the COMMON DOMESTIC GOOSE. The latter is too familiar to require description. It has been known to live over eighty years.

The WILD GOOSE or CRAVAT GOOSE, *A. Canadensis*, is from forty to forty-two inches in length. In its contour, especially about the neck, it seems to approach the Swans; the patch of white feathers on the neck contrasting with those of dark shade, has the appearance of a *cravat*. It breeds most abundantly in Labrador and high northern latitudes. In the Fur countries its arrival is anxiously looked for, and hailed with great joy. At Hudson's Bay, three thousand or more are sometimes killed and barreled up in a year.

The EGYPTIAN GOOSE, *Chenalopex*, (Gr. a goose or duck, Pliny,) *Ægyptiacus*, is a beautiful species, which passes over occasionally from Africa into Europe; it is particularly numerous in the island of Sicily. The upper part of the plumage is reddish brown; the under parts are buff, mingled with blackish lines. This Goose is figured on the monuments of the ancient Egyptians, and was regarded by them with veneration!

The GOOSANDERS or MERGANSERS, form a sub-family,—*Merganinæ*, including, according to Prince Bonaparte, two genera, *Mergus*, (Lat. a diver, from *mergo*, to dip,) the Smew, and *Merganser*, (Lat. from *mergo*, to dip, and *anser*, a goose,) the Goosander.

The SMEW or WHITE NUN, *Mergus albellus*, (Lat. from *albus*, white,) is found in the Arctic regions of both Continents; it is migratory in autumn, but especially in winter. Its food consists of small crustaceans, water insects, mollusks, and small fish. The nest is placed on the borders of rivers and lakes, and contains twelve whitish eggs. The Smew (when old) has upon the head a tufted crest of pure white; the edges of both mandibles of the beak, have saw-like teeth directed backwards; the point of the upper mandible is curved, and with the horny nail, forms a hook. (Plate IX. fig. 5.) The length of the Smew is fifteen or sixteen inches.

The GOOSANDER or JACKDAW, *Mergus merganser* (or *Castor*,) having also a saw-like and hooked bill, (Plate IX. fig. 5,) builds its nest among rolled pebbles on the banks of waters, or in bushes and hollow trees, and lays twelve or fourteen whitish eggs. The flesh of this and the preceding species, is rank, and by no means in request for the table. Its native abode corresponds with that of the Smew; it migrates southward on the approach of winter. The very old male has a large and thick tuft on the head; the plumage of the upper parts is deep black; the under parts, which in the Smew are white, are in the Goosander tinged with yellowish rose-color, (changing to white in stuffed specimens.) The length is twenty-six or twenty-eight inches.

SECOND FAMILY. DIVERS. (Short-winged.)

Colymbidæ, (Gr. *κόλυμβος*, *kolumbos*, a diver.)

These birds are more entirely aquatic than the Ducks, and remarkable for their powers of diving, and the great length of time which they can remain immersed. They have narrow, straight, and sharp-pointed beaks; the head is small; the legs, placed near the extremity of the body, are flattened sidewise, so as to present a thin edge before and behind; the toes are armed with broad, flat nails. In one genus, *Colymbus*, including the Loons,—the toes are united by a membrane, and there is a short tail; the two other genera, (including the Grebes,) have the toes divided midway to the base, and bordered with white oval membranes, and have no traces of a tail. Owing to the position of their feet, these birds are poor walkers, though extremely powerful and fleet swimmers and divers. They have short wings, and their ability to fly is consequently quite limited; but under the surface of the water, the wings are expanded and employed as fins. The thread-like, or downy plumage, is remarkably thick, and has a silvery gloss. The Divers' food varies with the situations which they frequent. It consists of fishes with their fry and spawn; crustaceans, water insects, &c., and occasionally vegetable substances. The Grebes are widely scattered over fresh waters; the Loons are confined to the oceans and coasts of temperate and arctic regions. These birds dive so instantaneously, that it is difficult to shoot them,—disappearing, as they do, at the first flash of the gun, and not returning to the surface within some two hundred yards, and then merely to raise the head for a moment and again disappear. The stomach of the Grebes, is generally found to contain a mass of their own feathers. These are probably conveyed thither in the bird's process of oiling its plumage, or, as has been said, "making its toilet." The largest, and finest species of Loon, is the GREAT LOON or DIVER, *Colymbus glacialis*, (Plate X, fig. 9.) This bird is thirty-two inches long,—the neck and head are black, glossed with purple or green; their upper parts black, marked with white spots, set in rows; the under parts pure white. The cry of the Great Diver is melancholy in its tone, resembling the howling of a wolf, and is said to portend rain. The flesh is dark, tough, and unpalatable. The eggs are two or three in number, of a deep olive color, spotted with brown, and about as large as those of a Goose.

The RED-THROATED LOON or SCAPE-GRACE, *Colymbus septen-*

trionalis, (Lat. northern,) is another species, breeding from Newfoundland northwardly. Length twenty-five inches.

The GREBES, (*Podiceps*,) have been variously placed by different naturalists. De Kay includes them with the COOTS, in a separate order, *Lobipedes*, (Lobe-footed,) and ranks them immediately *before* the Swimmers. We have followed Cuvier and others, in placing them with the Swimmers, and in the present family. Among the species are the HORNED GREBE, *P. cornutus*, (Lat. horned,) in length fifteen inches; common to Europe and America, and known by the names Dipper, Water-Witch, &c.; the CRESTED GREBE, *P. cristatus*, in length nineteen inches; commonly found in secluded ponds and lakes in the interior, but also seen on the sea-coast. It ranges from Mexico to 68° N. L.;—is found also in Europe; the RED-NECKED GREBE, *P. rubricollis*, (Lat. red-necked,) not quite so long as the preceding, and scarcely seen South of New York.

THIRD FAMILY. AUKS.

Alcidæ, (Lat. *alca*, alk or auk.)

The birds of this family, have a structure which pre-eminently adapts them to an aquatic life, and are, in their resorts and habits, exclusively maritime. The beak in these birds varies in length, and is more or less compressed; both mandibles are much curved and notched; the nostrils are almost entirely closed by a naked membrane; the feet small and entirely webbed; the legs short and placed far back, so that, in sitting, these birds assume an erect position; the tail has sixteen small feathers. In moving under water, the Auks make no use of their feet, but hold them out behind, as the Waders do theirs in flying, and use their short wings in the manner of fins, so that they may be said to *fly* beneath the surface. "Their movements, under water, precisely resemble those of the *Dyticidæ*, or Common Water-Beetles; the principal motion being more or less vertical, instead of horizontal as in the Grebes and Loons; they are, therefore, together with the distinct group of Penguins, the most characteristic divers of the class." Their food, obtained by diving, (an operation in which they are assisted by their wings as well as their feet,) consists of small fishes, crustaceans, and other marine animals. The Auks are frequently seen in immense numbers on rocky islets, and precipitate cliffs that overhang the sea, on the shelves and edges of which they lay their eggs, one only being deposited by each bird. The female, while sitting in an erect position, keeps the egg between her feet for the purpose of incubation. Many families gain their subsistence by procuring the eggs and young of these and similar birds. The storm-lashed and iron-bound coasts of Northern Europe and America, and the frozen islands of the Arctic Seas are the dreary homes of the Auks; some of them roam hundreds of miles out to sea.

The Penguins occupy, in the Southern Hemisphere, the place filled by the Auks or Puffins in the Northern. Their wings are very small,—mere rudiments, covered with an integument, resembling scales, and entirely powerless as organs of flight; but they not only aid the bird in its divings and evolutions under water, but also as a sort of front extremities when progressing on land. Being without the power of flight, and unable to run, this bird may be easily overtaken on land; but when it reaches the water, it has no difficulty in distancing its pursuers, swimming like a fish, and springing several feet over any obstacles which it meets in its course. The Penguins are peculiarly remarkable for having a kind of ball and socket union in the vertebræ, corresponding, in some degree, to what is seen in the reptiles.

The GREAT AUK, *Alca impennis,* (Lat. wingless,) is almost wingless, i. e. its wings are very small, entirely incapable of raising it in the air, but serving admirably as paddles to the bird when diving under water. The Lump-fish is said to be a special favorite of the Great Auk. Audubon says, "the egg is very large, measuring five inches in length and three in its greatest breadth; the shell is thick and rather rough to the touch; color yellowish white, with irregular lines and blotches of brownish black," which have been supposed to bear some resemblance to Chinese characters. Newfoundland is one of the breeding places of these Swimmers, and the Esquimaux who frequent that island are said to make clothing of their skins. The Great Auks are widely diffused in the northern hemisphere, but in high northern latitudes they "swarm." They may be seen on floating ice, but do not wander beyond soundings. The winter plumage, which begins to appear in autumn, "leaves the cheeks, throat, fore part and sides of the neck, white. In spring, the summer change begins to take place, and confines the white on the head to a large patch which extends in front and around the eyes; the rest of the head, the neck and upper plumage is deep black." The length of the Great Auk is about three feet.

The RAZOR-BILL, or BLACK-BILLED AUK, *A. torda,* has wings so far developed as to answer for the purpose of flight, though the bird uses them with great effect as oars, when swimming under water. Its length is about seventeen inches. These Auks breed from the Gulf of St. Lawrence to along the coast of

Labrador. Thousands of them are killed on that coast for the sake of the breast feathers, which are very warm and elastic. The eggs are about as large as a turkey's, being great in proportion to the size of the bird. Of these, incredible numbers are collected at Labrador and in its vicinity. The Razor-Bill is seen on the coast of New York State every autumn and winter; it is common in Europe.

The COMMON PUFFIN, or COULTER-NEB, *Fratercula Arctica*, or *Mormon fratercula*, (*Mormon Arcticus*, DeKay,) has a beak monstrously large, rivaling in its devélopment those of the Toucans and Hornbills, and from its enormous size and the sharpness of the edge, rendering this bird a formidable antagonist. This organ is shorter than the head, higher than its length, somewhat triangular in outline, and has its sides cut into furrows, (Plate IX. fig. 11.) The generic names applied to the Common Puffin, refer, in their signification, to its singularly grotesque appearance, with its short, thick-set form, its erect attitude, and above all, its extraordinary and brightly colored beak. It makes a burrow for itself on the lofty cliffs, but sometimes avoids this labor by occupying that of a rabbit which stands in awe of the formidable bill, and readily gives up his habitation. From the lofty cliff, the Puffin plunges fearlessly into the sea, and returns with its beak full of fish, which are secured by their heads, and lie in a row along the Puffin's bill. The length of this bird is from twelve to thirteen inches.

The LITTLE GUILLEMOT, *Uria alle*, or *Mergulus alle*, is from six to ten inches in length, sometimes, but rarely, seen on the coast of New York, its range being from 39° N. L. to the north pole. It is also called Sea-Dove, Sea-Pigeon, Pigeon Diver, or Ice-Bird. During the breeding season, it collects in vast numbers along the north and east coast of Baffin's Bay. Dr. Kane says it was not uncommon to kill more than a hundred in the course of a couple of hours. The long-sought and lamented Sir John Franklin killed and salted down so many of these birds as to augment his resources by nearly a two years' supply of food. "No other bird migrates in such numbers, or contributes so largely to the pleasures of the table." (Grinn. Arct. Exped.) The size of this bird compares well with that of a partridge; the feet are short, plunged into the feathers far back beyond the equilibrium of the body; it has three toes, all front and entirely webbed. While taking their food, consisting of small fish, crustaceans and medusæ, they can be approached so near as to be knocked down with poles and boat hooks. The whalers sometimes shoot them with dried peas. Upon the bare rock they lay, in company, each a single egg of a pale green, blotched with dark brown spots. So close are they together, that the birds, when sitting nearly upright, almost touch each other, covering the ledges of the rocks upon which their young are hatched, and from whence they take to the water in five or six weeks.

The PENGUINS, *Aptenodytes,* (Gr. *a,* priv., *ptenos,* winged; *dutes,* a diver,) seem to be among the Natatores what the Ostriches are among the strictly terrestrial birds. Swainson remarks that "the hind toe in the Penguins and Cormorants is placed almost as far forward as in the Swifts. In the Penguin the tarsus is so short as almost to be confounded with the sole of the foot, and is probably rested on the ground when the bird walks, just as in the bear and other plantigrade quadrupeds. The whole foot is remarkably flattened, as if to enable the bird to cover a greater breadth of ground." (Classification of birds, Vol. I.)

The bones are described as very hard, compact and heavy, having no aperture for the admission of air; but they contain, especially the bones of the extremities, a thin oily marrow. The sensations of these birds are by no means acute. One writer relates that he stumbled over a sleeping one and kicked it some yards without disturbing its rest. Another states that he left a number of these birds apparently lifeless, while he went in pursuit of others; but they afterwards got up and marched off with their usual gravity.

The habits of the Penguins are highly interesting, and have frequently been described. Their camps, towns, and rookeries, so called, are largely descanted upon by southern voyagers. Those at the Falkland Islands have attracted particular attention.* The rookeries are said to be designed with the utmost order and regularity, though they are the resort of different species. But in the midst of this apparent order, there seems to be a want of good government, the stronger species stealing the eggs of the weaker, if they be left unguarded. The King or Patagonian Penguin, *A. Patachonica,* (Plate X. fig. 9b,) is said to be the great-

* The rookeries at the Falkland Islands above referred to, sink into insignificance when compared with a settlement of the King Penguins recorded by Mr. G. Bennett, who saw at the north end of Macquarrie Island, in the South Pacific ocean, a colony of these birds which covered an extent of thirty or forty acres. He describes the number of Penguins collected together in this spot as immense, but observes that "it would be impossible to guess at it with any near approach to truth, as during the whole of the day and night, 30,000 or 40,000 are continually landing, and an equal number going to sea."

est thief of all. Three species are found in the Falkland Islands. Two of these, the King Penguin and the MACARONI, *A. chrysocome*, (Golden-haired or feathered,) deposit their eggs in these rookeries. The Jackass Penguin, *A. demersa*, (Lat. from *demergo*, to plunge in,) which is the third species, has its English name from the horrible brayings which it sets up at night. This makes its nest in burrows on downs or sandy plains, and does not appear to take invasion so quietly as the other species.

H. T. Cheever, in his "Island World of the Pacific," when referring to his landing on the Falkland Islands, says: "What was our surprise to find what we had thought a facing of white stones, to be innumerable Penguins, standing erect and in the rank and file of battle array, upon the declivity of the rocks, and occupying at least two acres, in dense columns, away back to the moss and grass. On every out-jutting angle or hollow, there was a dusky nest with a bird sitting upon it, and so unacquainted with man that we could climb up and lay hands upon them before they would move." He continues: "To those who have never seen a picture of a Penguin, it would be impossible to convey an idea, by description, of this odd amphibious creature. It has the head, bill, and two web-feet of a bird, and stands erect on land, sometimes two and a half and three feet in height. They have no wings nor proper feathers, but two fins or flippers, like the seal. Their motion on land is by successive hops in the most awkward manner conceivable. When going down a declivity, the center of gravity is often thrown too far forward, and away they tumble, and scramble, and roll over, until they get to the sea, in which they dive and swim with great swiftness. They are often seen singly, or two and three together, far out at sea. Their cry or bark is like the inarticulate human voice; and sounding, as it often does, from the surface of the ocean like the cry of a man in distress, it startles and appals one."

The largest species of the Patagonian Penguins is said to be four feet and a quarter in length, and to weigh forty pounds. When sitting or attempting to walk, they have been compared to a dog that has been taught to sit up and move in a minuet. Their short legs drive the body in progression from side to side, and were they not assisted by their flipper-like wings, they could scarcely move faster than a tortoise. This awkward make of the legs, which so disqualifies them for living on the land, admirably adapts them for life on the water, inasmuch as they serve for propellers, and being placed so far behind the moving body, and worked the more swiftly for being short, they push forward with great velocity; with their heads erect, and their fin-like wings hanging down as half arms, they "look like so many children with white aprons on." Hence they are said to "unite in themselves the qualities of men, fowls, and fishes! Like men, they are upright; like fowls, they are feathered; and like fish, they have fin-like instruments that beat the water before them and serve for all the purposes of swimming rather than flying." They are covered more warmly with feathers than any other bird, so that the sea seems entirely their element.

FOURTH FAMILY. PETRELS OR FULMARS.

Procellaridæ, (Lat. from *procella*, a storm.)

The form of the beak in the birds of this family is very remarkable; it appears to be constituted of several separate pieces soldered together. The upper mandible has the basal part separated from the tip by a deep, oblique furrow, and has on its summit a tube, (or two tubes united into one,) containing the nostrils; the point of this mandible takes the form of a curved and pointed claw or nail; the lower mandible is likewise seamed in a similar manner, and its tip is hooked downwards. (Plate IX. fig 10.)

The front toes are united by a membrane; the hind toe is reduced to a mere claw, which is elevated upon the tarsus and sometimes wanting. The wings are usually long, and the flight powerful.

The Petrels are eminently birds of the ocean, rarely approaching the land, except in the breeding period. Some of them appear to be almost always on the wing, following the course of ships for days together without alighting. Their food consists of small mollusks and crustaceans, and the oily particles which float upon the surface of the sea. In high latitudes, some of them feed with much voracity on the fat of slaughtered whales. Hence their flesh becomes apparently saturated with oil; and when alarmed, many of them occasionally eject fetid oil from their nostrils, as a defence. This family includes a number of species, about eight of which are found in America.

The COMMON FULMAR, *Procellaria glacialis*, (Lat. icy,) or *Fulmarus glacialis*, is considered the type of the true Petrels, having a stout, thick bill, with the upper mandible considerably hooked at the tip, and sulcated or furrowed; the lower mandible is straight and slightly truncated; the nostrils are united in a single tube; the legs of only moderate length. This bird is a native of the Polar regions, but is found, though in less numbers, in the Northern Seas of Europe and America. It is not uncommon off the coast from New York to Nova Scotia. The rocky St. Kilda, one of the western islands of Scotland, is the only place of annual resort for this bird in the British dominions. (Shelby.) It is from sixteen to eighteen inches in length; breeds in high latitudes, never coming to the coast except for the purposes of nesting, or when driven thither by gales. The bill, iris and feet are yellow; the head, neck, and lower parts pure white; the back and wings, of a grayish blue. Scoresby says: "The Fulmar is the constant companion of the whale-fisher. It joins his ship immediately on passing the Shetland Islands, and accompanies it through the trackless ocean to the highest accessible latitudes, ever keeping an eager watch for any thing thrown overboard; the smallest particle of fatty substance can scarcely escape it. It never dives but when incited to it by the appearence of a morsel of fat under water." Though like Mother Carey's Chicken, it follows in the wake of ships, its food is of a somewhat higher grade. being restricted to the garbage of the vessel, blubber, &c. This bird is the Mollemoke of Dr. Kane.

The Slender-Billed Fulmar, *P. tenuirostris*, is a species named by Audubon. Its length is eighteen inches and a half. It is common near Columbia river; is easily taken with a hook baited with pork, and during a gale is so tame as almost to allow itself to be taken with the hand.

The Southern Seas are visited by several species of Petrels. The largest, the Nelly or Break-Bones, *P. gigantea*, is a common bird, both in the inland channels and on the open sea. "In its habit and manner of flight," says Darwin in his Voyage of Adventure, "there is a very close resemblance to the Albatross, and, as with the latter bird, a person may watch it for hours together without seeing on what it feeds, so it is with this Petrel. The Break-Bone is, however, a rapacious bird, for it was observed by some of the officers of fort San Antonio, chasing a diver. The bird tried to escape both by diving and flying, but it was continually struck down, and at last killed by a blow on its head. At Port St. Julian, also, these Great Petrels were seen killing and devouring young gulls." These large Petrels are called by the sailors, Mother Carey's Geese.

The Shearwater, *Puffinus*, differs from the true Petrels by having a longer bill, and the tubular nostrils open, not by a common aperture, but by two distinct orifices.

The Wandering or Large Shearwater, *P. cinereus*, (ashy-colored,) is twenty inches in length, and of a sooty brown color. It is frequently seen off the shore from the Gulf of St. Lawrence to that of Mexico. According to Mr. Darwin, it is common to Cape Horn and the coast of Peru, as well as Europe. The flight of these *Wanderers* of the ocean is very rapid and long protracted. In calm weather they are fond of alighting on the water, in company with the Fulmars, and when at play among themselves, swim with great buoyancy and have a graceful appearance.

The Puffin, or Shearwater, *P. anglorum*, is a species that once largely inhabited a small islet near the southern part of the Isle of Man, but has of late deserted it. It is now abundant on the coast of South Wales. It has been found in the vicinity of Newfoundland. In the Orkney Islands it is called the Lyre, and is much valued, both on account of its serving as food, and for its feathers. This bird is described as standing nearly erect and flying with great rapidity. "It feeds on marine animal substances of all kinds, and when taken squirts out an oily fluid from its nostrils, in the manner of the Petrels." It breeds in burrows, laying one egg, which is white and about as large as that of the domestic fowl. The upper parts of the body are of a lustrous black; the under parts pure white; the sides of the neck speckled with black and white; length thirteen inches.

The Little Shearwater, *P. obscurus*, is of a brown color above; beneath, white; in length, ten or eleven inches. It is common to Europe and America; ranges northwardly from the coast of Mexico to that of New York.

The genus *Thalassidroma*, (Gr. *Thalassa*, the sea; *dromos*, a race,) including the smallest of the web-footed birds, has been separated from the rest of the Petrel group. They are of nocturnal or crepuscular habits, and seldom seen except in lowering or stormy weather, when they frequently follow in the track of ships. At other times and during clear weather, they are concealed in the holes of rocks and in burrows, and only come forth at night in search of food. Their flight equals in swiftness that of the Swallow tribe, which they resemble in size, color, and general appearance. They breed in the crevices of rocks or in burrows, like the rest of the family, laying but one egg, which is white and comparatively large.

The Stormy or Least Petrel, *T. pelagica*, (belonging to the sea,) or *P. pelagica*, is known to sailors under the name of Mother Carey's Chicken, and by them regarded as the precursor of a storm. This is the smallest of the Web-footed Birds, being only about six inches long. In the length of its wings and its swift flight, it is like the Chimney Swallow; in its plumage it is black with purple reflections, except the rump and a portion of the tail, which are white. It is met with on every part of the ocean, diving or swimming over the surface of the heavy rolling waves of the most tempestuous sea, quite at ease and in security. Long before seamen can discover any appearance of a storm, these birds, as if foreseeing and fearing its approach, flock together in large numbers, making a clamorous, piercing cry, thus warning the mariner of his danger. So oily is the Stormy Petrel said to be in its texture, that the inhabitants of the Faroe islands draw a wick through its body and use it as a lamp. A most singular peculiarity of this bird is its faculty of standing and even running on the surface of the water, which it does with the greatest facility. According to Buffon, it is from this practice that these birds are called *Petrels*, the name being derived from the Apostle Peter, who, as Sacred Scripture informs us, walked upon the water. This species is not observed to breed on the American coast, though it is not uncommon on the Banks of Newfoundland.

WILSON'S PETREL, or MOTHER CAREY'S CHICKEN, *T. Wilsonii*, is a little over seven inches in length; in the color of its plumage of a dark grayish brown, with some portions of white. It is less lively than the common Stormy Petrel.

The FORK-TAILED PETREL, *T. Leachii*, is eight inches in length; of similar plumage with Wilson's Petrel, but less active and does not breed so extensively on the American coast.

The genus *Diomedea*, (a proper name,) comprises, among other species, the ALBATROSS OF CHINA, *D. fuliginosa*, (Lat. sooty;) the YELLOW AND BLACK-BEAKED ALBATROSS, *D. chlororhyncos*, (Gr. yellow-beaked;) (this has been taken on the Pacific not far from Columbia river;) and the COMMON ALBATROSS, *D. exulans*, (Lat. wandering.) The beak in these birds is very strong, hard, long, and straight nearly to the end, where it suddenly curves. The upper mandible appears to be composed of many articulated pieces, furrowed on the sides and crooked at the point; the lower mandible is smooth and cut short; the wings are very long and narrow with the primary quill short and the secondaries long; the feet short; the three toes long and completely webbed.

Albatross is a word said to be corrupted "from the Portuguese Alcatraz, which was applied by the early navigators of that nation to cormorants and other large sea-birds."

The COMMON ALBATROSS, *D. exulans*, (Lat. wandering,) is the largest sea-bird known. The top of the head is a muddy gray, but the rest of the plumage is white, except a few of the wing-feathers, and several transverse black bands on the back. The range of these birds is very extensive. They are not confined to the Southern Ocean, as has been supposed, but are equally numerous in northern latitudes, (excepting, perhaps, the tropics.) From its often breeding with the Penguin, it has been supposed to have a peculiar affection for that amphibious creature, and a pleasure in its company. Their nests are seen together on uninhabited islands, where the ground slants to the sea. As if for mutual protection, the Albatross raises its nest on a hillock of heath, sticks, and long grass, about two feet high, and lays one egg; around this, the Penguins, in a circle, make their lower settlement in burrowed holes in the ground,—commonly, it is said, eight Penguins to one Albatross.

"The Albatross," says Cheever, "is the most beautiful and lovable object of the animate world which the adventurer meets with in all the South Pacific; when on the wing, it is the very ideal of beauty and grace. The capture of a whale a thousand miles from land, will bring them trooping from afar, as a carcass in Mexico or Louisiana, will the Turkey-Buzzards. I have watched them singly, keeping company with our ship, and have seen them gathered by hundreds when the cutting-in of a whale along side, allured them from a circuit of five hundred miles. They sit upon the water light and graceful as Swans, and feed on small marine animals, mucilaginous zoophytes, the spawn of fish, and blubber. Not unfrequently, they measure eleven feet from tip to tip of the outspread wings, and weigh from seventeen to eighteen pounds." Another voyager, (Ives,) mentions one shot off the Cape of Good Hope, "which measured seventeen feet and a half from wing to wing."

In the Arctic Exploring Voyage, Dr. McCormick met with one weighing twenty pounds, and having twelve feet stretch of wings. The Albatross does not seem to be a quarrelsome bird, but when attacked by its enemy, the Skua Gull, it seeks safety in flight. Sometimes, however, it does so by dipping its body in the water, its formidable bill appearing to repel its assailants. When it wishes to rise on the wing, "it has to tread water a long way, like a running Ostrich, before it can attain its due momentum and soar aloft; and when captured, and set at liberty in the ship, it can never, of itself, rise from the even surface of the deck, but we must toss the noble bird overboard, or lift him quite clear of the ship's rail, before he can raise his glorious pinions, and mount aloft in the air."

Billets of wood with inscriptions upon them, are often attached to these birds before setting them loose; in repeated instances, such birds have been captured in different and distant latitudes by other ships, and curious information has thus been communicated. "They are caught by baiting a hook with pork or blubber, and fastening a piece of wood near the bait, so that it may be kept floating, and letting it tow astern. Superstitious sailors sometimes ascribe the high winds and bad weather to their having killed an Albatross."

FIFTH FAMILY. GULLS.

Laridæ, (Gr. λάρος, *laros*, a mew or gull.)

These web-footed and well known sea-birds, are numerously dispersed over every quarter of the world, and, in some parts, are met with at certain seasons, in prodigious multitudes. They assemble together in rather promiscuous and straggling flocks, and greatly enliven the beach and rocky cliffs, by their irregular movements, while their shrill cries are often deadened by the noise of the waves, or nearly drowned in the roaring of the surge. Occasionally, taking a wide range over the ocean, they are seen by navigators many leagues distant from the land. They are all greedy and gluttonous, devouring, almost indiscriminately, whatever comes in their way, whether of fresh or putrid substances, until they are obliged to disgorge the contents of their overloaded stomachs; still, they can endure protracted hunger. The large kind of Gulls are most common in the cold climates of the North, where they breed and raise their young, feeding chiefly upon the remains of dead whales, which they find floating on the sea, among the ice, or driven on shore by the winds and waves. The True or Typical Gulls, (Larus.) are much more decidedly land birds than any other of the order. Those of the sub-genus *Xema* or Laughing Gulls, in particular, roam much inland; feed on insects and worms; build among herbage in low nests near the sea; lay eggs of an olive color, marked with large brown spots; and undergo seasonable changes of plumage; all of which may be said of the Plovers. To the Wading Birds, the Plovers especially, the Gulls, (*Larus*,) approach in their general form, in attitude, in the long and slender tarsus, with the hind toe small and set high up, (as in the Lapwing. *Vanellus*,) in the naked space above the heel, and even in the form of the beak, straight, slender, and swelling towards the tip,—and also in the internal structure.

We quote from Swainson some remarks, pointing out clearly the differences in the three sections into which the Gulls have been arranged, viz: FORK-TAILED GULLS, (*Rynchops*;) the THREE-TOED GULLS, (*Larus*,) and the FOUR-TOED GULLS, (*Lestris*.)

"The **Terns**, or Sea-Swallows, (*Sterna*,) constitute the fissirostral type; they have remarkably long wings, and slender bills; the tail is forked; and the plumage, generally, is of a delicate pearl-white, with more or less black upon the head; the species are numerous, and occur in both hemispheres. The extraordinary genus, *Rhynchops*,, or Skimmer, although possessing much of the general habits of the Terns, is eminently distinguished by the singular form of its bill, the upper mandible of which is considerably shorter than the under, and appears as if one-third of the length had been broken off; three species have been described, to which we add a fourth; they skim over the surface of the ocean with great swiftness, and scoop up small marine insects and other animals. The True or Typical Gull, (*Larus*,) are a numerous race, dispersed over every clime, and so closely resembling each other in plumage, that many of the species are even now but imperfectly understood; they are much like the Terns in general appearance, but the bill is stronger, and the upper mandible is much more curved towards the end; many are of larger size; and all, rapacious devourers of fish, and of every marine animal, dead or alive, which is cast upon the shore; they particularly abound in northern latitudes, but seem to range over the wide world of waters. The Parasitic Gulls, (*Lestris*,) are the raptorial representative in this family, and are almost confined to cold regions; they are known by their stronger conformation, their different shaped bill, and the rough scales upon their feet; these birds, like the frigate cormorants, derive their chief supply of food by robbing their more feeble congeners; they pursue the largest Gulls, and make them disgorge or relinquish their hard-earned prey. The Black-toed and the Arctic Gulls belong to this group, and both are occasionally seen on the northern shores of Britain."

Fork-tailed Gulls.

The **Black Skimmer**, *Rhynchops*, (Gr. *rhunchos*, beak; *ōps*, face;) *nigra*, (Lat. black.) This singularly endowed bird (referred to above) is dispersed in large flocks from Texas to New Jersey. It reaches the coast of New York State in May; breeds on sand beaches or islands; at night, ascending streams, sometimes to the distance of one hundred miles. The length of the male bird is twenty inches. The bill, for half its length, is a rich carmine, inclining to vermilion; and the feet are of the same color; the claws, black. The upper parts are a deep brownish black; the secondary quills, and four or five of the primaries, tipped with white; the tail-feathers of the male, are black, broadly margined with white, (in the female they are white ;) the under parts are white, with a roseate tinge. This bird is known under the names *Shearwater*, *Razor-bill*, *Cutwater*, *Skimmer*, *Floodgull*, and *Shippang*. Its eggs are three or four, white, blotched with shades of brown, laid in a slight hollow in the sand. Audubon says, "The flight of the Black Skimmer is perhaps more elegant than that of any water-bird with which I am acquainted. The great length of its narrow wings, its partially elongated forked tail, its thin body and extremely compressed bill, all appear contrived to assure it that buoyancy which one cannot but admire when he sees it on the wing. It is able to maintain itself in the heaviest gale; and I believe no instance has been recorded of any bird of this species having been forced inland by the most violent storm." These birds show much sagacity in finding their place of rendezvous in the morning, after having been scattered during the night in all directions in quest of food ; and evince great enmity to Crows and Turkey Buzzards, driving them as marauders from their breeding grounds. All possess great power and endurance in flight; their long forked tails and pointed wings, indicating both strength and swiftness.

Of the TERNS, twelve or more species might be enumerated. But we can only refer in particular to—the COMMON TERN, *Sterna hirundo*, (Lat. swallow,) found in abundance on the southern shores of Europe, and in many parts of Asia and Africa. This species, from fourteen to sixteen inches in length; is sometimes called the Big Tern,—in Massachusetts, the Mackerel Tern. It ranges on this Continent from the tropics to the Arctic circle.

The CAYENNE TERN, *S. Cayana*, is larger than the Common Tern, in its size and its robust tarsi, resembling the smaller Gulls. It breeds from Florida southwardly, but is met with from the intertropical regions to 55° N. L. Length from sixteen to nineteen inches.

—The NODDY TERN, *S. stolida*, (Lat. dull,) receives its common name from the breeding places of this species, one of the Tortugas Keys, called Noddy Key. The Sooty Terns, *S. fuliginosa*, breed on an island a few miles distant. The Noddy ranging from Florida southwardly, has been frequently celebrated by travelers who have crossed the equator. Its color is sooty brown; the bill, black; the crown, white; the tail, wedge-shaped and long. The Noddies form regular nests of twigs and dry grass, which they place on the bushes or low trees, but never on the ground. The female lays three eggs, of a reddish yellow color, spotted with dull red and purple. "When seized in the hand, the Noddy utters a rough cry, not unlike that of a young American Crow taken from the nest. On such occasions, it does not disgorge its food, like the Cayenne Tern and other species, although it bites severely, with quickly repeated movements of the bill, which, on missing the object aimed at, snaps like that of our larger Fly-Catchers." Length about sixteen inches.

—The SILVERY or LITTLE TERN, *S. argentea*, (Lat. silvery,) is closely allied to the *S. minuta*, (Lat. small or minute,) of Europe. The upper parts and tail, are a deep pearl gray; all beneath, silvery white. It is larger than the corresponding European species, and the entire upper parts, (with the tail,) are of a lighter shade. Length from nine to ten inches. The eggs are light yellowish white, with angular dark brown spots. The Silvery Tern breeds from Texas to Labrador.

THREE-TOED GULLS.

The GULLS, *Larus*, are represented by thirteen or more species on this continent. In these the hind toe is very small, and articulated high up on the tarsus; in one species entirely wanting.

The GREAT BLACK-BACKED GULL, *L. marinus*, is the largest Gull that is seen on the American coast, and described as exceedingly bold, voracious, and predatory in its habits. Its length is from twenty-eight to thirty inches; the expanse of wings about five feet and a half. It breeds on the coast, from Labrador northwardly; ranging in the winter, to New York, and migrating as far South as Florida. It is also common in many parts of the North of Europe, where it finds a home. Its nest is made of grass, rushes, and other materials, and contains three or four eggs, of an olive green, marked with very dark brown. Audubon remarks, "This bird must be of extraordinary longevity, as I have seen one that was kept in captivity more than thirty years." The back and wings are a deep bluish black; the quills, with black shafts, tipped with white; in the *summer*, the head and neck are pure white; in *winter*, the same parts are white, with brownish streaks.

The COMMON AMERICAN GULL, *L. zonorhyncus*, (Gr. zone or ring-billed,) has a mantle of bluish gray; the head, tail, and under parts, white; the outer quills are black, tipped with white. In the quills, however, the plumage changes with the age and season. It is popularly called the BROWN WINTER GULL,—a name referring to the plumage of the young, rather than of the adult. The ring on the bill is not always found. The length of this species is nineteen inches. It is allied to the *L. canus*, (Lat. gray,) or Gray Gull, of Europe, breeds from Maine northwardly,—and during the winter, is seen as far South as Mexico; sometimes it appears on the Pacific coast.

The Four-toed Parasitic Gulls. Jagers, or Skuas.

These birds all breed in high northern latitudes, spreading themselves into the interior on lakes and rivers; but in the winter are seen in temperate regions, and on this Continent as far South as Mexico. The bill is of moderate length, cylindrical, and hooked at the tip; the hook and tip, of separate pieces; the hind toe is small, and on a level with the others; the tail is even or rounded; the central pair of feathers very much lengthened. Of the several species, we can refer particularly only to

The Arctic Jager, *Lestris*, (Gr. a pillager,) *parasiticus*, which is seen in great numbers in the northern regions. Like the other Skuas, it obtains the greater part of its subsistence by pursuing and buffeting the peaceable Gulls, and compelling them to give up the produce of their toils. But the Jagers also feed on fish, insects, and worms. Temminck particularly mentions the *Janthina*, or Oceanic Snail, as forming a portion of their sustenance. "In truth, no animal substances seem to come amiss to them." The nests of these birds are composed of dry grass and mosses, and placed on unfrequented heaths, at some distance from the shore; the eggs are two, of a dark olive green, with irregular blotches of dark brown. Captains Parry and Ross speak of this bird as abundant at Baffin's Bay and in the islands of the Polar Sea. It is said, that it "is frequently met with inland, seeking its food along the water courses which occupy the bottom of ravines; differing in this respect from the Pomarine Jager, *L. pomarinus*, which is exclusively a Sea-bird." The length of the Arctic Jager is twenty-three inches. The plumage is "close, elastic, soft, and blended;" on the upper parts blackish gray; the neck and lower parts, white, the former tinged with yellow.

Sixth Family. Pelicans.

Pelecanidæ, (Gr. πελεκὰν, *pelecan*, a pelican.)

The Pelican family are characterised by having the hind toe united with the others in a single membrane, so that the whole four toes are webbed. The bill is, generally, longer than the head, strong, and sometimes compressed; the mandibles are dentate, (toothed;) the nostrils mere slits, the aperture to which is scarcely perceivable. With the exception of the Phaeton or Tropic-bird,—which, in many respects, agrees with the Gulls,—there is more or less of naked skin about the face and throat;

the skin of the throat is capable of being dilated; the wings are long and powerful; the feet short and robust; the tail consists of twelve, fourteen, twenty or twenty-four feathers.

Though their completely webbed feet seem to be perfect oars, peculiarly adapting these birds to an aquatic life, yet a very large part of them do not swim or dive at all, but perch on trees. They all fly well, and some, from the broad expanse of their wings, have uncommon powers of flight. Soaring far out over the ocean, when a fish first arrests their attention, they plunge down upon it, and instantly rise again into the air.

The birds of this family nestle and roost either on rocks or lofty trees; the eggs are encased with a soft, absorbent, chalky substance laid over the hard shell; the young are, at first, covered with long and flossy blackish down. They remain a great while in the nest, and when they leave it, are generally equal or superior to the adults in weight. The species are not very numerous, but are found in the seas and around the coasts of most parts of the globe. The plumage is usually black, (often glossed with metallic reflections,) and white.

This family may be arranged into the Pelicans proper, (*Totipalmes*, of Cuvier,) the Cormorants, the Darters, the Frigate Birds, the Gannets, and the Phaetons.

Pelicans proper.

The True Pelicans, *Pelecanus*, are large and heavy birds, with very long, rather narrow, and rounded wings; the tail is short, broad, rounded, with twenty to twenty-four feathers, which are broad, and abruptly pointed. A pouch which hangs under the lower mandible, is capable of containing a large quantity of water. It has been said by some writers, that this pouch "enables these birds to dispose of a superabundance of fish, which they take, either for their own use, or the nourishment of their young," and this has been the generally received idea. Audubon, however, who often noticed flocks of these birds, says "the idea that the Pelicans keep fish or water in their pouches to convey them to their young, is quite erroneous." He states, as the result of his observations, that the water is immediately forced out between the partially closed mandibles; and the fish, "unless larger than those on which they usually feed," are instantly swallowed, though afterwards *disgorged* for the benefit of the young. The Pelicans have long been celebrated as symbols of maternal love. Books of emblems have depicted this bird as tearing open the breast to nourish its young with its blood, but this representation is not well founded. The fact appears to be, that the bird, in the process of feeding its young, crushes the fish between its mandibles, and thus stains its white breast with drops of blood. The Pelicans are rarely seen more than sixty miles from land. They are gregarious, and numerous in Asia and Africa, as well as in Europe and America.

The COMMON WHITE PELICAN, *P. onocrotalus*, (Gr. *onokrotalos*, a pelican,) is an European species, with which that of *P. Americanus*, or the American White Pelican, very nearly agrees. The American, however, differs from the European bird in having a "long, thin, bony process in the upper mandible." "The male of the American species is sixty-one and three-fourths inches long; bill thirteen and three-fourths inches; expanse of wings one hundred and three inches." In this species, the feet and pouch are pale yellow, as are the long feathers on the breast, and the tuft on the back of the head.

The BROWN PELICAN, *P. fuscus*, (Lat. brown,) is, when mature, fifty-two inches in length; the expanse of wings is eighty inches. It is very abundant on the American coast as far northward as North Carolina; breeds on trees, and also on the ground; the pouch is usually from six to ten inches in depth, and will hold a gallon of water. This membrane is sometimes dried, and used for keeping snuff, gun-powder, and shot. The quantity of fish which the Brown Pelicans consume, is extremely large. They often times become so overburdened with food, that flight is difficult. Audubon examined one which had in the stomach upwards of a hundred small fishes; sometimes "he found in that organ a great number of live, blue colored worms, measuring about two and a half inches in length, and about the thickness of a Crow-quill." The bodies of these birds are greatly inflated by air-cells; their bones are very light; and they are hard to kill. The Black-headed Gull, which is abundant along the coast of Florida in spring and summer, closely watches the motions of these Pelicans, in order to seize the small fishes which in letting off the water from the bill, they sometimes allow to escape; for that purpose, the Gull alights on the Pelican's bill, or on his head, and seizes the prey when apparently just on the eve of deliverance,—the Pelican, meanwhile, exhibiting no symptoms of anneyance or anger.

The CORMORANTS are included in the genus *Phalacrocorax*, (Gr. *phalakros*, bald; *korax*, raven.) They are widely spread over many parts of the world, and every where remarkable for their voraciousness. The bill in these birds is about as long as the head, rather slender, nearly straight, and compressed towards the end, the upper mandible ending in a powerful hook; the sac under the throat is small, by no means comparing in size with that of the Pelicans proper; the nostrils are obliterated, but in youth open, (Aud.;) the wings of moderate size and broad; the tail of moderate length, very narrow and much rounded, having twelve or more strong shafted feathers. These birds differ from others of the family in being excellent divers. Their plumage is soft and generally blended, compact on the back and wings; usually of dark, but often rich colors, varying with age and the season of the year. They are capable of domestication and are trained to catch and bring in fish. The Chinese who use them for this purpose, put a ring around the neck as a hindrance to their devouring the fish. To increase the power of swallowing, it should be noted that the Cormorant has an additional bone peculiar to itself, on the back part of the head, called the xyphoid (sword-like) bone, which, moving with facility in each direction, by the action of the muscles attached to it, enlarges the opening of the gullet for the more easy passage of any unusually large fish.

The COMMON CORMORANT, *P. carbo*, (Lat. charcoal,) is spread over a considerable portion of Europe, especially the north. It is a common bird in England; in this country ranges in the winter and is plentiful as far south as New York; breeds in Newfoundland, Labrador and Baffin's Bay. It swims very low in the water; even in the sea its body is deeply immersed, so that little more than the head and neck can be seen above the surface; and most expertly does it dive after its fishy prey. It perches on trees, where it is occasionally known to build its nest, but it mostly selects rocky shores and islands, preferring, according to Selby, the summits, and not, like the Green Cormorant, the clefts or ledges. The nest is said to be composed entirely of a mass of sea-weed, frequently heaped up to the height of two feet, in which are deposited from three to five eggs, of a pale bluish-white, with a rough surface. Ravens and Peregrine Falcons have been observed to have nests on the same rocks with those of the Cormorant, and in some instances, close to them. This bird is sometimes three feet and four inches in length.

The GREEN CORMORANT or SHAG, *P. cristatus*, (Lat. crested,) does not perch on trees like the others. As illustrating the depth to which this bird dives, Mr. Yarrell says: "The Shag has been caught in a crab-pot fixed at twenty fathoms, or a hundred feet from the surface." The specific name is given to it from the crest or tuft of wide outspread feathers which appears in the spring on the back part of the head, and is capable of erection. The Shag is without the white feathers on the neck and thighs which are seen in the Common or Great Cormorant. The length is two feet, one or two inches.

The VIOLET GREEN CORMORANT, *P. resplendens*, (Lat. glittering,) is the most beautiful species which has been found within the limits of the United States. The gloss of its silky plumage suggested the specific name. This bird has been found in abundance near the Columbia river. The length of the female is two feet three inches.

THE DARTERS, OR SNAKE-BIRDS.

These birds, which are included in the genus *Plotus*, have bills longer than the head, slender, pointed, and finely serrated at the extremity; the tarsus is partly feathered above; the neck is much lengthened; the tail long, spreading and much rounded. The necks of these birds, often rapidly moved and bent, suggested the name of *Darter*, or *Snake Bird*. (Fig. on Chart.)

The AMERICAN ANHINGA, or SNAKE BIRD, *P. Anhinga*, is a common and constant resident from Florida to Georgia, and it passes up the Mississippi as far as Natchez. In the southern parts of Florida it is called the "Grecian Lady." This bird is seen only occasionally in the immediate vicinity of the sea, decidedly preferring rivers, small bays, or lagoons in the interior where the land is level and lies low. It is quite remarkable in its appearance and manners, often standing erect with the wings and tail spread out in the sunshine, and throwing its long slender neck and head, in every direction, by sudden jerks and bendings. Though adapted for protracted and powerful flight, as is shown by its form, long wings and large fan-like tail, this bird spends more than half its time by day in the water. On the approach of any danger, it sinks its body and swims with its head and neck only above the surface, when these parts, "from their form and peculiar sinuous motion, somewhat resemble the head and part of the body of a *snake*." The nest of the Snake Bird is found in different situations, sometimes in low bushes not more than eight or ten feet above the water; at others, on large and tall cypresses, overhanging the borders of rivers or other streams. The nest is of a circular form and two feet in diameter; the eggs of a sky-blue color.

FRIGATE BIRDS.

These birds, though in some respects nearly resembling the Cormorants, yet at the same time, differ from them in the very broad expanse of the wing, by which they are rendered the most powerful of the Swimming Birds. They also differ from the Cormorants in their feet, the webs of which are deeply notched, and in the form of the tail and beak. The tail is very long, deeply forked, and of twelve feathers; the bill is longer than the head, strong, and broader than high, except towards its curved extremity. The Frigate or man-of-war birds, seem particularly fond of the Flying fish, darting at it themselves when near the surface of the water, or obtaining it from other birds which they force to drop their prey. Often they sadly persecute the *Boobies.* Indeed, these birds are eminently raptorial. Ray speaks of their eagle eyes, vulturine claws, and kite-like glidings. Their immense extent of wing and dashing habits have obtained for them the name of the swiftest ships of war that sweep the seas.

The Frigate or Man-of-War Birds, are included in one genus and species, *Tachypetes,* (Gr. *tachus,* swift; *petáō,* to fly,) or *Fregata aquilus,* (Lat. from *aquila,* an eagle.) Their length is three feet five inches; the expanse of wings is eight feet; some accounts make it fourteen feet! Audubon says: "The Frigate Pelican is possessed of a power of flight which I conceive superior to that of perhaps any other bird." This bird is very common on the intertropical American coasts, and in the Atlantic and Pacific oceans, but always within reach of land. It resides constantly on and about the Florida Keys, where it breeds in vast numbers, on trees. Sometimes the nest is built on elevated rocky cliffs.

GANNETS, OR BOOBIES.

These have bills differing somewhat from those of the Frigate Bird, being long and resembling a lengthened cone which is very large at the base and compressed towards the slightly curved point; the edges of the mandibles are serrated; the hind toe is articulated to the inner surface of the tarsus, and all the four toes are united by a membrane; the wings are long; the power of flight is however not equal to that of the Frigate Bird; the tail is wedge-formed.

The COMMON GANNET of Europe, *Sula alba,* (Lat. white,) is sometimes called the SOLAN GOOSE. Its length is about thirty-four inches. The head and neck are of a buff color, all the rest of the plumage white, except the wing primaries, which are black. This species is also included among the birds of Madeira and South Africa.

The AMERICAN GANNET, *S. Americana*, is thirty-seven inches in length. Near the base of the upper mandible is "a sharp process and suture," which this bird can move in a small degree in swallowing a fish. This was formerly supposed to be identical with the European Gannet, but is now considered a distinct species.

The Booby Gannet, *S. fusca*, (Lat. tawny,) has the head, neck, and all the upper parts dusky brown; the under parts white; the face, bill and feet yellow. Its length is thirty-one inches.

The term *Booby* is more particularly applied to this species on account of the stupidity which it shows when assailed, calmly waiting to be knocked on the head, as these birds often do when sitting on shore, or when perching on the yard of a ship till the sailor climbs to their resting place, and takes them off with the hand. Notwithstanding all that has been said and written about the stupidity of this bird, its dullness may be questioned; it may not, like other birds, associate danger, certainly not at first, with the appearance of man; its wings are so long, and its legs so short, that when once at rest, it has difficulty in setting the former in motion, and when surprised has no resource but its beak, which is seldom feared by the aggressor. Audubon says: "I am unable to find a good reason for those who have chosen to call these birds *boobies*." It has been affirmed by many writers and eye-witnesses that this bird suffers greatly from the persecutions of the Frigate-bird, and the Lestris or Skua Gull, which force it to disgorge its food. All the old voyagers abound in entertaining stories relating to this subject, and it is hardly credible that *all* were *mistaken*. Audubon, however, says, "this *I* have never witnessed." The nest of the Booby is placed on the top of a bush at a height of four to ten feet; sometimes on ledges of rocks covered with herbage. It lays one egg, of a dull white color, about as large as that of a common hen. This bird ranges from Georgia southwardly, but is occasionally seen farther north. It is found in large numbers on Noddy island, one of the Tortugas, in company with the Noddies.

TROPIC BIRDS, OR PHAETONS.

These birds, (*Phaeton*,) are distinguished by two long slender tail feathers, and well known to navigators as the harbingers of the tropics. They are characterized by extraordinary length of wing and feeble feet; they are hence well formed for flight, and disport in the air far out at sea; on land they are seen perching on rocks and trees.

The COMMON TROPIC BIRD, *P. æthereus*, is somewhat larger than a partridge. The bill is red, with an angle under the lower mandible, as in the Gulls. The eyes are surrounded with black, which ends in a point towards the back of the head; three or four of the largest quill feathers, towards their ends, are black tipped with white; all the rest of the plumage is white, except the back, which is variegated with curved lines of black; the legs and feet are of a vermilion red. These birds are seldom seen but a few degrees north or south of either tropic. They glide along, most frequently without any motion of the wing, but at times, this smooth progression is interrupted by sudden jerks. When they perceive a ship, they never fail to sail around it as if to reconnoitre. They ordinarily return every evening to land to roost in the midst of the rocks where they place their nests. The long feathers of the tail are used by the inhabitants of the South Sea Islands as ornaments of dress.

What is the 2nd order of AQUATIC BIRDS? What is said of their distribution? What of the Grebe's foot? What is said of the feet of the other SWIMMERS? What of their motion on land and in the water? What is remarked of their plumage? What peculiarity of these birds is noticed by Cuvier? Have any of them very high powers of flight? What is said of their flesh? What are their resorts? How many families does this order include?

Give the leading characteristics of the 1st FAMILY. What is said of their nests and young? Upon what do they feed? How does their beak aid them in obtaining their food? Are they assisted by any other organ? What birds form the connecting link between the SWIMMERS and WADERS? Do the TRUE DUCKS include many species? What is said of the Shoveler or Spoon-bill? What sp. are referred to and what is said of each? What SUB-FAMILY is mentioned? How many genera does it include? Where are the Sea Ducks mostly found? What is said of their migration? Where do they make their nests? How do they cover their eggs? Which gen. includes peculiarly Marine Ducks? For what is the Eider D. remarkable? What is peculiar in its beak? What does Dr. Kane say of its appearance, &c.? What facts show the great numbers of these birds at the north? What is said of their nests? How many kinds of down and how do they differ? Illustrate the elasticity of the live down. Where are the King-Ducks found? What is said of their flesh and down? What is said of the Surf D.? What sp. is mentioned? Why is it called the Velvet D. and what is said of its down? Where is it very numerous? Where are the breeding places of the Canvas-backs? When are they seen on the coast of the U. S.? What else is said of them? What of the Spirit D.? Of the Long-tailed D.? Mention the different sp. of GEESE. Which of these is the origin of the Common Domestic Goose? What is said of the Egyptian Goose? What SUB-FAMILY is named? How many gen. has it? What is said of the Smew? What of the Merganser?

What is the 2nd FAMILY? Are they more or less aquatic than the Ducks? What is said of the beak, &c.? What of the Loons and Grebes? What is said of their ability to walk and fly? What of their power as swimmers and divers? How do they use their wings under water? What is said of their plumage and food? Which are Ocean birds? Which Freshwater? What of the diving of the Grebes? Of their stomach? Which is the largest of the Loons? What is said of it? Which are the other sp.? What is said of them?

Which is the 3rd FAMILY? What is said of their structure and habits? What characteristics are given? In moving under water do they use their feet? What insects do they resemble in such motion? Of what does their food consist, and how do they obtain it? Where are they seen in immense numbers? What is said of their eggs? In which hemisphere are the AUKS found? What birds fill their places in the Southern H.? What is said of their wings? Of their movements in water? For what are the PENGUINS peculiarly remarkable? What is said of the Great Auk? Of the Razor or Black-billed A.? Of the Common PUFFIN? Of the Little Guillemot? What relation do the Penguins sustain to the Swimming Birds? What does Swainson remark? What is said of the bones of the Penguins? What of their sensations? What of their habits? Which of their rookeries have attracted particular attention? Are they arranged with order? What is said of their extent? Describe the characteristics and habits of the Penguins, as given by Cheever.

Which is the 4th FAMILY? Give its characters. What are its habits? How many sp. in America? Which is the type of the TRUE PETRELS? In what localities is it found? To what class of persons is it a constant companion? What is this bird called by Dr. Kane? What other sp. are mentioned? Do they frequent the Southern Sea? Which is the largest? What does Darwin say of it? What names do sailors give these large P.? How does the Shearwater differ from the True Petrels? What is said of the Large S.? Of the Manx Puffin? Of the Little S.? What genus has been separated from the rest of this group? What is their size? Mention their habits? Which is the smallest of the Web-footed Birds? What do sailors call it? What is said of its plumage? What interesting particulars are given? What is said of Wilson's Petrel? Of the Forked-Tailed P.? What sp. of the Albatross is mentioned? What is said of the origin of the name? What is the size of the Common Albatross? Describe its plumage. What is its range? For what bird has it been supposed to have peculiar affection? What is said of its beauty and loveliness? What more is said of it?

Which is the 5th FAMILY? What is said of the distribution and habits of these birds? Where are the larger Gulls most common? Which of the Swimmers are most decidedly land-birds? What Gulls roam inland? What order of birds do the Gulls resemble? Into what sections are they arranged? What distinctions does Swainson make? Which of the Forked-Tailed Gulls are mentioned? What is said of the Black Skimmer? How many sp. of the TERN? What is said of the Common Tern? Of the Cayenne T.? Whence does the Noody T. derive its name? What is said of it? What of the Silvery T.? How many sp. of the Three-toed Gulls? What is said of the Great Black Backed G.? Of the Common G.? In what latitude do the Jagers breed? What characters are given? How does the Arctic Jager obtain its food? Upon what mollusk do they feed? What else is said of it?

What is the 6th FAMILY? How is it characterized? Do they swim or dive? What is said of their powers of flight? In what places do they build their nests? What is said of the eggs? Are the sp. numerous? Into what groups are they arranged? What is the size of the TRUE PELICANS? What use do they make of their pouch? What is Audubon's opinion relative to this subject? For what have the Pelicans been celebrated? How have they been depicted? Has this been done with good reason? What is said of the Common White P.? Of the Brown P.? What gen. includes the Cormorants? How do they differ from others of the family? What bone is peculiar to the C.? What is said of the Common C.? Of the Green C. or Shag? Which is the most beautiful sp. in the U. S.? What gen. includes the DARTERS or SNAKE BIRDS? What suggested the name? What is said of the Anhinga or American Snake B.? What group do the FRIGATE BIRDS resemble? How differ from them? Of what fish are they particularly fond? What other fish do they persecute? What does Ray say of them? Which is the only gen. and sp.? In what respects do the Gannets differ from the Frigate B.? What name is sometimes given to the Common G. of Europe? In what other regions is it found? What is said of the A. Gannet? Of the Booby? How are the TROPIC BIRDS distinguished and characterised? What is said of the Common Tropic B.?

GENERAL EXERCISE ON THE CHART.

What is the first division of Birds on the Chart? Into how many orders are the LAND BIRDS arranged? Name each, giving some peculiarity or characteristic. Name the SUB-ORDERS, and the forms or peculiarities of the bills upon which the divisions of Perching birds are based. Name and trace the families in each order. Which order is most numerous? Which the least numerous? Which contain the largest birds? Which the smallest? Which are the most beautiful? Which the most ordinary? How are the WATER BIRDS divided? What kind of feet have they? Which wade? Which swim? What is the form of each? Which the most awkward? Which most useful? How do the bills vary in all the different orders? How the toes, wings, legs and necks?

EXPLANATION OF PLATE XII.

Fig. 1st. Skeleton of a Tortoise, with the under part (sternum or plastron,) removed to show how the back-bone and ribs are expanded and united together, forming the carapace, dorsal plate, or buckler.

H. The three-branched shoulder. I. Humerus, between the shoulder-joint and the elbow. K. Ulna. L. Radius, both bones of the fore-arm. R. Femur, or the thigh bone. S. Tibia, the largest, and Fibula, the smallest bones of the leg.

Fig. 2d. Skeleton of a Bird, consisting of A. Cranium or Skull. B. Cervical vertebræ. C. The anchylosed or immovably fixed vertebræ of the back. D. The caudal vertebræ. E. Ribs. F. Breast-bone. G. Furcula, or merrythought. H. Clavicle, or collar-bone. H*. Scapula, or shoulder-bone. I. Humerus. K, L. Bones of the fore arm, ulna, and radius. M. Metacarpus of hand. N. Phalanges of fingers. R. Femur, or thigh-bone. o, o. Patella, or knee-pan. S. Leg, tibia and fibula. T, T. Os calcis, or heel-bone. U, U. Metatarsal bones. V, V. Metacarpal bones. O. Ilium. P. Pubis, and Q. Ischium, bones of the pelvis.

Fig. 3d. Skeleton of a Fish, showing the five sorts of fins, some of which are often absent. a, b, first and second dorsal fins on the ridge of the back, varying in number and form; c, the caudal or tail fin, as important to a fish as the rudder to a ship; d, anal fin, on the under part of the tail; e, one of the ventral fins which correspond to the hind feet of quadrupeds; f, one of the pectoral fins, which are analogous to the fore feet of quadrupeds, or the wings of birds.

Fig. 4th. Skeleton of a Frog, showing the absence (or mere rudiments,) of ribs, and its long hind limbs adapting it for sudden springs and long leaps.

Fig. 5th. Skeleton of a Boa-constrictor, consisting of skull, a; vertebral column, b; and ribs, c. The ribs, 304 in number, come forward in succession, like the feet of a caterpillar, and form 152 pairs of levers by which the animal moves from place to place. The jaws, d, have on each side, a double row of sharp, strong, close-set teeth, pointing backwards, thus giving a firm hold of its victims.

Fig. 6th. Skeleton of a Chameleon, showing how the toes and tail are adapted for clinging to the branches of trees, and that the trunk is mounted high upon the legs, forming in this respect an exception to most reptiles.

PL.XII.

1
H
I
K
L
S
R
2
A
H
B
C
I
L
H*
K
G
H
O
E
F
Q
M
D
o
o
P
N
R
S
T
T
U
U
V
V
3
a
b
c
d
e
f
4
5
a
d
b
c
6

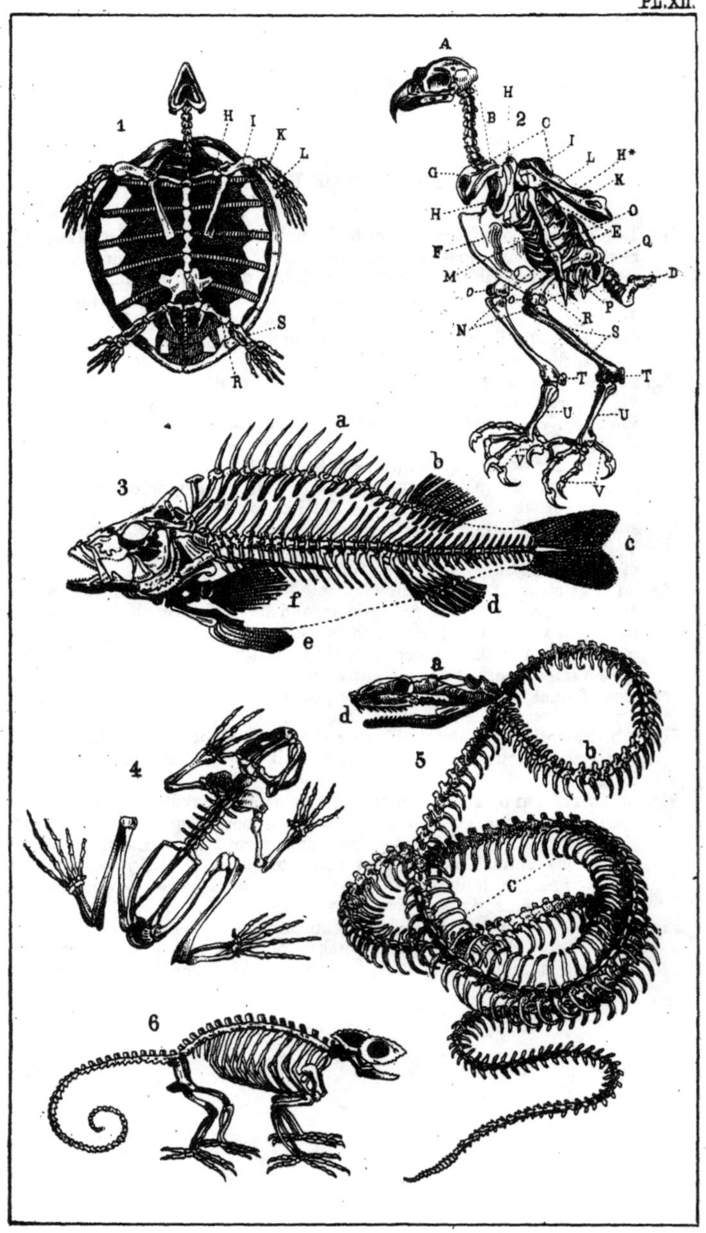

THIRD BRANCH OF ZOOLOGY.

HERPETOLOGY, (Gr. ἑρπητος, *herpetos*, a reptile, from ἑρπω, *herpō*, to creep.)

II. GRAND DIVISION OF THE VERTEBRATES. (Cold-blooded Animals.)

SECTION I.

THE second grand division of the Vertebrates, or Cold-blooded Animals, includes REPTILES and FISHES.

REPTILIA, (Lat. *reptilis*, a reptile, from *repo*, to creep.) REPTILES. The science which treats of Reptiles is called Herpetology, (or Erpetology.) They form one of the most remarkable of the vertebrate classes of the Animal Kingdom. They are highly characterised by the vertebral column, the articulations of which, in most recent adult forms, are spherically convex at one extremity and spherically concave at the other. The number of vertebræ varies exceedingly. As, for example, in the Surinam toad, (*Pipa,*) there are seven, and in the *Python* upwards of four hundred. The ribs also occur in various stages of development. A general survey of these creatures brings to view wonderful varieties of form and structure by which they are adapted to different localities. Some dwell on the land; others in the ocean. Many of them are found in rivers and morasses, and some are even arboreal in their habits, living amidst the foliage of the trees, intertwined with the branches, or flitting, with bird-like swiftness, from leaf to leaf or from branch to branch, in pursuit of their insect food. Though found in different latitudes, the hotter regions of the globe are the great nursery of the Reptiles; in tropical countries, they actually teem, swarming in sandy deserts, among dense and tangled brushwood, in humid forests, and extended pestilential swamps. They are termed COLD-BLOODED ANIMALS, their natural temperature being not much, if at all, above that of the atmosphere or water in which they dwell. Their power of producing animal heat is very limited, so that the system is at once affected by the lowering of the temperature of the medium which they inhabit. In our climate, and indeed in climates considerably nearer the meridian, they all undergo a state of torpidity, in some sheltered retreat, to which, as a refuge, their instinct directs them, and where they remain during the season of winter. Their blood, though cold, is red. In these, and in fact in all cold-blooded animals, the vital principle is much stronger than in those whose blood is warm. A frog has been kept alive forty days after having been subjected to the total privation of its lungs. The brain, which in reptiles is considerably inferior to that of birds, though superior to that of fishes, is not so essentially requisite to the exercise of their animal and vital functions as in the mammalia; for they continue to live and to execute voluntary movements for a considerable time after being deprived of it, and even after the loss of the head; their muscles also are strong and preserve their irritability for some time after life would appear to be extinct; their heart continues to pulsate for hours after it has been torn from the body. In the reptiles this organ is strikingly peculiar. In warm-blooded vertebrates it consists of two auricles and two ventricles; the left ventricle furnishing the system with blood, which, in the capillary vessels of the lungs has been acted on by the external atmosphere. In Reptiles the heart consists of but one ventricle and two auricles; and of these the right auricle receives the vitiated blood returned from the system to the heart; the left auricle receives the arterialized blood returned from the lungs; and both auricles convey their contents into the cavity of the ventricle. The vitiated and the arterialized blood thus become more or less mixed together; part of this mixed fluid is sent through the great arterial trunk, as a supply to the system, and part through the pulmonary arteries to be further oxygenated in the lungs, this ventricle having in itself the branching arteries both of the body and the lungs. Such is the circulation in the tortoises, lizards, and snakes. The blood of Reptiles is characterized by the possession of the largest globules to be found in the entire sub-kingdom of Vertebrates. Those in the tailed Batrachians, as the *Siren*, &c., are visible to the naked eye. As in fishes and in birds, these globules are elliptical in their outline, whereas in the Mammals, excepting the *Camelidæ*, they are circular.

The *Amphibia*, at an early stage of their existence, are furnished with gills, and, like fishes, respire in water; the gills, in due time, become obliterated, and lungs developed, as in the frog, the newt, &c., (Plate XIII.) Others, however, of this group, though they acquire lungs, never lose their gills, and are at the same time both aquatic and aerial in their respiration, or capable of breathing both in air and water. Such are the Proteus, the Siren, and the Axolotl. The former are called *Caducibranchiate*, (Lat. *caducus*, falling or perishable; *branchiæ*, gills;) the latter are termed *Perennibranchiate*, (Lat. *perennis*, durable; *branchiæ*, gills.) In the latter Amphibians, the ventricle receives blood from the auricles, and transmits it into an enlarged arterial vessel or bulb, which soon divides into separate branches, one being destined for each leaf of the gills essentially like those of a fish; here these arterial vessels sub-divide into five capillaries, and these at length, (as in fishes,) gradually pass into branchial veins which at last emerge into two vessels, and these unite to form the *aorta*, or great arterial trunk. Into this aorta, the blood purified in the gills, or branchiæ, is conveyed without being first sent back to the heart; and from this aorta, it is distributed throughout the system. But besides the branchial, these Amphibians have also a pulmonic, (Lat. *pulmo*, a lung,) circulation. By the pulmonary artery proceeding from the *aorta*, a portion of the blood which has already been partially oxygenated in the gills, is conveyed to the lungs, where it is still further purified. It is then sent through pulmonic veins to the left auricle, and from that to the ventricle, whence, mixing with the vitiated blood of the system, it is sent to the gills, and thence to the *aorta*, from which a portion again passes to the lungs, the rest to the system, and so on in a perpetual succession.

The Caducibranchiates, at the commencement of their existence, have only gills truly developed, and the circulation is branchial or fish-like. The lungs are, at this period, in a rudimentary state, and the pulmonary arteries exceedingly minute. In process of time, however, a new impetus is given to the pulmonary arteries and to the lungs, at the expense of the branchial arteries and the gills; as the former develop, the latter decrease, until at the last, the branchial apparatus *entirely perishes*, no trace of it being left; while certain vessels, which formed a junction between the branchial system of arteries and the pulmonic arteries, enlarge, and now add only to the pulmonic circulation. Thus the circulation in the frog and newt changes, by a wonderful transition, from that of a fish to that of a perfect reptile! while in the Proteus and its allies, it continues to be that of the fish, with the addition of a modified state of pulmonary circulation. We advert thus particularly to these peculiarities, because they present proof so striking of divine contrivance.

Reptiles have either four limbs, (Plate XII. figs. 1, 4, 6,) two, (see Chirotes or Bipes on the Chart,) or none, (Plate XII. fig. 5;) the ribs are sometimes very numerous, (Plate XII. fig. 5,) sometimes wanting or merely rudimentary, (Plate XII. fig. 4;) in the latter case, the ribs assist greatly in terrestrial locomotion. There is no true distinction between the chest and abdomen, no diaphragm or muscular expansion, dividing, as in quadrupeds, these two cavities. As the blood is of a low temperature, these animals need neither fur nor feathers for the retention of the vital heat. They are therefore covered either with horny plates, or with scales, or have the skin entirely naked. Their sight is in general extremely acute. On this sense they depend in their pursuit of food, and for their perception of the approach of enemies. The senses of taste, smell, and touch, in reptiles are comparatively feeble. With regard to hearing, considerable variation appears in different groups. In serpents, the sense is very acute, and they evidently derive pleasure from musical notes, a fact well understood by the serpent-charmers of the East. In lizards, also, the sense of hearing appears to be quick; in tortoises and in the Amphibians it is probably much more obtuse. In most cases the internal organs of hearing are entirely covered by the scaly investment of the head, or by the naked skin. In lizards generally, the tympanic, or drum-like membrane, is stretched over the external orifice of the ear, and is on a level with the scaly covering of the rest of the head; but in the crocodile, the external orifice, instead of being thus permanently closed, is provided with a firm, movable lid or operculum, by means of which the aperture may be either stopped or kept open. While basking on the margin of a river, or lying there in ambush for prey, the crocodile is able to raise the ear-lid, in order to listen attentively to every noise; but when he dives beneath the water, either for safety or to drown the victim he has seized, the entrance of water into the auditory cavities is prevented by the firm shutting of the lid, which accurately fits the orifice.

Reptiles are ordinarily produced from eggs. Many of them exhibit extraordinary fecundity. None of them, unless the Pythons be an exception, ever perform the process of incubation. They bury their eggs in the sand, deposit them in warm places of concealment, or leave them floating in the water exposed to the rays of the sun. In due time the young are hatched. In some Reptiles which produce eggs, as the Viper, the young is already formed and advanced within the egg at the time the parent deposits it. This reptile, in many northern and temperate regions, seems to represent the multitude of deadly snakes that infest the torrid regions ; while the harmless ringed snake takes the place of the huge Python of Bengal and Java. In the countries of the Polar circles, the snake, the lizard, the toad and the frog are never seen. The absence of the snails, insects and small animals upon which reptiles usually feed, excludes them from those dreary regions. The larger part of them are carnivorous; the Tortoises, however, are vegetable feeders. A few feed both on small animals, as slugs, insects, &c., and on leaves and fruits.

Reptiles probably number as many as two thousand species. They are either terrestrial, or aquatic, or both, and hibernate in temperate regions, passing nearly the whole winter in a state of lethargy. An extensive division of the Serpents have hollow fangs which they can erect at pleasure, when they open their mouths to bite, and these fangs have apertures, from which they eject into the wounds made by them, an active and deadly poison.

From the earliest times the forms and habits of Reptiles have attracted attention. They are found represented on the monuments of the ancient Egyptians, and numerous allusions to them are contained in the scriptures of the Old Testament. Nor should it be unnoticed that among the organic remains which the industry and science of inquiring minds have lately brought to light, none present forms more wonderful, or proportions more gigantic, than some of the Fossil Reptiles.

REPTILES are divided into four orders, viz.: (1) CHELONIA, Chelonians; (2) SAURIA, Saurians; (3) OPHIDIA, Ophidians; (4) AMPHIBIA, Amphibians.

These orders pass into each other by certain gradations of form, traceable in all, but most evidently so in the Saurians and Ophidians, all these gradations clearly pointing to the existence of one grand scheme, of which the parts respectively link together in admirable harmony.

REPTILES.

What is the 3d branch of ZOOLOGY? Of what does it treat? To what Grand Division of VERTEBRATES do they belong? What is said of the vertebral column? Of their form, habitat, &c.? Why are they called COLD-BLOODED animals? What is their condition during winter? What is said of the strength of the vital principle in Reptiles? How is this illustrated?

State what is peculiar in the heart of Reptiles. Explain the circulation of the blood in Tortoises, Lizards, &c. What striking facts are given respecting the blood of Reptiles? What is at first the condition of the Amphibia? What change do they undergo? Is this true of all of them? How do some of them differ? Explain the circulation in both divisions. How many limbs have Reptiles? Does their internal structure differ from that of Quadrupeds? What is said of their covering? What of their organs of sense? How are they usually produced? How do they dispose of their eggs? How is it with the VIPER? What is said of the diffusion of Reptiles? Upon what do they feed? How many species do they include? Name the ORDERS into which they are divided. Point out on the Chart some animals belonging to each of these orders.

SECTION II.

FIRST ORDER. CHELONIANS, or TORTOISES.

CHELONIA. (Gr. χελώνη, *chelonē*, a tortoise.)

The Tortoises may be arranged into the following divisions, viz: (1) LAND TORTOISES; (2) MARSH and RIVER TORTOISES; (3) MARINE TORTOISES or TURTLES.

They differ most widely from the general form of the class; for (1) They are clothed with a natural armor, not like that of the Armadillo, a simple, horny addition to the skin, but a part of the skeleton itself. The skeleton is, in fact, thrown to the outside of the body, so as to form an external bony envelope, covered with a horny or leathery sheathing, and enclosing, as in a box, the internal organs, and other parts of the bony frame-work which do not immediately enter into its composition. Tortoises may be likened to Frogs, so enveloped in horny armor as to be restrained by it from jumping.

The upper piece, or dorsal buckler, is termed the *carapace*. This is usually more or less arched, and consists of an expansion of the ribs into wide flat bones, all united firmly together, and also to the edge of the flattened spinal processes,—the whole forming a consolidated plate. (Plate XII. fig. 1.) To the margin of the plate thus formed, is added a third set of bones, regarded as representing the sternal ribs of the Crocodiles and other Lizards, and assisting to complete the circumference of the carapace.

The lower plate, or abdominal buckler, is termed the *plastron* or *sternon*. This, instead of being prolonged forwards into a keel to afford attachment to large muscles, as in the Birds, is extended sidewise for the protection of the body. It consists of nine bony portions, of which eight are in pairs; the ninth is single, and occupies the front part of the plastron.

(2) The Chelonians are also distinguished from other Reptiles by the entire absence of teeth. As a compensation for this, however, the jaws are cased in horny coverings with cutting edges, resembling the hooked beak of a Parrot, and with which they crop and mince the vegetable substances on which many of them feed. The armor in which the animals are encased, seems to be their most effectual defence. In the Land Tortoises the carapace is usually much arched and firmly united, so that, without injury, it can sustain a very great weight; the plastron in these also exhibits the highest degree of solidity, and is united to the carapace by an extended lateral surface. At the line of its union, it is sometimes slightly movable, but in most, it is fixed by an unyielding *suture*. Its front and hind margins are generally indented or notched, for the more easy egress of the neck and tail; but sometimes they simply end in a cross line; or, on the contrary, are prolonged into a point. In the Land Tortoises of the genus *Pyxis*, (Gr. *puxis*, a box,)—species *Arachnoides*, (Gr. spider-like,)—the Arachnoid Tortoises of India, the plastron is furnished with a transverse hinge, giving a power of motion to the front part, so that the animals can redraw their head and fore-limbs within the carapace, and close the plastron upon it. In another genus of the same group, *Cynixis*, of Guiana, the carapace, instead of being one solid whole, has the posterior portion distinct from the front portion, and movable, so as to close upon the hinder margin of the plastron, and shut in the hinder limbs and tail. In the aquatic species, the shell is generally more flattened, so as to present less resistance to the water. Some of them swim with considerable rapidity, and are much more active in their habits than others of the order. The shell of such has its parts less firmly united, and is, in some degree, flexible; it also affords much less complete protection to the body.

Notwithstanding the horny beak with which the jaws of Tortoises are in general furnished, the sense of taste is decidedly higher in these than in other Reptiles. The broad, thick, fleshy, and movable tongue, is provided with salivary glands, and nerves of taste, but is not capable of being protruded from the mouth. It is not an organ of taste merely, but filling out the entire cavity of the mouth, assists in the process of respiration; for "Turtles swallow the air they breathe."

The eyes are larger in proportion, and more movable in Tortoises than in other Reptiles. They have three eye-lids; two external, continued from the common skin of the head, and varying as to form in different genera; one, internal, resembling the nictitating membrane of birds; and moved by muscles appropriate to that office. The form of the pupil is round, as in birds. The iris is always colored, usually dark, but in some, red, or even milk-white.

Tortoises have no movable external ear, but in all are found the tympanic orifice and membrane which are wanting in Saurians, and the sense of hearing is consequently well developed. The sense of smell appears to be at a low degree. The nostrils open on the most anterior part of the upper jaw or mandible, and are close to each other. In the River Tortoises, and in the MATAMATA, a Marsh Tortoise, of South America, the nostrils are prolonged into a sort of flexible proboscis, which the animals can raise for the purpose of respiration, between the large, floating leaves of water plants, while they lurk with their bodies concealed below them, and immersed in the water,—lying in wait for their prey.

The males of the Tortoises are, in general, smaller than the females, and commonly distinguished by the plastron, which is slightly concave. Tortoises have a voice,—that is, they have, more or less, the faculty of uttering distinct sounds. They vary in their food according to the localities which they are accustomed to frequent. Some live on marine plants; others on small animals, in addition to vegetable food. They require but little nourishment, and can pass months, and even years, without eating. Turtles, in their growth, are exceedingly slow, coming to maturity the latest of all the Reptiles. But, at the same time, they are very long-lived. Land Tortoises have been known to live one hundred and twenty years, and some have even reached more than two hundred years.

According to Agassiz, their eggs, up to the seventh year, are of small size,—numerous, yet not distinguishable into sets; but with every succeeding year, there appears a larger and larger set of eggs; each set being made up of the usual number which the species lays, so that a Turtle of eleven years old, for the first time, contains mature eggs ready to be laid in the spring. The larger eggs always appear in regular sets, of a definite number, and these coincide with the number laid by that particular species at one time. Four sets can be readily distinguished; one of them mature eggs; another about half the size; a third still smaller; and the fourth smaller still, (about the size of a large pin's head;) below these, it is difficult to distinguish the difference in size. "Turtles," says Agassiz, "lay once a year;

therefore, it follows that an egg requires four years from the time there exists a marked difference among the eggs of different sizes, to acquire its full maturity."* Fresh-water Tortoises lay their eggs in moist ground, or in dryer places near the water; Marine Turtles lay theirs in hot sand; the Land Tortoises lay theirs upon dry ground. The time of the extrusion of the animal varies from six weeks to three or four months.

The divisions of the Chelonians vary, as made by different naturalists. Agassiz, who prefers names which have *priority of date*, divides the order TESTUDINATA, or CHELONIA, into two sub-orders.

I. AMYDÆ, with seven families.
  (1) *Testudinina*, Land Tortoises.
  (2) *Emydoidæ*, (3) *Cinosternoidæ*, (4) *Chelydroidæ*, (5) *Hydraspidæ*, (6) *Chelyoidæ*, (7) *Trionychidæ*, } River and Marsh Tortoises.

II. CHELONII, with two families.
  (1) *Chelonioidæ*, (2) *Sphargidæ*, Marine Turtles.

FIRST SUB-ORDER. AMYDÆ. DIGITATED.

This sub-order includes, besides the Land Tortoises, the Marsh and River Tortoises, intermediate in form between the Marine and Land Tortoises. The Marsh Tortoises proper, are sometimes arranged into two divisions, viz: (1) the *Pleurodera*, (Gr. *pleuron*, a side; *deirē*, neck,)—so named because the head is concealed, not by being drawn back in a straight line, but by the neck folded to one side of the opening of the shell. The arms are also incapable of being completely drawn within the carapace and plastron. Seven is said to be the number of the genera included in this division. None of the species belong to North America, but many to South America. (2) The *Cryptodera*, (Gr. *krupto*, to conceal; *deirē*, neck,) in which the cylindrical shaped neck can be folded back upon itself under the center of the forepart of the carapace. The pelvis is articulated to the internal surface of the carapace, while in respect to the plastron, it is free; this gives to many of the Marsh or Pond Tortoises a power to move the under portions of the osseous box; and this, accordingly, has a less degree of solidity; whereas, in the *Pleurodera*, the pelvis is firmly fixed to the roof of the carapace above, and to that of the plastron beneath. The species of this division are very numerous, and many are found both in North and South America.

* Contributions to the Natural History of the United States. Vol. I.

First Family. *Testudinina*, (Lat. *Testudo*, a tortoise.)

Land Tortoises. First Sub-Order, Amydæ.

The Tortoises of this family exhibit the greatest symmetry of form, and are, on various accounts, entitled to the first rank in the order. They are distinguished by their highly arched carapace, and still more, by their short, clubby feet, terminating in flat, spade-like nails. The outward armor is entirely ossified, and harder and thicker, in proportion to the animal's size, than in the Aquatic Tortoises. The shield is covered, externally, with epidermal scales, and the skin everywhere more or less protected with them; on the most exposed parts, they are thick and stiff, and form a continuous hard covering. The neck and legs are short, and can be drawn entirely within the shell. (See Radiated Tortoise on Chart.)

The Land Tortoises show nothing of the fierce dispositions exhibited by most of the other groups,—never attacking or making resistance, but resorting to the shield, and trusting to that alone for protection. Their feet, which, in shape, have some resemblance to those of the Elephant, are adapted to walking on solid ground only; when placed in water, these animals endeavor to walk, as if upon land, having no swimming motion. Their movement on dry ground is firmer and more steady, the weight being almost equally supported by both pair of limbs; and they can travel for a distance at a pace less slow than any other Tortoises. The fore feet have, usually, five toes, and the hind ones, four, which are furnished with short conical claws, well adapted for digging. The food of the Land Tortoises is entirely vegetable. They appear most fond of the succulent stems of plants and fleshy fruits. "I have often," says Agassiz. "seen our Gopher gnawing the stumps of cabbage, and apples falling from the trees in my garden, as the Squirrels do, holding them between their feet." The lungs are very much larger in the Land Tortoises than in any other family of the first sub-order. Their size is also, on the whole, larger than that of any other family of that division.

The Land Tortoises include four genera.

(1) *Cinixys*, of which there are two or three species found in Guiana. In these, the hind part of the carapace is not united to the front part, and is movable, so that the animals can shut in their hind limbs and tail.

(2) *Pyxis*. (Gr. *puxis*. a box.) This genera includes the Land Box Tortoises, having the front part of the plastron mova-

ble on a hinge, so that they can conceal the head and fore limbs within the carapace.

(3) *Homopus*, (Gr. *homos*, like each other; *pous*, foot,) including Land Tortoises which have the carapace and pastron immovable. They have but four nails on the fore feet, while the other genera have five. The VERMILION TORTOISE, *H. areolatus*, (Lat. divided into areas, or spaces,) is a species of this genus, found in Eastern Africa and Madagascar, and one of the smallest known of Land Tortoises, being seldom more than five inches in length.

(4) *Testudo*, having the carapace and plastron immovable. Of this genus, only one species is indigenous to North America, viz: *T. Carolina* or *T. Polyphemus*,—[*Xerobates*] (Gr. *xēros* from *xēra* land; *baino*, to go,) *Carolina*, Ag.,—the GOPHER TORTOISE, ranging from Florida to Georgia. It is from fifteen to seventeen inches in length. Its strength is so great that it can move easily with a man standing on its back. In habits, it is nocturnal; its flesh is excellent, and much sought after for the table. In sandy districts, it excavates holes in the ground, which much impede the movements of horsemen.

The ELEPHANTINE TORTOISE, *T. Indica*, (T. Elephantina,) *Cylindraspis*, (Gr. rolling or cylindrical shield,) *Indica*, is from three to four feet in length; and every way a huge animal. It sometimes has been known to weigh not far from three hundred pounds. The Elephant Tortoises are found in great numbers in the Galapagos islands, but, comparatively, not large,—averaging sixty pounds in weight. They are eagerly sought by crews of vessels. When captured, they serve for fresh meat, as they can be kept for a year in the hold of a ship without food or drink.

The GREEK OR EUROPEAN TORTOISE, *T. Græca*, is a well known species, found in the South of Europe.

The GEOMETRIC TORTOISE, *T. geometrica*, is a beautiful little creature, about six inches in length, found in South Africa and Madagascar. It has its specific name from the radiating lines of yellow, forming angular figures on the plates of the carapace.

The Charcoal Tortoise, *T. carbonaria*, is common in Brazil and other parts of South America. The carapace is deep black, and eighteen inches long. This kind is sold as a great delicacy in the markets at Caraccas.

The RADIATED TORTOISE, *T. radiata*, a handsome species, is a native of Madagascar, whence it is frequently taken to the Cape of Good Hope, and to the Mauritius and Bourbon isles. The carapace is hemispherical; the plates are simple and black, with a yellow central spot, whence diverge lines of the same color; the plates of the plastron are ornamented with black and yellow.

SECOND FAMILY. *Emydoidæ*, (Gr. εμύς, *emus*, a fresh-water tortoise.)

This is a most numerous family, including a large number of well known species, which present great varieties of habit, size, and structure. Dumeril and Bibron unite the *Emydoidæ* and *Chelyoidæ* into one family, under the name of ELODITES, referring the *Emydoidæ* to the division Cryptodera, and the *Chelyoidæ* to the division Pleurodera. Our limits do not allow us to detail minutely the distinctive peculiarities of this family, or to assign at large the differences existing in the various sub-divisions under which it is exhibited.

In these Tortoises, the box in which the animal is enclosed, is less thick and strong than in the Land Tortoises, not becoming completely ossified until late in life; its figure is oval, for the most part, broader behind than before. All the bony plates show great constancy and regularity of arrangement. The outside of the whole shield is covered with scales, as is, more or less, the skin of the head, neck, limbs, and tail; the tympanum is visible, and the eyelids are of equal height.

Nearly the whole of these are eminently MARSH TORTOISES. Almost all of them can withdraw and conceal their limbs within the carapace and plastron; but in the *Platysternon*, (Gr. broad-sternon,) the plastron or sternon from its width, suffices to conceal the limbs when folded; yet the head remains constantly exposed. The food of these Tortoises is both vegetable and animal. None of them catch active prey, or are in any way ferocious; when hard pressed, however, they defend themselves by biting. They lay their eggs on dry land, in holes which they dig with their hind legs; the land species, from two or three to five or seven; the water species, from ten or fifteen to twenty, thirty, or even more. (Agassiz.) The shell of the eggs is never brittle, but rather flexible, and less calcareous than in some other families.

The genera *Emys* and *Cistudo*, are prominent representatives of this family. In the *Emys*, the fore feet have five toes; the hind feet, four; the plastron is broad, immovable, solidly united to the carapace, and covered with twelve plates; the head is about the ordinary size; the tail long.

Emys Muhlenbergii, or Muhlenburg's Tortoise, is the smallest known species, about four inches in length; found in New Jersey and the eastern part of Pennsylvania.

E. rugosa, (Lat. wrinkled,) or *E. rubriventris*, (Lat. red-bellied,)—the *Ptychemys*, (Gr. *ptuche*, a fold; *emys*,) *rugosa*, of Agassiz,—is found as far South as Virginia, and North as far as the neighborhood of New York City. It is known by the name of the Red-bellied Terrapin, and as food is prized by epicures. Length from ten to seventeen inches. (Agassiz says fifteen inches.)

Emys concinna, (Lat. polished, beautiful,) is a handsome species, found in Georgia; in length, according to Agassiz, fifteen inches. This, and the preceding species, are the rarest of the family.

E. concentrica, is found both in North and South America. New York is said by DeKay to be its northern limit. It gives the preference to salt-water marshes, and is therefore called the SALT-WATER TERRAPIN. The flesh of this species is in particular request for the table. The plates of the carapace are olive-green, with concentric lines of brown, whence the specific name, *concentrica*. Sometimes it is called *E. palustris*. These Tortoises bury themselves in mud during the winter; they are then very fat, and taken in large numbers. The length is from seven to ten inches.

E. picta, (Lat. painted,) or *Chrysemys*, (Gr. gold-colored Emys,) *picta*, Ag.,—is probably the most *essentially aquatic* Tortoise of the entire family; in fact, it soon perishes, if removed from the water. The PAINTED EMYS (see Chart) is very common in the United States, but on account of the ill flavor of its flesh, never used for food. The plates of the carapace, which is considerably depressed, are of a deep brown color; the plastron is of a yellow or gold color. For the variety and beauty of its markings, this may be esteemed the handsomest of all the fresh-water Tortoises. It seems to enjoy much the rays of the sun, and sometimes floats in the water with the head just emerging from the shell, luxuriating in the genial temperature. It inhabits stagnant ponds or lakes, and is never found in rivers or running streams. Its length is from five to nine inches. When young, its contour is circular rather than oval. It is remarkably slow in coming to maturity, not laying eggs before it is seven years old.

E. guttata, (Lat. speckled,) the SPOTTED TORTOISE, is a small species found throughout the Union. Its length is about four inches. The carapace is black, or deep brownish black, with distant rounded yellow dots, occasionally with a few orange spots. It shows a preference for streams and ponds which have a muddy bottom. On a warm day, Speckled Tortoises may be seen basking in the sun, on a log or rock; but on the approach of any person, they slip suddenly into the water. This species is the *Nanemys*, (Gr. dwarf-emys,) *guttata*, of Agassiz.

The genus *Cistudo*, (Lat. a box,) includes TERRAPINS, or BOX TORTOISES. These, like the Emydes, have five toes on the fore feet and four on the hind ones. The head is very high; the plastron, broad, oval, and divided by a transverse hinge into two movable portions or valves, by means of which the whole body may be shut in. The beak of the upper jaw projects downward in the middle; the lower jaw is sharp pointed in front; the hind foot plantigrade. These Tortoises never take to the water from choice, and would be drowned if detained there. Indeed, they are so much on dry land as to be sometimes called Land Tortoises.

The CAROLINA TERRAPIN, or BOX TORTOISE, *C. Carolina*, or *C. Virginia*, (Agassiz,) is found in New England, also westward as far as Michigan, and southward as far as the Carolinas. In its general habits, in the vaulted form of its carapace, and in the structure of its feet, which are but slightly palmated, it appears to be a link between the Marsh and Land Tortoises. This species is not aquatic, preferring woods and dry places and living on vegetables and insects; occasionally, however, it is met with in swamps and moist places. Of all the Marsh Tortoises, it has the shortest and most convex carapace. The general color is dark brown with stars and blotches. The flesh is not much esteemed, but the eggs, which are about as large as a pigeon's, are thought to be excellent, and are much sought for. The length of this species is from five to seven inches.

BLANDING'S BOX TORTOISE, *C. Blandingii*, has a shell less raised than that of the Carolina Terrapin, and the lower jaw is hooked instead of the upper, as in the Carolina species. Its length is from seven to eight inches. This species was first accurately described and figured by Dr. Holbrook, in his valuable work on North American Herpetology. According to Agassiz, who deems this a "true Emys," the oldest name is *E. meleagris*.

The EUROPEAN BOX TORTOISE, *C. Europæa*, is widely diffused. It differs from the Carolina Terrapin in giving the preference to still waters, ponds and marshes, in the mud of which it delights to bury itself. This species is particularly fond of small fishes. These it kills previously to devouring them, but rejects the air sac, which rises and floats on the surface, so that the abundance or scarcity of these animals in any pool or sheet of water, is judged of by the numbers of these floating air sacs. The flesh of the European Box Tortoise, though not very delicate, is nevertheless eaten.

THIRD FAMILY. *Cinosternoidæ*, (Gr. κῖνεω, *kineō*, to move; στέρνον, *sternon.*) MUD and MUSK TORTOISES.

The Tortoises of this family have long and narrow bodies, covered by a shield which is entirely ossified. The marginal plates are twenty-four in number; the plastron is divided into three sections, and, "at least in the adult species, is made up of eight plates, there being no odd one, as in all the other families of the sub-order." (Ag.) The shield or carapace is covered with large horny scales; the head is large, elongated, and pyramydal in form; the mandibles are hooked, and in the under jaw covered with fleshy excrescences. In the female, the tail is short; in the males, thick and long. Their legs are slender, ill fitted for land travel, but easily carrying the body through the water over the bottom. As a family, they are dwarfish in their forms, the largest not being more than nine inches long, and the smallest not more than four inches. These Tortoises smell strongly of musk. Their disposition is a blending of shyness and ferocity. "They remind us," says Agassiz, "of the Insectivora among the Mammalia, the rapacious habits of which are in strange contrast with their size and feebleness." Their movements are abrupt and quick, but have little power; their food is chiefly animal; their habits aquatic, though sometimes they bask in the sun on the shore. They lay only from three to five eggs, having the shape of a lengthened ellipse with very blunt ends, and a glazed, shining surface, much smoother than that of the other turtles' eggs, and also quite thick and brittle.

This family, entirely American, is represented by the genera *Cinosternon* and *Sternothærus*, (Gr. *sternon; thairos*, a hinge.) The former has both the front and hind parts of the plastron movable upon an intermediate fixed position; the latter has the plastron solid, with the front part movable.

Cinosternon represents the MUD TORTOISES, or Cinosteroids. Several species are described.

C. Pennsylvanicum, or *Thyrosternum*, (Gr. *thurōn*, porch; *sternon*,) *Pennsylvanicum*, (Ag.) is very common in various parts of the United States. It inhabits ditches and muddy ponds, and often takes the hook. Its food consists of frogs and small fishes. The length is seven or eight inches.

Sternothærus, or *Ozotheca*, (Gr. *ozō*, to smell; *thēkē*, repository,) Ag., of sub-family *Ozothecoidæ*, includes the MUSK TORTOISES. Of the species the Musk Tortoise, *S. odoratus*, or *O. odorata*, Ag., is the most common and the smallest one known, being less than four inches in length. It is common in marshes and ditches from Maine to Florida. The carapace is gibbous or oblong, of a brownish color, with streaks of green. The color and marking are, however, not easily detected, as the animal is usually covered with mud, and an agglutination of water plants. It is sometimes called Mud Terrapin, and, on account of its disagreeable odor, Stink-Pot, and other names equally savory.

FOURTH FAMILY. *Chelydroidæ*, (Gr. *χελυδρος*, *cheludros*, a water-tortoise.)

SNAPPING TURTLES.

The body of these Tortoises is high in front and low behind, the upper surface is "like a shed-roof falling backwards, curved down on either side, lowest about the middle, less and less towards the ends." The carapace projects beyond the attached surface of the body, except at the neck, where it is joined with the plastron; the latter is not movable, of a cross-like shape, and covered with twelve plates. The head is very large and covered with small plates; the upper mandible is hooked; below the under mandible are two small wattle-like excrescences. The tail is extremely long, compressed and surmounted by a ridge of strong scales, as in the crocodile. The limbs are very robust, and the nails of the toes are strong, hooked and sharp. The head, though of great size, can be withdrawn within the carapace; but not so the tail and limbs. "The animal lives mostly in the water, but makes considerable passages over land. It does not, like the *Trionychidæ*, remain burrowed in the soft muddy bottom, but rather lies in wait for prey under shelving banks, or among the reeds and rushes."

This family is represented by the ALLIGATOR TORTOISE, or SNAPPING TURTLE, to which naturalists have given various names, among which are *Chelonura*, (Gr. *chelonē*, tortoise; *oura*, a tail,) *Serpentina*, (Say;) *Gypochelys*, (Gr. *gups*, a vulture; *chelus*, a tortoise,) *Serpentina*, (Agassiz.) Both internally and externally, it exhibits an approach to the alligator, and perhaps may be viewed as an intermediate link. When adult, it exceeds three feet in its total length. So great is the strength of its jaws that a large one has been seen to bite off a piece of plank more than an inch thick. It eats frogs, other aquatic reptiles, and even fish; it swims with celerity, and is prone to *snap* at every thing coming near it. Woe to the unwary duck or other animal that swims unguardedly within its reach. The Alligator Tortoise is a native of Carolina and the warmer districts of North

America. This Tortoise, according to De Kay, lays from sixty to seventy eggs, about the size of a small walnut.

FIFTH FAMILY. *Hydraspidæ*, (Gr. ὕδρα, *hudra*, a water-snake; ασπίς, *aspis*, a shield.)

This family includes four genera, viz.: *Platemys*, (Gr. *platus*, broad; *emus*, emys;) *Rhinemys*, (Gr. *rhin*, nose; *emus*, emys;) *Phrynops*, (Gr. *phrūnos*, a rubeta, or venomous toad; *ōps*, face;) *Hydraspis*. The whole are included by Wagler in one genus, *Platemys*. They have the head flattened and covered with a single delicate scale, or with a number of small irregular plates; the jaws are simple; two barbels appear under the chin; the carapace is very much flattened; the plastron is immovable; there are five claws on the fore feet, four on the hind. One species is found on the banks of the Macquarie River, (Australia;) other species are found in South America, living in marshes or else on the banks of rivers. So far as the head and neck are concerned, some of these animals, as the Chelodina of New Holland, appear more like a snake than a tortoise.

SIXTH FAMILY. *Chelyoidæ*, (Gr. χέλυς, *chelus*, a tortoise.)

The Tortoises of this family have a shield that is thick, completely ossified, and regularly divided into plates; the head is extraordinarily large, flat and triangular; the jaws are weak, neither pointed nor sharp edged, unfit for catching large active prey, or for tearing any tough vegetable or animal matter; the mouth is broad, but very close when its roof and floor are brought together, being well adapted for catching and swallowing minute animals; the legs are strong; the feet broad and compact, with long and sharp claws, the fore feet having five, the hind feet four. This family includes but one genus, *Chelys*. The only recognized species is the MATAMATA, *C. matamata*, (an aboriginal name,) found in Cayenne and Guiana, having the neck furnished with long cutaneous appendages, and two barbels on the chin. The head looks as if it had been *crushed*, and this, together with its fringes and skinny enfoldings, gives it a singularly grotesque appearance. Decidedly, it is the most remarkable of the Pond or Marsh Tortoises. When full grown, the Matamata is about two and a half feet in length.

SEVENTH FAMILY. *Trionychidæ*, (Gr. τρεῖς, *treis*, three; ὄνυξ, *onux*, nail.)

SOFT-SHELLED OR RIVER TORTOISES.

These Tortoises are distinguished by the complete absence of scales from the body, the shell being covered with a soft skin. The feet are broad, webbed, and move horizontally; of the toes, three on each foot are provided with nails or claws, whence the term *Trionyx*, (see derivation above.) The form is that of a flat orbicular disk, slightly elongated, with a long pointed head projecting upon a lengthened, slender neck. The structure of these animals is well adapted to life and motion in the water. They swim with great facility. In this process they are aided, not by their flattened and webbed feet alone, but by the loose and flexible skin of the body, forming a narrow flap or border around the edges of the shell, and performing the office of a fin. The soft carapace is generally dark colored, variegated with brown; but the plastron and all the under parts are pale, like the turbot. The plastron is not entirely ossified in the Trionyx proper, and is united to the carapace by cartilage. These Tortoises live mostly on the muddy bottom of shallow waters, burying themselves in the soft mud, leaving only the head, or a small part of it exposed. Sometimes they remain under water as long as half an hour, without coming to the surface to take breath. They are rarely seen on land, where, to them, locomotion is labored and unsteady. They lay from a dozen to twenty or more eggs, of a spherical form, having a thick but brittle shell, and about the size of a musket ball.

They feed upon fish, reptiles, and mollusks, especially Anadontas and Paludinas, fragments of which have been found in their intestines. Two genera represent this family.

(1) *Gymnopus*, (Gr. naked-footed,) or *Platypeltes*, (Gr. *platus*, broad; *peltē*, shield;) (2) *Cryptopus*, (Gr. *krupto*, to hide; *pous*, foot,) or *Aspidonectes*, (Gr. *aspis*, shield; *nectes*, a swimmer,) Agassiz. The type of the genus *Gymnopus*, or *Platypeltes*, is the *Trionyx ferox*, the species of this country earliest known to foreign naturalists. The *Tryonyx ferox*, or *P. ferox*, (Lat. fierce,) is found from Georgia to Western Louisiana. Though *fierce*, it is not very large. Agassiz says the largest tortoise which he ever saw or heard of belonging to this species, was one foot and a half in length. As the generic name, *Platypeltēs*, indicates, the carapace is broad. The great breadth of the cartilaginous circumference of the carapace, and the narrowness of

the plastron, are distinctive marks of this tortoise. In its native regions it reigns as a tyrant, producing great havoc among the finny tribes; it is very voracious, and eagerly seizes a hook baited with a fish.

The females visit the shore in May to lay their eggs. These are globular in shape and brittle; they are hatched in July. It is said that in its fierceness and voracity, this Tortoise will attack small quadrupeds, aquatic birds, and young alligators. The *Trionyx* of the Nile, *G. Ægyptiacus*, is much valued in Egypt, on account of the services it renders in devouring the eggs and young of the crocodile.

Tortoises of the genus *Cryptopus*, or *Aspidonectes*, have the plastron broad, and capable of closing up in front, so as to shut in the retracted head and limbs; in the rear they have a cartilaginous valve on each side, for shutting in the hind limbs.

A. spinifer, (Lat. thorn-bearing or prickly,) is a species common in Lake Champlain, and in most of our western rivers. Length fourteen inches.

Second Sub-Order. Chelonidæ, or Chelonii. Sea Turtles. Pinnated.

The entire structure of the Sea Turtles shows an express adaptation to aquatic habits. Not only is the carapace greatly flattened, but the limbs, in which the toes are not externally distinct, are likewise flattened and modified into large oars. On land, these animals shuffle along in the most awkward manner, and make, with toilsome efforts, only a slow progress; but they plough the waves, dive and ascend with admirable address and dexterity. They swim almost entirely by means of their front limbs; the other pair acting independently and being chiefly useful in aiding to balance the body, and guide the general course. They feed chiefly on marine plants. The gullet of these Turtles is lined with long cartilaginous processes, all tending towards the stomach. These appear designed to prevent the return of the food when the water which is swallowed in connection with it is regurgitated.

The Sea Turtles never resort to the shore, except to deposit their eggs. They lay them at night and in large companies. Those of most of the species are both nutritious, and agreeable to the taste. "American Sea Turtles," says Agassiz, "lay their eggs towards the end of May, or in the beginning of June. They lay a large number of them, about one hundred at a time, or even *more*, which they deposit on shore, in the dry sand. Their eggs are not large, in comparison to the size of the animal, and not perfectly spherical, their orbicular outline being more or less irregular." "I have no reason," he says, "to trust the reports that they lay eggs more than once a year." Other writers, however, say "the process is repeated *three times* a year." The eggs are almost unprotected by a shell, and hence it is necessary that the sand in which they are laid to be hatched by the heat of the sun, should be soft and movable. To obtain a suitable locality for their eggs, they often travel many hundred leagues. Ascension Island is to them a favorite place of resort. At the breeding season the Turtle-fishery is carried on. The flesh of the females is in the highest estimation, and at this season, it is supposed to possess its best quality. "The fishers suddenly advance from their watching places, and despatch the Turtles with clubs, or turn them quickly over upon their backs, in doing which, it is often necessary to use levers, several men at the same time combining their strength. A few skillful men, in the course of three hours, may turn over, and thus secure forty or fifty turtles." On the coast of Guiana, haul nets are employed for the capture of these creatures. In the Chinese and Indian seas, and also on the shores of Mozambique, boatmen take them by availing themselves of the natural powers and instincts of certain fishes, named *poissons pecheurs*, or Fish-fishers. The Turtles are usually met with in the warm latitudes of the ocean, and especially towards the torrid zone.

First Family. *Chelonioidæ.*

This family are characterized by having the carapace very broad, more or less depressed, of a somewhat heart-shaped outline, covered with horny scales, and bordered by a distinct marginal rim; also by having a flat nail on the thumb of each paddle. Three well defined genera are found along the coast of the United States; the greatest difference between them has relation to the structure of the mouth. Their food consists of aquatic plants, sea-weeds, and the like. In size, they much surpass the average size of the *Amydæ*; yet they are shy and inoffensive, not biting when hard pressed, but striking with their powerful flappers, and endeavoring to escape by quickening their speed. They lay their eggs at the end of May or beginning of June.

The Green or Esculent Turtle, *Chelonia mydas*, is considered the most important of the Turtles, its flesh being in great request as a luxury for the table, and as furnishing abundant and wholesome food to voyagers in tropical climates. It has twelve pair of scales of a greenish color, but they do not overlay each other, and are of no use in the arts. Green Turtles are very common in shallow parts of the sea near the islands and the shores of continents, within the tropics, where they may be seen in great numbers among the sea-weeds, grazing like a herd of cattle; occasionally coming to the surface to breathe, and sometimes remaining there, basking in the sunshine. They are often caught at sea in calm weather, a harpoon and line being used. The usual length is four or five feet, and the weight from four hundred to eight hundred pounds; but this Turtle has been known to "reach the length of eight feet, and a weight of fifteen hundred pounds." The flesh of the smaller ones is, however, the more highly esteemed.

The coast of Florida is one of the resorts for the females, which deposit each, every spring, between one hundred and two hundred eggs in the sand, where they hatch in about seven or eight weeks. But scarcely a thirtieth of this number gain the sea, or live a week after reaching it. Birds and beasts of prey thin the number of those hatched; and crocodiles and rapacious fishes are ready to seize upon such as escape destruction on land and gain the water.

The TORTOISE SHELL TURTLE, *Chelonia imbricata*, or *Eretmochelys*, (Gr. *eretmos*, an oar; *chelus*, a tortoise,) *imbricata*, (Ag.,) has the horny muzzle somewhat lengthened into a sharp point, and the lower jaw is received into a groove of the upper, so that the food can be cut as well as bruised by it. The shield has twelve pair of scales. They overlap each other, at least one-third of each lying over the one behind it; hence this species is named *imbricata*, (imbricated.) The plates increase only in front. As they enlarge there, "the older parts move backwards, where they are worn off by external mechanical agencies. This process goes on so fast that in a specimen of two feet in length, no trace of those primary scales which covered the whole shield, during the first year, could be found. This mode of growing and moulting, if we may call it so, is very similar to that in the human nail." (Ag.) The flesh of this Turtle has a disagreeable flavor, probably arising from the nature of its food. The animal is chiefly sought for the plates of the carapace, called "Tortoise Shell," and which are much thicker and stronger, as well as more clouded in color than those of any other species. The shell procured from the live Turtle is thought to be the finest. The epidermis, or outer shell, is said to change every year. The shell is removed from the bone of the carapace by presenting its convex surface to a glowing fire. The application of boiling water to the shell when removed, so mollifies it that it may be acted on like a soft mass, and by pressure in metallic moulds, made to assume a great variety of forms. A single Turtle yields about ten or twelve pounds of Tortoise Shell. When the stripped animal is set at liberty, the shell grows again; and hence it sometimes happens that in after years, the stripped Turtle is recaptured, and subjected to a second ordeal, but the shell in that case obtained is very thin. The Tortoise Shell Turtle never reaches so large a size as the Green or Esculent Turtle. Sometimes it is called the Hawk's-bill Turtle. (See Chart.) It is found in the warmer latitudes of the seas and coasts of this continent, and also in the seas of Asia.

Com. Wilkes, in his "Exploring Expedition," states that the chiefs in the Fejee Islands, keep Tortoise Shell Turtles in pens. Tortoise Shell "sometimes sells in Manilla for from two to three thousand dollars the picul, (one hundred and thirty-three English pounds.")

The Loggerhead Turtle, *C. caretta*, or *Thalassochelys*, (Gr. Sea-Turtle,) *Caouana*, differs from the tortoise shell Turtle, in having thirteen pairs of scales, and these not imbricated. The flesh is not much valued, though wholesome. The Loggerhead Turtle feeds upon fish and mollusks. It yields abundance of oil, for which alone it is sought. This species is numerous in the Mediterranean.

Second Family. *Sphargidæ*, (from Gr. σφαραγεω, *špharageo*, to roar loudly.)

These Sea-Turtles have the bony structure of the carapace covered with a thick layer of leathery skin, instead of plates or scales; the form may be compared to a flattened cone, with angular sides; the skeleton is light, and the shield narrow and small, compared with the size of the animal; in the full grown Turtle, the skin is quite smooth; but in the young is tuberculous. The paddles are without any distinct nails. These Turtles are the largest in size and lay a great number of eggs. Only one species is as yet recognized.

The Leathery Turtle, *Sphargis coriacea*, (Lat. leathery.) This has jaws of immense strength, the lower one being sharp edged, and turning up at a point which when the jaws are closed is received into a central indentation or notch of the upper jaw. The carapace is heart-shaped and has seven longitudinal ridges at equal distances from each other. (See Chart.) The eyes open almost vertically, which gives to the animal a strange aspect. This Turtle exceeds all others in size; the carapace is sometimes fifteen feet in circumference, and nearly seven feet in length. It sometimes weighs more than eighteen hundred pounds. Agassiz states he has seen those that weighed over a ton. The Leathery Turtle feeds upon marine animals, as well as plants. When aged, it is said to carry on its carapace "a world of parasites." This gigantic species is found in the Mediterranean, and in the Atlantic and Pacific oceans. It breeds on the Tortugas or Turtle islands, on the Bahama islands and Keys, and on the coast of Brazil, laying, on an average, about three hundred and fifty eggs, in two sets. A large sized specimen of this species, taken off Sandy Hook in 1816, is now in the American Museum, New York city. The Leathery Turtle was known to the ancient Greeks, and of its carapace the first lyre is supposed to have been formed; the seven ridges on the back suggesting the adoption of seven strings, which was the ancient number. Hence, it is sometimes called the LUTE TURTLE.

CHELONIANS.

Into what divisions may this Order be arranged? In what respects do they widely differ from the general form of the class? What is the shield of the back called? Of what does this consist? What in the TORTOISES represents the sternal or breast ribs of the lizards, &c.? What is the name of the lower plate of the armor? What is said of it? What is the second distinction between CHELONIANS and other reptiles? How are they compensated for the want of teeth? What is said of the armor of the LAND TORTOISES? What is peculiar in that of the genus *Pyxis?* In that of the genus *Cinixys?* How is it in the aquatic species? What is said of the organs of sense in Tortoises? What of their food? Of their powers of abstinence? Of their growth? Of their age? What does Agassiz say of their eggs? Into how many SUB-ORDERS does he divide the order Chelonia? Name the families included in AMYDÆ, or DIGITATED TORTOISES. Also those included in CHELONII or PINNATED, (finned,) T. What groups besides the Land Tortoises does the Sub-order Amydæ include? Into what two divisions are the strictly Marsh Tortoises sometimes arranged? Are the sp. of this Sub-order numerous? What is said of the forms of the Land Tortoises? What of their distinguishing characteristics? Of their disposition, habits, &c.? Name the sp. which are mentioned. Which of these are found in the U. S.? What is said of it? What can you say of the others?

Which is the 2nd and most numerous FAMILY? How does the box or armor of these T. differ from that of the Land T.? Where are the largest part of them found? How is it with the *Cistudo?* What gen. are prominent in this family? Which is the smallest sp. of Emys? Which is the largest? What is said of the SALT WATER TERRAPIN? Which is the most essentially aquatic? What is said of it? Which gen. includes the Box

TORTOISE? What is said of the CAROLINA TERRAPIN? Of the European Box T.?

Which is the 3rd FAMILY? Describe them. Of what does Agassiz say they remind us? What is said of their movements, food, &c.? On what continent are they all found? What genera represent this family? Which genus includes the MUD Tortoises? Which the MUSK T.?

What is the 4th FAMILY? Describe them. What turtle represents this family? How does it resemble the ALLIGATOR? What is said of its strength, food, &c.?

What is said of the 5th FAMILY? How many gen. does it include? Where are they found? What other reptiles do some or all of these resemble?

What is the 6th FAMILY? Name their characteristics. What gen. does it include? What is the only recognized sp.? What is said of it?

What is the 7th FAMILY? In what respects are these distinguished? For what element does the structure of these animals fit them? What helps them in swimming? In what waters are they usually found? Upon what do they feed? What gen. are mentioned? What is the type of the gen. *Platypeltes?* What is said of it? What of the *Trionyx* of the Nile? What is said of the plastron of *Aspidonectes?* What sp. of this gen. is mentioned?

What is the 2nd SUB-ORDER? How does the structure of the SEA TURTLES fit them for aquatic habits? What is said of their motion? Which limbs do they use in swimming? What is the chief use of the hind limb? On what do they feed? Are they ever found on shore? When do they lay their eggs and what is said of them? What is to them a favorite breeding place? At what time is the T. Fishery carried on? How are the Turtles secured? How do the Chinese boatmen take them? In what latitudes are the Sea T. found? What is the 1st FAMILY? How is it characterized? How many well defined gen. on the coast of the U. S.? What is said of their food, &c.? Which is the most important of the SEA TURTLE? Why? Where is it very common? How often caught at sea? What is said of the size and weight? What resort for the female is mentioned? What is said of their eggs? Do their young, when hatched, all reach the sea? How many pair of scales has the Tortoise-shell T.? Why are they said to be imbricated? At what part do the plates increase? What effect has this upon the older part? What does this mode of growing and moulting resemble? For what is this animal chiefly sought? Which shells are the best? How is the shell removed from the carapace? How is it made to assume various forms? Is the shell renewed upon the stripped animal? Where is this T. found? How does the Loggerhead T. differ from the Hawks-bill or Tortoise shell T.? For what is it sought?

What is the 2nd FAMILY? From what is the family name derived? What Sea T. does it include? What is said of the jaw, carapace, &c. of the LEATHERY T.? What of its size, weight, &c.? Where is it found? What breeding places are spoken of? Why is it sometimes called the LUTE T.?

How are the Chelonidæ or Testudinata usually divided, as given on the Chart? What is Prof. Agassiz' first division of them as there shown?

SECTION III.

Second Order. Saurians.

Sauria, (Gr. σαῦρος, *sauros*, a lizard.)

This order contains a numerous assemblage of animals remarkable for the differences in their size, which varies from a few inches to thirty feet; and not less so for differences in respect to strength, form and habit. All, however, agree in certain essential characters. Many species, generally of great dimensions, are known only in a fossil state.

The general contour of the body is lengthened; the skin is protected either by horny plates, by scales of various sizes and figures, or by granulations. The limbs are usually four in number; the toes armed with claws. The body always terminates in a tail, which is frequently of considerable length. The eyes are protected by eye-lids, except in certain instances; and in most species a tympanic membrane covers the orifice leading to the internal organs of hearing. The ribs, unlike those of the Tortoises, are distinct and movable; and there is a sternum or breast-bone, which is not found in serpents. The jaws are armed with teeth, as in snakes; but the bones of the jaws are firmly united together, and not separable into distinct parts, as in the latter animals.

The eggs of the Saurians have a hard calcareous shell; the young undergo no metamorphoses or changes like those of the newt and frog.

The tongue, in these animals, differs greatly in its form, and in the degree of freedom which it enjoys. In the Crocodile it is undeveloped and scarcely distinguishable from the general floor of the mouth, between the branches of the lower jaw. In other groups, it is broad, fleshy, and free only at its point; in the Chameleons, it is fleshy, cylindrical, and capable of being projected to a great distance, and then completely redrawn. In some genera, it is slender and deeply forked, like that of a snake, and when at rest drawn into a sheath; while in others, it is flat, very movable, and notched or forked at the tip. It is, in all, lubricated with a glutinous saliva, but does not appear to be endowed with a high sense of taste.

In most of the Saurians, the body is so remarkable for its length and cylindrical figure that, as Aristotle has observed, they resemble snakes with the addition of limbs. "Among all the reptiles," says M. Bibron, "these undoubtedly approach nearest to the Mammalia, both in the variety and rapidity of their different movements, especially if we compare their progression with that of Tortoises. There are, indeed, among the Saurians, species which enjoy many modes of progression; for they can creep, walk, run, climb, swim, dive, and even fly. Nevertheless, the elongated and heavy trunk of these Reptiles is not supported by the limbs without effort; they walk in general, with constraint and slowly, for the arms and thighs are short, slender, but slightly muscular, and directed outwardly; while the elbows and knees are too angular to support with ease the superincumbent weight. Still, however, notwithstanding this conformation, so faulty in appearance, (though not in reality,) they are capable of executing a great variety of movements, all bearing on progression."

The form of the tail, the length of the body, the structure of the toes, and the shape of the claws determine the character of the movements, and correspond with the general habits of the animal.

The hotter climates of the globe are the great nurseries of the Saurians. Persons who live in northern latitudes are ordinarily not likely to form any adequate idea of the variety of these creatures, which tenant their favorite abodes.

"In the intertropical latitudes, they obtrude themselves upon notice; they are in the common pathway, and even haunt the abodes of men; they swarm among the trees; they lie motionless upon the surface of the water, enjoying the hot rays of the sun; they cover banks and walls or crumbling ruins, and mingle their sparkling hues with those of the blooming vegetation amidst which they nestle."

Like the snake tribe, the Saurians moult their skin during the spring or summer, appearing afterwards in bright colors.

None of the Saurians are poisonous; none have poison fangs, though the ancients regarded many as venomous in the extreme. Of these an imaginary animal termed the Basilisk, (*Basilicus*,) was especially celebrated; a name which modern naturalists apply to a genus peculiar to South America.

Though the Lizard race do not possess the medical properties which have been ascribed to them, many of them, as the Iguanas, hold a high rank as articles of luxury for the table; and the flesh and eggs of the Teguixin or Monitor, a large species found in Brazil and other parts of South America, are esteemed for food.

The Saurians are divided into the following families, viz.: (1) *Crocodilidæ*, Crocodiles; (2) *Chamaeleonidæ*, Chameleons;

(3) *Geckotidæ*, Geckos; (4) *Iguanidæ*, Iguanas; (5) *Varanidæ*, Varans; (6) *Teidæ*, Teguixins; (7) *Lacertidæ*, Lizards; (8) *Chalcidæ*, Chalcides; (9) *Scincidæ*, Scinks.

First Family. Crocodiles.

Crocodilidæ, (Gr. *κροκόδειλος, krokodeilos*, a crocodile.)

This family includes three genera, closely related to each other, and agreeing in the general details of their structure. They are the largest of the Lizards found in America. They are called Alligators in the southern parts of the United States, and Caimans in the Antilles and South America. In Africa and Asia they are called Gavials. Those of America are distinguished by a broad and rounded snout; those of Africa by an elongated flat snout, and those of Asia by a pointed one like the beak of a bird. They differ from other reptiles in the tongue, which is thick, flat, and attached so much to the mouth that the ancients believed this member was altogether wanting. "To it, of all animals," wrote Herodotus, "nature has not given a tongue." The power of swimming is shown by the palmated feet, and by the lateral compression of the tail, which thus acts as a large and powerful fin; the tail is no doubt used as a weapon of defence, being armed with a serrated ridge of strong square scales. The lower jaw is rather longer than the upper, and both are armed with a single row of pointed teeth, the number of which does not vary with age, as in other animals. The Crocodiles are all inhabitants of the rivers and fresh waters of warm countries; but are most abundant in those latitudes which approach nearest to the equinoctial line. Their mode of feeding is very peculiar. They do not swallow their prey upon seizing it, nor is it ever eaten while fresh; but the victim is first drowned, and then conveyed to some hole at the edge of the water, where it is suffered to putrify before it is devoured. Their food consists principally of fish, crabs, and such other animals as they can catch. They seem to manifest an affinity to the Tortoises in the coverings of their bodies, being defended, like them, by plates or shields; with this difference, however, that in the Tortoises the plates are compactly united at the edges, while in Crocodiles they are sufficiently far apart to admit the free motion of all parts of the body and limbs. Owing to a peculiarity in the vertebræ of the neck, which bear upon each other by means of small false ribs, that render motion sidewise somewhat difficult, these creatures can not turn about with much facility, and may be avoided without difficulty, when on land they attempt the pursuit of man.

The Crocodiles of the Nile were regarded as sacred by the ancient Egyptians, and sometimes, when caught young, they were so tamed as to follow in the train of their religious processions. In some localities they have been killed of the length of thirty feet, (Swain.) Those of Egypt and Senegal, (Africa,) are less numerous, but more dangerous than those of this continent. No living species of this family is found in Europe, nor has any yet been detected in Australia, but remains have been discovered which indicate the former existence of this animal in territory now included in the British dominions.

Alligator (Lat. a binder) *Champsa,* (Gr. Crocodile, Hesiod.) ALLIGATORS. (See fig. on Chart.)

These have the head broad; the muzzle wide and rounded; the teeth of unequal length; the fourth or canine tooth of the lower jaw, (counting from the fore part of the jaw,) is the longest and is received into a corresponding cavity or pit in the upper jaw, when the mouth is closed, so that it is concealed. The hinder limbs are rounded and destitute of rigid scales; the webs between the toes are short.

The Alligators pursue fish with much dexterity, driving a shoal of them into a creek, and then getting into the midst of their prey and devouring them at pleasure. They also seize and feed upon dogs, frogs, pigs or other animals incautiously approaching too near to their lurking places. The usual method of capturing the Alligator is by baiting a large four-pointed hook and suffering it to float in the river. When the creature has swallowed the hook, he is hauled on shore and killed. Audubon gives an interesting account of the chase of a wounded Ibis by one of these animals. It had almost reached the terrified bird, "when," says he, "by pulling three triggers at once, we lodged the contents of our guns in the throat of the monster. Threshing furiously with his tail, and rolling his body in agony, the Alligator at last sunk to the mud; and the ibis, as if in gratitude, walked to our very feet, and then lying down, surrendered to us."

The principal species are the Caïman with bony eye-lids, *A. palpebrosus,* (Lat. from *palpebra,* an eye-lid,) found in Cayenne and Brazil, also in the Mississippi, as high as the Red River, in Carolina and Florida, and sometimes twenty feet in length.

The PIKE-NOSED ALLIGATOR, (see Chart.) *A. lucius,* (Lat. a pike,) found in the southern rivers of North America. In Louisiana, the Alligators of this species bury themselves in mud, where they become stiff, without being frozen. So intense is their lethargy, when the cold is severe, that they may be cut deeply without being roused. Their eggs are less in size than those of the Crocodile, being not much larger than a hen's. A peck of them are sometimes taken out of the place of deposit in the sand.

The SPECTACLED CAÏMAN, *A. sclerops*, (Gr. *skleros*, hard; *ops*, face,) is a native of Cayenne, Brazil, and Paraguay. It has its English name from a ridge across the forehead, and another before each eye, showing some resemblance to a pair of *spectacles*. The eggs are as large as those of a goose; usually about sixty are deposited. The Indians esteem them as food, and even relish the flesh of the Yacare, as the animal is called in Paraguay.

Crocodilus. The CROCODILE. (See Chart.)

This genus is distinguished from the preceding by the sudden narrowness of the muzzle behind the nostrils, which produces a large notch for the lodgment of the fourth tooth of the upper jaw, when the mouth is closed. The hinder margin of the leg is ornamented with a series of ridged scales, and the hind toes, especially the three outermost, are joined by webs to their point. The sublime description of the Leviathan in the book of Job, (chap. xii.,) evidently relates to the Crocodile. The most favorable season for catching the animal, is the winter, when it usually sleeps in sand banks, enjoying the warmth of the sun; or else, in the spring, while the female is "watching the sand islands, where she has buried her eggs." Sometimes it is harpooned, the coat of mail which protects the animal, being pierced by the weapon. The eggs of this formidable creature are but little larger than those of a goose. Many of them are destroyed yearly by birds of prey and quadrupeds, particularly the Ichneumon. Herodotus speaks of a bird called *Trochilus*, (supposed by some to be one of the Plovers,) which entered the jaws of this animal unmolested, and picked out, and devoured the *bdellæ*, suckers or gnats. These insects also infest the mouths of the Caïmans, of South America. Two species of this genus are found in this hemisphere, viz: the *C. rhombifer*, (Lat. *rhombus*, a rhomb; *fero*, to bear,) found in Cuba; and the SHARP-NOSED CROCODILE, *C. acutus*, (Lat. sharp,) found in St. Domingo and Martinique. The other species all belong to the Eastern Continent.

The COMMON CROCODILE, *C. vulgaris*, is found in the Nile, the Senegal, and the Ganges, and along the coast of Malabar.

The HELMETED CROCODILE, *C. galeatus*, (Lat. helmeted,) is found in Siam.

The TWO-RIDGED CROCODILE, *C. biporcatus*, (Lat. *bis*, twice; *porcatus*, ridged,) occurs in the Ganges, in the rivers of Pondicherry, and in those of Java.

The CUIRASSED CROCODILE, *C. cataphractus*, (Gr. *kataphractos*, mailed,) is found in the river Galba, near Sierra Leone, (Africa.)

The GAVIALS.

Gavialis. This genus is at once distinguished by the length and narrowness of the jaws, which are prolonged in a straight beak-like snout, armed with ranges of formidable teeth. Of this genus there is but one known species.

The GAVIAL OF THE GANGES, *Gavialis Gangeticus*, (see Chart,) one of the scourges of that celebrated river. The dying Hindoo, exposed upon its bank, and the dead body committed to its waters, become, not rarely, the food of this ferocious animal.

"In the living sub-genera of the Crocodilean family," observes Dr. Buckland, ('Bridgewater Treatise,' pp. 20,) "we see the elongated and slender beak of the Gavial of the Ganges, constructed to feed on fishes; while the shorter and stronger snout of the broad-nosed Alligators, gives them the power of seizing and devouring quadrupeds that come to the banks of rivers in hot countries. As there were scarcely any *mammalia* during the secondary periods, whilst the waters were abundantly stored with fishes, we might, *a priori*, expect that if any crocodilean forms had then existed, they would have most nearly resembled the Common Gavial; and we have hitherto only found those genera which have elongated beaks in formations anterior to, and including the chalk, while True Crocodiles, with a short and broad snout, like that of the Cäiman and the Alligator, appear for the first time in strata of the tertiary periods, in which remains of the *Mammalia* abound."

FOSSIL CROCODILES.

These have been found in the Eocene or early tertiary deposits of England. About seventy fossil members of the Crocodile family are known; but not many belong to the United States. In their structure, they conform most nearly to the Gavial of the Ganges.

The genus *Steneosaurus*, (Gr. *stenos*, narrow or straight; *sauros*, a lizard), affords the nearest link to the living species of the crocodile family.

The genus *Teleosaurus*, (Gr. *teleios*, perfect; *sauros*, a lizard,) resembles the living Saurians in the general contour of the head and jaws, but differs from them widely in the conformation of the muzzle, and the opening of the nose. Many species of the fossil Crocodiles were of enormous size, much larger than the living ones of the present day.

SUB-FAMILY. MARINE FOSSIL LIZARDS.

Enaliosauria, (Gr. ενάλιος, *enalios*, marine; σαύρος, *sauros*, a lizard.)

This group includes some very extraordinary fossil Saurians. Little else than the bones have been preserved, and from these alone the structure and habits are inferred. It is hence impossible to speak with certainty in regard to many parts of the living organization; while yet it is made quite clear, that in these extinct and gigantic reptiles, the extremities were flattened into fin-like flippers, connecting them with the CHELONIA, and, together with other peculiarities of their structure, pointing out the animals as exclusively aquatic. The Enaliosaurians "inhabited the seas of Europe during the Trias and Jura formations." We are not aware that any species have as yet been discovered in North America. The two genera, are,

1st, *Ichthyosaurus*, (Gr. *ichthus*, a fish; *sauros*, a lizard.) (See Chart.) This reptile is, according to Prof. Owen, a singular compound, in which the characters of the fish, the cetacea, and the bird are engrafted upon an essentially Saurine type of structure. Dr. Buckland, in his "Bridgewater Treatise," says of it, "It presents combinations of form and mechanical contrivances, which are now dispersed through various classes and orders of existing animals, but are no longer united in the same genus. Thus, in the same individual, the snout of a Porpoise is combined with the teeth of a Crocodile; the head of a Lizard with the vertebræ of a Fish; and the sternum of the *Ornithorhyncus* with the paddles of a Whale." The general outline of an *Ichthyosaurus* "must have most nearly resembled the modern Porpoise or Grampus. It had four broad feet or paddles, and terminated behind in a long and powerful tail." The structure of the skeleton is like that of a Saurian; but the vertebral column consists of more than a hundred vertebræ, each of which is hollow, and fashioned like those of fishes. The form of the sternal arch and the broad surfaces of the clavicles are adapted to give great strength to the chest, and enable the animal to breast the most disturbed waters. Dr. Buckland remarks, that "the bones composing the arch are combined nearly in the same manner as the Ornithorhyncus, of Australia, which seeks its food at the bottom of lakes and rivers, and is obliged, like the *Ichthyosaurus*, to be continually rising to the surface to breathe air." To this sternal arch the front paddles are articulated; they are nearly one-half larger than the posterior paddles, and in this part of the structure the cetaceous type appears to have been flattened. The bones of the head, the length of the muzzle, and the teeth, sometimes amounting to one hundred and eighty in number, present analogies to those of the Crocodile. The eyes, however, were extremely large, much larger than those of the latter animal, and we can easily imagine, glared ferociously as the monster darted towards its prey. Six different species have been enumerated. The commonest species, *I. tenuirostres*, (Lat. thin-beaked,) reaches the length of fourteen feet. The species *I. platyodon*, (Gr. *platus*, broad or large, and *odous*, tooth,) has been seen in specimens thirty feet long. The teeth are sometimes two and a half inches in length; and the orbit (of the eye) one foot in diameter. The vertebræ are one hundred and twenty in number. There is no evidence whatever that one species has succeeded, or been the result of the transmutation of a former species.

It should be added, that the first remains of the *Ichthyosaurus* were collected by a lady,—Miss Anning,—from the cliffs of Lyme Regis, Eng. The Ichthyosaurians are abundant throughout the Lias and Oolitic formations.

Plesiosaurus, (Gr. *plesios*, next; *sauros*, a lizard.) This genus was first described by Conybeare, in 1821. Its most remarkable character pertains to the vertebræ of the neck, which are from twenty to forty in number, (see Chart;)—more than in any other known animal. Conybeare conjectures, that as this creature breathed air, and had frequent need of respiration, it usually swam upon or near the surface of the water, arching back its long neck like the Swan, and plunging downwards at the fishes coming within its reach. Cuvier asserts,—"To the head of a Lizard, the *Plesiosaurus* united the tail of a Crocodile; a neck of enormous length, resembling the body of a Serpent; a trunk and tail having the proportions of an ordinary quadruped; the ribs of a Chameleon, and the paddles of a Whale." The greater length of its extremities would seem to indicate that movement on land was probably less difficult for this creature than for the *Ichthyosaurus*. It was, probably, in general, about ten feet long; though some species of this genus and the preceding one, must have exceeded twenty feet in length.

Prof. Owen enumerates no less than sixteen species of this extinct and most anomalous animal.

Pterodactylus, (Gr. *pteron*, wing; *daktulos*, a finger or toe,) The PTERODACTYLE, (see Chart,) The researches of geology have brought to light this Flying (fossil) Lizard, which received its name from Cuvier. The construction of the skeleton fully proves that it was capable of flying, or of skimming from one spot to another. The wings were, probably, much like those found in the Bat. The neck was very long and bird-like; the head large; the jaws armed with pointed teeth; and the tail very short. Six or seven species of this genus have been distinguished; one is almost the size of a Thrush; one of a Common Bat; and another considerably larger than the first. To these extinct reptiles, the little Dragons, (*Draco*,) have but a distant resemblance. The food of the smaller species consisted of insects, the larger preying upon the fishes, or the marsupials of their day. These very singular animals have only, within a comparatively short period, been admitted to a place among the Reptiles.

Iguanodon, (*Iguana*, and *odous*, a tooth.) This name has been given to the fossil remains of an extinct animal related to the Iguana; but which attained a far more enormous bulk. The bones were discovered by Dr. Mantell in the strata of Tilgate forest, Eng. The teeth are so much like those of the Iguana, as to show beyond question, its relation to this gigantic Saurian, which could not have been less than seventy feet in length. The teeth of the Iguanodon disclose some peculiar mechanical contrivances, fitting them for cropping tough vegetable food, such as that furnished by the plants found imbedded with it.

SAURIANS.

What is the SECOND ORDER OF REPTILES? Give the general characteristics of this order. What did Aristotle observe respecting the SAURIANS? In what respects do they come nearer the Mammalia than other reptiles? In what climates are they most numerous? At what season do they moult their skin? Are medicinal properties justly ascribed to them? Into how many families are they divided? What is the First Family? How many genera does this family include? How do they compare in size with the rest of the LIZARDS? Where are they called ALLIGATORS? Where CAIMANS? Where GAVIALS? How are these severally distinguished? In what particulars do they differ from other reptiles? What evinces their power of swimming? How are the jaws armed? Where are Crocodiles most abundant? What is peculiar in their mode of feeding? How are they related to the Tortoises? How did the ancient Egyptians regard them? Is any species found in Europe or Australia? What characters of the ALLIGATOR are given? What is said of its pursuit of fishes? How is it captured? What incident is related by Audubon? Mention the principal species. How does the gen. *Crocodilus* differ from the gen. *Alligator*?

Mention the species found in this Hemisphere. Also the other species and their localities. How is the gen. *Gavialis* at once distinguished? What is said of the Gavial of the Ganges? What is the remark of Dr. Buckland? What is said of Fossil Crocodiles? From what is the name Enaliosauria derived? Upon what is this group of marine Fossil Lizards founded? What two genera does it include? What does Prof. Owen say of the Ichthyosaurus? What does Dr. Buckland say respecting it? What further is said of it? How many species have been enumerated? Who first described the gen. Plesiosaurus? What is its most remarkable character? What was the conjecture of Conybeare? What else is said of this gen.? How many species have been enumerated? What is said of the Pterodactyle? What of the Iguanodon? Of the Chameleon? Illustrate this order from the Chart, tracing the gen. and families as there given.

SECTION IV.

Second Family. Chameleons.

Chamæleonidæ, (Gr. χαμαιλέων, chamaileon, a Chameleon, Chameleon-kind.)

These are a group of singular reptiles, not immediately related to any other family, but perhaps succeeding the Crocodiles as fitly as it would any other reptiles. It contains but a single genus, *Chamæleon;* the first peculiarity whereof consists in the absence of scales,—instead of which, the surface of the skin is covered with horny granulations of unequal size, but of symmetrical distribution; (2) the body is of a deep, compressed form, surmounted on the back by a sharp ridge; (3) the Parrot-like structure of the feet, (Plate XII. fig. 6,) longer in proportion than those of any other Saurian, having each five toes, divided into two opposing sets, one including two, and the other three, armed with five sharp claws, and connected together as far as the claws by the skin.

The internal organ of hearing is entirely hidden; the head is very large, and seems to be set upon the shoulders; the upper part usually showing an elevated crest or casque; and a ridged arch is over each of the large orbits to the muzzle. The mouth is very wide; the teeth are sharp, small, and three lobed. The eyes, though in themselves small, appear extremely minute; the whole of the ball, except the pupil, being covered with skin, forming a single circular eye-lid with a central dilatable aperture, The furrow between the ball of the eye and the edge of the orbit is very deep; and the eye-lid closely attached to the ball, moves with it. Each eye has the power of motion independent of its fellow!—so that we may see the axis of one directed upwards or backwards, while that of the other is in a contrary direction, giving a strange and most ludicrous aspect to the animal, in unison with its general contour and slow movements. In consequence of this independent motion of the eyes the animal when agitated, appears, in its movements, as if it were joined to another, with which it has no unity of purpose or action. For this reason, the Chameleon never goes into the water. He cannot swim; when in the water his power of concentration is lost; and he tumbles about as if in a state of intoxication. Moreover, he may be asleep on one side, and awake on the other!

As Cuvier observes, the only part of the Chameleon which moves with quickness, is its tongue. This organ is cylindrical and worm-like in shape, capable of being greatly elongated; it terminates in a fleshy tubercle, and is lubricated with a viscous saliva. When not in use, it can be withdrawn into the mouth, but is thrust forth with noiseless and arrow-like rapidity after insects, slugs, and the like, which come within its reach. On these the animal lives, and not "on air," as many of the ancients supposed. The gummy secretion at the tip of the tongue enables it to secure its food readily. When fully protruded, the tongue reaches to a distance equal to the length of the animal's body. "An insect on a leaf at an apparently hopeless distance, or a drop of water on a twig, disappear as if by enchantment, before the Chameleon," so marvelously rapid is the movement of its tongue; and here, doubtless, is the origin of the old idea relating to the *airiness* of its food. The structure of the grasping power of the tail in these animals can hardly fail to remind the beholder of the Spider Monkeys, and distinctly points to their arboreal habits; when they descend to the ground, their actions there appear strange and awkward. The females of this group dig a hole in the ground for the reception of their eggs, which they cover with earth and dry leaves.

In captivity, the Chameleons have little in their habits or manners that is pleasing or attractive. Like all the Lizard tribe, they are capable of enduring long-continued abstinence from food, and apparently without injury.

Fifteen species, mostly African, are described as belonging to the genus *Chamæleon*.

The Common Chameleon, *C. vulgaris*, is found in the south of Europe, as well as in Africa. This species, the emblem of hypocrisy and inconstancy, is the one so well known to the ancients, and respecting which so much has been said relative to its power of changing its form, and taking the color of near objects, and which was believed "to live on air."

The FORKED-NOSED CHAMELEON, *C. bifidus*, (Lat. divided into two parts,) is a very singular species, found in the Moluccas, India and Australia, having the top of the head flat, and the snout prolonged into two distinct branches.

The WARTY CHAMELEON, *C. verrucosus*, (Lat. warty,) is one of the largest species, averaging twenty inches in its total length. It is a native of Madagascar.

THIRD FAMILY. *Geckotidæ*. The GECKOS.

The Geckos are a numerous family, divided by Cuvier into seven sections, according to the structure of the toes, but bearing a strong resemblance to each other in their general characters, and are distinguished for their nocturnal habits. Their flattened form and broad head give them a peculiarly disagreeable appearance, which is increased by their sombre and rather toad-like hue : whence they have been subjected to the unfounded imputation of being venomous creatures, producing, by their touch, malignant disorders of the skin. Their limbs are short and the toes, which are nearly of equal size, are flattened and expanded on their under surface, either throughout the whole or a greater part of their length ; the dilated parts, or the disks, are often marked with regular but minute plates, so ranged as to produce a striated surface, and acting as suckers. The nails are sharp, hooked, and retractile, like those of a cat, so that their points may not become worn or blunted. The tongue is fleshy and broad, but short and capable of little protrusion, and notched at the tip, which alone is free. The eyes are large and full, with extremely small eye-lids, which, as in the Chameleon, form only a single membrane, leaving, however, a large aperture, and exposing the nictitating membrane. The pupil, as in the cat and other nocturnal animals, is linear when undilated, and contracts under the influence of light.

The orifices of the ears are placed on the sides of the head, the tympanum being considerably below the surface. The mouth is extremely wide ; the teeth are small, uniform, and implanted along the inner margin of the jaws ; the nostrils are placed laterally.

The skin is more or less covered with granulations or horny tubercles ; and in some species, it is extended along the sides and limbs into a kind of marginal fringe.

The voice of these reptiles is a sort of clucking cry, of which the term GECKO, uttered in a shrill tone, is an imitation.

Their food consists of insects and caterpillars, which they

often obtain by waiting in ambush for them, or by pursuing them into the holes and crevices to which they retreat for refuge. The imbricated suckers of the feet permit these reptiles to traverse ceilings, and suspend themselves on the under side of a leaf, while watching the movements of their prey.

Their sharp, hooked-like claws enable them to climb the bark of trees with perfect facility; to penetrate the cavities and clefts of rocks, and to ascend walls for the purpose of finding chinks or hollows in which they conceal themselves during the day, resting motionless, and affixed by the feet, with the back downwards.

We must not fail to notice the singular power which the Geckos have of reproducing the tail when it is lost by accident. Indeed, the tail appears to be *brittle;* and when broken off, it is soon replaced; but a swelling at the base of the reproduced member, marks its line of union.

These reptiles, though persecuted, seem partial to the habitations of men; attracted thither by the flies which swarm in the regions of their abode. It is useless to try to seize them. Their power of adhesion is instantly overcome in the case of danger; in their quick escape, not the slightest noise or rustle is heard, so that they vanish as if by magic. Mrs. Mason, of the Baptist mission in Burmah, says: "The first reptile that attracts the attention of new comers, is the Gecko, or House Lizard. They are every where; under the sides of tables and chairs; in the closets and book-cases, and among the food and clothing. They sometimes tumble from the roof upon the tables, but they usually come struggling with a centipede, or some other vermin in their mouths." So far from having any wish to destroy them, Mrs. Mason says their services were invaluable, the best "help" she had. "This harmless little creature," she continues, "is represented by English, French, and German authorities, as 'a species of poisonous lizard;' yet I have had them rest on the back of my hand, and hang suspended from my fingers, without the slightest disagreeable effect being produced." This is the animal mentioned in Prov. xxx., 28, correctly rendered by Jerome,

"The Gecko taketh hold with her hands,
And dwelleth in kings' palaces."

The Geckos are arranged, by some naturalists into seven genera, based upon the distinctive form of the toes and including about sixty species. They are found in Asia, Africa, America and Australia.

The COMMON GECKO, *G. verus*, (Lat. true,) was noticed by

Pliny and others of the ancients, under the name of Stellio, (Lat. a newt, or an animal having star-like spots upon its back.)

The BANDED GECKO, *Diplodactylus*, (Gr. *diplóos*, double; *dactulos*, finger,) *vittatus*, (Lat. banded or filleted,) is a singular species, found in Australia.

The LEAF-TAILED GECKO, *Phyllurus* (Gr. *phullon*, a leaf; *oura*, tail,) *platurus*, (Gr. *platus*, broad; *oura*, tail,) is a curious New Holland species, first described by Dr. Shaw, having a tail which is flattened horizontally in the shape of a leaf.

The WALL GECKO, *Platydactylus*, (Gr. broad-fingered,) *muralis*, (Lat. of a wall,) is a species common in southern Europe, where it attracts attention by its power of ascending smooth perpendicular walls. It is this species which is called by the Italians, *Tarantola*, or *Tarantula*.

The LEAF-FINGERED GECKO, *Phyllodactyla*, (Gr. leaf-fingered,) *tuberculosus*, (Lat. pimpled or tuberculated,) is found in California.

The SMOOTH GECKO, *G. lævis*, (Lat. smooth,) or *Platydactulus theconyx*, (Gr. *thēke*, a bag or sheath; *onux*, a nail,) is a native of South America and the Caribbee Islands. Specimens of this species, in which the tail has been broken off and replaced by another of imperfect growth, are seen in cabinets.

FOURTH FAMILY. *Iguanidæ*. The IGUANAS, or Thick-tongued Lizards.

These form a very numerous group of reptiles, of which the genus *Iguana*, (aboriginal name,) may be considered the type. The whole have been comprised, (see Chart,) in forty-six genera and one hundred and fifty species. Further discoveries, together with modifications of former classifications, have increased the number of genera to over fifty. Of the entire number of species belonging to this family, about one hundred are natives of America. North America possesses a considerable number, but not more than three species are found within the limits of the United States.

In all the genera of the Iguanas, the body is covered with horny plates or scales, often keeled, spinous or tuberculated, but never investing bony centres or rings. Nearly all have a horny ridge or crest along the middle of the back and tail. The teeth vary in their mode of attachment, but are never rooted or fixed in sockets. The tongue is of moderate size and free at the extremity only; it is thick, fleshy and spongy or velvety on its surface, never cylindrical, nor playing in a sheath. The eyes are protected with movable eyelids. The fingers are free, distinct, and all furnished with claws. The auditory orifice is usually visible, and often surrounded with pointed scales.

The senses of sight and hearing in the Iguanas, appear to be well developed; taste they seem to have in a fair degree, but not smell; the touch is moderate.

MM. Dumeril and Bibron divide these reptiles into two sub-families; (1) the *Pleurodonta*, (Gr. *pleuron*, side; *odous*, tooth,) having the teeth palatine, or in a sort of furrow running along the jaw bones and to which they adhere simply by their inner surface. All the genera are American, with the exception of one genus, *Brachylophus*, (Gr. *brachus*, short; *lophos*, crest,) found in India.

(2) The *Acrodonta*, (Gr. *akros*, the highest part or summit; *odous*, tooth,) having the teeth soldered to the ridge or upper edge of the jaws, of which they appear to be a continuation, and from which they rise. Our space allows us to notice but a few of the genera and species.

I. *Iguana*, distinguished by having a long flap or fold of skin under the throat, on the part nearest the chin, somewhat like a dewlap, and by having two series of palatine teeth, a long compressed tail, and a dentated crest along the back.

The animals of this genus are arboreal in their habits, but often, however, visit the ground, and occasionally take to the water, in which they swim with ease and rapidity. They are easily tamed, though they retain a degree of fierceness, and will often attempt to bite. The female visits the sea shore, or the borders of rivers, in order to deposit her eggs in the sand.

The incessant destruction of these creatures for the sake of their flesh, has rendered them exceedingly scarce in localities where they were once abundant. Their eggs are much esteemed. When attacked, they seldom attempt to escape, but gaze at their assailants, inflating their throats prodigiously, and assuming as formidable an air as possible, They show themselves to be very tenacious of life, and are generally killed by plunging a sharp instrument into the brain. A well known species inhabiting South America and the West India Islands, is the *Iguana tuberculata*, (Lat. having tubercles,) often reaching five feet in length, and sometimes measuring even six; the sides of the neck are covered with *tubercles*, whence the specific name. The general color of this species is green, more or less tinged with olive; or yellowish, marbled with a brighter tint; the tail is ringed with dusky black. It is fierce in its aspect and disposition. On account of the excellence of its flesh, the animal has also the specific name *sapidissima*, (Lat. most savory ;) it does not, however, suit "some constitutions."

II. *Anolius*, or *Anolis*. This genus is distinguished by an expansion of the skin on the last joint but one, (or the penultimate joint,) of the toes; by the possession of two rows of palatine teeth; by the absence of pores from the thighs. In some species, both the back and tail are without a ridge or crest; in others, a crest consisting of minute scales runs along the middle line of the back, and sometimes along the tail.

Like the Chameleons, the animals of this genus have the power of changing their color. They are smaller in size, the largest being not more than a foot in length; climb the branches of trees with great facility; and even rest upon the leaves, secured by the disks with which their toes are provided. The males are said to make a barking noise like that of a small dog, and to curl the tail over the back while running. In these animals, as well as those of the preceding genus, the middle parts of the body and the tail are more slender and fragile than the other portions, so that they often suffer a break, followed, however, by a reproduction and consequent deformity. One species is found in the United States, viz.: the Carolina Anolis, *A. Carolinensis*. It is very abundant in the southern sections of the Union, where it is known as the Green Lizard or Chameleon. This is a very beautiful animal, of a light golden green above and greenish white beneath; the throat pouch, when inflated with air, is of a vermilion color. It keeps about gardens, and often, in search of flies, enters the windows of houses, and can even walk upon glass by means of the disks of the toes.

The Great Crested Anolis, *A. velifer*, (Lat. sail-bearing,) is one of the species which have upon the back a sail-like crest.

III. *Basilicus*, (Gr. *basilikos*, a kinglet.) This genus varies from the *Iguana*, in the absence of femoral pores, and in having a more contracted dewlap. A triangular fold of thin skin, sustained by a cartilage, and rising vertically from the middle longitudinal line of the back of the head, gives a singular aspect to the animals of this genus, which appear as if crowned with a raised hood or pointed cap. An elevated, serrated ridge or crest of scales passes along the middle of the back and tail, in the males of one or two of the species, supported by bony appendages, and presenting the appearance of a continuous fin.

The Mitred Basilisk, *B. mitratus*, (Lat. mitred,) found in Mexico and regions further south, receives its name from the conspicuous pointed hood or crest on the occiput or hind part of the head. It should be noted that the Basilisk of modern naturalists, is not to be confounded with the Malignant Basilisk, or serpent of the African deserts, pictured by the fancy of poets, whose very glance the ancients believed to be fatal to all who came within its influence. The true Basilisk or Cockatrice is, notwithstanding its formidable appearance, a perfectly harmless reptile, possessing great activity and seeking its insect food among the trees.

IV. *Amblyrhyncus*, (Gr. *amblus*, blunt; *rhunchos*, a beak or muzzle,) is an anomalous genus found in the volcanic Galapagos islands, so noted for their peculiar forms. The head is short and has a blunt muzzle; the scales of the body are not tuberculated; the skin of the throat is dilatable, but not formed into a dewlap; a high crest appears upon the back and tail. Two species are found in the Galapagos islands, one terrestrial and burrowing under ground, *A. subcristatus*, (Lat. somewhat crested;) the other marine, *A. cristatus*, (Lat. crested,) living exclusively on the rock-bound sea, feeding on sea-weed, and seldom found at much distance from the shore. "It is of a dirty black color; stupid and sluggish in its movements. The limbs and strong claws are admirably adapted for crawling over the rugged and fissured masses of lava which every where form the coast; on the black rocks, six or seven of these hideous reptiles may oftentimes be seen basking in the sun." (Darwin's "Voyages of the Adventure and the Beagle.")

V. *Tropidolepis*, (Gr. *tropis*, a keel; *lepis*, a scale,) is a genus confined to North America, and embracing ten species. The Lizards which it includes have rough *carinated* (keel-like) scales on the back and sides, while those of the other parts are imbricated. The body is depressed and oblong in shape; the head short, depressed and rounded in front; the neck contracted and smooth beneath; the thighs have a series of distinct pores, but there is no crest either on the back or tail.

The Brown Swift, *T. undulatus*, (Lat. varied with waves,) is found within the Atlantic states as far north as New York, and also in the Western States. It is often seen running along fences or among trees, particularly in hilly or sandy districts, abounding in pine trees, among which it seeks its insect food; and hence is called the Pine or Fence Lizard. This little creature, from five to eight inches long, is venomous in its aspect, but really harmless. Like the Chameleon, it changes its color. It is very active, and therefore called Swift.

VI. *Phrynosoma*, (Gr. *phrunos*, a toad; *sōma*, body.) (Holbrook.)

The genus *Agama* formerly included both *Tropidolepis* and

Phrynosoma, but as now restricted, it contains no American species. The genus Phrynosoma is closely allied to the preceding. It includes several species inhabiting Texas, Mexico and California. The short, squat, nearly orbicular body, the feeble limbs, the long spines fringing the hind part of the head, and the shorter ones scattered along the back, give the animal quite a singular appearance. The species which are most numerous are *P. cornuta*, (Lat. horned,) or *spinosa*, (Lat. spiny ;) *P. orbiculare*, (Lat. orbicular.) They are named Tapayaxan, or Horned Frog, from their fancied resemblance to the latter animal. These species feed upon insects, which they take by stealing upon them imperceptibly: they have the strange habit of feigning death when handled or even approached.

SECOND SUB-FAMILY, Acrodonta, without palatine teeth, and the greater part without any external auditory orifice. All the species are found in the old world. We barely notice some of the more prominent genera.

I. *Draco*, (a dragon,) including eight or nine species found in India, Java, Sumatra, etc.

These Lizards are of small size, and at once distinguished from all other Saurians, by the possession of a pair of parachute appendages, formed by the horizontal extension of the wings of the sides, and resembling those of a butterfly. They are the only living representatives of the fabulous dragons of olden time, celebrated in romance and fable. The "wings" can be folded up or expanded at will, but they can not be made to strike the air, and raise the animal after the manner of a bird or bat ; they, however assist this little *dragon*, only a few inches in length, in fluttering from branch to branch in search of insects, or when, like the Pteromys, or Flying Squirrel, it shoots from tree to tree. One of the most common species is the *D. Daudini*, of Bibron, or *D. volans*, (Lat. flying,) of Gmelin, found in Java.

II. *Stellio*, (Lat. a newt or stellion,) is a genus characterized chiefly by having the tail encircled with rings of large scales that are often spinous. It furnishes the only European representative of the present family, viz.: *S. vulgaris*, the COMMON STELLION.

III. *Grammatophora*, (Gr. *Grammata*, letters ; *phoreo*, I carry,) so called from a fancied resemblance of the tubercles of the neck to letters. The back is without a crest, but has cross-rows of large scales. Some have a fold across the throat. One species is the *G. muricata*, the MURICATED LIZARD.

FIFTH FAMILY. VARANIDÆ. VARANS.

The Varans are worthy of particular attention on account of the light which they shed upon the organization of certain fossil Saurians. They are also interesting on account of the size of some of the species, which is inferior only to that of Crocodiles.

These reptiles are covered with non-imbricated tubercles; *i. e.*, they do not overlay each other, like tiles on a roof. These are set in the skin, rounded (except on the under part of the body, where they are angular in shape,) and arranged in circular bands or rings. The body is elongated, rounded, and without dorsal crests; the toes are distinct, very long, and armed with strong claws. The tail is more or less compressed, and at least twice as long as the body; the tongue is fleshy and very extensible, being, when fully protruded, twice as long as the head; it is of a slender figure, and deeply forked at the tip, like the tongue of a snake.

The Varans are divided into two distinct groups, viz: (1) the eminently Terrestrial group, which have the tail nearly conical in shape, and which dwell far from the water, in desert and sandy places; (2) the Aquatic group, consisting of those which inhabit the banks of rivers and lakes. In this latter group, the tail is compressed laterally, and surmounted by a ridge, formed by two series of flattened scales. In these the tail is an important organ of progression in the water; they lash it rapidly and powerfully from side to side, and thus propel themselves along with great celerity, cleaving the water like an arrow. The body, in consequence of the air with which the lungs are filled, floats on the surface, and is directed by this powerful organ, at once a rudder and an oar.

The motions of these animals on land, are quick and active. It is not certain that any of them are arboreal, or able to climb trees, but they can scramble up rocks and craggy precipices. They run with facility; but owing to the length of the tail and manner in which they work it from side to side, pressing, at the same time, against the ground, their movements are sinuous, like those of a serpent; and they can spring upon their prey.

The pupil of the eye is circular, and yet many are said to be nocturnal in their habits; others, however, are undoubtedly diurnal.

The food of the Varans consists of the larger kinds of insects, such as locusts, crickets, and beetles,—of birds, eggs, and small mammalia. It is said, "they unite themselves in packs on the borders of lakes and rivers, to attack quadrupeds which unsuspectingly approach to quench their thirst." M. Dumeril quotes Latour as saying that he had "seen them hunt down a young deer which was crossing a river, and succeed in drowning him;" and, on one occasion, had "found a bone of the thigh of a sheep in the stomach of one of these animals which he dissected."

No evidence exists that they ever attempt to injure man unless previously molested by him.

Such are the animals which, in certain parts of their organization, bear the closest resemblance to the extinct Saurians. If the habits attributed to these Varans bear any relation to those of the Saurians now swept from the earth, then "we might have in those annihilated giants, no bad representatives of the dragons of our wildest legends."

The species of this family are not numerous, though widely distributed. But one belongs to the North America, viz: the *Mexican Heloderma, Heloderma horrida*, one of the Aquatic Varans. In Mexico, the belief is general, but erroneous, that the bite of this species is fatal. Others are found in Asia, Africa, and Oceanica. Only two species of Terrestrial Varans are known; one is peculiar to the island of Timor, (*V. Timoriensis;*) the other is

The Desert Varan of Egypt, *V. arenarius*, the Ouaran-el-hard of the Arabs,—about three feet in its total length. It is less active than the aquatic species, and especially than that inhabiting the Nile.

The Varan of the Nile, *V. Niloticus* or *Monitor Niloticus*, Nilotic monitor, is a noted aquatic species, attaining the length of five or six feet, and common in the Nile. It was held in great veneration by the ancient Egyptians, probably, says Cuvier, because it destroyed the eggs of the Crocodile, of the approach of which it is said to warn persons by a hissing noise, and hence was called *monitor*. There are several conspicuous fossil Saurians, some of which seem to be allied to the Varans, and which are represented in the Cretaceous (Lat. *creta*, chalk) system of the United States,

(1) The *Geosaurus*, (Gr. *gē*, the earth, *sauros*, a saurian.) This name was given to this fossil by Cuvier, not in reference to its habits as a living animal, as it was no doubt aquatic, but in "allusion to the earth,—the *Ge* (Γῆ) of the Greeks, as the fabled mother of the Giants." Remains of this animal were first obtained from the white lias, at Manheim, Franconia. According to DeKay, remains have also been found in the marl of the green sand in New Jersey, and named *G. Mitchelli*, after the late Dr.

Samuel L. Mitchell, (Ann. of the Lyc. of New York, Vol. III.) Cuvier judged from the remains, that the animal was intermediate between the extinct Enaliosauria, or Sea-Lizards, and the living ones. The length of this fossil species is estimated at from fifteen to twenty-five feet. (2) The animal of Maestritcht, *Mosasaurus*, (Lat. *mosa*, the Latin name of Maestricht, and Gr. *sauros*, a lizard,) named by Conybeare from a fine specimen obtained from Maestricht, at the time of its capture by the French army. Specimens of this fossil, *M. maximiliani* or *M. major* have been obtained from New Jersey and the banks of the Yellow Stone River. DeKay gives the length, from fourteen to fifteen feet; but Dr. Buckland judges the animal to have been twenty-five or twenty-eight long, (see his "Bridgewater Treatise,") and so constructed as to "possess the power of moving in the sea with sufficient velocity to overtake and capture such large and powerful fishes as, from the enormous size of its teeth and jaws, we may conclude it was intended to devour."

SIXTH FAMILY. *Teidæ*. TEGUIXINS.

The *Lacertidæ* have been arranged, by M. Dumeril, into two divisions, viz: (1) Pleodonta, (Gr. *pleos*, full, not hollow; *odous*, *odontos*, a tooth,) distinguished by having solid and rooted teeth; (2) Coelodonta, (Gr. *koilos*, hollow,) which have the teeth hollowed by a sort of canal, and but slightly adherent to the bones of the jaws. The latter are peculiar to the Old World; the Pleodonta are confined to this continent, and none are included in the family *Teidæ*, which have the head-plate horny, and the scales small and granular, and sometimes with large plates. This division is clearly separable from the Helioderms, of Mexico, which have the head shields and scales of the body tubercular and the teeth groved within the ridge of the jaw.

The present family of Lizards includes twelve genera, which may be divided into two groups, the one with the tail compressed or flattened vertically; the other with rounded tail; or they may be divided into those in which the front has the cross-folds, with six-sided scales between; (2) those in which the throat has a collar of large shields.

Those which have compressed tails, show a marked resemblance to the Crocodiles, which is increased by their great size. The tail is flattened somewhat like an oar, and the surface being increased by caudal crests, these animals are able to move in the water, which they inhabit, with nearly or quite as much facility as the Crocodiles.

The species *Crocodilurus*, (Gr. Crocodile-tailed,) *lacertinus*, is nearly six feet in length; inhabiting the waters of Guiana and Brazil. This is sometimes called the *Tupinambis*.

The *Teius Teguixin*, or *Teguixin Monitor*, of Gray, *Tupinambis Monitor*, (Daudin,) is the true Tupinambis, the *Sauvegarde*, (the Safeguard,) of Cuvier. This is one of the most noted species. In their habits, the SAFEGUARDS are highly aquatic. They are, indeed, able to run with great swiftness along the ground, and they dig for themselves burrows or hiding places in the earth, but when hard pressed, are sure to take to the water. They are found in South America, and reach from four to six feet in length. Sometimes they are seen as long as eight feet.

D'Ayara states, that "they feed on fruits and insects," and that "they also eat serpents, toads, young chicks and eggs." He also relates that "they are fond of honey; and in order to obtain it without injury from the bees, they come forward at intervals, and as they run away, each time, give the hive a blow with their tail, until, by repeated attacks, they weary out the industrious insects, and drive them from their home."

The *Thorictes*, (Gr. from *thorax*, coat of mail,) *dracæna*, is a very large species found in Guiana, and, in some instances, being almost seven feet in length, of which the tail occupied five feet. This, and the species *Crocodilurus lacertinus*, were formerly included in the genus *Ada*, divided into the two genera by M. Bibron.

The genus *Ameiva* includes six species, some of which have the tail more rounded or conical, and two plates on the throat. These are more terrestrial or arboreal in their habits.

The genus *Cnemidophorus*, (Gr. *knēmis*, a greave or leggin; *phoreo*, I carry,) is interesting as including the only representative of the family in North America. This is the *C. sex-lineatus*, (Lat. six-lined,) which is abundant in the Southern States, and as far North as North-eastern Maryland. It is easily known from the other Lizards by the six yellow lines along the back, and the long tail. When pursued, it runs with almost incredible swiftness; climbing trees with great facility, but not leaping from branch to branch, like the Green Lizard, *Anolis Carolinensis*.

Acrantus, (Gr. *akrantos*, imperfect,) is a large South American genus, which has but four toes visible on the hind feet.

SEVENTH FAMILY. *Lacertidæ*, (Lat. lacerta, lizard.)

SLENDER-TONGUED LIZARDS.

This family includes the Coelodonta, already defined; and which are found in the Old World. No true Lizard has yet been discovered either in Australia or the Polynesian islands.

In many respects, these and the American Teidæ agree. The body is rounded and elongated; the tail generally exceeds the body in length, and is always well developed; the head is pyramidal, flattened above, and covered with plates; the tympanum is distinct, and sometimes externally apparent; the feet have each four or five separate toes, armed with hooked claws; the eyes have the nictitating membrane in addition to the ordinary eyelids; the mouth is very wide, and its edges are covered with large (labial) plates; the teeth hollowed and placed in a groove within the ridge of the jaw.

The True Lizards inhabit all the warm countries of the Old World, and some of those which are considered temperate; but in the latter, they pass the winter in a torpid state. When excited by the heat of the sun, they are extremely active and vivacious,—the most so indeed of all the Saurians. It is, how ever, only by sudden darts, and for short distances, that they perform their movements. If these animals do not soon gain their burrows, or hiding places, they become fatigued, and fall an easy prey to their enemies. Hence, they never undertake long excursions from their native spot, or from the retreat which they have chosen. In their course over the ground, or when making their way among tangled herbage, the movement of their bodies is serpentine. They help themselves onward not simply by their limbs, but also by the body, and especially the tail. The latter is so brittle, that it breaks off easily, but it is soon renewed; the renewed part being clearly distinguishable by a difference of coloring from the rest, and the vertebræ, instead of being hard and bony, are cartilaginous.

Although quite inoffensive, Lizards defend themselves with much energy when attacked, and bite more sharply than might be supposed. The larger part of them feed upon insects; though some of them prey upon small animals, such as mice or frogs.

The typical genus, *Lacerta*, contains species which are widely spread over Europe and Africa, and remarkable for their brilliant colors, as well as their quick movements. The Lizards of this genus are easily distinguished by the throat collar of broad scales; the tongue is long and forked; the scales of the tail are disposed in rings; a minute plate of bone above protects the orbits of the eyes; a long row of pores runs down each thigh; the palate is toothed.

The EYED LIZARD, *L. ocellata*, of Southern Europe, attains to about sixteen inches in length. Its ground color is a bright glossy green, ornamented with round eye-shaped spots of gold and blue, and with rings and irregular markings of black. It is very bold and resolute; when attacked by a dog, it fastens itself on the muzzle of its enemy, and will suffer itself to be killed before it will let go its hold. The female lays seven or eight oblong eggs.

The GREEN LIZARD. *L. viridis*, is an elegant species, but in size is much less than the preceding. It is readily tamed, and taught to come to the hand for food; will lie coiled in the hand without attempting to escape; on account of its beauty and gracefulness, it is often kept in cages furnished with an inner compartment filled with dried moss or bran; amidst which it buries itself in order to pass the winter. It seldom bites; and, indeed, its bite is said to be "a pinch scarcely to be felt."

The SAND LIZARD, *L. agilis*, is considerably larger than the Green Lizard; is a native of England and most parts of the Continent of Europe. Its general color is a sandy brown, spotted with black on the sides, each spot having a white or yellowish dot in the center. Unlike the Green Lizard, it is impatient of confinement, and soon pines to death, never growing familiar. It is sometimes a foot long, measuring from the nose to the extremity of the tail. The female buries her eggs in the sand, and leaves them to be hatched by the heat of the sun.

The VIVIPAROUS or SCALY LIZARD, *Zootoca*, (Gr. *zōos*, living; *tikto*, to bring forth,) *vivipara*. This species of the sub-genus *Zootoca* is characterised by the palate being toothless. This is also found in Great Britain and in the Continent of Europe, and is "a pretty, active little creature, frequenting dry, sunny banks, thickets and copses." It seldom exceeds five or six inches in length, and is very gentle and harmless. It differs from the preceding species in producing living young. The eggs are hatched before exclusion, and not deposited in the sand; hence, the term applied to it, "viviparous." This Lizard ordinarily produces four or five young, often seen in company with the mother, and for sometime, probably, guided by her, but lively and alert, and capable of procuring their own food. This species presents various markings, but in most, the upper parts are of a greenish or olive brown, with lines of dark brown on the back and side; the under parts orange, spotted with black, or, in the female, pale gray, with a tinge of green.

The genus *Ophiops*, (Gr. *ophis*, serpent; *ōps*, eye,) is principally distinguished by having no eye-lids, or merely rudimentary ones, like the Serpents.

The species *Ophiops elegans*, is of an olive color above, with two lines of yellow on each side of the body, having two rows of black spots between them.

EIGHTH FAMILY, *Chalcidæ*. CHALCIDES.

This and the succeeding family of Skinks, each conduct to the Ophidia, or Snakes. These two families have, therefore, sometimes been regarded as constituting an intermediate order between the Saurians and Ophidia, and termed Saurophidia, or Lizard-Snakes. Some of the genera of the present family are, by Cuvier, classed with the Snakes, as they are without limbs, and resemble the latter in other respects in their structure.

The animals of this family are readily distinguishable by the arrangement of the scales or markings of the skin, and by the lateral furrow found in many species. Some of them are furnished with four legs; others with but two; while another portion of the family are entirely serpent-like in their appearance, in consequence of the absence of these members outside of the skin. The trunk of the body blends with the head and tail, without any distinct lines of division, and is covered with scales which, instead of being imbricated like those of fishes, are arranged in whorls or rings enclosing the body. Where the scales are absent, furrows in the hardened skin exhibit similar markings. The teeth are not implanted in the jaws, but appended along the margin or internal edge,—thus showing the true pleurodont character; the tongue is free, but not very extensible; it is broad, and covered with papillæ, and is notched at the front; the ears are apparent externally in some species, while others present no such indications. The eyes are generally small and slightly developed. Some species have movable eyelids; in others, these organs are not movable; while a few have the entire ball of the eye covered by the skin.

The Reptiles are confined chiefly to Africa and America. Mexico, California, and the Southern parts of the Union have quite a number of species, some fifty of which have been described and arranged in sixteen or more genera. They have been divided into two sub-families, according as the skin is covered with scales, or destitute of them. (1) Ptychopleura, (Gr.

ptuchē, a fold; *pleura*, side,) distinguished by a fold of the skin upon the side; (2) Glyptoderma, (Gr. *gluptos*, graved or carved; *derma*, skin,) distinguished by square or card-like divisions, sometimes colored, and then, like mosaic work, extending in regular order over the skin. The first sub-family have scales arranged in the manner described above. All have a fold or furrow on each side, and are in possession of eye-lids.

Of these may be mentioned the Ophisauros, (Gr. serpent-lizard,) found in North America, having, with the head of a lizard, the body of a snake, and the snake-like manners which such a form involves. It is called the Glass-snake, from the fact that the body is very brittle, and may be broken by a slight blow.

Two species exist in the United States, viz: (1) *O. ventralis*, which is limited to the Southern or South-Eastern States; (2) *O. lineatus*, which is met with in the South-West, and as far North as Michigan.

II. Genus *Pseudopus*, (Gr. *pseudos*, false; *pous*, foot,) including reptiles which, in their form and movements, resemble snakes,—having no front limbs, and hind limbs which are mere scaly, undivided appendages.

The Scheltopusik, (*P. Pallasii*, Cuv.,) is so named by the natives of the desert of Naryn, near the Volga, (Russia.) It is a native, not of Europe only, but of Africa and Asia. This reptile is eighteen inches long; of a reddish yellow or chestnut color, clouded with black. It frequents wooded valleys and gives chase to small Lizards, which, together with insects, constitute its food. Being of a quiet and inoffensive disposition, it is, when captured, sometimes kept alive in rooms. It is recorded, however, that on one occasion, one of these reptiles so kept, got access to a nest of young birds, which it soon demolished, and, no doubt, fully enjoyed.

III. Genus *Chalcides*, includes species chiefly found in South America, having both fore and hind limbs, but in a rudimentary condition. The front pair terminate in three or four scaly tubercles; the posterior pair are represented by two slender spines; the tongue is arrow-like in figure, with a sharp, two-cleft point; the surface is covered with large, flat, imbricated papillæ, resembling in form and arrangement, the scales of a fish. Four species are described. One, (*C. Schlegii*,) a native of Java; the others are found in Guiana, Columbia, and Chili. These reptiles have no external ear, by which they are distinguished from the following genus.

IV. Genus *Chamaesaura*, (Gr. *chamai*, on the ground; *saura*, a lizard,) which has an outward auditory cavity, and the rudimentary limbs without any sub-division, or but one toe on each foot. The only species is the *C. anguina*, of the Cape of Good Hope, having the head covered with many side shields or scales, and the cylindrical and elongated body covered with elongate, keeled scales.

V. Genus *Saurophis*, (Gr. Lizard-Serpent,)—includes reptiles with more highly developed extremities, each foot having four toes. The only species known, is the *S. tetradactylus*, which inhabits the southern part of Africa.

The other genera of this sub-family have four toes on each foot, of these we can only refer to the genus.

VI. *Gerrhonotus*, (Gr. shield-back,) of which there are eight species, seven inhabiting Mexico, and one California. In these reptiles, the thighs are destitute of the pores. They produce their young alive; and in their habits, closely resemble the Lizards.

VII. *Zonurus*, (Gr. *zonē*, belt; *oura*, tail,) is a genus in which the limbs are four and robust; the feet each furnished with five toes; the tail is short, and the head triangular and flattened; the scales of the back and sides are square, in a close cross series.

The Cordyle Lizard, *Z. griseus*, (Bibron,) or *Cordylus* (Gr. a knotty club,) *griseus*, (Cuvier,) is an example of this genus. It is a native of South Africa, where it is common.

VIII. *Tachydromus*, (Gr. swift runner,) is a genus found in Cochin China, China, Borneo, and Java, distinguished by having keeled ventral shields; and the throat with keeled scales. It has, like the preceding, four limbs, but they are less robust; five toes, but three not fully developed, and a greatly elongated form.

The Tachydrome, T. *sexlinĕatŭs*, (Lat. six-lined,) receives its specific name from having three lines extending longitudinally on each side.

Second sub-class. Glyptoderma, (Gr. carved-skin.) This division nearly corresponds with the family *Amphisbaenidæ*, of some authors. The lateral furrow peculiar to these reptiles, is faintly seen in the more typical Chalcides referred to above. Most of the species have been classed by some with the *Ophidia*, which they greatly resemble. From the latter, however, they are distinguished by their Saurian head and tongue; and by having the vertebræ united by fibrous or thread-like cartilage.

I. Genus *Amphisbaena*, Double Walkers, so called from the strong resemblance between the front and hind extremities of the membranous body, the head, tail, and intermediate part being of the same circumference. Appearing to have a tail at each end, they are supposed to be capable of progression in either direction.

M. Bibron enumerates ten species, of which two are natives of Africa, the rest of America.

The DUSKY AMPHISBAENA, *A. fuliginosa*, and the WHITE AMPHISBAENA, *A. alba*, are species measuring nearly two feet in length; found in Brazil and Cayenne. They bore the ground like worms, and, it is said, move either way with equal facility. They are often found in the earthy habitations of the Termite-Ants, which they follow through their winding galleries, for the purpose of feeding on them. The flesh of these creatures, dried and reduced to a fine powder, is sometimes administered as an infallible remedy in cases of broken bones, or dislocated joints; on the inference, that as it has the power of uniting its own body, if cut in two, and of healing, in so marvelous a manner, amputation in itself, it has at least the power of curing a simple fracture in another!

II. Genus *Chirotes*, (Gr. from *cheir*, hand,) has no hind limbs, but has a pair of short front limbs placed near the head, and what is remarkable among Saurians, each having five fingers, or at least four fingers or toes, armed with claws, and a tubercle representing the fifth. The possession of a sternum distinguishes these reptiles from the Amphisbaena. The body is snake-like; the head, neck and trunk, are of equal circumference. Only one species is known;

The CHANNELED CHIROTES or BIPED, *C. canaliculatus*, (Lat. channeled,)—eight or ten inches in length, a native of Mexico, and extending to the eastern side of the Rocky Mountains. The eyes are almost imperceptible, covered with transparent skin, but destitute of eye-lids. Its upper surface is yellow; the under, white, and the whole body covered with little square compartments, disposed circularly. In the absence of hind feet, while the front ones are present, this creature presents a strong resemblance to the Siren, a genus of the Batrachians.

NINTH FAMILY. SCINKS, (or SKINKS,) or LEPIDOSAURIANS.

Scincidæ, (Gr. σκίγκος, *skinkos*, a kind of lizard.)

We come now to the last family of the Saurians, which, to the general characters of the order, join many distinguishing peculiarities. They have the head covered with large plates, which have angular and regular shapes. These render them distinguishable from all the other families of the order, except the True Lizards and the Chalcidians, which, as we have seen, possess them also. The rest of the body is invested with scales, of greater or less magnitude, and of variable forms; but always arranged in a *quincunx* or five-fold order, and overlaying each other like the tiles of a house, as we see in large scaled fishes, as, for instance, the Carp. The scales of the under parts and sides, are nearly of the same size and shape as those of the back. This distinguishes them from the True Lizards, in which the ventral scales are much larger than those of the back, with the outlines angular. There are no lateral furrows or folds of skin extending along the flanks; this again separates them from the Chalcidians. The tongue is free, fleshy, notched at its point, without a sheath, and covered either altogether, or in part, with papillæ. The whole surface of the scales being generally smooth and polished, many of these reptiles glide easily into small crevices; and they creep by giving a tortuous and snake-like motion to the trunk and tail. The limbs vary in different groups, being four, two, or none; when present, they are short.

The Skinks include about a hundred species, variously distributed over the globe. The largest number of species is found in Australia, which has nearly forty peculiar to itself. Asia claims the next largest number; then comes Africa, and afterwards America. Europe numbers scarcely more than six or eight species. Five species are found within the limits of the United States.

The Skinks have been arranged into three sub-families, distinguished from each other by peculiarities relating to the eyes.

I. Saurophthalmia, (Gr. *sauros*, a saurian; *ophthalmos*, an eye.) The members of this sub-family have movable eye-lids, which can be brought together so as to entirely cover the eye. Most of them have four feet, but some have two, while others appear to have none. All are without femoral pores. The lowest form of this group is the *Acontias*, (Gr. a serpent that darts from a tree on its prey.) Of this, only one species is known, the *Acontius meleagris*, found at the Cape of Good Hope. Though much like a serpent in the absence of feet, and of a tympanic orifice, it possesses most of the characters of the Skinks. The tongue, as in the Blind or Slow Worm, is flat, and like an arrow-head, with scarcely any notch at the tip. The eyes are very minute, and there is only a single eye-lid, which proceeds from the lower part of the orbit. The scales are smooth and imbricated.

II. Genus *Anguis*, (Lat. a snake,) is probably the best representative of the Serpentine or Footless Skinks. This is characterised by a cylindrical aad snake-like body and tail, as well as by the absence of limbs. The eyes, as in the preceding species, are very minute. Only one species is known, viz: the SLOW

WORM or BLIND WORM, *A. fragilis*, (Lat. brittle.) When irritated or alarmed, the Slow Worm, by a forcible contraction of all the muscles of the body, becomes perfectly stiff, and then breaks in two, with the slightest blow, or upon an attempt to bend it. Hence, Linnæus applied to it the term *fragilis*. This beautiful and harmless reptile, is found in various parts of Europe, appearing early in the spring, and going into winter quarters in October. It feeds on insects, earth-worms and slugs; being particularly partial to the latter. The Slow Worm is said to shed its skin like the true snakes. The female produces her young alive, in July or August, or at least lays from ten to sixteen eggs, from which the young soon escape; development having considerably advanced previous to the deposit. The general color is yellowish brown or yellowish gray, with lines and spots of black; the under parts are white, with whitish reticulations or net-work.

III. *Tropidophorus*, (Gr. *tropis*, carina or keel; *phoreo*, to carry.) In this genus there are four strong limbs, each with five compressed toes. The body is fusiform or spindle-shaped. The scales upon the body are thick and striated, but rounded on the muzzle; the tail has four spinose keels above, but is smooth on the sides. The species *T. Cocincinensis*, is a native of Cochin China. It is, on the upper parts of the body, of a color inclining to olive, or a yellow brown. The neck is banded, the color being brown; with marks of a much deeper shade, representing a succession of figures like the letter X; spots of deep brown also appear on the tail; and a row of whitish points along the lower parts of the sides.

IV. Genus *Seps*, (Gr. a small serpent.) exhibits a form somewhat snake-like, it being much elongated, but still provided with four limbs. These, however, are very small and weak, and have toes of unequal length. This is represented by a single species, *S. chalcides*. This curious reptile is a native of Southern Europe, and, except in the possession of limbs, resembles the Slow Worm. Like that reptile, it brings forth its young alive, and feeds on insects, earth worms and slugs. It is said to be perfectly harmless, though some suppose it to be a poisonous animal. It spends the winter in its ground-burrow, but emerges again in the spring, and lives during summer in sunny spots covered with herbage and underwood.

V. *Tetradactylus*, (Gr. four toed.) This genus has four toes on each foot.

VI. *Champsodactylus*, (Gr. with crocodile-toes,) has five toes in front, and four behind.

VII. Heteropus, (Gr. with unlike feet,) has four toes in front and five behind.

VIII. Trachysaurus, (Gr. rough lizard,) has five toes to each foot. The Rough Skink, *T. rugosus*, of New Holland, attains to a very large size, and is very singular in its appearance.

IX. Plestiodon, is a genus found in the United States, also having five toes to each foot; the fore feet short and scaly, with five sharp nails; the hind feet larger, with long slender toes, also furnished with nails. The species *P. fasciatus*, (Lat. banded,) was formerly *Scincus fasciatus*, (as in the Chart.) The body has five yellow lines upon it, from which the specific name is derived; the color above is bluish black. The length is from six to eight inches. This reptile is common in the Southern parts of New York State, and has been seen as far North as Massachusetts. It is found in Japan. The species *P. erythrocephalus*, (Gr. red headed,) is twelve inches long, and found from Pennsylvania to Florida. One species, *P. Americanus*, which is the largest, is said to attain the length of twenty-five inches. In the Southern States, they are called Scorpions, and regarded as poisonous, but not justly. The larger species are capable of inflicting a severe bite. The smaller ones are found about old logs, and sometimes under the bark of trees. A species of this genus is found in Egypt.

X. *Scincus*. This genus, as now restricted, includes but one species, *S. officinalis*, peculiar to Northern and Western Africa and Syria, having the tongue notched and scaly, the teeth conical and blunt, and two rows on the palate; the muzzle is wedge-shaped; the scales are smooth and shining, like those of a fish. The limbs are four, with five toes on each foot. The tail is conical and pointed. The upper parts are usually yellow, or of a silvery gray, mingled with brown and blackish. The under parts are, generally, of a silvery white. It is termed by Bruce, El Adda. In ancient times, it was regarded as an efficacious remedy in various diseases, especially those of an eruptive nature. According to Pliny, it was useful for curing wounds made by poisoned arrows, and, at the present day, it is kept by the druggists of Southern Europe; though its reputation has greatly waned. "It runs with considerable rapidity, and, when alarmed, it buries itself in the sand with singular quickness, burrowing, in a few moments, a gallery of many feet in depth. When caught, it struggles to escape, but neither attempts to bite, nor to defend itself with its claws."

II. Sub-family *Opiophthalmoi*, (Gr. serpent-eyes,) including Skinks, in which the eyes, like those of Serpents, are either without eye-lids, or else have them in the form of a narrow ring, partly or entirely surrounding the eye. Most of the species are found in New Holland. The genus *Hysteropus*, (Gr. *husteropus*, hind-footed,) (*Bipes*, Cuv.,) is without fore feet, and the hind ones are but short, flattened appendages, without any division into toes.

III. Sub-family *Typhlopthalmoi*, (Gr. blind-eyes,) includes Skinks which are entirely blind, having eyes so minute as to be completely rudimentary. Of this division, there are but two genera, each with a single species, viz: the *Dibamus*, of New Guinea, with hind oar-like feet; and *Typline Cuvierii*, of South Africa, without any feet whatever.

Which is the SECOND FAMILY of the SAURIAN REPTILES? What is its only gen.? What characteristics are given? What is said of the organ of hearing? Of the mouth, teeth, &c.? Describe the eyes. What peculiar power has each eye? How does this animal therefore appear when agitated? How does he act when in the water? Which is the only part of the C. that moves quickly? Describe the tongue and its uses. In what respect is the Chameleon like the Spider-Monkeys? What sp. are referred to?

What is the THIRD FAMILY? Give its characters. What is the origin of the name Gecko? On what does the animal feed? How is it able to traverse ceilings and the under side of leaves? How does it use its sharp hook-like claws? What singular power does it possess? Repeat the remarks of Mrs. Mason. What is the number of sp.? Where are they found? Which are referred to?

What is the FOURTH FAMILY? Which is the typical gen. of this family? What is said of the number of gen. and sp. which it includes? Give their general characteristics. Into what two SUB-FAMILIES have they been arranged? What is said of the gen. *Iguana?* What sp. of it are referred to? What is said of the gen. *Anolis?* What sp. are mentioned? What is said of the gen. *Basilicus?* What of the MITRED BASILISK? What of the gen. *Amblyrhyncus?* What sp. are named? What of the *Tropidolepis?* What of the BROWN SWIFT? What of the gen. Phrynosoma? Which are the most numerous sp.? Where are all the sp. of the SECOND SUB-FAMILY found? Which is the first gen. mentioned? What is said of it? What other gen. is mentioned?

What is the FIFTH FAMILY? Why do the VARANS deserve particular attention? Describe them. Into what groups are they divided? Give their peculiarities. Are they nocturnal? Upon what do they feed? What is the remark of Latours? What sp. belong to N. A.? What is said of it? What two sp. of Terrestrial Varans are known? What is said of the V. OF THE NILE? What FOSSIL Saurians allied to the Varans are mentioned?

What is the SIXTH FAMILY? Into what divisions have the LACERTIDÆ been arranged? Which of these is confined to this continent, and included in the Teidæ? Does it include the Helodorma of Mexico? How many gen. and groups does it contain? Which group resembles Crocodiles? What is said of the *Tupinambis?* Give the other name. What is said of the TEGUIXIN MONITOR? What very large sp. is found in Guiana? What representative of this family is found in N. A.? How is it known from the other Lizards?

What is the SEVENTH FAMILY? To which Continent are they confined? Describe the habits of the TRUE LIZARDS. What is said of the gen. *Lacerta?* Describe the *Eyed-Lizard.* The GREEN L. What of the LAND L.? Of the VIVIPAROUS or SCALY L.?

What of the EIGHTH FAMILY? To what order is this and the succeeding family closely related? How have they been regarded? How are the Chalcides readily distinguished? What further characters are given? To what regions are these Reptiles confined? Are any found in the U. S.? How many sp. have been described? Name the sub-families into which they have been arranged? What gen. is found in N. A.? Why is it called the GLASS-SNAKE? What sp. in the U. S.? State particulars in regard to the other gen. and sp.

What is the NINTH FAMILY? Give the peculiarities of the family? How many sp.? Name their localities. How many sub-families? How distinguished from each other? Describe the gen. and sp. Name the first sub-family. What is said of the second sub-family? What of the third?

SECTION V.

THIRD ORDER.—OPHIDIANS OR SERPENTS.

Ophidia. (Gr. ὄφις, *ophis*, a serpent.)

The Ophidians are particularly distinguished by the total absence of external limbs in a majority of the species, or else the limbs are so rudimentary as to be discoverable only by dissection, or on very close examination: so that, as Cuvier remarks, they are more truly deserving of the name of Reptiles than any other order.

They possess an elongated form, with which is conjoined not only great flexibility, but amazing strength. Their upper surface is covered with narrow and somewhat pointed scales, of small or moderate size, imbricated or disposed like tiles; these are called *squamæ.* The under surface is covered with broad transverse scales or plates, called *scuta,* of which the hind edge in one overlays the front edge in the other. The top of the head is also usually covered with plates. The whole of the delicate and pellucid outward membrane which covers the scales, is shed entire, and renewed once a year, or perhaps oftener; it is sometimes called the *slough.* All serpents pass the winter, or coldest part of the year, in a torpid state; when coming out from this state, the skin is cast or exuviated; it is first detached around the head, and is pushed off gradually, being turned inside out, like the finger of a glove. This rejection of the slough was to many of the ancients, a sign of a renovated state of existence; they regarded these reptiles as leading a protracted life of annually renewed vigor and beauty. The internal framework or skeleton of serpents is extremely simple, consisting of the skull, the vertebral column, and the ribs, (Plate XII. fig. 5.) The breast bone is wanting; so also are the bones of the hips, and of the limbs, excepting where the hind pair exist in the form of hook-like stylets, as in the Boas. A reference to the plate of the skeleton just referred to, must satisfy any one of its elegance, and also suggest the idea of its flexibility, which an examination of its parts will fully confirm.

The vertebral column consists of a series of bones united to each other by beautiful ball-and-socket joints; the head of each separate vertebra being received into a deep cup-like cavity of the one succeeding it. The whole of the spine is, in reality, a chain of these joints, firmly locked together, each movable to such extent as is consistent with the safety of the spinal cord.

Serpents are capable of twisting themselves in the most extraordinary manner; but their pliability consists less in the mobility of each joint separately, however great this be, than in the number of joints into which the vertebral column is divided. Two ribs, one on either side, arise from each of the distinct bones of that column. Its bones are exceedingly numerous, being always more than a hundred, and in some species amounting to more than three hundred.

The ribs, forming a large portion of a circle, embrace nearly the whole circumference of the body; and to these reptiles are the efficient agents of locomotion. They each severally play on a convex protuberance of the respective vertebræ, and are acted upon by powerful muscles, which move them backwards and forwards. Instead of being attached at their extremity to a breast bone, as is the case in the Mammals and Lizards, each pair, by means of a slender cartilage, is connected with one of the *scuta*, or shield-like plates of the under surface. The ribs may be likened to the limbs of the millipede or thousand-legged worm; they support the weight of the snake, and in its progression, work like the legs of that insect.

Although destitute of limbs, yet some serpents are capable of rapid advances. On the surface of the ground, their progression is made in two ways. The ordinary movement, when the body is straightened out entirely in contact with the ground, is by a succession of short steps, taken by the numerous ribs, as is seen in the millipede ; the ribs moving in pairs, and each pair, as advanced, carrying forward with themselves, the scales to which their extremities are fixed, and which serve as so many points of resistance to a backward movement. But the reptile makes a more rapid progress by throwing the body into large curvatures, the fore part being fixed, and the rest brought up by the action of the muscles and ribs; the hind part of the body being next fixed, and the fore part thrown forwards, and so on, alternately.

The most rapid movements are probably made when the entire body is gathered up into one vertical loop, like a bent spring, the head and tail being more or less approximated ; the sudden straightening of this loop or spring, with the tail as the fulcrum or point of resistance, may enable the animal to spring forward at one operation, to a distance greater than the length of the body.

The exceeding flexibility of their bodies enables many species to climb trees in pursuit of their prey, and on these, some are habitually found ; others are constant inhabitants of the water.

The bones of the face, excepting in a few species, have a high degree of mobility, and on account of their peculiarities, deserve special notice.

The lower jaw is not directly articulated with the upper, as in other animals, but connected with it by two bones which are movable upon each other. The extremities of the lower jaw also, instead of being anchylosed, or immovably fixed, are connected by an extensile ligament. Thus, unlike what is seen in the Mammals, where the bones forming the jaws and face are firmly locked together, those of the snake have no connection but that which is made by ligaments and skin. Hence the serpent tribe can swallow their food undivided, and many times larger in bulk than the circumference of the body. In this process they are aided by the expansibility of the skin, the gullet, and the stomach ; but something additional was needed to complete their capability of swallowing such enormous masses of food ; how shall such masses be made to pass through the jaws? Here is a difficulty which, without some peculiarity of structure must have been insuperable, but which, in the case of these reptiles, is fully met by the attachment of the lower jaw to bones movable upon each other, and allowing of a sort of *natural dislocation,* so that the jaw gives way in the act of swallowing, and recovers itself when the prey is fairly engulphed. The head of the snake may hence be pointed to as exhibiting one of the beautiful and striking instances of marked adaptation and harmony, so extensively apparent in the works of God.

Serpents can hardly be said to have two lungs, one of the two being generally abortive, or merely rudimentary.

All of them have teeth, but these serve only to retain their food, and are not adapted to mastication. In the harmless snakes, (or rather those which are not poisonous,) the upper and under jaws are furnished with a number of small, but very sharp teeth, pointing backwards, (Plate XII. fig. 5;) the palate is also armed with two similar rows, so that there are six lines of teeth in the mouth. The venomous species not only have the jaws very small and freely movable upon a bony peduncle or footstalk; but each branch of the upper jaw has a long, recurved, pointed tooth, traversed by a canal or tube, leading from a large gland situated beneath the eye. The fluid secreted by the gland passes through this tube into the bottom of the wound which the poison-fang inflicts. When not called into use, the poison-fangs lie concealed along the roof of the mouth; but when about to bite, the snake raises them up, and in the act of biting, compresses the poison-glands, by means of a peculiar muscle for that purpose, and thus instils a few drops of the deadly fluid into the puncture. These large fangs are, in truth, the only teeth in the upper *jaw*; the others above are arranged in two rows on each side, along the bones of the palate. The branches forming the lower jaw are slender, and but partially furnished with teeth.

It is common to hear persons speak of the *sting* of the serpent; but from the explanation here given, it will be noticed that properly speaking, the serpent has no sting; the fatal wound is produced by a *bite*.

Most species of venomous serpents are ovo-viviparous, i. e., the young are hatched before exclusion and born alive, whence the general name of Vipers—a contraction of *Vivipares*, (born alive)—though with a few of this division, whether they be so born or not, seems a matter dependent on the latitude and the mean temperature of the region in which they dwell. Some of the non-venomous or harmless snakes are also ovo-viviparous; the others are oviparous. The eggs are often more than thirty in number, rounded and agglutinated in bead-like rows, by a sort of mucous substance. The shells of the egg in oviparous serpents, although cretaceous, are soft like the eggs of the common hen when she has not enough calcareous matter in her food, called soft eggs. Their color is ordinarily yellowish or grayish white. The Creator has in this as in other instances, beneficently provided against the increase of dangerous animals, by assigning a small number of young to the venomous species, while many of the harmless kinds are extremely prolific. The females often take care of their young for a time. On the approach of danger, they have been seen to receive the whole family into their throats, and when it has passed, to restore them again to the open air.

The voice or hiss of serpents, which is often exerted, is more or less loud and piercing. It is the expression of anger or impatience; the warning of an attack, or the signal of defiance.

Their senses exhibit different degrees of development. They cannot be said to possess that of touch in a high degree, though they have what is sufficient to regulate their progression, and indicate the kind of surfaces with which their bodies are brought into contact.

The tongue is soft, fleshy, bifid (or divided into two branches) at its extremity, and working in a sheath. It is never venomous, as is commonly supposed. As an organ of taste, it cannot be very susceptible. The prey is swallowed entire, and under circumstances which afford little or no opportunity for the exercise of taste by the tongue.

The sense of smell, judging from the structure and habits of these reptiles, cannot be very acute.

The eyes are generally very small, not protected by movable eye-lids, nor by a nictitating membrane, so that they always appear to be fixed or on the watch. It is remarkable that the transparent cornea seems to form part of the skin and epidermis, with which it is detached at each moult of the reptile. Vision, excepting for a time previous to a change of the skin, when it is evidently less perfect, appears sufficiently acute in reptiles of the present order.

Serpents sometimes grow from a length of twelve or fourteen inches, when they are first excluded, to that of twenty-five or thirty feet, and attain to a great age. They are extremely tenacious of life, often surviving very severe wounds. Instances have occurred in which the head, severed from the body, has, after a considerable time, not only retained vitality, but bitten with fury.

The popular opinion that serpents are capable of exercising a power of fascination over their intended victims, is perhaps not well founded. The most eminent ornithologists refer the effects produced upon birds by the presence of these reptiles, to the fear amounting to terror which is thus occasioned, and to the instinctive solicitude for their young, which induces them to approach these reptiles too nearly for their own safety. The serpent tribes are indeed very generally regarded with feelings of horror and aversion, for which it would not be difficult to account; and yet to some these reptiles have furnished objects of religious veneration. The ancient Mexicans adored the Boa, and in the blindness of their superstition, sought to propitiate it with human victims. Among the bronze relics of the Egyptians, is a figure of the Cobra, with expanded hood, which was probably regarded as the image of a divinity, or one of the household gods. Figures of the Hindoo Chrisna sometimes present him entwined by a large Cobra, which is fixing its poisoned fangs in the heel; and again they represent him as crushing the head of the Serpent, while he triumphantly tears the creature from his body. The origin of these emblems cannot well be doubted; they, in all probability, spring from traditions related to the great prophetic promise of scripture, Gen. iii. 15. "The serpent stands as an emblem of the principle of evil to be ultimately destroyed with the poison of death itself, by the seed of the woman."

The divisions of the present order have been variously given by systematic naturalists. We like the arrangement of Mr. J. E. Gray, who divides the order into five families, viz.: (1) *Colubridæ;* (2) *Boidæ;* (3) *Hydridæ;* (4) *Viperidæ;* (5) *Crotalidæ.* (On the Chart the reptiles of the last two families are, for convenience, arranged among the *Boidæ.*)

Between three and four hundred species are enumerated, of which about one-fifth are venomous. But few species of *Ophidia* have been found in a fossil state.

FIRST FAMILY. *Colubridæ.* (Lat. *coluber*, a serpent.)

This family of the Ophidians includes snakes the larger portion of which are harmless, but few being provided with poisonous fangs. They are distributed over the globe, and are more numerous, considered both as individuals and species, than any other family of the order.

Dr. Gray arranges them, together with the *Boidæ* and *Hydridæ*, into the sub-order *Colubrina*, of which he gives the following definition: "Jaws strong, both toothed, sometimes with fangs in front or grooved teeth behind. Head moderate or indistinct; crown often covered with regular shields." The section *Colubridæ* have the belly covered with broad scales; the tail conical and tapering, and rarely compressed; the nostrils are open and placed at the side of the muzzle, near the top. The head is usually covered with large regular plates, the variations of which as to number and shape, afford good specific distinctions. The Colubridæ are mostly oviparous and carnivorous.

The leading genera of this family as found so numerously in the United States, are *Coluber* and *Tropidonotus*, (Gr. *tropis*, a keel; *nōtos*, back.)

The genus *Coluber* includes most of the larger familiar species which have smooth scales, without the *keel*, or longitudinal ridge along the center, which appears in the *Tropidonotus*. The body is usually slender and cylindrical. The snakes of this genus are rarely seen in water; they deposit their eggs in decayed wood, sand, or other localities.

Of the well known species found in the United States are (1) the CHAIN SNAKE, *C. getulus*, from its quick movements also called the racer; length four to six feet; (2) the MILK SNAKE, *C. eximius*, (Lat. select or distinguished,) sometimes, from its chestnut colored spots and light colored ground, called the Chequered Adder; also named the House Snake; it is not unfrequently found in dairies and cellars in which milk is kept, which it is said to seek with avidity; length two to five feet; (3) the GRASS SNAKE, or Green Snake, (see Chart,) *C. vernalis*, (Lat. vernal,) is found from Massachusetts to Pennsylvania, has been numerous in the marshes about Salina and Cayuga, (N. Y.;) length from twelve to twenty-four inches; (4) the RINGED SNAKE, *C. punctatus*, (Lat. dotted,) (see Chart;) it emits a disagreeable odor; occurs from Maine to Louisiana, under rocks and the bark of decayed trees; length twelve to eighteen inches; (5) BLACK SNAKE, *C. constrictor*, (Lat. one who binds together); abundant in all parts of the land and from three to six feet in length. It climbs trees with great facility, and moves very rapidly over the ground. This, as well as the species above mentioned, is called the Racer, on account of its pursuit of terror-stricken persons fleeing before it, an enemy it were wiser resolutely to face. Its climbing power renders it formidable to birds and young squirrels in their nests. It has been supposed to exert a fascinating influence over birds; we have already intimated that the unusual actings of a bird in its presence *may be* occasioned chiefly by the danger threatening its brood, which the reptile might devour at a single meal; (6) *C. Alleghaniensis*, is a larger species of Black Snake, from five to eight feet in length, said, however, to be much more gentle than the other, seldom showing any disposition to bite, which the Common Black Snake is very likely to do. Both of these species, and indeed all the Colubrines in North America, are non-venomous and harmless. The Black Snakes and some of the other larger species are exceedingly bold and resolute, and defend themselves obstinately when attacked. They even engage in deadly conflict with the Rattle Snakes; and owing to their superior agility, are generally victors, evading the poisonous thrusts of their antagonists, and seizing the opportunity to strangle them in their folds, like the Boa or Python; indeed, the specific name, *Constrictor*, is given in allusion to the mode in which the Black Snake kills its prey.

The snakes of the genus *Tropidonotus*, differ from the True Colubrines, in possessing the power of flattening or depressing the body. This enables them to swim well, and hence, they are all more or less aquatic. They are generally viviparous, the eggs being developed previous to exclusion.

Of this genus, (1) the familiar STRIPED or GARTER-SNAKE, *T. tænia*, (Lat. a ribbon,) or *T. sirtalis*, (two to five feet long,) is the typical representative. Though frequently found about the water, or in marshy places, it is as often on high dry ground, and has been noticed at an elevation of two thousand feet above tide water. When irritated, without the means of escape, it raises its scales so as to give the body a roughened appearance; and under such circumstances, it will bite, leaving a troublesome, though not dangerous wound. Its fecundity is so great, that in one instance, it is said, eighty-one young, each over nine inches in length, were taken from a single female; (2) the Water-Snake, *T. sipedon*, also called the Water-Adder, sometimes the Moccasin-Snake, and erroneously thought to be poisonous, is found rather abundantly in the Northern States, and also, to some extent, in the Middle States. This Snake is from two to five feet in length; it moves in the water with great ease, and may be said to live in it habitually. In the Southern States, its place is supplied by the beautiful GREEN-SNAKE, *Leptophis*, (Gr. *leptos*, thin; *ophis*, snake,) *æstivus*, about two feet in length. Another species, *L. saurita*, is found in the Northern, and, to some extent, in Western States. It is the Ribbon-Snake, or the Little Garter-Snake, as it is called in New York, of a chocolate brown color, gentle but very nimble, climbing trees with facility. Length from one two feet. These are the only species of the genus *Leptophis*. Both have long and slender bodies, carinate scales, and very long tails.

The SAND-SNAKE, *Psammophis*, (Gr. sand-snake,) *flagelliformis*, (Lat. of whip-form,) is a long, slender, and exceedingly swift species, found in South Carolina and Florida,—its tail having unimbricated scales, and being one-fourth its length.

The DIAMOND-SNAKE, *Coronella Sayi*, is conspicuous for its

minute white specks, scattered over a dark ground. It is one of the Snakes that often engage in successful conflict with the Rattle-Snake.

The HARLEQUIN OR SCARLET-SNAKE, *Elaps fulvus*, (Lat. red or tawny,) is found in Carolina, Louisiana, and Upper Missouri. It is distinguished by having a fang permanently fixed on each side of the upper jaw, with which may be connected a rudimentary poisonous gland; but the animal is considered entirely harmless. The head is scarcely larger than the body; the length twenty inches.

The RED-SNAKE, *Calamaria amœna*, (Lat. delightful to the eye,) is a beautiful little serpent, of a reddish brown color, from six to twelve inches in length, and found in the Eastern and Middle States, under stones and logs. It has a small head, smooth scales, and a short, abrupt tail.

The HOG-NOSED-SNAKE, *Heterodon*, (Gr. different or unequal teeth,) *platyrhinos*, (Gr. broad-nosed,) is two feet in length,—called also the Buckwheat-nose, (from a fancied resemblance between that grain and its rostral plate,)—and also the Deaf Adder and Yellow Viper. This species is well known, throughout the United States, and *H. niger*, (Lat. black,) is known as the BLACK VIPER, about three feet in length, and found in Georgia and Tennessee. Both present a formidable appearance, from flattening the head and whole body when irritated; but are entirely harmless. Passing over many colubrine species found in the United States and elsewhere, we name the *C. quadrilineatus*, (Lat. four lined,) which is the largest of the European Serpents, often attaining to six feet in length, and found in Spain and Italy. This formidable, though not venomous snake, is probably the Boa, of Pliny.

We must also not omit to notice an African genus, *Deirodon*, (Gr. *deirē*, neck; *odous*, tooth,)—which strikingly illustrates the special adaptations to particular uses and ends, which are presented in the animal kingdom. This snake is said to live almost entirely on the eggs of birds, and for this its entire organization seems expressly designed. The *mouth*, when full grown, has no teeth whatever, so that the egg is readily received into the open jaw, and there is no hazard of its being prematurely broken. The inferior spinous processes of the seven or eight lower cervical vertebræ, shoot forward with the gullet or æsophagus, where they are overspread with a layer of hard cement, and made to resemble long, sharp teeth. The eggs, in their descent, press against these teeth, and are sawed open lengthwise,—then crushed by the contraction of the gullet, and carried into the stomach; the shell, as well as its contents, subserving the purposes of food.

SECOND FAMILY. *Boidæ.* BOAS AND PYTHONS.

This family includes species which, although not venomous, are exceedingly terrific on account of their gigantic size and amazing strength. In these, the ventral shields are narrow, transverse, and often six sided; the pupil is oblong and erect, excepting in the genus *Tortrix.* But perhaps the most marked peculiarity, is their possession,—contrary to the general rule in serpents,—of hooked-like claws, connected internally with a series of bones, representing, though imperfectly, those of the lower limbs. The tail is prehensile, and can be firmly twined around any object. The Boas are natives of South America; the Pythons of Asia and South Africa. Some serpents, kindred to these, are also seen in Australia.

Imagination finds it difficult to picture more formidable objects than the reptiles of this group; and yet, if we can credit the statements of ancient writers, serpents far more terrific than these, were once found in the Eastern Continent. Livy refers to one which "had its lair on the banks of the Bagradas, near Utica, and swallowed many of the Roman soldiers in the army of Regulus," and which was finally killed by stones discharged from military engines. The skin, afterwards taken to Rome by Regulus, it is said, "measured one hundred and twenty-three feet!" This, however, may be an exaggeration, or the term "feet," is, perhaps, to be understood in a more limited sense than that which we assign to it.

The Boas, properly so called, sometimes reach the length of forty feet. In their entwining folds, acting with the combined energy of thousands of muscles for crushing their victims, they possess a power which no man or animal can successfully resist. To climb, to swim, to dart along the ground, are endowments of these powerful reptiles, and they avail themselves of each as occasion requires.

In the Boas, the head is covered with small scales to the muzzle; and the scuta of the tail are undivided. In the Pythons, there are plates over the anterior part of the head, and the scuta of the tail are divided.

The EMPEROR BOA, *Boa constrictor*, (see fig. on Chart,) is characterised by a broad chain extending along the back, and consisting alternately, of large, blackish, and somewhat hexagonal marks, and of pale, oval dashes or spots. The epithet

"Emperor," given to this Boa, indicates the religious veneration with which it was regarded anciently by the natives in Mexico and South America. It is more terrestrial in its habits than the Anaconda; resorting to dry places, among bushes, trees, and rocks. It climbs trees with great facility, and hangs suspended from them by its prehensile tail, ready to drop upon and crush any unfortunate creature that may pass beneath. Its length is from thirty to thirty-five feet.

The ANACONDA, *Boa scytale,* (Gr. *skutale,* a club or rod,) or *Eunectes,* (Gr. a good swimmer,) *murinus,* (Lat. from *mus,* mouse.) This species is of a brownish color, with a double series of roundish, black blotches running along the back. The spots on the sides are annulated and ocellated, the disks being white, surrounded by blackish rings. The trivial specific name, *murinus,* is given to it, because it is said "to lie in wait for mice." These, together with fish, frogs, etc., are truly "small game" to this creature, which *constricts,* and swallows down whole Sheep, Peccaries, Agoutis, etc. When the prey is dead, this, and the other Boas, thrust out their tongue, vibrating in token of their desire of food; the jaws and throat become lubricated with saliva, as a preparation for swallowing the enormous meal. The position of the mass in the alimentary tube indicates the completion of the process. When gorged with food, the animal is for some time torpid and defenceless, and may easily be killed. Occasionally, it is destroyed by shooting, lassoing, etc. The thick skin is frequently tanned, and converted into leather for boots and saddles. The Anaconda is said to attain the length of from thirty to forty feet; but the common specimens seen in museums and menageries, rarely exceed ten or fifteen. Among the other species, are

The ABOMA, *Boa cenchris,* (Gr. spotted,)—found in South America and the West Indies, is one of the largest of the family, sometimes attaining a gigantic size. It is of a yellowish color, with a row of large brown rings, running the whole length of the back, and variable spots on the sides.

The BOJOBI of the Brazilians, or the GREEN BOA, *B. canina,* or *Xiphosoma,* (Gr. sword-body,) *caninum,* (Lat. dog-like,) having a muzzle which shows some resemblance to that of a dog.

The CORAL-SNAKE, *Tortrix corallinus,* found as far North as Florida, and often kept tame in houses, belongs to this family.

The PYTHONS are natives of East India and its islands, and of Southern Africa. Two species are distinguished by placing their eggs in a group, and covering them with their bodies. One of these, is

The TIGER PYTHON or ROCK-SNAKE, *Python*, (Gr. *puthōn*,) *tigris*, a native of India and Java, and elegantly marked. It is said to be as large as the largest Boa, but more slender, and greatly to be feared. Stories are told of the tiger falling a prey to this formidable reptile.

The RETICULATED PYTHON, *P. reticulatus*, is found in Hindostan, Ceylon, and Java. It is said to increase until it reaches thirty feet in length, and can "manage a buffalo," crushing it in its huge folds. It is one of the most brilliant species of the entire family, "the whole body being covered with a gay lacing of gold and black."

The Pythons, in the British Zoological Gardens, "are fed with rabbits, which they destroy by winding round and crushing them; they are then easily swallowed; the expansive power of the jaws permitting a very small specimen to manage such animals."

THIRD FAMILY. MARINE SNAKES.

Hydridæ, (Gr. ὕδρα, *hudra*, a water-snake.)

The truly Aquatic or Marine Snakes, are all confined to the intertropical regions. They are mostly found in the seas and rivers of the East Indies. These singular reptiles, excepting that they are destitute of fins, are not unlike the eel, particularly in the form of the tail, which is expanded in a vertical direction, and flattened laterally, so as to act the part of a paddle. Some species, however, have conical tails, and these are thought to live in fresh water. In the Indian Seas, numbers of these snakes collect together, forming shoals, which may be seen swimming about in pursuit of fishes and other prey. It is very seldom that the true sea snakes visit the land. Sometimes they coil themselves up on the shore, where they lay their eggs. It is supposed that they live on sea-weed. They are often found asleep on the surface of the sea, when they are easily caught, as they are unable to descend without throwing themselves on their backs, probably for the purpose of expelling the air from their capacious lungs. They are frequently thrown ashore in the surf, to the terror of the natives. Occasionally, they are carried up rivers by the tide; but they cannot long live in fresh water. The fishermen of the Eastern seas, often take them in their nets, and greatly dread them on account of the poison of their bite.

The species are said to be, without exception, venomous. Dr. Cantor, who was in the service of the East India Company, and had favorable opportunities of studying the peculiarities of these serpents, captured by him in fishing nets, refers to the case of a British officer, who "died within an hour or two after the bite of a serpent caught at sea;" and also to numerous experiments of his own, "in which fowl, fish, and other animals, invariably died within a few minutes after the bite had been inflicted." We refer to these facts, because it has been stated, that "the Marine Serpents are harmless."

Rev. John Williams, in his "Narrative of Missionary Enterprises in the South Sea Islands," says: "That in the Samoa group are water snakes, some of them beautifully marked with longitudinal stripes of yellow and black, and others with rings alternately white and black." He adds, "the natives esteem both the Land and Sea Snakes as good food."

The MARINE SNAKES, in common with the BOIDÆ, have narrow, elongated scales on the belly, nearly resembling those on the back; the ventral shields are narrow, hexagonal, or band-like,—the eyes and nostrils look upwards, the latter usually placed in the middle of a shield, with a slit or groove on its outer edge; the fangs are of moderate size, and intermixed with the maxillary teeth; the pupil is small and round.

Of the species with compressed teeth, or true Marine Snakes, are the TWO-COLORED PELAMYS, *Pelamys bicolor*, with hexagonal scales, found in the Pacific Ocean; and the BANDED SEA-SNAKE, *Chersydrus*, (Gr. *chersudros*, an amphibious serpent,) *fasciatus*, (Lat. banded,) or *C. granulatus*, found in meadows. (For figure of which see Chart.)

FOURTH FAMILY. VIPERS. Sub-order VIPERINA, (venomous snakes.)

Viperidæ, (Lat. *vipera*, a viper.)

This family contains nine genera and twenty species, found chiefly in Asia and Africa; none of them have been discovered on the American continent. Unlike the Colubrine Snakes, these have few if any teeth in the upper jaw; but they have, in common with the Crotalidæ, glands secreting a poisonous fluid, which, on occasion, they discharge through their fangs in front. These glands are connected with muscles which are capable of exerting a powerful compression, and thus of ejecting the venom with great force into a wound. The shields of the muzzle in this family, are broad and band-like; the scales keeled, except in the genus *Acanthophis*, (Gr. spiny-serpent;) the tail is short and tapering.

The Common Viper, of Europe, *Vipera berus* or *V. communis*, is greatly feared, though its venom is said not to be as virulent as that of the kindred reptiles found in hotter regions of the globe. It does not often happen that death follows the bite of this species in the case of human beings. Ammonia or hartshorn, given internally, and fomentations applied to the part, to be gently rubbed afterwards with oil, are the remedies usually employed. To persons laboring under general debility, or to children of weak and irritable constitution, especially if the reptile be in full energy, during the heat of summer, the bite of the Common Viper is known to prove fatal. The surest remedies for its bite, are the immediate removal of the poison by suction, washing, excision of the part, &c. The Viper, as already intimated, brings forth its numerous young alive. These, though but a few inches in length, crawl about, and are as fierce as the parent,—throwing themselves into an attitude of defence when molested, and hissing with anger. Mice, lizards, and nestling birds, are the food of this species.

The Esping, of Sweden, or *Aspic*, of England, *V. chersia*, is perhaps only a variety of the Common Viper,—but is even more virulent; seldom, however, more than six inches in length. The rapid reproduction of the Common Viper, renders ineffectual the many efforts which are made for its extermination.

The genus *Naja* contains the Hooded or Spectacled Serpents, *Cobra de Capello*,—characterized by having the head covered with large plates, and the skin of the upper part of the back dilatable, or capable of such expansion as to form a sort of hood, impressed with a mark somewhat like a pair of spectacles. (See Chart.) Their bite is deadly in the extreme. They are found in Ceylon, from six to fifteen feet in length. The *hood* and *spectacles* show themselves when the reptile is enraged and preparing for an attack. The extension of the membranous skin serves as a warning to those who are within reach of the animal. The *Naja tripudians*, (Lat. dancing,) *N. larvata*, (Lat. frightened, distracted,) are species of India. To *N. tripudians*, the Portuguese originally gave the name of *Cobra de Capello*. The *Naja haje*, (see Chart,) is an African species, and indubitably the one which the ancients have described under the title of the Asp, or Aspis of Egypt, or of Cleopatra. The Najahs of South Africa are said, when irritated, to expel poison from the points of their fangs, and are supposed to have the power of ejecting the poison to a distance.

The Cobras are the serpents upon which the serpent charmers in India and Egypt chiefly practice their arts, and which are often taught to dance to their rude music. It should be noted with reference to the contest in the presence of Pharaoh between Moses and Aaron and the magicians of Egypt, (Exodus viii., 9—12,) that it is stated, on good authority, the modern Egyptian jugglers possess the power of throwing the *N. haje* into a state of catalepsy, and rendering it still and immovable, in other words, changing it into a rod, by pressing the nape with the fingers.

Dr. Cantor has brought to notice a new genus of snakes, nearly allied to the Cobras, called *Hamadrydas*, (Gr. *hama*, together with; *drus*, an oak or any tree,) which has a few maxillary teeth beyond the poison fangs, thus connecting the venomous serpents to the harmless, that have a complete row of maxillary teeth. According to Dr. C., the Hamadrydas feeds upon other serpents. It is said to be from eight to twelve feet in length, and exceedingly fierce, not merely ready to defend itself, which is all the common Cobra does unless greatly provoked, but quick to attack and to pursue when opposed. Its poison is a "pellucid tasteless fluid, in consistence like a thin solution of gum arabic in water," and reddening litmus paper, like that of other members of the family.

The Cerastes or Horn Snake, (see Chart,) is a native of Egypt and Lybia, and characterized by having a group of elevated horn-like scales over each eye. Its general length is about two feet. The color is sandy red, with irregular brownish markings; hence it cannot easily be distinguished from the sands of the desert, in which it dwells; so that it may be trodden upon unsuspectingly by man or cattle, and inflict a poisonous wound before its presence is perceived. It moves with great rapidity and in all directions, forwards, backwards, and sidewise, which makes it the more dangerous.

Another most deadly snake, called the Death Adder, and Black Snake, *Acanthophis tortor*, (Lat. torturer,) is found in Australia. The small woods and sandy heaths around Botany Bay are largely infested with it, and every where it is greatly dreaded by the colonists, on account of the mortal wounds which it inflicts. It is hideous in its aspect and thick in proportion to its length, which is two or three feet. The genus *Acanthophis* links the Viperine group to the Rattle snakes; the tail terminating in a single horny spine, instead of being invested as in the Rattle Snakes, with dry scaly pieces, resembling so many bell-like appendages, and forming a rattle. The head is covered in front with large plates; the scuta, or plates beneath the tail, are double.

Fifth Family. Rattle Snakes.

Crotalidæ, (Gr. κρόταλον, *krotalon,* a rattle or bell.)

The Crotaline group, including Rattle Snakes, exhibit the following characters: the face has a large pit on each side; the head is large behind, crown flat, covered with scales or small shields; the jaws weak, the upper with long fangs in front, and no teeth; the belly is covered with broad band-like shields, and there are no spurs or rudimentary feet. The species are all more or less venomous; a part of the family are viviparous, the rest ovo-viviparous.

The *Crotalus,* or Rattle Snake, is the type genus of this family. All the species are distinguished from the others of the group by the presence of a rattle at the end of the tail, (see Chart.) This consists of a number of joints of a horny texture, loosely joined together, so that when rapidly vibrated, they make a distinct whirring noise, which has been compared to that of peas shaken about in a dry bladder, or to the sound produced by the locust, and is heard at some distance. The rattles vary in number according to age; the basal bell or rattle is the last formed and the largest; one is erroneously said to be added every year. The head is covered with scales, but in one subgenus with plates.

Rattle Snakes, especially when irritated, exhale a disgusting odor. The peccary is said to destroy and devour them, as does the common hog also; but horses and dogs avoid them. They are sluggish and inactive in their movements; and, though highly venomous, seldom attempt to inflict any injury upon man unless molested by him. His approach calls forth the noise of the rattle, which usually precedes any blow. These reptiles never ascend trees, always capturing their prey upon the ground. Usually they rest coiled spirally, in paths, or clear spots in the woods, waiting for their prey, upon which they dart, when within the proper distance. In mid-winter and during hard frosts these serpents intertwine themselves together, in ball-like masses, and become totally torpid. At that season they may be handled without danger. They eat indifferently all kinds of birds, but not frogs, to which the Black Snake is so partial. Their food also consists of small animals, such as rabbits, squirrels, rats, &c., and sometimes even dogs are killed by them. Two or three species of the genus *Crotalus* are found in North America.

The Common or Banded Rattle Snake, *C. durissus,* is dispersed abundantly throughout the United States, though rarely met with north of the parallel of 45°. It is particularly numerous in the region of the Alleghany mountains. The length is not often more than four feet.

The DIAMOND RATTLE SNAKE, *C. adamanteus*, (Lat. adamantine,) is more formidable than the preceding species. This is found on the coast of the States south of North Carolina. It has been known to exceed eight feet in length, and in thickness to equal that of a stout man's leg. Those of this species are seen much about water, and are therefore called WATER RATTLES, to distinguish them from the common species, which keeps on dry land.

The CASCAVELLA, *C. horridus*, is common in South America. The kindred genus *Crotalophorus*, (Gr. *krotalon*, a rattle; *phoreo*, to carry,) has several species in North America, which are usually termed Ground Rattle Snakes. These have the head covered with shields, and the rattles very small. The Miliary or LITTLE CAROLINA RATTLE SNAKE, *C. miliarius*, though but twelve or fourteen inches in length, is dreaded on account of the intensity of its venom.

The MASSASAGUA, *C. Kirtlandi*, is another species found in northern Ohio and Michigan. Length twenty-seven inches.

The BOIQUIRA, or as the natives term it, the Queen of Serpents, is found in Brazil.

The MINIMARU, or Jergon, *Lachesis* (Gr. name of one of the *Parcæ*, or Fates) *picta*, is a species found in Peru, having the head heart-shaped and covered with scales, and a thick upper lip. It haunts the higher forests, while in those lower down its place is filled by its no less fearful relative, the BUSH-MASTER, *L. rhombeata*, (Prince Max,) which is from six to eight feet in length. The genus is characterized by double scuta beneath the tail, which ends in a short horny point.

The COPPERHEADS, (genus *Trigonocephalus*, triangular-headed,) are perhaps even more to be dreaded than the Rattle Snakes, since they are equally venomous and give no warning of their presence.

The COPPER-HEAD, *T. contortrix*, in length two to three feet, is most extensively distributed. Damp meadows are its favorite resorts, where it shows itself to the peril of persons who are engaged in mowing or passing through them. It sometimes finds its way into damp cellars, where, however, it makes itself useful by destroying rats and mice. This and the Banded Rattle Snake are the only really venomous kinds found in the middle and northern States.

The WATER MOCCASIN, *T. piscivorus*, (Lat. *piscis*, fish; *voro*, to devour,) is the pest of southern plantations. This species, like the harmless water snakes of the Middle States, may be seen lying over bushes which overhang the water, into which they plunge on the slightest alarm.

The LANCE-HEADED VIPER, *T. lanceolatus*, is abundantly distributed through several of the West India Islands. It attains six or seven feet, and sometimes even nine feet in length, and is greatly dreaded. It is said to be remarkable for its activity, and to abound among the sugar plantations, in which many of the laborers fall victims to its bite.

What is the 3rd ORDER of REPTILES? How are they particularly distinguished? What does Cuv. say of them? What is said of their form? What of their covering? Describe the upper and lower scales. State the difference in their arrangement and give their respective names. At what time and in what manner do they change their skin? What are the parts of the skeleton? Describe the vertebral column. How do you account for the flexibility of Serpents? What is said of the number of their bones? What are their agents of locomotion? How? To what may the ribs be compared? What is said of their various modes of progression? Why do the bones of the face deserve especial notice? How is the lower jaw articulated? How are its extremities connected? How are these things related to the serpent's swallowing its food? Have they two lungs? What is said of their teeth? What is peculiar to the venomous species? Describe the action of the poison fang? Does the serpent sting? What is said of the young of venomous serpents? What of the non-venomous? How is the goodness of the Creator herein displayed? Do serpents care for their young? What is said of their voice or hiss? What of their senses? What of their growth and tenacity of life? What of their powers of fascination? How were they regarded by the ancient Mexicans, Egyptians and Hindoos? Into how many families does Mr. J. E. Gray divide this order? In what family are the last two included on the Chart? How many sp. are enumerated? What proportion is venomous? Have any been found in a fossil state?

What is the 1st FAMILY? What does it include? What family compose the SUB-ORDER COLUBRINA? What are the characters of the Colubridæ? What its leading gen.? What well known sp. of the gen. *Coluber* is found in the U. S.? Repeat what is said of them severally. How does the gen. *Tropidonotus* differ from the true Colubrines? State what is said of the two sp. mentioned. What of the other gen. and sp. of this family?

What is the 2nd FAMILY? Why are the sp. of this family very terrific? What is said of their shields? Name a more marked peculiarity. Is the tail prehensile? Where are the BOAS found? Where the PYTHONS? What statements are made by Livy? What is said of the size, &c., of the Boa? Repeat what is said of the EMPEROR BOA. Of the ANACONDA. Of the ABOMA. Of the BOJOBI or GREEN BOA. Of the CORAL SNAKE. What two sp. of Pythons are mentioned? What is said of them? How are the P. in Zoological gardens of London fed?

What is the 3rd FAMILY? Where are they mostly found? In what respects do they resemble the Eel? Do true SEA SNAKES visit the land? Are they easily caught? Can they live in fresh water? Are they venomous? What is said by Dr. Cantor? What by the Rev. I. Williams? What kind of scales have the Marine Snakes? What species of true Marine Snakes are mentioned?

What is the 4th FAMILY? How many gen. does it contain? Are any found on this continent? What is said of their teeth and shields? What is said of the bite of the COMMON VIPER of Europe? What remedies for this bite are mentioned? What is said of the young, &c., of the Viper? What of the Esping or Aspic? What gen. contain the HOODED or SPECTACLED SERPENTS? How characterized? What is said of their bite? When do the hood and spectacles appear? What sp. are mentioned? Which is the ASP OF CLEOPATRA? What is said of the powers of jugglers? What new gen. has Dr. Cantor brought into notice? What is said of the Cerastes or Horned Snake? What of the Death Adder? What link does this form?

What is the 5th FAMILY? By what characteristics are they distinguished? What is the type of this family? How is it distinguished from the others? Describe the Rattle Snake. What is said of their habits and movements? What is their condition in mid-winter? What is said of their food? Mention the sp. found in N. A., also those of S. A.

What gen. includes the Copper-heads? Why are they to be especially dreaded? What sp. is found in the Northern and Middle States? What are its favorite resorts? What is said of the Water Moccasin? What of the Lance-headed Viper?

Name and trace those figured or mentioned on the Chart.

SECTION VI.

FOURTH ORDER. AMPHIBIA, (Gr. 'αμφιβιος, *amphibios*, having a double life.)

The Reptiles of which we have treated in the preceding sections, are covered with plates, shields, or scales. In those of the present order, the skin is naked, smooth, and often moist, or lubricated with a fluid secretion, which, in some cases, is acrid and apt to irritate the skin of the persons who handle it. As in snakes and lizards, the skin is frequently shed; in some species in shreds; in others entire. In some, as in the frog, the blood, through the delicate vessels of the skin, as well as in the vessels of the lungs and gills, undergoes those changes which are necessary for the maintenance of animal life. This cutaneous respiration can, however, take place only while the skin is kept moist; the same remark may be made respecting the gills of fishes, of tadpoles, and various crustacea. To meet this exigency, the Reptiles have a skin which is capable of secreting a fluid by which it is preserved in a humid condition. The healthy action of the skin, co-operating with that of the lungs, is really essential to their existence. Dr. Townson, of England, in his tracts on the "Respiration of the Amphibia," states from actual experiment, that a frog, when placed on blotting paper well soaked with water, absorbed nearly its own weight of the fluid in the short time of an hour and a half; and, it is believed, these reptiles never discharge it, except when they are disturbed or pursued, and then only to lighten their bodies and facilitate their escape. The form of the Amphibians is variable. Besides the naked skin, we may mention, as general characters of the present order, that the skull is united to the column of vertebræ by two *condyles*, (Gr. *kondulos*, a protuberance or knot on the end of a bone,) situated on the back of the head; the teeth are generally numerous, of small and equal size and close set; the toes usually unfurnished with claws; the ribs either wanting, (Plate XII. fig. 4,) or rudimentary and not attached to the breast-bone; and the animals are oviparous, the eggs having soft, not calcareous shells. A change of form and habit, as we have stated in our general description of the Reptiles, occurs in many, which begin their existence with branchiæ, or gills, that afterwards become obliterated; while in others, the branchiæ continue throughout their lives. Several prominent naturalists, including Prof. Agassiz, are inclined to separate the Amphibians from the Class of Reptiles, regarding them as possessing the distinguishing characters upon which classes are founded; but as this point seems not perfectly settled, and they have usually been numbered with the Reptiles, they are so arranged on the Chart.

The Amphibians may be divided into two sub-orders.

FIRST SUB-ORDER. CADUCIBRANCHIATA, (Lat. *caducus*, perishable; *branchiæ*, gills.)

The distinguishing characteristic of this sub-order is that the Amphibia which it includes, commence life with gills for the aëration of the blood, i. e., the air effects, through the medium of the gills, a change corresponding to the arterialization of the blood through the medium of the lungs, in other animals. The gills, however, are possessed only in the early or tadpole state; they become gradually obliterated, and lungs are developed.

This sub-order includes five families, viz.: (1) *Cæciliidæ*, Cæcilia; (2) *Ranidæ*, Frogs; (3) *Bufoidæ*, Toads; (4) *Salamandridæ*, Salamanders; (5) *Amphiumidæ*, Menopoma, &c.

First Family. *Cæciliidæ*, (Lat. *cæcilia*, from *cæcus*, blind;) *Apodous*, (or footless.)

The Reptiles of this family, Cuvier, following Linnæus, placed in his third and last family of Ophidians, calling them naked serpents, and observing that those who placed it among the Batrachians, "did not know whether the form underwent a metamorphosis or not." Müller, however, has proved that the Cæcilia has, at a very early period, gills, which are soon lost. The name Cæcilia was given to these Reptiles on account of their supposed blindness. The eyes are, in fact, exceedingly small, and nearly hidden under the skin. Cuvier asserts that in some species the eyes are wanting altogether. The Cæcilia are named by Dumeril, *Ophiosomata*, (Gr. *ophis*, a snake; *sōma*, a body.) They have a snake-like body, destitute of limbs, and with vertebræ resembling those of fishes, short ribs, and no sternum or breast-bone. They are undoubtedly to be regarded as a connecting link between the Ophidians and the Amphibia. Their skin is smooth, viscous, and marked with a regular series of ring-like furrows; and the scales, which are very minute, are not to be found, except by an examination of the substance of the skin itself. The head is depressed; the tongue is thick, rounded and velvety; the skull united to the vertebræ by two tubercles or condyles, as in the other Batrachians, whereas there is only one in snakes. There are both maxillary and palatal teeth. In their intestines, Cuvier says there is to be found "a quantity of vegetable matter, vegetable earth and sand." Nine species of this singular group are described as belonging to Asia and America. These Reptiles are ovoviviparous, producing their young alive, to the number of six or seven. Not much is known respecting their general habits; they bury themselves in the soft mud of marshy places, piercing through it, in a worm-like manner, often to the depth of many feet; they creep slowly on the ground, and, when in water, swim like an eel, striking to the right and left with their tail.

The Ringed Cæcilia, *C. annulata*, (Lat. ringed,) which is figured on the Chart, is an inhabitant of Brazil, Cayenne, and Surinam. It is remarkable for the bluntness of the tail, the distinctness of the rings, extending from the head over its whole length to the tail, and for the position of the false nostrils, below and a little before each eye.

ANOURA, OR TAILLESS AMPHIBIA.

This group comprises the Frogs, the Toads, with their allied forms, constituting, in the whole, a numerous assemblage. In these animals, the form of the body is short and broad. During the tadpole state, there are no limbs, but a long compressed tail, is their organ of locomotion, (Plate XIII. figs. 1, 2, 3, 4;) in this state it is called a tadpole ; subsequently four limbs are developed, (figs. 5 and 6,) and the tail disappears, (fig. 8.) The skull is very short and broad. Ribs are wanting ; the seven or eight anterior vertebræ only are distinct; the tympanic orifice is open ; the breathing is at first effected by gills, and afterwards by lungs. Warm and temperate, but moist climates are the localities most favorable to the Anourous Amphibia.

SECOND FAMILY. *Ranidæ,* (Lat. *rana,* a frog.) FROGS.

In this family of tailless Batrachians, the posterior legs are long and formed for leaping ; the hind toes are webbed ; teeth are found both on the upper jaw and on the palate ; the mouth is wide ; the tongue folded back, broad, soft, fleshy, and notched ; the eyes are prominent, and they are protected by a movable membrane well adapted to guard them against those injuries to which, from the Frogs' mode of life, they would be peculiarly liable. In the tailless Frogs, which are nocturnal in their habits, the pupil is linear. The ears are extremely small, yet by the answers which the Frogs make to each other, even at a great distance, by croaking, they show that they have powers of hearing which meet their wants. To enter into a more minute description of these harmless, and in gardens, highly useful Reptiles, seems unnecessary. All are familiar with their croak, their mode of leaping and swimming, their bright eyes and their coloring. The Frogs, like other Reptiles, pass the colder months of the year in a state of torpor, buried deep in the mud at the bottom of ponds or sluggish streams, and so mingled together as to form almost a continuous mass. In the spring they emerge, when they begin their singing, which has some meaning besides mere noise, each male frog having a different note from his neighbor.

The Ranidæ are more or less accustomed to dwell in the water or its neighborhood, voraciously consuming the larger insects, and especially slugs, which are a favorite food. It is, therefore, the opinion of some, that instead of being wantonly and cruelly destroyed, they ought to be protected. The Frog seizes its food with great rapidity, using its tongue for that purpose, which, being quite as long as the animal's body, darts at prey with arrow-like speed, and it is swallowed entire, secured by the glutinous adhesive secretion which lubricates the extremity of that organ. This rapid swallowing seems to indicate that the taste is not very acute. The focal axis of a Frog's eye is precisely as far distant as the length of each Frog's tongue, and at that angle these animals catch their prey.

The eggs of Frogs are gelatinous and numerous; some naturalists represent the number as thirteen hundred, and even as many as fourteen hundred. The black points discernible in the eggs are the germs of the Tadpole, or immature young. The development is rapid, but few days elapsing, in some places, before the young is hatched, though where the climate is less mild it is not hatched before the expiration of a month or more. The tadpole state is quickly passed, and the metamorphosis becomes complete. With the disappearance of the gills and tail, the habits of the animal are changed; atmospheric air now becomes the sole element of respiration. While yet tadpoles, they were the prey of fishes; now they become the prey of the weasel, the snake, and various kinds of water-fowl, which feed eagerly upon them. Very few out of every thousand that are hatched survive the summer. Frogs are capable of being tamed, and instances are related of their visiting houses regularly at the hour of meal-time, and partaking of offered food. A story is related by Mr. Bell of one which had such strong partiality for warmth, that during the winter seasons, he "regularly and contrary to the cold-blooded tendency of his nature, came out of his hole in the evening and directly made for the hearth in front of a good kitchen fire, where he would continue to bask and enjoy himself till the family retired to rest . . . frequently nestling under the warm fur of the cat, whilst the cat appeared extremely jealous of interrupting the comforts and convenience of the frog."

Besides the change of form in the Frogs, and the power of the naked skin to act upon the air in such a way as to fulfil, in a great degree, the office of lungs, and the fact that aërated water may be made to subserve this process of cutaneous respiration; besides, also, their power of long abstinence from food, their hybernation, and their age, as great as thirty-six years, in the case of the tailless species, startling stories are told of their issuing forth alive from the heart of trees, or the solid rock, after the confinement of centuries. The experiments of Dr. Buckland, however, favor the idea that frogs and toads cannot live more than one or two years, completely excluded from air and destitute of food.

Sixteen genera of these Amphibia have been enumerated, three of which are found in North America. It is worthy of being noted that Frogs were introduced into Ireland from England, as late as the year 1696, by Dr. Gwythers, a fellow of the University of Dublin, (and we add here that still more recently were *snakes* imported into that country.)

The Bull Frog, or Croaker, *Rana pipiens*, (Lat. chirping,) is peculiar to North America, and found throughout the Union. It is very large, the body being from six to twelve inches long, and a half pound or more in weight. Sometimes they have been known to "measure two feet from one extended extremity to the other." This species are noted for their bull-like bellowings, which may be heard to a great distance. They are voracious and predatory, devouring insects, fish, and even snakes. Their hind legs are used for food and are "excellent eating." These Frogs are sometimes reared specially for the table.

The Marsh Frog, *R. palustris*, (Lat. marshy,) is one of our most beautiful frogs, and extremely active. Length, three inches.

The Edible, or Green Frog, *R. esculenta*, found in Continental Europe and in parts of Asia and Africa, is essentially aquatic, inhabiting either running or stagnant streams. The croak of the male in the summer months, where the numbers are large, is said to be almost intolerable. The meat of these frogs is described as delicate and well flavored. In Vienna, (Austria,) they are considerably used, being preserved for eating, and fattened in "froggeries."

The Wood Frog, *R. silvatica*, (Lat. woody,) is found from Massachusetts to Virginia; it is in length two and a half inches. It may be at once known in the woods by its wonderful and rapidly repeated leaps, which render its capture very difficult. To this nearly corresponds the Red or Common European Frog, *R. temporaria*, (Lat. temporary or changeable.)

In the West Indies and South America is a species *Cystignathus* (Gr. vesicated jaw,) *ocellatus*, called Bull Frog, and distinguished by the entire absence of a web on the hind feet, which can clear a wall five feet in height. The palm of the hand is provided with quite large tubercles; that at the base of the inner finger is the largest of all. The first phalanges are marked beneath by similar tubercles. Small tubercles also appear under the articulations of the first and second phalanges, except under the inner toe. In the Antilles these frogs are reared in a state of domestication, and said to become familiar.

The genus *Ceratophrys*, (Gr. horned eye-lid,) includes frogs of beautiful colors, found in South America, having a granular or tuberculous skin, with the edge of the upper eye-lid prolonged to a point, resembling a horn. The species *C. granosa*, (Lat. full of grains,) is figured on the Chart.

The Painted Frog, *Discoglossus*, (Gr. orb-like tongue,) *pictus*, is somewhat remarkable for its rounded tongue, and the markings of its skin.

The Thimble Frogs, *Dactylethræ*, are a peculiar kind of Frogs found in Africa, and deriving their name from having some of their toes enveloped at their tips by a conical horny claw or cap.

The Tree Frogs, (*Hyladæ*,) are arboreal in their habits, and capable of leaping, like birds, among the branches. They are described as beautiful, both in form and coloring. The foot of these Frogs differs in its structure from that of the other animals that make their home in trees; it is not a grasping organ, nor is it furnished with claws for clinging, but has suckers somewhat like those we have described as belonging to the Gecko. The enlarged and rounded tip of each finger, both of the fore and hind paws, has an apparatus consisting of a little cushion moist with a thick glutinous fluid, and applying itself so closely to the surface it touches, as to support the animal's weight. It disengages or fixes its fingers at will. The cushioned apparatus is like an air-pump at the extremities of the fingers, giving the animal the ability to walk on the ceiling, on the polished surface of a mirror, even to suspend itself by one finger, if so disposed. Tree Frogs are numerous in some of the Southern and Western States. The frogs of the genus *Hyla*, (Gr. *hule*, a wood,) differ from the common frog, (*Rana*,) in the greater length of the hind legs, and also in the male having a membranous sac under the throat, which is distended during their hoarse and oft-repeated croaking. So alert are these Frogs that they have been known to clear an interval of twelve feet in descending from one branch to another. Their leaps are also made with much address and precision, indicating great distinctness and power of vision.

It must not be supposed that these Frogs pass *all* their lives on trees. On the contrary, like others of the race, they are at first aquatic animals, and when adult, visit the water to deposit their eggs, which is generally done in April. They also hybernate in the mud at the bottom of lakes and marshes. At this time their croakings are so loud and discordant, that "they might be taken for the cry of a pack of hounds in full chase;" and in the stillness of night, "the din of their united voices may frequently be heard at the distance of a league, especially on the approach of rain." After the young are hatched by the heat of the sun, as in the case of the Common Frog, they continue in the tadpole state about two months, swimming in the water and feeding upon insects and worms. When the tail and gills have disappeared, they, with unerring instinct, take to the woods. The Tree Frog lives about thirty-six years. It is slow in attaining its full growth, which does not take place until the fourth year; nor does it breed before this period. Its fine green color is not perpetual; after the breeding season, the animal becomes of a reddish brown, which soon changes to gray, mottled with reddish; the color next assumed is blue, and this again changes to green, which is the summer tint. The agreeable colors and sprightliness of this frog, occasion it, not unfrequently, to be kept in cages.

The Northern Tree Frog, *H. versicolor*, (Lat. changing color,) is spread over a large part of the United States. This Frog is particularly clamorous in rainy weather. Dr. DeKay says he has been assured "that it possesses ventriloquial powers in no inconsiderable degree." It appears to assimilate its color to that of the tree on which it rests. This species is very similar to *H. viridis*, of Europe, (see Chart.)

Third Family. *Bufoidæ*, (Lat. *bufo*, a toad.) Toads.

In the Toads, the tailless Batrachian structure has its highest development. Cuvier distinguishes them as having an inflated body, a warty or tuberculous skin, and a tumor of variable size behind each eye, consisting of a gland from the pores of which exude an unctuous and offensive fluid. They have no teeth; the hind limbs do not much exceed in length the fore pair. They crawl rather than leap, and after passing from the tadpole state, retire from the neighborhood of water to dry situations. Their saliva has been supposed to be poisonous; but this is a mistake. There are, however, glands on the skin of the back and sides, that give out a fluid which in some species is acrid, capable of producing irritation in a very sensitive skin, and probably intended for the defence of the Toad against the attacks of carnivorous animals.

Toads are nocturnal in their habits, evening and night being the principal season of their activity, and their favorite slugs then also creeping abroad. They hibernate in holes in the ground, the interstices of walls, or other similar retreats, in the spring emerging from their state of rest and seeking the water in order to deposit their eggs, in the form of strings of jelly three or four feet long, with a double row of black dots, when their loud croak may be heard at a considerable distance. They are two or three weeks later than the Frogs in depositing their eggs, after which process they return to the land. In August, the tadpoles, having completed their transformation, leave their native element for the land, dispersing themselves in all directions.

Of the genus *Bufo*, about twenty species are enumerated; several are found in the United States, but only one is seen in the northern parts of the Union. These have simple toes and a distinct tympanum.

The Common American Toad, *B. Americanus*, (*B. vulgaris*, Storer,) is about three inches in length. It is furnished with a sac for holding the water which it obtains through the skin. The skin, shed at certain intervals, is, according to Mr. Bell, swallowed as soon as it is detached.

The Natter Jack, or Running Toad, *B. calamita*, is a species of toad of a yellowish brown color, with a bright yellow line running down the middle of the back. It never leaps, nor does it crawl with the usual toad-pace, but its motions are more like running. This species is found in Ireland.

The Bahia Toad, *Phryniscus nigricans*, is a species noticed by Mr. Darwin, at Bahia Blanca. He graphically says of it, "If we imagine, first, that it had been steeped in the blackest ink, and then, when dry, allowed to crawl over a board freshly painted with the brighest vermilion, so as to color the soles of its feet and parts of its stomach, a good idea of its appearance will be gained." Instead of being nocturnal, like other toads, it crawls about during the heat of the day, over dry sand-hillocks and arid plains.

The Mitred Toad, *B. margaritifer*, (Lat. pearl-bearing,) receives its name from the peculiar conformation of the head. It is an American species.

The Surinam Toad, (see Chart,) *Pipa Surinamensis* or *monstrosa*, is from six to eight inches in length, and four to five in breadth. It has a large and triangular head, is without teeth or a tongue; its tympanum is concealed beneath the skin; its eyes are small and placed near the margin of the upper jaw. The skin is of dirty brown color, thickly studded with reddish tubercles. The general uncouthness of its appearance is increased by a phenomenon almost unexampled in the animal kingdom. The female has the back pitted with a great number of small cells, and in these the male carefully places the eggs which she has deposited. When this has been done, she repairs to the water: the skin of the back now swells; the pits deepen, and in due time the Tadpoles appear; on the back they pass the Tadpole state, and do not emerge till they have lost their tail and their limbs are developed. The female then returns to the land. This Toad is not unfrequently found in houses. The Pipa is not restricted to Surinam, but is found in various parts of South America.

TAILED AMPHIBIA. URODELA. (Gr. οὐρά, *oura*, a tail; δῆλος, *dēlos*, manifest.)

This division is one of peculiar interest to the naturalist, as well from the variety of forms which it includes, as from the successive changes which these forms exhibit. They are characterised by their *permanent* tail, their rudimentary ribs, the possession of four or two limbs, the absence of a breast bone, the simple lungs, the teeth in both jaws, and the want of an external ear. These amphibia are widely dispersed over the northern temperate portions of both continents. North America and Japan possess the largest variety of forms.

FOURTH FAMILY. *Salamandridæ*, (Gr. σαλαμάνδρα, *salamandra*, a salamander.)

NEWTS.

Of these, some species are terrestrial, visiting the water only in the breeding season; others make it their permanent or nearly permanent abode. The Tadpoles, or young of the Newts, undergo a transformation, essentially resembling that of the Toads and Frogs, with this difference, however, that the tail merely changes its form, and is never lost. In their general appearance, they resemble Lizards. The jaws have minute teeth, and a double row also extends down the palate.

LAND NEWTS. (Occasionally found in water.)

The Land Newts, included in the genus *Salamandra*, have the tail, when the animals are adult, round and tapering; on each side of the head is a gland, similar to that of Toads; the tongue is short and thick, enlarged above, and attached by a slender root in the center. The breathing in the first aquatic or tadpole state, is by external gills; afterwards atmospheric by lungs.

These reptiles frequent humid places, and take up their abode in the soft ground, among decayed trees in wooded districts, in ditches and shady spots, and in caves and old crumbling buildings. They are sluggish and slow in their movements, not often quitting their retreat, except during rainy weather, and at night. The courage in danger for which they have been renowned, is nothing more than stupidity. Flies, worms, slugs, &c., constitute their food. Their size varies from two to seven inches, and they also show great varieties of color. For example, we have the Yellow-bellied Salamander, with the upper parts reddish brown; the Violet-colored; the Red-backed; the Slate-colored, (with orange beneath;) the Salmon-colored; the Blotched, (gray, with large bluish-black blotches;) the Yellow, (spotted with black,) otherwise the Long-tailed Salamander; the Granulated, (greenish above, varied with gray and brown beneath;) the Red; the Scarlet; the Black; the Spectacled Salamander, &c. The number of species is very large, even as found in the United States, and this general reference to them must suffice.

They are said to pass the winter in a kind of underground burrow, numbers assembling together, and intertwining themselves for the sake of mutual warmth. Like other reptiles, they shed their cuticle; they are ovoviviparous, forty or fifty being the produce at the same time, of a single female. Though tenacious of life, a little salt or vinegar thrown on the Salamander, produces convulsions and death. From some species, there exudes a milky or glutinous secretion, which is occasionally projected several inches; it is acrid, and of a powerful odor, and is described as fatal to small animals. In this fact, we perhaps have the origin of Pliny's statement, that the Salamander "infects with its poison, the vegetables of a vast extent of country, and even spreads death around, like a pestilence." This, and the ancient stories of its being a body of ice, and uninjured by the strongest heat, and of its having a deadly bite, are now regarded as utterly groundless. It may be, however, that larger and more formidable species formerly existed.

The Triton is distinguished by its fish-like tongue, which is attached more or less at its borders, and has only the front extremity free. Of this genus, there are several species found in brooks and marshy places,—varying in length from three to six inches. The Tiger Triton, *T. tigrinus*, is of a bluish-black color, with numerous irregular blotches over the head, body, tail, and extremities. The length is from six to seven inches, including the tail, which is longer than the body. Specimens of this Newt have been obtained from the vicinity of Oneida Lake, (N. Y.)

The largest Water Newt of England, is the CRESTED TRITON, *T. cristatus*, about six inches in length, and which feeds on aquatic insects, the Tadpole of the Frog, and even the smaller species of Water Newts. The manner in which the female deposits her eggs, is very singular. When present, she chooses the leaf of the smart-weed, (*Polygonum Persicaria*,) as the place for the deposit. "She first applies her head to the edges of a leaf, and turns it with her snout in such a way that the lower surface of the leaf is turned towards her breast; then, with her fore-paws, she passes the turned leaf beneath her body, seizes it with her hind-paws, and conducts it beneath the vent, folding it, at the same time, and forming with it an angle, the opening of which is directed towards the tail. The egg, in escaping from the vent, would thus pass through the middle of the angle formed by the leaf, but the Salamander (or Newt) stops it in its fall by her hind feet, shuts up this angle with them, and thus forms in the leaf a fold in which the egg is held. Still, on the removal of the feet, the egg would fall to the bottom of the water; but the careful parent, before she quits the leaf, folds it so firmly with her hind feet, that the gluten with which the envelop of the egg is surrounded, spreads from the pressure on the two internal surfaces of the leaf, and prevents the folds from opening." (See Plate XIII. fig. 9, which represents the Triton on the leaf; also, for the figures in the same plate, representing the animal as it appears in the transition from the tadpole to the perfect state, and the explanations of these figures, as attached to the plate.)

AQUATIC NEWTS.

These are distinguished by having the tail flattened on the sides, and by the absence of glands from the sides of the head. The body is covered with watery excrescences.

These reptiles spend nearly all their lives in water. They are remarkable for the facility with which they successfully reproduce their tail when it is cut off.

The males, during the breeding season, are distinguished by a high membranous crest upon the back, and another one along the upper side of the tail. The limbs are short and feeble, and progression in water is effected by the paddle-like action of the tail. It should be remarked, that as the Land and Water Newts are, some of them, at least, closely alike in anatomical structure, some naturalists reject this division, and have introduced other distinctive terms.

FIFTH FAMILY. *Amphiumidæ*, or *Menopomidæ*.

These tailed Batrachians are nearly all found in North America. The gills, after a short time, suddenly disappear, leaving orifices upon the neck, and the respiration is performed by the lungs alone. They are not known to undergo any transformations. Among them are included the genus *Menopoma*, characterised by having a robust and flattened body, with the head distinct from the neck, and the skin wrinkled into numerous folds. The tail is broad and much compressed, and the soles of the feet have a marginal fold of skin, qualifying the animal for rapid movement in the water. It rarely leaves that element except at night. These creatures, sometimes twenty or twenty-four inches in length, are extremely voracious, feeding on insects, fish, and in some instances, on small mammals. They are of a slate or blackish color. As they bite at a hook, they are sometimes caught by the angler, to his disappointment, and perhaps terror, though, as their teeth are very small, they are not capable of doing him any serious injury.

The GROUND PUPPY, ALLEGHANY HELL-BENDER, *M. Alleghaniensis*, is a species found in the Alleghany river, coming from the Mississippi waters. Its tail is nearly as long as its body.

The AMPHIUMA has an eel-shaped body, with the head and neck continuous. The limbs are exceedingly minute, and divided in one species into two, in another, into three jointless toes.

The THREE-TOED AMPHIUMA, *A. tidactylum*, (three-toed,) is found in Alabama and Arkansas. The largest member of the present family *Megalobatrachus*, (Gr. great batrachian,) is found in Japan. Specimens of it have been seen more than three feet in length, and weighing eighteen pounds or upwards.

SECOND SUB-ORDER. PERENNIBRANCHIATA. (Lat. *perennis*, enduring; *branchiæ*, gills.)

The name of this sub-order is applied to Amphibians, respecting some of which diversity of opinion has been entertained, and still exists. Though they acquire lungs, at least rudimentary ones, respiration is aquatic by means of gills which are external and persistent,—the animals continuing in a perfect tadpole state, by an arrest of development. These singular forms the Chart arranges in one family.

FAMILY. *Proteidæ*, (Gr. *Πρωτεύς*, *Proteus*, a proper name.)

The name which DeKay gives to these reptiles, is *Sirenidæ*, (Gr. siren.) More recently they have been included in a sub-order of the *Urodelan Batrachians*, termed *Trematodera*, a name referring to the perforations or apertures on the side of the neck, which remain through life.

The first genus we shall notice, is *Proteus*, of which there is a single species, *P. anguinus*, (Lat. snaky.) Few reptiles have excited more interest than this curious species,—an apparent link between the amphibia and fishes. Its branchiæ are not, indeed, covered as in fishes, but are exposed, presenting the form of a beautiful pink tuft on each side of the head; the body is eel-like, as are all its movements; the tail is compressed; the eyes are rudimentary,—with small black dots under the skin, (as has been discovered by dissection;) the jaws are furnished with minute teeth. The limbs are very small and feeble, and, in fact, almost useless; the toes are three on the front, and two on each hind limb. The skin is smooth and delicate.

The Proteus dwells in the subterraneous waters of the great cavern of Adelsburg or "Grotto of the Maddalena," situated near the main road from Trieste to Vienna, (Austria.) "These subterranean waters communicate with, and supply a small lake in the celebrated cavern; and it is in this lake, where no sunlight ever enters, inclosed by barriers of piled up rock, deep in the bowels of the earth, that the Proteus is found, reposing in the soft mud, precipitated by the fluid, and lining the rocky basin." At Sittich, which is about thirty miles from the cavern, it is also noticed, though rarely, being "thrown up by water from a subterranean cavity."

According to Sir Humphrey Davy, the Protei are seldom found in dry seasons, but are often abundant after great rains. The length of a moderate-sized one, is about a foot; the thickness varies from that "of a quill, to that of the thumb." The nature of its food is not certainly known; though its numerous teeth would indicate it to be carnivorous. The skin is of a pale flesh color, but when exposed to the light, it approaches olive brown. The light appears to act upon it with a power that is too stimulating, and the animals, when exposed to it, creep under any object that may shelter them from its influence. In the mysterious nature of the Proteus, and its singular dwelling-place, how manifest is the hand of the Creator, assigning to every thing the bounds of its habitation, and so organizing every thing, that it shall accomplish its allotted destiny.

The *Menobranchus*, (Gr. enduring gills,) or *Phanerobranchus*, (Gr. manifest gills,) of Fitzinger, is clearly allied to the Proteus. It has a body moderately elongated; the tail is deep and flattened at the sides; the head is flat and large. There are two rows of small conical teeth in the upper jaw, and one row beneath. The branchial tufts are large. The toes are four on each foot. The species called the BANDED PROTEUS, or Big Witch Lizard, *M. lateralis,* (*Necturus lateralis,*) is of a brownish color, with blackish spots, and often a dark lateral line. It is stouter and longer than the Proteus, the length varying from one to three feet. It is found in the great northern lakes of this continent. A spotted species, *M. maculatus*, is found in Lake Champlain and Lake George; and a third species, *M. punctatus*, having more uniform markings than the others, lives in Santee River, South Carolina. The Menobranchus has been found in the Erie Canal, (N. Y.) The animal appears to move slowly in the water, but the powerful tail must render it able to move with much celerity. Though its flesh is white, and perhaps savory, it is looked upon by fishermen with disgust and aversion.

The term *Siren* represents eel-like animals, having three gill-tufts on each side, and utterly destitute of hinder limbs. The front limbs are feeble; the toes are four in number, small and clawless. The lower jaw has teeth, the upper none; but there are ranges of teeth on the palate. The eye is very small, and the ear hidden from view. The gills in these animals have less external development than in the Menobranchus; but, on the other hand, the lungs perform their part more completely.

The LACERTINE SIREN, (*S. lacertina,*) is probably the largest species, reaching the size of three feet. It is black above; dusky beneath; and found in the muddy marshy grounds of South Carolina and Florida.

The AXOLOTL, *Siredon pisciformis*, of Mexico, is common to the lakes in the vicinity of the city of Mexico. It is found in the coldest mountain waters. The length of the Axolotl is eight or ten inches; its general color green, spotted with black. The flesh resembles that of an eel, and is considered quite agreeable. The gills and gill-openings of this animal are highly developed; a continuous flap extends across the throat; the tail is compressed and fin-like.

Respecting this creature naturalists have been much perplexed. Many doubt whether the gills be permanent, and are of the opinion that the Axolotl is nothing more than the tadpole of a large species of Salamander, or else of some species, the perfect form of which is yet to be discovered.

Fossil remains of Amphibia have been found both in Europe and America. Traces of batrachian foot-marks are thought to be discernible in the new red sand-stone of Massachusetts and Connecticut; and in the coal measures of Westmoreland county, Pennsylvania, are found foot-prints which are regarded as those of airbreathing vertebrates.

What is the 4th Order of Reptiles? Describe their skin. How is it exuviated? With what furnished? What kind of changes do the vessels of the skin undergo? What is necessary in order to their action? How is the skin moistened? What is said of its healthy action? What is the remark of Dr. Townson? What is said of the form of the *Amphibians?* What other characteristics are mentioned besides their naked skin? What changes of form and habit do they undergo? How do some naturalists regard them? Into how many orders are they divided? What is the 1st? Name its distinguishing characteristics. How many families does it include? What is the 1st Family? Where did Cuv. place it? What has been proved by Müller? Why was the name Cæcilia given to these reptiles? How are they characterized? What is said of their habits? What sp. is mentioned? What is said of it?

What do the Tailless Amphibians include? What characters of this group are given? Name the 2nd Family. Give its characters. In what condition do they spend the colder months? Upon what do they feed? How do they secure their food? What is said of their eggs? What changes do these animals undergo? Can they be tamed? How long do they probably live without air or food? At what time were they introduced into Ireland? What Frog peculiar to N. A. is mentioned? What is said of it? What other sp. are named? Give some account of them. Describe the Tree Frogs. In what respects do they differ from common frogs? Do they never leave the trees? What is said of their young, &c.? What sp. are mentioned? What is the next Family? How does Cuv. distinguish them? Is their saliva poisonous? What is said of their skin? Of their habits? How many sp. of them? Which are named? State what is said of them? How are the Tailed Amphibia characterized? What is the 4th Family of Amphibians? What division is made of the Newts? Describe the Land Newts? What places do they frequent? What is said of their size and color? Of the number of species? What else is said of them? How are the Tritons distinguished? In what places are the sp. found? What is said of their size? Describe the Tiger Triton? Which is the largest Water Newt of England? In what manner does the female deposit her eggs? What are the characteristics and habits of the Aquatic Newts? What is the 5th Family? Where are they nearly all found? Do they undergo any transformation? How is the gen. Menopoma characterized? What further is said of these animals? What sp. is referred to? Describe the Amphiuma? What sp. are mentioned and what is said of them? What is the 2nd Sub-order? Repeat what is said of it. In what family does the Chart include these Amphibians? Has it received any other name? What genus is first noticed? Can you give its character? What is said of its places of abode? What does Sir Humphrey Davy remark respecting it? What is its size, &c.? Give the characters of the genus Menobranchus? What is said of the Banded Proteus? What other sp. are mentioned? Describe the animals included in the generic term Siren. Which is the largest sp.? What is said of the Axolotl? What of Fossil Amphibians?

Name and trace the Amphibians figured on the Chart.

PL. XIII.

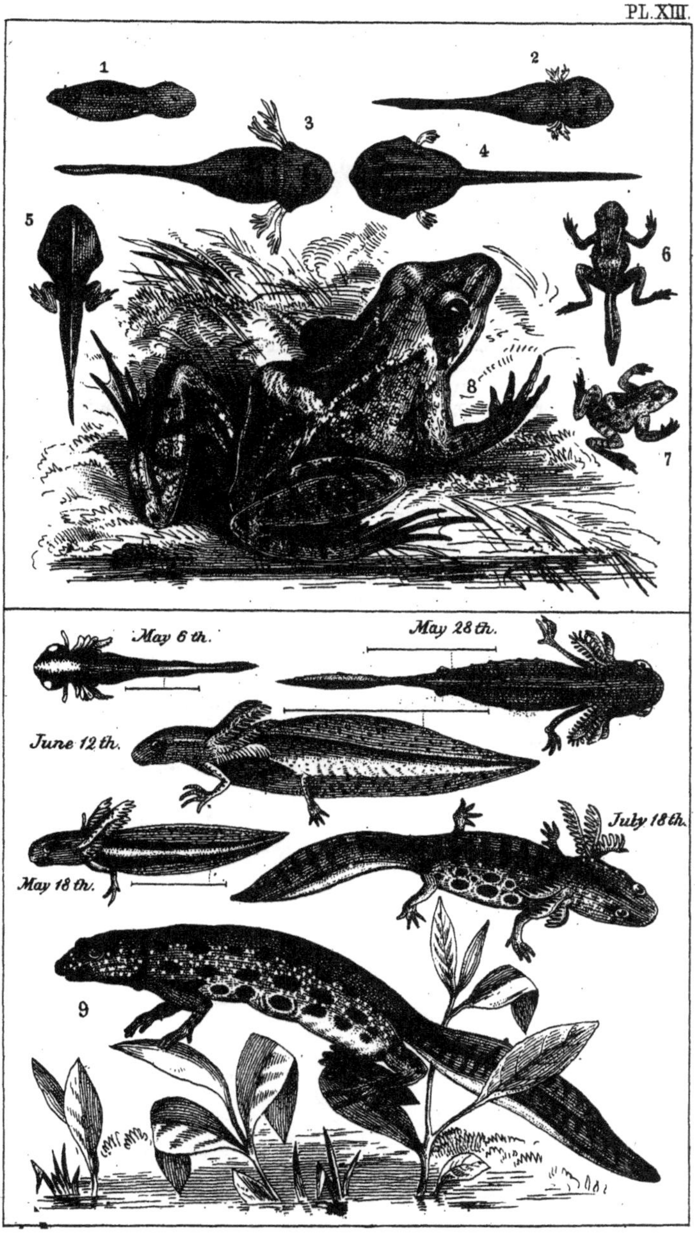

1
2
3
4
5
6
8
7

May 6th.
May 28th.
June 12th.
July 18th.
May 18th.

9

EXPLANATION OF PLATE XIII.

THE FROG IN ITS DIFFERENT STAGES.

1. Just emerged from the egg, a tadpole, fishlike creature.
2. Gills, in branching tufts, on each side.
3. The blood is seen to course through the gill filaments.
4. The gills begin to disappear, the eyes are formed, and the little tadpole or pollywog begins to devour vegetable matter with voracity.
5. Ceases to respire water and the hind legs begin to show themselves.
6. The fore legs appear and the tail is being absorbed.
7. It breathes by lungs, the tail has disappeared, and the legs are perfected.
8. The full grown Frog, living upon insects, mice, birds, &c..

GREAT WATER NEWT IN DIFFERENT STATES.

May 6th. The young Salamander Tadpole or Newt just escaped from the egg.

May 18th. The fore feet have lengthened, and the eyes are perceived.

May 28th. The hind feet begin to appear, and the fore feet are well developed.

June 12th. The hind feet almost developed, and lungs extend half way down the trunk.

July 18th. It has arrived at the maturity of its tadpole state, and after the 27th respires atmospheric air, having attained its perfect state.

9. The Great Water Newt depositing its eggs on the leaves of the smart-weed, or amphibious knot-weed, (*polygonum persicaria.*) She is folding the leaves over to protect the eggs.

FOURTH BRANCH OF ZOOLOGY.

ICHTHYOLOGY. (Gr. ιχθὺς, *ichthus*, a fish; λόγος, *logos*, a discourse.)

Class, *Pisces*. FISHES. COLD-BLOODED VERTEBRATES.

SECTION VII.

WE come now to that part of Zoology which treats of Fishes, their structure and form, their habits and uses, and their classification. The Fishes, as a class, possess a greater number of species than any other of the primary divisions of the Vertebrates; and, indeed, the species not improbably exceed in number those of the Mammals, Birds, and Reptiles taken together. Our limits will therefore allow us to do nothing more than present a general view of this part of the Animal Kingdom, followed by succinct notices of the orders and families.

The most prominent characteristics of Fishes are (1) that they generally have cold red blood; (2) they breathe by gills instead of lungs; (3) they have a two-chambered heart; (4) they use fins as organs of progression; (5) they have the skin naked or covered with scales of varied structure; (6) they are almost incredibly prolific.

The blood, generally cold, assumes the temperature of the surrounding element. It should be stated, however, that in some of the swift Oceanic Fishes of the Mackerel family, such as the Tunny and the Bonito, the blood is found to be 10° higher than that of the surface of the sea, even within the tropics. The blood-disks are sometimes circular, and sometimes oval; they are larger than those of the mammalia and birds, smaller than those of reptiles, especially the amphibia. The gills consist of bony or cartilaginous spines, usually placed parallel with each other like the teeth of a comb, but sometimes arranged in bunches. These organs are analogous to lungs in terrestrial animals, being adapted to extract from the air contained in the water, the oxygen which is needed for the renewal of the blood. The breathing apparatus formed by the gills is double in form, placed on each side of the neck. Most commonly it consists of several series of laminæ, or membranous plates, fixed upon slender arches of bone. Over these thin membranous plates branch out innumerable blood vessels, whose walls are so thin as to permit the fluid contained in them to absorb the oxygen with which they are brought in contact when the fish takes in water through the mouth. In order to carry off the water when deprived of its oxygen, and to bring in fresh portions to be successively respired, a constant current is produced over the surface of the gills, by the action of the fish while taking in water at the mouth, and throwing it out on each side, behind the gills, through orifices which it has for the purpose, called the gill-openings. The apparatus for breathing is protected by large bony plates, or opercular bones, making up the chief portion of the sides of the head. These are four in number, and are termed the *operculum*, the *sub-operculum*, the *pre-operculum*, and the *inter-operculum*. The *first* of these covers the gills. The *branchiostegous rays*, often mentioned in descriptions, are situated under the opercular bones. In the Sharks, *Squalidæ*, (Lat. *squalus*, a sea-fish,) and the Rays, *Raidæ*, (Lat. *raia*, a ray,) the gills are attached at their outer margin, with a separate orifice to each, through which the water escapes. The orifices, usually five in number, are, with the mouth and nostrils, on the under surface, and completely hid when the fish is laid on its belly.

The heart consists of but one auricle and one ventricle. The blood collected from the venous system, is accumulated in the single auricle, thence it is sent into the ventricle, and this drives it into the gills where it is changed from venous to arterial blood, and thence circulated through the body in arteries, aided by the contraction of the surrounding muscular fibres. Hence it will be perceived the heart never contains any but venous blood, the arterial first proceeding from the gills.

Most of the bony fishes have a membranous bladder, commonly of a lengthened form, placed along the body between the spine and the bowels, and having a structural relation to the lungs in the higher Vertebrates. This is filled with air, and well known as the air-bladder, or swimming-bladder. Whenever possessed, it aids more or less the process of respiration. It also serves another important purpose, which is to enable the fish to vary its specific gravity, and thus float at any desired elevation in the water. In appearance it varies; sometimes, as in the Sea-Porcupines or Hedge-Hog fishes, (*Diodon*, double toothed,) and their allies, it is two-lobed; in the Electric Eels and the Carp fishes, *Cyprinidæ*, (Gr. *kuprinos*, a carp,) it is divided by a transverse partition, which in the latter allows of inter-communication, through a narrow orifice. In the (*Pimelodus catus*,) or Common Cat-fish, Family *Siluridæ*, (Gr. *silouros*, a sheat-fish,) it is divided into four cavities or compartments. In many species there are closed or blind tubular processes proceeding from various parts of the surface; in others the bladder is sub-divided into as many irregular cells; all this showing it to possess the rudimentary remains of the lungs of air-breathing animals. In marine fishes the bladder usually contains a gas having in it a greater proportion of oxygen than of atmospheric air, while in those of fresh water, nitrogen predominates. The species which are without the air-bladder, or have it only in a rudimentary state, are generally Ground Fishes, keeping close to the bottom. Sometimes the possession of an air-bladder exposes fishes to danger; Gurnards, *Triglidæ*, (Gr. *trigla*,) and Conger Eels, (*Anguilla conger*,) at times appear to distend the air-bladder so much that it loses its elasticity or power of contraction, or as fishermen say, these fishes "blow themselves," becoming unable to sink or to make their escape. The Sea-Porcupine, (Plate XIV. fig. 7,) has the habit of filling its body with air, and of floating helplessly in this condition at the surface; but in the case of this fish, the air is taken, not into the bladder, but into the huge stomach.

A writer speaks "of a gentleman of his acquaintance who had a Gold-fish which swam about for more than two months, with its belly upwards, appearing perfectly healthy and lively," and who attributed this change in the natural position of the fish to an enlargement or defect in the air-bladder.

Water is the well known sphere of life and motion to fishes. The Flying fishes, *Exocœtus volitans*, (Plate XIV. fig. 3,) and some of the Gurnards can indeed raise themselves into the air and keep their position there for a few seconds; some of the Frog-fishes, *Lophidæ*, (Gr. *lophos*, a neck or crest,) and Eels, *Anguillidæ*, (Lat. *anguilla*, an eel,) can crawl upon the exposed mud or sand in the interval occurring between the ebb and flow of the tide; and the Anabassidæ, or Climbing Perches leave the water in order to obtain food; but these are only deviations from a general rule. Some inhabit fresh water only; some only the sea; others can exist in both, either by periodical migration, or at pleasure. Near the city of Bristol, (Eng.,) the Eels are known annually to ascend the trees whose branches hang into a pond, pass over to the opposite branches and drop into an adjoining stream, and thus migrate to far distant waters. The trees at such times appeared to be quite alive with the eels. (Gosse.) Eels descend rivers to spawn in the brackish waters of estuaries. Salmon ascend rocky rivers from the sea, leaping cascades and overcoming various obstacles to deposit their eggs in fresh water.

The form of the fishes is decidedly the one best adapted to facilitate progression through such a medium as water, being commonly that of a spindle, swelling in the middle and tapering to each extremity. To this, however, there are exceptions. The Skates, *Raiidæ*, and Flat-fishes, *Planidæ*, are flattened horizontally; the Chaetodons, or Hair-Tooths, *Chaetodontidæ*, and the famed Dories, *Zeina*, (Gr. from *Zeus*, Jupiter,) a branch of the *Scombridæ*, (Lat. *scomber*, a mackerel,) are flattened vertically; some are of a globe-like form, as the Diodon and the Sun-fish, *Orthagoriscus*, (Gr. a sucking-pig;) some of serpent-like form, as the Eels and Lampreys, *Petromyzon*, (Gr. stone-sucker;) and some, as the Ribbon-fishes, *Cepolidæ*, resemble in length and thinness the fabric after which they are named.

The organs of motion in this class are fins. These have the form of a delicate membrane, investing a series of bony or cartilaginous rays, and which is more or less transparent. These rays are slender bones, consisting, in some cases, of a single piece, stiff and spinous; in other instances, they are made up of several pieces jointed together, and hence flexible; the latter are frequently divided each into two or more branches at the tip. The bony character of the fin rays affords a basis for two of the orders, viz.: ACANTHOPTERYGII, Spiny-finned Fishes, and MALACOPTERYGII, Soft-finned Fishes.

The fins of Fishes are of five kinds, which have received their names from their position upon the body, viz.: (1) the *dorsal* or back fins, (Plate XII. fig. 3a,) usually single, but sometimes divided into two or three fins, at varying distances from each other; (2) the *caudal* or tail fins, (c,) which in the true fishes are vertical, but in the fish-like mammalia are horizontal; (3) the *anal* or vent fins, (d;) (4) the *pectoral* or breast fins, (f;) (5) the *ventral* or belly fins, (e.)

The pectorals and ventrals are arranged in pairs, and correspond to the fore and hind limbs in other vertebrate animals; the pectorals, for instance, representing the wings in birds; the ventrals the feet. The dorsal or medial fin aids in keeping the body in a perpendicular position in the water; scarcely any fishes are without this fin, many have two dorsal fins, and a few, as the

Haddock and Cod, have three dorsals. The anal fin corresponds to the dorsal, but is placed beneath the body, just behind the vent. The principal instrument of motion is the tail fin. In those fishes which swim most swiftly, the tail is forked, each division being pointed, as are also the pectorals; while in those of less active or sluggish habits, the tail fin, as well as the pectorals, is commonly short, even or rounded. The rapid and powerful strokes of this fin, given obliquely right and left upon the water, urge the fish rapidly on in a straight course. The pectorals and ventrals do not appear to be much used for communicating motion; their chief office is to balance the body, or for turning, and for rising and sinking in the water.

Fins without distinguishable rays, or in which the rays are covered with a mass of fatty matter, or else entirely absent, are called *adipose*. A fin of this description is found on the back of the Brook Trout, (*Salmo fontinalis*,) in the rear of the main dorsal fin.

The bones are less dense and compact in their structure than those of the other Vertebrates, yet some of the Spiny-rayed Fishes possess considerable hardness. In the third group, CHONDROPTERYGII, or CARTILAGINOUS FISHES, (see Chart,) which includes the formidable Rays and Sharks, the skeleton is composed of gristle or cartilage instead of bone. Some of the species of this order seem, however, to make an approach to the osseous divisions. This is especially true, (1) of the SPOONBILL, *Polydon reticulatus*, an extraordinary fish, two feet or more long, found in the Mississippi, known at once by its snout, which is excessively prolonged, very flat and lanceolate, and in length nearly equal to the whole body; (2) the COMMON or SHORT-NOSED STURGEON, *Acipenser* (Lat. a sturgeon,) *brevirostris*, which has the body covered by hard bony tubercles; (3) the *Chimæridæ*, or SEA MONSTERS, so named from the fantastic shape of their heads, which have a singular hoe-shaped appendage, tipped with spines and somewhat-like a crest, upon their snout, (see Chart.) The body of one species, *Chimæra borealis*, which looks almost as much like a reptile as a fish, terminates gradually in a long slender filament.

The cone-shaped cavities of the *vertebræ*, or joints of the spine, are in the Fishes filled with a jelly-like substance, continued through the whole spine, by means of a hole pierced through the center of each vertebral joint. Though the tubular perforation is usually small, yet in many of the gristly or cartilaginous fishes, it is of so great a diameter as to reduce the vertebræ to mere cartilaginous rings.

Connected with the vertebræ above and below are spinous

processes for the attachment of muscles. Within the cavity of the belly the inferior processes are absent, but are replaced by lateral ones, to which the ribs are attached. These are usually numerous, slender and flexible bones, each of which sends off a branch of almost equal length and thinness. Some species, as the Herring and Pilchard, *Clupidæ*, (from *clupea*, a shad or herring,) send off thread-like branches from each of the vertebræ, so that the bodies of these fishes seem to be filled with long and slender bones.

The form of the skull varies much in the different orders, but generally it consists of pieces corresponding to those which form the head in other vertebrates. The line of distinction between the head and body it is difficult to draw, in consequence of the entire absence of a neck.

Teeth are very numerous, sometimes being found in almost all the bones of the mouth. They are usually simple spines, curved backwards, but the form is often much modified. The teeth of the voracious SHARKS, for example, are flat and lancet-like, the cutting edges being notched like a saw. In some species of these terrible fish they are so numerous that upon opening the mouth "the eye sees nothing but a forest of pointed teeth, any one of which, if detached, would be sufficient to inflict a most severe wound." In the Sharks of the genera *Pristis*, (Gr. Saw-fish,) and *Mustelus*, (Hound-fish,) the teeth differ, being flat, blunt, and tesselated. It is a remarkable provision that in some species the teeth are arranged in series of rows of which the outer one only is in use, the others remaining flat in the mouth until called into exercise by the injury or destruction of the outer row. The front teeth of the FLOUNDERS, (*Platessa*,) are compressed plates; the WRASSES, *Labridæ*, (from Gr. *labros*, greedy,) have flat grinding teeth; the SHEEP'S-HEADS, (*Sargus*, or *Sparus ovis*,) have the grinding surface convex; the Gilt-heads, *Chrysophrys*, (Sea-Breams,) have round, flat grinding teeth, arranged like the stones of a pavement, and often with strong pointed canines in front, able to crush and grind to powder the shells of the crustaceans and mollusks upon which they feed; the beautiful Chaetodons of warm climates, have, as the name denotes, teeth which resemble bristles; the Perches have teeth on the upper and lower jaw, slender, minute, numerous and closely set; the bold and fierce Pikes have teeth scarcely less formidable in size, form and sharpness than the canines of carnivorous animals.

The number as well as the form of the teeth greatly varies. While the Pike, the Perch, the Cat-fish, and many others have the mouth crowded with almost numberless teeth, the Carp and the Roach, *Cyprinidæ*, have only a few strong ones in the throat and a single flat one above; and the Sturgeon, the Pipe-fish, *Syngnathus*, and the Sand-launce, *Ammocætes*, (Gr. sand-bedded,) are entirely toothless.

The skin is either naked or covered with scales which appear in various states of development, as true imbricated scales, as isolated scales, as spiny bristles, hard, bony, enamelled plates. Most of the fishes have on their sides a longitudinal row of scales, in each of which is a perforation. These perforations were formerly thought to secrete mucus; but Prof. Agassiz has proved them to be the openings of tubes, which together with similar tubes opening on the skull, penetrate all parts of the frame and freely admit water, which serves to counterbalance the external pressure.

The colors of Fishes, including as they do all shades and lustres, are not surpassed even by those of the Birds. The hues are, however, evanescent, disappearing immediately after death. The effect of fear in changing the color of the human hair is well known. From the statement of a writer in the "New Sporting Magazine," it would appear to have a similar effect upon some Fishes, particularly Trout.

The brain is small and the face has not much expression; the tongue is mostly cartilaginous, and sometimes covered with teeth. This, connected with the fact that the food is almost always swallowed whole as soon as it is seized, seems to warrant the inference that Fishes have not acute taste. The sense of smell they probably possess in considerable perfection, the olfactory nerves being very large, and distributed over a great extent of surface. There is no external ear, nor even an auditory orifice, yet there is a complex internal apparatus of large size for the reception of sounds.

The eyes are distinguished by their almost immovable position; the cornea is flat, but the crystalline lens is perfectly spherical; the latter is familiar in the form of a white globule in a boiled fish, the transparency being destroyed by heat; eye-lids are not present. From the density of the watery medium inhabited by Fishes, a large number of the rays of light are absorbed and lost in passing through it; hence the eyes of fishes are very large, so as to collect as many of the remaining rays as possible; they are also of brilliant hues. These hues are owing to a membrane called the *choroid*, spread around the back of the eye, composed, to a large extent, of highly reflecting microscopic crystals. The eyes of some species gleam like quadrupeds.

Generally they are placed opposite to each other on the two sides of the head, so as to look sidewise, but in species that habitually live in deep water, they are placed on the top of the head, and look upward. In one genus of Sharks, called, on this account Hammer-heads, the head is enormously widened, so as to present two long lateral processes, at the extremities of which the eyes are placed; the shape of the head, much resembling that of the hammer used in caulking ships, (see Hammer-headed Shark on the Chart.) We must not omit here to notice the fact that a species of blind fish, *Amblyopsis*, (Gr. *amblus*, dim. *opsis*, vision,) *spelæus*, (Gr. *spēlaion*, a cave,) has been found in the Mammoth Cave in Kentucky. The Saw-fishes have the snout prolonged into a straight bony blade, along the edges of which are set pointed teeth directed outward.

Some of the species are endowed with a property quite peculiar to this class of Vertebrates. This is the power of giving electric shocks, at will, to other creatures, possessed by the Torpedo, (Plate XIV. fig. 13;) and the *Gymnotus*, or electric eel. The electric organs consist of numerous six sided cells containing a number of delicate membranous plates, separated from each other by a transparent jelly-like fluid. In the Torpedo, the plates are placed vertically, and form two masses one on each side of the head; in the Gymnotus, they are horizontal and form four sub-organs, one pair on each side of the body.

The organs of voice are, in fish, entirely wanting. The Cat-fish, *Pimelodus*, is, however, said to make a peculiar sound by the vibration of its *cirri*, or barbels, (Plate XIV. fig. 8.) The Weak Fish, *Otolithus regalis*, makes a peculiar and seemingly abdominal grunting when caught, as does also the Black Drum, *Pogonias chromis*.

The food of fishes is, for the most part, animal. Some browse the sea-weeds that wave around the rocks of the coast, and others nibble the soft parts of fresh water vegetation; but the great majority are carnivorous. The soft-bodied animals of the sea, such as the *Actiniæ*, the *Medusæ*, the *Annelidæ*, and the naked *Mollusca*, afford food to multitudes; others are furnished with strong teeth to grind down the newly formed parts of coral, and devour the living polyps; and a large number feed greedily on Star-fishes, *Crustacea*, and the Shelled *Mollusca*. In fresh water, worms, leeches, and the larvæ of insects satisfy the appetite of many. Besides these sources of supply, Fishes everywhere feed upon Fishes, the larger upon the smaller. Their voracity is extremely great, no limit to their appetite appearing but the actual capacity of the stomach.* Some, as the Trout, act the part of *tyrants* over their fellows.

Fishes are almost incredibly prolific. One species, the Blenny, produces its young alive, sometimes two or three hundred at a time, and able to provide for their own support. A species of viviparous fish, but two inches in length, inclusive of the caudal fin, and containing twenty-two perfect fish, has also been found in a canal connecting with Lake Pontchartrain. Larger specimens have been received by Agassiz from Lake Erie, and also from California. But, generally, the continuation of the race is accomplished by means of eggs, called, in the aggregate, *spawn;* and before exclusion, *roe.* The eggs are deposited in various places, on sticks, stones, grass, furrows in the sand, etc. In rare cases, as the Goby of the Mediterranean, some North American *Cyprinidæ* or Carps, and the HASSARS, *Callichthys*, of Demerara, a nest for the reception of the spawn is built, consisting of a single pile of stones, or else, as in the last named fishes, a more complicated structure of grass and sticks. The *Stickleback*, (*Gasterosteus*,) forms of sea-weed and common coralline, pear-shaped nests, which hang from the rocks, variously intermingled with each other. The Shark, instead of depositing almost innumerable eggs in a season, like the Cod or the Herring, produces two eggs, of a square or oblong form, (see fig. on Chart,) the coat of which is composed of a tough, horny and semi-transparent case; each corner is prolonged into a tendril, of which the two which are next to the tail of the enclosed fish, are stronger and more prehensile than the other pair. The use of these tendrils appears to be their entanglement among the stalks of sea-weeds, and the consequent mooring of the egg in a situation of protection and comparative security. The part of the skin near the head, is weaker and more easily broken than any other part,—a provision for the easy exclusion of the animal, which occurs before the entire absorption of the yolk of the egg,—the remainder being attached to the body of the young fish, enclosed in a capsule, which for a while it carries about. The position of the animal while within the egg, is, with the head, doubled back towards the tail,—one very unfavorable for the process of breathing by internal gills. But as a provision

* "At a lecture delivered before the Zoological Society of Dublin, Dr. Houston exhibited, as 'a fair sample of a fish's breakfast,' a Frog-fish two feet and a half long, in the stomach of which was a Cod-fish, two feet in length; the Cod's stomach contained the bodies of two Whitings of ordinary size; and the Whitings, in their turn, held the half digested remains of many smaller fishes, too much broken up to be identified."

for this emergency, on each side a filament, of the substance of the gills, projects from the gill-opening, containing vessels in which the blood is exposed to the action of the water. These processes are gradually absorbed after the fish is excluded, until which the internal gills are incapable of respiration. This presents an analogy with the Frogs and Newts. We advert to it thus particularly as impressively manifesting the Divine benevolence when the object of so much contrivance and care, is the dreaded and hated SHARK! The horny cases just referred to are frequently found on the sea-shore, and are called Sea-Purses, Mermaids'-Purses, &c. Some species, as the PENNY DOG, *Galeus vulgaris*, and the SMOOTH HOUND, (*Mustelus lævis*,) bring forth their young alive, without any capsule or covering at all.

Some species, as the Pipe-fishes, (*Syngnathus*,) are ovo-viviparous. What is very singular, the *male* Pipe-fish is provided with a pouch, into which he receives the spawn as it is deposited by his mate, and in which he carries it about until the young are hatched. And, as if to make the resemblance to the Marsupials complete, the young are in the habit of retiring for shelter into the paternal pouch, for sometime after they are able to leave it and roam at their own pleasure. It is somewhat remarkable, that it is the male, generally, who assumes the care of the eggs, and the construction of the nest. Instances are not wanting, of striking parental devotedness and foresight.

The following Table of Mr. Harmar, (Phil. Trans.,) shows the different degrees of fecundity in different species of fishes.

NAME OF FISH.	Weight of Fish.		Weight of spawn.	Number of Eggs.
	ozs.	drs.	grs.	
Carp, ...	25	5	2.571	205,109
Cod, ...			12.540	3,686,760
Flounder, ...	24	4	2.200	1,357,400
Herring, ...	5	10	480	36,960
Mackerel, ...	18		1.223	546,681
Perch, ...	8	9	765½	28,323
Pike, ...	56	4	5.100½	49,304
Roach, ...	10	6½	361	81,586
Smelt, ...	2	0	149½	32,278
Sole, ...	14	8	542½	100,362
Tench, ...	40	0		383.252

But far more productive than these, is the Salmon, (*Salmo Salar*,) for "the ovarium of one female, has been known to produce 20,000,000 eggs!"

Prof. Dana estimates the greatest number of eggs in the Thick-lipped Grey Mullet, (*Mugil Chelo*,) to be 13,000,000; in the Cod-fish, (*Gadus Morrhua*,) 11,000,000; in the Turbot, (*Pleuronectes maximus*,) 9,000,000; in the Plaice, (*P. platessa*,) 6,000,000; in the Carp, (*Cyprinus carpio*,) 600,000 to 700,000; in the Perch, (*Perca fluviatilis*,) 71,000.

It has been estimated that the progeny of a single Herring, if allowed to multiply, undisturbed for thirty years, would not only be sufficient to meet every demand for this fish, but become even inconveniently numerous; and that, too, notwithstanding hardly one among the millions of young Herrings comes to maturity, in consequence of the ravages made by rapacious fish, and by other means. Although so extensively used, the supply of this fish is always found equal to the demand. The same might be said of the Cod, the Mackerel, the Tench, &c.

The longevity of fishes seems to be undoubted, however it be true that few reach their natural term of years. Pike and Carp kept in fish ponds, have been known to live to a great age. A Pike taken in Prussia, in 1754, bore a ring which testified to its having been placed in the pond two hundred and sixty-seven years before; how old it was when put in was unknown. Carp, it is clearly shown, have attained the age of a century. Buffon speaks of one that was one hundred and fifty years old.

"Cartilaginous fishes," says Swainson, "continue to grow all their lives; and as many of these, particularly the Rays, habitually live in the deep recesses of the ocean, and thus seldom run the risk of being captured by man, we may probably attribute their enormous and almost incredible size to their great age." It is thought to be "a rare thing for a fish to die of natural decay." But owing to the ravages made among them, the actual average of life is with fishes of comparatively short duration. They are capable of enduring great extremes of temperature,—a fact which may be regarded as indicating their low place in the scale of organization. Experiments have shown that several species of fresh-water fish can live many days in water so hot that the hand could not be held in it a single minute. Eels have been alive in hot springs, in which the temperature is pretty regularly 113° Fahr. But such cases are far less wonderful than that recorded by Humboldt and Bonpland, who "saw living fishes, apparently in health and vigor, thrown up from the bottom of a volcano, with water and hot vapor that raised the thermometer to 210° Fahr.,—a heat only two degrees less than that of boiling water!"

On the other hand, the cold of freezing does not always destroy the life of fishes. Eels and Perch are conveyed from place to place in a frozen state, which revive on being thawed. So is it, according to Dr. Richardson, with the Grey Sucking Carp. Gold fishes which have been thawed out of a solid body of ice, have completely revived.

Species which live near the surface of the water, have less tenacity of life than those which seek the deep waters. Mackerel, Salmon, Trout, and Herrings, of the former kind, die almost as soon as they are taken out of the water; while Carp, Eels, Tenches, Skates, and the Flat-fishes, which live near the bottom, and have a low standard of respiration, and a high degree of muscular irritability, with less necessity for oxygen, sustain life for sometime after they are taken out of the water, and their flesh continues good for several days.

The Eels, *Muraenidæ*, and the Blade-eels, *Ophidiadæ*, show extraordinary tenacity of life; even removal of the skin, and the division of the body into parts, not immediately producing death.

The flesh of the larger part of fishes is useful for food; though that of some species is somewhat indigestible. Fish of fresh-water are more generally edible than those of the sea; but as a whole, are not so savory. Fishes are also valuable for other purposes: some for the oil which they yield; the air-bladder of the Sturgeon furnishes the isinglass of commerce; the roes of the Sturgeon, Pike and some other fish, furnish caviar; the shagreen skin of some *Placoids* is employed for polishing and for making ornamental coverings. The bones are used for fish-hooks and other purposes.

The first scientific CLASSIFICATION of Fishes is that of Artedi, which was made in 1738; the next, that of Linnæus, made between twenty and thirty years later.

The Classification of Cuvier is generally adopted, and the one to which the Chart most nearly conforms. The entire class of Fishes is first divided into OSSEOUS and CARTILAGINOUS.

The OSSEOUS FISHES are sub-divided into SPINE-RAYED and SOFT-RAYED.

They are also arranged into three sections,—I. *Pectinibranchii*, which have the *branchiæ*, or gills, in continuous, comb-like ridges, and include all the ordinary and typical fishes. This section is comprehended in two orders.

I. ACANTHOPTERYGII, (Gr. spine-rayed,)—distinguished by having the anterior part of the dorsal, anal, and ventral fins, furnished with simple, spiny rays. The Perches, Mullets, Gurnards, Mackerels, &c., belong to this order.

II. MALACOPTERYGII (Gr. soft-rayed,) having all the fin rays soft and flexible, with the exception, sometimes, of the first ray of the dorsal and pectoral fins. There are also three Sub-orders, founded either upon the position of certain fins, or their absence, as (1) the *Abdominales*, in which the ventral fins are situated far behind the pectorals, as in the Carp, Tench, Bream, Dace, Roach, Pike, Salmon, etc.; (2) the *Sub-brachials*, or *Sub-brachiati*, (terms derived from the Latin *sub*, under; *brachialis*, armlet, or *brachium*, arm,) in which the ventral fins are immediately beneath the pectoral fins or armlets, (or even a little before them,)—as in the Codfish, Haddock and Whiting. To this group also belong the Flat-fishes,—such as the Plaice, Flounder, Turbot, Sole, etc.; (3) the *Apodes*, (Gr. footless,) including the Eels, which receive this name from their possessing no ventral fins.

The other sections of the Osseous division are, (2) *Lophobranchia*, (Gr. tuft-gills,) including Bony Fishes which have the gills in tufts, (not pectinated,) and arranged in pairs along the branchial arches; (3) *Plectognathi*, those in which the bones of the head are closely combined, including the *Gymnodontidæ* or Naked-toothed Fishes, the *Balistidæ*, or File-fishes, and the *Ostracionidæ*, or Trunk-fishes.

The Cartilaginous Fishes are sub-divided (1) into those which have the gills free, *Eleutheropomi*, (Gr. free-covers or *operculæ*.) The gills in these, are pectinate or comb-like, and there is only a single gill opening. Of these, the Sturgeons furnish an example; (2) those with fixed gills, (*Branchiis fixis*,) and which have more than one gill opening on each side,—including the Sharks, (*Squalidæ*,) the Rays, (*Raiidæ*,) and the Stone-Suckers, (*Petromyzonidæ*.)

The division *Plagiostomi* includes those Cartilaginous Fishes which have on the under side of the face, and at a greater or less distance from the extremity, the broad *transverse mouth*,—such as the Sharks and the Rays. The gills in all are fixed with five or six gill-openings.

The *Cyclostomi*, (Gr. with circular mouths,) are those Cartilaginous Fishes which breathe by a "series of cells," the gills not being comb-shaped fringes, but forming sacs or pouches by the union of two opposite ones along their edges. Here are found the Stone-Suckers, so called because the animal applies its circular lip to the surface of a stone or other solid body in the water, and drawing in the piston-like tongue, produces a vacuum in the mouth, while the pressure of the super-incumbent body of the water causes the lip to adhere to the stone with immense tenacity, until, by the protrusion of the tongue, the vacuum is voluntarily destroyed.

The lowest and most anomalous of all the species of Fishes, is the BRANCHIOSTOMA or LANCELET, (*Amphioxus lanceolatus,*) usually about two inches in length, and generally distributed throughout the seas of Europe and North Africa. So unique is the structure of this minute creature, that, on the Chart, it is dissevered from the Stone-Suckers. "A vertebrated animal without a brain, a fish with the respiratory system of a mollusk, and the circulatory system almost of an Annelide,"—presents a combination of characters which has challenged its right to a place among the Vertebrates, and seems to justify its separation from the Lampreys, with which some naturalists have ranked it.

The Myxines, or GLUTINOUS HAGS, of the most Northern and Southern seas, are almost equally strange in form and structure, having been classed by Linnæus and other writers, among the WORMS.

Their place is filled in the higher parts of the Southern hemisphere, by the equally curious and nearly allied genus, *Heptatrema.**

The *Lepidosiren,* (Gr. scaly-siren,) is the *connecting link* between FISHES and Reptiles, being so dubious in its organization, that its true position is disputed. Most naturalists of Continental Europe consider it to be a reptile, while Prof. Owen confidently maintains its claim to a place among the Fishes. If assigned to the Reptiles, its position would be as a *fourth* order of the Batrachians.

Prince Bonaparte divides the Fishes into four orders, viz: *Acanthopterygii, Malacopterygii, Plectognathi,* and *Cartilaginei.* His arrangement is by many highly esteemed.

Agassiz names the orders of Fishes from their scales, (his classification being applicable to the *fossil* as well as the living forms,) viz:

(1) CTENOIDS, (from Gr. *ktenos,* a comb,) in which the scales consist of plates whose posterior or free margin is pectinated, or comb-like, as in the Perch, Bass, Pumpkin-seed, &c.

---

* The name *Heptatrema* (meaning seven apertures or perforations) was given to this genus by Dumeril. It is found, however, that the number of apertures varies. Mr. C. Girard describes one of these fishes of the Southern Hemisphere, as having *fourteen* breathing holes. (See "*U. S. Naval Astronomical Expedition,*" published at Washington, D. C., 1855.) He adopts the generic name of Müller, *Bdellostoma,* (Gr. *Bdello,* I suck; *stoma,* mouth,) founded on the structure of the mouth, and calls the specimen which he figures, *Bdellostoma polytrema,* (Gr. many perforations.)

(2) Cycloids, (from Gr. *kuklos*, a circle,) those whose scales are entire, as in the Salmon, Trout, Shiners, &c.

(3) Ganoids, (from Gr. *ganos*, splendor,) having scales of an angular form, composed of horny or bony substances, covered with a thick coat of enamel, so that they become teeth-like in their structure, as in the Gar-pike.

(4) Placoids, (from Gr. *plax*, a plate or slab,) thus named from the irregularity which the scaly coverings exhibit, so that the skin resembles shagreen as in the Sharks, Rays, &c.

What is the fourth branch of Zoology? Of what does it treat? What is said of the number of fishes? What are their most general characteristics? What is said of their blood? Describe the breathing apparatus and the circulation of the blood. What have they analogous to the lungs in higher vertebrates? What purposes does it subserve? State its variation in different fishes. Of what kind are the fishes which have not the air-bladder? How does its possession expose to danger? What habit has the Sea-Porcupine? What is related of the Gold-fish? What fishes sometimes leave their proper element? Relate facts respecting the migration of Eels. What is the usual form of fishes? What exceptions are mentioned? What are the organs of motion? Of what do they consist? What orders are based upon differences in their structure? Of how many kinds are the fins? Name them, and describe their uses. What additional fin is mentioned? How do the bones compare in structure with those of other vertebrates? What forms the skeleton of the Cartilaginous Fishes? What is said of the vertebræ of fishes? What of the bones of the Herring and Pilchard? What of the head? Describe the teeth, with their variations as to form and number? What is said of the skin? What of its colorings? Of the senses of taste, smell, and hearing? Of their eyes? What causes their brilliant hues? Where are the eyes placed? Are these organs wanting in any species of fish? What power is peculiar to some of this class? Describe the electrical organs. Have they a voice? Do any give forth sound? Describe their food. Are any fish viviparous? What are the eggs in the mass called? What before exclusion? Where are they deposited? What fish construct nests? Are any ovo-viviparous? What singular facts are mentioned in regard to the male fishes? What of the Shark's eggs? Which is the most prolific fish? Give other instances from the table. What facts show the longevity of fishes? What their capacity to endure heat and cold? Which species show the greatest tenacity of life? What is said of their flesh? Explain the classification of Fishes as given on the Chart. Which is the most anomalous of all the species? What other strange fish are mentioned? Which connect the Fish and Reptiles? Into how many orders are the Fishes arranged by Prince Bonaparte? Give their name, and also the orders of Agassiz and Müller?

Which of Agassiz' orders of fish figured on the Chart has a hetero-circal or uneven tail or caudal fin? How has this order been otherwise divided? Upon what are these divisions based? Name the prominent fish in each. Trace each family upon the Chart, giving some prominent characteristic of each. Name some of the fish in each family, giving both the common, generic, and specific names when they are mentioned. Which is the most numerous family? Which is the smallest? Which most useful for food?

NOTE.

Müller's classification of fishes (made in 1846) is the most recent, but it has since been somewhat modified. This gives eleven orders, viz: 1. DERMOPTERI, in which he includes the Amphioxidæ or Lancelets, placed in the sub-order *Pharyngobranchii;* and the Myxinoidei, (Myxines,) and Petromyzontidæ, (Stone-Suckers,) in the sub-order *Marsipobranchii;* II. MALACOPTERI, with sub-orders *Apodes*, *Abdominales;* III. *Pharyngognathi*, with sub-orders *Malacopterygii*, (including Scomberesocidæ, and *Acanthopterygii*, (including Chromidæ, Cyclo-Labridæ, Cteno-Labridæ ;) IV. ANACANTHINI, with sub-orders *Apodes*, (including Ophididæ,) and *Thoracici*, (including Gadidæ, Pleuronectidæ, and Echineidæ;) V. ACANTHOPTERI; VI. PLECTOGNATHI; VII. LOPHOBRANCHII; VIII. GANOIDEI; IX. PROTOPTERI, which includes the one family Sirenoidei, made up of *Lepidosirēn paradoxa*, of Brazil, and *Lepidosirēn* or *Protopterus annectens*, of the Gambia River, Africa; X. HOLOCEPHALI, (including Chimæroidei and Edaphontidæ;) XI. Plagiostomi.

Agassiz, in the "Essay on Classification," contained in his recently published work, thus remarks,—"I am satisfied that the differences which exist between the Selachians, (the Skates, Sharks and Chimæræ,) are of the same kind as those which distinguish the Amphibians from the Reptiles proper, and justify their separation, as a class, from the Fishes proper. I consider also the Cyclostomes as a distinct class, for similar reasons; but I am still doubtful whether the Ganoids should be separated also from the ordinary Fishes. This, however, cannot be decided until their embryological development has been thoroughly investigated, though I have already collected data which favor this view of the case. Should this expectation be realized, the branch of Vertebrata would contain the following classes:—

1st class: Myzontes; with two orders, Myxinoids and Cyclostomes.

2d class: Fishes proper; with two orders, Ctenoids and Cycloids.

3d class: Ganoids; with three orders, Cœlacanths, Acipenseroids and Sauroids; and doubtful, the Siluroids, Plectognaths and Lophobranches.

4th class: Selachians; with three orders, Chimæræ, Galeodes and Batides.

5th class: Amphibians; with three orders, Cæciliæ, Ichthyodi and Anura.

6th class: Reptiles; with four orders, Serpentes, Saurii, Rhizodontes and Testudinata.

7th class: Birds; with four orders, Natatores, Grallæ, Rasores, Insessores, (including Scansores and Accipitres.)

8th class: Mammalia; with three orders Marsupialia, Herbivora and Carnivora."

SECTION VIII.

Osseous Fishes

First Order. Acanthopterygii, (Gr. 'ἄκανθα, *akantha*, a spine; πτέρον, *pteron*, wing.)

The Ctenoids, or Acanthopterygiians, including three-fourths of all known fishes, are almost all marine. They are ornamented with hard, shining, tooth-like scales of beautiful colors. The spiny fins of most of them are constructed for long continued motion.

FAMILIES.

(1). *Percidæ*, (Gr. *perkē*, a kind of fish.)

The Perches. These comprise one-seventh of all spine-rayed Fishes. Most of them, including the Gropers or Mailed Perches, are marine, but the typical species, the Common Perch, of Europe, (*Perca fluviatilis*,) with two separated dorsal fins, the rays of the first spinous, of the second flexible, is found only in fresh water. To this nearly corresponds the Yellow Perch, *P. flavescens*, the most conspicuous of the numerous North American species, and found both in salt water and fresh. In 1825, Yellow Perch were transported from Skaneateles to Otisco Lake and Onondaga Lake, and appeared to thrive after the transfer. They are common in ponds and streams, and in all the great lakes. The eggs of the Perch are of the size of a poppy seed, and joined together by a viscid substance, in long strings. Among the most remarkable fishes of this group are those included in the genus *Polynemus*, (Gr. *polus*, many; *nēma*, a thread or filament,) distinguished by having the ventral fins inserted farther back than the pectorals, and also for having numerous long flexible filaments placed near the latter fins, from four to ten on each side, and sometimes twice the length of the body. Interest attaches to these fishes from the fact that the bladders of several species yield pure isinglass. The Suleah Fish of India, *P. sele*, is the one from which it is said to have been first procured. Shoals of this species are found in the estuaries of the river Ganges; they are three to four feet in length, and eight to ten inches in depth. The species *P. Americanus* has several rays attached to the pectoral fin. It is about a foot in length, and of a silvery color. All the fishes of this family agree in the toothed or comb-like edges of the scales, and in having serrated or spined gill-covers, and fins destitute of scales.

The ROCK-FISH, or STRIPED-BASS, *Labrax lineatus*, is a fish that has the tongue covered with teeth or prickles; the opercula are somewhat different from those of the Perches proper, but in other respects it closely resembles them. It is much esteemed, especially when taken in autumn. These fishes are brought into market, (dead,) during the winter, and sold in great numbers. Like the shad, with which they are taken, they run from the salt waters into the fresh for the purpose of spawning. (For other genera and species of this numerous family, see Chart.)

(2) *Triglidæ*, (Gr. *trigla*, a surmullet.)

The GURNARDS, or MAILED CHEEKS. These have enormous pectoral fins, yet live near the shore. The name Gurnard is supposed to be derived from the French word *gronder*, to grumble or emit sounds. Cuvier called them "Fishes with hard cheeks," referring to the encasement of the head and face in a solid buckler of bone, which is their most obvious character. Like other bottom fish, they live for some time out of the water.

The PIPER, *Trigla lyra*, is rather an uncommon European species. The RED GURNARD, *T. cuculus*, a well flavored fish, occurs on the coast of the United States. Other American species are found in the genus *Prionotus*, which closely resembles Trigla. Among the Gurnards is included the Flying-fish, *Dactylopterus* (Gr. finger-wings) *volitans*, (Lat. flying,) having very large pectoral fins divided into two portions and serving as wings; but it is not so good a flyer as the Flying-fish, *Exocœtus*, belonging to the Pike family. To this family also belongs the genus *Gasterosteus*, (Gr. bony-belly,) including quite small, but very active and voracious fish, found in both fresh and salt water, and popularly called STICKLEBACKS. They are only from one to two and a half inches long, but so elastic is the Three-spined Stickleback, (represented half the usual size, Plate XIV. fig. 2,) that it leaps nine times its length, in perpendicular height, from the water. Its extraordinary voracity is shown by the fact that it has been known to devour in five hours, seventy-four young dace, and on the following day, sixty-two; (some dace, it should be mentioned, are exceedingly small.) Several species of Sticklebacks are found in the waters of the United States. In some parts of England, these fish are so numerous as to be used for manure. They are so pugnacious as to destroy each other; and yet some of them manifest great care in building and watching their nests.

(3) *Scænidæ*, (Gr. *skiaina*, a sea-fish.)

The *Maigres*. These resemble Perches, but live in the sea, and attain to a great size. They make a sort of purring sound. The air-bladder is long, tapering, and fringed along each side, giving it a singular appearance; the head is generally enlarged with cavernous swellings. Among the American fishes of this family are the WEAK-FISH, *Otolithus regalis*, abundant on the Atlantic coast, and an excellent salt-water fish. It is called Salt-water Trout on the southern shores. The RED-BASS, or Sea-bass, *Corvina ocellata*, is taken off the coast as far north as Long Island Sound, and for food, is highly prized. The LAFAYETTE, or CHUB, *Leiostomus obliquus*, abounds on the coasts of the Middle States, and is in some estimation for food. A species, *L. xanthurus*, (Gr. yellow-tail,) known as the YELLOW JACK, or YELLOW TAIL, is found off the coast of South Carolina. The KING-FISH, *Umbrina nebulosa*, distinguished from others of the family by a cirrus or tuft on the under jaw, is thought by many to be the best fish which appears in the New York market. The Drum, *Pogonias chromis*, (Gr. *pogonias*, bearded; *chromis*, a fish,) is a large and fine flavored fish.

(4) *Sparidæ*, (from Lat. *sparus*, gilt-head.)

The SEA BREAMS. These fishes have flat grinding teeth, sometimes strong pointed canines in front. The common Gilt-head, *chrysophris*, (Gr. *chrusophrus*,) *aurata*, can crush such thick stony shells as the Periwinkles, Whelks and Turbos, (or Tops;) a more voracious fish is scarcely known. The famed SHEEPSHEAD, *Sargus ovis*, abundant on the coast, and much esteemed for food, is of this family.

(5) *Maenidæ*, (Gr. *mainē*, a small sea-fish.)

The MENDOLES. This is comparatively a small family, (not mentioned on the Chart.) The common Mendole, *Maena vulgaris*, is considered so utterly worthless that the name is used at Venice as a term of derision. A West India species decomposes with remarkable rapidity, the flesh becoming soft almost immediately after it is dead. The species of the genus *Smaris*, (Gr. a sea-fish,) are sought for in the Mediterranean. One species is called the KING-FISHER of the Sea, *S. alcedo*, in allusion to its beautiful tints.

(6) *Chaetodontidæ*, (Gr. *chaitē*, hair or bristle; *odous*, tooth.)

The *Chaetodons*. These are thus named because their teeth are so long, fine and slender as to resemble the bristles of a brush. Cuvier called them *Squamipennes*, to express the manner in which their fleshy fins are covered with scales, like the rest of the body, which is flat or round, and thin, with long bat-like fins. The scales reflect the most brilliant hues. The species are numerous and tropical. The ARCHERS, *Toxotes*, (Plate XIV. fig. 4,) eject water out of the tubular mouth with such precision as to bring down any insect within their reach. In Java, they are kept in glass vessels for amusement, an insect being suspended by a thread above for the fish to shoot at.

(7) *Anabassidæ*, (from Gr. *anabaino*, to ascend.) CLIMBING PERCHES.

This family includes but one genus, *Anabas*, fishes whose respiratory organs are so constructed as to enable them to sustain life for a space of time out of water, by having small apertures or some receptacle where they can preserve sufficient water to moisten their gills. There is but one species, *A. scandens*, (Lat. climbing.) When a pond is dried up in which these Perches are found, it is said they are guided by a remarkable instinct in traveling towards the nearest water. Swainson says the Climbing Perch "quits the water and ascends the roots of the mangrove trees, (in East India,) an effort it accomplishes by using its ventral fins as little feet." (These fish are by some naturalists included in the family *Labyrinthibranchiæ*, a name referring to the vascular membrane, folded together in a number of laminæ, and occupying the upper part of the front branchial arches, and which serves to retain water for moistening the gills during the travels of these fishes on the land.)

(8) *Scombridæ*, (Lat. *scomber*, a mackerel.) This includes the Mackerels, an important as well as numerous family of almost entirely marine fish, found in all seas. Many are pelagic, (roving far from land.) They are taken in such quantities as to prove them to be inexhaustible. They live near the surface and are among the fishes which quickly decompose. The TUNNY, *Thynnus*, of the Mediterranean, is from three to four and even fifteen feet in length. Fried in cutlets, this fish resembles veal, the flavor being quite as much like that of flesh as of fish. The BONITA, (*Scomber pelamys*,) found on our coast, is a species of Tunny, which in the tropics pursues the Flying-fish.

The SWORD-FISH, *Xiphias gladius*, the largest of the order, being from twelve to fifteen feet long, has the beak lengthened into a long, powerful weapon, which it sometimes drives with such violence as to penetrate to a great depth into the timbers of ships. The PILOT FISH, *Naucratēs ductor*, follows vessels, and thus acts as a guide to the Sharks. Among the other American fishes of this family found on our coast, are the following; the SPANISH MACKEREL, *Cymbium*, (Gr. *kumbion*, a small bowl;) the CRAB-EATER, *Elacate*, (Gr. a distaff;) the CAROLINA LICHIA,

Lichia Carolina; the TRACHINOTE, *Trachinotus;* the YELLOW and SPOTTED MACKERELS, *Caranx crysos*, (Gr. yellow or golden,) and *C. punctatus*, (Lat. spotted;) the HAIR-FINNED BLEPHARIS, (*Blepharis crinitus*,) which is the *Zeus crinitus* of Akerly, kindred to the well known Dory or John Dory, *Z. faber*, an European species much esteemed by epicures, and of which strange things have been often recited; the SHINER, *Vomer*, (Lat. a plough-share;) the SERIOLE, *Seriola*, (Lat. a small jar;) the BLUE FISH, or GREEN FISH, *Temnodon*, (Gr. *temno*, to cut; *odous*, tooth; so named from the very sharp teeth,) sometimes called HORSE MACKEREL; the principal species, *T. saltator*, being the TAILOR, or SKIP-JACK of the more southern waters, "twenty of which," it was formerly said, "would fill a barrel;" the BOTTLE-HEADED DOLPHIN, *Coryphæna*, (Gr. *korus*, a helmet; *phaino*, I display,) *globiceps*, (globe-headed.) The fishes of this genus, including the fishes generally known as dolphins, and celebrated for their beauty, are, however, rarely found off the coasts, being mostly inhabitants of mid-ocean. The species *C. hippuris*, is famed for its beautiful play of colors when dying. The Dolphins are conspicuous enemies of the Flying-fish. Other fishes of this family are, the *Lampugus*, a rare and exceedingly beautiful fish, and the Harvest-fish, *Rhombus*.

(9) *Cepolidæ*, or *Tæniadæ*, (Lat. *tænia*, a ribbon.)

This is a small family of fishes allied to the Mackerels. They are chiefly distinguished by an elongated, flattened shape; their general appearance being that of a bright silver ribbon. They have the popular names of Ribbon-fish, Lath or Deal fish. The body is not thicker, except in the middle, than a sword. Most of the species inhabit the Mediterranean. The eleven-rayed Band-fish, *Cepola rubescens*, (Lat. turning red,) is seen on the coasts of England; it displays brilliant colors; sometimes is called Fire-flame and Red-ribbon. As showing the appropriateness of the name Ribbon-fish, it is related, that a specimen of this species, "though nineteen and a half inches in length, having been carefully *folded up like a ribbon*, passed to Belfast, (Ireland,) in a franked letter of the ordinary size and legal weight, viz., less than an ounce." (Magazine of Nat. Hist.) The Silvery Hair-Tail, *Trichiurus*, (Gr. hair-tail,) *lepturus*, (Gr. thin-tail,) having a tapering tail, ending in a filament, is found off the coast of the U. S.

(10) *Teuthidæ*, (Gr. *Teuthis*, a kind of fish.)

This is another, not numerous family, sometimes called Lancet-fish, resembling the Mackerels in appearance and some other respects, but peculiar for the cutting spines in each side of the tail, and a horizontal spine before the dorsal fin. They have but one row of teeth, and are among the small number of the class that feed entirely on vegetable substances. The DOCTOR-FISH, *Acanthurus*, (Gr. spine-tail,) *cæruleus*, (Lat. dark blue,) has caudal lancets which are short, hard, and glassy, and are enclosed in a yellow membranous sheath. It is common on the coasts of the West India islands, South Carolina and Florida. The SURGEON-FISH, *A. phlebotomus*, (Gr. vein-cutting,) is another species found off the coasts of the United States. The caudal lancets of these fishes are analogous to the horns of ruminating animals, and to be regarded rather as defensive than offensive weapons.

(11) *Atherinidæ*, (Gr. *atherina*, from *ather*, a thorn.)

This is a family sometimes included in the *Mugilidæ*, called Silver-sides, from the silvery band on the side; the two dorsal fins are far apart; the anterior one spinous. The genus *Atherina* is represented by several species of small fishes. Silver-sides used to be caught in New York Harbor, and sold for bait, under the name of Anchovies and Sea-smelts. These small fishes were formerly supposed to be all included in the genus *Atherina*, but Mr. Charles Girard, (see "United States Astronomical Expedition,") has, within a few years, proposed three additional genera, of which several species are found in S. A.

(12) *Mugilidæ*, (Lat. *mugil*, a mullet.)

The Mullets are lengthened, and often cylindrical in form, with a somewhat projecting snout, and an extremely small mouth, placed beneath. They inhabit both salt and fresh water; indeed, a change from salt to fresh water seems necessary to them. A number of species of the genus *Mugil* are found in the United States. The Common Mullet, *M. albula*, is throughout the greater part of the year, taken in large numbers on our Southern coast. These fishes have not been considered carnivorous; but the shells obtained from the stomach of one of them by Mr. Thompson, of Belfast, (Ireland,) filled a large sized cup. They swim in large shoals near the surface; Gosse says that the Grey Mullets, *M. capito*, assemble to feed every evening at a certain knocking, and are the only fish with which he is acquainted, that select for food nothing that has life, except that they sometimes swallow the Sand-Worm.

(13) The *Gobidæ*, GOBIES, and *Blennidæ*, BLENNIES, are, on the Chart, included in one family, both having flexible and slender-dorsal spinous rays. They have no swimming bladder. Some of the Gobies proper, have no visible scales. This numerous family of small unimportant fishes, have not even beauty to

recommend them. They are soft to the touch, being invested with a mucous slime; hence the generic name, *Blenna.* (Gr. mucus.)

The Wolf-Fish, or Sea-Cat, *Anarrhicas lupus*, is much larger than others of this family, being a formidable, voracious fish, from three to eight feet in length, with a broad cat-like face, and a grinning mouth, bristling with stout, sharp teeth, so strong as to crush the hardest shells, and even stone.

The genus *Zoarces*, (with three American species,) has the dorsal, anal, and caudal fins united.

The Butter-fish, *Gunnellus mucronatus*, (pointed,) has a long compressed body, and the ventral fins rudimentary.

The German Dragonet, *Callionymus lyra*, is an exception to others of the family, in having beautiful colors. It has no visible scales.

In Italy, Blennies are fried in numbers, like Sprats in England, and eaten by the poorer classes. Some species of the genus, *Zoarces*, and probably others, produce their young alive by dozens. The Shanny, *Pholis*, (Gr. a scale,) deposits its eggs on the roofs or sides of cavities in rocks, near the low water mucks, and being of a bright amber color, with a polished surface, it appears as if paved with round stones. The *Physis*, of the Mediterranean, forms a nest of sea-weed in which to deposit its spawn, and attends upon the young.

(14) *Lophidæ*, (Gr. *lophos*, a crest.)

These are distinguished by the lengthening of the carpal bone, by which, as on an arm, the pectoral fin is supported. The family includes some of the most singular looking fishes in the entire class, such as Frog, or Toad Fishes, &c., grotesque and reptile-like; without scales; hiding themselves in the mud, and attracting their prey by agitating the filamentary processes on the head. The feet-like pectoral fins assist them to crawl on the bottom of the sea, and also upon land. These, on account of the soft and yielding nature of the skeleton, were formerly classed with other Cartilaginous or Soft-rayed Fishes, (see *Lophius Americanus*, *Squalidæ*, on the Chart;) but Cuvier demonstrated its fibrous structure, and fixed its position among the bony fishes.

The *Antennarius*, (Lat. from *antenna*,) is found in tropical seas. It is said to crawl about the fields for two or three days at a time. So tenacious of life are the fishes of this genus, that they have been transported alive from tropical seas to Holland, and sold "for twelve ducats a piece." Their voracity is great,—in fact, they seem to be mostly mouth and stomach. On the coast of Scotland, these "Sea-Devils" are met with, four and even five feet long. The Mouse-fish and Toad-fish are small species of this family.

(15) *Labridæ*, (Lat. *labrum*, a lip,)—a family deriving its name from the fleshy lips appended to the jaws. It has been divided into two sections, *Cyclo-labridæ*, having cycloid-scales, and *Cteno-labridæ*, having a dorsal fin supported in front by spines. This family includes the WRASSES or ROCK-FISHES,—numerous small fishes of brilliant orange and blue color, arranged in stripes with wavy lines. The genus *Ctenolabrus* is represented by the NIBBER or COMMON BERGALL, *C. cæruleus*, found on the coast from New Jersey northward. On account of its prevailing color, it is also called the BLUE-FISH, BLUE PERCH, CUNNER or CONNER, and CHOGSET; the last mentioned name being derived from the Mohegan tongue. The flesh is insipid and watery. The *Tautoga Americana*, the COMMON BLACK-FISH or TAUTOG, (in the Mohegan dialect,) much valued for the table,—is found on the coast between Massachusetts and Chesapeake Bay.

SECOND ORDER. MALACOPTERYGII, (Gr, *μάλάκὸς*, *malakos*, soft; *πτερὸν*, *pteron*, wing.

The CYCLOIDS. These fishes are a step lower in organization. Soft-fins or rays distinguish them from those of the preceding order. The genera and species are less numerous, but as furnishing food for man, the order is the most important of all, including such fish as Salmon, Pike, Herring, Cod, Carp, Turbot, Halibut, &c. The order comprises all the Ground Fishes,—those which are restricted to fresh waters, and lie in wait for their prey.

SUB-ORDER. ABDOMINALES.

The fishes of this sub-order have the ventrals behind the pectoral fins, and not attached to the humeral or shoulder-bone.

---

(16) *Siluridæ*, (Gr. *silouros* from *seio*, I move; *oura*, the tail.) This is a family represented by the Cat-fish, Bull-pouts, Bull-heads, and Horned-pouts, (*Pimelodus*,)—the last name being derived from the fleshy filaments, (*cirri* or barbels,) floating from the mouth. These cirri are supposed to aid them in obtaining food, while groping in the mud. They are without scales, and covered with a slimy coat of mucus. Some South American species have large angular, bony plates, and are, therefore, said to be mailed. The Oceanic Cat-fish has only six barbels.

The Sheat-fish or Sly Silure, (see Chart,) *Silurus glanis*, (Gr. a kind of shad,) is the only species of Europe, and perhaps the largest of European fresh-water fishes,—attaining the length of ten, twelve, or even fifteen feet.

One species, *Silurus electricus*, (the *Malapterurus electricus*, of later writers,) an inhabitant of the Nile and of the rivers of Central Africa, has electric properties similar, or intermediate to those of the Torpedo and Gymnotus, though the organs are of much finer texture.

Of the American fresh-water forms the most noted are included in the genus *Pimelodus*, (Gr. *Pimelē*, fat,) distinguished by having an adipose dorsal fin. One species, found in the Mississippi, has been known to weigh one hundred pounds. The genus *Noturus*, (Gr. back-tail,) includes the Stone Cat-fish. It has its generic name from having the back fin confluent with the tail-fin.

The *Pimelodus cyclopum*, (Humboldt.) of South America, inhabits the highest regions in which fish are known to live, occurring at Quito, 16,000 feet above the level of the sea. They are found in subterranean lakes, and sometimes are ejected from the craters of the Cotapaxi and Tunguaraga volcanoes.

In this family are included the BLIND FISHES, *Amblyopsis spelæus*, (DeKay,) of the Mammoth Cave, (Kentucky,) in which the eyes are invisible, or appear in a rudimentary state, on the dissection of the fish. It is said "they are acutely sensitive to sounds, as well as to undulations produced by other causes in the water." (Silliman's Journal, second series, Vol. XVII.)

(17) *Cyprinidæ*, (Gr. *kuprinos*, a carp.) This family includes by far the greater part of fresh-water fishes, though the flesh of not very many is valuable for food. Few of them are found in tropical waters. The Carps have no teeth in the mouth, but they appear in various kinds upon the posterior branchial arch, (or pharyngeal bone.)

The species *Cyprinus Carpio*, (Lat. a carp,) is highly prized for food. It is particularly abundant in Europe, and has been naturalized in waters of the United States, especially in the Hudson River.*

The GOLD-FISH, *Cyprinus auratus*, of our parlors, so conspic-

* These fish were first successfully introduced by H. Robinson, of Newburgh. The spawn is deposited among the grass along the sides of the ponds or rivers which they inhabit. These fishes reach the size of three or four inches the first year, and sometimes become quite large, though the size varies considerably. The Breams are from five to seven inches long; the Chubsuckers from seven to twelve; the Suckers from seven to eighteen. They, together with the Dace, Sheepshead, Killi-fish, Red-fin, &c., are found in the waters of New York.

uous among fresh-water fish for the beauty and variations of their colors, are of this family. The true home of these fish is a lake in China, whence they have been taken, and introduced to other countries. When kept in globes, care should be taken not to give them more food than they can eat at a time, as the unconsumed portion, dissolving in the water, may affect their breathing. The eggs should be removed to another vessel, or else the fish will eat them.

The GUDGEONS, *C. gobio*, appear to delight in slow rivers, and swim together in shoals. They seize the bait with avidity, and hence afford excellent amusement to anglers.

The SLIMY TENCH, *Tinca vulgaris*, is common in lakes of the European continent, and sometimes found in ornamental waters and ponds, but is seldom found in rivers, being fond of still and muddy waters. It is considered a very prolific fish, and of quick growth. The Tench ranks among the most useful fresh-water fish of Europe.

The BLEAK, *Cyprinus alburnus* or *Alburnus lucidus*, is another European species, from the scales of which is chiefly obtained the silvery matter used in the preparation of artificial pearls.

The VARIEGATED or CARP BREAM, *Abramis versicolor*, is a savory fish, sometimes called the Yellow-bellied Perch and Wind-Fish; found in the Connecticut and Hudson Rivers, and in other waters. When a light breeze ruffles the water, thousands of these fish are sometimes seen darting to the surface. Near Peekskill, N. Y., it is called the Dace, from its resemblance to Dace of Europe, *C. leuciscus*.

The BARBEL, of Europe, *Barbus*, (Lat. from *barba*, a beard;) *vulgaris*,—named from the cirri or barbs attached to its mouth,—frequents the deep and still parts of rivers,—is very numerous in the Thames, Eng. Its flesh is coarse and unsavory, and held in little estimation.

Several species of Dace are found in North America. Among them are the Black-nosed Dace, *Leuciscus atronasus*, (Lat. black-nosed;) the SPAWN-EATER, *L. Hudsonius*, supposed by fishermen to live entirely on the spawn of other fishes,—first described by DeWitt Clinton, formerly governor of the State of New York; the SHINER, *L. chrysopterus*, (Gr. yellow-finned;) the SILVERY DACE, *L. argenteus*, found in Massachusetts; the PIGMY DACE, *L. pygmæus*, which is only an inch long. Other species are sometimes quite small.

The SUCKERS, *Catastomus*, (Gr. *kata*, against; *stoma*, mouth,) embrace many species known by their very fleshy lips, which can be applied to any object like a *sucker*. Different names are applied to the several species, such as Mullet, Buffalo-fish, Red-Horse, &c., &c.

The species of *Cyprinidæ* are extremely numerous in American waters, and many, no doubt, are yet to be described. But we must not omit to notice two singular species; the first, *Cyprinodon umbra*, remarkable as being one of the inhabitants of the subterranean lakes in Austria, where darkness perpetually reigns; the second, the FOUR-EYED LOACH, *Anableps tetrophthalmus*, (Gr. four-eyed,) found in the Brazilian rivers. "It is," says Mr. Edwards, in his "Voyage up the Amazon," "always seen swimming with the nose above the surface of the water, and propelling itself by sudden starts. The eye of this fish has two pupils, although but one crystalline and one vitreous humor and but one retina. It is the popular belief that, as it swims, two of its eyes are adapted to the water, and two to the air."

(18) *Esocidæ*, (Gr. *isox*, a kind of pike.) The PIKES, (*Esox*,) are the most voracious and destructive of all fresh-water fish. Their lengthened form enables them to live in shallow waters, and even when considerably large they sometimes are found in small brooks. The Trout alone can compete with these fishes, and not often are both found in the same waters. Lacepedé calls them the Sharks of our ponds and rivers. Only one species, *E. lucius*, is found in Europe; sometimes attaining a length of nineteen feet, and a weight of seventy pounds. A skeleton of one has been preserved at Manheim, which weighed three hundred and fifty pounds, and was probably between two hundred and three hundred years old. The species are numerous in the waters of this Continent. Those of Lake Erie and other Northern lakes, as the Muskalonge or Muskellunge, *E. nobilior*, *E. estor*, are very large. The more Southern species are smaller. The American species form two divisions; one of which has the opercular or gill covers entirely scaly, and dark reticulated markings; the other having scales only on the upper half of the gill covers, and marked with light spots on a dark ground.

The COMMON PICKEREL, *E. reticulatus*, abounds throughout the Eastern and Middle States, and in the waters of Ohio.

The Banded Gar-fish, *Belone truncata*, has very minute, soft scales, and the upper part of the body is of a beautiful transparent sea-green.

The *Scomberesox*, or Bill-fish, (*S. Storeri*, Mass. Report,) has a broad silvery band on the body; and is, hence, sometimes called the Silver-Gar. Both these fish have the head and snout very much elongated; the Bill-fish has the dorsal and anal fins divided into finlets, as in the Mackerel,—hence, the name *Scom-*

beresox, or *Mackerel-Pike*. The two last named genera are sometimes united in a separate family, (Scomberesocidæ.) The Flying Fishes of tropical seas, *Exocœtus*, (Gr. Ἐξωκοιτος, *exokoitos*, (sleeping out of the sea,)—were so named because believed by the ancients to sleep on the beach. They have the specific name *volitans*, (Lat. flying,) from having the pectoral fins so enlarged as to resemble wings, (Plate XIV. fig. 3.;) when in the air they move so rapidly as to resemble birds more than fish. They fly straight forward, remaining out of the water thirty seconds or more at a time. Two or three hundred of them are sometimes seen together.

(19) *Fistularidæ*, (Lat. from *fistula*, a pipe or tube.) This family includes the Pipe-mouthed and Trumpet fishes having tubular mouths, which, it is thought, they use in drawing up their food, like a syringe. The genus *Fistularia*, (the tobacco-pipe Fish,) has several species on our coast. The Trumpet-fish, *Centriscus*, besides the tubular snout, has a short compressed body, of which the head forms the larger portion.

(20) *Salmonidæ*, (Lat. *salmo*, a salmon.)

This is the TROUT family, inhabiting both fresh and salt water, and the most completely toothed of all the fishes. They agree with the Herring family in the structure of the upper jaw, and are distinguished by having a small fatty fin behind the true dorsal fin. Their flesh is unrivaled; all the members of the family are eagerly sought for by anglers, from the salmon or lake trout, the mackerel trout, the white fish of the large lakes, and the Bass of Otsego, to the small frost-fish or smelt caught in Lake Champlain, through holes in the ice, to which the fish rush in crowds to breathe the fresh air. Different causes have been assigned for the various shades of color in the flesh of Salmon. Such as live upon fresh water shrimps and other small crustaceans, are said to be the brightest; those feeding upon aquatic vegetables dull, and the darkest of all.

DeKay thinks it doubtful whether any trout feed on vegetables. Those of ponds are externally dark colored; those in clear streams with sandy bottoms, are bright; and those in salt, brackish streams are not only bright externally, but have the flesh more of the Salmon color. The most conspicuous species is the *Salmon salar*, (Lat. a kind of trout,) which is the true Salmon found on the northern shores of both Europe and America, and ascending the rivers in summer; sometimes attaining a weight of fifty pounds or more. It is not only valuable for food, but the women of the Tungooses, in Siberia, tan the skin so as to render it flexible, for the purposes of clothing.

The BROOK TROUT, *S. fontinalis*, is met with from Maine to

the southern parts of Virginia. This seldom exceeds four pounds in weight. Dr. Mitchell speaks of a Salmon of Lake Huron, which weighed one hundred and twenty pounds, but the Lake Salmon are not now often found to exceed eighty pounds. To the prolific nature of the Salmon we have already referred. These delicious fish were formerly quite abundant; indeed, it is not many years since they were, in Massachusetts, a perfect drug. We have read of a boy who was apprenticed in Newburyport, with the special condition in his indentures, that he should not be obliged to eat salmon more than three times a week.

(21) *Clupeidæ*, (Lat. *clupea*, a river fish or shad.)

The fishes of this family, including Herrings, Pilchards, Sprats, Sardines, Anchovies and Shad, are among those esteemed as most useful and indispensable. Both the maxillaries and intermaxillaries are employed to form the margin of the upper jaw. These fish are exceedingly abundant. Four hundred thousand Anchovies are said to have been taken at one haul, on the coast of Sardinia. These latter fish are preserved with salt, after removing the head and intestines. They are about the size of the little finger, and used as a condiment.

Herrings, (*Clupea harengus*,) are now supposed to live in the vicinity of the places where they are caught, approaching the shore to spawn in such numbers that the water is filled with loose scales rubbed off in the crowd. The Herring fishery along the coasts of Europe and America, gives employment in summer to many thousands of people. The consumption in Europe alone of two thousand millions of these fish, annually, does not seem to decrease their numbers. They are valuable in commerce, either pickled or smoked. To prepare the Red or Smoked Herring, the fish are sprinkled with salt, and lie about six days in heaps on a brick or stone floor. Rods are then passed through the gills, care being taken not to have them touch each other. These rods are suspended in tiers, in ovens, holding from ten to twelve thousand, where the herrings are smoked for a month with hard wood, and after being cooled, are packed for market. The Emperor Charles V, in 1556, erected a monument, and ate a herring over the grave of a fisherman of Zealand, who had improved the art of pickling herring. Several species of Herring are caught on the coasts, and in the rivers of the Atlantic States. Some idea of the extent of the herring-fishery in Maine may be obtained from the fact that at Treat's island, there were in five days, caught, salted, and stored up for smoking, what would make or pack five thousand boxes. Some kinds of Herring are used instead of guano, for enriching land.

SARDINES are a small species of Herring, much prized as a relish. From forty to fifty thousand are often taken at a single haul in the Mediterranean, the Baltic and the Atlantic. The American Shad, *Alosa præstabilis*, (Lat. excellent,) or *A. sapidissima*, (Lat. most savory,) is a beautiful and savory fish which enters our rivers between January and May, the time varying with the latitude, passing to a considerable distance from the mouths of the rivers in order to spawn. They descend the Hudson river during the latter part of May, when they are called *Back Shad*, and are lean and scarcely fit to eat. Shad are caught in large seines, and in gill-nets attached to long spears, and often set in from seven to ten fathoms of water. They are taken in large numbers in the Hudson and other rivers, and more especially in Chesapeake and Delaware Bays. In abundant seasons they are sold at from six to ten dollars per hundred, and packed away in salt. This species is of much finer flavor than the COMMON SHAD, *A. vulgaris*, of Europe.

The ALEWIVES, *A. tyrannus*, appear in great numbers in Chesapeake Bay, from March to May. In New York waters, they appear with the shad, about the first of April, but not in numbers sufficiently large to form a separate fishery. They are numerous on the coast of Massachusetts, and very good food.

The MOSSBONKER, *A. menhadan*, is valuable and largely used for manure, and in some places as bait for mackerel, cod, and halibut. This fish also has the names, *Bony-fish*, *Hard-head*, and *Menhadan*, "the last being the name given by the Manhattans." It is dry, full of bones, and without flavor, and therefore is seldom eaten.

SUB-BRACHIALS.

These are distinguished by having the ventral fins under the pectorals, and the pelvis immediately attached to the bones of the shoulder.

(22) The *Gadidæ*, or COD FISH family, have an elongated body, covered with soft scales not extending on the head. The genus *Morrhua* represents the true Cod. The best known species is the *Morrhua vulgaris*, found in the seas of Europe as far south as Gibraltar, and in those of America as far as Newfoundland; its maximum size is sixty or seventy pounds. The species commonly found through the whole year off the coast of the United States, and going into deep water in the spring, is *M. Americana*, from one to three feet in length. Occasionally it attains an immense size. Specimens are sometimes taken which

weigh seventy or eighty pounds. Dr. Storer speaks of one which reached the enormous weight of one hundred and seven pounds; a cod of fifty pounds, however, is thought to be very large. The Cod-fishery, it is well known, is extensively followed in the Eastern States, particularly Massachusetts; giving employment to a large number of persons and requiring an amount of tonnage which ranks only second to that employed in the whale fishery. Fishing vessels of all nations are found off the Banks of Newfoundland. Cod fish are taken with hooks or seines sunk to a considerable depth in the sea. The months of May and June are the season for securing them. They are preserved by simply salting them green, or they are salted and then dried. The oil, *oleum jecori*, from the liver of the cod, is quite largely used as a medicine, and considered to be highly valuable, especially in pulmonary complaints. The roe is also extensively used as bait for herrings. Other fishes of this family are the Power Cod, *M. minuta*, from four to eight inches long; the Tom Cod, or Frost-fish, *M. pruinosa*, (Lat. frosty,) a savory fish, and caught in large quantities; the Haddock, *M. æglefinus*, nearly as common in our market as the Cod, but inferior in size and as an article of food; the Whiting, *Merlangus vulgaris*, an European species; (the name Whiting is also applied to the American species, *M. albidus*, Lat. whitish;) the Burbot, *Lota vulgaris*, is considerably esteemed; the Cusk, *Brosmius vulgaris*, (Storer;) the Hake or Codling, *Phycis Americanus*. These are all equally palatable with the Common Cod; the Coal-fish, *M. carbonarius*, (Lat. from *carbo*, a coal,) ranging on both shores of the Atlantic; *M. purpureus*, (Lat. purple-colored,) abundant on the shores of New England and sometimes on that of New York, and known under the name of *Pollack*.

(23) *Planidæ*, or *Pleuronectidæ*, (Gr. *pleuronectes*, side-swimmer,) the Flat-fish Family.

These, from their want of symmetry, really stand alone among the Vertebrates. The eyes are both on one side of the head, usually one above the other, and often varying in size. The upper surface of these fishes resembles the ground in which they lie in wait for their prey; the under surface, from being never exposed to the action of light, is white. The upper and white surface are really to be regarded as the *two sides*, right and left, so that instead of being depressed, it is compressed, or flattened vertically, like the Chaetodons, though the latter, like other fishes, swim with the back uppermost, notwithstanding their thinness; but the Turbot swims or grovels along the bottom upon its side, the colored side, right or left, being uppermost. To this the term Pleuronectes refers. DeKay designates the FLAT-FISH, having the eyes and colored surface on the right, as dextral species, and the FLOUNDERS, which have the eyes and colored surface on the left, as sinistral species. Of the latter is the OBLONG FLOUNDER, *Platessa oblonga*, (Plate XIV. fig. 5,) found on our coast, and from fifteen to twenty inches in length. The Turbot, *Rhombus maximus*, is considered the best of European fishes. The SPOTTED or WATERY TURBOT, *Pleuronectes maculatus*, or *Rhombus aquosus*, (Storer,) is found on our coast and sometimes called the English Turbot, but is distinguished from that fish by the absence of the numerous tubercles on the colored side, which characterize the latter.

The Halibut, (*Hippoglossus vulgaris*,) has a longer body and sharper teeth than others of the family. Sometimes it reaches a great size. Dr. Storer speaks of one that weighed six hundred pounds, though a Halibut weighing two hundred pounds is considered large. The fins are regarded by epicures as a very choice part of this fish. There are several species of Flat-fish, *Platessa*, most of which are prized for food. The FLOOK, *P. flesus*, and the DAB, *P. limanda*, are European species. The COMMON SOLE, *Achirus mollis*, is found abundantly on our Atlantic coast.

(24) *Cyclopteridæ*, (Gr. circular or cup-shaped fins.) The LUMP-FISHES, or LUMP-SUCKERS.

These are a small family, having the ventral fins so united as to form a sort of cup-shaped disk, with a funnel-shaped cavity in the center, by which they adhere firmly to any solid object. The body is rough, being covered with very bony tubercles. They are called SUCKERS on account of a curious sort of sucking disk, by means of which they adhere to the rocks of the bottom, or to any other substance. The skeleton is so soft that some members of the family are said to dissolve after death into a mucilaginous jelly, in which hardly any trace of bone remains. These fish are represented by three genera: *Lepidogaster*, *Lumpus*, and *Liparis*, the two latter having American species. The LUMP-SUCKER, *Lumpus anglorum*, or *Cyclopterus cæruleus*, is called in Scotland, the COCK-PADDLE. In England it has the name of SEA OWL, as well as Lump-fish and Lump-sucker. Its appearance is remarkably grotesque. The ventral unite with the pectoral fins, and form a single disk. Some of the family have two disks, one formed by the pectorals, the other by the ventrals; hence these fish have been called *Discoboli*, (Gr. throwers of the DISCUS, or quoit.) They are now sometimes included with the Blennies. Pennant says that one of the Lump-fishes thrown into a pail of water, adhered so firmly to the bottom that the pail was lifted by taking hold of the tail of the fish.

(25) *Echeneidæ*, (Gr. *echo*, to hold; *nēus*, a ship.)

This family is represented by the genus *Echeneis*, the name referring to the flattened disk of cartilaginous plates, covering the top of the head, and enabling the fish to attach itself to other bodies.

The COMMON SUCKER, *E. remora*, (Lat. delay,) is found throughout the Atlantic Ocean. It has sometimes been taken from the bottom of vessels in the harbor of New York. One species, the WHITE-TAILED REMORA, *E. albicauda*, (Lat. white-tail,) is called the SHARK-SUCKER, from being frequently found attached to that fish.

SUB-ORDER APODES, (Gr. footless.)

These are without ventral fins.

(26) *Anguillidæ*, (Lat. an eel,) or *Muraenidæ*, (Gr. *muraina*, a kind of fish.)

This is the Eel family, which have long, snake-like bodies and small scales so imbedded in the soft, slimy skin, as to be scarcely perceptible. The dorsal, anal, and caudal fins are united, and the rays so delicate as to be with difficulty enumerated. They have been estimated to be as many as three hundred and twenty, or three hundred and forty. During the season of its activity the eel is a voracious feeder. Conger or Sea Eels, *Anguilla conger*, (Lat. sea-eel,) or *Conger occidentalis*, (Lat. western,) are larger than the Common Eels. Yarrell, in his British Fishes, says that "specimens of Conger Eels weighing eighty-six pounds, one hundred and four pounds, and even one hundred and thirty pounds, have been recorded, some of them measuring more than ten feet long and eighteen inches in circumference." (See fig. on Chart.)

"The ancient Romans reared these fish with great care, in consecrated ponds, and they even decorated them with jewels. Six thousand were served up at one entertainment given to Cæsar when he entered upon his dictatorship." The branchial pouches of Eels enable them to crawl and remain some time out of water, and thus they can move from one place to another in search of food, being hardly inferior to any other fish in the power of enduring abstinence from their native element. They are strongly susceptible of magnetic or galvanic influence. Their eggs are so diminutive as to escape observation, which may have given rise to the notion that these fish are viviparous.

The ELECTRIC EEL, *Gymnotus* (Gr. *gumnos*, naked ; *nōtos*, back,) *electricus*, has no tail fin, and the scales are imperceptible. It is sometimes five or six feet long. By its electric shocks, it knocks down men and horses, and by repeating its discharges is able to kill them. It can be obtained only after its electric power has been exhausted by successive shocks. The Indians of South America drive wild horses into the muddy ponds in which these Eels abound, in order to secure them. Two specimens, taken in the waters of the Amazon, have been sent to Professor Henry, of the Smithsonian Institute. The Gymnotus, (see Chart,) and the Torpedo are able either to emit or withhold this electric power.

(N. B. The Electric Eels are sometimes separated from the Common Eels, and formed into the family *Gymnotidæ*.)

SUB-ORDER LOPHOBRANCHII, OR LOPHOBRANCHIA. (Gr. tuft-gills.)

The fishes of this sub-order are characterized by having the gills in small tufts along the branchial arches, instead of being comb-like. In this and the following sub-order the internal skeleton is but partly ossified.

---

(27) *Syngnathidæ*, (Gr. *sùn*, together ; *gnathon*, jaw.) PIPE-FISHES.

These fishes have the body covered with angular, bony plates, so arranged that the body itself is many sided. The gill-covers are large, but soldered down for the greatest part of their edge, leaving only a small orifice for the discharge of the water which has been respired. The male pipe-fish, *Syngnathus*, has a pouch or pocket in which he receives the eggs as they are laid. In this he also carries the young for some time. Some species are without pouches, but have indentations on the abdomen where the eggs are placed.

The SEA-HORSE, *Hippocampus*, (Gr. a sea-horse,) has eyes which move independently, and is the only fish known to have a prehensile tail. It is found in the Hudson river from five to six inches in length. When dried this fish curls up and in form resembles a horse. (See fig. on Chart.)

The SHORT-NOSED SEA-HORSE, *H. brevirostris*, is found on the coasts of Great Britain. It is about five inches long ; sometimes it is found coiled up in oyster shells.

SUB-ORDER PLECTOGNATHI. (Gr. plaited or twisted jaws.)

This sub-order is distinguished by the interior union of some of the bones of the head.

(28) *Gymnodontidæ*, (Gr. naked-teeth.) BALLOON and GLOBE-FISHES.

These can scarcely be said to have real teeth; but the jaws are covered with enamel so divided into plates as to answer the purpose of teeth.

In the SEA PORCUPINE, *Diodon*, (Gr. two teeth,) each jaw has a single piece; hence the generic name. The form of this fish is somewhat cubical; it has the singular property of puffing itself up into a globular ball, (Plate XIV. fig. 7,) and in this shape floating on the surface. The length varies in different species, from two to seven inches. In the Puffer or Balloon-fish, *Tetraodon*, (Gr. four teeth,) the suture in the middle of each jaw gives it the appearance of four teeth. Like the Diodon it can inflate and contract itself at pleasure. When it inflates itself the formidable spines with which the body is covered, become erected. Its flesh is unwholesome if not poisonous. The Puffer can bite severely, and can emit water in self-defence; its spines are also an effectual guard, but the most curious thing about it is that when handled, it emits a beautiful red excretion, which stains ivory and paper a permanent carmine red.

The small GLOBE-FISH, *Acanthosoma* (Gr. spiny body) *carinatum*, (Lat. ridged,) is armed with spines and susceptible of inflation, (Plate XIV. fig. 6.) It is quite small, being only one inch in length. The color is of olive brown above, silvery beneath. The GLOBE FISH, *T. lævigatus*, (Lat. smoothed or polished,) is from one to two feet in length. The Common Puffer, *T. turgidus*, (Lat. swollen,) is from six to twelve inches long. A species of electrical Globe-fish, *T. lineatus*, is found in the Nile. The SUN FISH, or MOON FISH, *Orthagoriscus*, (Gr. a sucking pig,) appears as if the fins were set in or near the head, and the tail abruptly cut off, so that its aspect is most singular. The Sun-fishes are without spines, and have not the power of inflation. A species found on the coast of France weighs over three hundred pounds.

Balistidæ, (Gr. from *balista*, a military engine resembling a stringed bow.) FILE-FISHES.

These are fishes of a less grotesque appearance than the Diodons and Tetraodons, found most largely in the still waters of tropical seas. The body is compressed and has a lengthened conical or pyramidal snout, ending in a small mouth having distinct teeth in both jaws. The skin is roughened with scaly granulations or prickles; in the typical forms there are two dorsal fins; in others, the front dorsal fin is sometimes represented by a single spine. The ventral fins are often wanting or else indis-

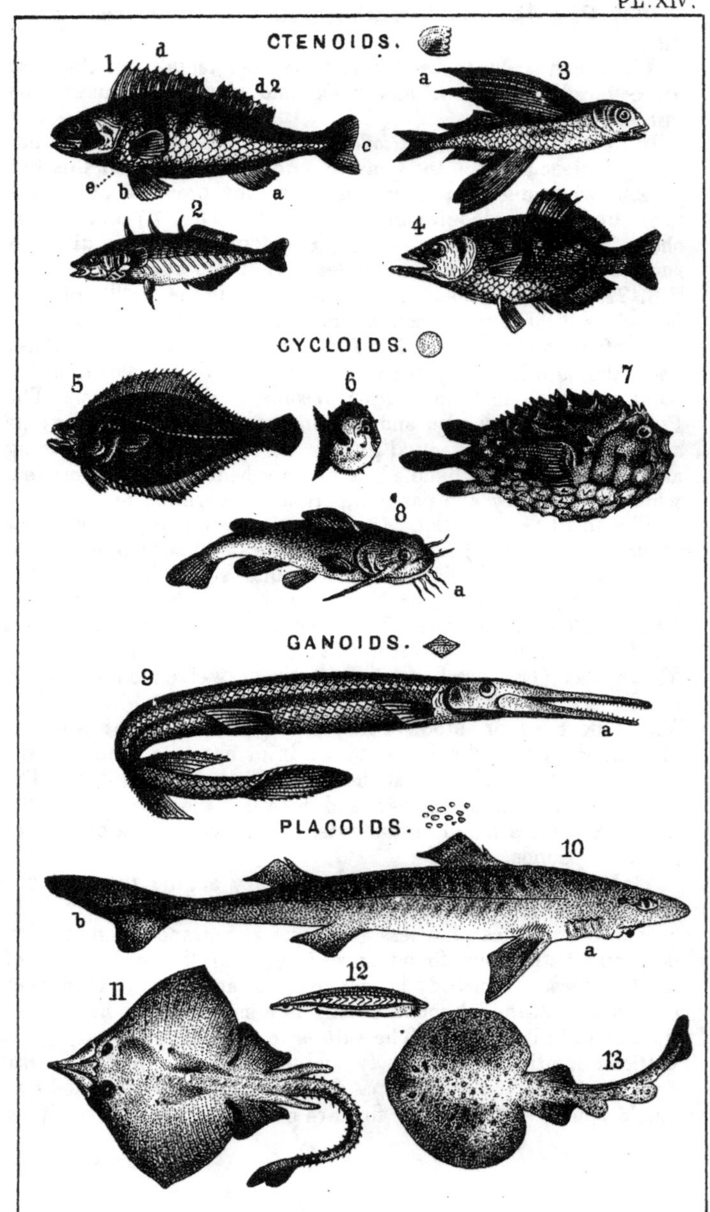
CTENOIDS.
1
d
d2
c
e
b
a
3
a
2
4
CYCLOIDS.
5
6
7
8
a
GANOIDS.
9
a
PLACOIDS.
10
b
a
11
12
13

EXPLANATION OF PLATE XIV.

CTENOIDS, (Toothed scales.)

Fig. 1. Perch; a, anal fin; b, ventral fin; c, caudal fin; d, dorsal fin; d2, second dorsal fin; e, pectoral fin.
Fig. 2. Three-spined Stickleback.
Fig. 3. Flying-fish; a, long wing-like pectoral fins.
Fig. 4. Archer-fish, with a mouth fitted for shooting insects.

CYCLOIDS, (Round scales.)

Fig. 5. Plaice, showing the eyes both placed on one side, as is usual in the Flat-fishes.
Fig. 6. Short Head-fish.
Fig. 7. Porcupine or Balloon-fish.
Fig. 8. Bull-pout, or Cat-fish; a, barbels or cirri.

GANOIDS, (Enameled scales.)

Fig. 9. Buffalo Bony Pike, or Gar fish; a, long narrow jaws, covered on the inside with rasp-like teeth; a row borders the edge, of bony pointed ones.

PLACOIDS, (Flat-scales.)

Fig. 10. Spiny Dog-fish; a, branchial openings on each side, in the place of gills for breathing; b, the heterocercal tail, as in all Sharks and Sturgeons; the back bone runs to a point above the tail, which is placed below like a triangular rudder.
Fig. 11. Ray; body flattened out like a dish; tail long and slender, with the dorsal fins upon it; pectoral fins large, uniting with the snout in front.
Fig. 12. Amphioxus, or Lancelet, the lowest form of Vertebrates.
Fig. 13. Torpedo, or Numb-fish.

tinct. One of the dorsal fins is fronted with a strong bony spine. The bones or rays of this fin are so contrived as to act in concert for suddenly elevating it at the pleasure of the fish. It is a singular fact that if the foremost or largest ray be pressed ever so hard, it will not stir; and yet if the last or least ray be pressed very slightly, the other two immediately fall down with it, just as a cross-bow is let off by pulling down the trigger. To this peculiarity there is a reference in the name of the typical genus *Balistes*. The fishes of this genus are covered with large and hard rhomboidal scales. The DUSKY BALISTES, *B. fuliginosus*, (Lat. dusky,) twelve inches long, is found off the coast of the United States. The MASSACHUSETTS FILE-FISH, *Monocanthus*, (Gr. a single spine,) has very small scales, and a single large spine in place of the first dorsal fin. Its length is from three to five inches. The LONG-TAILED UNICORN-FISH, *Aluteres cuspicauda*, (Lat. pointed or spear-tailed,) has the skin covered with small and almost invisible granules. Length from six to nine inches.

(30) *Ostracionidæ*, (Gr. *ostrakion*, a shell or covering.) TRUNK FISHES.

These are a group of singular fishes, found principally in the American and Indian seas. They are enveloped in a bony crest or covering, so united as to form an inflexible shield, leaving only the tail, fins, mouth, and a small part of the gill-openings capable of motion, passing through openings in the Armadillo-like shield. There are no ventral fins, and but one dorsal. These fish have little flesh, but a large liver, abounding in oil. The surface, in some species, is armed with spines. There is but one North American genus, *Lactophrys*. The species *L. camelinus*, (camel-like,) has the back elevated into a spine, and is three and a half inches in length.

GANOIDS.

These are characterised by having the scales bony, and covered, externally, with enamel, generally angular and continuous. Most of them are extinct species.

(31) *Sauridæ*. This name has been employed by Agassiz to designate the fishes of this group, which also comprises the *Polypterus*, (Gr. *polus*, much or many; *pteron*, fin,) of the Nile. This latter (fresh-water) fish is usually about eighteen inches in length, and partakes both of the osseous and cartilaginous kinds; but is thought by some to be "most nearly allied to those species of the genus *Esox*, which are furnished with large, long, and bony scales." Its color is sea-green. It is called by the Egyp-

tians, *Bichir*, and is said to be one the best of the Nilotic fishes for the table. The back has a long row of finlets. (See Chart.)

The Alligator-Gar, *Lepidosteus*, (Gr. *lepis*, scale; *osteon*, bone,) is confined to North America. The scales are smooth and of adamantine hardness; the upper jaws consist of many pieces. (Plate XIV. fig. 9.)

The Buffalo Bony Pike, *L. bison*, is sometimes three feet in length.

The Flat-nosed Bony Pike, *L. platyrhynchus*, (Gr. broad-snout,) is two feet in length, and found in Florida and the Western rivers.

Placoids. (Gr. πλάξ, *plax*, a plate or tablet.) Plate-like scales.

Cartilaginous Fishes. Chondropterygii, (Gr. cartilage-winged.)

The skeleton in these fishes is not entirely destitute of calcareous matter, but this is arranged in separate grains, and does not form fibres or plates. The gelatinous substance, which in other fishes, fills the intervals of the vertebræ, and communicates from one to other by a small hole, forms in several genera of this division, a continuous cord, which perforates them all.

First Order. Eleutheropomi, (Gr. ελεύθεϱος, *eleutheros*, free; πωμα,, *pōma*, cover.)

The fishes of this order have pectinated or comb-like gills, which are free, as in ordinary fishes, with one large external opening on each side, furnished with a strong operculum or cover; they are without rays; the upper jaw, formed by the palatial bone, is firmly united to the maxillary; the intermaxillary bone is rudimentary.

(32) *Chimæridæ*, (Gr. *chimaira*, fabulous monster,) Sea-Monsters.

These are so called from the fantastic shape of the head, which has a singular hoe-shaped appendage, tipped with spines upon the snout. The second dorsal fin extends to the tip of the tail, which is drawn out into a long slender filament. The eggs are large, coriaceous, and have flattened hairy margins; these are esteemed by the Norwegians, who use them mixed up with their pastry. The only species of the genus *Chimæra*, viz: *C. monstrosa*, is abundant in the Arctic seas. (See Chart.)

(33) *Sturionidæ*. The Sturgeons.

These fishes have the body covered by hard bony tubercles or plates. The mouth, situated beneath the head, is small and toothless; it is placed on a sort of foot of three joints, by means of which it can be protruded and retracted at pleasure. On its under surface, as in most cartilaginous fishes, are several cirri, beard or worm-like appendages, which hang down in front. It is so much like India-rubber, that boys put pieces of it in their balls, to make them bound. The body is long and tapering, ending in a tail unequally forked, the upper lobe being considerably the longer. Sturgeons live on small fishes and worms. They grow to a great size, many of them measuring more than twenty feet long, and some weighing more than two thousand pounds. The roe is remarkable for its quantity of eggs, containing sometimes one hundred and fifty millions, and weighing one-fourth of the whole fish. It in fact constitutes its chief value, as from it *caviar*,—so much prized, is furnished. For preparing it, the roes, taken out and placed in tubs, are cleansed with water; the fibrous parts, by which the eggs are connected, being removed, the spawn is rinsed in white wine or vinegar, and spread to dry. It is then put into a vessel and salted, being crushed down at the same time with the hands, and afterwards inclosed in linen bags to drain off the moisture. Lastly, it is packed in tubs, pierced in the bottom, that any remaining moisture may yet drain off, and closed down for domestic use or exportation. Sometimes it is said to be preserved, after having been salted and seasoned, by being rolled up into large balls, and immersed in vessels of oil; or the rolls are inclosed in wax, so that the air may be more effectually excluded. (Gosse.) The flesh of the Sturgeon is another article of considerable commerce. It is smoked or broiled in slices, and pickled, and in this form exported. So fat and unpalatable, (as some regard it,) it was deemed by the ancient Romans one of the most sumptuous dishes; and at all "great dinner-parties, this fish was always carried by servants decked with garlands and flowers, and attended by a band of musicians. On the Hudson River, it is called "Albany Beef," from its frequent exposure in the markets of that city. The swimming bladder of the Sturgeon is also profitable. If cut open and washed, and its silvery glutinous skin be exposed for some hours to the heat of the sun, and separated from the external skin, it furnishes the best isinglass, the value and uses of which are well known. They migrate during the early summer months, deposit their spawn, and return again to the sea. Those of North America are almost fresh-water fish.

The chief species are the COMMON STURGEON, *Acipenser*, (Lat. a sturgeon,) *sturio*, of the seas and rivers of Europe; the Br-

LUGA OF ISINGLASS STURGEON, *A. huso*, of the Caspian Sea, probably the largest species, sometimes weighing, it is said, three thousand pounds, and from which the caviar of commerce is made in great quantities; the STERLET, *A. ruthenus*, of the Mediterranean, Black and Caspian seas, which is said to yield caviar of a very superior sort; the LAKE STURGEON, *A. rubicundus*, (Lat. ruddy,) four feet long, of a yellowish red on the back, and olivaceous red on the sides, found in Lakes Ontario and Erie, and the upper lakes; the Sharp-nosed Sturgeon, *A. oxyrhincus*, (Gr. sharp-nosed,) seven feet long, and found in the rivers of the United States. A species of the genus *Scaphirhyncus*, (Gr. boat-nosed,) viz: the SHOVEL-FISH, *S. platyrhynchus*, (Gr. broad-nosed,) is found in the Mississippi river. Of the genus *Polyodon*, (Gr. many toothed;) is the Spoon-bill, *P. folium*, (Lat. a leaf,) (previously referred to,) and also an inhabitant of the Mississippi. This fish has an enormous gill-cover, with a large branchial aperture, nearly like that of the generality of fishes; and it is also furnished with an air-bladder: hence, though placed next to the Sharks, Swainson appears to doubt the propriety of such a position of it. It has a snout greatly extended, much dilated, and, together with the head, nearly as long as the body; the tail is highly heterocercal, and the skin entirely naked.

SECOND ORDER. PLAGIOSTOMA. (Gr. πλάγῖος, *plagios*, transverse; στόμα, *stoma*, mouth.) Gills not free.

The fishes of this order have a cartilaginous cranium, in which the parts are not separately discernible. The cartilaginous, teeth-bearing jaws are attached to the skull, also by cartilages. The gills are fixed by their external edges, with five small external openings on each side. The face is prolonged in front; and on its under side is situated the broad transverse mouth; the ventrals and pectorals, soft and fleshy, like the other fins, are always present; the pectorals, in the male, having long appendages on their internal margins. The covering of these fishes consists of shagreen, or of plates variously modified. The swimming-bladder is wanting; the teeth are placed on the roof of the mouth and the lower jaw. The order includes two families, *Squalidæ*, (Sharks,) and *Raiidæ*, (Rays.)

(34) *Squalidæ*, (Lat. *squalus*, a kind of sea-fish.) The SHARKS. This dreaded family of fishes are distinguished by having the branchial openings lateral, the eye-lids free, the pectoral fins distinct from the head, the body slender, and somewhat spindle-

shaped. The mouth is generally placed far beneath the end of the nose; and the upper part of the tail is longer than the lower. These fish are generally of a large size, sometimes almost gigantic. They are carnivorous, and very voracious. Some of them are universally dreaded on account of their ferocity, their appetite for human flesh, their strength, and the formidable array of teeth with which their mouth is furnished. These are triangular, finely serrated, and exceedingly sharp, lying quite flat in the mouth; but when seizing their prey, are raised by the action of muscles by which they are joined to the jaw. To this, and the singular method in which these formidable creatures are continued, we referred, however, in the general description of the Fishes. The most useful part of these fishes is the liver, from which oil is obtained; a Shark twenty feet in length, yielding about two barrels. The rough skin is used for polishing ivory and wood, and for making thongs, &c., for carriages; converted into shagreen, it serves for covering small cases and boxes. The flesh is not eatable, being coarse, and of a disagreeable flavor.

The WHITE SHARK, *Carcharias*, (Gr. marine-dog,) *vulgaris* or *Squalus Carcharias*, found in tropical seas, has been known to cut a man's body in twain at a single snap; and it is stated that human bodies have been found entire in the stomachs of these terrible monsters. It is suggested, that their insatiate voracity may result from the great quantity of gastric juice with which they are supplied, causing them to digest with great rapidity, and from the tape and other worms which abound in their intestines. Their sense of smell is acute, so that they discover their victims at a distance; and they follow in the wake of ships for the purpose of devouring whatever may be thrown or fall from them into the sea. The White Shark is said to measure, sometimes, thirty feet in length, and to exceed one thousand pounds in weight.

The THRESHER SHARK, *Carcharias vulpes*, has the upper part of the tail nearly as long as the body, or even longer. The tail is its principal organ of defence; it literally threshes its enemies. Sometimes it is called the Fox-Shark, and the SWINGLE-TAIL. This species, which is from twelve to fifteen feet in length, is found on the coasts of North America as far North as Nova Scotia.

The SMALL BLUE SHARK, *C. obscurus*, from two to six feet in length, is frequently taken on our coast.

The MACKEREL PORBEAGLE OR MACKEREL SHARK, *Lamna*, (Gr. a plate,) *punctata*, has a pyramidal snout, with the nostrils under the base, and the gill-openings all in front of the pectorals; there are no pectoral orifices. Its general color is a dark slate. The oil of its liver is much esteemed by curriers. The length is from four to ten feet. The surface under the lens, exhibits numerous minute plates; to this its generic name has reference.

The HOUND-FISH, *Mustelus canis*, has blunt teeth, forming a closely compacted pavement in each jaw, with temporal orifices. The lower lobe of the tail-fin is short. Its length is from two to four feet.

The BASKING SHARK, *Selachus maximus*, has the gill-openings all before the pectorals, long, and nearly surrounding the neck; it has no air-holes behind the eyes; the teeth are small, of various forms, but generally conical. This species is over thirty feet in length. It is said its liver will yield eight barrels of oil. It has the popular name of *Basking Shark*, from its habit of continuing for some time in one place. It is sluggish, inactive, and less fierce than the other species, and inhabits the Northern seas, but is occasionally seen off our coast.

The SMALL SPOTTED DOG-FISH, *Scyllium canicula*, has a prominent and slightly pointed jaw, with the nostrils pierced near the mouth, and a cylindrical shaped body. It keeps near the bottom of the water, and feeds on fish and small crustaceans. This, and the larger Dog-fish, *Scyllium catulus*, are found on the British and French coasts. The larger species is three or four feet in length, and does much damage to the fisheries on account of its voracious habits. In Scotland, these fish are said to form no inconsiderable part of the food of the poor. The species *S. catulus*, is sometimes called the Rock-Shark. This, or a similar species, is found on the coast of the United States.

The SPINY DOG-FISH, *Spinax acanthias*, (Storer,) is easily recognized by the spiracles or air-holes which are placed, one on each side of the temple, just behind the eye. It has a sharp, strong spine in front of each of the two dorsal fins. (See Plate XIV. fig. 10.) Its teeth are in several rows, small and cutting. The color is slate; the length from one to three feet. This species is very numerous about Cape Cod, where they are much sought for the oil which they furnish. Of the immense numbers of them found in tropical seas, some idea may be formed from the fact, that in the single harbor of Kingston, (Jamaica,) from one hundred to one hundred and fifty thousand are destroyed annually. Twenty thousand, it is said, have been taken in a seine at one time.

The ANGEL-FISH, *Squatina angelus*, seems, in its form, to unite together the Sharks and the Rays. Swainson includes it with the Rays, remarking that this and the species *S. Dumerili*, found on the American coast, have the two dorsals and the caudal fin in shape and situation the same as what is seen in the Torpedoes. The length of the American species is from three to four feet. It is said to have acquired the name of *Angel-fish* from its extended pectoral fins having the appearance of wings; and it is called MONK-FISH, because its rounded head appears as if enveloped in a hood.

The SAW-FISH, *Pristis*, (Gr. *pristis*, a saw,) has the body flattened in front, with the gill-openings beneath, as in the Rays; but they are chiefly distinguished by a very long snout, which is in form like the blade of a two-edged sword, and armed on each side with pointed bony spines. This saw-like weapon, the fish often buries in the flesh of the whale and other marine animals. The Saw-fish is sometimes included with the Rays.

(35) *Raiidæ*, the RAYS.

These are a family of fishes which have the body flattened as in the Saw-fish, and the pectorals greatly enlarged, as in the Angel-fish, both which, in their structure, seem to approach the present group. In the Rays, the pectorals are very broad and continuous with the head, sometimes stretching out in front of it in the form of lobes, so that these fishes present an appearance disk-like, or more or less rhomboidal, the snout forming one corner, and the projecting tail another; the other two corners being the angles of the pectoral fins; the ventrals, in the males, have appendages like those of the Sharks; the dorsal fins, two, sometimes three in number, are small, and placed far back on the slender tail. The eyes are on the upper surface, as are also the temporal spiracles; the mouth, the nostrils, and the gill-openings are placed in the under surface, and thus concealed from view. The mouth is small and set with numerous teeth, which are placed in close array, like paving stones. As in many of the Sharks, the eyes have a nictitating membrane or skin which can be drawn over the eye at pleasure, and serves as an eye-lid. The young of the Rays are enveloped at birth in capsules of a thin horny or leathery substance to which filaments are attached. The prolongations of the angles of the envelope give it some resemblance in shape to a hand-barrow. But the most distinguishing peculiarity of the Rays is their barbs or prickles, varying in length, according to the size of the fish, by which they are able to tear the flesh and inflict severe wounds. These fishes are strictly ground feeders, groveling along on the soft muddy bottom, and moving with a peculiar undulating action of the pectoral fins.

Some of the species of the tropical seas grow to a great size and are proportionally ferocious.

The RAYS PROPER, *Raia*, include several species, some of which are found on our coast, such as the CLEAR-NOSED RAY, *R. Diaphanes*, (Gr. clear,) from one to three feet long, caught with cod-fish, and sometimes eaten; the PRICKLY RAY, *R. Americana*, from one to two feet long, (Plate XIV. fig. 11;) the SPOTTED RAY, *R. ocellata*. When captured, this species whips its tail about with great activity, and hence has the name of *Whip Ray;* the HEDGE-HOG RAY, *R. erinaceus*, length about eighteen inches; the PRICKLY-STING RAY, *Pastinaca*, (Lat. sting-ray,) *hastata*, (Lat. from *hasta*, a spear,) having two or more spines or barbs in the tail, which is longer than the body; the whole length is from five to eight feet; this species is numbered among the edible rays; the SMOOTH SKATE, *Raia lævis*, in length from two to four feet; the THORNBACK, *R. Clavata*, (Lat. knotted or thorned,) has large and numerous spinous tubercles.

The EAGLE RAYS, *Cephaloptera*, (Gr. head wings,) often grow to an enormous size, specimens having been seen twenty-five feet in length and thirty in breadth. One was taken at Barbadoes a few years ago, which weighed thirty-five hundred pounds and required seven pair of oxen to draw it on shore! (Kirby.) The Eagle Rays are nearly or quite as dangerous to man as the Sharks. They are known to fishermen under the name of "Devil Fish." The species *C. vampirus*, the OCEANIC VAMPIRE, is from sixteen to eighteen feet in length. It is very powerful, sometimes seizing the cables of small vessels at anchor, and drawing the vessel for several miles, with great velocity. Passing by some other divisions, we must refer to the

ELECTRIC RAYS, or TORPEDOES, *Torpedinidæ*, fishes which have long been celebrated for their electrical powers, while their shape is so singular that they look more like gigantic tadpoles than fish, (Plate XIV. fig. 13.) The head is entirely surrounded by the pectoral fins, which give to it in some species, a completely circular appearance; the tail is thick, fleshy, and only moderately long, terminated by a distinct, large, and triangular fin. The electrical organs constitute a pair of galvanic batteries, arranged in the form of perpendicular hexagonal columns, placed on each side of the head and gills, the small cells being filled with mucus. These fishes are less powerfully electrical than the *Gymnotus*, but can benumb the arm of a person touching one of them.

THIRD ORDER. CYCLOSTOMI, (Gr. κύκλος, *kuklos*, a circle; στόμα, *stoma*, a mouth.)

The fishes of this order have already been referred to as having sac or purse-shaped gills. These are fixed and open outwards by several apertures. The mouth consists of a circular fleshy lip, with a cartilaginous ring supporting it; this peculiarity gives name to the order.

(36) *Petromyzonidæ*, (Gr. stone-suckers.) These have lengthened, cylindrical, eel or worm-shaped bodies, destitute, both of pectoral and ventral fins, but having foldings of skin above and below, serving the purpose of dorsal, caudal and anal fins, though without any supporting rays.

The SEA LAMPREY, *Petromyzon Americanus*. In making its furrow, preparatory to spawning, it uses its sucker-like mouth; with it separately removing stones of large size, and thus quickly constructing a large furrow. It is from two to three feet in length.

The MUD LAMPREYS, *Ammocœtes*, (Gr. sand-bedded,) include several species which differ from the true Lampreys chiefly in the form of the mouth, which is not suctorial, but composed by the projecting upper lip, the lower one being transverse.

These fish are found in large numbers, in sand or mud flats. They are from three to four inches in length, varying in thickness "from that of an earth worm to that of a swan's quill." They are dug up from a depth of four or five inches below the level of the water, and used as bait for fishes.

The MYXINOIDS or Glutinous Hags, *Myxynoidei*, approach the lowest form of the Vertebrates, and by Linnæus and other writers, are classed with the Worms. These curious animals are eel-shaped, and measure, when full grown, about a foot and a half. The head is scarcely distinguishable from the body, and is obliquely truncated in front, ending in a large round mouth, the frame-work of which is a membranous, maxillary ring, furnished above with a single tooth. The tongue has in each end two rows of strong teeth. The Hag has no eyes; the branchial openings are two in number; the skin is covered with slime, furnished from a row of pores on each side of the belly. An obscure fin runs along the hinder portion of the back; and is continued round the compressed tail. The color is of a dark bluish-brown above, and whitish beneath.

The HAG or BORER, *M. glutinosa*, is found in the northern seas of Europe. It does mischief by entering the mouths of fishes caught in the lines of the fishermen and eating up all the fleshy parts of their bodies, leaving only the skin and bones. The name Borer is given to it, because, as is said, it pierces a small aperture, and thus makes its way into the body of Cod, or other fishes which it attacks. (See general account of Fishes.)

FOURTH ORDER. BRANCHIOSTOMA, (Gr. *branchia*, gills; *stoma*, mouth.)

The term here used to designate an order, is sometimes employed as *generic*, and the name *Amphioxidæ*, (37,) given to the family of which it constitutes the sole genus. Müller, (see his classification,) ranks it in the sub-order PHARYNGOBRANCHII, (Gr. throat-gills,) a name referring to the position of the branchial sac. This extraordinary animal, at an early period thought to be a mollusk, was first discovered on the British coast, during the latter part of the last century. Müller is no doubt correct in saying, "it is evidently a vertebrated animal and a fish," though it has more the aspect of a worm than a fish. Yarrel in his "History of British Fishes," calls it the LANCELET, *Amphioxus lanceolatus*, (Plate XIV. fig. 12.) It has a naked skin and no fin except the dorsal, which extends over the entire length of the back. The mouth is entirely inferior, elongated or circular, the margins having a row of filaments.

The vertebræ are reduced to a single, cartilaginous column or thread, flexible, transparent, and scarcely to be distinguished from the horny pen enveloped in the flesh of some of the Cuttle-fish. There is no trace of a brain, and the heart "presents entirely the form and distribution of blood vessels and extends over wide spaces," characters of themselves sufficient to distinguish the *Branchiostoma* from all other fishes. The blood is white. Müller considers it connected with the Cyclostomatous fishes through its dorsal chord and the absence of jaws; but as inferior to them in not having a distinct brain and in the peculiarities of its respiratory system. The Lancelet is only about two inches in length, and lives in sandy ground at a depth of between ten and twenty fathoms of water. It probably buries itself in the sand. Other curious particulars could be given relating to this lowest of the Vertebrates, did our limits permit, but here we close our account of the Fishes and of the sub-kingdom VERTEBRATES, to which they belong.

It is difficult to state with accuracy the number of species belonging respectively to the several families of Fishes, as new researches, made from time to time, vary the assigned numbers. The following tabular view is given, however, as an approximation to a true account:

1	**Percidæ,**	**600 species.**	19	**Fistularidæ,**	**20 species.**
2	**Triglidæ,**	260 "	20	**Salmonidæ,**	132 "
3	Scienidæ,	250 "	21	Clupidæ,	180 '
4	Sparidæ,	240 "	22	Gadidæ,	110 "
5	Maenidæ,	61 "	23	Pleuronectidæ,	150 "
6	Chaetodontidæ,	150 "	24	Cyclopteridæ,	40 "
7	Anabassidæ,	1 "	25	Echeneidæ,	20 "
8	Scomberidæ,	400 "	26	Anguillidæ,	175 "
9	Cepolidæ,	34 "	27	Syngnathidæ,	100 "
10	Teuthidæ,	80 "	28	Gymnodontidæ.	100 "
11	Atherinidæ,	50 "	29	Balistidæ,	110 "
12	Mugilidæ,	80 "	30	Ostracionidæ,	30 "
13	Gobidæ,	400 "	31	Sauridæ,	
14	Lophidæ,	50 "	32	Chimæridæ,	
15	Labridæ,	500 "	33	Sturionidæ,	24 "
16	Siluridæ,	400 "	34	Squalidæ,	114 "
17	Cyprinidæ,	723 "	35	Raiidæ,	130 "
18	Esocidæ,	120 "	36	Amphioxidæ,	

What is the 1st ORDER of OSSEOUS OR BONY FISHES? What other name has it? How is it characterized? What is the number of the PERCH Family? Are they marine or fresh water fish? Which is the typical species? What other sp. closely resemble it? What is said of the Striped Bass? Which are among the most remarkable fishes of this group? What is said of them. In what respects are all the perches alike? How are the GURNARDS characterized? Whence is the name derived? What did Cuv. call them? What sp. are mentioned? What two remarkable fish are found in this family? State what is said of them. What is the 3d FAMILY? What is said of them? What of the 4th and 5th Families? Why are the CHAETODONS so called? What does Cuv. name them? Why? What is peculiar in the Archers? What gen. includes the CLIMBING PERCHES? What is said of them? What is the next Family? What is said of their numbers? What is said of the Tunny, Bonito, and Sword-fish? Name the other sp. What is said of the Bottle-headed Dolphin? Which is the 9th FAMILY? Why are they called Ribbon-fishes? What fact shows the name to be appropriate? What sp. is found on our coast? What is the next Family? What fishes do they resemble? What is said of the Doctor-fish? What of the Surgeon-fish? What is the 11th FAMILY? State what is said of it. Name the next Family. Repeat what is said of them. In what respect are the Blennies and Gobies alike? What other characters are given? State what is said of the sp. referred to. How are the LOPHIDÆ distinguished? What singular looking fishes does this family include? What is said of the Antennarius? Why are the Labridæ so named? What two sections do they include? What is said of the Wrasses or Rock-fishes? What sp. are mentioned?

Name the 2d ORDER of FISHES. What is said of their organization? How distinguished from the preceding order? Why is it esteemed important? What SUB-ORDER is first mentioned? How characterized? What is the 16th FAMILY? What fishes represent it? What is said of the Horn-pouts? Of the Sheat-fish or Sly-Silure? What is peculiar to the *S. electricus?* What is said of the gen. *pimelodus?* What large sp. is referred to? What remarkable S. American sp. is mentioned? What is said of the

BLIND-FISHES? What is the 17th FAMILY? How numerous are they? Are they found in Tropical waters? What is said of their teeth? Where is the sp. *Cyprinus carpio* particularly abundant? Who first introduced it into American waters? What is said of the Gold-fish, the Gudgeon, and the Slimy Tench? What fish furnishes the silvery matter for artificial pearls? Name other fishes of this family. What remarkable sp. are found in Austria and Brazil? How are the PIKES characterized? What does Lacepede call them? What is the only European sp. and what is said of it? Are the American sp. numerous? How are they divided? What sp. are mentioned? What is said of the BILL-FISH? What of the FLYING-FISH or *Exocœtus*? What family includes the Pipe and Trumpet fishes? What is said of them? Which is the 20th Family? How are they characterized? How regarded by anglers? What is said of their color? Which is the most conspicuous sp.? What is said of the BROOK TROUT? What of their numbers in former times? What is the 21st FAMILY? How are they esteemed? What is said of their abundance? What of the HERRING fishery? How are the Red or Smoked Herring prepared? For what are the H. used besides food? What is said of SARDINES? What of SHAD, &c.?

How are the SUB-BRACHIALS distinguished? Which is the 22nd FAMILY? Which is the best known sp.? Which is common on the U. S. coast? What is said of it? What other sp. are mentioned? Which is the next FAMILY? In what respect do they stand alone among vertebrates? Describe them. How does DeKay designate the FLAT-FISH? How the FLOUNDER? What sp. of Turbot are mentioned? What is said of the HALIBUT? What is the 24th FAMILY? How are they characterized? Why called SUCKERS? What genera represent them? What is said of the Lump-sucker? What does Pennant state? What is the next Family? What gen. represents it? To what does the name refer? What sp. are mentioned? What is said of them? What is the next SUB-ORDER? Name the 26th Family? How are they characterized? What is said of the number of fin rays? Which sp. are the larger? What does Yarrell say? What enables these fish to remain out of water? What is said of their eggs? Repeat what is said of the Electric Eel? What is the next SUB-ORDER? How is it characterized? What is the 27th FAMILY? Give their general characters? What is remarkable in the male PIPE-FISH? What in the SEA-HORSE? What is the next SUB-ORDER? How distinguished? What is the 28th FAMILY? Have they real teeth? What have they instead? What is said of the SEA PORCUPINE? What of the Puffer, or Balloon-fish? What of the small Globe-fish? What other Globe-fishes are named? What is said of the Sun-fish? Mention the next Family? What is said of it? What is said of the Trunk-fishes? How are the GANOIDS characterized? What is the 31st Family? What is said of the Polypterus of the Nile? What other sp. of this family are mentioned?

What is said of the PLACOIDS or CARTILAGINOUS FISHES? Name the 1st ORDER? How is it characterized? What is the 32d Family? What is said of it? What is the next Family? What is said of their covering? What of their mouth and body? Of their size? In what respect is the roe remarkable? What valued food does it furnish? How is it prepared?

From what part of the Sturgeon is isinglass obtained, and how is it prepared? What sp. of Sturgeon are mentioned and what is said of them? What is the 2nd ORDER of CARTILAGINOUS FISHES? How is it characterized? What is the 34th FAMILY? What is said of their size, voracity, &c.? What is said of the White Shark? Of the Thresher S.? What other sp. are mentioned? What is said of the Spiny Dog-fish and the Spotted Dog-fish? Of the Angel-fish and Saw-fish? What is the 35th FAMILY? Give their characters? What sp. of RAYS PROPER are mentioned? What is said of the EAGLE RAYS? Of the Electric Ray? What is the 3d ORDER of CARTILAGINOUS FISHES? What peculiarity gives name to the order? What is the 36th FAMILY? What is said of this FAMILY? What of the Sea Lamprey? What of the Mud Lamprey? Of the Myxinoids or Glutinous Hags? What is the 4th ORDER of CARTILAGINOUS FISHES? How is the term sometimes employed? What Family does it include? How does Müller rank it? What is it called by Yarrell? Describe this fish.

Name and trace from the Chart some of the largest fish, giving a sketch of each from the book. Which is the largest of the Mackerel Family? Which next? Which is the largest of the Flat-fishes? In what do they differ from all other fish? Which of the Placoids is the largest? How do the caudal fins of this order differ from those of other orders? What families present the most singular forms? Which are nearly round? Which long and slender? Which three or four sided? Describe the Sun-fish and Chimæra from the Chart. Which fish crawl upon land or climb trees, and in what orders are they found

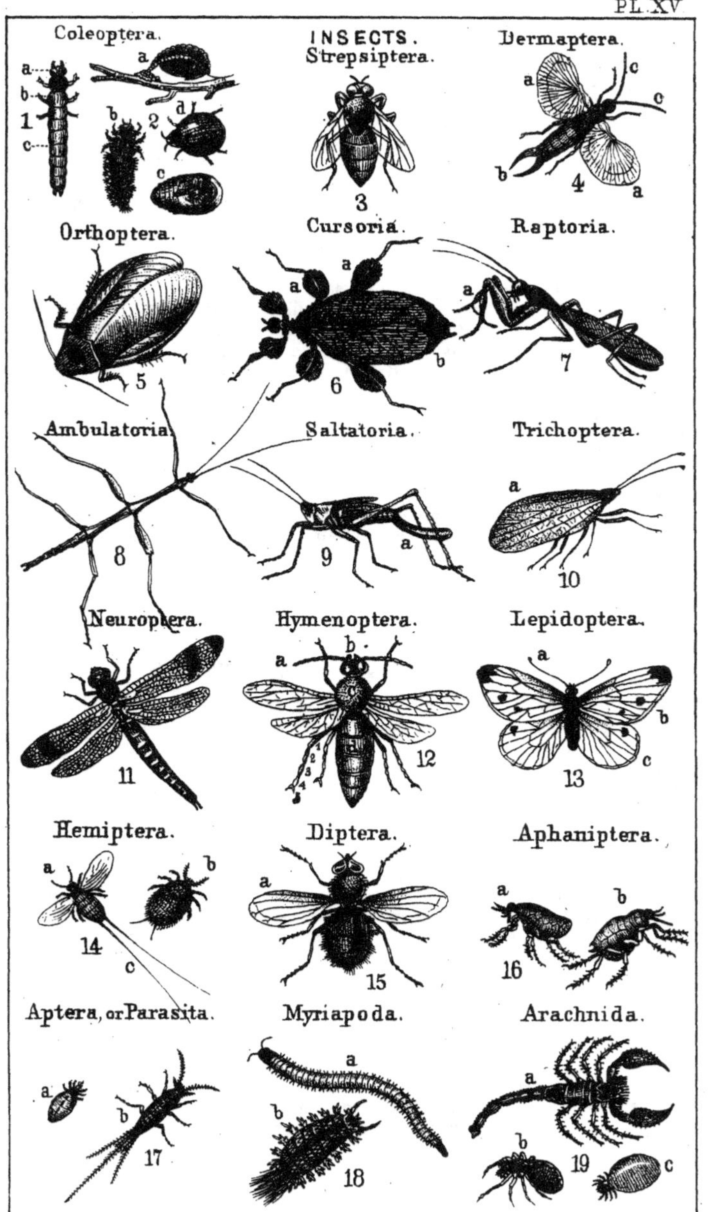
INSECTS.
Coleoptera.
Strepsiptera.
Dermaptera.
Orthoptera.
Cursoria.
Raptoria.
Ambulatoria.
Saltatoria.
Trichoptera.
Neuroptera.
Hymenoptera.
Lepidoptera.
Hemiptera.
Diptera.
Aphaniptera.
Aptera, or Parasita.
Myriapoda.
Arachnida.

EXPLANATION OF PLATE XV.

INSECTS WITH BITING JAWS.

Fig. 1. Larva of a beetle, usually consisting of thirteen segments; a, the head; b, the three segments of the thorax, to which the legs and wings are attached; c, the nine segments of the abdomen.

Fig. 2. The Lady-bird or Lady-bug, *Coccinella septempunctata;* a, pupa with the outer case, b, larva; c, pupa; d, complete insect.

Fig. 3. Wasp-fly, *Stylops.*

Fig. 4. Earwig, *Forficula;* a, the large posterior wings; b, anal forceps; c, the antennæ with fourteen joints.

Figures 5, 6, 7, 8, 9, belong to the four sections of the Order Orthoptera.

Fig. 5. Cockroach, *Blatta orientalis.*

Fig. 6. Walking-leaf Insect, *Phyllium siccifolium;* a, foliaceous expansions upon the feet; b, the true wings, far exceeding in size the wing covers.

Fig. 7. Praying Mantis, *Mantis religiosa*, named from the attitude they assume while waiting to grasp their prey with their raptorial feet, a.

Fig. 8. Walking-stick, Walking-beetle or Spectre, *Spectrum femoratum*, without wings or wing covers, can scarcely be distinguished from the branch on which they rest.

Fig. 9. Katydid, *Platyphyllum concavum;* a, the curved ovipositor, about one-fourth of an inch long.

Fig. 10. Caddis-fly, *Phryganea;* a, front wings, fibrous with branching nerves; the hind wings are largest, but folded when at rest.

Fig. 11. Dragon-fly or Darning-needle, *Libellula;* the compound eyes, nearly cover the entire head, and contain about 12,000 lenses.

Fig. 12. Hornet, *Vespa crabro;* a, antennæ; b, head; c, thorax; d, abdomen; 1, the coxa or hip joined to the body; 2, trochanter, or second joint of the leg; 3, femur or thigh; 4, tibia or shank; 5, tarsus, composed of five or less joints, and terminated by two hooked claws.

WITH MOUTHS FOR SUCKING, PUMPING, OR PIERCING.

Fig. 13. Cabbage Butterfly, *Pontia brassica;* a, the knob at the end of the antennæ, distinguishing butterflies from moths or millers, whose antennæ are feathery or saw-like; b, anterior wings; c, posterior wings.

Fig. 14. Cochineal Insect, *Coccus cacti;* a, the male, with red body, white wings and two setæ or bristles c, at the apex of the abdomen; b, the female, without wings, having shorter antennæ, and a beak of which the male is destitute.

Fig. 15. Blow or Blue-bottle Fly, *Musca vomitoria;* a, the only pair of wings, which are transparent and without scales or dust. There are no wing covers, as the lower wings are reduced to two small knobbed threads, called halterers or poisers.

Fig. 16a. Flea, *Pulex irritans;* b, Jigger or Chigoe, *P. penetrans*, feet long, bristly, and adapted for leaping.

Fig. 17a. Louse, *Pediculus;* b, *Lepisma.*

Fig. 18a. Thousand-legged Worm or Millipede, *Iulus terrestris*, has usually two pair of feet to each segment, each foot ending in a claw. The feet in different sp. vary from 12 to 300 pair. b, Brush-tailed Centipede.

Fig. 19a. Scorpion, *Scorpio afer;* the extended tail-like abdomen ending with the sting; b, Spider, *Arachnida;* no distinct head, eyes from two to eight, neither wings, antennæ, or upper lip; c, Tick *Ixodes.*

SECOND SUB-KINGDOM.

ARTICULATES. (Lat. *articulus*, a joint.)

SECTION I.

In leaving the Vertebrated Animals, we, in the descending scale, first come to the class of ARTICULATES, (*articulata.*) These rank first among the INVERTEBRATED Group, or those animals that are destitute of a back bone. They are so named because the different parts of their body are composed of movable pieces *articulated* or jointed to each other. They deviate from the Molluscous animals in generally possessing a skeleton; but the skeleton, unlike that of the Vertebrates, is exterior instead of interior, being composed of a series of rings, protecting the internal parts, and serving as points of attachment for muscles. Though exhibiting considerable diversity of character among themselves, the Articulates are usually provided with a skin, which is either soft, as in the leech and earth-worm, or horny and crustaceous, as in the crab and craw-fish. Some families are destitute of feet, but the greater portion possess these members. When limbs are present, they are never fewer than six. Articulated animals have the trunk of the body, for the most part, long, cylindrical, and divided transversely into segments. In the lowest of the series, where there are no appendages for locomotion, and all the movements are effected by the body itself, as in the common worm,—the segments appear to be perfectly simple, but, as ascending in the scale, we observe that gradually, the segments develop lateral organs, which are of kinds quite various, according to the character of the animal.

In many of the *Annelidans*, and in the *Myriopoda* or Centipede tribe, especially, the articulated character of these animals is conspicuous, the segments being numerous, and all of nearly equal size, and each possessing a short pair of legs, which are themselves also jointed. In the Crustaceans and the *Arachnida* or Spiders,—the divisions are reduced to eight or ten in number; and in the Insects, to six. Where the design is to lighten the body, there the segments are reduced in number and size, as in the insects and the crabs; in the Annelidans, as the Earth-worms and Lug-worms, we find the number of segments increased.

The animals of this class are active; hence, their skeletons are light and thin. The muscles or organs of motion are attached to the interior of the skeleton; but as this is hard and unyielding, it is necessary that it undergo a process of exuviation, which occurs in all the Articulate animals, going on through all the stages of their existence. Phosphate of Lime, which enters into the bones of the vertebrate classes, constitutes also the material out of which the skeletons of a majority of these animals are formed. Considering their size, it may be doubted whether any other animals possess so large an amount of muscular power as the Articulates. The bulk of their bodies is really made up, in great part, by the muscles which move them. Throughout the animal kingdom, the muscular power corresponds with the amount of respiratory action, and the development of animal heat; in various forms of the Articulates, this law is remarkably displayed.

The strongest resemblance to each other, exhibited by these animals, exists in the nervous system. The brain is extremely small. Two nervous cords, surrounding the æsophagus or gullet, run along the centre of the lower surface of the animal; these cords are studded, at regular intervals, with knots or *ganglia*, forming so many centers from which the nerves pass off to the different segments.

The head also has its ganglia, in which the cords terminate anteriorly. In cases where the members are not distributed along the entire body, but limited to one part, as in Insects, Arachnidans, and the higher Crustaceans, there is a corresponding concentration of the ganglia in that particular part; indicating by its degree, the elevation of the animal in the series.

The organs of sense are very imperfectly developed, and in some instances, entirely wanting, excepting that of sight.

No organ of smell has been discovered, unless it be assigned to the antennæ. Some naturalists have described organs of hearing in the insects, while others regard the antennæ as instruments for the exercise of that sense, and also of feeling.

The digestive apparatus is, for the most part, in accordance with the carnivorous habits of the Articulates. Where animal food is eaten, the process of digestion is less complicated than where vegetable food is used.

The lengthened form of these animals impresses its character upon their digestive, and also upon their circulating apparatus.

In most of the Articulates, the blood moves forward in one or more large dorsal arterial vessels, from which side branches are given off, terminating in various trunks that convey the blood backwards to the dorsal vessel. The blood is more highly organized, has a deeper color, and contains a larger quantity of corpuscles and fibrin than in either the Radiates or Mollusks.

Respiration is accomplished by organs which, in all cases, are perfectly symmetrical in those of this class which, like the Crustaceans, habitually live in water by means of branchiæ or gills; in others, by means of *tracheæ* or air-tubes, which receive air by certain lateral openings, called *Stigmata*, (Gr. dots or marks.) In rare instances, there exist cellular cavities, analogous to lungs.

The ARTICULATES may be arranged into the following classes: I. INSECTA, Insects; II. MYRIAPODA, Thousand-legged Worms, &c.; III. ARACHNIDA, Spiders, &c.; IV. CRUSTACEA, Crabs and Lobsters; V. CIRRHOPODA, Barnacles; VI. ANNELIDA, Annelidans or Worms. (Plates XV. and XVI.)

FIFTH PART. ENTOMOLOGY. (Gr. ἔντομα, *entoma*, an insect; λογος, *logos*, a discourse.)

First class of ARTICULATES. *Insecta*, (Lat. from *inseco*, to cut into.) INSECTS.

The name given to this class refers to the divided structure of the body or trunk of the animals which it includes. This is generally composed of thirteen (sometimes fourteen) sections, of which one forms the head; three the intermediate thorax, and nine the abdomen. The head of a perfect insect has usually three pair of jointed appendages. The first pair are called *antennæ* or feelers. They are affixed to the sides of the head for the most part, between the eyes and the mouth, and have from one to sixty joints or articulations.

All true insects have six jointed or articulated legs, attached to the thorax, and, usually, two or four wings, situated upon its three rings. (Plate XV. fig. 12.) The abdomen, which is furnished with many rings, contains the digestive organs. The breathing process is accomplished by means of spiracles or pores, on the side of each ring, for admitting the air, which is thus made to permeate the whole body.

Insects have a circulating apparatus, of which the central organ, corresponding to the heart or the aorta in the higher orders of animals, is a vessel or tube running beneath the skin of the back, from which the white and cold blood is distributed in tissued channels or veins. The corpuscles have forms like those which are found in animals of the superior grades. The nervous system consists of a symmetrical arrangement of nervous threads in two lines, situated on the face of the abdomen, and connected by knots or *ganglia*, at every ring of the body.

The mouth of insects, although made up of the same essential parts, has these modified into two principal forms of structure, one of which is adapted to chew, and the other to suck food. The former are named mandibulate, (from Lat. *mando*, to chew;) the latter, haustellate, (from Lat. *haustellum*, a sucker.) In the order Hymenoptera, however, biting mandibles are united with sucking jaws. (Plate XV. fig. 12b.)

But the most striking peculiarities of insects relate to the changes or metamorphoses which they undergo during their stages of growth, corresponding, in some degree, with the developments made in other animals, yet differing from them in being stationary at certain periods. (Plate XV. fig. 2.)

By far the largest part are oviparous. The eggs are generally oval, but they are seen in other forms,—sometimes round and sometimes cylindrical. Some are smooth and shining; others are beautifully sculptured. They vary as to color, but white and green predominate.

The Flesh-Fly, *Musca carnaria*, is ovoviviparous, the eggs being hatched within the body.

The *larva* state of insects commences when the egg is converted into a footless worm, resembling the higher Entozoa, or the inferior Annelidans, in its organization, and continues until the wings begin to appear. The term *larva*, (a *mask*,) was originally adopted by Linnæus, who regarded insects, while under this form, as *masked*. It is applicable to the young of all insects. In the Scaly-winged Insects or Butterflies, (*Lepidoptera*,) and most of the Sheath-winged Insects or Beetles, (*Coleoptera*,) the larva, at the time of its escape from the egg, has the rudiments of three pair of legs upon the thorax,—though these are little more than simple claws, except in the Carnivorous Beetles. The soft, white larvæ of the Beetles are called *Grubs;* those of certain Flies or Two-winged Insects, (*Diptera*,) are called *Maggots;* those of Butterflies, Moths, and Millers are termed *Caterpillars*. The young of the HEMIPTERA, including Bugs, *Cicadæ*, Plant-Lice, &c.; and of the ORTHOPTERA, including the True Locusts, Crickets, Cockroaches, &c., do not emerge from the shell until they have a close resemblance to the parents in every thing, excepting wings; and they can hardly be regarded as having the characteristics of real *larvæ*.

In the larval state, insects eat most voraciously,—indeed, their entire energy seems to center in the eating process. Their growth is great, and often rapid. The comparative weight of that remarkable insect, the Great Moth, *Cossus ligniperda,* to that of the young one that has just crept out of the egg is, as 72,000 to 1,—an increase of seventy-two thousand times! This insect occupies three years in attaining to its perfect state. The Maggots of Flesh-Flies are said to increase in weight two hundred times in twenty-four hours. Caterpillars, in the same time, consume three times their weight of food.

Larvæ are subject to moultings, or changes of the skin; the number varying with the species. This moulting is most strikingly exhibited in the Silk-worm, *Bombyx mori,* which casts its outer skin five times in a month. While undergoing this process, the larva does not eat, but it absorbs the fat beneath the outer skin, which favors casting it off.

The larval state is the one in which insects continue the longest, varying, however, in duration, from hours to months and years.

The Caterpillars of several Butterflies and Moths, live in large societies, in habitations or tents, sometimes of a pyramidal form, and which are constructed by their united skill.

When the worm has fixed itself in some suitable and secure retreat, the *pupa* is formed, and encased in the last skin, which, in two winged insects, becomes more rigid; or else a new and beautiful case is made,—a robe of silk, impervious to water, being laboriously woven from a single thread, which is formed and spun from the juices of the body,—impressively illustrating the instinctive power of the insect as related to its successive developments.

The name of the third state, *pupa,* (child or doll,) refers to the swathed appearance of most insects during its continuance, it resembling, in miniature, a child trussed up in swaddling clothes.

This state has two modifications. (1) That of those which, in general form, resemble their larvæ; (2) That of those which are entirely unlike their larvæ. Of the first kind are the Hemiptera, &c.,—which have the pupa somewhat incomplete, and possess rudimental wings; also those which have an incomplete pupa, and are also without wings, as Lice, *Pediculus,* which, and the Spring-tails, *Podura,* together with some other wingless insects, undergo no metamorphosis, coming forth from the egg almost in the condition in which they remain all their lives.

Of the second kind, are those which undergo a complete metamorphosis. These include those in which no trace of a future insect can be perceived, as in the Fly, *Musca*, and others of the dipterous or two-winged insects; those in which the thorax and abdomen are distinct, and enclosed in a horny case, as in the Butterflies; and those in which the parts are covered by a membrane, but distinct, as in the order Hymenoptera, and some of the Two-winged Insects.

In the Coleoptera, Lepidoptera, Hymenoptera, Diptera, and some of the Neuroptera, the pupa state is one of complete inactivity as to all manifestations of animal life, while yet the interior formative processes are carried on with extraordinary energy. In the egg, the development in the case of these insects, was only carried far enough to enable the larvæ to come forth, and to obtain their own food. In the pupa state, it is continued at the expense of the nutriment which they had collected and stored up within their bodies, so that the passage into the pupa state might almost be compared to a second entering into the egg.

Of those which are not, like the Silk-worm, protected by a cocoon, some suspend themselves by their hind extremity; others, as the Butterfly, *Papilio*, attach themselves, with the head above, and a thread around the body to keep it in its position. Some of these hanging pupæ exhibit bright colors and golden spots, whence the name *Chrysalis*, (from Gr. *chrusos*, gold.)

In the Ant-tribes, the Neuters do not acquire wings. Some of them, which are two or three times as large as the rest, and somewhat differently formed, are named "soldiers," it being their special office to defend the nest, rather than to nurture young; and in the White Ants, *Termites*, the "soldiers" appear to be pupæ arrested in their development, while the "workers" have the characters of permanent larvæ.

The period of inactivity in the pupa state, greatly varies in duration; some insects remaining inactive for years or months, while others pass through that state in a few days or hours, and reach the fourth or last stage, when the insect is called *Imago*, (Lat. image.) Now, having laid aside its *mask*, and cast off its *swaddling* bands, it becomes a proper *image* or representative of its species. Whenever an insect is spoken of without the restricting terms *larva* and *pupa*, it should be remembered, the *imago* state is meant. In this state, the three principal parts of head, thorax, and abdomen, are distinctly perceptible; the insect now eats much less food than when in its first state. Some, indeed, live so short a time as to need no food, as the Silk-worm and the

Gad-flies. The May-fly or Day-fly, (*Ephemera,*) commits its eggs to the water, and dies in a few hours, though including its larva and pupa states, it had previously lived two or three years. The Butterfly needs only a little honey; the Fly daintily sips its food, while the larvæ of both eat most voraciously.

Some insects are able to endure abstinence from food for a long time. The Ant-lion, *Myrmeleon,* (Gr. *murmēx,* ant; *leon,* lion,) can remain for six months, uninjured, without food, though daily devouring an insect of its own size when it can be obtained; Beetles have been known to live two or three years without food of any kind. Most insects feed themselves, but the young of those which live in societies, and continue longer than most others in the adult state, as the Bees, Wasps, Ants, &c., are fed by the older ones, which also store up food for future use.

Most insects are extremely prolific. They are, as we have seen, produced from eggs laid by the female; though there is one remarkable exception to this rule in the *Aphis* or Plant-lice, (order Hemiptera,) which increases by a process of gemmation or budding, somewhat after the manner of the Polypi,—females being thrown off at once for several generations, of which each has the power to multiply its kind in the same way,—even to the seventh or ninth generation; when eggs are again laid, and the gemmating or budding process is again renewed. According to calculations based upon observation, the whole brood in a season from a single Aphis, will amount to the immense number of 1,000,000,000,000,000,000! but the insect is extremely feeble; "the touch destroys it; the winds, rains, and cold, sweep off its numbers by hundreds of thousands." (Emmons.)

The Queen-Bee, *Apis mellifica,* (Lat. honey-bee,) lays fifty thousand eggs; the female White-Ant, *Termes bellicosa,* has an abdomen fifteen hundred or two thousand times as large as the rest of the body, and lays eighty thousand eggs in twenty-four hours, and forty or fifty millions in a year.

Insects usually deposit their eggs where the young larvæ may find appropriate food. Thus, the Silk-worm places hers on the leaves of the Mulberry, *Morus multicaulis,* (Lat. many-stalked mulberry.) The Hessian-fly, *Cecidomya destructor,* deposits its eggs upon the young leaf of the wheat, where it joins the stem or straw (culm) near the earth; while the Wheat-fly, *C. tritici,* places hers in the wheat-head; the Gad or Horse-Fly, *Oestrus,* (Gr. *oistros,* a gad-fly,) *equi,* (Lat. of a horse,) deposits hers in hundreds upon the hairs of the horse. Ichneumon-Flies, *Ichneumonidæ,* deposit theirs in or upon the bodies of Caterpillars and other larvæ, by means of a sharp and strong abdominal tube or ovipositor (egg-placer) of great length. The larva of the *Pimpla lunator*, according to Prof. Emmons, sometimes, in company with the *Sirex*, deposits its eggs in young maple trees, introducing the ovipositor into the wood, sometimes to the depth of three inches.

There is indeed scarcely any organized substance upon which insects are not adapted to prey. Growing vegetables and living animals are alike subject to their attacks,—these, when dead, also supply with food many kinds of insects; and even when such substances are decomposed or much decayed, they furnish nutriment to particular species. Hence, though sustaining much damage by the injury which the insects do to plants and trees, man also derives important benefit from them, by their removal of putrid substances, the noxious exhalations of which would poison the air, and thus detract greatly from his health and comfort. They are frequently useful to plants in bringing the pollen to the pistils, and thus effecting the continuance of the species in cases where it could not be done except by extraneous methods. Large Grasshoppers are in the Levant, dried and consumed for food; some savage nations eat the large grubs which are found in rotten wood. The Great-Moth, *Cossus*, which the ancients esteemed as a delicacy, was a larva of some kind; and a species kindred to this one is at this day eaten in Brazil. Ants are also eaten by the natives in that country. While attending to these uses of insects, we may also refer to that which is made of the *Cantharides* or Blistering Flies,—to that beautiful dyeing material, cochineal, furnished by insects of the genus *Coccus*,—to the galls formed on oak trees by the genus *Cynips*, and which are employed in the arts; to the art of Caprification or causing figs to ripen by suspending upon the trees branches of the wild Fig-tree, (*Caprificus*,) which is infested by an insect that pierces the fruit and hastens its maturity; and to *manna*, used as an agreeable food in the East, which, though not directly produced by insects, is made to flow from the *Tamarisk mannifera*, (manna bearing,) by the puncture of a small species of *Coccus*. The destruction of the larvæ of some insects by those of others, is in some instances, actually enormous, so that the undue multiplication of insects, which might result from the very great number of their eggs, and from their rapid growth, is counteracted not only by the influence of the many beasts, birds, reptiles and fishes, which feed upon them, but also by the numerous onsets which insects make upon each other. In these, they sometimes show considerable contrivance, availing themselves of traps, excavated in the sand, by which they secure their prey, as in the case of the Ant-lion, an insect, in its perfect state, resembling the

Dragon-fly. If by any means any poor, unwary insect, found in the neighborhood of the Ant-lion larva, seems likely to escape, jets of sand arrest his progress, and carry him to the bottom of the pit-fall, where he is instantly seized; his juices sucked out, and the body jerked out of the den, which, if injured, is soon repaired, and ready for another victim. A plan quite similar to this is also adopted by the larva of a Fly, (*Leptis vermileo.*)

The locomotive powers of insects are unsurpassed by those of any other animals. These are peculiarly conspicuous in the Dragon-flies, Termites, Bees and Ants. Even the Swallow is unable, in this respect, to match the Dragon-fly or Darning-needle, which can elude its pursuer by flying backwards and forwards, right and left, without turning its body. Its twenty-four thousand eyes guard it against surprise, by enabling it to see in all directions. The wings of Musquitoes are said to vibrate three thousand times a minute.

The organs of sense in Insects have a high degree of development. This is more particularly true of the sight. Of the two kinds of eyes found in adult insects, compound and simple, the latter, termed *ocelli*, (eyelets,) and *stemmata*, (stems,) are alone present in the larvæ, though these are sometimes entirely without visual organs. In perfect insects, the eyes are compound, that is they consist of many eyes, each of which is perfect in itself, having the proper humors and lenses necessary for the exercise of vision. In addition to the compound eye, which often fills up the largest part of the head, Insects sometimes have simple eyes upon the forehead, generally three in number, set in the form of a triangle, which are suited to view only such objects as are near. The compound eye is immovable, round, oval, or kidney-shaped, and examined under the microscope, appears reticulated, this appearance being occasioned by the hexagonal lines which bound each eye or lens. The number of lenses, each fitted for vision in its own sphere, is almost incredibly great. The number in the Dragon-Fly has already been mentioned; that of the common Fly is 4,000; of the Butterfly, from 6,000 to 30,000; and of the Mordella Beetle, 25,000; while that of the ant has but fifty lenses.

There seems no doubt that insects have the sense of hearing, for though the precise organ which is subservient to it has not been fully ascertained, there is abundant evidence that they are guided and influenced by sounds, one of the most striking instances of which is that the male of some Insects, such as *Cicadæ*, Crickets, &c., emit peculiar sounds, which attract the females to them. A nocturnal butterfly, *Acherontia atropos*, produces a plaintive cry which is said to proceed from the head. These sounds are produced entirely by mechanical means, and cannot be considered vocal.

It is thought by some naturalists that the organ of hearing is situated in the base of the antennæ. These are supposed to be also the chief organs of touch. (Plate XV. fig. 4c.)

Insects seem to possess the sense of smell. The Flesh-fly deposits eggs in the thick fleshy petals of the Carrion-flower, (*Stapelia*,) deceived by its odor, which resembles tainted meat.

Many insects, particularly the Coleoptera, which include the Snap-Bug, *Elater*, and the Fire-fly, *Lampyris*, are luminous at night. Several North American species of the *Sphinx*, or Hawk Moth, seem to be phosphorescent, by dim candle-light, or when shaded from direct light. When the light is extinguished, nothing appears, however, excepting a peculiar reflection.

Insects are essentially terrestrial, but many, as the Whirligigs or Water-Fleas, (*Gyrinus*,) swim on fresh water; and some, as the Skippers, (*Hydrometridæ*,) walk with the body raised above it, the tips of their feet touching the surface; and a genus *Halobates*, (Gr. *hals*, the sea; *baino*, to go,) is seen in the Southern Atlantic, far out from the land.

The Insects are divided by Kirby and Spence, into twelve orders, as presented on the Chart, viz.:

1. Coleoptera.	7. Hymenoptera.
2. Strepsiptera.	8. Lepidoptera.
3. Dermaptera.	9. Hemiptera.
4. Orthoptera.	10. Diptera.
5. Tricoptera.	11. Aphaniptera.
6. Neuroptera.	12. Aptera.

First Order. Coleoptera, (Gr. κολεὸς, *koleos*, a sheath; πτερὸν, *pteron*, a wing.) Beetles. Mouth mandibulate.

These insects are almost incredibly numerous, between seventy and eighty thousand species being found in the cabinets of collectors. In the Royal Museum at Berlin, Prussia, is a single collection of forty thousand species. The singular forms and brilliant colors of many of these insects, the size of their bodies, the solid texture of their integuments, which renders preservation comparatively easy, and the nature of their habits, which affords every facility for their capture, have combined to render Beetles objects of peculiar attention and interest to entomologists.

The upper wings of these Insects are horny or leathery, and shield or sheathe the lower ones; the metamorphosis is perfect, the pupa being torpid; the mouth is mandibulate or chewing. The wing-cases are called *elytra*, (Gr. coverings,) and are unsuited for flight. Many of these Insects, particularly in the larva state, are quite injurious to vegetation; but at the same time, they are, as a whole, very useful in diminishing the numbers of other noxious or destructive insects, and in removing fungous and offensive matters.

(1) The TIGER BEETLES, *Cicindelidæ*, (from gen. *Cicindēla*, Lat. a glow-worm,) so called on account of their fierceness and voracity, are found in sandy localities and dusty roads. They feed upon other insects, are good runners, and fly with facility. Those of the genus *Cicindēla* are the most numerous.

2. GROUND BEETLES, *Carabidæ*, (from gen. *Carăbus*, a crab, i. e., crab-like,) are those which are commonly found under stones and rubbish, and generally, but not always, nocturnal. They are predaceous, feeding upon insects and larvæ. The colors are black, with blue and purple hues.

The CATERPILLAR HUNTERS, *Calosoma*, (Gr. beautiful body,) include species having colors in which green and blue predominate. They are found in trees, and lessen the number of injurious insects which infest them.

3. The DIVING BEETLES, *Dyticidæ*, (from gen. *Dyticus*, a diver,) are large hardy insects, sometimes seen in water bordered with ice. They feed upon minute fish, larvæ and worms.

4. The LADY BIRDS, or Lady Bugs, *Coccinellidæ*, (gen. *Coccinella*, from Gr. *kokkos*, a berry, i. e., berry-like,) are well known, small, hemispherical insects, having bright colors and often marked with spots. (Plate XV. fig. 1d.) They feast on gourd-like plants, such as melons and pumpkins, but are of great service, both in their larva and perfect state, in destroying the Plant-lice. The larva is of a long oval shape, with a pointed tail; of a black color, with red and white specks, and a rough surface, (Plate XV. fig. 2b.) It changes to a short, blackish, oval chrysalis, or pupa spotted with red, (a, c,) and which gives birth to its beautiful inmate in May or June, (d.) The eggs of these insects may be seen upon the under surface of leaves, in a cluster of thirty or forty, placed in contact and gummed by one end to the leaf. These hatch within a few days.

5. The WATER-LOVERS, *Hydrophilidæ*, (gen. *Hydrophilus*, Gr. water-lover,) are found in ponds and ditches, or in stagnant waters, which they seem to prefer.

6. CARRION BEETLES, *Silphidæ*, (from Gr. *silphē*, a cockroach, i. e., cockroach-like.) These include the *Sexton Beetle*, *Necrophorus*, (Gr. *nekros*, a dead body; *phoreo*, I carry.) This is

about an inch in length, of a black hue, and extremely fetid. It is noted on account of its finding the carcases of small animals, such as mice, rats, birds, frogs, &c., shortly after death, burying them by working the earth from beneath them, and afterwards covering them. In these dead animals, the Sexton Beetle deposits its eggs.

7. DUNG BEETLES, *Geotrupidæ*, (Gr. *ge*, the earth; *trupaō*, to bore.) These, with other similar families, are, in their larva state, incapable of much locomotion, and generally live in the ground.

8. SCAVENGER BEETLES, *Scarabæidæ*, (gen. *Scarabæus*, from Gr. *Skarabos*, a beetle or scarabee.) These Beetles use the flat shield of their heads for working in the ground and in the dung upon which they feed. One species labor in pairs, the one beetle pushing their ball backwards with the hind feet, and the other walking up the ball on the opposite side, thus making it roll. The *Copris* rolls together a small ball which it immediately buries.

9. STAG BEETLES, *Lucanidæ*, (gen. *Lucanus*,) include some very large sized beetles, distinguished by having the antennæ terminated by a large jointed club. The males of *Lucanus cervus* have singular horns affixed to the head and thorax. (Plate II. fig. 8.)

10. GIANT BEETLES, *Dynastidæ*, (gen. *Dynastes*, Gr. a ruler.) These include some of the largest of the order. The males have horns or tubercles arising from the head or thorax. A most remarkable species is the Hercules Beetle, *Dynastes Hercules*, found in South America, measuring sometimes not less than five or six inches in length, having a horn of enormous length in proportion to the body, proceeding from the upper part of the thorax. Its larva continues about six years, and is three or four inches long. It is sometimes eaten fried, and esteemed a luxury.

11. ROSE BEETLES, *Cetoniidæ*, (of which *Cetonia*, is a prominent genus,) form an extensive group, including several which are distinguished for their brilliant colors. The common ROSE CHAFER, *C. aurata*, may be cited as an example, found in roses and upon the flowers of the privet, an insect nearly an inch long, of a shining green color above, and copper-red beneath, with white marks in the elytra. In its larva state, it feeds upon moist rotten wood, and is often met with under ground, in ants' nests.

12. SPRINGING BEETLES, *Elateridæ*, (gen. *Elater*, a charioteer.) These have a strong spine situated beneath the thorax, which fits at pleasure into a small cavity on the upper part of the abdomen;

thus enabling the insect, when laid upon its back, to spring up with great force and agility, in order to regain its position. Their larvæ are known in New England and New York, by the name of *Wire-worms*, and are injurious to corn and herbaceous roots. One species, *Elater noctilucus*, (Lat. shining by night.) is one of the most brilliant of the fire-flies which inhabit South America and the West India islands. In Cuba, ladies use these phosphorescent insects as ornaments for the hair. (See Chart.)

13. Wood-Borers, *Ptinidæ*, (genus *Ptinus*,) is a rather numerous family of insects, of small size, oval form and destructive habits. They are of obscure colors, and counterfeit death by withdrawing their head and antennæ, and contracting their legs. The Wood-Borers are found in old houses, which their larvæ perforate in every direction; also among furniture, books, &c.

The Death-Watch, *Anobium*, (Gr. *anō*, I end; *bios*, life,) *tessellatum*, (tesselated or checkered,) is of this family. It strikes its jaws upon the wood in which it has its abode, so as to imitate the ticking of a watch. The generic name we suppose to refer to the superstitious notion that when its beating is heard it is a sign that some person in the house will die within a year, and hence is derived the name *Death Watch*. (See Chart.)

14. Fire-flies, or Glow-worms, *Lampyridæ*, (leading genus *Lampyris*, Gr. *lampuris*, a glow-worm.) have a lengthened, depressed body, and flexible elytra. In some species the females are wingless, and in others they have only short elytra. They prey, in the larva state, upon the bodies of snails, and not upon plants. When alarmed, they draw in their antennæ and legs, and remain motionless, as if dead. The common Glow-worm, seen in the Middle States of the Union, is the female of the species *Photuris*, (Gr. *phōs*, light; *oura*, tail.) *versicolor*, (Lat. of changeable color.)

15. Corn and Nut Weevils, *Curculionidæ*, (*Curculio*, a corn-worm or weevil.) This family of Snouted Coleopterous insects includes the Diamond Beetles and other splendidly colored species, as well as the Corn or Grain Weevils. The Nut Weevil, *Balaninus*, (Gr. from *balanos*, acorn or nut,) *nucum*, (Lat. of nuts,) see Chart, is often found in the Chinquapin nut, and sometimes renders worthless almost the entire crop, which, in a short time, become wormy.

16. Cockchafers, *Melolonthidæ*, (leading genus *Melolontha*,) are well known and destructive insects. An instance is given of a farmer whose crops were completely destroyed by the larvæ of

the common Cockchafer, "of which eighty bushels were gathered up."

The genus Phyllophaga, (Gr. *phullon*, a leaf; *phago*, to eat,) includes several species, which are furnished with strong jaws for cutting the leaves of plants. They are injurious both in the larva and the perfect state; in the former, eating the roots of grass, &c., and in the latter, the tender leaves of fruit and other trees. Formerly they were included in the genus *Melolontha*. They are well known by the name of Horn Bugs, though their more appropriate name is May Beetles.

17. Pea Bugs, Wheat Weevils, &c., *Bruchidæ*, (genus *Bruchus*, Gr. *brouchos*, a locust—locust-like.) The Pea-bug, *Bruchus*, is a small hairy insect, gray and rather egg-shaped, which deposits its eggs in the pea-pod in its early state, and in which they are hatched. Multitudes of the larvæ are destroyed in preparing green peas for the table. The *Calandra granaria*, or Corn Weevil, of Europe, is a species that has been introduced into this country from Europe, in samples of grain, to which it is very hurtful. Linnæus calls it *Curculio granaria*.

18. Blister Beetles, *Cantharidæ*, (Gr. *Kantharis*.) Among these are the *C. vesicatoria*, (Lat. from *vesica*, a blister,) of a beautiful changeable or metallic green color, about three-quarters of an inch in length, and well known for its medical uses. In Spain, Portugal, and Italy, these insects are abundant. Potato vines and other plants are, in mid-summer, often infested by insects allied to the Spanish-flies.

N. B. The above account includes all the families of Beetles to which the Chart refers, though but a small part of the entire number.

Second Order. Strepsiptera, (Gr. στρεπτὸς, *streptos*, twisted; πτερὸν, *pteron*, wing.)

This order of insects is named by Latreille, Rhipiptera, (Gr. fan-wings.) They have the front wings replaced by a kind of twisted halterers; the posterior are large and folded like a fan. (Plate XV. fig. 3.) The tarsi have from two to four articulations. The mouth is armed with two slender acute jaws wide apart, and two pointed palpi, or feelers. The order includes a limited number of insects, arranged in the two genera *Xenos*, (Gr. a guest or stranger,) and *Stylops*, (Gr. *stulos*, a stylos or graver; *ōps*, face.) The larvæ are vermiform, and have six feet. The pupæ are inactive. They are all small, mite-like creatures, the largest not being a quarter of an inch in length. The larvæ are parasitic on the bodies of the wasps and bees, where they lose their feet and become larvæ of a different form—an instance of retrograde metamorphosis. The perfect insects are very short-lived, but very active. They were first observed by Kirby.

THIRD ORDER. DERMAPTERA, (Gr. δέρμα, *derma*, skin; πτερὸν, *pteron*, wing.)

The insects of this order are by some included among the ORTHOPTERA to which in the organs of the mouth they correspond, and which they resemble also in being active and in feeding during the pupa state. But they differ from them in the structure of the wings, which fold both longitudinally and transversely to bring them under the elytra, (wing-covers.) This order includes the EAR WIGS, (*Forficula*,) Plate XV. fig. 4, which live in damp places and feed on vegetable food. These insects have the tarsi three-jointed; their antennæ are long and slender and made up of many articulations. The Ear-wig sits over her eggs and assiduously watches the young when they appear.

FOURTH ORDER. ORTHOPTERA, (Gr. ὀρθὸς, *orthos*, straight; πτερὸν, *pteron*, wing.)

In this order the metamorphosis is imperfect, the elytra, or wing covers are coriaceous and veined, with the inner margins overlapping; in some cases the wings are wanting, or so small as to be entirely useless; the mouth is mandibulate, (with jaws,) and this organ and the thorax are much like those of the Beetles. The body is generally long; the head vertical, and the antennæ slender. The feet are well developed; but though some are very active, others are remarkably slow in their movements. The order includes (1) COCKROACHES, *Blattidæ*, (genus *Blatta*, Plate XV. fig. 5,) hiding by day and seeking food by night, and in tropical countries extremely troublesome. Scalding or fumigating them in their hiding places is one of the best methods of exterminating them. They sometimes even penetrate the brick walls of buildings, destroying both animal and vegetable substances; (2) the PRAYING INSECTS, *Mantidæ*, (Gr. *Mantis*, a prophet.) which use their fore legs as arms and hands, and when waiting for their prey, raise their feet as if in supplication, (Plate XV. fig. 7,) whence their name. They eat other insects, are great fighters, and when confined will eat each other. The smaller kinds of these insects are seen occasionally in New England and New York; (3) SPECTRES, *Phasmidæ*, (Gr. *phasma*, a spectre or apparition,) These have the wings somewhat undeveloped or entirely absent. They eat leaves, live upon trees, and present some very curious forms. Some are called walking-sticks, from their resemblance to a stick. One species, found in the Moluccas, is ten inches long. *Phyllium*, (Plate XV. fig. 8,) is a genus that has wings which look like a leaf, whence the name, which means a *Walking-Leaf*. One or two of these remarkable insects are met with in New York and in some of the Eastern States; (4) the CRICKETS, *Achetidæ*, (*acheta*, a chirper,) which, although they present a general likeness to the Grasshoppers, differ from them in their habits, being entirely terrestrial, and having, more or less, the power of burrowing. They appear to live both upon vegetable and animal food, which they search for at night. The Crickets are good runners, but do not fly as well as the Grasshoppers; (5) LOCUSTS, *Locustidæ*, (Lat. *locusta*, a locust.) The abdomen of the female has a sharp, flattened ovipositor; the males make a loud stridulation, or whizzing, by means of their upper wings. These insects are quite arboreal in their habits, and from the green color of many of them, they are hardly perceptible among the foliage. They sometimes appear in great numbers; (6) Grasshoppers, *Acridiidæ*, (Gr. *akris*, a locust.) The female is without an ovipositor. The males of these insects make their peculiar noise by rubbing their hind-thighs against the wing-covers. To this family belongs the KATYDID, *Platyphyllum*, (Gr. broad-leaf,) *concavum*, (Lat. concave or hollow,)—ranked among the Grasshoppers. This singular insect is of a grass-green color, and derives its name from the notes which it sends forth. It reaches its perfect state in September, depositing its row of eggs upon the twigs of the trees in which it dwells. (See Plate XV. 9, and explanations.)

This order has been divided into four sections, founded on differences of habit arising from the peculiar construction of the organs of locomotion, (Plate XV. figs. 5, 6, 7, 8, and 9.) (1) The Runners, (Orthoptera cursoria;) (2) the Graspers, (Orthoptera raptoria;) (3) the Walkers, (Orthoptera ambulatoria;) (4) the Jumpers, (Orthoptera saltatoria.) The RUNNERS include the Cockroaches; the GRASPERS, the Praying Insects; the WALKERS, the Walking Sticks, &c.; the JUMPERS, the Grasshoppers and Locusts.

FIFTH ORDER. TRICHOPTERA, (Gr. θρὶξ, *thrix*, hair, πτερον, *pteron*, a wing.)

The genus *Phryganea*, (Plate XV. fig. 10,) which is the only one of this order, is by some joined with the genera *Hydropsyche* and *Limnophilus*, to form the family Phryganeidæ, and referred to the order Neuroptera. The insects of this order have four membranous and reticulated wings; the posterior pair are the larger; the front pair are generally hairy,—hence the name of the order. The name of Caddis-flies has been given to these insects which come from the various species of case-worms. The larvæ are inactive, residing in water, in a *case* formed of bits of shells or sticks, or of sand or saw-dust. The pupa is inactive. The Caddis-fly is often used as a fish bait.

SIXTH ORDER, NEUROPTERA, (Gr. νεῦρον, *neuron*, nerve; πτερὸν, *pteron*, a wing.)

This order of mandibulate insects exhibits a considerable variety of characters. According to Westwood, it includes twelve families. It is estimated to include not far from a thousand species. These insects have usually four reticulated nervures, (wings with horny divisions, thin, and lace-like.) The wings are of unequal size; instead of the hind-wings, there are sometimes only pedicles or stems. The antennæ are usually short and bristly. The pupæ are sometimes active and sometimes torpid; the larvæ are six-footed and very active, mostly predaceous, and either terrestrial or aquatic.

The DRAGON-FLIES, *Libellulidæ*, (genus *Libellula*,) include nearly two hundred known species. To these we have already referred. While on the wing, they deposit their eggs in water and in it pass both their larva and pupa state, gliding through it, or crawling about in the mud at the bottom. The hinder part of the body has several leaf-like processes, which can be drawn together or opened at pleasure. These close the opening of a cavity having very muscular sides. When the Dragon-fly wishes to move rapidly, it opens this cavity, which thus becomes filled with water; then by contracting the walls of the cavity, it throws out the water forcibly, like a stream from a syringe; aided by the re-action produced by the jet against the surrounding fluid, the creature shoots forward, with its legs closely packed along the sides. The pupa is no less active, fierce and voracious than the larva, differing from it only in having upon the thorax the rudiments of wings, which in the perfect insect are so admirable for their firmness, transparency and gloss. Even after it has reached the imago state, its ferocious manners still continue. It has even been known to devour its own body, when confined and deprived of musquitoes and the other insects upon which it usually feeds.

The ANT-LIONS, *Myrmeleonidæ*, (genus *Myrmeleon*, Gr. ant-lion,) are distributed throughout the world. These are terrestrial, spider-like in their appearance, and short and thick, having mandibles strongly toothed on the inside, so that the insect may suck the juices of its victims, and so constructed that it can hold its food firmly, though unable to chew it. To the curious devices which the larva of this insect employs for entrapping its prey, we have already alluded.

The Ant-lions have been found under the limestone ledges of Schoharie, and the larvæ have also been seen beneath such ledges near Burlington, Vt. (Emmons.)

The MAY-FLIES, *Ephemeridæ*, or Ephemeral-flies, are so named from the Greek word *ephemeros*, (diurnal,) in allusion to the extreme brevity of their existence. Their larvæ live in the water; they take refuge under stones, and in the earth and mud, feeding upon its slime. In their perfect state these insects generally live but a few hours, taking no nourishment; but if the sexes be kept apart, it is said, they will live from one to three weeks. Sometimes they issue forth in such numbers that "the ground is covered by their bodies when they die, to such a thickness as to make it worth while to cart them away as manure. The swarms of one species with white wings have been so abundant as to resemble a fall of snow."

The TERMITES, *Termitidæ*, which include the genus *Termes*, (Gr. *terma*, an end,) are distinguished by wings having few transverse nervures or horny divisions, and folding horizontally; the tarsi are four-jointed; the antennæ short and moniliform; the body is white and oblong in shape.

The head of the White *Ants*, as they are called, though differing from the true ants, is large and rounded; and besides the ordinary compound eyes, they have three ocelli or simple eyes, situated on the upper surface; the antennæ are long, and composed of about eighteen joints.

The Termites are chiefly confined to the tropics, though some few species extend into the temperate regions. Swainson, Kirby and Spence, and other writers who have observed the operations of these ants, either in Africa or South America, present many interesting particulars respecting their wonderful economy and habits. These insects unite in societies, composed each of an immense number of individuals. In the warmer regions, the ravages of some species are often fearfully great. A species discovered by Latreille at Bordeaux, (Fr.,) frequently attack the wood work of houses, in which they form innumerable galleries, all leading to a central point. In building, they avoid piercing the surface of the wood-work; and hence it appears sound, when the slightest touch is sometimes sufficient to cause it to fall to pieces.

One of the largest and best known species is the *Termes bellicosus*, or Warlike Ant, (see Chart,) found on the coast of Africa. These Ants build conical nests or edifices, sometimes of enormous size, nearly as hard as stone, and very commonly twelve feet in height, (see Chart.) They are often quite numerous, appearing almost like huts of savages; and Mr. Cummings says, "are of the greatest service to the hunter, enabling him to conceal himself with facility on the otherwise open plain."

The male and female, or King and Queen, have their *royal* chamber near the center of the hillock, and never leave it. They are both perfect insects, but the wings which they once had are lost soon after their admission to their place of abode. To the almost numberless eggs dropped by the Queen-mother, we have already referred. In times of scarcity, the Hottentots feast upon these eggs, which they call rice, on account of their resemblance to that grain. They usually wash them, and cook them with a small quantity of water, declaring that they are savory and nourishing. When they find out a place where the nests are numerous, it is said they soon become fat from eating the eggs, even when previously much reduced by hunger. "Sometimes they will get half a bushel out of a single nest."

The larvæ, in their full grown state, are perhaps a quarter of an inch in length. They are far the most numerous and the workers of the colony, building, foraging and nursing. The soldiers or fighters are comparatively few, not more than one to a hundred of laborers; but they are many times larger, and armed with sharper and more formidable jaws. They appear as defenders when the nest is assailed, and will even attack the assailants, biting with considerable force. The species *T. frontalis*, of South America, works galleries in logs and stumps of trees, and in the ground also, plastering them with a hard mixture of clay.

SEVENTH ORDER. HYMENOPTERA, (Gr. ὑμὴν, *humēn*, a membrane; πτερον, *pteron*, a wing.)

In the insects of this order, inferior in numbers only to the Beetles, the nervures, or veins of the wings, form the basis of numerous sub-divisions. The wings differ from those of the Neuroptera in being of a less delicate construction, and having fewer nervures. The mandibles are distinct, but better fitted for imbibing nourishment by suction than by mastication; the body is of a hard consistence; the antennæ are variable, but for the most part slender, showing twelve articulations in the male, and thirteen in the female; the tarsi are generally pentamerous or five-jointed. These insects are also peculiarly distinguished by the prolongation of the body in the case of the females, into an organ which in some is a sting, in others an *ovipositor*, or instrument for depositing the eggs—usually having the power of boring a hollow for their reception. The Hymenoptera are among the most remarkable of the class, for their instinctive faculties, their social qualities and habits, and their powers of locomotion. The Bees, the Wasps, the Ants, the Saw-flies, the Ichneumons and the Gall-flies have, from the remotest periods, been objects of attention to the observers of nature.

The order is sometimes arranged into two sections, viz.: the TEREBRANTIA, in which the female has a saw or borer for the deposition of eggs; and the ACULEATA, in which the abdomen of the females and neuters is possessed of a sting, which is connected with a poison reservoir. The former section includes seven families; the latter seventeen.

TEREBRANTIA—BORERS.

The SAW-FLIES, *Tenthredinidæ*, (Gr. *Tenthrēdōn*, from *tentho*, to gnaw,) are the only ones of the order which have feet. The larvæ feed upon leaves or vegetable matter. The ovipositor of the female appears to combine the properties of a saw and file. The Saw-fly, *Tenthredo*, is also named *Cimbex ulmi*, (Lat. of an elm,) because it inhabits the Elm.

The WOOD-WASPS, or Horn-Tails, *Uroceridæ*, (from Gr. *oura*, a tail; *keras*, a horn,) are a family of insects which often do great mischief to fruit trees and also to forest trees, especially resinous ones. The females have an ovipositor in the form of a slender horn, consisting of five pieces—two outside grooved and forming a hollow tube; the other and inner pieces are needles, with which the trunks of trees are pierced to make a place of deposit for the eggs. The grub-like larvæ burrow in the green solid matter of trees and eat the wood.

The ICHNEUMON-FLIES, *Ichneumonidæ*, (Gr. ichneumōn,) have narrow bodies and rather long antennæ; the feet are long and adapted for running, and the ovipositor is straight. These insects fly and move about in a restless manner, keeping their antennæ in a constant vibratory motion. They perform a useful part in preventing an undue multiplication of Moths and Butterflies, upon the larvæ of which these flies deposit their eggs, but through so small an opening as not to check the growth of the larvæ. When the larva passes into the pupa state, the eggs of the ichneumon hatch, and the progeny feed upon it, so that, in the end, instead of a butterfly, there comes forth a brood of ichneumons. This is a very numerous family, including thousands of species.

The GALL-FLIES, *Cynipidæ*, (genus *Cynips*,) are a small family of insects, the larvæ of which are parasitic in plants, where they cause the excrescences called *galls*.

The family *Evaniidæ*, (genus *Evania*,) includes the AMERICAN HATCHET WASP, *Pelecinus* (Lat. a hatchet) *politrurator*, (Lat. a furbisher or polisher,) (see Chart,) which is seen by the road sides in the fall of the year, flying slowly, as if borne down by its long and slender abdomen.

The SNAKE WASPS, *Ophidion*, (from Gr. *ophis*, a serpent,) (see Chart,) of which there are several species, have the abdomen three times as long as the thorax, and the antennæ nearly the length of the insect, which is about one inch. This wasp is seen late in the summer or the beginning of autumn, hovering over brambles, &c., looking after caterpillars as a place of deposit for its eggs. The genus *Evania* is parasitic in ship Cockroaches.

ACULEATA-STINGERS.

The Spider Wasps, *Sphecidæ*, (genus *Sphex*, Gr. a wasp,) have an elongated body; the abdomen is attached by a long, slender peduncle, (see fig. on Chart,) and armed with a sting. These wasps are extremely active and difficult to capture. In the perfect state, they suck the fluids of flowers, but the larvæ are furnished with animal food by the adult.

The WASPS, *Vespidæ*, (genus *Vespa*, Lat. a wasp,) like the bees, include males, females and workers. Like the bees also, they are social and dwell in small communities, though there are some solitary species, among which no neuters are found. The Wasps and Hornets are natural paper makers, societies of them living, during summer, in nests divided into hexagonal cells, opening downwards, formed of paper-like material, which is impervious to water. During a season, two or three broods are raised successively in the same set of cells. The nests may be seen on trees, sometimes from twelve to sixteen inches in diameter. The small "Yellow Jackets," as they are termed, build under ground.

The Paper-Wasp, *Polistes* (Gr. the founder or chief of a state,) *fuscata*, (Lat. swarthy,) either fastens its comb to the branch of a tree, or to the shelving parts of a house.

The Paste-board Wasps, *Chartergus*, (Gr. paper-work,) make their nests of a solid and rather thick paste-board. Their structures have been seen in Pennsylvania, but are more common in South America.

The Common Hornet, (Plate XV. fig. 12,) *V. crabro*, (Lat. a hornet,) is considerably larger and more formidable than the Wasp, building its nests in decaying hollow trees, or beneath their roots, and in timber yards, or under the eaves of barns, etc. Its sting, as is well known, often produces serious consequences.

The Large American Hornet, *V. maculata*, (Lat. spotted,) often enters houses to catch flies.

The Ants, *Formicidæ*, (*formica*, an ant,) a well known and interesting family, to be distinguished, however, from the *White* Ants already described, as belonging to another order. In addition to the males and females, which form a small part in any community of ants, and which are alone furnished with wings, there are *neuters*, or workers, by which the labors are chiefly performed, not only constructing the nests, but feeding and taking care of the young grubs. These alone survive the winter, in our climate remaining torpid during that season; but it is otherwise with them in the torrid zones. There they are active, night and day, during the entire year; to these the words of inspiration, (Proverbs vi.,) have particular reference,—so indefatigable is their industry—that to them the indolent and inactive may well be pointed for lessons of instruction. "Go to the ant, thou sluggard, consider her ways and be wise."

The Red Ants, *Formica rubra*, construct their nests upon the branches of trees. These are said to be the only ones which feed upon their own species. Extremely lively representations have been given of the wars sometimes carried on between two or three Ant-cities, equal in size and population, and situated at about one hundred paces from each other.

Of this ant a minute species, *M. domestica*, is found in companies, either under stones, or else in old galls upon oak shrubs, which they enter by the opening that is made when the *Cynips* leaves. A few of these, which have large heads, appear to be the workers. These ants often swarm in houses.

The SLAVE-MAKING or Rufescent Ants, *F. rufescens*, make war upon other ants, for the sole purpose of procuring slaves to labor for them. Most of the slave dealers are reddish, while those who are captured to become their servants are black. Besides adults, however, larvæ and pupæ are seized, and brought up by their captors, commencing their labors when they reach their perfect state; yet their masters do some part of the work. According to Westwood, the large Yellow Ant of the United States, makes slaves of the Black Ants.

Certain Ants, called COW-KEEPERS, are very fond of the liquid matter which is given out by the Aphides, or Plant-lice, and actually attend upon these "Honey-flies," as Swainson calls them, for the purpose of obtaining it. They even have the power of making them yield it at their pleasure, by patting the abdomen of the *Aphis* alternately on each side; and thus they "milk their cows." They are called "Cow-keepers," for the reason that they sometimes seem to claim a right to the *Aphides* inhabiting a particular branch or stalk, and resist the approach of strangers. To rescue the "Cows" from their rivals, they will take the *Aphides* into their mouths, keep guard around them, sometimes enclose a certain number in a tube of earth, or other materials near their nests, so that they may be always at hand to supply them with the desired food. The most remarkable Cow-keeper is the Yellow Ant, *F. flava*, of Gould, which secures within the common nest, a large number of Honey-flies of the species *Aphis radicum*, (Lat. of roots,) which derives its food chiefly from the roots of grass and other plants. The Yellow Ants, it is said, bestow upon these little creatures care and solicitude equal to that which they give to their own offspring. In India, the honey-like secretion which the *Aphides* cast upon the ground, is so abundant in quantity that the natives collect it when dry, and sell it in the country bazaars as a sweetmeat. The honey, it is said, may be kept for seven or eight years, without losing its sweetness. In Brazil, the insects not only furnish ants with milk, but, ruminant-like, have horns growing out of their heads; and hence are called the "cattle" of the ants.

The DRIVER ANTS, of South Africa, according to the observations of Dr. T. S. Savage, an American missionary to that region, include in their communities, *Neuters*, *Soldiers*, *Workers*, and *Carriers*. These do not construct nests, but live temporarily in crevices, sometimes "ranging about in vast armies," and when they enter houses, causing rats, lizards, &c., and even man himself to flee. They travel at night or in cloudy weather, as the direct rays of the sun are almost immediately fatal to them. "I know of no insect," says Dr. Savage, "more ferocious and determined upon victory. It may literally be said they are against everything, and everything against them. 'Conquer or die,' is their motto." They are useful in keeping down the more rapid increase of other noxious insects, and also in consuming much dead animal matter.

The SOLITARY BEES, *Andrenidæ*, consist only of males and females. The species of the genus *Andrena*, are quite numerous. They make their appearance in the early spring and summer months, and have very much the appearance of Hive-bees. The females collect pollen from the stamens of flowers, rather by means of the general hairiness of the body than with the posterior tarsi. They burrow in the ground in sandy districts, especially, if exposed to the sun, often to a considerable depth.

Of the Bees proper, *Apidæ*, (Lat. *apis*, a bee,) there are several groups, differing from each other, to some extent, in their qualities and habits. The HUMBLE (Bumble) BEES, *Bombus*, (from Gr. *bombos*, a humming or buzzing.) construct their nests under ground, in fields and pastures. The females, which are unlimited as to number, assist the neuters in working. The larger females alone survive the winter, and in the first fine days of spring, construct their cells, and rear a brood of workers, which, in due time, assist in the construction of new cells. The honey which these bees collect, is of an inferior kind, and their wax is not so clean, or so capable of fusion as that of the True Honey-Bees.

The MASON BEES, *Megachile*, (Gr. great lips or jaws,) *muraria*, (Lat. from *murus*, a wall,) build their cells by agglutinating grains of sand and gravel.

The UPHOLSTER BEES, *M. papaveris*, (Lat of a poppy,) line the holes which they excavate for their young in the earth, with an elegant coating of leaves or flowers, preferring, for this purpose, the brilliant scarlet furnished by the leaves of the wild poppy. The species *M. centuncularis*, (from Lat. *centunculus*, patchwork,) coat their dwelling with the leaves of trees.

The CARPENTER BEES, *Xyclopa*, (Gr. *Xulon*, wood; *kopto*, to cut,) bore with great labor out of solid wood, long cylindrical tubes, and divide them into various cells, in which the young are placed with a quantity of pollen-paste.

The VIOLET CARPENTER BEE, *X. violacea*, is common about Paris, and in the gardens of Southern Europe. Among these bees, the females perform all the labor; the males have no stings. The species *X. victima*, is found in the United States, and bores in the lower surface of white-pine structures.

But the most important, and, indeed, the most interesting of the family, is the Common Hive Bee, *Apis mellifica*, (Lat. *mel*, honey; *facio*, to make.) The Hive includes three kinds; the Female; the Male or Drone; and the Worker, (see Chart for figures showing the relative size, &c., of each.) The bees collect honey, pollen, and propolis, feeding their young with the former two, and using the latter for filling up crevices in their cells, and for the needed repairs. The *wax* is secreted by the workers, and appears between the segments of the lower side of the abdomen, in the form of small scales.

Every hive is under the government of the Queen Bee. She is lady paramount, and suffers no other queen to share her dominion. At the swarming season, the old queen becomes so sadly disturbed by the encroachments of the young queens, that she rushes forth from the hive, attended by a large body of her subjects; thus, the first swarm is formed. In seven or eight days afterwards, the queen next in age departs, also taking with her a supply of subjects. When all the swarms have left the original hive, the remaining queens fight until one gains the throne. The Queen Bee lays about eighteen thousand eggs. About eight hundred of these prove males or drones, and four or five queens; the remainder are workers. The cells are six-sided. Those in which the drones are hatched, are much larger than the cells of the ordinary working bees. The royal cells are much larger than any others, and are of an oval shape. When a worker larva is placed in a royal cell, and fed in a royal manner, it imbibes the "principles of royalty," and becomes a queen accordingly. This practice is resorted to if the Queen Bee die, and there be no other queen to take her place.

The form of the cells is such as to afford the greatest space and strength, with the least amount of material. How the bees are enabled to give them this form, unless by a divinely implanted instinct, it is difficult to tell. Three figures will admit the junction of their sides without vacant spaces between them, viz: the square, the equilateral triangle, and the hexagon, the last being the strongest and most convenient. And this is the very form in which the bees build their cells. The bottom of each cell, on one side, meets three on the other, and is supported by the divisions between them; and it is formed by three plates that meet at an angle, which profound mathematical investigation demonstrated to be the very angle which combines the greatest strength with the least material. Kirby and Spence say,—"Maraldi calculated that the great angles were 109° 28′, and the smaller ones 70° 32′; and König calculated that they ought to be 109° 26′, and 70° 34′, to obtain the greatest strength with any given amount of material." But subsequent examination showed that the bees were right and König wrong.

EIGHTH ORDER. LEPIDOPTERA. (Gr. λεπις, *lepis*, a scale; πτερον, *pteron*, a wing.)

These insects, comprehending, perhaps, one-fourth or one-sixth of the entire tribe, have a suctorial mouth and rudimentary mandibles. Their metamorphosis is complete. The beautiful Butterflies are the representatives of this order, and also of all those winged visitants that flit about our lamps during the evenings of summer; the one are diurnal; the other nocturnal. They all have four membranous wings, usually covered with minute scales; the mouth is suctorial, consisting of a tubular thread-like organ, which, when not in use, is rolled into a compact spiral coil; their bodies are soft and covered with hair; the feet are pentamerous, (have the tarsi five-jointed;) generally, they are hairy and of equal length; though sometimes the front pair are so small as to be of no use in walking. The Lepidoptera feed upon the juice of flowers, but, in the perfect state, they sometimes need none. They may be arranged into three great divisions: (1) the Butterflies Proper, *Papilionidæ*, (Papilio,) which have thread-like antennæ and bear a knob, (Plate XV. fig. 13;) (2) the *Sphingidæ*, (*Sphinx*,) or the Hawk-Moths, which have the antennæ, thick in the middle, and at the tip often hooked; (3) the Moths (in general) having the antennæ somewhat naked, of bristle form, or else feathered on the sides.

I. The Butterflies Proper include at least three hundred species, sometimes most gorgeously colored, of which large diurnal ones are found in the United States. The Butterfly, ***P. turnus***, is one of the most common species; in its markings and forms, resembling the ***P. machaon***, or Swallow-tailed Butterfly of Europe. (See Chart.)

CABBAGE BUTTERFLIES, *Pontia*, (Gr. a sea-green surface,) *Brassica*, (Lat. cabbage.) These are common and destructive in our gardens. (Plate XV. fig. 13.) The eggs are yellowish and laid on the under side of cabbages, turnips and radishes; the pale green worms come out in about a week, and attain their full size of an inch and a half, in three weeks.

The HAIR STREAKS, genus *Thecla*, derive their name from the

26

delicate, straight, or zig-zag lines on the under side of the rings. Some species frequent hedges, others the oak and ash trees.

The *Nymphalidæ*, (*Nympha*, a nymph,) include many beautiful Butterflies, called Red and White ADMIRALS, PAINTED LADIES, FRITTELARIES, the front legs of which appear incomplete, but in their ability for strong flight, they are more than compensated for the deficiency of their feet. The genus *Vanessa* includes many species, of which V. Io, PEACOCK BUTTERFLY, is pictured on the Chart.

The TORTOISE-SHELL BUTTERFLY, *V. urticæ*, (Lat. of a nettle,) nearly resembles the Peacock Butterfly. The Caterpillers live in societies, changing their skins frequently, and constructing a new tent on another part of the plant at each moult, until the last, when each individual feeds by itself, and the society is dissolved.

The SKIPPERS, *Hesperiidæ*, (leading genus *Hesperia*,) have the four hind shanks furnished with two pairs of spurs. They have a jerking kind of flight, from which their popular name is derived; and in many respects, they approach the moths.

The TITYRUS SKIPPER, *Eudamus tityrus*, often strips the locust tree of its foliage. It forms its habitation of the leaves of that tree bound together by silken threads, and also feeds upon its leaves.

II. The Hawk-Moths, *Sphingidæ*, (leading genus *Sphinx*,) are also named Humming-Birds, being capable of flying for a long time, and of poising themselves in the air, like the Humming-Bird. (See Chart for figures of *Sphinx ligustri*, or Privet-Hawk-Moth, in the larva, pupa, and imago or perfect state.) Many beautiful species of Hawk-Moths are seen on fine summer evenings. The *Philampelus*, (Gr. vine-lover,) *Satellitia*, is of this family, (for figure of which see Chart.)

III. MOTHS.

The TIGER-MOTHS, *Arctiidæ*, (*Arctia*,) have the feelers and tongue usually short and thick, and the antennæ doubly feathered; both the Caterpillars and Moths are downy. They fly only at night. The family includes different genera, but we can only name the American Tiger-Moth, *A. virgo*, of a pink red color, with two central, triangular spots, and other markings; and the GREAT-TIGER MOTH, *A caja*, (see Chart.) an English insect, but one that is represented in the *A. Americana*, which it closely resembles. The latter has the base of the fore wings marked with white branching spots, which partly resemble a cross; the wing beyond the middle is also marked with a white irregular cross, something like an X.

The SILK-WORMS, *Bombycidæ*, (leading genus *Bombyx*,) represent some of the largest and most beautiful species of nocturnal Butterflies; among which is the *Attacus luna*, or GREEN EMPEROR MOTH, which is about five inches in the expanse of wings, (see Chart;) the Caterpillar is also of a bluish green color; when in motion, three inches in length, and in feeding, preferring the leaves of the hickory. For figures of the *Bombyx mori*, or Silk-Worm in its different stages, see Chart. The larvæ have sixteen feet,—feed upon leaves; and spin the silken cocoon out of a single thread, with the assistance of a gummy matter, which soon hardens. Other species than the *Bombyx mori*, (Lat. of the mulberry,) are reared for the silk, and more of it might be obtained, if warm water dissolved the gum of the cocoon, as it does in the true Silk-Worm. The *Tineidæ*, (*Tinea*,) are the smallest Moths in the section. These infest woolens, furs, etc. The best way to protect such articles against these Moths, is to put them together with tobacco-leaves, camphor, or turpentine, in a tight bag early in the spring, before the eggs of this insect are laid.

The HONEY-COMB MOTH, *T. cerella*, (Lat. from *cera*, wax,) is notorious for its depredations upon the wax of the Bee-Hive.

The LEAF-ROLLERS, *Tortricidæ*, (*Tortrix*, i. e., twister or roller,) comprehend many species of insects, the larvæ of which do great damage to the fruit of apple and the foliage of forest trees. The larvæ of the *Carpocapsa*, (Gr. fruit-eater,) *pomonella*, (Lat. from *pomum*, fruit, apples, &c.,) known as the Apple-Worm, came to this country with the apple, and this worm has become naturalized among us.

NINTH ORDER. HEMIPTERA, (Gr. ᾑμισυς, *hēmisus*, half; πτερόν, *pteron*, wing.)

This order is distinguished by having the rostrum or jaw compounded, i. e., formed for piercing and sucking. The insects which it includes, live upon vegetables and animal juices, those feeding upon vegetables being the most numerous. The name Hemiptera, first used by Linnæus, refers to a characteristic of some of the order in having a thickening on the basal part of the anterior wings, while the other part is thin and transparent. Others apply to it the term *Rhynchota*, (Gr. *rhunchos*, beak or gape,) having reference to the character of the mouth. The metamorphosis in this order is only semi-complete, both the larva and pupa being active, and, at all times, taking food.

The order includes two sections, viz: Homoptera, (Gr. like-wings,) and Heteroptera, (Gr. different wings,)—in the first of which the wings are of a uniform; in the second, of a varied texture. (To the section Homoptera, Latreille gives the name Hemiptera, while Leach calls it Omoptera.)—Westwood divides the Homoptera, which he considers a distinct order, into three sections, viz: Trimera, (Gr. three parts,) Dimera, (two parts,) Monomera, (one part or division,)—these terms having reference to the divisions of the tarsi; and the Heteroptera into two sections, Hydrocorisa, (residents of water,) and Aurocorisa, (residents of air.)

Of the Homopterous division, having the four wings all of a firm membraneous texture, are the BARK-LICE, or SCALE-INSECTS, *Coccidæ,* (typ. gen. Coccus.) (Plate XV. fig. 14.) Of these there are several species,—found on the leaves and bark of different plants. The *Coccus cacti* (of the cactus,) is, on account of its beautiful crimson color, used as a coloring-matter. It is a native of Mexico, and feeds upon a particular kind of Cactus, called Indian-fig, and extensively cultivated for the express purpose of rearing it. The annual amount of Cochineal exported to Europe is, according to Humboldt, eight hundred thousand pounds; and it requires about seventy thousand insects to make a pound. Lac or Shell-lac, employed for making sealing-wax, is the product of a species of *Coccus.* The Mealy-Bug, *Coccus adonidum,* (of Adonises or flowers,) found in hot-houses, is reddish, but covered with a white, powder-like substance. The *Coccidæ* belong to Westwood's Monomera.

The PLANT-LICE, or Vine-fretters, *Aphidæ,* (leading genus *Aphis,*) infest the roots of vegetables, (often doing them great injury,) and also the leaves of most plants, such as roses, asters, apples, pears, peaches, cabbages, &c.,—each plant having its own peculiar species. Their bodies are soft, of an oval form and have upon the abdomen two tufts or pores. The females are usually wingless, but not always. The upper wings corresponding to the wing-covers in the Hemiptera Proper, are the larger and used for flight or as aids in leaping. To the prolific powers of the Plant-lice reference has already been made. A young leaf that curls, or that has an unhealthy appearance, is probably infested with these lice. Fumes of tobacco, turpentine or sulphur, are a remedy against them, and also against the Mealy-Bugs. The *Aphidæ* belong to the *Dimera.*

The WOOL-FLY, or the Apple-tree-blight, *Eriosoma,* (Gr. *erion,* wool; *soma,* body,) is an insect of a woolly appearance; without wings, but wafted from tree to tree by its cotton envelope. Its microscopic eggs,—covered with the same soft, downy substance as the body, are found in the crotches and chinks of trees, where they hatch, and produce the Apple-tree-blight. The wounds of this insect produce warts and excrescences on the surface of the trees, and finally result in its death.

The Jumping Plant-lice, *Psyllidæ*, (*Psylla*, Gr. *Psúlla*, a gnat,) are similar to the Plant-lice, but more active. These are dimerous; they have ten articulated antennæ, and the females have an ovipositor.

The Lantern-flies, *Fulgoridæ*, (*Fulgora*, Lat. from fulgeo, to shine,) include a number of trimerous species of bright colors and large size. Many of them have a curious prolongation of the forehead, sometimes nearly as large as the rest of the body. Whether they are luminous or not, is a point not positively settled. They probably give out light at particular seasons. The species *F. candelaria*, (Lat. from *candela*, a candle,) is yellow, and the elytra black, marked with yellow spots. It is said a Chinese edict exists against young ladies keeping Lantern-flies.

The Harvest-Flies, *Cicadidæ*, (*Cicada*,) are distinguished by their robust body, their large and triangular head, with three stemmata, their prominent eyes; the antennæ short and thin, with six articulations, and, usually, by large transparent wings. The Harvest-flies are trimerous. They have long been particularly noticed on account of the noise made by the male, differing in different species. The species which has attracted most attention, is *C. septendecim*, (Lat. seventeen,) the Seventeen Years Locust, which often does very great damage to trees. The female, with her ovipositor, inserts her eggs in their tender branches, which causes them to die, so that the tops of the forests, sometimes, on this account, look as if they had been scorched by fire. Miss M. A. Morris has ascertained that trees also suffer much from the *larvæ* of these locusts, which penetrate six inches under ground, and reach the roots. She says further, that the larvæ are destroyed by those miners, the Moles.

The Dog-day Harvest-fly, *C. canicularis*, (Lat. from *canicula*, the dog-star,)—according to the observation of Mr. Harris, has, for many years in succession, been regularly heard at Cambridge, on the twenty-fifth day of July, between the hours of ten in the forenoon and two in the afternoon. Its body is thicker and proportionably shorter than that of the Seventeen Years Locust, but its habits are quite similar.

The Heteroptera, which have the upper wings partly thick, and partly thin, include several families. Among these are the

Notonectidæ, (Gr. back-swimmers,) the True Water-Bugs, (Hydrocorisa,)—named from their habit of swimming with the back below. These, from the peculiar appearance of the body, are sometimes called Boat flies. The hind feet are long and fringed, held out when at rest, like a pair of oars, and used like them in swimming. The larvæ and pupæ differ from the perfect insect only in their smaller size and the absence of wings.

Hydrométridæ, WATER-MEASURERS, or SKIPPERS, (Aurocorisa.) These live on the surface of standing or running waters, and sometimes move with great rapidity.

The boat-shaped insects of the genus *Hydrometra*, (Gr. *hudōr*, water; *metron*, measure,) are furnished with fore feet suited to locomotion. These move over the water rather slowly. Their larvæ have the abdomen extremely small, which is also true of the Oceanic Halobates, (Gr. *hals*, the sea; *baino*, to go,) which seems to confirm the general idea of Agassiz, (see our account of the Turtles,) that fresh-water forms are of higher grade than the marine.

Reduviidæ, (Genus *Reduvius*.) These are another family of SKIPPERS, which are active and predaceous, their strong beak or rostrum enabling them to pierce insects that have a covering tolerably hard. The puncture which they make is said to be rather poisonous.

Cimicidæ, or Land Bugs, (Aurocorisa,) include the Bed-bug, *Cimex lectularia*, (see Chart,) so odious, and so widely spread. It is said this bug was "unknown in England until after the Great Fire of London, in 1666, when it was introduced in the fir-timber imported for rebuilding the city." Westwood, however, asserts, it was known there as early as 1503.

Coreidæ, (genus *Corus*.) These are small, elongated bugs, found in small fruits, some of them of a red and yellow color bordering the elytra and upper surface. One species is the SQUASH-BUG common on the leaves of the squash and pumpkin, which lays its eggs about the last of June. It should be crushed with the foot before that time.

Scutelleridæ, (genus *Scutellaria*.) 1. These insects derive their family name from having the scutellum, (dimi of *scutum*, a shield,) so large as to cover the abdomen and wings. These are the bugs of unpleasant smell, found on strawberry-vines and other berries. Some of them are above the medium size of insects, and not a few are clothed in bright colors. Those of the genus *Pentatoma*, (Gr. five sections,) are among the most common. Like others of the family, they secrete an ill-scented fluid. The antennæ are divided into five joints,—whence the name.

TENTH ORDER. DIPTERA, (Gr. δις, *dis*, twice or two; πτερον, *pteron*, wing.)

These insects are distinguished by the possession of only two wings and a pair of small knobbed appendages, (as in the common fly and the musquito,) called *halterers* or poisers. The wings are membranous, and without any covering, except a few hair-like scales, which, in some species, appear at the base. They are never folded upon themselves, remaining expanded when at rest, as in the insects of the preceding order. Their nervation is quite different from that exhibited in the other orders. The mouth is suctorial, and in many, has a fleshy proboscis, that encloses lancets capable of penetrating flesh, or the softer parts of vegetables. In a few genera, as the *Oestrus* or Gad-fly, the mouth is closed. These insects are all small; but what is wanting in size is made up in numbers. They are every where, and also are attendants upon man,—sometimes to his great annoyance; but it should be remembered, they are highly useful in cleansing the earth's surface of impurities, both animal and vegetable. In this order the transformations are imperfect. The pupæ sometimes take the incomplete form,—having the limbs visible, and without a cocoon. The larvæ are white and fleshy, cylindrical in shape, and without feet. They are seen in carrion and in galls; or in living caterpillars; and sometimes among vegetables pickled with vinegar, and in the brine of salt works. We can only refer to some of the more conspicuous families.

1. *Culicidæ*, represented by the genus *Culex*, (Lat. a gnat.) This family includes the numerous Gnats and Musquitoes, *C. pipiens*, (Lat. peeping,) distinguished by the tufted antennæ of the males. The pupæ of these are active; the larvæ are inhabitants of water; hence, these insects are abundant, chiefly in damp situations. Mankind are attacked by the female gnats alone; the lancets of the mouth being in the males fewer and weaker. These insects lay two or three hundred eggs in stagnant water, joined together so as to form a little raft floating upon the water, where they hatch in about three days, producing small greenish worms, that in fifteen days become the wrigglers of open rain-water casks and stagnant pools, breathing through the tail, and darting first one way, and then another. From this pupa state, they emerge as full-grown Musquitoes, Gnats, Midges, &c., breathing through openings in the sides, and ready to pierce the flesh, suck the blood, and instil their inflammatory poison into the wounds made by their pointed proboscis;—four or five generations may be

produced in a summer. A few small fish kept in a cistern or open water cask, destroy the larvæ of the gnats as fast as they hatch, and prove a sure defence against these annoying insects, so far as this source is concerned.

In warm climates, these insects are a serious trouble ; it there becomes indispensable to protect beds against them at night by a netting of gauze, called a *Musquito-bar.*

2. *Tipulidæ,* (Lat. *tipula,* a water-spinner,) known as the DADDY-LONG-LEGS. These, in their slender body and feet, considerably resemble the gnats. Their antennæ have, usually, from fourteen to sixteen articulations. Among them are found the insects which do the most serious injury to the crops of the farmer. These are represented in the genus *Cecidomyia,* including the HESSIAN-FLY, *C. destructor,* (Lat. destroyer;) the WHEAT-FLY, the pest of wheat-fields, *C. tritici,* (Lat of wheat;) the WILLOW-FLY, *C. salicis,* (Lat. of a willow,)—found in a reddish gall upon low willow-bushes.

The WHEAT MIDGE PARASITE, *Platygaster,* (Gr. broad-belly,) *tipulæ,* (Lat. of the *tipula,*)—a minute fly, somewhat resembling the Winged Ant,—performs the part of a public benefactor, by depositing its eggs in the larvæ of the Wheat-midge,—(a single egg in each,) and thus preventing the development of great multitudes of them in the perfect form, though, like some other benefactors, it has been charged with committing the very injuries which it has instrumentally limited. There have been collected in Europe twenty thousand species of insects preying on wheat.

3. *Muscidæ,* (Lat. *musca,* a fly.) This is a well known and numerous family, as may be inferred from the fact that not much short of eighteen hundred species are described as existing in Europe alone, which is probably not half the entire number. Meigen, a German, described six hundred species which he collected in a distance of ten miles circumference. The type of the family, is the common House-fly, *Musca domestica,*) but great diversity is exhibited in the habits of different species. Among the various kinds, are the FLESH-FLIES, *Sarcophaga,* (Gr. flesh-feeding;) the CHEESE-FLIES, *Prophila,* (Gr. very fond,) *casei,* (Lat. of cheese,)— the larvæ of which skippers infest cheese; and the species, *P. petasionis,* (Lat. of gammon,) are found in smoked hams.

PLAGUE-FLY. During the prevalence of the Yellow Fever in Norfolk, Va., not very long since, the PLAGUE-FLY, as it is called, made its appearance there in large numbers. This is a flat insect, with black back and red belly, and has very large wings.

Its presence, during the time of pestilence, is regarded as a good omen, it being supposed to devour the malaria.

BOT-FLIES.

4. *Oestridæ*, (Lat. *oestrus*, a gad-bee,) The flies of this family, the larvæ of which are known by the name of *bots*, infest different quadrupeds, and a species fouud in Peru, assails man himself. The horse licks them off his coat; they are then hatched by the warmth and moisture of the mouth, and conveyed to the stomach; sometimes they are laid in the skin of the ox, Antelope, etc., and on the head of sheep. They are called *gastric*, *cutaneous*, and *cervical*, according to the place in which they breed. From the *Oestrus bovis*, (Lat. of an ox,) oxen run to the water for protection. The *Oestrus tarandi* deposits its eggs under the skin of the Rein Deer. The presence of these insects occasions much annoyance and terror to these and other animals, upon whom the larvæ are deposited.

GAD-FLIES.

5. *Tabanidæ*, (Lat. *tabanus*, an ox-fly, or gad-fly.) This family includes the largests insects of the order, having prominent eyes, and a mouth which, in the female, has six, and the male, four piercers. Many of the perfect insects are greedy of flesh and insects,—and some are so even in the larva state. They often become a great pest to cattle. In Africa, it is said, even the lion is afraid of them. The males of these insects draw their nourishment from flowers; the females alone are blood-suckers.

ELEVENTH ORDER. APHANIPTERA. (Gr. αφανὴς, *aphanes*, not manifest; πτεϱὸν, *pteron*, wing.)

This order includes the tribe of Fleas, *Pulicidæ*, (Lat. *pulex*, a flea,) having no proper wings, but simply two scales on each side. All of them are very minute in size, and similar in their habits. In their perfect state, they are parasitic. The larvæ of Fleas, (*Pulex irritans*, (Lat. provoking,) issue from the egg in the form of very small worms, that attain their full size in about twelve days, and feed upon animal matter. In the silken cocoon which they weave for themselves, they pass in quiet the pupa state. The CHIGOE, JIGGER, &c., *P. penetrans*, (Lat. piercing,) is numerous in the West Indies and South America. It often buries itself deeply in the skin, both of men and animals, depositing an immense number of eggs, which, when hatched, are extremely irritating, and sometimes produce ulcers and death.

Against these insects, wormwood is a remedy. For this order Latreille proposed the name SIPHONOSTOMA, (Gr. siphon, or sucker-mouth.)

TWELFTH ORDER. APTERA, (Gr. ἄπτερος, *apteros*, wingless.)

These wingless insects Latreille arranges into two orders,—(1) THYSANAURA, (Gr. *thusanoi*, hairs; *oura*, tail,)—which includes the SUGAR-LICE, &c., *Lepismidæ*, (from *Lepisma*, a scale,)—so named from their minute silver colored scales. They have a row of movable appendages, resembling false legs, on each side of the abdomen, which is terminated by long jointed hairs or bristles; and also the SPRING-TAILS, *Poduridæ*, (from *Podura*, Gr. *pous*, a foot; *oura*, a tail,) that have the abdomen lengthened into a forked tail, by which they are enabled to make surprising leaps. Some species are found on trees or among moss; others beneath stones, or, at the time of a thaw, they are sometimes seen hopping about on the snow.

(2) PARASITA, which includes the different kinds of lice, *Pediculidæ*, (from Lat. *Pediculus*, a louse,)—almost entirely destitute of eyes, most prolific and most disgusting;—their very *name* presents a warning against a want of *cleanliness;* also Bird-lice, *Nirmidæ*, which infest birds, not feeding upon blood, but obtaining their food from the feathers, in which they are found.

Which is the SECOND SUB-KINGDOM? Why are they so named? How do they differ from the Vertebrates and Mollusks? What is said of their skin, limbs and body? In which division is the articulated character of these animals most conspicuous? How is it in the Crustaceans, Spiders, &c.? What is remarked of their muscles? Of what material is the skeleton of most of them formed? What is said of their muscular power? In what respects do these animals most resemble each other? Describe it. What is said of their senses? Of their digestive apparatus, &c.? Into how many classes may they be arranged?

What is the FIFTH PART OF ZOOLOGY? What is the FIRST CLASS OF ARTICULATES? To what does the name INSECTS refer? Of how many sections is the body usually composed? What is said of the antennæ? How many legs and wings have true insects? What contains the digestive organs? How is the breathing accomplished? Describe the circulation and nervous system. What is said of the mouth? What forms the most striking peculiarity? What is said of the eggs? When does the larva state commence? What is said of it as related to different insects? What is the third state? Name its modifications. Describe it as presented in the different orders. What is the insect in its last or perfect state called? How does it differ from the insects as existing in the other states? Can insects long abstain from food? What facts illustrate their prolific nature? Where do they deposit their eggs? In what respects are they beneficial, and in what injurious? What is said of their locomotive powers? Of their organs of sense? Are they all terrestrial?

What is the FIRST ORDER OF INSECTS? What is said of its number? Of its various forms, colors, &c.? What characters of this order are given? What is said of the Tiger Beetles? Give particulars respecting the other families named? What is the SECOND ORDER? What characters are given? In what two genera are its insects included? What is said of them? What is the THIRD ORDER? In what other order are these insects sometimes included? Why? How do they differ from it? Describe the EARWIGS. What is the FOURTH ORDER? Give its characters. Name the sections into which the order has been divided. Upon what are they founded? Describe the families referred to. What is the FIFTH ORDER? What genera does it include? Describe the wings of these insects. Why called Caddis-flies? What is the SIXTH ORDER? What characters are given? How many families is it said to include? What is said of the DRAGON-FLY? Of the ANT-LION? Of the May-flies? State particulars respecting the TERMITES. What is the SEVENTH ORDER? What characteristics can you give? In what respects are these insects remarkable? What two sections does this order include? What is the chief peculiarity of each? What is said of the SAW-FLIES? Describe the different species of WASPS. Of ANTS and BEES. What is the EIGHTH ORDER? What are its leading characteristics? What division of insects does it include? What are its leading characteristics? Give some account of the Butterflies. Of the Hawk-Moths and Moths Proper. What is the NINTH ORDER? How is it distinguished? What two sections does it include? How are the Homoptera characterised? What is said of the Bark-lice? Of the Plant-lice? Of the Harvest-flies or Cicadidæ? Of the Tree-hoppers? How are the Heteroptera characterised? Describe the Families mentioned. What is the TENTH ORDER? How are these insects distinguished, &c.? Describe the families referred to. What does the ELEVENTH ORDER include? What is said of them? What is the TWELFTH ORDER? How does Latreille arrange it? What is said of the Sugar-lice? Of the Spring-tails? Of the Lice-Tribe?

SECTION II.

SECOND CLASS. MYRIAPODA, (Gr. *μυρίος*, *murios*, innumerable; *πούς*, *pous*, a foot.)

The Articulates of this class occupy a position between insects and worms. They agree with the Annelidans in the lengthened extension of their trunk, in the similarity of the segments from one end of the body to the other, and in their cylindrical form. They, however, have more complete eyes than any of the Worms; and in their breathing apparatus and other parts of their organization, are more like the Insects. From the latter they differ in the absence of wings, and in having the body divided into a series of segments, each of which is provided with a pair of legs.

The Class is divided into two orders: I. CHILOPODA; II. CHILOGNATHA.

In both orders, the first segment, or head, is furnished with numerous eyes on each side, and also with a pair of jointed antennæ; the mouth is fitted for mastication, being provided with a pair of powerful cutting jaws; in the centipede and its allies it has also a pair of appendages formed by a metamorphosis of the legs of the first segment of the body. These are adapted not only to hold and tear its prey, but to convey poison into the wounds thus made, the poison being ejected through a minute aperture near their points, (Carpenter.) The covering of these animals is firm and of a horny character. The number of feet varies from twelve pair to upwards of three hundred.

The muscular apparatus consists of a series of distinct muscles for moving the segments and legs. When the young is hatched, it consists of but few segments, but these increase in number until it is fully grown, by the sub-division of the last segment but one. The first number of segments is eight or nine; but they continue to increase until the number is sixty or seventy. The larva has no legs, these organs not appearing until after the first exuviation of the skin. During their growth, the Myriapoda have considerable power to reproduce lost portions of their body, such as the legs and antennæ, but this power is lost when their development ceases. The bite of these animals is said to be more injurious than that of scorpions, but not often fatal. Ammonia is the best remedy.

FIRST ORDER. CHILOPODA, (Gr. χεῖλος, *cheilos*, lip; πόυς, *pous*, a foot; i. e., lip formed from foot.)

CENTIPEDES.

This Order contains sixteen genera, including about one hundred species, and arranged into four families. The name *Scolopendridæ*, was formerly given to it, but is now appropriated to one of the families of which the leading genus is *Scolopendra.*

The CENTIPEDE, *Scolopendra*, (Gr. *centipede*,) has four pair of eyes, a flattened body containing, with the head, twenty-two segments, and one pair of legs to each segment. Under the second lip, which is formed by the second pair of dilated feet, and terminates in a sharp hook, is an opening through which a poisonous fluid is thrown out. These animals are nocturnal; and in the West India Islands and the hot parts of this continent, they are formidable pests. They often find their way into beds, in the most cleanly houses. Their bite is extremely painful when first given, and is followed by local inflammation and fever. The Centipedes of this genus live upon animal matter, and run rapidly. They grow to be five or six, and even twelve inches in length.

The ELECTRIC CENTIPEDE, *Geophilus*, (Gr. loving the ground,) *electricus*, possesses electrical properties, giving out at night a light nearly equal to that of the glow-worm. Some species of this genus will live a day or two in water, and for the space of two weeks, parts of the body will stir after being separated.

The CENTIPEDE, *Scutigera*, (Lat. shield-bearing,) *coleoptrata*, is widely diffused on the Eastern continent. It is found in the United States, to which it is supposed to have been introduced in shipping.

Other prominent species are the LONG-HORNED CENTIPEDE, (see Chart,) and the Brush-tailed Centipede, (Plate XV. fig. 18b.)

SECOND ORDER. CHILOGNATHA, (Gr. *χεῖλος*, *cheilos*, a lip; *γνάθος*, *gnathos*, a jaw; i. e., lip formed from the jaw.)

MILLIPEDES.

This order includes the Millipedes, which have two pair of feet, (Plate XV. fig. 18a,) attached to each of the numerous segments, and usually terminated by a simple claw. They are nearly allied to the Centipedes, but the body, instead of being flattened, is often cylindrical. These animals move slowly; when disturbed or at rest, they roll themselves up into the form of a ball. Their eyes are composed of numerous hexagonal lenses, as in the insect tribes. The spiracles or breathing holes are situated behind each pair of feet. Besides these, there are outlets for odoriferous glands, situated on the sides. The Millipedes usually feed upon putrescent matter. They are included in fourteen genera, with about eighty species, embraced in six families.

The GALLY-WORM, or THOUSAND-LEGGED WORM, *Iulus terrestris*, (see Chart,) has about forty segments, to which are attached innumerable feet, in pairs or fours. When disturbed, this worm gives forth a fluid of a very disagreeable odor, from the orifices on the sides of the body.

There are five species of the genus *Iulus*. The BORDERED IULUS, *I. marginatus*, (Lat. bordered,) is about three inches long, blackish, with a rufous border on the segments. This is common in the United States.

The PILL CENTIPEDE, *Glomeris*, (a ball,) is a myriapode resembling the wood louse in its form, and its habit of rolling itself into a ball.

THIRD CLASS. ARACHNIDA, (Gr. ἀράχνη, *arachne*, a spider.)

These animals, including Spiders, Mites, and Scorpions, are separated from Insects on account of their external form, structure and habits. They differ from Insects in having no antennæ, in the eyes, which are in most species eight, and even when two in number, are never placed on the side of the head; in the legs, which are usually eight, though in some species six, and in others ten in number; in the breathing apparatus, consisting of radiated wind-pipes, communicating with a sort of gills inclosed in pouches in the lower part of the abdomen.

The skin of the Arachnida is in general rather leathery and horny; like the bones of the larger animals, giving support to the soft parts, and attachment to the muscles, the legs being externally united to a common breast plate, from which they radiate.

The greater portion of these animals are carnivorous, and furnished with organs adapted to their predatory life.

Nerve-knots, or *ganglia*, make up the nervous system of the Arachnida. These are uniform in their composition, and more concentrated than in the Insects.

The organ of hearing in these animals is not known; though it is certain that they hear. The eyes of Spiders and Scorpions, externally formed in exactly the same manner, "are smooth, glittering, and without divisions; and are as much dispersed as those that are disposed at random over the body. The Wolf-Spider, which catches its prey by leaping on it, has its eyes placed in the same manner."

Male spiders are uniformly much smaller than the females, being often not one-fourth as large.

The female spider lays nearly one thousand eggs in a season. These are soft and compressible, before they are laid, lying in the ovarium, or egg-bag, within the spider's body, squeezed together in a flat manner, but when laid, assuming a round form. The eggs are excluded unlike those of birds, from a cavity just behind the breast. Here there is a hook-like organ which the spider can move in such a manner as to direct each egg to the exact spot in the nest cup where it would have it placed. The sense of touch in this organ must be very acute, as by touch alone it can be guided, the eyes being so situated in the upper part of the head, that they cannot be brought within sight of the nest.

Latreille arranges the Arachnida into two orders:—

I. *Pulmonaria*, (Lat. *pulmo*, a lung,) which have pulmonary sacs or air-pipes for respiration, similar to those of Insects, and from six to twelve eyelets. These include the COMMON SPIDERS,

Araneidæ, (Lat. *araneida*, a spider,) usually having eight feet, (Plate XV. fig. 19b.) The palpi or feelers resemble small feet, without a claw at the tip. The frontal ones are terminated by a movable hook curving downwards, having on the under side a slit for the emission of a poisonous fluid which is secreted in a gland of the preceding joint. Though much is said of the effects of spider-bites, "there is still wanting evidence on which to rest the charge of poisoning man by biting him," even against spiders of tropical climates. Sometimes, however, the bite of the larger ones produces unpleasant inflammation. At the same time, people "have been known to eat them with bread, as a great delicacy."

The most remarkable office of spiders is that of weaving their webs, by means of a silken thread drawn from fleshy warts situated on the abdomen, four to six in number, containing thousands of openings, from each of which descends a thread, so thin as to be invisible to the naked eye until all are formed into a common thread. One set of warts or spinnerets is employed in producing threads which are glutinous, while another set produces those which are smooth. This may be shown by throwing some dust upon a spider's web like that of the GARDEN SPIDER, *Epira* (Gr. *peirō*, to affix,) *diadema*, which weaves one of the strongest, when it will be found to adhere to those which are spirally arranged, but not to those which radiate from the center, which are the stronger ones. Their webs have been manufactured into stockings and gloves; to obtain one pound of spider's silk, however, the webs of six hundred thousand spiders would be needed.

A curious thing in the natural history of spiders is their power of reproducing their limbs after they have been broken off; in such cases it is never a part of a leg which is reproduced; but if a part of a leg be removed, it proceeds to throw off the residue, and after the next moult, the missing limb again appears.

The Mason, or TRAP-DOOR SPIDER, *Mygale*, (Gr. *mugalē*, a mouse-spider,) *cæmentaria*, constructs a sort of tube in which it dwells and lies in wait for such animals as come within reach. Some of the holes or tubes are closed by a trap door. The largest species is found in South America.

The *Lycosa* (Gr. *lukos*, a kind of spider,) *tarantula*, the TARANTULA of Italy, is the poisonous species the bite of which, it has been supposed, could be cured by music. Some species of the same genus are found in the United States.

The *Pedipalpi*, (Lat. feelers to the foot,) differ from the Spiders proper, chiefly in the great development of the palpi or feelers, which form long arms, ending in a pincer-like claw.

The Scorpions, *Scorpionidæ,* form the larger part of this division. These have a jointed, tail-like extension of the abdomen, ending in a curved spur. (Plate XV. fig. 19a.) They are found in temperate as well as tropical regions, living under stones, in damp places, and even in houses. They are particularly fond of the eggs of spiders and insects. Their sting is said to become increasingly poisonous as the animal grows older.

The generality of Scorpions, as *Scorpio Europæus*, have six eyes; but there are some of the most formidable kind, as *Scorpio afer*, the African Scorpion, which have eight.

Second Order. Trachearia, (Gr. τραχεια, *tracheia*, a windpipe.)

This includes those forms of the class which have two or four eyes, and breathe by means of trachial tubes or air-pipes, similar to those of Insects. These include, (1) the various forms of Mites, *Acaridæ*, such as the Cheese-Mites, *Acarus*, the Itch-Mite, *A. scabiei*, (of itch,) named from the cutaneous disease of which it is the origin; the Sugar-Mite, *A. saccharinum*, found in the brown sugar of commerce; the Red Spider, *A. tellarius*, the pest of hot houses and green houses, &c. Camphor and sulphur are the best remedies for removing these minute, and some of them almost microscopic animals. These are not now considered as ranking among insects, differing from them as they do, in structure, and having in most cases, like spiders, eight feet, while no insect has more than six feet. (2) Ticks, *Ricinia*, while no insect has more than six feet. (2) Ticks, *Ricinia*, (Lat. *ricinus*, a tick,) embracing the genus *Ixodes*, (Gr. sticky,) (Pl. XV., fig. 19.) some species of which are free, and others parasitic. The latter are without eyes. They are well known from attacking sheep, cows, horses, dogs, and even tortoises, burying their suckers so deeply in the skin that they cannot be removed without tearing the flesh. They deposit a prodigious quantity of eggs, which are discharged from the mouth. (3) Shepherd Spiders, or Harvest-men, *Phalangidæ*, genus *Phalangium*, (Lat. a spider,) of which the greater part live upon the ground, on plants, or at the roots of trees, and are very active; others, which are less active, hide themselves between stones, or in mosses. Their legs are long and slender, the tarsi consisting of more than fifty joints. These spider-like creatures are known as Harry-long-legs. (4) The Sea Spiders, *Nymphonidæ*, are also included in this order, though sometimes referred to the class *Crustacea*,. Our limits do not allow us to enumerate *all* the families, or to give any further particulars respecting those which are mentioned.

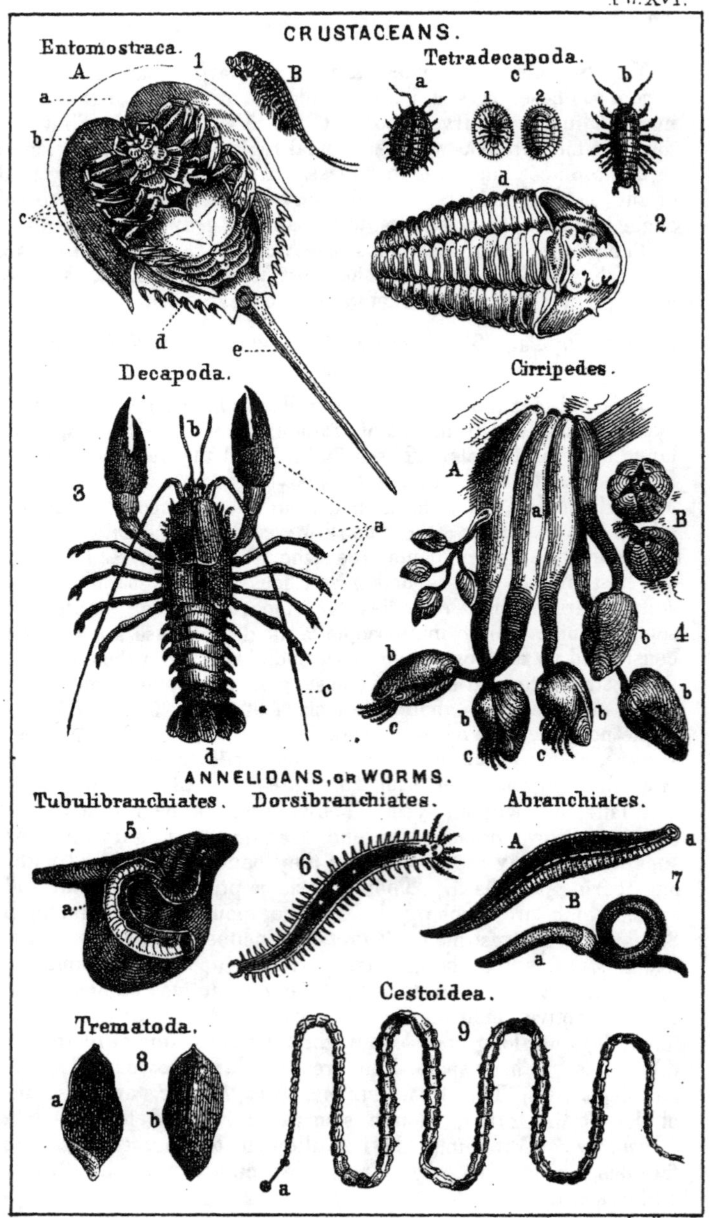
CRUSTACEANS.
Entomostraca.
Tetradecapoda.
Decapoda.
Cirripedes.
ANNELIDANS, OR WORMS.
Tubulibranchiates.
Dorsibranchiates.
Abranchiates.
Cestoidea.
Trematoda.

EXPLANATION OF PLATE XVI.

CRUSTACEANS.

Fig. 1. A. King Crab, Mollucca Crab or Horse Foot; a, the opening through which the animal emerges when casting off the old shell; b, small feet in front of the mouth, considered by some naturalists as *antennæ*, or feelers; c, the other five pair of true feet surrounding the mouth, used for walking and mastication; d, triangular shield covering the body to which the five pair of false feet or swimming legs are attached; e, long pointed tail.

B, Branchipus stagnalis, in road-side ditches and cart-wheel ruts.

Fig. 2a. Sow Bug, or Wood-louse; b, Limnoria terebrans; c, Fluvicola; 1, under side; 2, upper side, d, Trilobite, found only in a fossil state.

Fig. 3. Craw-Fish, or Fresh-water Lobster; (*Astacus fluviatilis;*) a, five pair of true feet; b, inner antennæ, supposed to be the organs of hearing; c, outer and longer antennæ, organs for smelling; d, five broad caudal plates, by which the animal is assisted to dart about so rapidly.

Fig. 4. A, Duck Barnacles, Pedunculated or Stalked Cirripedes; a, peduncle or stalk by which they are attached to submerged substances after the third moult; b, shell of five or more valves; c, six pair of feathery cirri for obtaining its food; B, Acorn Barnacles, (Sessile Cirripedes,) composed of six valves, with an operculum or cover of four pieces, between which the curly tentaculæ protrude.

ANNELIDANS, or WORMS.

Fig. 5. Vermilia, (a,) lives in an irregular twisted tube, attached by its whole length to shells, stones, &c.

Fig. 6. Sea Centipede, or Nereis; organs and gills distributed throughout the body; each of its numerous feet has two tubercles, two bundles of bristles, and a cirrus above and beneath.

RED-BLOODED WORMS.

Fig. 7. A. Medicinal Leech; a, the mouth, with three sharp teeth disposed in a triangle; no distinct head; moves by the adhesion and detachment of the sucking disks at each extremity.

B. Earth or Angle Worm; eight retractile bristles on the under side of each ring, assisting in their contractions and dilations, enabling it to creep at a pretty good pace; no distinct head, but the fore part, (a,) acts as a sort of awl in penetrating the earth.

ENTOZOA, INTESTINAL, or WHITE-BLOODED WORMS.

Fig. 8. Fluke, Distoma; a, upper side; b, under side; an inch long; two eyes; two suckers; infests the liver of animals. In sheep it produces or aggravates the disease called the rot.

Fig. 9. Tape worm; flat or ribbon-like: sometimes 60 or 100 feet long, with 500 divisions, each of which adheres to the intestine by a strong sucker, and may become a separate and perfect animal. The segments diminish in size so as to form a thin neck with a small globose head, (a,) mouth very indistinct.

SIXTH BRANCH OF ZOOLOGY.

CRUSTACEOLOGY. (Lat. *crustacea;* Gr. *λογος*, *logos*, a discourse.)

CLASS CRUSTACEA. (Lat. from *crusta*, a shell or crust.)

SECTION I.

THIS Class includes animals, some of which dwell on land, others in fresh or salt water, and which are covered with a soft shell or crust. They are oviparous, and divided into segments or rings, articulated into each other, to the inside of which their muscles are attached. The outer covering generally possesses a considerable degree of hardness, containing no small proportion of carbonate of lime. Its solidity varies; sometimes it is membranous.

The way in which the animals free themselves from the old shell is quite singular; they generally manage to get out of it without occasioning the least change in its form. When the shell is first stripped off, the surface of their bodies is extremely soft; and it is some time before the substance which has been exuded from the pores on the surface of their skin, acquires a hard consistence.

They generally have a distinct heart and a circulatory system or blood vessels, but no internal skeleton, properly so called. They breathe by means of gills or branchial plates, or else by the skin. The breathing apparatus is adapted to aquatic rather than aërial respiration. In those genera in which the head is not separated from the thorax, the shield protects the whole of the thorax. Other genera have the head distinct from the body, which is divided into seven segments, to the lower sides of which the feet are attached; these, for the most part, have a tail, consisting of many segments. The limbs vary from six to fourteen, each having six articulations. The two front limbs, and sometimes even three on each side, are provided with pincers at other times they are terminated by simple hooks, and not unfrequently, they have appendages which fit them for swimming. There are two mandibles, a lip below, and from three to five pairs of jaws; these small, leg-shaped appendages are not adapted to locomotion, but being placed near the mouths, assist in the operation of feeding.

The eyes vary in number, usually being compound, seated on peduncles, which are sometimes movable, and at others fixed. Some of the parasitic species are destitute of eyes in their perfect state, though possessing them when young and able to swim about. The Crustacea have the senses of taste and hearing, probably also have that of smelling, though the precise location of its organ has not been ascertained. Some of them have the power of emitting light in the dark. Others are able not only to detach one of their limbs when seized upon by an adversary, but can reproduce the severed limb. This, however, is always of a less size than the others, until it has once or twice changed its crust.

The Crustaceans include five orders.

First Order. Malacostraca, (Gr. μαλακὸς, *malakos*, soft; ὄστρακον, *ostrakon*, shell;) or Decapoda, (Lat. ten-footed.)

This Order is divided into two sections. (1) *Brachyoura*, (Gr. short-tailed,) including the Crabs, the species of which are various in size, color, and modes of living, having bodies covered by an external skeleton or calcareous crust, with ten articulated limbs, adapted for swimming and for walking also, (with oblique steps,) and breathing by gills. They have two pincer-like claws, and jointed antennæ, and throw off their crust annually, at the end of spring. When they have lost a pincer or foot, it is reproduced with the new shell, and also at other times. When their legs or claws become injured or are touched with a hot iron, they themselves cast them off. The material out of which the new shell is hardened, is furnished by two calcareous concretions, called Crabs' eyes, found in summer on both sides of the stomach. These animals live on dead bodies, putrid flesh, and all descriptions of worms and insects found in water. Our references to species must be confined chiefly to those noticed on the Chart. The Edible Crab, of England and

Western Europe, *Cancer pagurus*, (Gr. *pagouros*, a crab,) sometimes attains a large size, weighing ten or twelve pounds. It casts its shell between Christmas and Easter.

The COMMON EDIBLE CRAB of the United States, *Lupa hastata*, (Lat. from *hasta*, a spear,) is of smaller size. In forty-eight hours after the old shell is cast, it is renewed and firmly consolidated. In the interval, these Crabs are termed "Soft-shelled," and eagerly sought after. They are abundant in the muddy shores of bays and inlets.

LONG-TAILED CRUSTACEANS.

(2) The *Macroura*, (Gr. long-tailed,) are so named from the large and well developed tail, ending in a fan-shaped fin, which assists them in swimming. They walk well, but are best adapted for swimming, shooting backwards through the action of the abdomen and the tail fin. The antennæ are usually long, the first pair not being received into a cavity as in the Crabs.

The *Palinurus*, is a large crab, shaped much like a lobster, but is more cylindrical with none of the feet cheliform or claw-like. It is armed with a very hard crust, and is in general use when in season, as an article of food. Prof. Dana, (Silliman's Journal.) speaks of two gigantic species of this genus, *P. vulgaris*, of the Mediterranean, (see Chart,) and *P. lalandii*, of the Cape of Good Hope, each a foot and a half long, independent of the antennæ. The COMMON PRAWN, *Palæmon vulgaris*, found near our coasts or at the mouths of rivers, is nearly allied to the Prawn of England, *P. serratus*, (Lat. saw-shaped ;) these are esteemed a great delicacy ; also allied to the species *P. squilla*, (a prawn,) of the European continent. The Common Prawn is not more than an inch and a half in length; but the RIVER Prawn of the Carolinas and Florida, *P. fluviatilis*, is seven or eight inches long.

The RIVER CRAY OR CRAW FISH, *Astacus*, (Gr. *astakos*, a kind of lobster,) *fluviatilis*, (Plate XVI. fig. 3, with explanation,) may be regarded as the Fresh Water Lobster. It is found in the fresh waters of Europe and the north of Asia, placing itself in holes of the banks, or under stones, where it lies in wait for small mollusks, little fishes, the larvæ of insects and decomposed animal substances upon which it subsists. It is said that it will live for upwards of twenty years, and becomes large in proportion to its age. The eggs, when laid, are collected under the lower part of the body or tail ; the young, which at birth are very soft, take refuge under the tail of the parent for some days.

Several species of the River Lobster are found in the United States.

The LOBSTER, *Astacus marinus*, is very abundant and of great commercial value. Good sized ones are four and a half inches long, from the tip of the head to the end of the back shell. The pincers of one of the tail claws are furnished with knobs, and those of the other claw are serrated. With the former it keeps firm hold of the stalks of sub-marine plants, and with the latter, it cuts and minces its food very dexterously. The fecundity of the Lobster is very great. Dr. Baster says that "he counted 12,444 eggs under the tail of one female lobster, besides those that remained in the body unprotruded." In a boiled lobster they are bright red and called the coral. Lobsters are very voracious, and are caught at night in pots or traps made of twigs baited with garbage, (refuse flesh, entrails, &c.,) or in nets let down into the sea, the place being marked by a buoy. Sometimes they are taken by torch light, with a pair of tongs or forceps of wood. Their eyes are placed so that they can see in every direction. When alarmed, they spring to a surprising distance. Usually they weigh one or two pounds, but sometimes four or even six.

The COMMON LOBSTER of this country, *Homarus*, (Astacus,) *Americanus*, (see Chart,) is, however, much larger, averaging in weight four pounds, and sometimes reaching the weight of fifteen, twenty, and even thirty pounds. The Common Shrimp of Europe, *Crangon*, (Gr. *krangōn*, a shrimp.) *vulgaris*, is closely allied to the shrimp of our own country; it is of a pale greenish color, about an inch and a half or an inch and three-fourths in length. The *C. septemspinosus*, (Lat. seven-spined,) is known by the name of Bait shrimp, and extensively used. It is found from Florida to the Arctic regions.

(3) *Anomoura*, (Gr. anomalous tail.) This is a section intermediate to the two preceding, including crabs having the front part of the body crustaceous; the lower part soft and rolled upon itself. They are in the habit of resorting to the dead shell of a univalve mollusk, which is exchanged for a larger one as they increase in size, and seem to prefer the shells of the *Trochoidæ*, (see Chart.) Hence they are called HERMIT CRABS.

SECOND ORDER. TETRADECAPODA, (Gr. fourteen-footed.)

This includes several families of small Crustaceans, some of them marine or fresh-water species; some of them terrestrial or parasitic, which, from the number of their feet, may be referred to the present order. We follow Prof. Dana in placing here (1) the WOOD-LICE, *Oniscidæ*, (Gr. *oniskos*. a wood-louse,)—sometimes referred to the order *Isopoda*, (Gr. equal-footed.) These have fourteen slender feet, adapted to walking, and the first pair of antennæ rudimentary; the second pair being alone complete and conspicuous. The species *Oniscus asellus*, (Lat. a chee-slip or sow-insect,) (Plate XVI. fig. 2a,) is the COMMON SOW-BUG, found under stones and decaying wood. The fourteen feet gradually increase in size from the front; the antennæ have eight articulations. The Sow-bug feeds upon decomposed vegetables. It carries its eggs in a sac beneath the body. The color is dusky brown above; beneath greyish white. The *Porcellio*, (Lat. a sow-bug or wood-louse,) *spinicornis*, (Lat. spiny horns,) is very similar to the *Oniscus*, but its antennæ have only seven articulations. The third joint of the antennæ is armed with an acute *spine*. This also has the popular name of Sow-bug, and is found in similar situations with the preceding. The species *P. granulatus*, (Lat. granulated,) or Hog-louse, is black and unspotted. It is roughened with small elevated grains,—whence the specific name.

To the Sow-bugs are similar, in most respects, those of the genus *Armadillo*, which, from their habit of rolling themselves into a ball, are known by the name of PILL-BUGS. These are of dull lead color, with three lines of large yellowish spots on the upper part.

The WHALE-LOUSE, *Cyamus ceti*, (see Chart,) referred by Latreille to the order LÆMODIPODA, (Gr. throat or jaw-footed,)—the *Oniscus ceti*, of Linn., has at least *twelve* feet, of which eight are perfect, and the others in the form of slender, jointed appendages. It attaches itself to whales, and occasionally to tunnies and other large fish.

(2) The SAND, or BEACH FLEAS, *Gammaridæ*,—sometimes referred to the order AMPHIPODA, (so named from having two kinds of feet, cheliform or claw-like, and simple.) The family includes several genera and species. *Orchestria*, (from Gr. *orchecomai*, to leap,) *longicornis*, (Lat. long horned,) is a species having the lower antennæ longer than the body, and the four front feet terminated in a compressed claw. They subsist upon dead animal substances. They are found on the shores of Long Island, where, to conceal themselves, they dig holes in the sand. Other Sand or Beach Fleas are included in the genus *Talitrus*. Of this is the species *T. quadrifidus*, (Lat. four-cleft,) which have a body composed of thirteen segments, exclusive of the head; and the antennæ shorter than the body. The tail has

three appendages terminating in four spines,—whence the specific name. These are of a dark horn color, and frequently found hidden under stones and sea-weed.

The genus *Gammarus* includes Fresh-water Shrimps, which are very active, and common in running streams. They may often be found under stones and pieces of wood. These have the last joint of the antennæ composed of numerous minute ones; the upper antennæ are as long as the lower, and sometimes longer, with four articulations, the last ending in a bristle,—the lower antennæ have five articulations; the tail has small, bundle-like spines.

We can barely name the parasitic *Cymothoids*, (genus *Cymothoa*,) which are fourteen footed,—formerly arranged with the Isopoda, most of which attach themselves to the mouths and gills of fishes,—and of which the *Seriolis* has been thought to present, at first sight, a resemblance to the extinct form of the Trilobites.

Ligia is another genus of the present order, having an oval, oblong body, with transverse segments, and two short appendages at the end of the tail. The two outer antennæ are quite conspicuous. (See figure of *L. oceanica*, Plate XVI. fig. 2b.)

Limnoria is another marine genus, which has the head nearly as large as the first segment; the tail has six distinct rings with two appendages on each side. The species *L. terebrans*, (Lat. boring,) can roll themselves up into a ball. These, and the *Ligia oceanica*, both in great numbers sometimes attack the timbers of ships, docks, etc., and soon render them useless.

Third Order. Entomostraca, (Gr. εντομα, *entoma*, an insect; ὄστρακον, *ostrakōn*, a shell, i. e., shell insects.)

This term is applied to Crustaceans for the most part inhabiting fresh-water. In these, the nervous knots which supply the place of the brain, consist of one or two globules merely. The heart assumes the form of a long vessel. The gills are composed of hair-like processes, forming a portion of the feet, or of a certain number among them, and sometimes the mandibles and upper jaws. The number of feet varies, and in some genera is said to be over a hundred. Nearly all have a shell, consisting of one or two pieces, generally almost membranous and transparent,—the coverings are like those of the insects, rather horny than calcareous. The antennæ, varying much in form and number, serve in many species for swimming.

Dr. Baird says, most of them are "essentially carnivorous."

In this he discerns a decided fitness, as tending to prevent the hurtful effects of putrid air that might attend the decomposition of the amazing number of these animals abounding in ponds and ditches. These Crustaceans, however, in their turn, become the prey of other animals. They form a considerable part of the food of fishes; and it is thought that the quality of some of the fresh-water fishes, of which a species of trout may be particularly mentioned, may, in some degree, depend upon the abundance of this portion of their food. Among the genera belonging to this order, we refer first to the *Cyclops*, (see Chart,) (Gr. circular or rounded eye,)—a fresh-water genus, in which the body is pear-shaped, and the upper, or larger pair of antennæ, are employed as aids to locomotion. Species of these may be seen jerking themselves along in springs and stagnant waters. When they lose part of an antenna, it reappears as, in the case of some others of the class, at the time of the next moult. They are carnivorous, and when without other food, even eat up their own young. Some of the kindred marine species appear to be phosphorescent. These minute Crustaceans are very prolific. They are tenacious of life, reviving after having been frozen, though they soon die when removed from water and dried. Many of them furnish food to the water larvæ of insects.

(2) *Daphnia*, (Gr. a laurel-berry,)—the ARBORESCENT WATER-FLEA. In this genus the body is enclosed in a bivalve shell, though the head is exposed, having a compound and somewhat movable eye. These Crustaceans are found in stagnant waters in company with the Cyclops, which they resemble in their movements. They are sometimes so numerous in water as to give it "a muddy hue, like the red dust of iron, or as if blood had been mixed with it." On the back of the shell is seen, at certain seasons, a black saddle-shaped appendage, containing two eggs, from which, in the spring, the species are reproduced.

(3) *Cypris*, (see Chart.) This is likewise enclosed in a bivalve shell, with a dorsal hinge. The antennæ are four,—the second pair large, and fitted to aid in swimming. Many species may be seen in summer-time swimming about in stagnant pools, and they often show beautiful variations of color.

(4) *Limulus* or *Polyphemus*,—this is sometimes referred to the order XIPHOSURA, (Gr. sword-tail,)—a name referring to the long, hard, and sharp tail-spine of this creature, which, in some places, is used for pointing spears. The body is covered with a large carapace shield, (Plate XVI. fig. 1;) is rounded in front, having the hind part smallest, with spines on the sides, and deep notches behind; the gill-feet are appended to the abdomen. Six feet,

strongly articulated and adapted to walking, are attached to the thorax. The common name of these Crustaceans, is the King-Crab or Horse-foot. The first name refers to its size, the last to its shape. They feed on animal substances, and are gathered as food for hogs and poultry, and also used as manure. Lamarck calls them giant branchiopods, in allusion to the gigantic stature of some of the species. The color is of a uniform dark brown. To this order we assign the TRILOBITES, (sometimes arranged in a separate order,)—fossil animals, the knowledge of which is limited to the shell or crust. (Plate XVI. fig 2d.) Feet have not been found in connexion with their remains, so that it cannot be certainly known whether or not they possessed these members. Agassiz remarks, "there is an incompleteness and want of development in the form of their body that strongly reminds us of the embryo among the Crabs." Their food is supposed to have been small water animals; their habitat the vicinity of coasts in shallow waters, where they lived gregariously in vast numbers. We here also place *Fluvicola Herricki*, a singular Crustaceous animal which has been found adhering to rocks in and near the water of West Canada Creek. "It is detached with considerable difficulty, and when so detached, partially rolls itself up." (DeKay.) The locality in which they are found, is noted for fossils and petrefactions; and, as De Kay intimates, it is a singular coincidence that it should furnish animals so strongly resembling the extinct trilobites, see Plate XVI. (fig. 1 and 2c,)—which presents figures of some of these animals that were found in Clinton, Oneida county, N. Y., in a ravine a little North of Hamilton College. They seem to be allied to the present order.

FOURTH ORDER. CIRRIPEDES, (Lat. *cirri*, ringlets or tufts; *pedes*, feet.)

These animals were ranked by the earlier naturalists among the Mollusks, and they certainly possess many characters in common with some of those animals, yet exhibit greater symmetry of form. The body is prolonged, and from each side proceed long and slender feet, curving together into a kind of curl,—whence the name *Cirripedes*, curl or tuft footed. They are inclosed in a shell, which is more or less conical. These animals are subject to a metamorphosis, the young having two valves like the bivalve Mollusks, and capable of swimming about until they become permanently affixed. In this state, they are able to protrude the limbs from the fore part of the shell, the front pair being of considerable size, and furnished with a sucker and hooks for attachment to submarine substances. The six hind pair of limbs are used for swimming. The shell is not made up of simple layers, as in the Mollusks, but is traversed by a complex series of canals, through which nourishment is conveyed.

The Cirripedes are divided into two principal groups,—the *pedunculated* and the *sessile*, both of which are widely distributed by ships, floating wood, sea-weed, mollusks, turtles, whales, etc.

I. Campylosomata, (Gr. *καμπύλος*, *kampulos*, curved; *σῶμα*, *soma*, body.) The division contains the pedunculated forms, that is, those which are furnished with stems, (Plate XVI. fig. 4A,) by which they attach themselves to wood or other objects,—among them are the *Anatifa*, (Lepas,) Common Barnacle, consisting of five pieces, of which two are large valves, somewhat like those of a muscle; two smaller are articulated to those near the point; and one unites the valve along the back edge; and thus they envelop the whole of the mantle. Barnacles often adhere to the bottoms of ships in such numbers as to impede their sailing.

II. Acamptosoma, (Gr. ἄκαμπτος, *akamptos*, uncurved; σῶμα, body.) This section includes the sessile or unpedunculated forms. It is represented by the *Balanus*, (Lat. acorn,) or Acorn-shell, (Plate XVI. fig. 4B,) so named from its resemblance to the acorn, it being short and conical in form. The mouth is protected by an operculum, consisting of two or more valves. These animals are found in great numbers on rocks and piers along the coast. The species *B. psittacus*, (a parrot,) is quite large; it is eaten by the natives of Chili. The *Coronula* attaches itself to the backs of whales, imbedding itself in the skin.

Fifth Order. Rotatoria, or Rotifera. Wheel-bearing Animalcules.

This order includes animalcules not to be distinctly perceived, except with the microscope. They receive their name from peculiarities of structure, and are wonderfully minute,—some of them being less than the five-hundredth part of an inch in length. Nearly all of them are aquatic in their habits; their bodies are transparent; hence, their general structure can, with the help of the microscope, be easily recognised. They have usually an elongated form, similar on the two sides; and at the front extremity are one or two rows of vibratile *cilia*, usually arranged in a circular manner, which, when in motion, appear like revolving *wheels*. The posterior extremity is prolonged into a tail, possessing three joints, each of which has a pair of prongs or points. The circular arrangement of the *cilia* forms what are called the *wheels*. By the successive vibration of these, the appearance of a continual rotation is produced; and their action creates rapid currents in the surrounding fluid, by which the supply of food is obtained,—consisting of other animalcules of still smaller size, and less complex structure. Between the wheels, the head is occasionally protruded, bearing two red spots, supposed to be eyes; on its under surface there is a projecting tubular spike, which is believed to act as a syphon conveying water into the general cavity to aid perspiration. The vital power of some species is extraordinarily great, they having been known to revive after being kept in dry sand for four years. The wheel-animalcules do not propagate by spontaneous division, but by eggs inconceivably minute, so that they can be raised in the air with vapor, and transported in every direction. Much diversity of opinion has existed in relation to the proper classification of these animalcules, of the wonderful structure and variety which the microscope has made such interesting revelations; but the lengthened form of their bodies, the location of the mouth and eyes at one extremity; the occasional appearance of cross or transverse lines shadowing forth a division into segments; and especially the character of the nervous system, so far as it can be ascertained, are among the proofs that they should have a place with the Articulates. With these, Dr. Grant was one of the first to place them. Leydig proposed to call them CILIATED CRUSTACEANS. We follow the suggestion of Prof. Dana in placing them next the Cirripedes.

The common species, *Rotifa vulgaris*, is remarkable for the two circles of vibratile cilia or vibrillæ, referred to above, and for the posterior forceps or pincers. One species (*Melicertă ringens*) has the power to withdraw itself into an outward case; and has the vibratile cilia distributed into four divisions.

SEVENTH BRANCH OF ZOOLOGY.

HELMINTHOLOGY. (Gr. ῞ελμινς, *helmins*, a worm; λογος, a discourse.)

Class Annelidans or Worms.

This lowest division of the Articulates is arranged by Cuvier and other naturalists, into *two* sections; the one embracing the class Annelidans, or Red-blooded Worms, and ranked with the Articulates; the other, including the Intestinal or White-blooded Worms, is ranked by them with the Radiates. Agassiz considers the nervous system of the latter Worms, which has been made a ground of their separation, though somewhat different, as yet *essentially* the same with that of the Articulates. We follow him as well as other distinguished naturalists, in placing *all* the Worms in this latter class.

First Division. Annelida, (Lat. *annulus*, a ring.) Red-blooded Worms.

These always have their bodies formed of a great number of small rings nearly equal in size, varying in number from twenty or thirty to more than five hundred, according to the length of the animals. Their skin is soft and pliable; and their bodies, not having any external skeleton, are also soft, and in general more or less cylindrical. The head is usually distinct, furnished with two or four eyes; the sides have attached to them feet, or rather bristle-like projections, which are used for locomotion, and vary widely in different species. Most of the annelidans are marine; but some live in fresh water.

FIRST ORDER. TUBULIBRANCHIA, (*Tubulicolæ*, dwelling in tubes, Cuvier.) SEDENTARY ANNELIDANS.

These are characterized by having their branchiæ in the form of plumes, or of small tree-like figures, attached to the head or fore part of the body. Nearly all inhabit tubes, which are calcareous, sandy, or membranous. The order may be arranged into two families.

First Family *Serpulidæ*, (Lat. *serpula*, a small snake.)

Worms, the tubes of which are calcareous and singularly twisted. They have the branchial tufts separated into two distinct parts by a pendunculated operculum, or else protected by a solid one when they are drawn into the shell.

1. *Serpula*. This genus includes worms which adhere to stones, shells, and other sub-marine substances. The branchiæ are of a beautiful red, or variegated with yellow and violet, and used in taking the minute living objects upon which the worm subsists. They are found in the Mediterranean and European seas. This genus has been estimated to embrace sixty or more recent and fossil species. (Fig. on Chart.)

2. *Vermilia*,—Worms so named from the red line on each side of the ridge which appears upon the back. (Plate XVI. fig. 5.)

3. *Ditrupa*,—Worms free, living in a tubular shell, open at both ends, with twenty-two branchiæ, in two sets, and feathered with a row of cilia. These are nearly allied to the *Serpula*.

SECOND FAMILY, *Amphitritidæ*.

Worms which have around the mouth numerous thread-like tentacles; and tubes formed by a mucous secretion to which are attached fragments of shells, etc.

1. *Amphitrite*. These have the thread or straw-like processes in the form of a comb or that of a crown.

2. *Sabella*,—Worms about the size of a finger, living in tubes composed of sand, clay or fine mud. The plumes are highly brilliant and delicate, sometimes of a rich orange color.

3. *Terebella*,—Worms living in tubes of similar composition with the preceding, having on the neck arborescent, not fan-shaped gills.

Second Order. Dorsibranchiata, (Lat. *dorsum*, back; *branchiæ*, gills.)

These Annelidans have their organs, and especially their branchiæ, distributed nearly equally along the whole or a part of the body. All the species are aquatic and worm-like, swimming with facility and active in crawling. The head is distinct from the trunk; they are furnished with two pair of rudimentary eyes. The order includes the Sea-Mice and the Sea-Centipedes, arranged into several families or groups.

First Family. *Aphroditidæ*, (Gr. from *Aphroditē*, Venus.) Sea-Mice.

These include species oval in form, some of which are superbly colored. Usually they have two pairs of jaws. The gills are concealed under two rows of scales covering the back, and hidden by a kind of flocky down or tow, from which issue brilliant spines or bristles. The species *Aphrodite aculeata*, (Lat. prickly,) is six or eight inches long, and two or three inches wide. Cuvier says that these Sea-mice do not yield in beauty either to the plumage of the Humming-birds, or to the most brilliant precious stones.

Second Family. *Eunicidæ.*

These are represented by the genus *Eunice.* This is furnished with tuft-like gills, and has the trunk armed with three pairs of horny jaws. Each of the feet has two cirri and a bundle of bristles; there are two tentacles on the head, above the mouth, and two on the neck. The Gigantic Eunice, *E. gigantea*, found in the seas around the Antilles, is sometimes upwards of four feet in length, being the largest annelidan known.

Third Order. Abranchiata, (Lat. *a*, priv. or without; *branchiæ*, gills.)

The Worms of this order are without branchiæ, respiration being accomplished by means of the skin. The order includes two principal groups, of which the one is terrestrial, the other aquatic.

FIRST FAMILY. *Lumbricidæ*, (Lat. *lumbricus*, an earth-worm.)

The EARTH or ANGLE WORM has a body composed entirely of numerous rings; is of a reddish or bluish hue, and of a shining aspect. It secretes a viscous or glutinous substance which protects the body and greatly facilitates its progress through the earth. This worm is enabled to creep at a good pace, by contracting and dilating its rings, the retractile bristles on the under side of each ring assisting locomotion. The fore part of the head in earth-worms acts as an awl in penetrating the earth, which they loosen, enrich, and prepare for the labors of the farmer by admitting the air and water. By their castings, which so annoy the gardener, they, in a few years, cover a barren waste with vegetable, or rather animal mould. Not improbably every particle of earth in old pastures has passed through the intestines of worms. They are known as coming to the surface in wet weather and at night. The power of reproducing mutilated parts is very great in this entire family, of which more than twenty species have been described. The eggs are in capsules, or membranous cocoons. Each egg produces two worms. The species *Lumbricus terrestris* (Plate XVI. fig. 7b,) attains nearly a foot in length, and has a hundred and twenty rings.

SECOND FAMILY. *Hirudinidæ*, (Lat. *hirudo*, a leech.) LEECHES.

These include various genera, both marine and fresh water. All are without limbs or bristles, but have a sucker at each end of the body, which enables them to move about and to adhere to living bodies, penetrating the skin, by means of their three jaws and teeth, and drawing the blood, upon which they were formerly supposed to subsist.* Two species of Leech are almost exclusively medicinal; the GREEN LEECH, *Hirudo officinalis*, and the BROWN LEECH, (spotted underneath,) *H. medicinalis*. Other species are, however, sometimes used. Fresh-water leeches soon die after having been removed from the water. Many leeches have eight eyes. There are several marine species which attach themselves to Torpedoes, Turtles and

* It is very remarkable that blood is not the natural food of the Leech; and that the fluid which it so greedily swallows, does not pass into the intestines, but remains in the stomach for many months, and what is still more curious, it does not coagulate during the whole of that time, as it would do in an hour if exposed to the air, but continues to retain its fluidity. (Gosse.)

Fishes, particularly the Skate. Leeches are so much used for medicinal purposes that methods have been adopted for cultivating them. Some enterprising Frenchmen have recently leased marshes in Ireland, and sowed them broad cast with leeches, in the hope of thus deriving large profits. The value of those annually used in France, is estimated at from one to one and a half million of dollars. The species *H. geobdella*, (Gr. earth-leech,) frequently leaves the water to pursue earth-worms.

FOURTH ORDER. ENTOZOA, (Gr. εντος, *entos*, within; ζῶον, *zōon*, an animal.)

This order includes the various minute animals which are produced and developed *within* other living beings. They are exceedingly various in form and organization, having but one character in which they mostly agree, viz.: that they are parasitic, living within and at the expense of the bodies of other animals.

Some species, both in their appearance and internal structure, so closely resemble individuals placed in other classes, that they can be said to differ from them only in respect to the localities in which they are found. They have been discovered in all the Mammalia, from man down to the Cetacea; and they are even more numerous in Birds, Reptiles, and Fishes than in the Mammals. The invertebrated animals have also parasites peculiar to themselves. They have been found in Insects, Mollusks, and even the Acalephs. They fix themselves, according to the species, in various parts of the bodies which they infest, such as the intestines, brain, liver, kidneys, muscles, blood, and bones. In some cases, the same species are found in water, as well as within animals.

FIRST SUB-ORDER. NEMATOIDEA, (Gr. νῆμα, *nema*, a thread; ειδος, *eidos*, form.) ROUND WORMS.

These are the highest in organization of the Entozoa, having a round, long, and elastic body, and a complicated structure, there being a true intestinal canal. The mouth, by its varieties, affords generic characters; the females are longer than the males, and for the most part oviparous. They have been divided into eleven genera. We have room to notice only (1) those of the genus *Ascaris*, which include the COMMON ROUND WORM, *A. lumbricoides*, so named from its general likeness to the *Lumbricus*, or Earth-Worm. This occurs in the hog and ox as well as in man, and chiefly inhabits the small intestines. The male is smaller and more abundant than the female. This worm is white, from six to twelve or fifteen inches long. It is frequently fatal to children, in which it penetrates to the stomach, and even to the mouth. Five hundred have sometimes been passed from a child in the course of seven or eight days; also Pin-Worms or Thread Worms, *A. vermicularis*, (Lat. from *vermiculus*, a little worm.) These are very minute, the male seldom exceeding two lines, and the female five lines in length, and being proportionally slender.* They dwell in the large intestines, sometimes in immense numbers and producing great irritation.

2. *Filaria*, (including the GUINEA WORMS,) of which three species inhabit the human body. Some are found in various animals including insects and their larvæ. These are long worms, smooth and thread-like, and of a somewhat rigid texture. The GUINEA WORM, *F. Medinensis*, occurs in Arabia, Upper Egypt, Guinea, the West Indies, and other hot climates. It is generally white, but sometimes of a brown color. The length varies from six inches to twelve feet, and it is about as thick as the string of a violin. It infests the muscles and subcutaneous tissues, principally of the lower limbs; sometimes it locates itself about the eye and under the tongue. Occasionally it makes its way to the surface of a skin, creating a pustule or sore, when it may be taken hold of and cautiously and gradually extracted. If broken off, however, the part remaining enclosed produces inflammation, and may render amputation indispensable. Within the tropics, people sometimes seem to be affected by it almost epidemically, nearly half the men in a regiment of soldiers, having at the same time been attacked by it. It seems that it may exist under the skin many months or even a year without being detected. The Guinea worm is said to be sometimes seen swimming in the waters of the countries which it inhabits.

Species of the *Filaria* have recently been found in the blood of dogs. The HAIR-WORMS, *Gordius aquaticus*, (see Chart,) are nematoids found in free water, or as internal parasites of insects. The latter swallow the eggs of the Hair Worm, after they have been deposited in water; and "in this position the egg is hatched, producing the *Gordius*, which becomes impregnated, and escapes from the insect into waters where it deposits its eggs." It has erroneously been supposed to be developed from a horse hair.

* The line here referred to is the twelfth part of a French inch.

Second Sub-Order. Acanthocephala, (Gr. from ακανθα, *akantha*, a thorn; κεφαλη, *kephale*, a head.) Hooked Worms.

This order contains but one genus, *Echinorhynchus*, (Gr. *echinos*, a hedge-hog; *rhunchos*, beak,) with numerous species. The generic name refers to the chief character, which is a straight, round trunk, armed with rows of recurved tooth-like hooks. These Worms are generally found in the intestinal canal; sometimes in the neck under the skin. They occur in all vertebrates except man, sometimes boring through the intestines and passing into other parts of the body. The species *E. gigas*, which is from three to fifteen inches long, infests hogs, particularly such as have been shut up to be fattened.

Third Sub-Order. Trematoda, (Gr. from τρῆμα, *trēma*, a foramen or hole.) Fluke Worms.

These worms have a soft and rounded or flattened body. The head is indistinct, with a suctorial foramen; one or more suctorial pores appear on the surface of the body, furnishing the basis of their subdivision into genera. They have no intestinal canal. The *Fasciola*, (Lat. a small bandage,) *hepaticum*, (Gr. *hepaticos*, of the liver,) is a representative of this group,—a worm that infests the liver, gall-bladder, and sometimes the contiguous veins or ducts; is frequently found in numerous ruminant and other animals. It is particularly common in sheep, in the disease called the *rot*. This worm has sometimes been found in the gall-bladder of man. Its shape is considerably like that of a melon seed, (Plate XVI. fig. 8.) These worms have two pores, one in front, the other ventral; hence they are sometimes denoted by the generic term, *Distoma*, (Gr. two-mouthed.)

Some Trematods occur in birds and fishes.

Fourth Sub-Order. Cestoidea, (Gr. from κεστος, *kestos*, a band; ειδος, *eidos*, form.) Tape Worms.

Of these worms eight genera have been described. The head varies greatly in the different genera; generally it has two or four pits or suctorial orifices, and sometimes four retractile tentacles. There is no trace of an intestinal canal, unless it be connected with vessels proceeding from the suckers. Two genera contain species that infest the human body:

1. *Bothriocephalus*, (Gr. *bothros*, a groove; *kephale*, head.) This is a long, flat, jointed worm, with two longitudinal grooves,

one on each side of the head; infesting birds, fishes, and reptiles. The species *B. latus* is common in the intestines of man, in Switzerland, Russia, parts of France, &c. It is distinguished from the *Tænia* by the form of its segments, which are broader than they are long, and by the openings of the ovaries, which are beneath instead of at the sides.

2. *Tænia*, (Lat. a band.) This genus also has the body flat, long, and articulated, but the articulations are so small and indistinct for some distance from the head, (Plate XVI. fig. 9a,) that its existence was for a long time unknown, and it was supposed the worm obtained its nourishment through the lateral pores. The head is round, with four suckers forming a square about the mouth.

The Common Tape Worm, *T. solium*, inhabits the human intestines, but not with equal frequency in all countries. It is, however, more widely distributed than the *B. latus*. The length to which this worm attains is considerable, but it may be difficult to assign its limit. Sometimes it is twenty feet and even more, in length. One species, *T. cateniformis*, (Lat. chain-like,) about an inch long, infests the cat.

Fifth Sub-Order. Cystica, (Gr. κύστις, *kustis*, a bladder.)

These worms are either flat or round, terminating behind in a transparent cyst or bladder filled with a perfectly clear fluid; the head is retractile and provided with two or four pits, or four suckers and a circle of small hooks, or with four unarmed tentacles. This is the lowest group of the class, nothing is known of its nutritive and some other organs. They are represented by the Hydatida, or Hydatid, which consists of a globular bag, composed of condensed albuminous matter, of a laminated or plate-like texture, and containing a clear and colorless fluid.

The young are developed between the layers of the parent cyst, and thrown off internally or externally, according to the species. Some have doubted whether it be an animal. Its structure resembles that of the lowest forms of *Algæ*, or Sea-weed, as the Red Snow, (*Protocœcus nivalis*,) of the arctic regions. Acephalocysts have been found in almost every structure and cavity of the human body. Some species live in the brain and spinal cord of sheep, and in the brain of oxen, giving rise to the disease called "staggers."

What is the position of the MYRIAPODA? In what respects do they agree with the ANNELIDANS, and how do they differ? In what respects do they resemble and how differ from Insects? Into how many ORDERS are they divided? Describe their characteristics. Into how many families is the 1st ORDER arranged? Name and describe the CENTIPEDES referred to. Also those of the 2nd ORDER. What is the 3d class? Why separated from Insects? State how they differ. Describe the characteristics and habits of SPIDERS. Which is the 1st ORDER? Repeat what is said of the Common S. What is said of the Mason or Trap-door S.? Of the Tarantula? How do the Pedipalpi differ from the Spiders proper? What family forms the largest part of this division? What is said of them? Which is the 2nd ORDER? What forms does it include? Repeat what is said of the Mites. Of the Ticks. Of the Shepherd Spider. What other sp. are referred to?

What is the 6th BRANCH of ZOOLOGY? Describe their characteristics and habits? How many orders do they include? What is the 1st ORDER? Give the characters, &c., of the Crabs, or Short-tailed Crustaceans. What sp. are mentioned and what is said of them? What is said of the Long-tailed or Second Section? What is said of the Shrimps? Of the River Prawn? Of the Cray or Craw-fish? Of the Lobster, *Astacus marinus?* How does the Common American Lobster compare with it in size? What sp. of Shrimps are mentioned? What Crabs are included in the 3rd Section, and what is said of them? What is the 2nd ORDER? What families does it include? Which is first mentioned? What is said of it? Name and describe the sp. referred to? What is the 2d family mentioned? What gen. and sp. are named? What is said of them? What is the 3rd ORDER? What are its characteristics, &c.? What is said of the Cyclops? Of the Daphnia? Of the Cypris? What is said of the King Crab, or Horse-foot? What of the Trilobite? What singular Crustacean is next spoken of? What is the 4th ORDER? How ranked by the earlier Naturalists? Give their characters, &c. Into what two groups are they divided? What is said of the Barnacles? Of the Acorn-shells? What is the 5th ORDER? What is said of them? Name the sp. referred to.

What is the 7th BRANCH of ZOOLOGY? How was the class Annelidans arranged by Cuvier and others? How again by other naturalists? Give the characters of Red-blooded Worms. Which is the 1st ORDER? How distinguished? What Family is first named? How distinguished? What gen. are mentioned? What is said of them? What Family is next mentioned? Repeat what is said of it. Also of the 2nd Family. What is the 3d ORDER? What leading character is noticed? How is the Order divided? What is said of the Earth Worms? What of the Leeches? What is the 5th ORDER? Give the general account. What is the 1st Sub-order? Repeat what is said of Round Worms. What is said of Hooked Worms? What of the Fluke? Of the Tape-worm? What of the Cystica, or Hydatids? Name the White and Red-blooded worms found upon the Chart? By what forms are they distinguished? Where do they live?

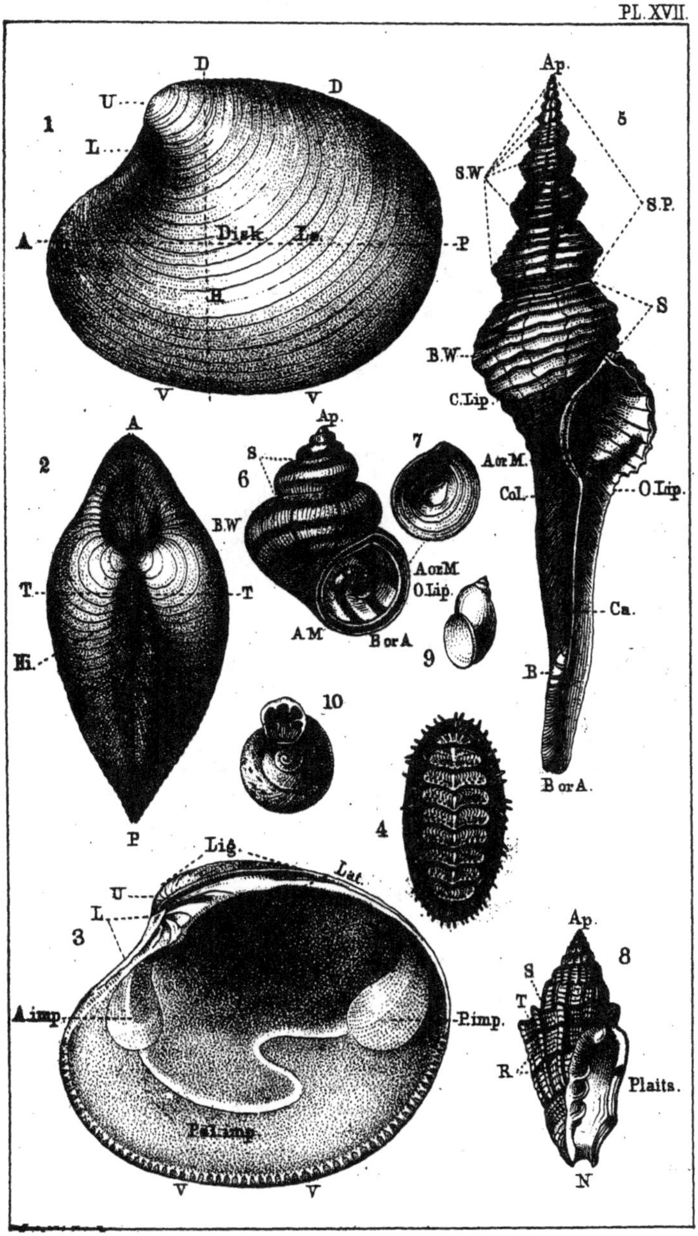

PL. XVII.

EXPLANATION OF PLATE XVII.

1, 2, 3. Round or Hard Clam, *Venus Mercenaria*, showing the different parts of a bivalve shell.

U. Umbones or bosses. The swelling part of bivalve shells near the beaks. The highest points of the beaks are the summits.

L. Lunule, a crescent-like mark or spot near the anterior or posterior slopes in bivalve shells, sometimes called areola.

D. D. Dorsal or superior border, near the bosses or beaks.

V, V. Ventral or inferior border, or border lip, at the base of the shell, opposite the beaks.

A. Anterior or oral extremity, the part in which the ligament is not placed. In a univalve it is the greatest distance from the apex.

P. Posterior or anal extremity, that side of the bosses containing the ligament.

Le. Length in bivalves is taken horizontally, or from the posterior to the an terior margin; in univalves it is taken perpendicularly or from the apex to the base.

Disk, the middle part of the valves.

H. Heighth. T. Thickness, through the shell from disk to disk.

Lig. Ligament, an external substance, uniting the two valves, and which in fact is the true hinge; the internal or cartilaginous part is often continued between the teeth.

The hinge is composed of the ligament, the cartilage and the teeth.

C. Cardinal teeth, i. e. the serratures or dentations beneath the bosses.

Lat. Lateral teeth, at the sides of the cardinal teeth.

A. imp. Anterior muscular impression. P. imp. Posterior muscular impression; these indented marks upon the shell show where the adductor muscles are attached.

Pal. imp. The Pallial or marginal impression formed by the mantle of the animal.

---

4. A Multivalve Shell, one composed of many pieces, as the *Chiton*.

5. Fusus. A spindle-shaped UNIVALVE SHELL, showing the different parts.

Ap. Apex, or posterior part of a univalve shell, the point or nucleus of a shell, the top of Limpets and all univalves, and the bosses or beaks of bivalves.

Sp. The Spire includes all the volutions except the body whorl.

S. W. Spiral whorls; each complete turn is termed a whorl or volution.

B. W. Body or basal whorl, is the last and usually much the largest.

S. Suture, the line where the whorls of spiral shells meet or fit into each other. When grooved or furrowed it is said to be canaliculated.

Col. Columella or Pillar, the internal support round which the whorls wind.

C. Lip. Columella, inner or pillar lip, folds over the lower part of the columella.

O. Lip. The outer lip is the external edge or termination of the last whorl.

A. or M. Aperture, mouth or front, from which the body can protrude.

Ca. Canal, groove, or furrow in the beak as in *Fusus*, *Murex*, &c.; in the Buccinum, Harpa, &c., it is only a notch, as in fig. 8.

B. Beak, or rostrum, the continuation of the body whorl.

B. or A. The base or anterior part.

---

6. A Turbinated Shell, (*Paludina vivipara*,) with the young shells.

7. The OPERCULUM, (door or cover,) closing the mouth, found in nearly all predaceous univalves, and always attached to the foot of the live animal.

8. *Tiara*, showing a turreted shell, with plaits or folds on the pillar; S, striæ; N. notch at the base, R. ribs, and T. tubercles.

9. *Physa*, showing the reverse or sinistral aperture.

10. Snail, (*Helix anastoma depressum*,) showing the reflexed lip and the teeth.

EIGHTH BRANCH OF ZOOLOGY.

MALACOLOGY. (Gr. μαλακὸς, *malakos*, soft; λόγος, *logos*, discourse.)

THIS is the science of the structure and habits of soft animals, or Mollusks. Many of these, from the number, variety, and beauty of their shells, invited attention at an early period, under the name of CONCHOLOGY, (Gr. *konchē*, a shell; *logos*, a discourse;) but in order to a natural classification, and a knowledge of the habits of the class, it was found that the entire animal must be known; hence, Conchology has been merged in MALACOLOGY.

THIRD SUB-KINGDOM. MOLLUSCA, (Lat. *mollis*, soft.) MOLLUSKS.

The Mollusks are, as a whole, inferior to the Articulates in their organization and faculties, but yet are superior to the Radiates, thus ranking as the third series in the Animal Kingdom.

In their external form, they are exceedingly various. Their internal parts are always soft, fleshy, moist, and cold; although a small number of them have some solid internal pieces intended for the protection of certain organs. The nervous system, instead of being developed in the form of a spinal cord, is composed of ganglia and nerves, which are dispersed, more or less irregularly, in different parts of the body. A few species have organs analogous to the ear; many are furnished with eyes; but it is not certain that they possess any sense of smell. Many of them appear to have no other organs than those subservient to touch and taste. The sense of feeling is probably most acute in the tentacula. The organs of sense and locomotion are generally arranged with symmetry. The muscles are attached to the skin; and by the alternate elongation and contraction of certain parts, the animals crawl on the ground, swim on the water, and lay hold of objects; but, as their limbs are not supported by bones or other solid parts, their motions are usually very slow. They are never furnished with feet arranged in series on each side of the body, as in the vertebrates and insects; but many possess a fleshy tongue-like appendage, called a foot, which is used, in some cases, either for progression, as in the snail, or adhesion, as in the limpet and chiton, (see Chart, and Plate XVII.) The location of this organ, on the lower part of the body of univalve Mollusks, (see figures of Harpa, Buccinum, and a Haliotes, on the Chart,) suggested the distinctive term for the order, GASTEROPODA. The organs of respiration are always distinct, and present the form of gills, i. e., blood-vessels dividing into parallel branches, which are brought into contact with the air contained either in the atmosphere or in the water.

The blood of the Mollusks is white, bluish, or limpid. There is always a heart in them, but it is singularly placed; indeed, some of them seem to have several hearts. In no other animals is the circulation more unequal; but always, however, there are blood-cavities into which the blood-vessels open, and from which other vessels arise and diffuse again the blood into the organs. The stomach is sometimes simple; sometimes divided into several parts; there is always a large liver.

In some, the sexes are separated; and in others united; all of them produce eggs,—which, in some cases, are deposited externally; in others, hatched within, so that Mollusks are either oviparous or ovoviviparous. The young of all have, from the first, nearly the form which they present when mature.

The soft, and usually sensitive skin, frequently forms plaits or folds enveloping the body either wholly or in part. The portion of covering thus formed, is termed the mantle. It is often almost entirely free, presenting two large laminæ or lobes, which cover the rest of the animal, as in the Cypræa; or the two laminæ unite so as to form a kind of tube, as in the Solen or Razor-Shell. Sometimes the mantle forms a sort of disk, of which the margins only are free; or it surrounds the body in the form of a bag.

In a large number of Mollusks, the soft skin is protected by a sort of calcareous crust, which is secreted from the mantle, in deposits of successive layers, composed of a kind of glutinous substance, mixed with carbonate of lime,—differing, as Prof. Dana has shown in his admirable work, (see Narrative of Exploring Expedition,) from the Polyps, (Radiates,)—in which the limestone portions form a part of the animal, and are not mere excretory matter, resembling shell. Sometimes the whole shell appears to be horny, but most commonly the calcareous portion predominates, and the inner surface is more compact than the other.

In some cases, the shell is internal, or lodged in the skin, but, generally, it is external, and affords complete protection to the animal.

Mollusks which, like the Cuttle-fish, (see Chart,) have no outer shell, are said to be *naked;* those having a shell, are called *Testaceous* or *Conchiferous.* The shell varies in form, the shape being determined by the animal itself. Sometimes it resembles a shield that covers the back of the Mollusk, but more frequently it is like a conical tube spirally twisted; or it may be composed of two distinct pieces united by a joint; hence, the arrangement of these animals into Univalves and Bivalves. The first, or the MOLLUSCA CEPHALATA, have a distinct head, bearing lips or jaws, and are furnished with eyes and tentacula; the Bivalves, or the MOLLUSCA ACEPHALA, have a more simple organization. These have no distinct head, and are destitute of jaws, and other hard parts of a mouth. The shells are often ornamented with colors variously disposed, the animals themselves being furnished with the materials for beautifying as well as constructing their outward coverings. The skin is full of pores, containing colored fluids, which, penetrating the calcareous substance before it hardens, form its variegated tints. The regularity of the markings is admirable. It is accounted for by the fact, that the pores containing the colored matter are arranged in the skin of Mollusks with undeviating order, as the spots upon the leopard, or the stripes upon the tiger. When the liquid exudes, it stains the shell; and the uniformity of pattern in the shell results from the order in which the pores are placed in the mantle. The numerous spines or digitations found in many of the shells, (see Murex and Pteroceras, on the Chart,) are formed by the prolongations of the mantle bearing upon its edges the material for this calcareous deposit.

The parts of a univalve shell, are (1) the *body* or lower part; (2) the spire or tapering portion; (3) the *turns* or *whorls*; (when the lower whorls of the spire are pressed into the body whorl or turn, they are said to be *retuse;*) (4) the *suture* or line of junction of the turns; (5) the *columellar* or *pillar*, the axis of the shell; (6) the *mouth* or *aperture* with its *peristone* or margin, which may be complete or not, and may be described as forming an *outer lip*, and an inner *lip*; (7) the lid or *operculum*, (from *operior*, to cover,)—the plate or door with which some species close the *aperture*. The spiral turns may be smooth, or variously marked with striæ, laminæ, ribs, nodosites, or spines, the markings being longitudinal or transverse. In its natural position, the mouth is beneath and forward, the spire pointing backwards and to the right side. Some shells have the mouth on the left side and are called *sinistral;* those of the ordinary form have the mouth on the right side, and are called *dextral.*

Bivalve shells are composed of two pieces, kept together by a sort of hinge. When the two valves are equal, the shell is said to be *equivalve;* when unequal, *unequivalve.* They may be round, elliptical, ovate, linear, or of various forms. The more or less prominent part of the valve at the joint, is the *umbo.* When the umbo is nearly in the middle, the shell is said to be *isomeral* or *equilateral;* when not, *anisomeral* or *inequilateral.* The *hinge* may be plain, but it generally presents various prominences, called *teeth,* with *depressions,* the teeth of one valve filling the depressions of another. The valves are farther kept together by an elastic fibrous *ligament,* which tends to throw them open. They are brought near to each other by a pair of strong *muscles* extended internally from one valve to the other, and leaving strong *impressions* on the inner surface. The *teeth* are distinguished into *cardinal* or *central,* and *lateral.* The *surface* may be convex in various degrees; concentrically striate, laminate, or rugose, or radiated from the umbones (or bosses) with striæ, ridges, grooves, ribs or spine. In the natural position, the hinge is uppermost on the back; that end of the shell to which the ligament is nearest, is above, and is called the *posterior* end; the other or lower, toward which is the head of the animal, is the *anterior* end; the thin edges of the valves are their *ventral margins.* On the inner surface of the valves are seen the impressions made by the muscles, and that left by the mantle. (See figures of Plate XVII. together with the explanations of the same.)

Some shells, as the Pearl Oyster, *Avicula margaritifera,* the *Pinna* and the *Modiola,* (see Chart,) fix themselves by silky filaments called a byssus; some by a sort of cement, as the Oyster, (*Ostrea;*) others by forming a vacuum, as the *Patella* or Limpet, and still others attach themselves to rocks by the same substance as that of which the shells are made, as the *Vermetus.* The shells, which by any of these means are rendered stationary, are called *fixed* shells, subsisting upon the little animals which are brought near by the motion of the water; the other shells are called *free.* Mollusks are also (1) terrestrial. These feed on vegetables, have always four tentacula, and their eyes placed at the tip of these organs; (2) *fluviatile* or fresh-water shells, that have only two tentacula, which are flat, and have eyes at the base; (3) Marine, which are most numerous, most beautiful, and most highly prized. We find the shell and the habits of the Mollusks wisely adapted to the situations which they occupy.

Some that belong to rapid streams, have an exceedingly hard and substantial shell, adapted to contend with the most boisterous elements. Others, by their very levity, are enabled to float on the surface of the water, and offering no resistance, are carried along on the surface of the waves. The *Pinna* anchors itself by its *byssus* to rocks, and thus is secure against all dangers. Others, as the *Nautilus* and Argonaut, by adding to the weight of their bark, can descend and seek a shelter in the ocean's bed. There are numerous and beautiful contrivances for their preservation. Breaches will, however, sometimes be made in the outer coverings; but these they have the power to repair, by exuding a calcareous matter similar to that with which the shell was first constructed. They are peculiarly abundant in warm climates; being larger and more brilliantly colored, the greater the light and heat to which they are subjected. Including the soft, naked species, as well as those protected by a hard calcareous shell, it is believed the number of species will not fall short of ten or twelve thousand, and this, exclusive of the fossil species, which are thought to be still more numerous. It is said there are scarcely eight hundred living shells found in the Mediterranean, or on the French shores of the Atlantic Ocean, but more than twelve hundred fossil shells have been found in that stratum of limestone in which the city of Paris is built, and of which such extensive deposits exist in the neighborhood. "In that single stratum is found, at this day, one third more fossil shells than live on the whole extent of the French shores." (Agassiz.)

We had designed to follow this general description of the Mollusks, with explanations of the various sub-divisions, after the manner adopted in the preceding sub-kingdoms; but already this volume has swelled far beyond the limits originally assigned to it. Hence, we are constrained to close our account of shells here, referring to the Plates found in this work, and to the Chart, with its numerous figures and explanations, for further illustrations; also to Manuals of Conchology, already published, until, if circumstances should hereafter warrant, we may be able to prepare a volume in which the Articulates, Mollusks, and Radiates, shall be presented in a manner corresponding in fullness to the view herein presented of the VERTEBRATES.

What is the Eighth Branch of Zoology? Of what does it treat? Under what name was the science of shells formerly known? Why was it changed? Which is the THIRD SUB-KINGDOM? What is said of their external form? Of the internal parts? What of the nervous system? Of their organs of sense? To what are the muscles attached? How are the animals enabled to crawl, swim, &c.? Why are their motions very slow? What is said of the organs of respiration? Of the blood, stomach, &c.? What is meant by the mantle? How does it vary? How is the skin in many Mollusks protected? Is the shell always external? What are those called which have no shell? What those which have a shell? How is the shape of the shell determined? Whence the divisions into UNIVALVES and Bivalves? What other names have these divisions received? What is said of them? What is said of the coloring of shells? Name the parts of a Univalve Shell, &c. Give the explanation of Bivalve shells as furnished by the text and Plate XVII? What are *fixed* shells? What are *free* shells? What other division of the Mollusks are given? What is said of the adaptation of the shell and habits of Mollusks to their respective situations? In what climate are they most abundant and most brilliantly colored? What is said of the number of species?

EXPLANATION OF PLATE XVIII.

Class Echinoderms.

Fig. 1. Sea-Slug or Sea-Cucumber, *Cucumaria frondosa*, a Holothurian eaten by the Chinese; a, the branching tentacula which surround and fringe the mouth; b, the five rows of perforations for the sucker-like feet; c, the vent.

Fig. 2. Sea-egg or Sea-urchin; the vent at the apex. The mouth is beneath and central. The large tubercles in wide rows support the large spines; the small perforations are for the passage of the sucker-like feet which assist in locomotion.

Fig. 3. Five-finger, radiated Star-fish, or Cross-fish, *Asterias (Uraster) sanguinolenta;* a, the eye spots at the extremity of each ray.

Fig. 4. *Asterias (Uraster) rubens;* a, both mouth and vent on the under side.

Fig. 5. Pentagonal Star-fish, *Asterias tesselata.*

Fig. 6. Medusa's-head, *Gorgonocephalus*, or *Euryale*, having five arms, which in some individuals branch off into 5,000 filaments.

Fig. 7. *Pentacrinus Briareus*,—a, shows the head and arms; b, the upper part, half the natural size, with the arms entwined around the plated integument of the abdominal cavity, which terminates above by a sort of proboscis.

Fig. 8. Stone Lily, Lily Encrinite, *Encrinus liliiformis.*

Class Acaleph.

Fig. 9. *Rhizostoma Cuvieri;* a common jelly-fish of European seas; a, pedunculated mass depending from four roots in the center.

Fig. 10. *Lymnorea triedra;* a, side view; b, view from above; c, eight finely divided appendages surrounding the long center proboscis.

Fig. 11. *Rhodophysa helianthus;* a, the short bladder-like body; b, rib-like gelatinous bodies, from which filamentous processes, c, depend.

Fig. 12. *Callianira triploptera;* a, wing-shaped appendages fringed with a double row of vibrating cilia; b, a pair of long, branched, tentacular-like appendages; c, tubular body.

Class Phytozoa, or Polyps.

Fig. 13. Sea Anemone, *Actinia.*

Fig. 14. *Hydra fusca*, a common fresh water polyp, highly magnified.

Fig. 15. *Caryophyllea*, with two series of numerous tentacula putting forth from multiradiate cells.

Fig. 16. Warty Gorgonia, or Warted Sea-Fan, *Gorgonia verrucosa;* a, represents the sucker-like feet; b, the same magnified.

Fig. 17. Sea-pen, *Pennatula grisea.* The branches put out on each side of the central axis, like the barbs of a feather; on these branches the Polyps are situated, and by their tentacula its course seems to be directed.

Fig. 18. *Madrepora abratanoides;* a, twelve tentacles protruding from deep cells.

Fig. 19. Thick tentacled Fungia, swarming with numerous tentacula, all belonging to one animal.

Fig. 20. *Fungia patellaris*, Mushroom coral.

Class Protozoa.

Fig. 21. *Rhizopods*, Low forms allied to mollusks by their shells, as Foraminifera, &c.

Fig. 22. Polygastrica, (Animalcules, or microscopic animals.)

Fig. 23. Sponges *Spongia;* a and b, sponges of commerce; c, tube sponge.

PL. XVIII.

ECHINODERMS.
ACALEPHS.
PHYTOZOA, OR POLYPS.
PORIFERA, OR PROTOZOA.

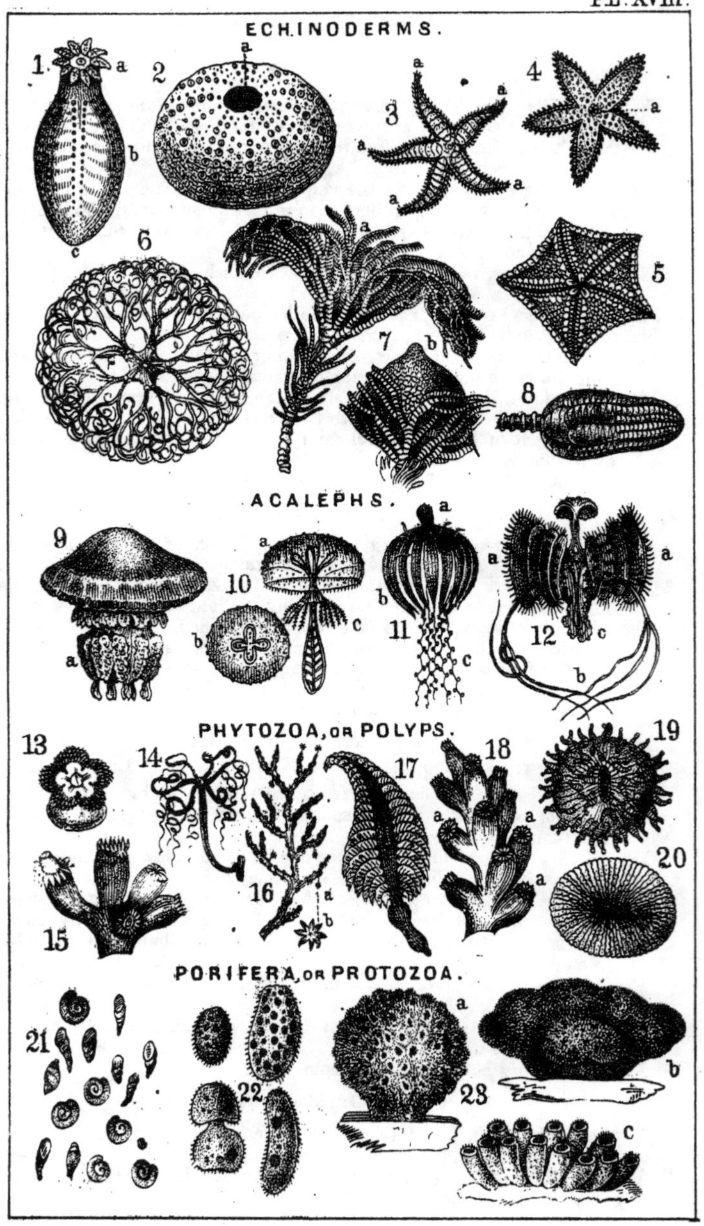

FOURTH SUB-KINGDOM.

RADIATES. (Lat. *radius*, a ray.)

THIS includes animals so named from the arrangement of the parts round an axis somewhat as in plants; hence they were called by Cuvier, ZOOPHYTES, Plant-like animals. These introduce us to the

NINTH BRANCH OF ZOOLOGY.

ACTINOLOGY, (Gr. ακτὶν, *aktin*, a ray; λογός, *logos*, a discourse.

RAY-LIKE ANIMALS.

The terms which are here employed are not, by any means, equally applicable to all the beings included in this last sub-division of the Animal Kingdom. While in some of them the radiated arrangement of the parts is very easily seen, in others it can be traced only by a close microscopic examination. Agassiz says it can be perceived in all with "*sufficient* observation." Such are the differences of form and in the degree of organization among the Radiates, that instead of undertaking to present additional characters, describing them as a whole, we shall at once proceed to consider the four classes into which they are divided.

I. ECHINODERMS, (Gr. εχῖνος, *echinos*, a sea-urchin; δέρμα, *derma*, skin.)

These are all marine, and characterized by having a well organized skin, under which or attached to which are often found plates of solid matter, forming a kind of skeleton, and sometimes joined together like the stones of a pavement. They have a digestive and vascular system. A circular nervous system has been detected in many of the species. Some of them have the bodies raised upon a footstalk, the base of which is fixed; others have no pedicle, and can move freely, though slowly along the bottom. Of the latter kind the Star-fish, *Asterias*, is an example, which drags its slow pace along with the aid of more than eight hundred sucker-like feet. In these, at the extremity of the rays are dots, which, according to Ehrenburg, are eyes. The Echinoderms usually have the mouth armed with hard bony teeth. Both the mouth and the gullet are, in the Star-fish, extremely dilatable, and admirably fitted for securing the Crustaceans and Shell-fish upon which they feed. When the prey is over large, the gullet, together with part of the stomach, can be protruded and turned over so as to draw the desired food into the cavity. In this way, shell-fish are swallowed whole by these animals; living specimens have sometimes been taken out of the cavity.

The sudden and violent dismemberment by which many of the Echinoderms save their central disk at the expense of their rays or arms, is a striking peculiarity; the length of time during which these severed parts still continue to be endowed with motion, is also remarkable. These animals have great power of reproducing lost parts, but they do not seem to be able to increase individuals by gemmation, as in the Acalephs and *Protozoa*.

FIRST ORDER. HOLOTHURIDEA, (Gr. ὁλοθούριον, *holothourion.*) SEA-SLUGS, or SEA-CUCUMBERS.

These animals have a more or less lengthened and sometimes worm-like body; it is soft, leathery and contractile, with earthy matter deposited about the mouth, but without any outer shell. When irritated, the contractile power of the body seems to be increased. The general form is quite variable. Over the body are spread numerous pores for the secretion of mucus, and perforations for the exit of the sucker-like feet. When not generally distributed, these latter organs are arranged in five rows or furrows, representing what are fancifully termed *ambulacra* or avenues. The mouth is bordered with tentacles which can be withdrawn. Mariners have named these animals Sea-Cucumbers, from the resemblance of their form and prickly surface to the garden cucumber. Of the genus *Cucumaria*, there are several species, differing in quality, known among the Malays under the name of Trepang, or *Biche-de-mer*. Commodore Wilkes, (Exploring Expedition,) says "the most esteemed kinds are found on the reefs in water, one or two fathoms deep, and are caught by diving. The natives also fish the *biche-de-mer* on rocky coral bottom by the light of the moon or of torches, for the animals keep themselves drawn up in holes by day, and come forth by night to feed, when they may be taken in great numbers. The motions of the animal resemble those of a caterpillar, and it feeds by *suction*, drawing in with its food much fine coral and some small shells.

The prepared article finds a ready sale in the China market, where it is used as an ingredient in soups. When brought on shore, the animals are placed in bins where they remain the next day; the entrails are then removed; the larger kinds are cut along the under part for three or four inches to make them dry more rapidly; then thrown into pots of the form of sugar boilers, containing from one hundred to one hundred and fifty gallons, and boiled fifteen or twenty minutes, after which they are thoroughly dried in a large building. This process makes the slug lose two-thirds of its weight and bulk. When cured it resembles smoked sausage. The *biche-de-mer* is sometimes taken to Canton, but more usually to Manilla, whence it is shipped to China."

Sometimes it is as much as two feet in length, and from seven to eight inches in circumference; but a span long, and two or three inches in circumference is the ordinary size.

SECOND ORDER. ECHINIDEA, (Gr. εχῖνος, *echinos*, a sea-urchin; εἶδος, *eidos*, form.) SEA-URCHINS OR SEA-EGGS.

These furnish complete examples of the type of radiated animals. Their hard covering and habits of living in the sand have preserved them in many rocky strata, so that in their fossil as well as living forms, they are objects of interest to the naturalist. Their usually oval or circular form has suggested for them the name of Sea-eggs. Their more or less rounded form, without any arms, distinguishes them from other Echinoderms. Within their integument calcareous matter is deposited, forming a series of regular plates, studded with tubercles which are jointed with spines varying in form and size, according to the genus to which they belong. These spines "have a beautiful microscopic structure, being composed of cells which are arranged around a common center, almost in the same manner as the zones of wood in a tree." The plates run in vertical rows or avenues, twenty in number, two of which are wide and two narrow, alternately, the tubercles of the wide pair supporting the larger spines; the narrow ones have vertical rows of minute perforations, which allow the passage of the sucker-like feet, that in addition to the spines, perform the office of locomotion. The ovarium-holes are situated on the apex, and from these the eggs are extruded. The forms of this order vary from that of the CAKE-URCHINS, *Scutella*, (Lat. a salver,) slightly convex above, to the sub-globular *Echinus*, (Plate XVIII. fig. 2.) This, with other genera having large spines, is found on the bottom of the sea; while the Cake-Urchins, having short, bristly spines, burrow in sand.

The HEART-URCHIN, *Spatangus*, (Gr. *spatangos*, a kind of sea-urchin,) has a thin, delicate shell, of a lengthened, gibbous form, with the vent posterior and placed upon the upper surface, (see Chart.) The structure of the mouth in the Sea-urchin deserves special notice. It is formed of ten series of hard plates, furnished with teeth which are moved by very distinct muscles, and put in motion by a complicated nervous system; armed also with five jaws so arranged as to correspond with the ten series of plates. So powerful are these jaws that they are able to crush shell-fishes and the hardest bodies. According to Agassiz, a more complicated organization is scarcely to be found in the Animal Kingdom. The mechanism of these five jaws, with their singular array of arched teeth, Aristotle compared to a lantern; hence the *Echinus* has been called "Aristotle's Lantern."

Professor Forbes informs us that "in a moderate sized urchin, there are sixty-two rows of pores in each of the ten avenues; and as there are three pairs of pores in each row, the total number of pores is 3720; but as each sucker occupies a pair of pores in each row, the total number of suckers is 1860." He says also that "there are above three hundred plates of one kind, and nearly as many of another, all dovetailing together with the greatest nicety and regularity, bearing on their surface above four thousand spines, each spine perfect in itself, and of a complicated structure, and having a free movement in its socket." "Truly," he adds, "the skill of the Great Architect of Nature is not less displayed in the construction of a sea-urchin, than in the building of a world."

THIRD ORDER. ASTERIDEA, (Gr. αστὴρ, *aster*, a star; εἶδος, *eidos*, a form.) STAR-FISHES.

This order of Radiates is distinguished by having the body more or less lobed, and the lobes chaneled beneath for cirrhi, which act as suckers, and are organs of motion. In some genera the arms, instead of having lateral cirrhi or filaments, separate into branches. The *Euryale* or *Gorgonocephalus*, (Gr.

gorgon-headed,) (see Chart,) is remarkable for its five arms dividing into pairs of branches, which terminate in curled filaments, and form a sort of net work. So numerous are these branches, it is said, they may number eight thousand in one individual.

The STAR-FISH, *Asterias*, has almost innumerable perforations through which the cupping-glass feet protrude, enabling the animal to crawl up a surface as smooth as glass, and also assisting it to hold its prey. The rays are so much enlarged as to become, on that account, less flexible, and not so well fitted for locomotion, but for this the animal is compensated by numerous perforations.

The SCUTELLATED STAR-FISH, *A. scutellata*, has an angular body, the lobes or rays of which are short, not exceeding the diameter of the disk; other species have a body furnished with elongated rays, whose diameter exceeds the diameter of the disk. The genus *Ophiura*, (Gr. *ophis*, a serpent; *oura*, a tail,) in its long, slender arms, shows a resemblance to the tail of a serpent, (see Chart.) These arms are flexible; and by giving them a waving motion, the animal is able to swim. A number of species of Star-fish are found on our coast,—some of which have proved very destructive to the oyster beds. The Providence (R. I.) Journal, in reference to some in that vicinity, says, "It is impossible to estimate the injury that has been done by the Star-fish; probably not less than twenty thousand bushels of oysters have been destroyed,—and unless it disappears, oystermen will hardly be willing to plant their beds."

FOURTH ORDER. CRINOIDEA, (Gr. κρῖνον, *krinon*, a lily; εἶδος, *eidos*, form.) ENCRINITES.

This order includes species by far the larger portion of which are extinct, but which are found abundantly in limestone or the lower rocks. In the fossil state, they are known by the name of Encrinites,—a name suggested by the stony stem, and a crown of rays bending in sigmoid curves, like the Greek letter sigma, (σ,) resembling "the stalk and elegant bell-shaped blossom of a liliaceous flower." The living species are very rare. The form of the body is oval or cup-like, protecting the internal soft parts, and composed of numerous plates. The arms are five or more in number, simple or branched, with lateral jointed appendages, and situated around the upper margin of the body, the mouth being placed between them. It is when the arms are closed that some of the species assume a lily-like appearance. (See *Encrinus liliiformis*, on the Chart.) The joints composing the rounded stem have perforations in their cavities by which they can be strung as beads, "which," says Dr. Buckland, (Bridgewater Treatise,) "caused them in ancient times to be used as rosaries." In the northern parts of England, they are still called "St. Cuthbert's beads." They are also known by the name of Wheel Stones.

The *Pentacrinus Europæus*, found in the Irish coast, is now considered to be the young of the Rosy-feathered Star-fish, *Comatula*. (See Chart.)

The *P. Briareus*, (see Chart,) is a fossil species, having great length of stem and numerous side-arms,—whence the specific name. It is frequently found in contact with masses of drifted wood. (See Pl. XVIII.)

The Pear-shaped Encrinite, *Apiocrinites*, (Gr. *apion*, a pear,) *rotundus*, is so named from the form of its body. The figures on the Chart represent it with its expanded and closed arms. It is fixed by a jointed peduncle. The surfaces of the joints of the vertebral column are striated with rays.

The Pentacrinites have pentagonal stems, and are found in the more recent strata. Besides the species found in the Bay of Cork, above referred to, a larger one, *Holopus*, (Gr. *holos*, whole; *pous*, foot,) *rangei*, is found in the West Indian seas.

Second Class. Acalephs, (Gr. ακαλήφη, *akalēphē*, a nettle.)

This class derives its name from the stinging power possessed by a large portion of the animals composing it, and sometimes in so high a degree as to be a terror to bathers. They are known by the names Sea-Nettles, Medusæ, Sea-Jellies, &c. Having the power of free motion, they float in all seas, especially those of the warmer latitudes. They are not like the Echinoderms, enclosed in a thick integument; but, on the other hand, one of their most striking peculiarities is their extreme softness. Some of them attain considerable size, but almost all are without any internal support or skeleton. Their soft tissues give them the appearance of a mass of jelly,—a mere net work of animal filaments, the intermediate spaces of which are filled up with sea-water. Hence, although some of the largest reach the size of two feet, and a weight of fifty or sixty pounds, yet, when they are dried, the pounds become grains. Many of the Medusæ are extremely beautiful, reflecting the prismatic rays. DeBlainville represents their form as "nearly always circular, sometimes discoidal or spheroidal, but most frequently hemispherical." They quite commonly have a form like that of our umbrellas;

the central part being thickest, and the under surface concave. In some cases, cirri or tentacles, varying in length, form, and number, are attached to the circumference. Some of the tentacles have a colored spot (thought to be an eye) at their base. It is quite certain that they are sensible to light. Some of the smaller Medusæ have been "known to shun a bright light, and to sink into deep water to avoid it." The chief seat of touch seems to be in the tentacula or cirri, which are also capable of wonderful expansion and retraction. Many of these Medusæ make no sign when wounded in the umbrella or disk. The food, consisting of small fishes and marine animals, is conveyed to the mouth, not by the tentacles and cirri alone, but also probably by contractions in the disk. These animals appear, in most cases, to be bisexual, i. e., the two sexes are often united in the same individual. Like some other lower organizations, they have the power of producing their offspring by gemmation. "Fancy," says Prof. Forbes, "an Elephant with a number of little Elephants sprouting from his shoulders and thighs,—bunches of tusked monsters hanging epaulette-fashion from his flanks in every state of advancement. The comparison seems grotesque and absurd, but it really expresses what occurs among our Naked-eyed Medusæ."

The phosphorescent or luminous appearance of the sea, and which is shown in Southern latitudes with such brilliancy and beauty, is to be chiefly ascribed to multitudes of these animals.

First Order. Pulmonigrades. Lat. *pulmo*, a lung; *gradior*, to advance.) Medusæ or Jelly-fish.

The name of this order above given, refers to the contractile and expansive power of the umbrella-shaped disk belonging to the animal which it includes, and which in the exercise of this power, resembles the breathing lungs. The order is also, with reference to the umbrella or disk, named Discophora, (disk-bearing.) As illustrating this order, we simply refer to the genera and species noticed on the Chart, with the figures and explanations there given.

Second Order. Physograda. (Gr. φυσάω, *phusaō*, to inflate; Lat. *gradior*, to advance;) or Siphonophora, (Gr. σιφων, *siphon*, a pipe or sucker, φορέω, *phoreō*, to bear.)

This order is composed of animals which are supported and capable of moving in the water, by the possession of one or more bladders, which they inflate with air at will. With reference to this means of support, Cuvier called them Hydrostatic Acalephs. When several air-bladders exist, instead of a single large one, they are usually affixed to the same stalk, like currants upon the stem, and rise out of the tentacular apparatus.

The Portuguese Man-of-War, *Physalis*, (Gr. water-bubble,) *pelagica*, (*Physalia pelagica*. Linn.) This noted Acaleph (for which see Chart,) has a very large air-vessel, beneath which the digestive apparatus is arranged. The sac is surrounded by a sort of crest or sail, which is usually elevated entirely above the water, when the animal is floating at the surface. It has a small orifice at each end, from which the air can be expelled when the animal wishes to sink, and it is distended when it wishes to rise. From the under side of the air-sac hangs a mass of flask-shaped tentacular appendages terminated by suckers, and sometimes hanging down like fish-lines, to an extent of fifteen or sixteen feet. This creature possesses an active stinging power, and is also very contractile, so that it is able to draw up its prey. It would seem the short suckers are attached to the bodies of the entrapped animals, and that the *Physalis* derives its nourishment by imbibing their juices through the pores of its numerous cirri.

The *Physophora*, (Gr. bladder-bearer,) *Muzonema*, has two series of vesicles, and very numerous tentacles and filaments. (See Chart.)

The Rhizophysa, (Gr. root-bladder,) has a very contractile body with an aid-bladder at one extremity, and is provided throughout its length with tentacular appendages covered or mingled with filaments. (See Chart for figure of *R. filiformis*.)

The *Apolemia Urania* has an elongated worm-like body possessing in the fore part two rows of numerous swimming organs, and behind solid scaly organs, between which come forth tentacle-like cirri furnished with vermiform suckers.

The *Rhodophysa*, (Gr. *rhodon*, a rose; *phusa*, a bladder,) *helianthus*, (Gr. sun-flower,) has a short cylindrical and fleshy body, swollen above into an air-bladder, and having below a number of gelatinous rib-like formations with appended filaments. (See Chart and Pl. XVIII. fig. 11.)

THIRD ORDER. CILIOGRADA. (Lat. *cilia*, eye-lashes, or vibratile hairs; *gradior*, to advance,) or CTENOPHORA, (Gr. κτεις, *kleis*, a comb; φορέω, to bear.)

This order is named from its flat phosphorescent vibrillæ or rows of cilia, arranged lengthwise upon the surface of the body, and by means of which it is propelled through the water. It is supposed by some that the cilia are organs of breathing as well as of locomotion. The genus *Beroe* varies in form from globular to cylindrical; the cilia also vary in length and position. The species *B. ovata*, (Lat. egg-shaped,) exhibits the greatest celerity in the movement of its delicate organs, and the most beautiful variety of colors, (see Chart,) as these organs play to and fro in the rays of the sun. The oval-shaped body is open at the large end, transparent, and of a firm gelatinous consistence; easily contracting and widening, but always open and expanded when in motion. The species *Beroe*, (Cydippe,) *pileus*, (Lat. a cap or hat,) has a regular body divided into eight sections by rows of cilia. From an internal cavity issue a pair of long retractile appendages, (see Chart,) furnished with vibratory cilia. These beautiful forms are said not often to exceed three inches and a half in length, and two inches and a half in the transverse diameter. *Callianyra triploptera*, (Gr. triple-winged,) seems in its structure and general character, not far removed from Beröe. It has two pair of winged-shaped appendages, fringed with a double row of filaments upon the edges. (Plate XVIII. fig. 12.)

The GIRDLE OF VENUS, *Cestum Veneris*, has a ribbon-shaped body, (see Chart,) sometimes six or eight feet long, its breadth not being as many inches. The margins are fringed with beautifully colored phosphorescent cilia, which at night gives it the appearance of a band of flame in motion along the water. Some naturalists add a

FOURTH ORDER. CIRRIGRADA. (Lat. *cirri*, locks or tufts; *gradior*, to advance.)

In this order the form is discoidal, and there is an internal discoidal skeleton distinguishing these animals from the Pulmonigrades. They are named from the cirri which are attached to the disk upon which the organs are arranged. Some of the cirri are tubular and furnished with suckers. The *Velella*, (Lamarck,) in addition to the oval, sub-cartilaginous skeleton, is surmounted by a vertical and oblique crest. This form is widely diffused. The animals are met with far out at sea, often in considerable masses. Sailors are said to fry and eat them.

THIRD CLASS. PHYTOZOA, OF ZOOPHYTA. (Gr. φυτον, *phuton*, a plant; ζῶον, *zōon.*) PLANT-LIKE ANIMALS.

The larger part of the animals composing this class are marine,—some species are fixed to the soil. None of them are properly free, swimming animals, although some of them can move at will from their location. Portions of them are unconnected with others,—independent and single; others are joined in large societies, having the base of the stems in union. Some have no hard-support; others secrete a stony skeleton, termed *corallum*, (coral,) thus constructing the well-known coral reefs and islands,—modifying "the shades of the ocean's depth, and forming whole mountain ranges." The corals cannot properly be regarded as the shells with which the animals cover themselves, the hard parts being, in reality, a part of the internal structure. Prof. Dana, in his magnificent and standard work on the Zoophytes,—says "the corallum is entirely concealed within the polyp, as completely as the skull of an animal beneath its fleshy covering. All corals are more or less cellular, and through the cellules, the animal tissues extend." In some instances, however, the coral is exposed, i. e., when the increase occurs from a terminal secretion upon a separate stem. As the stem increases in length above, the part below dies. This increase and disappearance of vitality above and below are common; and thus are formed the huge masses of coral. The most common species engaged in the production of coral banks, are the *Meandrina labyrinthica*, (Brain Coral,) the *Caryophyllia*, *Madrepora*, *Porites*, and *Astrea*, especially the latter. (See Chart.) A solid dome of the Astrea, twelve feet in diameter, has, according to Prof. Dana, a living exterior not more than a half or three-fourths of an inch in thickness. "It is a well-known fact," he says, (Silliman's Journal, Jan., 1837,) that corals cannot grow above the surface of the water; and that reef-building corals cannot grow at a greater depth than from ten to twelve fathoms, or above the surface of the water at low tide; therefore a coral reef cannot be more than sixty or seventy feet thick." As a condition of coral growth, "the sea-water must be pure and transparent. Corals will not grow, therefore, on muddy shores, or in water upon the bottom of which sediment is deposited."

Some of the Polyps increase both by eggs and buds; but not all of the latter can in their turns produce eggs. They seize their food with their tentacles; their whole surface is covered with vibratory cilia so exceedingly minute as to be discernible only through the aid of the microscope. These minute hairs are perpetually in motion, producing a continuous current of water. They are not under the animal's control from the period of its escape from the egg. Even in the egg, and when the animal is at rest, these cilia are in motion,—their action wafting small portions of organic matter to the mouth of the animal, itself incapable of going after other food, and thus supplying its wants. By means of their tentacles, however, Polyps are able to seize upon larger prey.

FIRST ORDER. ACTINOIDA, (ACTINOIDS,) (Gr. ἀκτίν, *aktin*, a ray, Ray-like,) SEA-ANEMONES and CORALS.

The name given to these animals refers to the radiated disposition of the tentacles, which, when expanded, sometimes resemble the petals of a flower; when contracted, the mass assumes a lemon-like shape. The order includes, however, not these actiniæ or flower-shaped genera alone, which do not secrete a coral, but others also, which are coralligenous.

The exterior surface of the Actinoids is either fleshy or leathery,—slimy, and exceedingly sensitive. The mouth, is simple and bordered with tentacles,—it is situated above, while in the Star-fish, it is beneath. Each of the tentacles is a tube, the walls of which are formed of longitudinal muscular fibres. By the contraction of these fibres, the animal can shorten the tentacles in all directions. Around the entire tube are circular fibres. These pull the tentacles in succession, so as to elongate it to three or four times its usual length, thus enabling the animal to seize its larger prey. The interior cavity or stomach, is a simple sac, which the animal contracts or shuts at pleasure. The digestive power is great and rapid, commencing to act as soon as the food is within the cavity. Fish, Crabs, and Shell-fish, are speedily assimilated, the harder parts being ejected in the course of ten or twelve hours; and the juices produced by the influence of the walls of the stomach, are diffused in the lower cavity, into which the water which came in with the food is also poured. The Actinoids have "no blood, no vessels, no respiration proper, though the contact of water produces a sort of respiration." (Ag.)

The reproduction is both by division and by eggs. At first, the young have but five or ten tentacles, but these steadily increase until they become almost innumerable, though uniformly "multiples of five." Lost parts, especially the tentacles, are soon replaced. If the body be cut into several parts, each may survive and become a complete animal. Actiniæ can endure water hot enough to blister the hand; they can be thawed out alive after having been frozen; though to dip them in fresh-water is said to kill them immediately. "A strong light incommodes, noise startles them, and they are affected by odors." Recorded facts show that the duration of life in these inferior forms, is often quite considerable. Gosse, in a note found in "Life in its Lower forms," speaks of one that was living in 1856, which attained the age of thirty-five years.

The Actiniæ are found in every sea—some suspended from the walls of sub-marine cliffs; others covering the more exposed sides of rocks with a flower-like tapestry, and some confining themselves to the smooth sands, on the surface of which they spread out their tentacula, and even slowly withdraw under the sand when danger threatens. Some of them have a stinging quality. Many are used for food in tropical countries, on the coasts of which they are more numerous than in cold climates.

The order ACTINOIDA is divided by Prof. Dana into the Sub-orders, 1. ACTINARIA; 2. ALCYONARIA. We refer first to genera and species figured on the Chart belonging to the first (non-coralligenous) family, *Actinidæ.*

The *Iluanthus* (Gr. mud-flower,) *Scoticus*, has a round mouth surrounded by numerous filiform tentacula. The body tapers to a point which is probably buried in the soft mud in which it lives.

The SUN-FLOWER ANEMONE, *Actinia helianthus*, has the mouth encircled by tubulous tentacula, giving it somewhat the appearance of a sun-flower.

The PURPLE SEA ANEMONE, or ANIMAL-FLOWER, *A. equina*, has a soft skin, finely striated, usually of a beautiful purple, often clouded with green. The tentacula, which number one hundred, vary much in color.

The WHITE SEA-ANEMONE, *A. plumosa*, (*A. dianthus*,) Plate XVIII. fig. 13, is four or more inches broad; it has the margins of the mouth expanded into lobes, all furnished with innumerable tentacula. There is an inner row of these, still larger.

The LARGE LEATHERY SEA-ANEMONE, *A. senilis*, (*A. crassicornis*, thick horned,) is three inches broad, with a leathery, unequal envelope, of an orange color. The tentacula are in two ranges, usually marked with a rose colored ring. Its abode is commonly in the sand. This species occasionally masters and swallows a victim even much larger than itself. Dr. Johnston, in his "History of British Zoophytes," thus remarks: "I had once brought to me a specimen of *Actinia crassicornis*, that might have been originally two inches in diameter, and that had somehow contrived to swallow a valve of the great scallop, *Pecten maximus*, of the size of an ordinary saucer. The shell fixed within the stomach, was so placed as to divide it completely into two halves, so that the body stretched tensely over, had become thin and flattened, like a pancake. All communication between the inferior portion of the stomach and the mouth was of course prevented, yet, instead of emaciating and dying of an atrophy, the animal had availed itself of what undoubtedly had been a very untoward accident, to increase its enjoyments and its chance of double fare. A new mouth, furnished with two rows of numerous tentacula, was opened upon what had been the base, and led to the under stomach; the individual had, indeed, become a sort of Siamese twin, but with greater intimacy and extent in its unions."

The Sea Anemone, *Edwardsia vestita*, (Lat. clothed,) is one of the last discovered species, named after a distinguished naturalist. It forms for itself a shell or *clothing*, into which it can retire at pleasure, or when in shallow water the tide recedes, leaving it exposed to the air. (Plate II. fig. 3.)

The genus *Lucernaria*, (Lat. the plant *verbascum*,) includes animal flowers which are bell-shaped, free or fixed to sea-weeds by a narrow disk or stalk, from which they expand to a broad, eight-sided disk, in the center of which is a quadrangular mouth, and at each angle a bundle of tentacula; surrounding the mouth are festoons of ovaries. The largest are about an inch in height. They are of various colors, but usually pink. The species figured on the Chart is *L. auricula*, (Lat. the ear-lap.)

The other genera of the sub-order noticed on the Chart are nearly all coralligenous. The forms which the corals assume are extremely various, such as those of trees, shrubs, leaves, obelisks, domes, etc. Their substance consists principally of carbonate of lime. The surface is usually "covered with radiated cells, each of which marks the position of one of the polyps; and when alive, animals appear like plants on every part of the Zoophyte." The frame work or skeleton is called Polyparium, or Polypary, (Polyp-structure.) It should be remembered it is made, not as the bee constructs its cell, but by secretions of the animal tissues, increasing without the consciousness of the polyp, in the same manner as the bones and other structures in the higher orders of animals.

Family Astræidæ. These Actinoids have the corolla calcareous with marginal tentacula, excavated cells, and circumscribed stars. Of these are the *Astræa ananas*, Pine-Apple Coral, (see Chart.) Prof. Dana says: "calculating the number of polyps that are united in a single astræa dome of twelve feet in diameter, each covering a square inch, we find it exceeding 100,000; and in a Porites of the same dimensions, in which the animals are under a line in breadth, the number exceeds five and a half millions. There are here, consequently, five and a half millions of mouths and stomachs to a single zoophyte, contributing together to the growth of the mass, by eating, and growing and budding, and connected with one another by their lateral tissues and an imperfect cellular or lacunal communication."

The BRAIN CORAL, *Meandrina labyrinthica*, (see Chart,) is of this family. Recent species belong to the South Atlantic and Indian Oceans. Fine specimens have been received from Bermuda.

Family *Fungidæ*, (Lat. *fungus*, a mushroom.) These have the tentacles short and scattered; when aggregate, the disks are confluent. The surface of the coral is without proper cells and stellate. There are nine recent species, mostly from the Indian seas, and as many fossil ones. One species is the MUSHROOM CORAL, (see Chart,) *Fungia patellaris*, (*F. fungites*, Linn.,) a circular coral with radiating plates like the under surface of some mushrooms. The Thick-tentacled Fungia, *F. crassitentacula*, shows the animal on the external surface, (Plate XVIII. figs. 19 and 20,) with the protruding tentacles.

Family *Caryophyllidæ*, (genus *Caryophyllia*.) These have the radiating cells or plates striated externally and collected into a solid conical polyparium fixed at the base, (Plate XVIII. fig. 15.) See Chart for figure of *C. cyathus*, (Gr. *kuathos*, a cup or ladle.) The *Oculina*, a white coral, is of this family.

Family *Zoanthidæ*, (Animal flowers.) These have the exterior of a somewhat leathery consistence, and short marginal tentacles, in two or three species, in the midst of which the mouth is situated. For the Animal-flower, *Zoanthus Solanderi*, see Chart. This family is not coralligenous.

Family *Madreporidæ*, (Fr. *madre*, spotted, and *pore*, from Gr. *poros*.) These have deep cells extending to the center of the corallum, which is very porous and fixed. The tentacles are twelve. The species *Madrepora abrotanoides*, (Gr. southern wood-like,) Lamarck, or *M. muricata*, Linn., is an example—see Chart and Plate XVIII. fig. 15. (The figure on the Plate gives this species in magnified, that on the Chart in diminished size.)

The Chart names several other Madreporic corals, living and fossil, which we have not room to notice particularly. The fossil genus *Catenipora*, (chain coral,) is found in transition rocks. The animal is unknown, contained in tubular cells united laterally in a calcareous polypary, of a conical form. (For figure of *C. escharoides* lattice-like, see Chart.)

Sub-Order Alcyonaria.

Family *Pennatulidæ.* Sea-Pens. This interesting family of Corals is represented on the Chart by *Pennatula phosphorea'* which has a stony axis and is free or has the base sunk in the mud. From the axis a series of lateral branches passes off on each side, resembling the barbs of a feather, whence the generic name. On these branches the polyps are situated. This species, when disturbed, emits a phosphorescent light. By the movement of the eight tentacles, the animal seems to have power to direct its course.

Family *Gorgonidæ.* Sea-fans. These have the polyp mass rooted and tree-like, consisting of a central axis backed with a polypiferous crust. The axis is horny or fasciculate, but not calcareous. These include many beautiful fixed corals. Some *Gorgonias* found on the Atlantic coast, when stripped, have the appearance of Whale-bone. The Warty Sea-fan, *G. verrucosa,* is somewhat fan-shaped, and when dry, backed with a white warted crust. The species *G. flabellum,* (Lat. a small fan,) is reticulated with the branches inwardly compressed. It is found in the warm seas of India and America, and three feet in length.

The Red Coral, *Coralium rubrum,* (see Chart,) has the entire stem converted into a stony axis; the flesh is external, and in this alone are the polyp cells. This species is branched, one foot high, varying from a deep red to a beautiful rose color. It takes a high polish, and is employed for purposes of ornament.

Our limits render it necessary to omit details of other families of Corals, some of which are found in American seas; also many particulars respecting the wonders wrought by the Coral-insects. The statements of Prof. Dana contained in our general remarks, preclude the idea that the coral islands, of which so much has been said, are exclusively the work of these insects. In many instances the coral extends to a much greater depth than these animals are known to live; in other instances it presents a surface considerably elevated; but corals "cannot grow above the surface." Com. Wilkes, (of U. S. Exploring Expedition,) and others, regard them as in part at least, of volcanic origin. Some subterranean movement must have lifted these islands from the bed of the ocean; the coral being in some instances not less than eight or nine thousand feet high. Agassiz, it is said, has for the first time succeeded in preserving alive in this country, some coral insects. They were kept in water, carefully and frequently changed. Lady Wortley, (see her "Travels in the United States,"

1851,) says she "hardly dared to breathe while looking at them, for fear she should blow away their precious lives; it was most interesting to watch them as they flung about what seemed their fire-like white arms, like microscopic opera dancers or windmills; and most curious to mark their complexions and contortions, all the twistings and twirlings, and flingings and wringings of these curious little creatures."

Third Order. Hydroids, (Gr. ʽυδρα, hydra.)

This order contains animals, some of which have, and others have not the coralline matter. In these the internal cavity is tubular; some of them have the power of moving from place to place. The coral or hard material varies in different genera, sometimes merely showing itself in calcareous granules diffused through the body.

The eggs of the Hydroids hang in bunches externally from the lower end of the upper cavity, in graceful forms and sometimes beautifully colored. Prof. Dana divides this order into four families. 1. *Hydridæ*, not coralligenous; 2. *Sertularidæ*, with corneous coralla; 3. *Campanularidæ*, with corneous bell-shaped corolla; 4. *Tubularidæ*, with coralla tubular and corneous.

I. *Hydridæ*. It will be noticed that hitherto our attention has been given to marine radiates. This family comprises minute fresh water polyps which are soft and naked. Numbers of them are seen often in stagnant pools and ditches, clustering upon aquatic plants, etc. Their structure consists of a fleshy tube, the aperture of which serves as a mouth to receive and exclude food; it is bordered with from six to eighteen extremely flexible, thread-like arms. These are so fine as to appear like gossamer, and may be stretched six or eight inches. The Chart names the Green Polyp, *H. viridis*, (Plate XVIII. fig. 14,) and the Yellow Polyp, *H. fusca*. When in search of prey, the Hydra permits its arms to float loosely through the water, and thus succeeds in obtaining a supply of food. If any of the minute crustaceans or aquatic insects but touch one of these tentacles, the thread is suddenly "thrown into cork-screw coils," other threads are also coiled around the victim, and it is soon borne quite motionless, to the mouth. At the bottom of the fleshy sac is a saucer-shaped body, "in the center of which is a small, oval, solid body, bearing on its summit a calcareous dart, pointed at the extremity, and bifid, or sagittate at its base." The darts are thrust out with force, ejecting, as is thought, a subtle poison at the same time;

and hence "worms and the larvæ of insects die suddenly from the touch of these gelatinous threads."

The Hydridæ increase by minute gemmules or buds developed from the common substance of the body; but unlike some of the Zoophytes, the point of union becomes more and more tender, and they are finally detached. These polyps may be artificially increased. If the body be divided, each segment will become a new animal, and even "a small portion of the skin soon grows into a polyp. If cut off pieces be placed in contact, and pushed together with a gentle force, they will unite and form a single one." One species, it is said, "can be actually turned inside out like a glove, and yet perform all the functions of life as before, though that which was the coat of the stomach is now the skin of the body, and *vice versa.*" (Gosse.) This power of reproduction gave to this polyp the name of *Hydra*, in allusion to the fabled monster whose heads sprouted as often as they were cut off by Hercules.

FOURTH CLASS. PROTOZOA, (Gr. πρῶτος, *prōtos*, first, or lowest; ζῶον, *zōon*, animal.) INFUSORIA. FORAMINIFERA. RHIZOPODS.

The above several names are applied to minute animals which have been observed and studied since the discovery and improvement of the microscope. Leeuwenhoek, in 1675, first observed them in standing water, though he was not then certain as to their animal nature. They are termed PROTOZOA, because regarded as the first manifestations of animal life. Being found in infusions of vegetable and animal substances, they are called INFUSORIES, though *infusion* is not essential to their production. Some of them are minute shells, consisting of one or more chambers, united by a small perforation or *foramen*, and are hence named FORAMINIFERA. The term RHIZOPODS, (Gr. *rhiza*, a root; *pous*, a foot,) is also applied to these latter animals.

The Infusories have been divided into the "illoricated" and the "loricated," the former composed of a single, homogeneous, soft substance, analogous to fine membrane and unshielded or naked; the later covered by a siliceous or calcareous lorica, or external shield, in which the animal is enclosed. The pellucid membrane of the animal contains, according to Ehrenberg, a long curved intestine, with numerous globular bodies suspended to it somewhat like grapes. These he regarded as so many stomachs, and therefore called the animals POLYGASTRICA, (many-stomached,) dividing them into two legions: 1. ENTERODELA, (with the intestines apparent,) and 2. ANENTERA, (without intestines,) each legion including both naked and coated species.

These Polygastrica seem to be universally diffused, one set of forms inhabiting salt water, another fresh. Every mineral fount has its peculiar inhabitant. They are found with the red snow of the Alps and the poles, and in the waters of hot springs. In a word, wherever organic matter exists in a decomposing state, there they abound, "acting as scavengers in devouring in the state of comminution and decay, those particles of decomposing vegetable matter, which, if left to be diffused throughout the atmosphere, might be productive of the most pernicious malaria."

Of the Enterodela an example is had in *Bursaria truncatella.* This is found in ditch water, and is so large as to be seen by the naked eye, resembling an egg in shape, with one end deeply hollowed; and from one-fourth to one-third of a line long. (N.B. The line employed in Natural History is the twelfth part of a *French* inch.) Of the Anentera we name of the genus Monas, the species *M. crepusculum,* the TWILIGHT MONAD, one of the most minute and most simple of all the living beings made known by the most powerful microscopes, resembling a mere ciliated cell, and in size only the twelve thousandth part of an inch. If a few stalks of hay be tied together and suspended in a jar of water, the contents remaining untouched, the second day after, there will appear a sort of scum on the surface of the water, that has become turbid and slightly tinged with green. When a minute drop of this liquid is examined with a microscope having a magnifying power of about two hundred diameters, the water is found to swarm with immense multitudes of minute round or oval atoms, which move rapidly with a gliding action. These are MONADS.

All infusions of vegetable and animal substances are found to be speedily filled with animals resembling these or others of the genus *Vibrio,* the latter bearing some likeness to an eel, (now placed among the Intestinal Worms.) Upon these minute creatures strong poisons seem to have no immediate effect, though a few drops of alcohol suffice to strike dead the five millions of living beings found in a barrel of vinegar.

All the Infusoria seem to be provided with a mouth, generally terminal, but sometimes placed near the middle of the body. The breathing organs, so far as known, are simple openings. The sense of feeling perhaps has its appropriate organs in the mouth and the vibratile cilia by which it is surrounded. The eyes are supposed to be the dark red or black stigmas which the microscope reveals as situated in front, on the upper side. Most of the Polygastrica have a single stigma; some, as the *Distigma*, have two. They never sleep, and are most tenacious of life. The reproduction occurs by spontaneous division and by gemmation or budding. The division goes on with wonderful rapidity, either transversely or lengthwise, each half forming an independent animal. Ehrenberg asserts that the *Hydatina seta*, increased in twelve days to sixteen millions, and another species in four days to one hundred and seventy billions.

The Infusoria are generally colorless and translucent; but some are green, some yellow, and a few red. The colored species give their peculiar tinge to the water. The shape is globular, oval, spindle-like, cylindrical, or vermiform. Some are continually changing their form, as those of the genus *Amœba* or *Proteus*, belonging to the ANENTERA. This "consists of a mass of clear jelly-like matter, with a few granules, two or three supposed stomachs, and a contractile bladder," and from its power of changing its form, has "long been celebrated among naturalists."

The term FORAMINIFERA, or RHIZOPODS, is restricted to animals of low organization, "consisting of a slimy, transparent jelly, invested with a hard, usually calcareous shell." They are found in sea-sand, and amongst marine refuse dredged up from deep water. Owing to the spiral form of many of their shells, these creatures were long erroneously regarded as mollusks, and as allied to the *Nautilus*. Ehrenberg thought them to be allied to the Bryozoa or Moss-corals, (minute animals aggregated in great numbers like the coralligenous Zoophytes, having a distinct stomach, and an intestine curved upon itself, with an outlet near the mouth, "the tentacles of which are covered with vibrillæ, and covered with a membranous, horny, or calcareous tube; now referred to the Tunicates or lowest class of Mollusks.") But the true position of the Foraminifera is probably between the Amæba on the one hand, and the sponges on the other. The Foraminifera in the calcareous shell, present various appearances. Sometimes they are comparatively large and conspicuous; at others so small that their existence can be shown only by means of high magnifying powers. Through the foramina, long delicate processes of the soft animal, termed *pseudopodia*, (or false feet-like,) are protruded. These are probably used to some extent, "for tactile, prehensile, and locomotive purposes, or for the imbibition of nutritive fluid."

The Foraminifera have peculiar interest for the geologist. Recent strata owe their origin to the long continued accumulation of these minute atoms. White chalk rocks are mainly composed of them. Below these their numbers decrease. They exist, however, in every formation from the Silurian to the Tertiary. In most countries Silicious infusorial shells abound in salt-marshes and the superficial marls which are associated with peat. The fertilizing power of the guano is in part attributable to the silicious shells of infusorial Diatoms with which it is filled. Their remains constitute the Berg-mehl, or Mountain meal, which in Swedish Lapland is, in seasons of scarcity, used, mixed with flour, for sustaining life. With the silicious shields of the Diatoms are also found the calcareous shields of Foraminifera.

Prof. Bailey has described a bed of infusorial earth at West Point, N. Y. Thirteen or more similar deposits have been discovered in other states, sometimes fifteen feet in thickness. These American fossils are mostly found under banks of peat. The forms of this country are similar in character to those of Europe.

The true position of Sponges, *Spongia*, is not easily fixed. Like many of the *Polypifera*, they have a firm horny or stony skeleton, immersed in a soft gelatinous living mass. If they belong to the Animal Kingdom, they are at the very lowest point, showing no sensation when pierced, torn, burnt, or acted on by acids; so that in respect to sensitiveness, they are surpassed by some kinds of plants. The species of *Spongia* are very numerous, one hundred and fifty having been described by Lamarck. Their forms are also exceedingly various, (see Chart and Plate XVIII. fig. 23a, b, c.) They are mostly marine, though some are found in fresh water. The best known species and the one seen in shops, *S. officinalis*, is found attached to rocks in the Mediterranean, and gathered by divers. The cup-shaped sponge, *S. usitissima*, is found in American seas.

Agassiz, in his recent work, expresses the decided opinion that the division of the Animal Kingdom called PROTOZOA, differing from all other animals in producing no eggs, does not exist in nature; and that the beings which have been referred to it, might be divided and scattered, partly among plants and partly among animals. It would however be premature to suppress the entire class until further results have been attained.

What is the 4th and last Sub-Kingdom? What does it include? What is the 9th branch of Zoology? How is this term derived? Is the radiated form equally manifest in all? Into how many classes are the Radiates divided? Name the first. Give the general characteristics of the Class. What is the 1st Order? What are its leading characters? Why have the animals been called Sea-cucumbers? What does Com. Wilkes say of them? How large are they? What is the 2nd Order? Give its characters. How do the forms vary? What is said of the Cake and Heart Urchins? What of the mouth? Repeat the remark of Prof. Forbes. What is the 3rd Order? How distinguished? What is said of the Gorgon-headed Star-fish? Of Asterias? Of Ophiura? To what mollusks are Star-fish peculiarly destructive? What is the 4th Order? Are there many living sp. of this order? What are the fossils called? What is said of the body, arms, &c.? What is said of the joints of the rounded stem? For what were they anciently used? What are they still called? What sp. are mentioned and what is said of them? What is the 2nd Class of Radiates? Why is it so called? What are their common names? Give their characters and habits. Which is the 1st Order? To what does the name refer? Has it any other name? What gen. and sp. are noticed on the Chart? What does it say of them? What is the 2nd Order? What animals does it include? What did Cuv. call them? What is said of the Portuguese Man of War? What other Radiates of this order are mentioned? What is said of them? What is the 3rd Order? Why so named? What are the uses of the cilia? What sp. are mentioned? What is said of them? What 4th Order is added by some naturalists? What is said of it? What is the 3rd Class? Give the substance of what is said respecting it. Which is the 1st Order? Describe their characteristics, &c. Into what Sub-orders does Prof. D. divide it? What sp. of the family Actinidæ are mentioned? Give particulars respecting them. What is said of the forms assumed by Corals, &c.? What 2nd family is mentioned? What does Prof. D. say of the Astræa dome, &c.? What 3rd family is named? What is said of it? Name the other families of this Sub-order. What families of the 2nd Sub-order are mentioned? What is said of them? Are the Coral islands entirely the work of the Coral insects? What has been said of these islands by Com. Wilkes and others? Have the Coral insects been kept alive? Repeat the extract from Lady Wortley's Travels. What is the 2nd Order? What is said of it? Into what families does Prof. D. divide it? Give the substance of what is said of the Hydridæ. What is the 4th Class? What several names have been applied to these animals? Why termed Protozŏa? How have the Infusories been divided? Of what are the former composed? How are the latter protected? Why did Ehrenberg name them Polygastrica? Into what two legions did he divide them? What example is given of the first? What gen. of the second is mentioned? What is said of the Twilight Monad? Of the Vibrio? What are the breathing organs of the Infusoria? What is said of their senses, reproduction, &c.? To what is the term Foraminifera restricted? What is said of them? What of the Sponges? Give a general view of these animals as presented on the Chart.

CONCLUSION.

We thus terminate our sketch of "Nature in Living Forms." The reader or student of this volume, while surveying these "Forms," in their structure, their organs, and their variations from a common type, beginning with MAN, who stands in high pre-eminence at the head, and passing downward until reaching the group we have last contemplated, in which are discoverable but the faintest traces of animal existence, and marking, too, in his progress the adaptation of all to the stations and offices for which they were severally designed, has been presented with developments of a plan of being, beautiful and harmonious in its various parts and gradations, essentially unaffected by lapse of time or changes of locality and climate, and having its origin in the far-reaching intelligence and unbounded goodness of the Infinite Creator.

We have in this volume but attempted to lift from the face of "Nature" a portion of its mysterious veil. May those who are induced to examine the present work, not only derive from it motives for still further research in the interesting department of Natural History, but also for increased diligence and fidelity in performing their own incumbent offices as parts of the wondrous whole, cultivating towards each other feelings of true affection and brotherhood, and above all, binding themselves by cords of supreme love and obedience to the one common Father, extolling

"Him first, Him midst, and without end."

INDEX.

www.ingramcontent.com/pod-product-compliance
Lightning Source LLC
LaVergne TN
LVHW021048110826
845150LV00001B/13

* 9 7 8 1 4 2 5 5 6 8 9 0 0 *